Medicinal Plants Used in Traditional Persian Medicine

This book is dedicated to Prof. Dr Mehdi Mohaghegh who devoted his life's work to the history of Persian Medicine

Medicinal Plants Used in Traditional Persian Medicine

Edited by

Hartwig Schulz

Consulting and Project Management for Medicinal and Aromatic Plants, Stahnsdorf, Germany

Seyed Ahmad Emami[1,2]

[1]Department of Pharmacognosy, School of Pharmacy, Mashhad University of Medical Sciences, Mashhad, Iran
[2]Department of Traditional Medicine, School of Pharmacy, Mashhad University of Medical Sciences, Mashhad, Iran

Farsad Nadjafi[1,2]

[1]Shahid Beheshti University, Medicinal Plants and Drugs Research Institute, Tehran, Iran
[2]MateriaMed GmbH, Vienna, Austria

CABI

CABI is a trading name of CAB International

CABI
Nosworthy Way
Wallingford
Oxfordshire OX10 8DE
UK

Tel: +44 (0)1491 832111
E-mail: info@cabi.org
Website: www.cabi.org

CABI
200 Portland Street
Boston
MA 02114
USA

Tel: +1 (617)682-9015
E-mail: cabi-nao@cabi.org

A catalogue record for this book is available from the British Library, London, UK.

ISBN-13: 9781800621657 (hardback)
9781800621664 (ePDF)
9781800621671 (ePub)

DOI: 10.1079/9781800621671.0000

Commissioning Editor: Rebecca Stubbs
Editorial Assistant: Emma McCann
Production Editor: James Bishop

Typeset by Straive, Pondicherry, India
Printed and bound in the UK by Severn, Gloucester

Contents

Contributors

Mamak Hashemi Habib Abadi, Department of Persian Medicine, School of Medicine, Hamadan University of Medical Sciences, Hamadan, Iran.

Amir Hossein Abdolghaffari, Department of Toxicology & Pharmacology, Faculty of Pharmacy, Tehran Medical Sciences, Islamic Azad University, Tehran, Iran; and GI Pharmacology Interest Group (GPIG), Universal Scientific Education and Research Network (USERN), Tehran, Iran.

Maryam Akaberi, Department of Pharmacognosy, School of Pharmacy, Mashhad University of Medical Sciences, Mashhad, Iran.

Seyyed Ahmad Askari, Traditional Medicine and Materia Medica Research Center, Shahid Beheshti University of Medical Sciences, Tehran, Iran.

Sayyedeh Fatemeh Askari, Cardiovascular Diseases Research Center, Birjand University of Medical Sciences, Birjand, Iran; and Department of Pharmacognosy and Traditional Pharmacy, School of Pharmacy, Birjand University of Medical Sciences, Birjand, Iran.

Zahra Ayati, Department of Traditional Pharmacy, School of Pharmacy, Mashhad University of Medical Sciences, Mashhad, Iran; and NICM Health Research Institute, Western Sydney University, Westmead, NSW 2145, Australia.

Sajad Azad, Department of Traditional Pharmacy, School of Pharmacy, Mashhad University of Medical Sciences, Mashhad, Iran.

Parmis Badr, Phytopharmaceutical Technology and Traditional Medicine Incubator, School of Pharmacy, Shiraz University of Medical Sciences, Shiraz, Iran; and Pharmaceutical Sciences Research Center, Shiraz University of Medical Sciences, Shiraz, Iran.

Hamed Baharara, Department of Clinical Pharmacy, School of Pharmacy, Mashhad University of Medical Sciences, Mashhad, Iran.

Roodabeh Bahramsoltani, Department of Traditional Pharmacy, School of Persian Medicine, Tehran University of Medical Sciences, Tehran, Iran; and PhytoPharmacology Interest Group, Universal Scientific Education and Research Network, Tehran, Iran.

Zahra Boghrati, Department of Traditional Pharmacy, School of Pharmacy, Mashhad University of Medical Sciences, Mashhad, Iran.

Mina Borhani, Department of Pharmacognosy, School of Pharmacy, Mashhad University of Medical Sciences, Mashhad, Iran.

Seyed Ahmad Emami, Department of Traditional Medicine, School of Pharmacy, Mashhad University of Medical Sciences, Mashhad, Iran; and Department of Pharmacognosy, School of Pharmacy, Mashhad University of Medical Sciences, Mashhad, Iran.

Fatemeh Etemadpour, Department of Phytopharmaceuticals (Traditional Pharmacy), Shiraz University of Medical Sciences, Shiraz, Iran.

Seyed Majid Ghazanfari, Iranian Research Center on Healthy Aging (IRCHA), Sabzevar University of Medical Sciences, Sabzevar, Iran.

Shakila Hajizadeh, Department of Traditional Pharmacy, School of Traditional Medicine, Shahid Beheshti University of Medical Sciences, Tehran, Iran.

Bahia Namavar Jahromi, Infertility Research Center, Shiraz University of Medical Sciences, Shiraz, Iran; and Department of Obstetrics & Gynecology, School of Medicine, Shiraz University of Medical Sciences, Shiraz, Iran.

Lida Jarahi, Department of Community Medicine, Faculty of Medicine, Mashhad University of Medical Sciences, Mashhad, Iran.

Mohammad Reza Kanani, Medicinal Plants and Drugs Research Institute, Department of Botany, Shahid Beheshti University, Tehran, Iran.

Seyedeh Mahnaz Karimi, Department of Traditional Pharmacy, School of Persian Medicine, Tehran University of Medical Sciences, Tehran, Iran.

Abdolali Mohagheghzadeh, Department of Phytopharmaceuticals (Traditional Pharmacy), Faculty of Pharmacy, Shiraz University of Medical Sciences, Shiraz, Iran; and Pharmaceutical Sciences Research Center, Shiraz University of Medical Sciences, Shiraz, Iran.

Leila Mohtashami, Department of Pharmacognosy, School of Pharmacy, Mashhad University of Medical Sciences, Mashhad, Iran.

Saeideh Momtaz, Medicinal Plants Research Center, Institute of Medicinal Plants, Karaj, Iran; and Toxicology and Diseases Group, Institute of Pharmaceutical Sciences, Tehran, University of Medical Sciences, Tehran, Iran; and Gastrointestinal Pharmacology Interest Group, Universal Scientific Education and Research Network, Tehran, Iran.

Ghazaleh Mosleh, Phytopharmaceutical Technology and Traditional Medicine Incubator, School of Pharmacy, Shiraz University of Medical Sciences, Shiraz, Iran.

Farsad Nadjafi, Shahid Beheshti University, Medicinal Plants and Drugs Research Institute, Tehran, Iran; and MateriaMed GmbH, Josefstädterstrasse 43–45, Vienna, Austria.

Motahare Nayebzadeh, Department of Phytopharmaceuticals (Traditional Pharmacy), Shiraz University of Medical Sciences, Shiraz, Iran.

Maryam Nikoosokhan, Science and Technology Park of the University of Tehran, College of Agriculture & Natural Resources, University of Tehran, Karaj, Iran.

Roja Rahimi, Department of Traditional Pharmacy, School of Persian Medicine, Tehran University of Medical Sciences, Tehran, Iran; and PhytoPharmacology Interest Group, Universal Scientific Education and Research Network, Tehran, Iran.

Hartwig Schulz, Consulting and Project Management for Medicinal and Aromatic Plants, Waltraudstraße 4, Stahnsdorf, Germany.

Zahra Taghipour, Department of Traditional Pharmacy, School of Persian Medicine, Tehran University of Medical Sciences, Tehran, Iran.

Mojgan Tansaz, Department of Traditional Medicine, School of Traditional Medicine, Shahid Beheshti University of Medical Science, Tehran, Iran.

Meysam Zaeri, Department of Phytopharmaceuticals (Traditional Pharmacy), Faculty of Pharmacy, Shiraz University of Medical Sciences, Shiraz, Iran.

Elaheh Zibaee, Department of Traditional Pharmacy, School of Pharmacy, Mashhad University of Medical Sciences, Mashhad, Iran.

About the Editors

Hartwig Schulz (PhD) studied chemistry in Bonn and Kiel (Germany). After completing his studies, he carried out several years of research in the field of human nutrition (vitamin E, phytosterins, plant lipids) at the University of Kiel, Germany. Then, as head of an analytical department, he worked 10 years for one of the largest perfume and flavor manufacturers worldwide (Symrise). From 1996 to 2019, he was the director and professor of a research institute under the responsibility of the Federal Ministry of Agriculture in Germany. Hartwig Schulz is author of numerous scientific articles and book chapters and as the president of the German Society for Quality Research (DGQ), he was also engaged for 8 years in food research on national and international levels. Since his retirement, he is working as a consultant, especially in the field of "medicinal and aromatic plants", and is also supporting various development aid projects on a voluntary basis.

Seyed Ahmad Emami (PharmD, PhD) graduated in Pharmacy from Mashhad University of Medical Sciences (1993, Mashhad, Iran) and received his PhD (2001) in Pharmacognosy from Isfahan University of Medical Sciences. He is currently serving as a full professor in the Department of Pharmacognosy and the Department of Traditional Medicine, School of Pharmacy, Mashhad University of Medical Sciences in Iran. With his long years' experience in the field of pharmacognosy of medicinal plants occurring in Iran, he has contributed significant advances on all aspects of medicinal plants used in Traditional Persian Medicine. From 1998 to the present he has supervised numerous students during their academic education and performed research projects in different fields such as isolation and identification of new valuable substances in various medicinal plants endemic to Iran. Furthermore, he studied the Islamic culture and civilization. He has published his scientific results in numerous international journals (203 articles in English and French, and 35 articles in Persian and Arabic) and books (26 books in English and French, and 25 books in Persian and Arabic).

Farsad Nadjafi holds a PhD in Agroecology and has been an academic member of the Medicinal Plants and Drugs Research Institute of Shahid Beheshti University in Tehran since 2006, being the director of this research institute for 2 years. His specialty is domestication and cultivation of valuable and endemic wild medicinal plants of Iran and also ethnobotanical study of these species. He is currently the manager and owner of MateriaMed GmbH in Vienna, a company that is active in the field of cultivation and processing of medicinal plants. He has published numerous scientific articles in the field of cultivation and domestication of medicinal plants, has many contacts to institutes working in the field of Traditional Persian Medicine, and is reviewing the relevant Persian/Arabic literature. When selecting the plant species, priority will be given to (endemic) plant species that have been more or less unknown in Western medicine up to now.

Preface

Traditional Iranian Medicine (TIM) or Traditional Persian Medicine (TPM) encompasses the extensive knowledge and practices used in Persia from ancient times to the present day for the diagnosis, prevention, and elimination of diseases. It is based solely on practical experience and observations passed down from generation to generation. This book presents the latest scientific findings, challenges, and applications in this exciting field, providing insight into the treasures of the Arabic and Persian medical systems, as well as the interactions with Ayurveda in India, Traditional Chinese Medicine (TCM), and Western traditional medicine (the so-called "monastic medicine") that interacted with or emerged from Islamic medicine in the Middle Ages. As a result of the wide geographical spread and extensive travels within their territories, as well as the exchanges with communities in the Middle East, India, and North Africa, Muslim authors described the individual importance of medicinal plants and gradually developed their botanical systematics.

Iran has a long history of great importance in the field of traditional medicine practices; this knowledge heritage mainly goes back to the time of Babylonian–Assyrian civilization, and every generation added its individual experience and new elements to this "cultural database". Even nowadays, numerous medicinal plants are still used in Iran as curatives for various types of health problems. A great part of this traditional knowledge has not yet been taken into account by scientific research to a sufficient extent, and it therefore seems advisable to verify the effectiveness of traditional treatments with the help of new scientific tools available today, especially in those cases where no or only few supporting data are available in the literature.

In this context, extensive literature studies were carried out, referring in particular to various ethnobotanical studies. Those plant species relevant in this context as well as their individual traditional uses in traditional medicine and the form of application are described in detail. The individual ethnomedical practices, which have so far been made transparent through extensive research, derive from the experience of local people who have developed remedies against a wide range of diseases handing down the knowledge from generation to generation over the millennia. These traditional medicinal practices today represent an important source of inspiration in the development of new phytopharmaceutical medicines and new therapeutic strategies. The World Health Organization (WHO) has therefore repeatedly called for increased attention to be paid to the scientific validation of traditional medicines, as they are capable of making a very important contribution to public health in different parts of the world.

It should also be pointed out here that this process of processing and preserving ethnomedical knowledge still offers considerable potential in Iran as well. This was especially observed during the coronavirus pandemic in the early 2020s, where several research groups were able to determine antiviral efficacy against SARS CoV-2 virus infection in some Iranian plant endemic species.

The first records of medicinal plants date back more than 5000 years to the time of the Sumerians in southern Mesopotamia, who prepared lists with plant species known to them, such as myrrh or opium, on clay tablets. Ancient Chinese and Egyptian papyrus writings dating back to 3000 BC also reveal the medicinal use of plants at that time. The "Ebers Papyrus" from ancient Egypt, for example, provides extensive information on more than 850 herbal medicines, including aloe, mandrake, cannabis, garlic, castor oil, and juniper.

In present days, herbal medicines are experiencing a renaissance and are extensively used for therapeutic purposes. A broad spectrum of applications opens up for phytopharmaceuticals, ranging from self-therapy of sleep, stress, and mood disorders to digestive complaints, colds, and viral diseases. Herbal medicines are also used to treat asthma, skin eczema, migraine, menopausal symptoms, chronic fatigue, and irritable bowel syndrome, as well as in cancer therapy.

Manuscripts of Traditional Persian/Iranian Medicine (TPM/TIM), written by Persian scholars especially in the Middle Ages, not only represent a treasure of traditional medical and pharmaceutical knowledge, but also provide a collection of ingenious studies regarding the use of medicinal plants.

In writing this book, we have been guided mainly by Avicenna (AD 980–1037) and his important medical work, *Canon of Medicine* (Latin: *Canon medicinae*, Arabic: *al-Qānun fi al-Tibb*). Ibn Sinā (انیس نبا), as he was called in Persia, is considered in medical literature to be one of the most important physicians, astronomers, philosophers, and writers of the Islamic Golden Age, but is also honored as the father of early modern medicine. Of his 450 works, about 240 have survived, 40 of which relate to medical aspects.

Due to the steady increase in knowledge in modern medicine, especially in the Western world, the methods of traditional medical systems have been increasingly pushed into the background in recent decades and the therapeutic potential of evidence-based medicine could therefore not always be fully exploited. As in TCM and the Indian Ayurveda science, various studies have therefore been initiated in TPM/TIM to objectively test the performance of traditional holistic therapy methods by using modern methods of analysis.

In response to the increased demand in the Western world for methods and products of traditional medical systems, for some selected "prosperity diseases" such as Alzheimer's disease, various cardiovascular diseases, and diabetes but also selected skin disorders and various infection diseases, this book compares the therapeutic approaches developed already by Avicenna with the methods of modern medicine practiced today. In this context, the focus of this comparison is mainly based on the concept of traditional medical systems using "food as medicine". Although it is generally known today that in the Western world a more balanced nutrition would help to reduce the comparatively high number of "prosperity diseases", it is usually not easy to implement the necessary dietary changes in practice. Therefore, it is sometimes necessary to assist this process by consumption of valuable dietary supplements and in connection with appropriate conventional medical treatments.

Knowing that the introduction of new phytopharmaceuticals based on plant species endemic for Persia into the European and US markets is a time-consuming and sometimes costly process, the authors describe the individual advantages to increase awareness for these "new" medicinal plants and their special usages in individual European countries, especially in the field of alternative and complementary medicine.

Furthermore, the individual historical connections of TPM/TIM with the other well-known traditional medical systems as well as the ethnobotany and collection of Persian medicinal plants and the production of medicinal products and tea mixtures are explained in more detail.

The editors are very grateful to all chapter authors who provided their time and energy, contributing with their special knowledge to the different facets of Iranian medicinal plants. Furthermore, they would like to thank the CABI publishing house for the very pleasant cooperation and the always competent support.

Hartwig Schulz
Seyed Ahmad Emami
Farsad Nadjafi

1 The History of Persian Medicine At a Glance, Emphasizing the Life and Works of Avicenna

Seyed Ahmad Emami[1,2]*

[1]*Department of Traditional Medicine, School of Pharmacy, Mashhad University of Medical Sciences, Mashhad, Iran;* [2]*Department of Pharmacognosy, School of Pharmacy, Mashhad University of Medical Sciences, Mashhad, Iran*

Abstract

This chapter gives a detailed overview of the history of Traditional Persian Medicine (TPM) and, in this context, specifically describes the biography of Ibn Sinā (also known as Avicenna). This famous philosopher and physician was born near Bukhara in Central Asia (Uzbekistan) in AD 980 and died in Iran in 1037. In his youth he acquired extensive knowledge in physics, mathematics, logic, and metaphysics, and began studying medicine at the age of 13. His outstanding medical knowledge earned him the attention of the Sultan of Bukhara, whom he successfully treated for a serious infection. During his busy life, he managed to write nearly 131 books, including the famous *Qānun fi aṭ-Ṭibb* (*Canon of Medicine*), which was first translated into Latin in the 12th century and became the standard textbook not only in Asia but also in the medical schools of Europe for several centuries. Avicenna originally wrote the book in Arabic, but it was later translated into several other languages, including Persian, Chinese, Hebrew, German, French, and English. In addition, the *Canon* contains descriptions of about 800 mineral, herbal, and animal-based medicinal materials, which until today are an important basis for TPM.

History of Medicine in Iran

Historical and geographical position of Iran

Iran is a relatively high land that covers a major part of the Iranian Plateau. The country is located between 25° and 45° N (latitude) and 40° and 63° E (longitude). The time difference between the easternmost and westernmost parts of Iran is 1 h and 18 min. The area of Iran is about 3.7% of the area of the Asian continent and 1.1% of the area of the world's drylands (Badiʿi, 1999). Iran is bounded by the republics of Armenia and Azerbaijan, the Caspian Sea, and the Republic of Turkmenistan from the north; Afghanistan and Pakistan from the east; the Oman Sea and the Persian Gulf from the south; and the republics of Iraq and Turkey from the west. In terms of plant groups, the land of Iran can be divided into six vegetation zones: (i) the region of northern moist broadleaf forests, which are located between

*Email: emamia@mums.ac.ir

 Medicinal Plants Used in Traditional Persian Medicine (eds. H. Schulz, Seyed Ahmad Emami and Farsad Nadjafi)
DOI: 10.1079/9781800621671.0001

the ridges of Talish and Alborz in the parts overlooking the Caspian Sea; (ii) the region of western broadleaf forests, which includes all the forests of the Zagros region and encompasses Kurdistan, Kermanshah, Lorestan, and Bakhtiari; (iii) the mountainous steppe region, which includes the mountains of North Khorasan and Azerbaijan; (iv) the dry region of the plateau rivers (foothills), which covers a large part of the country; (v) the dry region of the desert rivers, which includes all the deserts and sand fields as well as internal sand dunes of great extent and covers the eastern and southern desert areas; and finally (vi) the forest region of the southern coast, which is located on the edge of the Iranian Plateau (Rahnamai, 2019).

The plateau of Iran is one of the habitats of ancient humans and the place of one of the ancient civilizations of the world, where many historical monuments from past eras remain in its various areas. The special geographical position of the Iranian Plateau, which connects Central Asia and the Far East to the Near East and Europe, turned this region into one of the most important centers of waves of migrations and settlements and various wars and developments since the prehistoric era. This plateau has been a place of passage, invasion, and life of different tribes. Asian, African, and European physical characteristics in Iranian people, and their speaking in different languages and dialects are signs of the mixing of the people of this land. Apparently, the oldest site related to the Old Paleolithic period that has been found in Iran so far is located in Khorasan and in the bed of the Kashaf Roud (a river), where the stone tools found there are estimated to be about 800,000 years old (Smith, 1968). The history of "Iran", which means "land of Aryans", begins with the migration of a group of Aryan peoples (Indian and Iranian) into the Iranian Plateau. The Aryans themselves were Indo-European peoples who lived in Central Asia after separating from their fellow races in the 2nd millennium BC. According to some researches, Aryan immigrants entered the western regions of this plateau around 1000–800 BC when the Iranian Plateau was in the Iron Age (Young, 1967; Mallory, 1989). Later in Iran/Persia, various dynasties including Medes (670 to 550 BC), Achaemenid Empire (546 to 330 BC), Seleucid Empire (312 to 129 BC), Arsacid Empire (238 BC to AD 224), and Sassanid Empire (AD 220 to 651) ruled. Shortly after the the assassination of Yazdegerd III, the last Sassanid king who was killed while fleeing (AD 651), Persia almost officially became part of the Islamic Caliphate and entered its Islamic history. In the post-Islam era, the various dynasties included the Tahirids (AD 821 to 873), Saffarids (AD 861 to 1002), Samanids (AD 819 to 1004), Buyids (AD 934 to 1055), Ghaznavids (AD 963 to 1187), Seljuk Empire (AD 1037 to 1194), Khwarezmian (AD 1071 to 1231), Ilkhanate (AD 1256 to 1335), Timurids Empire (AD 1370 to 1506), Safavids (AD 1501 to 1736), Afsharids (AD 1736 to 1796), Zands (AD 1751 to 1794), and Qajars (AD 1794 to 1925). The Pahlavi dynasty ruled from AD 1925 to 1979 and with the fall of the last dynasty, the Iranian monarchy was dismantled and turned into the Islamic Republic (Zarrinkoob, 2012).

Medicine in ancient Persia

There is not much information left about the medicine of ancient Persia, but it seems that there were regular medical institutions in the vast territory of Achaemenid. One of the famous sages and physicians of Persia during the Achaemenid period was Jamasp, who is mentioned in *History of Balʿami* (*Tārikh-i Balʿami* in Persian) as the author of books on geometry and medicine (Balaʿmi, 2016). In addition, as another instance, according to Xenophon, Cyrus the Great (Kurush Kabir in Persian) realized the importance of people's health and its effect on improving the individual and social situation (Xenophon, 2001). He tried to improve the situation of medicine and treatment, provided medical facilities and free medicine, and paid the cost of treatment from the general budget of the country (Fig. 1.1).

In addition to local doctors, foreign doctors were also in the service of Achaemenid kings. During the Achaemenid era, special

Fig. 1.1. The Tomb of Cyrus (Persian: آرامگاه کوروش بزرگ) is the final resting place of Cyrus the Great, the founder of the ancient Achaemenid Empire. The mausoleum is located in Pasargadae, an archaeological site in the Fars Province of Iran. (Image extracted from p. 621 of *A journey from London to Persepolis; including wanderings in Daghestan, Georgia, Armenia*, by John Ussher, *c.*1865 CE. Original held and digitized by the British Library (accession number: HMNTS 10076.g.6.), https://commons.wikimedia.org/wiki/File:USSHER(1865)_p621_TOMB_OF_CYRUS,_MURGHAB.jpg.)

attention was paid to health matters, of which water and food hygiene is an example. Achaemenid kings tried to use safe and boiled water in a clean container during their travels and campaigns. The Persian king Darius I attached great importance to medical knowledge (Fig. 1.2) and encouraged scientific doctors. According to some available documents, Darius appointed one of his generals known as "Udjahorresnet" to repair the medical school located in Sais city (Lloyd, 1982).

In an inscription left by him and kept in the Vatican today, the mentioned general said: "I chose the sons of famous people as students and provided the necessary books and equipment for them to study medicine. The Emperor did this because he knew the superiority of this art to restore health to a sick man" (Posener, 1936). The mentioned document is actually considered an example in which the existence of medical schools during the Achaemenid era can be understood.

A point that needs attention is that although Greece is considered the cradle of medical knowledge and Hippocrates is considered the father of medicine, the Eastern world, especially ancient Persia, has had a significant impact on the foundation of Greek medicine. Hippocrates was born in 460 BC on the island of Kos and died in 375 BC. According to Hippocrates, every disease has a cause, and the basis of his work is accurate observations and recording of the patient's condition. Hippocratic medicine is based on the theory of four humors, the principle of which entered Persia with the religious thoughts of the Aryans, and the Persian physicians developed and elevated it. The principle of this theory is based on the hypothesis of a small world (Kahjahān in Persian) and its similarity with a big world (Mahjahān in Persian). In this way of thinking, the human being represents a mirror of the whole view of the world. The theory of four natures is also based on the similarity of different human temperaments with the

Fig. 1.2. The relief stone of Darius the Great in the Behistun Inscription, a large rock-relief carved at Mount Behistun, near the city of Kermanshah in Iran. (By Surenae, CC BY-SA 4.0, https://commons.wikimedia.org/w/index.php?curid=125510181.)

four elements of nature. The similarity between these two worlds is explained in the 13th part of *Bondahesh*. Apparently, the concept of four humors, *Akhlat Arbaʿah* in Arabic and Persian, which is the basis of Hippocratic medicine, came from ancient Persia to Greece.

The mentioned transfer took place after Alexander's attack on Persia and during the Seleucid Empire period. During the Arsacid Empire era, Persian medicine was a mixture of Greek and Zoroastrian methods of medicine. Among the famous physicians of the mentioned era, Azonax, who mastered both medicine and manual mechanics, should be especially noted, and also King Phraates I (Farhad in Persian), who lived in the middle of the 4th century AD, should be mentioned. One of the other researchers of medical knowledge during the Arsacid period is Mithridates (Mehrdad in Persian) VI, the founder of the school of Mithridatism, which means getting used to poisons. He was suspicious of his fellow men and even of his mother, and for fear of being poisoned, he took small amounts of arsenic daily, steadily increasing the dose. In this way he was able to better resist the deadly quantities of this poison. He also conducted many studies on various other poisons, and his reports can therefore be considered today as the first foundations of immunology.

During the Sassanid Empire era, medical teachings were a mixture of Greek, Indian, and Zoroastrian schools of medicine. During this period, the "GundeShapur" University and Hospital was founded by King Shapur I in the northwest of Khuzestan between Shushtar and Shush (at the site of today's village of Shahabad) that reached its peak under Khosrow I, also known as Anushirvan ("with the immortal soul"), who ruled from AD 531 to 579. In the mentioned university, Greek, Syriac, Indian, and Persian professors were engaged in teaching and translated various books into Pahlavi. At first, Greek medicine was taught in Syriac in this university, but apparently from the time of Anushirvan, its official language was changed to Pahlavi, and efforts were made to translate books related to various sciences from different languages into Pahlavi. In addition to Syriac and Greek scholars, a number of Indian doctors also lived in GundeShapur School, who translated some books from Indian into Persian language. In addition to teaching the medicine of different nations, Persian and Zoroastrian medicine was also taught and researched at this university. Although in GundeShapur medical knowledge was given special attention, philosophy, theology, mathematics, botany, and pharmacology were not neglected either. Thus, many books of philosophy and other intellectual and experimental sciences were translated and reached the Islamic world in this way through the masters of this school. The city of GundeShapur was conquered by Muslims in the 12th century AD and had activities until two centuries later, but then that university was no longer of interest and its top professors went to Baghdad, where they boosted Baghdad University and translated scientific works into Arabic. The translation of the mentioned books is due to the valuable and continuous efforts of some famous translators such as John Grammaticus, Aaron Alexandrian, Sergius of Rechina, Issa ibn Sahar Boxt, Sabet ibn Qorrah, Qusta ibn Luqa of Heliopolis, Sahl ibn Shapur, Shapur ibn Sahl, Ibn Serapion, al-Kendy, and members of some families such as the Boxtishu (the servant of Christ), Massarjuyah, Hunayn, Karkhy, Massuyah, and Sinān families who came generally from the graduates of this university. Regarding the medical formations of ancient Persia, it should be said that Persians considered Fereydon (Triteh) as the first doctor. In ancient Persia, the head of healthcare was called "Dorostbod̲", which is probably the original form of the said word "Dord̲sotbod̲" who was also the king's special physician and the head of GundeShapur Hospital and was considered the head of physical medicine in the country. In this context it is important to mention that all doctors, both physical and spiritual doctors, were under the great presidency of Mōbadān Mōbad (head of the clergy). During the Sassanid Empire era, in addition to GundeShapur, there were public and charity hospitals in other cities of Persia, where patients and people who were blind, crippled, disabled, and strangers were hospitalized and treated.

There were three classes of doctors in hospitals as follows: (i) divine healers who were called "psychiatrists" and were considered the highest class of doctors; (ii) herbal medicine specialists; and (iii) legal doctors who were the official doctors of the country and trusted by the health system. Of course, the fourth category, called kard pezeshk (A doctor who works with a knife=surgeon), worked under the supervision of the last category. In ancient Persia, cesarean section (*Rostamineh*) was performed, which Ferdowsi also refers to in *Shahnameh* (Ferdowsi, 1987). Also, a mixture of marijuana and wine was used to induce anesthesia. The ancient Persians already had extensive research on poisons and their antidotes, but there is only little information left about the medicine that was used in serious and very difficult cases. In ancient Persia, many medicinal plants such as agarwood (*Aquilaria malaccensis* Lam.), asafoetida (*Ferula assa-foetida* L.), barberry (*Berberis integerrima* Bunge), black mustard (*Brassica nigra* L.), cinnamon (*Cinnamomum verum* J. Presl), citron (*Citrus medica* L.), cumin (*Cuminum cyminum* L.), damask rose (*Rosa* × *damascena* Mill.), frankincense (*Boswellia sacra* Flück.), gum tragacanth (*Astragalus tragacantha* Ucria), haoma (*Ephedra* spp.), hemp (*Cannabis sativa* L.), labdanum (*Cistus creticus* subsp. *creticus*), myrtle (*Myrtus communis* L.), narcissus (*Narcissus tazetta* L.), olive (*Olea europaea* L.), pepper (*Piper nigrum* L.), pomegranate (*Punica granatum* L.), saffron (*Crocus sativus* L.), Syrian rue (*Peganum harmala* L.), true indigo (*Indigofera tinctoria* Gouan), violet (*Viola odorata* L.), and white water-lily (*Nymphaea alba* L.) were used; and also aromatherapy with aromatic drugs such as Arabian balsam (*Commiphora gileadensis* (L.) C. Chr.), camphor (*Cinnamomum camphora* (L.) J. Presl), and sandalwood (*Santalum album* L.) was practiced. Furthermore, some minerals such as alum (*Alumen*), lapis (*Lapis lazuli*), nushadir salt (*Sal ammoniacum*), and sulfur are among the drugs used in ancient Persian medicine (Emami *et al.*, 2017).

Since there are some adaptations and similarities between medicine in ancient Persia after Islam and ancient Greek medicine, it is necessary to take a brief look at the history of medicine in ancient Greece before dealing with the former.

Medicine in ancient Greece

The founder of Greek medicine is considered to be Aesculapius, who was probably a semi-historical, semi-legendary character with a divine aspect. Hippocrates (460–375 BC), a Greek physician who is generally considered to be the founder of medical knowledge, had a deep influence on Roman medicine as well as common medicine in the Middle Ages (Fig. 1.3) (Garrison, 1966; Nuland, 1988). However, Hippocrates owes his fame to his eloquent works known as *The Hippocratic Corpus*, which contain some 70 textbooks, lectures, research notes, and philosophical essays on various subjects in medicine, in no particular order. Notable among the treatises of the *Corpus* are *The Hippocratic Oath*, *The Book of Prognostics*, *On*

Fig. 1.3. Asclepius (center) arrives in Kos and is greeted by Hippocrates (left) and a citizen (right), mosaic, 2nd–3rd century AD. (Mosaic Arrival of Asclepius, Archaeological Museum of Kos; blackbird (GR) by Tedmek – CC BY-SA 3.0, https://commons.wikimedia.org/w/index.php?curid=15847465.)

Regimen in Acute Diseases, *Aphorisms*, *On Airs, Waters and Places*, *Instruments of Reduction*, *On The Sacred Disease*, and others (Singer and Underwood, 1962; Rutkow, 1993; Iniesta, 2011), but the prescriptions, prescription methods, and drug regimens invented by him also play a major role in his famous work. For example, he recommended asparagus (*Asparagus officinalis* L.) and garlic (*Allium sativum* L.) as a diuretic, poppy (*Papaver somniferum* L.) to induce sleep, and willow (*Salix alba* L.) to relieve pain and fever. The interesting point is that the big plane tree that still casts a shadow on the grave of this great scientist in the state of Larissa in the north of Greece is the one under which he already taught medicine during his lifetime.

The first written treatise in the field of medicinal botany parallel to the rise of Greek civilization was presented by Theophrastus (371–287 BC) (Fig. 1.4). This treatise was entitled *De historia plantarum* and originally consisted of ten books, nine of which are extant.

The work is arranged into a system whereby plants are classified according to their modes of generation, their localities, their sizes, and according to their practical uses such as foods, juices, herbs, etc. (Long, 1842). The first book deals with the individual parts of plants; the second book with the reproduction of plant species and the times and manner of sowing; the third, fourth, and fifth books are devoted to trees, their types, their locations, and their practical applications; the sixth book deals with shrubs and spiny plants; the seventh book describes various herbs; the eighth book deals with plants that produce edible seeds; and the ninth book is focused on plants that produce useful juices, gums, resins, etc. (Long, 1842). Although this work shows the valuable effort of the author in presenting an important subject with the necessary scientific coherence, it also presents the limited boundaries of the author's work, which itself is caused by the limited information and knowledge available in the field of medicinal

Fig. 1.4. The frontispiece to an illustrated 1644 edition of *De historia plantarum* by the ancient Greek scholar Theophrastus. (By Henricus Laurentius (editor), http://www.abocamuseum.it/uk/bibliothecaantiqua/Book_View.asp?Id_Book=161&Display=P&From=S&Id_page=98935, Public Domain, https://commons.wikimedia.org/w/index.php?curid=1294802.)

plants at that time. However, in this book, in addition to the interesting information about different medicinal plants, for the first time, ferns that were used as anthelmintic are mentioned. Also, in this book, for the first time, the differences between round pepper (*Piper nigrum* L.) and long pepper (*Piper longum* L.) are stated. Ironically, reading this book is not very easy, because the names of the plants mentioned in this book are completely different from their common and modern names, and this makes it difficult to recognize them.

One of the most important books in the field of medicinal plants is *De materia medica* written by Pedanius Dioscorides. He lived in the 1st century AD (AD 40–90) and was a surgeon in Nero's army. During his military trips, he investigated the plants and animals of some regions such as Italy, Greece, Asia Minor, Spain, and France, and collected a significant number of minerals, plants, and animal samples, and at every possible opportunity, he researched the medicinal properties of the collected samples. The mentioned work was edited in five volumes, which contain all the information related to medical knowledge at that time and the description of 600 plant samples (Fig. 1.5).

In this work, an attempt has been made to classify medicinal plants, which, of course, is not based on the letters of the alphabet, but based on the common features of the plants and their similarities. The descriptions are very accurate and presented in a very simple and unambiguous language, and the preparation methods are also stated accurately and correctly. Dioscorides' frequent travels as an army doctor provided him with an opportunity to study and record all the uses and benefits of foreign plants. In fact, almost until the end of the 16th century AD, his writings were considered a reference and standard work in the field of pharmacology, and this shows the extent of his knowledge about the subject under discussion. *De materia medica* was translated into Arabic for the first time in the second half of the 3rd century of Hijri (9th century AD)[1] by Istfan bin Basil and Hunayn ibn Ishaq. The aforementioned translation was published in 1952 in the city of Tétouan, Morocco. Another translation of it was made by Mehran bin Mansour in the 12th century AD. Fortunately, the recent translation with the correction and research of Seyed Mahmoud Tabatabai was published in 2011 by the Tehran University of Medical Sciences publishing house in five volumes and at the same time the Persian translation by his pen was published by the same publisher in five volumes. Centuries later, Dioscorides' works were used by Muslim scholars who influenced the great universities of their day, especially Montpellier, the most famous botanical center in Europe. In the medieval period, *De materia medica* was circulated in Greek, as well as Latin and Arabic translation (Osbaldeston and Wood, 2000).

While being reproduced in manuscript form through the centuries, it was often supplemented with commentary and minor additions from Arabic and Indian sources. Ibn al-Bayṭār's commentary on Dioscorides' *De materia medica*, entitled *Tafsīr Kitāb Diāsqūrīdūs*, has been used by scholars to identify many of the flora mentioned by Dioscorides (Ibn al-Bayṭār, 1989). The importance of Dioscorides' works made many other writers try to expand the topics mentioned by him or to criticize his writings. The topics of the written books are often lexical, but undoubtedly the most complete book in this field is *Commentarii in sex libros Pedacii Dioscoridis Anazarbei de medica materia* written by the Italian scientist Pietro Andrea Mattioli (AD 1501–1577) in the 16th century AD. His book replaced Dioscorides' book as an accepted work. Mattioli criticized and analyzed the statements of Dioscorides and by accurately naming each plant, he tried to solve the inadequacies and ambiguities in his book and to reclassify the plants in it. In his work, he recorded new plants such as sunflower (*Helianthus annuus*) and added a number of foreign plants such as horse chestnut (*Aesculus hippocastanum*) to this collection. Some other Greek and Roman scholars in this field include Scribonius Largus (AD 1–50), Gaius Plinius Secundus or Pliny the Elder (AD 23–79), and Oribasius (AD 320–403). Today, Largus is known for the use of words such as "anodyne" (analgesic) and "epispastic" (blister-opening drug), which

Fig. 1.5. Illustrated page from the Vienna *Dioscorides codex* (*Codex medicus Graecus* of the Austrian National Library; unknown author). (Public Domain, https://commons.wikimedia.org/w/index.php?curid=745191.)

are common in the medical vocabulary. It is said that his activity was only in the field of medicinal botany. Although Oribasius played an effective role in the field of knowledge of medicinal plants, Pliny contributed the most in this field by writing his multi-volume work entitled *Naturalis historia* (*Natural History*) (Fig. 1.6). The work is divided into 37 books, organized into ten volumes. These cover topics including astronomy, mathematics, geography, ethnography, anthropology, human physiology, zoology, botany, agriculture, horticulture, pharmacology, mining, mineralogy, sculpture,

Fig. 1.6. The *Natural History* of Pliny in a mid-12th-century manuscript from the Abbaye de Saint Vincent, Le Mans, France. (By PHGCOM, photographed at Musee de Cluny, Public Domain, https://commons.wikimedia.org/w/index.php?curid=6834497.)

art, and precious stones. In this work, he has carefully described the knowledge and scientific thinking of his time. The result of his work, which was a collection of theories and views of about 400 researchers of that time, today is considered as one of the most important reference books of pharmacology in the ancient world.

Galen or Claudius Galenus (AD 129–216) is a well-known person in the medical community (Fig. 1.7) and certainly his fame as a physician is more than that of Dioscorides. The physician, who came from Pergamum, classified drugs based on a kind of thermal scale and in this way provided a suitable factor for identifying the right drug for each disease. Galen's theory of the physiology of the circulatory system remained unchallenged until 1242, when Ibn al-Nafis wrote his book *Commentary on Anatomy in Avicenna's Canon* (*Sharh Tashrih al-Qanun liʾ Ibn Sinā*), in which he reported his discovery of pulmonary circulation (West, 1985). Galen may have produced more work than any author in antiquity, and may have written as many as 500 treatises (McClellan and Dorn, 2006). In AD 191, a fire in the Temple of Peace destroyed many of his works, in particular treatises on philosophy (Houston, 2003). In the Abbasid period (after AD 750) Arab Muslims began to be interested in Greek scientific and medical texts for the first time, and had some of Galen's texts translated into Arabic, often by Syrian Christian scholars. As a result, some texts of Galen exist only in Arabic translation while others exist only in Medieval Latin translations of the Arabic. Various attempts have been made to classify Galen's vast output. The most complete compendium of Galen's writings is the one compiled and translated by Karl Gottlob Kühn of Leipzig between 1821 and 1833 (Kotrc and Walters, 1979). This collection consists of 122 of Galen's treatises, translated from the original Greek into Latin (the text is presented in both languages).

The history of Islamic medicine

In the history of medicine, Islamic medicine or Arabic medicine refers to medicine developed

Fig. 1.7. Portrait of Galen (an 18th-century engraving by Georg P. Busch). (Public Domain, https://commons.wikimedia.org/w/index.php?curid=40855568.)

in the medieval Islamic civilization and written in Arabic, the lingua franca of the Islamic civilization. Despite these names, a significant number of scientists during this period were not Arab. Some consider the label "Arab-Islamic" as historically inaccurate, arguing that this label does not appreciate the rich diversity of Eastern scholars who have contributed to Islamic science in this era. Latin translations of Arabic medical works had a significant influence on the development of the modern medicine, as did the

Arabic texts chronicling the medical works of earlier cultures.

In the world of Islam, Muslim physicians or other medicinal practitioners found a very congenital atmosphere to conduct medical research and teach medicine, which they inherited from Greece and ancient Persia. During the Abbāsid era between the 8th and 10th centuries AD, great physicians of the medical school of GundeShapur, who were mostly Persians, translated the ancient medical books from Greek, Syriac, Pahlavi Persian, Sanskrit, and other languages into Arabic. They included some famous translators such as John Grammaticus, Aaron Alexandrian, Sergius of Rechina, Issa ibn Sahar Boxt, Sabet ibn Qorrah, Qusta ibn Luqa of Heliopolis, Sahl ibn Shapur, Shapur ibn Sahl, Ibn Serapion, al-Kendy (Nadjmabādi, 1995; Ṣafā, 1997; Daffāʿ, 1998; Ibn Abi Oṣṣaybaʿh, 1998; Sezgin, 2001; Tadjbakhsh, 2003; Tayarani-Najaran *et al.*, 2014), and members of some families such as the Boxtishu (the servant of Christ) family, Massarjuyah family, Hunayn family, Karkhy family, Massuyah family, and Sinān family (Al-Aʿlouchi, 1967; Nadjmabādi, 1995; Ṣafā, 1997; Daffāʿ, 1998; Ibn Abi Oṣṣaybaʿh, 1998; Sezgin, 2001; Tadjbakhsh, 2003; Tayarani-Najaran *et al.*, 2014).

After translation of medical books of the ancient world known as the "Greek medicine", but owing its development also to the ancient Persian medicine, the Muslim physicians scrutinized these works. They found the defects in these books, removed the faults, and perfected them through experience, revision, and practice.

In the meantime, the term "Islamic medicine" has become established for this medicine. Although most of the books are written in Arabic and were therefore long referred to in the West as "Arabic medicine", the term "Islamic medicine" has now become accepted for it, referring to the four founders of this scientific discipline, namely Ali ibn Rabban Ṭabari, Rhazes, Ali ibn Aʿbbās Majussi Ahwāzi Arjāni (Hali Abbās), and Avicenna. Medicine in other Islamic countries also owes its existence and perfection to these well-known physicians. Therefore, Islamic medicine today is also synonymous with Traditional Iranian Medicine (TIM) and Traditional Persian Medicine (TPM). Islamic Spain also made valuable contributions to medicine by producing scientists such as Abu al-Qāssem al-Zahrawi.

The history of Islamic medicine in Iran

Abu al-Hassan Ali ibn Rabban Ṭabari (AD 807–861) was the first great Islamic physician practicing in Persia. His main work is *Firdaws al-Ḥikmah* (*The Paradise of Wisdom*) that contains seven parts, 30 essays, and 360 chapters (Fig. 1.8a). In general this book is a complete collection of medicine and pharmacy (Nadjmabādi, 1995; Daffāʿ, 1998; Ibn Abi Oṣṣaybaʿh, 1998; Sezgin, 2001; Tadjbakhsh 2003). It was published in 1928 in Berlin (Aṭ-Ṭabari, 1928).

Abu Bakr Mohammad ibn Zakariyyāʾ al-Rāzī, also known as Rhazes (AH 251–313/AD 865–925), a renowned Persian physician, philosopher, and chemist (Fig. 1.9), wrote about 250 books and treatises (Nadjmabādi, 1992, 1995; Ṣafā, 1997; Daffāʿ, 1998; Ibn Abi Oṣṣaybaʿh, 1998; Sezgin, 2001; Tadjbakhsh, 2001, 2003).

Al-Ḥāwi (*The Continens*) is well known as Rāzi's most important and most complete book on which he worked for 15 years (Fig. 1.8b). The book was translated into Latin in 1279 by Faraj ibn Salem (Farrgut) and was reprinted five times in Europe between 1488 and 1542. The Arabic text of *al-Ḥāwi* was published in Heydarabad, India in the 7th decade of the 20th century (Rāzi, 1955–1971). Among other famous medicinal books of Rāzi one can mention (Nadjmabādi, 1992, 1995; Ṣafā, 1997; Daffāʿ, 1998; Ibn Abi Oṣṣaybaʿh, 1998; Sezgin, 2001; Tadjbakhsh, 2001, 2003):

- *Man lā Yaḥḍuruhu aṭ-Ṭabib* (*For One in Need of a Physician*), a medical advisory for the general public. Rāzi was probably the first Persian doctor to deliberately write a home medical manual (remedial) directed at the general public. The book covers 36 chapters.
- *Al-Manṣouri* (*Book for al-Mansur*) that contains ten chapters. In *al-Manṣouri*, Rāzi presented a description of the

Fig. 1.8. Six important books of Islamic traditional medicine of Iran before the Mongol attack: (a) *Firdaws al-Ḥikmah*; (b) Al-Ḥāwi; (c) *Hedāyat al-Motaʿallemin fi aṭ-Ṭibb*; (d) *Al-Abniyah aʿn Ḥaqāyeq al-Adwiyah*; (e) *Al-Maliki*; (f) *Zakhirah Khārazmshāhi*. (By Milad Iranshahi.)

identification of tempers, anatomy, hygiene, orthopedics, wounds and sores, and bites, as well as a complete course of therapeutics. This book was translated into several European languages and was published many times.

- *Al-Jodari wa al-Ḥaṣbah* (*The Smallpox and The Measles*), which was the first book on differential diagnosis of the smallpox and the measles. It was reprinted more than 40 times in Europe. *Al-Jodari wa al-Ḥaṣbah* has been translated into Persian by M. Nadjmabādi and published by Tehran University Publications in 1992 (Rāzi, 1992).
- *Al-Morshed* (*The Guide*), which comprises 29 chapters and and adapts one of the writings of Hippocrates.

Some other of his medical books are aṭ-Ṭibb al-Mlouki (The Royal Medicine), Bur al-Sāʿah (Medical Emergencies), al-Taqseem wa al-Tashjir (Divisions and the Branches), al-Qarābādin al-Kabir (The Great Pharmacopoeia), and al-Shukuk al'a Jālinus (Doubts about some opinions of Galen). Rhazes was the most important specialist in clinical and practical medicine in the Islamic world.

One of Rhazes' contemporary physicians was Abu Manṣour Ḥassan ibn Nooḥ Qamari Bukhāri (died in AD 997). He had a

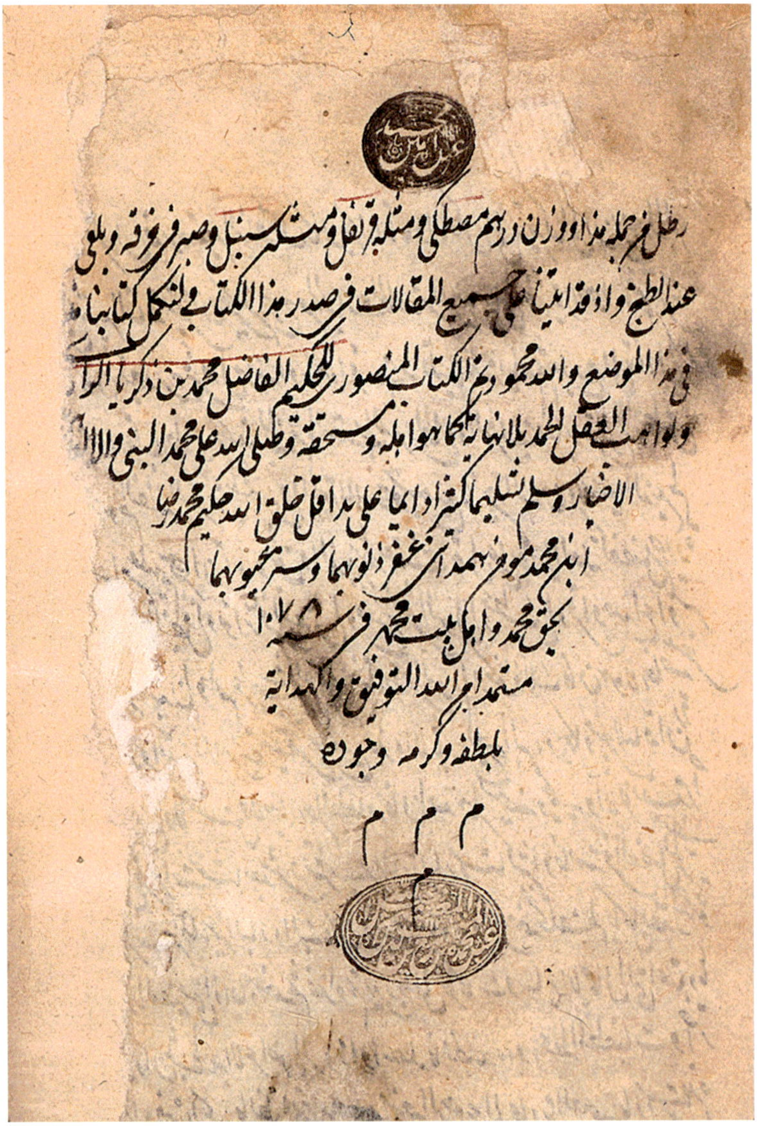

Fig. 1.9. Cover of the famous *Razi's Book of Medicine (Al-Manṣouri)*. (As reproduced in the frontispiece to McCarthy's *Freedom and Fulfillment* (Boston: Twayne, 1980) by Abu Bakr al-Razi – transferred from http://en.wikipedia.org/wiki/Image:Colophon-Razi's_Book_of_medicine_for_Mansur.jpg, Public Domain, https://commons.wikimedia.org/w/index.php?curid=673628.)

prominent position to the Samanid kings. His books include: *Ghenā wa Menā* (*Wealth and Wishes*), *E^clat al-E^clal* (*The Cause of Diseases*), *Moālejat Manṣouri* (*Manṣouri's Treatments*), *Maqālahh fi al-Boḥran* (*An Article on Crisis*), *Resālah fi A^clāj Amrāḍ al-Ṣadr* (*A Treatise on Curing Chest Diseases*), *Resālah fi Ḥommiyāt al-Dāyereh* (*A Treatise on Intermittent Fevers*), *Maqālah fi al-Maraz al-Estesqsā* (*A Treatise on Dropsy*), and *al-Tanwir* (*Enlightening*). He wrote the latter book in Arabic and Persian (Nadjmabādi, 1995; Daffā^c, 1998; Ibn Abi Oṣṣayba^ch, 1998; Sezgin, 2001).

One of the renowned physicians who was the student of Abu al-Qāssem Moqāne^ci (a Rhazes student) is Abu Bakr Rab^ci ibn

Ahmad Akhaweyni Bukhari (died in AD 983). Akhaweyni dedicated his whole lifetime to medicine (Nadjmabād, 1995; Tadjbakhsh, 2001, 2003). He recorded his medicinal attempts in *Hedāyat al-Mota'allemin fi aṭ-Ṭibb* (*An Educational Guide for Medicinal Students*). The book was written in an eloquent Persian language and contains three parts. The first part comprises 51 chapters on elements, tempers, humors, simple and compound organs and also descriptions on functions, souls, foods and drinks, physical movement, rest, sleep, and others. The second part, written in 130 chapters, applies pathology cap-a-pie. In the third part, containing 19 chapters (Fig. 1.8c), Akhaweyni introduces various types of fevers and pulses. This book was published by Ferdowsi University of Mashhad, Iran in 1965 (Akhaweyni Bukhāri, 1965).

Contemporary with Akhaweyni was Moafaq al-Din Abu Manṣour Ali Herawi, who was the author of *al-Abniyah a'n Ḥaqāyeq al-Adwiyah* (*Basics of the Realities of Drugs*). It is the most ancient book written in Persian language (Nadjmabādi, 1995; Tadjbakhsh 2001, 2003) and contains the Persian names of simple drugs listing the individual properties in alphabetic order. In addition, a total of 584 mineral, plant, and animal drugs are described in more detail (Fig. 1.8d). A manuscript of this book, written by Assadi Ṭussi, is available in Vienna National Library. The book was later published by Tehran University (Al-Herawi, 1992).

The most noted Muslim physician after Rhazes was Ali ibn A'bbās Majussi Ahwāzi Arjāni, known as Haly Abbās (AD 930–994). He is the author of the valuable book *Kāmil al-Ṣinā'ah aṭ-Ṭibbiyyah* (*Complete Book of the Medical Art*) or *al-Maliki* (Nadjmabādi, 1995; Daffā', 1998; Ibn Abi Oṣṣayba'h, 1998; Sezgin, 2001; Tadjbakhsh, 2001, 2003).

The book *al-Maliki* is divided into two parts. Each part contains ten discourses that cover the complete course of medicine. The first ten deal with the theory of medicine and its divisions, as well as types of tempers, elements, humors, anatomy, physiology, general principles of hygiene, diseases and their divisions, types of pulses, kinds of fevers, symptom of diseases cap-a-pie, and subjects on the period and consequences of diseases. The second ten contain topics on health and hygiene care, introductions to all kinds of therapeutic methods, treatment of different types of fevers, dermatologic ailments, all kinds of bites and poisonings, headaches and psychological diseases, respiratory diseases, heart diseases, gastrointestinal diseases, genitourinary diseases, a complete course on surgery and orthopedics, and finally a course on pharmacology and pharmaceutics (Fig. 1.8e). The Latin translation has been published three times in Europe and the Arabic text was printed in Bulāq, Egypt (Ahwāzi Arjāni, 1877).

Shaykh al-Ra'is (The Chief Principal) Abu Ali Ḥussain ibn Abdullah ibn Sinā, known as Avicenna (AD 980–1037), was the most prestigious scholar of Persia and the Islamic world.

Some of Avicenna's contemporary physicians were (Nadjmabādi, 1995; Daffā', 1998; Ibn Abi Oṣṣayba'h, 1998; Sezgin, 2001; Tadjbakhsh, 2001, 2003):

1. Abu al-Khayr Ḥassan ibn Sowār, known as Ibn Khammār (AD 943–1049), who wrote books such as *al-Ḥawāṣel fi al-Ṭibb* (*The Consequences in Medicine*) and *Imteḥān al-Aṭebā* (*Trial of Physicians*).

2. Abu Ali Ahmad ibn Yaqub ibn Moskuyeh (AD 937–1030), who wrote the books *al-James* (*The Comprehensive*) and *al-Adwiyah al-Mofradah* (*Simple Drugs*).

3. Abu Rayḥān Mohammad ibn Ahmad al-Biruni (AD 973–1048), the author of *aṣ-Ṣaydanah fi al-Ṭibb* (*Book of Pharmacy*) where he presents 799 simple mineral, herbal, and animal drugs. There exists an ancient Persian translation of this book, which has been published. Of the two texts of *aṣ-Ṣaydana*, one corrected by the late A'bbās Zaryāb Khooei has been printed in Iran (Biruni, 1991) and the other by the late Hakim Muhammad Sa'eid and Ranā Ehsānallahi was printed in Pakistan (Biruni, 1973).

4. Abu al-Qāssem Abd al-Raḥmān ibn Ali ibn Ahmad ibn Abi Ṣādeq (AD 995–1078) is another physician contemporary with Avicenna, being his student and the master of Jorjāni. Ibn Abi Ṣādeq wrote *Sharḥ Foṣul al-Aboqrāt* (*An Exposition to the Hippocrates' Articles*), *Sharḥ Masāel Ḥonain* (*An Explanation to Ḥunain's Problems*), *Sharḥ Taqdameh* *al Ma^crefah al-Boqrāt* (*An Explanation to Hippocrates' Prognosis*), and *al-Adwiyah al-Qalbiyah* (*Cardiac Medicines*).

The most important celebrated physician after Avicenna was Seyyed Ismāʿil Jorjāni (AD 1041–1136). Jorjāni wrote valuable books on medicine during his lifetime (Nadjmabādi, 1995; Daffāʿ, 1998; Sezgin, 2001; Tadjbakhsh, 2001, 2003) but his most significant book, which is considered to be the most detailed medical book in Persia, is *Zakhirah Khārazmshāhi* (*The Treasure of Khārazmshāh*). This book contains nine main books and two appendices on simple and compound drugs (Fig. 1.8f): the first book of *Zakhirah* is about medical science, identifying types of different humors and temperaments, and also general aspects of anatomy; the second book discusses health and diseases and also types of pulses, sweating, urine, and feces; the third book includes a complete course on maintaining health; the fourth book is allocated to ways of diagnosing diseases and disease duration; the fifth book is on identifying different types of fevers and methods to cure them; the sixth book is assigned to methods of curing diseases cap-a-pie; the seventh book describes types of inflammation, wounds, and fractures, and the ways to treat them; the eighth book describes various cosmetic applications; the ninth book is assigned to various types of poisons, antidotes, bites, Poisonings, and their individual treatments; and the final section explains simple and compound drugs in detail.

In fact, the mentioned book is an encyclopedia, fraught with pure Persian medicopharmaceutical terms. The complete text of *Zakhirah* was photoprinted in 1977 by the Iranian Culture Foundation (Jorjāni, 1977). Due to its importance, the book was also translated into Hebrew and Turkish.

Jorjāni wrote a summary of the *Zakhirah* named *al-Aghraḍ aṭ-Ṭibbiah wa al-Mabāhethi al-Aʿlāiah* (*Medical Goals and Aʿlāiyeh's Discussions*). This book provides an introduction to medicine, deals with the treatment of various diseases, and finally describes simple and compound medicines. A photoprint of *al-Aghrāḍ aṭ-Ṭibbiah* was published in 1966 by the Iranian Cultural Foundation and fortunately edited by H. Tadjbakhsh and published by Tehran University Press (Jorjāni, 2005–2006).

The third book of Jorjāni in medicine is named *Khofi Aʿlāii* (*A very compact book that fits in Ala ad-Din Atsiz's boot*), which is an abbreviated medical text consisting of two parts. The first part includes the theoretical aspects of medicine and the second provides scientific medical knowledge comprising seven articles. The book was lithographed in Kanpur, India in 1891. It was also published with valuable footnotes and descriptions (Jorjāni, 1998).

The fourth book of Jorjāni, which is called *Yādegār* (*The Memorial*), is an extract text and codified in five parts. *Yādegār* has been edited by M. Mohaghegh and published by the Institute of Islamic Studies in Tehran (Jorjāni, 2003).

One of Jorjāni's books is *Zobdat aṭ-Ṭibb* (*Selected Topics in Medicine*). This book was written in Arabic, and its context was ordered in numerous tables.

Other famous physicians of the Islamic civilization before the Mongol invasion will not be discussed further here, but at least their names and works will be briefly mentioned (Nadjmabādi, 1995; Daffāʿ, 1998; Ibn Abi Oṣṣaybaʿh, 1998; Sezgin, 2001; Tadjbakhsh, 2001, 2003):

1. Abu al-Ḥassan Ahmad ibn Mohammad Ṭabari (AD 932–995), the author of *al-Muʿālejāt al-Boqrātiyah* (*Hippocratic Treatments*).
2. Ḥakim Maysari (died in AD 981), the composer of *Dāneshnāmeh Dar Pezeshki* (*The Handbook of Medicine*), which is a versified medical textbook in Persian, published in Tehran (Ḥakim Maysari, 2015).
3. Abu Ali Ahmad ibn Abd al-Raḥmān ibn Manduyeh Iṣfaḥāni (died in AD 1020), the author of books such as *Nahāyat al-Ekhtesār* (*The Utmost of Conciseness*) and *al-Kāfi* (*The Sufficient*).
4. Abusahl Maṣiḥi (AD 971–1012), author of the book *al-Meah fi aṣ-Ṣenaʿāh aṭ-Ṭibbiyah* (*The Hundred Chapters in Medicine*).
5. Abu al-Faraj Ali ibn Ḥussein ibn Hendowi Qomi Rāzi (died in AD 1029), author of *Meftāḥ aṭ-Ṭibb* (*A Key to the Medicine*).

6. Sharaf al-Din Mohammad ibn Yussef Ilāqi (killed in AD 1041), writer of the book *al-Foṣul al-Ilāqyah* (*Ilāqyian Articles*).
7. Abu al-Ḥassan Ali ibn Ahmad ibn Hobal Tabrizi (AD 1121–1213), composer of the book *al-Mokhtār fi aṭ-Ṭibb* (*The Selected in Medicine*) that has been published in Ḥeyderabad, India, in four volumes.
8. Badr al-Din Mohammad ibn Bahram Qalānessi Samarqandi (late 12th to early 13th century AD), author of the book *al-Aqrābādin* (*The Pharmacopoeia*) in 49 chapters, which is a very complete work on various compound drugs.

The Mongol war began in AD 1219 with the attack of Changiz to Utrar, then spread to the Mavāra an-Nahr (Transoxiana) and Khorasan, Iraq, and Ajam Iraq (part of Persia comprising Tehran, Isfahan, and Arak). It caused the destruction to many cities, the murder of many inhabitants, and the colapse of many Islamic civilizations including ancient Persia. This cruel attack also resulted in a destructive shock to the scientific development of medicine.

One of the famous physicians after the Mongol attack was Abu Hāmed Mohammad ibn Ali Samarqandi (died in AD 1221). He was the author of the books *al-Asbāb wa al-Aʿlāmāt* (*Etiology and Symptomology*) and *al-Faraq Bayn al-Amrāḍ al-Moshkelah* (*Differences Between Difficult Diseases*) (Daffāʿ, 1998; Tadjbakhsh, 2001). Some of the great physicians from the post-Mongol era are mentioned below:

1. Qotb al-Din Mahmud ibn Masuʿd Kazeruni Shirāzi (AD 1236–1311), one of the greatest sage physicians who was the author of the book *al-Toḥfat al-Saʿdiah* (*An Offering to Saʿd*), which is the commentary on the *Canon*.
2. Shams al-Din Mohammad ibn Mahmud Āmoli (died in AD 1352) was a great scientist of his time. He wrote the books *Nafāis al-Fonoun* (*The Precious Arts*) and *Sharh Kolyāt Qānon ibn Sinā* (*Commentary on Generalities of the Canon*).
3. Mahmud ibn Oʿmar Chaghmini Khārazmi (died in AD 1344) is the author of *Qānonchah* (*The Small Canon*) (Fig. 1.10a) (Chaghmini Khārazmi, 1422).

Some of the great physicians who emerged in Persia during the Timurid era are discussed below:

1. Ali ibn Ḥussein Ansāri, known as Zeyn al-Attār (died AD 1403), has left his book *Ekhtiyārāt Badiʿi* (*Badiʿi Selections*) (Fig. 1.10b) that contains two articles (Ansāri, 1992).
2. One of the famous physicians in that time is Yusuf ibn Mohammad Herawi, also known as Yusufi, who lived in the court of Ẓaheyr al-Din Bāber, and his son, Homāyoun. He impressed several treatises in medicine. An eight-treatise collection of Yusufi, namely *Ṭibb Yusufi* (*Yusufi's Medicine*), has been published in Lahore, India.
3. Another famous physician of this age is Mohammad ibn Yusuf Herawi who wrote the worthwhile book entitled *Baḥr al-Javāher* (*Sea of Gems*), which is a dictionary of medical words and was published in the Qajarid era.
4. Another great physician of this period of time is Bahā al-Dowlah Seyyed Mohammad ibn Qāssem Nourbakhsh (AD 1455–1509). He is the author of *Kholasata al-Tajāreb* (*Summary of Experiences*) in Persian, which includes 28 chapters and subjects such as general health, children and elderly health, cosmetics, types of different fevers, simpleton, leprosy, frosting, contusion, poisoning, a course of disease treatment cap-a-pie, compound drugs, and finally weights usually used in pharmacology.

Some of the great physicians who emerged during the Safavid era are the following:

1. Ḥakim Moẓaffar ibn Mohammad Ḥussein Shafāee Iṣfahāni (died in AD 1555) is one of the famous physicians in the reign of Safavid. His famous book entitled *Ṭibb Shafāee* (*Shafāee's Medicine*) was translated by Angelus de St. Joseph (French) to Latin and was reprinted in the Netherlands with the new title *Pharmacopeia Persica*.
2. Sharaf ad-Din Ḥassan ibn Ḥakim Molāye Iṣfahāni (AD 1558–1627), famed as Hakim Shafāii, the author of *Qarābādin Shafāee* (*Shafāee's Pharmacopoeia*).
3. Mir Mohammad Akbar, known as Shāh Arzāni, who lived in the time of Urang Zib in India. He left the books *Ṭibb Akbari* (*Akbari's*

Fig. 1.10. Four important books of Islamic traditional medicine of Iran after the Mongol attack: (a) *Qānonchah*; (b) *Ekhtiyārāt Badiʿi*; (c) *Toḥfata al-Momenin*; (d) *Makhzan al-Adwiyah*. (By Milad Iranshahi.)

Medicine), *Mizān aṭ-Ṭibb* (*The Scale of Medicine*), and *Qarābādin Qāderi* (*Qāderi's Pharmacopeia*), which were all written in Persian language.

4. Seyyed Mohammad Momen ibn Mohammad Zamān Ḥusseini Tunekāboni was a famous physician who lived during the Safavid reign. He was the chief physician of Shah Suleymān Safavi and the author of *Toḥfata al-Momenin known as Toḥfeḥ Ḥakim Momen* (*Ḥakim Mo'men's Gift*). This book is written in Persian and contains five diagnoses and two orders (Fig. 1.10c). The first diagnosis concerns the explanation of different medical views and the designation of the degrees of the medicines; the second diagnosis refers to the explanation of names and expressions; the third and most detailed part of the book contains monographs of drugs in alphabetical order; the fourth diagnosis discusses the treatment of poisoning; and the fifth diagnosis deals with weights. The first order offers methods for preparation of simple drugs and the second is about preparation of compound drugs (Tunekāboni, 2007). *Toḥfeh Ḥakim Momen*, has been published in Iran and India several times.

Some of the great physicians who emerged in Persia in the Afsharid age are:

1. Ḥakim Mirzā Mohammad Hāshem ibn Mohammad Hādi Aʿlawi Shirāzi known as Aʿlawi Khān (died in AD 1695) lived in the Afsharid era. He wrote the books *al-Qarābādin Kabir* (*The Grand Pharmacopeia*) and *Majmaʿ al-Jawāmeʿ* (*Union of Comprehensives*) written in Persian language and published by Ḥakim Aʿqili Aʿlawi Khorāsāni Shirāzi (Wāseṭi, 1988).

2. Ḥakim Seyed Mohammad Ḥussein ibn Mohammad Hādi Aʿqili Aʿlawi Khorāsāni Shirāzi was one of the renowned physicians in the reign of the Afsharids. He was the author of *Makhzan al-Adwiyah* (*Drug Treasure*), which is a comprehensive book describing several simple drugs (Fig. 1.10d). *Makhzan al-Adwiyah* discusses predecessors' experiences about types of natural drugs. The book is ordered alphabetically. It was published numerous times in Iran and India (Wāseṭi, 1988; Aʿqili Aʿlawi Khorāsāni Shirāzi, 2014). He also wrote another book focused on general medicine entitled *Kholāsat al-Ḥikmah* (*Essence of Wisdom*).

We can consider Ḥakim Aʿqili as the last Islamic physician, although a few physicians like Mirzā Ahmad Tonekāboni, Mirzā Bābā Tabib Shirazi, and others emerged after him, but at this time Persia established cultural connection with Europeans and educated people and officials were influenced and intimidated by Western civilization and new European medicine. The ancient Islamic medicine was then neglected in Persia and also in other Islamic countries. Finally, European medicine gradually pushed aside the TPM. Fortunately, the body of Islamic medicine managed to continue on the Indian subcontinent. Some of the compiled books that verify the permanency of Islamic medicine on the Indian subcontinent are: *Moḥiṭ Aʿẓam* (*The Grand Ocean*) and *Excir Aʿẓam* (*The Grand Elixir*) written by Ḥakim Mohammad Aʿẓam Khān (Ḥakim Mohammad Aʿẓam Khān, 2019); *Tadhkarata al-Hind* (*A Collection of Indian Drugs*) written by Ḥakim Reḍa Ali Khān Dakani (Wāseṭi, 1988); and also *Kanz al-Adwiyah* (*The Treasure of Drugs*) written by Ḥakim Mohammad Najm al-Ghani Khān (Wāseṭi, 1988). Finally, the founding of Hamdard Great Institute in the Indian subcontinent is an example of the permanence of Islamic medicine in India.

Avicenna

Shaykh al-Ra'is (The Chief Principal) Abu Ali Ḥussein bin Aʿbdullah bin Ḥasan bin Ali bin Sinā (Avicenna) (AD 980–1037) has a very high position in medicine and philosophy. This claim derives primarily from his following books: *ash-Shifā* (*The Book of Healing*), *al-Ishārāt wa al-Tanbihāt* (*Remarks and Admonitions*), *an-Naajāt* (*Book of Salvation*), *Uʿyun al-Ḥikmah* (*Principles of Wisdom*), and *Dāneshnāmeh-e-Aʿlāii* (*Aʿlāii's Encyclopaedia*).

Biography

Avicenna was born in AD 980. His father Aʿbdullah was from Balkh and during the reign of Noah bin Manṣour (Noah II), one of the Samanid kings, he moved from Balkh, became a civil servant in a village named Khormaithan, and married a girl named Setarah from a nearby village named Afshnah. Some time later, his father moved to Bukhara where he left his son to tutors to learn the Quran and literature. Ibn Sinā learned jurisprudence and polemics from a teacher named Ismāil Zāhid and logic and geometry from Abu Ābdullah Nāteli. The young Ibn Sinā soon surpassed his masters in these techniques. After Nāteli left Bukhara, Ibn Sinā began to research and study in the divine and natural sciences; shortly after, the desire to learn medicine appeared in him, and he carefully read what previous doctors had written and soon achieved great success in this field. He achieved so much that the great scientists of medicine turned to him and studied under him. Ibn Sinā, while engaged in medicine and treatment of patients, did not neglect other sciences and spent one-and-a-half years re-studying logic and philosophy. After that, he turned to theology and started studying the book *Metaphysics* by Aristotle, but despite reading this book 40 times, he could not understand the author's intention until he accidentally came across the book of *The Aims of Aristotle's Metaphysics* written by Abu Naṣr Fārābi, and the difficulties of the mentioned text became easy. At that time, he had just turned 17 years old. When Ibn Sinā was 18 years old, Noah II fell seriously ill and the doctors were unable to treat him. Ibn Sinā treated Noah II and was allowed to study in his precious library in return for this service. After some time, Ibn Sinā's father passed away, and times changed his life. He went from Bukhara to Gorgānj, Khārazm. He lived in this country for some time, becoming close to the ruler, and he managed to write several books in this city. After that he went to Ray and entered the service of Majd ad-Dawlah, one of the rulers of the Buyid dynasty who was suffering from a mental illness, and he cured him; and from there he went to Qazvin and then to Hamedān. He stayed in this city for a long time, and it was in this city that he entered the service of Shams ad-Dawlah Daylami. At the same time, he wrote the *Al-Qānun fi aṭ-Ṭibb* (*The Canon of Medicine*) and started the compilation of the great book *Ash-Shifā* (*Book of Healing*) at the request of Abu Oʿbaid Bozjāni. When Shams ad-Dawlah passed away and his son succeeded him, Ibn Sinā did not accept his ministry and soon after he was accused of having correspondence with the ruler of Iṣfahān and because of this he was imprisoned. He spent 4 months in prison and wrote three books there. After being released from prison, Ibn Sinā was in Hamedān for some time, then he secretly left Hamedān and went to Iṣfahān. Aʿlā ad-Dawlah, the ruler of Iṣfahān, received him warmly and respected him very much. He became his companion. In this city, the master completed the *Ash-Shifā*, and in 1038, during a trip to Hamedān with Aʿlā ad-Dawlah, he fell ill and died.

Principal works

Many works of Ibn Sinā have been left behind or could be attributed to him; a comprehensive list of his oeuvre can be found below. This list includes 131 original scripts written by Ibn Sinā and 111 works attributed to him. Unfortunately, some of his books have been lost, but the names of some of the most important works left by him are as follows (Mahdavi, 1954).

Philosophy

- *Ash-Shifā* (*Book of Healing*): This book is the most detailed philosophical work of Ibn Sinā that has survived so far (Fig. 1.11a). *Ash-Shifā* is the most important and comprehensive work of Ibn Sinā and represents an encyclopedia in Arabic language in which humanities, summaries of opinions of eminent philosophers of ancient Greece, and commentators of the School of Alexandria, Platonists, and Neoplatonists were expressed and subjected to scientific analysis and criticism. This book consists of four sections: logic (nine subjects, 38 articles, and 248 chapters), natural sciences (eight subjects, 35 articles, and 206 chapters), mathematics (four subjects: geometry consisting of 15 articles; astronomy consisting of

Fig. 1.11. Four important books of Avicenna: (a) *Ash-Shifā*; (b) *An-Naajāt*; (c) *Al-Ishārāt wa al-Tanbihāt*; (d) *Al-Qānun fi aṭ-Ṭibb*. (By Milad Iranshahi.)

14 articles and 79 chapters; arithmetic consisting of four articles; and music consisting of six articles and 16 chapters), and theology (one subject, ten articles, and 59 chapters) (Ibn Sinā, 1984). According to Jawzjāni's report, in Hamedān Avicenna prepared the list of the main titles of *ash-Shifā* materials in two days and without referring to any book, and then devoted himself to the work of writing it, until all the natural sciences (except the books of animals and plants) and theology and part of the logic were written; then he went from Hamedān to Iṣfahān with several people, including his disciple Jawzjāni, where he devoted himself to completing the *ash-Shifā* book and wrote the sections on logic and mathematics; on the way to Sabon, he finally wrote the rest of *ash-Shifā*, namely the books on plants and animals as well as *an-Naajāt*.

- *An-Naajāt* (*Book of Salvation*): This book is a summary of *ash-Shifā*, which was written according to the common tradition of summary writing in that era (Fig. 1.11b). This book also includes four types of logic (149 chapters), natural sciences (six articles and 47 chapters), mathematics (four subjects: geometry consisting of ten chapters; arithmetic; almagest (in astronomy) consisting of nine chapters; and music) and theology (two articles and 58 chapters) (Ibn Sinā, 1985).
- *Al-Ishārāt wa al-Tanbihāt* (*Remarks and Admonitions*): This book, written at the end of Ibn Sinā's life, is his most mature and fruitful book in the field of philosophy and includes his last views (Fig. 1.11c). This book also has four sections: logic, natural sciences, theology, and mysticism, which are placed instead of mathematics (Ibn Sinā, 1982).
- *Al-Mabada* and *Al-Maᶜād* (*Origin and Resurrection*): In this book, in three articles, Ibn Sinā talks about *Wājib al-Wujud* (God) and His attributes, the order and nature of the grace of existence from the first to the last (and lowest) being, and finally the immortality of the soul, its happiness, and misery in the afterlife, and the topics he considers to be the fruit of two metaphysical and natural sciences.
- *At-Taᶜliqāt* (*Commentaries*): As the name suggests, this book contains Ibn Sinā's commentaries on the philosophical works of earlier philosophers and on his own works, and sometimes they are also answers to the questions that were put to him. Logical, natural, theological (= metaphysical), and mathematical topics are discussed in this book beside other various topics.
- *Al-Mobāḥethāt* (*Discussions*): This book is also the result of written dialogues between Ibn Sinā and some of his students (especially Bahmanyār and Abu Manṣour bin Zailah), which, like *at-Taᶜliqāt*, covers various topics.
- *Uᶜyun al-Ḥikmah* (*Principles of Wisdom*): This book presents the main topics of logic, natural sciences, and theology in a compact and concise manner.
- *Dāneshnāmeh-e-Aᶜlāii* (*Aᶜlāii's Encyclopaedia*): Ibn Sinā wrote this encyclopedia at the request of and for Aᶜlā ad-Dawlah Kākuyah. This book is apparently the first philosophical book written in Persian language after the establishment of Islamic philosophy. The encyclopedia includes four sections: logic, natural sciences, mathematics, and theology.
- *Al-Enṣāf* (*The Fairness*): According to Ibn Sinā, this book contains nearly 28,000 problems. The manuscript was destroyed during Masoᶜud Ghaznavi's attack on Iṣfahān. Ibn Sinā wanted to rewrite it, but apparently he did not succeed in doing so. Parts of this book were found by Aᶜbd al-Raḥmān al-Badawi and included in the book of *Arasṭu end al-Aᶜrab* (*Aristotle from the Perspective of the Arabs*).
- *Risalah fi Aqsam al-Uᶜloom al-Aᶜqliyyah* (*Treatise on the Types of Philosophical Sciences*): In this treatise, Ibn Sinā divides philosophy into theoretical and practical, then divides each of them into types, defines them concisely, and points to their topics, issues, and branches.

- *Risalah al-AḍḤāwiyyah fi Amr Al-Maʿād* (*Treatise al-AḍḤāwiyyah about the Truth of the Resurrection*): Avicenna wrote this treatise for a person named Abu Bakr bin Muḥammad, nicknamed al-Sheikh al-Amin, and in it, about the truth of the resurrection, various opinions about it, invalidation of false opinions, the truth of man (i.e. the soul) and its survival, he discussed the necessity of resurrection and the hereafter. In this treatise, Ibn Sinā considered physical resurrection impossible and interpreted individual Quranic verses that indicate physical resurrection.
- *Risalah al-Ḥudood* (*A Treatise on Limits*): In this treatise, he defined the concepts and philosophical terms.
- The Story of *Ḥayy ibn Yaqẓān*: This work is a mystical treatise by Ibn Sinā, and a Persian translation and commentary has been found on it, attributed to his student Jawzjāni.
- *Manteq al-Mashrqeen* (*Logic of the Orientalists*): This book is the logic section of *Ḥikmahqtol al-Mashrqeen* (*Philosophy of the Orientalists*) and in the introduction Ibn Sinā says that he wishes to include in it the most important contents of true knowledge, which only those who are very intellectual and opinionated and who benefit from correct guesses can find it. Ibn Sinā says that he wrote this book only for those who are like him, and for the people of philosophy, he considered *Kitāb ash-Shifā* to be sufficient, and even beyond their needs. In this way, it is clear that Ibn Sinā wanted to present a special school of thought in this book, other than what he had presented in *ash-Shifā*; but unfortunately, except for the logic part, nothing of it has reached us, and this amount is not much different from the logic of *ash-Shifā*.

Medicine

Avicenna also wrote 61 books and treatises on medical science, but it is generally agreed that his book *Al-Qānun fi aṭ-Ṭibb* (*The Canon of Medicine*) is Avicenna's masterpiece (Fig. 1.11d). It represents a source book of medicine in the Eastern and Western worlds. The *Canon* contains five main books each divided into some arts, tuitions, sentences, and chapters:

- The first book of the *Canon* discusses the concept of medicine, particularly the medicine extent and its subjects, and also topics around humors, tempers, elements, organs, spirits, functions, and powers. Themes on diseases and their etiology, hygiene, and finally general guides to treatment are also mentioned.
- The second book is assigned to simple drugs and includes about 800 mineral, herbal, and animal-based medicinal materials. The drugs are ordered alphabetically (Abjad), and in each monograph, the manner, characteristics, the best type (quality) of drugs, nature, application, properties, and indication are mentioned.
- The third book of the *Canon* diseases are treated in 22 arts. Each art comprises several articles. In fact, this part acts as a complete review of pathology.
- The fourth book offers ways to cure common diseases such as fevers and edema, and it also includes orthopedics, toxicology, and cosmetic and hygienic products.
- The fifth and final book that is allocated to compounded drugs is called *Qarābādin* (*The Pharmacopoeia*) and represents properties and recipes to make all kinds of pills, mixtures, powders, syrups, suppositories, tablets, and others (Ibn Sinā, 1981–1997).

There have been numerous expositions of the whole *Canon* or its parts and it has been summarized many times. The book has been translated into European, Hebrew, and Persian languages and it has been reprinted frequently.

- *Al-Adwiyah al-Qalbiyah* (*Cardiac Drugs*) includes an introduction and 16 chapters.
- *Al-Orjozah fi aṭ-Ṭibb (A Poetical Treatise in Medicine)* has 1338 lines of poetry.

- *Al-Orjozah fi Tadbir aṣ-Ṣeḥah fi al-Foṣoul al-Arbaʿah* (A Poetical Treatise in Health Measures in Four Seasons) describes various health measures during the four seasons in 1338 lines of poetry.
- *An-Nabḍ* (*The Pulse*) consists of nine chapters written in Persian about various topics related to blood vessels.

Students of Avicenna

Ibn Sinā trained knowledgeable and capable students such as Abu Uʿbaid Jawzjāni, Abul Ḥasan Bahmanyār, Abu Manṣour bin Zailah Iṣfahāni, and Abu Aʿbdullah Muhammad bin Ahmad al-Maʿṣoumi, the most prominent of whom are listed below (Moṭṭahari, 2020):

1. Abul Ḥasan Bahmanyār bin Marzbān is a very famous student of Ibn Sinā who, among other works, wrote the book of *at-Taḥṣil* (*The Obtain*). He was descended from Zoroastrians of Azerbaijan, but later he became a Muslim. One of Ibn Sinā's books called *al-Mabāḥeth* (*The Topics*) contains the master's answers to his questions.
2. Abu Uʿbaidullah Aʿbdulwāḥid bin Muhammad Jawzjāni was in his service continuously from AD 1013 until the death of Ibn Sinā, and after his death he collected and compiled his works. During Avicenna's ministry in the service of Shams ad-Dawlah, he asked Bu Ali to explain Aristotle's books. Jawzjāni has added the mathematics part of *an-Naajāt* and the mathematics and music parts of *Dāneshnāmeh-e-Aʿlāii* in his own style. His death has been considered 10 years after the death of Ibn Sinā in AD 1047.
3. Abu Aʿbdullah Muhammad bin Ahmad al-Maʿṣoumi is another famous student of Ibn Sinā, who wrote *Risālah al-Eʿshq* (*A Treatise on Love*) in his name. When the scientific dispute between Avicenna and Abu Riḥān Biruni was interrupted due to some of Abu Riḥān's words, al-Maʿṣoumi took over the continuation of the debate with Abu Riḥān. Al-Maʿṣoumi's material along with Ibn Sinā's answers and Biruni's questions and answers have been published in the book of *al-Asalah wa al-Ajwabah* (*Questions and Answers*) in Iran and Lebanon.
4. Sheikh Ali Nasāi Khorasāni is another student of Ibn Sinā whom Nāṣer Khosrow mentions in his travelogue.
5. Abu Manṣour Ḥussein bin Ṭahir bin Zailah Iṣfahāni was one of the Persian scholars of the 10th and 11th centuries and one of Ibn Sinā's students. According to the knowledge of his time, he was particularly skilled in mathematics. He died in AD 1049.

Acknowledgment

I would like to thank Professor Dr. H. Schulz and Dr. F. Najafi for their careful proofreading and scholarly editing of this chapter. I am also grateful to my dear colleague Dr. M. Iranshahi for providing beautiful and attractive photos of the important books of Islamic traditional medicine of Iran.

Note

[1] In the following text, the year numbers mostly based on the Islamic calendar (anno Hegirae, AH) have been converted to the Gregorian calendar (anno domini, AD) for consistency.

References

Ahwāzi Arjāni, A.A. (1877) *Kāmil al-Ṣināʿah aṭ-Ṭibbiyyah (The Perfect Art of the Medicine)*, ed. I. al-Dassuqi, 2 vols. Saʿādat Press, Bulāq, Cairo (in Arabic).
Akhaweyni Bukhāri, R.A. (1965) *Hedāyat al-Motaʿallemin fi aṭ-Ṭibb (An Educational Guide for Medical Students)*, ed. J. Matini. Ferdowsi University Press, Mashhad, Iran (in Persian).
Al-Aʿlouchi, A.H. (1967) *History of Medicine in Iraq*. Asaad Press, Baghdad, pp. 16–27 (in Arabic).

Al-Herawi, A.R. (1992) *Al-Abniyah aʿn Ḥaqāyeq al-Adwiyah* (*Basics of Realities on Drugs*), ed. A. Bahmanyār. Tehran University Publications, Tehran (in Persian).
Ansāri, A.Ḥ. (1992) *Ekhtiyārāt Badiʿi* (*Badiʿi Selections*), ed. M.T. Mir. Pakhshe Rāzi Private Joint Stock Co., Tehran (in Persian).
Aʿqili Aʿlawi Khorāsāni Shirāzi, S.M.H. (2014) *Makhzan al-Adwiyah* (*Drug Treasure*), ed. M.R. Shams Ardakani, R. Rahimi and F. Farjadmand. Sabz Ārang Publisher, Tehran (in Persian).
Aṭ-Ṭabari, M.A. (1928) *Firdaws al-Ḥikmah* (*Paradise of Wisdom*), ed. M.Z. Ṣiddiqi. The E.G.W. Gibb Memorial Trust, Buch-und Kunstdruckerei Sonne GmbH, Berlin (in Arabic).
Badiʿi, R. (1999) *Detailed Geography of Iran*, Vol. 1. Iqbāl Publisher, Tehran, p. 19 (in Persian).
Balaʿmi, M. (2016) *Tarikh Balaʿmi* (*History of Balʿami*), 1st edn, ed. M.T. Bahār and M. Parvin Gonābādi. Hermes Publishing House, Tehran, p. 606 (in Persian).
Biruni, M.A. (1973) *Kitāb al-Ṣaydanah*, ed. Saʿīd, H.M. Hamdard. National Foundation, Karachi.
Biruni, M.A. (1991) *Aṣ-Ṣaydanah* (*Pharmacy and Materia Medica*), ed. A. Zaryāb Khooei. Iran University Press, Tehran (in Arabic).
Chaghmini Khārazmi, M.Oʿ. (1422) *Qānonchah* (*The Small Canon*), ed. I. Nāẓem. Almaeʿe Publisher, Tehran (in Arabic).
Daffāʿ, A.A. (1998) *Leaders of Medicine in Islamic Civilization*. al-Risālah Publishers, Beirut, pp. 164–171, 175–178, 222–226, 238–242, 254–280 (in Arabic).
Emami, S.A., Iranshahi, M. and Iranshahi, M. (2017) *Herbal Pharmacopoeia*. Behnashr and Mashhad University of Medical Sciences Publications, Mashhad, Iran, p. 28 (in Persian).
Ferdowsi, A. (1987) *The Shahnameh*, Vol. 1, ed. D. Khaleghi-Motlagh. Persian Heritage Foundation under the Imprint of Bibliotheca Persica, New York, pp. 265–270 (in Persian).
Garrison, F.H. (1966) *History of Medicine*. W.B. Saunders Company, Philadelphia, Pennsylvania, pp. 92–93.
Ḥakim Maysari (2015) *Dāneshnāmeh Dar Pezeshki* (*The Handbook of Medicine*), ed. B. Zanjāni. Tehran University Publications, Tehran (in Persian).
Ḥakim Mohammad Aʿẓam Khān (2019) *Excir Aʿẓam* (*The Grand Elixir*), 8 vols. Safir Ardehal Publisher, Tehran (in Persian).
Houston, G.W. (2003) Galen, his books, and the Horrea Piperataria at Rome. In: *Memoirs of the American Academy in Rome*, Vol. 48. University of Michigan Press, Ann Arbor, Michigan, pp. 45–51.
Ibn Abi Oṣṣayba ʿh, A.Q. (1998) *Oʿyun al-Anbā fi Ṭabaqāt al–Aṭebbā* (*The Sources of News on the Classification of the Physicians*), ed. M.B. Oʿyun al-Soud. Dar al-Kotob al-Iʿlmiyah, Beirut, pp. 294–295, 379–392, 401–421 (in Arabic).
Ibn al-Bayṭār, A.A. (1989) *Tafsīr Kitāb Diāsqūrīdūs* (*Commentary on Dioscorides' De materia medica*), ed. Ibrahim ben Marād. Dār al-Gharb al-Islāmi, Beirut (in Arabic).
Ibn Sinā, H.A. (1981–1997) *Al–Qānun fi aṭ–Ṭibb (The Canon of Medicine)*, 5 vols. Jāmiʿa Hamdard, New Delhi (in Arabic).
Ibn Sinā, H.A. (1982) *Al-Ishārāt wa al-Tanbihāt* (*Remarks and Admonitions*). Reprinted by Ayatollah *Maraʿshi* Najafi Library Publications, Qom, Iran (in Arabic).
Ibn Sinā, H.A. (1984) *Ash-Shifā* (*The Book of Healing*), ed. I. Madkour *et al.*, 9 vols. Reprinted by Ayatollah *Maraʿshi* Najafi Library Publications, Qom, Iran (in Arabic).
Ibn Sinā, H.A. (1985) *An-Naajāt* (*Book of Salvation*), ed. M.T. Daneshpazhouh. Tehran University Publications, Tehran (in Arabic).
Iniesta, I. (2011) Hippocratic corpus. *BMJ* 342, d688. DOI: 10.1136/bmj.d688.
Jorjāni, S.I. (1977) *Zakhirah Khārazmshāhi* (*Treasure of Khārazmshāh*). Photoprint of the manuscript dated AD 1206, edited with introduction by A.A. Saeedi Sirjani. The Iranian Culture Foundation, Tehran (in Persian).
Jorjāni, S.I. (1998) *Khofi Aʿlāiy* (A very compact book that fits in Ala ad-Din Atsiz's boot). Eṭelāʿt Institute, Tehran (in Persian).
Jorjāni, S.I. (2003) *Yādegār* (*The Memorial*), ed. M. Mohaghegh. The Institute of Islamic Studies, Tehran–McGill Universities, Tehran (in Persian).
Jorjāni, S.I. (2005–2006) *Al-Aghrāḍ aṭ-Ṭibbiah wa al-Mabāhethi al-Aʿlāiyah (Medical Goals and Aʿlāiy's Discussions)*, ed. H. Tadjbakhsh, 2 vols. Tehran University Publications, Tehran (in Persian).
Kotrc, R.F. and Walters, K.R. (1979) A bibliography of the Galenic Corpus. A newly researched list and arrangement of the titles of the treatises extant in Greek, Latin, and Arabic. *Transactions and Studies of the College of Physicians in Philadelphia* 1(4), 256–304.
Lloyd, A.B. (1982) The inscription of Udjahorresnet. A collaborator's testament. *Journal of Egyptian Archaeology* 68, 166–180.

Long, G. (ed.) (1842) Theophrastus. *Penny Cyclopaedia of the Society for the Diffusion of Useful Knowledge* 24, 332–334.
Mahdavi, Y. (1954) *Bibliography of Ibn Sinā*. Tehran University Publication, Tehran (in Persian).
Mallory, J.P. (1989) *In Search of the Indo-Europeans*. Thames and Hudson, London, pp. 49–50.
McClellan, J.E. III and Dorn, H. (2006) *Science and Technology in World History: An Introduction*. Johns Hopkins University Press, Baltimore, Maryland.
Moṭṭahari, M. (2020) *Mutual Services of Islam and Iran*, 60th edn. Sadra Publishing House, Tehran, p. 493 (in Persian).
Nadjmabādi, M. (1992) *Mohammad ibn Zakaryā Rāzi, Iranian Physician, Chemist, Philosopher.* Razi University Publications, Kermanshah, Iran (in Persian).
Nadjmabādi, M. (1995) *History of Medicine in Iran During the Islamic Era*. Tehran University Publications, Tehran, pp. 324–640, 719–742 (in Persian).
Nuland, S.B. (1988) *Doctors: The Biography of Medicine.* Knopf, New York, p. 5.
Osbaldeston, T.A. and Wood, R.P.A. (2000) *Dioscorides Materia Medica*. IBIDIS, Johannesburg, South Africa.
Posener, G. (1936) La première domination Perse en Egypte. *Bibliothèque d'étude de l'institut français d'archéologie orientale 11* XIII, p. 206. L'Institut Français d'Archeologie Orientale, Le Caire, France.
Rahnamai, M.T. (2019) *Iran, Its Environmental and Natural Resources*. Mahkāmeh Publisher, Tehran, p. 19 (in Persian).
Rāzi, M.Z. (1955–1971) *Al–Ḥāwi fi aṭ–Ṭibb* (*Continens*), ed. Aʿbdul Muʿid Khān, 22 vols. Oʿthaniah Oriental Publications Bureau, Oʿthaniah University, Hyderabad, India (in Arabic).
Rāzi, M.Z. (1992) *Al-Jodary wa al-Ḥaṣbah* (le *Livre sur la Viriole et la Rougole*), translated and commented by M. Nadjmabādi. Tehran University Publications, Tehran (in Arab and Persian).
Rutkow, I.M. (1993) *Surgery: An Illustrated History*. Elsevier Science Health Science Divison, London and Southampton, UK, p. 27.
Ṣafā, Z. (1997) *History of Rational Sciences in Islamic Civilization*, Vol. 1. Tehran University Publications, Tehran, pp. 165–179, 206–271 (in Persian).
Sezgin, F. (2001) *Geschichte des Arabischen Schrifttums*, Band III: *Medizin-Pharmazie, Zoologie-Tierheilkunde*, translated into Persian by K. Jahandari. Ministry of Islamic Culture and Guidance Publications and Printing Organisation, Tehran (in Persian).
Singer, C. and Underwood, E. A. (1962) *A Short History of Medicine*. Oxford University Press, New York and Oxford, p. 27.
Smith, P.E.L. (1968) *Paleolithic Archaeology in Iran*. The American Institute of Iranian Studies Monograph No. 1. The University Museum, University of Pennsylvania, Philadelphia, Pennsylvania, p. 15.
Tadjbakhsh, H. (2001) *History of Veterinary Medicine and Medicine of Iran*, Vol. 2. Tehran University Publications, Tehran, pp. 284–295, 301–312, 317–332 (in Persian).
Tadjbakhsh, H. (2003) *History of Human and Veterinary Medicine in Iran*. Fondation Merieux, Lyon, France, pp. 127–135, 139, 145–156, 162–174.
Tayarani-Najaran, Z., Tayarani-Najaran, F. and Emami, S.A. (2014) The history of Islamic medicine at a glance. In: Watson, R.R., Preedy, V.R. and Zibadi, S. (eds) *Polyphenols in Human Health and Disease*, Vol. 1. Elsevier, Amsterdam, pp. 17–26.
Tunekāboni, M.M. (2007) *Toḥfata al-Momenin known as Toḥfeḥ Ḥakim Momen* (*Ḥakim Mo'men's Gift*), ed. R. Rahimi, M.R. Shams Ardekani and F. Farjadmand. Shahid Beheshti University of Medical Sciences and Shahr Publications, Tehran (in Persian).
Wāseṭi, N. (1988) *A History of Iran Pakistan Medical Relations.* Iran Pakistan Institute of Persian Studies, Rawalpindi, Pakistan, pp. 25, 34–35, 58, 62–72, 79–82, 138–143 (in Persian).
West, J.B. (1985) Ibn al-Nafis, the pulmonary circulation, and the Islamic golden age. *Journal of Applied Physiology* 105(6), 1877–1880.
Xenophon (2001) *The Education of Cyrus*, tr. W. Ambler. Cornell University Press, Ithaca, New York.
Young, T.C. Jr (1967) The Iranian migration into the Zagros. *Iran* 5, 11–34.
Zarrinkoob, A. (2012) *Ruzegaran* (*The Ages*). Elm Publisher, Tehran (in Persian).

2 Botany, Collection, and Cultivation of Some Medicinal Plants used in Traditional Medicine of Iran

Farsad Nadjafi[1,2*] and Mohammad Reza Kanani[3]
[1]Shahid Beheshti University, Medicinal Plants and Drugs Research Institute, Tehran, Iran; [2]MateriaMed GmbH, Vienna, Austria; [3]Medicinal Plants and Drugs Research Institute of Shahid Beheshti University, Department of Botany, Tehran, Iran

Abstract
The cultivation and collection of mdicinal and aromatic plants have a long tradition in Iran. This chapter presents extensive information on the botany, taxonomy, and individual occurrence of some selected medicinal plants that are dealt with in more detail in the following chapters. In addition, important notes on the cultivation of the most significant species and the sustainable collection of numerous aromatic and medicinal plants in Iran are described in detail.

Botany, Ecology, and Taxonomy of Medicinal Plants

In determining Persian and Arabic names, two references, Mozaffarian (2008) and Ghahreman and Okhovvat (2008), are used.

Vachellia nilotica (L.) P.J.H. Hurter & Mabb

Synonyms: *Acacia nilotica* (L.) Willd. ex Delile, *Gumifera nilotica* (L.) Raf., *Mimosa nilotica* L.

Family: *Fabaceae* (*Leguminosae*).

Arabic name: *Sanṭ*.

Persian name: *Karat*.

English name: Egyptian acacia, Gum arabic tree, Babul, Thorn mimosa (Fig. 2.1).

Distribution

Africa: Algeria, Angola, Benin, Botswana, Burkina Faso, Cameroon, Cape Verde, Chad, Djibouti, Egypt, Eritrea, Ethiopia, Gambia, Ghana, Guinea, Guinea-Bissau, Kenya, Libya, Malawi, Mali, Madagascar, Mauritania, Mauritius, Mozambique, Namibia, Niger, Nigeria, Senegal, Somalia, Sudan, Tanzania, Togo, Uganda, Zambia, Zimbabwe. *Asia*: Bangladesh, India, Iran, Iraq, Myanmar, Nepal, Pakistan, Saudi Arabia, Sri Lanka, Yemen. *Americas*: Argentina, Bahamas, Bolivia, Cuba, Ecuador, Jamaica, Mexico, Peru, Puerto Rico, Trinidad and Tobago.

*Email: md@materiamed.at

 Medicinal Plants Used in Traditional Persian Medicine (eds. H. Schulz, Seyed Ahmad Emami and Farsad Nadjafi)
DOI: 10.1079/9781800621671.0002

Fig. 2.1. *Acacia nilotica* subsp. *indica*. (By Dinesh Valke from Thane, India, Nalla Tumma Telugu: నల్లతుమ్మ, CC BY-SA 2.0, https://commons.wikimedia.org/w/index.php?curid=51605008.)

Oceania and Australasia: Australia, Galápagos, New Guinea.

Morphology

Perennial plant, shrub, or tree 2.5–10 m high with a dense spherical crown, stems and branches usually dark to black-colored, branchlets purple-brown, bark thin, rough, fissured, shortly or densely pubescent, with lenticels, exuding a reddish low-quality gum. Thorns thin, straight, light gray, usually in 3–12 pairs, 5–7.5 cm long in young trees, mature trees commonly without thorns. Leaves bipinnate, 3–4 cm long, often with 1–2 petiole glands and other glands between all or only the uppermost pinnae, pinnae 2–11 pairs, with 7–25 pairs of leaflets (1.5–7 mm long) per pinnae, tomentose, rachis with a gland at the bottom of the last pair of pinnulae. Peduncles clustered at nodes of leafy and leafless branchlets. Flowers in globulous heads 1.2–1.5 cm in diameter, golden-yellow, set up either axillary or whorly on peduncles 2–3 cm long located at the end of the branches. Pods straight or lightly curved, 5–15 cm long, 0.5–1.2 cm wide, strongly constricted, white-gray, thick, and softly tomentose, fleshy when young, indehiscent, becoming black and hard at maturity. Seeds deep blackish-brown, smooth, subcircular, compressed, areole 6–7 mm long, 4.5–5 mm wide (Engler and Prantl, 1897–1915; Liogier, 1985–1997; Rechinger, 1963–2001).

Senegalia senegal (L.) Britton, N.L. Britton & P. Wilson, Sci. Surv. Porto Rico & Virgin Islands 6: 538 (1930)

Synonyms: *Acacia senegal* (L.) Willd., *A. cufodontii* Chiov., *A. glaucophylla* Brenan, *A. kinionge* Brenan, *A. oliveri* Vatke, *A. oxysprion* Chiov., *A. platyosprion* Chiov., *A. pseudoglauca* Chiov., *A. pseudoglaucophylla* (Chiov.) Chiov., *A. rostrata* Sim, *A. rupestris* Stocks, *A. rupestris* Stocks ex Boiss., *A. spinosa* Marloth & Engl., *A. thomasiae* Brenan, *A. trispinosa* Stokes, *A. unispinosa* (Fiori) Chiov., *A. verek* Guill. & Perr., *A. virchowiana* Vatke & Hildebrandt, *A. volkii* Suess., *Mimosa senegal* L., *M. senegalensis* Houtt., *Senegalia oliveri* (Vatke) Kyal. & Boatwr., *S. retusa* Raf.

Family: *Fabaceae* (*Leguminosae*).

Arabic name: *Qathād.*

Persian name: *Qivand, Qhatād, Hashāb.*

Common name: Senegal-gum, Kher, Three-thorn acacia (Fig. 2.2).

Distribution

Africa: Angola, Burkina Faso, Cameroon, Chad, Ivory Coast, Egypt, Ethiopia, Gambia, Ghana, Congo, Kenya, Mali, Mauritania, Mozambique, Namibia, Niger, Nigeria, Rwanda, Senegal, Somalia, Sudan, Tanzania, Uganda. *Asia*: India, Oman, Pakistan. *Americas*: Puerto Rico.

Morphology

Low branching shrub or small tree to 7 m high (maximum 15 m). The tree flowers during the rainy season and loses its leaves during the dry season. When water is available only at great depth, deeply penetrating tap roots can develop, and the tree will grow considerably larger than normal. The bark of *A. senegal* is yellow to brown and smooth in young trees, becoming dark gray, gnarled, and cracked on older trees. It is armed with recurved prickles just below the nodes, in pairs or in groups of 3. The leaves are 3.5–8 cm long, with 3–8 pairs of pinnae. The rachis (main stem of the compound leaf) sometimes bears prickles. The pinnae are 1.5–2.5 cm long with 5–25 pairs of leaflets per pinna. The leaflets are 1–9 mm long, 0.5–3 mm wide, elongated-oval shape, sparsely hairy or smooth on both surfaces. The flowers are borne on spikes 3–8.5 cm long. The hairy fruits are yellow-brown, flat, and papery, 1.8–9 cm long, 1.2–3.4 cm wide, 2 mm thick, and open on maturity to discharge the seeds (Engler and Prantl, 1897–1915; Liogier, 1985–1997).

***Adiantum capillus-veneris* L., Sp. Pl. [Linnaeus] 2: 1096 (1753)**

Synonyms: *Adiantum affine* Willd., *A. africanum* R. Br., *A. capillaceum* Dulac, *A. capillus* Sw., *A. coriandrifolium* Lam., *A. cuneifolium* Stokes, *A. emarginatum* Bory, *A. emarginatum* Bory ex Willd., *A. fontanum* Salisb., *A. formosum* J. Rémy, *A. formosum* J. Rémy ex Gay, *A. fumarioides* Willd., *A. lingii* Ching, *A. maderense* Lowe, *A. marginatum* Schrad., *A. michelii*

Fig. 2.2. *Senegalia senegal*. (By NehaVindhya, CC BY-SA 3.0, https://commons.wikimedia.org/w/index.php?curid=14951148.)

Christ, *A. modestum* Underw., *A. moritzianum* Link, *A. paradiseae* Baker, *A. pseudocapillus* Fée, *A. remyanum* Esp. Bustos, *A. remyanum* Espinosa, *A. repandum* Tausch, *A. rimicola* Sloss., *A. schaffneri* E. Fourn., *A. subemarginatum* Christ, *A. tenerum* Buchanan, *A. trifidum* Willd. ex Bolle, *A. visianii* Schloss & Vuk.

Family: *Pteridaceae*.

Arabic name: *Shaʿr el arḍ*, *Shaʿr el kenzir*.

Persian name: *Paresiavoshān*.

English name: Venus's hair fern (Fig. 2.3).

Distribution

Africa: Algeria, Angola, South Africa, Chad, Comoros, Djibouti, Egypt, Ethiopia, Kenya, Lesotho, Libya, Madagascar, Malawi, Mali, Mauritania, Mauritius, Morocco, Mozambique, Namibia, Niger, Nigeria, Somalia, Sudan, Tanzania, Tunisia, Uganda, Zambia, Zimbabwe.

Fig. 2.3. *Adiantum capillus-veneris* L. (By tato grasso, CC BY-SA 2.5, https://commons.wikimedia.org/w/index.php?curid=1451833.)

Asia: Afghanistan, Bangladesh, Cambodia, China, Georgia, India, Iran, Iraq, Japan, Kazakhstan, Korea, Laos, Lebanon, Myanmar, Nepal, Oman, Pakistan, Palestine, Philippines, Saudi Arabia, Sri Lanka, Syria, Taiwan, Tajikistan, Thailand, Turkey, Yemen. *Europe*: Albania, Bulgaria, Cyprus, France, Great Britain, Greece, Ireland, Italy, Portugal, Spain, *Americas*: Bahamas, Brazil, Chile, Cuba, Guatemala, Haiti, Honduras, Jamaica, Mexico, Peru, Puerto Rico, USA, Venezuela. *Oceania and Australasia*: Fiji, Samoa, Tonga.

Morphology

Epilithic or terrestrial plant, 10–40 cm. Rhizomes creeping, slender, scales dense, lanceolate, brown, margins entire. Fronds 2-pinnate, closely spaced or remote; stipe castaneous-black, glossy, distally glabrous, 6–25 cm × 8–16 cm, base cuneate, apex acute; rachises, costae, and stalks same color as stipes, slightly zigzag, color passing into lamina base; pinnae from second pair upward all similar but progressively smaller; pinnules 2–4 pairs per ultimate pinna, obliquely ascending, alternate; stalk castaneous-black, 1–2 mm, slender; blade 12–20 mm × 10–15 mm, thinly herbaceous, dark brown-green or green, glabrous, base cuneate, upper margin rounded, 2–4-lobed or divided into twiglike segments; sterile pinnules with apex obtuse, with marginal teeth broadly erose or triangular; fertile segments with apex truncate, slightly depressed or straight, entire or with erose teeth on both sides; terminal pinnules flabellate, usually larger, base narrowly cuneate, stalks up to 1 cm. Veins multidichotomously forked, reaching margins, visible on both surfaces. Sori 3–10 per pinnule, on apices of lobes of upper margin. Perispore thickly granular (Tristram, 1884; Post, 1896; Rechinger, 1963–2001; Davis, 1965–1988; Castroviejo *et al.*, 1980, Liogier, 1985–1997; Sell and Murrell, 2006).

Ajuga chamaepitys **(L.) Scherb., Pl. Verticill. Unilab. Gen. Sp.: 24 (1774)**

Synonyms: *Ajuga rechingerae* Bilik, *A. chamaepitys* Guss, *Bugula arvensis* Gray, *B. chamaepithys*

(L.) Scop., *B. chamaepitys* (L.) Kuntze, *Chamaepitys vulgaris* Link, *Teucrium chamaepitys* L.

Family: *Lamiaceae* (*Labiatae*).

Arabic name: *Kamāfiṭus*.

Persian name: *Labdisi hamedani*.

English name: Yellow bugle (Fig. 2.4).

Distribution

Africa: Algeria, Egypt, Morocco, Tunisia. *Asia*: Caucasus, Iran, Iraq, Kazakhstan, Lebanon, Russia, Saudi Arabia, Syria, Transcaucasia, Turkey, Turkmenistan. *Europe*: Albania, Belgium, Bulgaria, Cyprus, France, Germany, Greece, Hungary, Italy, Poland, Portugal, Romania, Spain, Switzerland, Ukraine.

Morphology

Perennial, biennial, or annual herb; stems prostrate or ascending, glabrous to densely lanate-villous, 5–30 cm. Cauline leaves cuneate to cuneate-oblong, 3(–5)-partite with linear segments 0.5–3(–4) mm wide, the segments sometimes 3-fid, hirsute, villous, lanate or subglabrous. Bracts similar to the leaves. Verticillasters 2–4-flowered. Calyx 4–6 mm, the teeth as long as or shorter than the tube, teeth 0.5–1(–2) × tube. Corolla yellow sometimes drying pinkish, rarely entirely purple, 8–30 mm. Stamens exserted; filaments hairy. Nutlets 2.5–4 mm (Engler and Prantl, 1897–1915; Bobrov *et al.*, 1934–1964; Rechinger, 1963–2001; Davis, 1965–1988; Assadi *et al.*, 1989).

***Alpinia officinarum* Hance, J. Linn. Soc., Bot. 13: 6 (1871)**

Synonyms: *Languas officinarum* (Hance) Farw.

Family: *Zingiberaceae*.

Arabic name: *Kholanjān*.

Persian name: *Kholanjān*.

English name: Chinese ginger.

Fig. 2.4. *Ajuga chamaepitys*. (By Saphon A, CC BY-SA 3.0, https://commons.wikimedia.org/w/index.php?curid=2082847.)

Distribution

Asia: Cambodia, China, Myanmar, Vietnam.

Morphology

Rhizomes elongate, terete. Pseudostems 35–115 cm. Leaves sessile; ligule lanceolate, membranous, entire, 2–3(–5) cm; leaf blade linear, 20–30 cm × 1.2–2.5 cm, glabrous, apex caudate, base attenuate. Racemes erect, 6–10 cm; rachis tomentose; bracteoles small, ≤ 1 cm. Pedicel 1–3 mm. Calyx 7–10 cm, puberulent, apex 3-toothed. Corolla tube slightly shorter than calyx; lobes oblong, 1.5 cm, central one hoodlike. Labellum ovate, white with red streaks, 2 cm. Filament 1 cm; anther 6 mm. Ovary tomentose. Capsule red, globose, 1 cm in diameter (eFloras, 2008).

Althaea officinalis **L., Sp. Pl. 2: 686 (1753)**

Synonyms: *Althaea balearica* J.J. Rodr., *A. Kragujevacensis* Pancic ex Diklic & Stevan., *A. micrantha* Borbás, *A. pulchra* Klotzsch, *A. sublobata* Stokes, *A. taurinensis* DC., *A. villosa* Blatt., *A. villosoides* Blatt., *A. vulgaris* Bubani, *Malva maritima* Salisb., *M. officinalis* (L.) K.F. Schimp. & Spenn.

Family: *Malvaceae*.

Arabic name: *Khaṭmi bostāni*.

Persian name: *Charme giāh*.

English name: Common marshmallow (Fig. 2.5).

Distribution

Africa: Algeria, Tunisia. *Asia*: Afghanistan, Caucasus, Iran, Iraq, Kazakhstan, Kyrgyzstan, Lebanon, Pakistan, Russia, Syria, Tajikistan, Transcaucasia, Turkey, Turkmenistan, Uzbekistan. *Europe*: Albania, Belarus, Belgium, Bulgaria, Denmark, France, Germany, Great Britain, Greece, Hungary, Ireland, Italy, Netherlands, Poland, Portugal, Romania, Spain, Switzerland, Ukraine. *Americas*: USA.

Fig. 2.5. *Althaea officinalis* L. (CC BY SA 4.0, https://commons.wikimedia.org/wiki/File:Guimauve_officinale_dans_les_pr%C3%A9s_sal%C3%A9s_de_la_plage_de_Donnant.JPG#/media/File:Guimauve_officinale_dans_les_prés_salés_de_la_plage_de_Donnant.JPG.)

Morphology

Perennial herb, erect, up to 2 m, densely gray-pubescent; velutinous indumentum of densely stellate hairs. Leaves triangular-ovate or rhomboid-triangular, acute, crenate-serrate petiole 1–4 cm, stellate tomentose; leaf blade ovate-orbicular or cordate, 3–8 cm × 1.5–6 cm, 3-lobed or not lobed, papery, densely stellate tomentose, base cordate or rounded, margin bluntly dentate, apex acute. Flowers solitary or clustered in axillary and terminal inflorescences, borne in fascicles in the leaf axils, peduncle of axillary inflorescences shorter than the subtending leaf. Epicalyx lobes 9, linear-lanceolate, 4 mm, densely stellate strigose. Calyx persistent, cup-shaped, 5-parted, longer than epicalyx, densely stellate hirsute, lobes lanceolate. Sepals ovate, acute, curved over the fruit. Petals white to pinkish white, 1.5–2 cm, obovate-oblong. Anthers purplish-red.

Staminal column c8mm. Ovary 15–25-loculed. Mericarps rugose, stellate-pilose, 8 mm in diameter, enclosed by calyx, puberulent. Seeds reniform (Boissier, 1867–1888; Bobrov *et al.*, 1934–1964; Rechinger, 1963–2001; Davis, 1965–1988; Castroviejo *et al.*, 1980; Sell and Murrell, 2006).

Prunus amygdalus (L.) Batsch

Synonyms: *Prunus dulcis* (Mill.) D.A. Webb, *P. stocksiana* (Boiss.) Brandis, *P. stocksiana* (Boiss.) Burkill, *P. korshinskyi* Hand.-Mazz., *P. ramonensis* (Danin) Eisenman, *P. communis* Fritsch, *P. communis* (L.) Arcang., *Druparia amygdalus* (Batsch) Clairv., *Amygdalus stocksiana* Boiss., *A. sativa* Mill., *A. microphylla* (Post) Browicz, *A. fragilis* F. Heller, *A. fragilis* Borkh., *A. elata* Salisb., *A. dulcis* Mill., *A. decipiens* Poit. & Turpin, *A. communis* L., *A. amygdalus* (Batsch) Frye & Rigg, *A. amygdalina* Oken ex M. Roem., *A. amara* Duhamel, *A. amara* (C.F. Ludw.) Hayne.

Family: *Rosaceae*.

Arabic name: *Lawz*.

Persian name: *Bādām*.

English name: Almond (Fig. 2.6).

Distribution

Africa: Algeria, Morocco, Tunisia. *Asia*: Afghanistan, Iran, Iraq, Kazakhstan, Kyrgyzstan, Lebanon, Libya, Pakistan, Syria, Tajikistan, Transcaucasus, Turkey, Turkmenistan, Uzbekistan. *Europe*: Bulgaria, Cyprus, France, Great Britain, Greece, Hungary, Italy, Portugal, Spain, Switzerland. *Americas*: USA.

Fig. 2.6. *Prunus amygdalus* (L.) Batsch. (By Manfred Heyde, CC BY-SA 3.0-migrated, https://upload.wikimedia.org/wikipedia/commons/5/53/Almond_Prunus_dulcis.jpg.)

Morphology

Tree or shrub, 2–6(–8) m tall, branches erect or horizontally spreading, spiny but young shoots glabrous, previous year's branchlets brownish, grayish-brown to grayish-black with age. Leaves variable, 2–9(–12) cm × 1.5–3 cm, elliptic, ovate-lanceolate or oblong-lanceolate, crenate-serrate, glabrous; petioles 10–20(–30) mm, dentate-serrate, glandular, entirely glabrous or pubescent, leathery, with 2–4 nectaries at apex or at base of leaf blade. Flowers solitary or pairs opening before leaves, up to 4 cm diameter; pedicel 3–5 mm, 4–10 mm in fruit, glabrous; hypanthium broadly campanulate or cylindrical, outside glabrous, 6–8 mm × 2–5 mm; sepals tomentose at least on the margin, broadly oblong to broadly lanceolate, margin pubescent, outside glabrous, apex obtuse, 5–6 mm; petals *c.*20 mm, white or pinkish, obovate-oblong to oblong, 1.2–2 cm × 0.5–1.2 cm; stamens unequal; ovary tomentose; style longer than stamens. Fruit 3–6 cm, ovoid-oblong to oblong-ovoid, compressed, densely pubescent, olive-green; mesocarp thin, coriaceous, eventually splitting at maturity; endocarp yellowish-white to brown, broadly ellipsoid or shortly oblong, ovoid, asymmetric, 2–3(–4) cm, stone pitted and ± grooved, keeled, dorsal suture generally straight, base obliquely truncate to orbicular-truncate. Seed sweet or bitter (Rechinger, 1963–2001; Davis, 1965–1988; Castroviejo *et al.*, 1980; Assadi *et al.*, 1989; Sell and Murrell, 2006).

Aristolochia bottae **Jaub. & Spach, Ill. Pl. Orient. 1(10): 173, t. 98 (1844)**

Synonyms: *Aristolochia maurorum* subsp. *bottae* (Jaub. & Spach) Bornm., *A. maurorum* var. *latifolia* Boiss.

Family: *Aristolochiaceae.*

Arabic name: *Aristolochia.*

Persian name: *Zaravand-chopoghak.*

English name: Dutchman's pipe, Birthwort.

Distribution

Asia: Iran, Iraq, Lebanon, Palestine, Syria, Transcaucasia, Turkey.

Morphology

Rootstock cylindrical, 80–100 cm long and 2 cm thick. Stems erect, 20–40 cm, hirtellous, branched. Leaves lanceolate to linear-lanceolate, acute to obtuse, often 2–8(–10) cm × 1.5–6 cm, puberulous, mucronate, base widely subhastate, oblong-spathulate, widely divaricate with tips ± incurved; petiole 2–10 mm. Peduncles 2.5–7 cm. Perianth 4–8 cm long, glabrous to puberulous, pale brown or greenish-brown outside; utricle ovoid to ovoid-oblong, 8–20 mm × 7–18 mm; tube 4–9 mm broad at base, gradually dilated above to 6–12 mm broad; limb ovate, acute, or obtuse, rarely retuse, 1.8–4.5 cm × 2–3.5 cm, hairy at throat, obtuse, auricles rounded. Ovary tomentose. Capsule obovoid, 4–7 cm × 2–3.5 cm (Tristram, 1884; Post, 1896; Rechinger, 1963–2001; Davis, 1965–1988; Ghahreman, 1978).

Aristolochia fontanesii **Boiss. & Reut., Pugill. Pl. Afr. Bor. Hispan.: 108 (1852)**

Synonyms: *Aristolochia atlantica* Pomel, *A. attica* Orph. ex H. Duch., *A. longa* L., *A. multinervis* Pomel, *A. pallida* Salzm. ex Ball, *A. rotunda* Bové, *A. rotunda* Bové ex Duch.

Family: *Aristolochiaceae.*

Arabic name: *Zarāvand ṭawil.*

Persian name: *Zarāvande darāz.*

English name: Long aristolochia (Fig. 2.7).

Distribution

Africa: Algeria. *Europe*: Albania.

Morphology

Usually branched, glabrescent, 20–80 cm; tuber cylindrical. Leaves 2–6(–9) cm,

Fig. 2.7. *Aristolochia fontanesii* Boiss. & Reut. (By Said Zidat, CC BY-SA 4.0, https://upload.wikimedia.org/wikipedia/commons/3/36/Aristolochia_fontanesii.jpg?uselang=de.)

ovate-triangular; petioles up to 1(1.5) cm. Flowers 3–5(–6) cm, brownish or yellowish-green, limb brownish-purple; pedicels about equaling petioles. Capsule 1.5–3 cm, globose to pyriform (Tutin *et al.*, 1964).

Aristolochia rotunda L., Sp. Pl.: 962 (1753)

Synonyms: This species has no synonyms.

Family: *Aristolochiaceae*.

Arabic name: *Zarāwand modaḥraj*.

Persian name: *Zarāvande gerd*, *Nakhode alwand*.

English name: Snakeroot (Fig. 2.8).

Fig. 2.8. *Aristolochia rotunda* L. (By Hectonichus, CC BY-SA 3.0, https://upload.wikimedia.org/wikipedia/commons/0/0b/*Aristolochiaceae*_-_Aristolochia_rotunda.JPG.)

Distribution

Europe: Albania, Bulgaria, France, Great Britain, Greece, Italy, Spain, Switzerland, Turkey.

Morphology

Rootstock globose or ovoid, often irregular, simple or rarely branched, glabrescent; stems 20–60 cm tall. Tuber ovoid or globose. Leaves 3–8(–10) cm × 3–7 cm, cordate-ovate or ovate-orbicular, amplexicaul. Petiole subsessile, not more than 0.4 cm. Perianth 3.5 cm, yellowish or brownish-pink outside. Flowers (2–)3–5 cm, yellow, limb dark brown; utricle 4–7 mm × 3–4 mm; tube straight, 2–3 mm broad at base, scarcely dilated above; limb as long as tube, oblong, to 5 mm broad, obtuse, dark brown, with purple striations in throat; ovary puberulous to pubescent; capsule globose, 1–1.5 cm in diameter (Davis, 1965–1988; Sell and Murrell, 2006).

Artemisia absinthium L., Sp. Pl.: 848 (1753)

Synonyms: *Absinthium bipedale* Gilib., *A. majus* Garsault, *A. officinale* Brot., *A. vulgare* Lam., *Artemisia absinthia* St.-Lag., *A. albida* Willd. ex Ledeb., *A. baldaccii* Degen, *A. doonense* Royle, *A. inodora* Mill., *A. pendula* Salisb., *A. rehan* Chiov., *A. rhaetica* Brügger.

Family: *Asteraceae* (*Compositae*).

Arabic name: *Afsantin*.

Persian name: *Afsantin*.

English name: Absinthe (Fig. 2.9).

Distribution

Africa: Algeria, Egypt, Eritrea, Ethiopia, Libya, Morocco, Tunisia, Western Sahara. *Asia*: Afghanistan, Borneo, Cambodia, Caucasus, China, India, Indonesia, Iran, Iraq, Japan, Kazakhstan, Kyrgyzstan, Lebanon, Mongolia, Myanmar, Pakistan, Philippines, Siberia, Syria, Taiwan, Tajikistan, Thailand, Transcaucasia, Turkey, Turkmenistan, Uzbekistan, Vietnam. *Europe*: Albania, Austria, Belarus, Belgium, Bulgaria, Denmark, Finland, France, Germany, Great Britain, Greece, Hungary, Ireland, Italy, Netherlands, Norway, Poland, Portugal, Romania, Russia, Spain, Sweden, Switzerland, Ukraine. *Americas*: Argentina, Bolivia, Brazil, Canada, Chile, Colombia, Cuba, Dominican Republic,

Fig. 2.9. *Artemisia absinthium* L. (By Valerie 75, CC BY-SA 3.0-migrated, https://upload.wikimedia.org/wikipedia/commons/9/9c/Artemisia_absinthium_001.jpg.)

Ecuador, Guatemala, Haiti, Honduras, Mexico, Paraguay, Peru, Puerto Rico, USA, Uruguay. *Oceania and Australasia*: Australia, New Zealand.

Morphology

Aromatic perennial herb, suffrutescent, more or less gray sericeous or puberulent, 40–150 cm tall. Stems 1–3 cm. Leaves 2–3-pinnatisect or 5-lobed, lobes ± oblong, subacute, petiolate, upper leaves 4–6 cm × 2–4 cm; leaf blade ovate-elliptic or ovate, 8–12 cm × 7–9 cm, 2- or 3-pinnatisect; lobes lanceolate or linear-lanceolate, usually obtuse. Inflorescence narrow or broad panicle; bracts similar to leaves but smaller. Flowering stems erect, sericeous at least above, up to 1 m; outer flowers filiform, female, inner hermaphrodite, fertile, corolla yellowish, glabrous, marginal female florets 15–25; corolla yellow, glabrous, obliquely 2-toothed; disk florets 30–70(–90), bisexual; capitula hemispherical, depressed-globose, many-flowered, 3–5(–6) mm broad, shortly paniculate inflorescence, nodding; involucre globose or subglobose, 2–3(–4) mm in diameter; outer bracts oblong, herbaceous; inner ovate, herbaceous, with a wide scarious margin. Receptacle hemispheric, conspicuously pilose or hairy. Achenes oblong, 0.8–1 mm, with apical corona or without (Bobrov *et al.*, 1934–1964; Rechinger, 1963–2001; Tutin *et al.*, 1964; Davis, 1965–1988; Assadi *et al.*, 1989; Sell and Murrell, 2006).

***Artemisia montana* (Nakai) Pamp., Nuovo Giorn. Bot. Ital., Ser. 2, 36: 461 (1930)**

Synonyms: *Artemisia gigantea* Kitam., *A. gigantea* f. *montana* (Nakai) H. Hara, *A. vulgaris* var. *montana* Nakai, *A. vulgaris* subsp. *latiloba* Nakai, *A. montana* var. *montana*.

Family: *Asteraceae* (*Compositae*).

Arabic name: A species of *Berenjāsf*.

Persian name: A species of *Dermaneh*.

Common name: For various species in the genus include Mugwort, Wormwood, Sagebrush, Sage, Suttle wood, Artemisia (Fig. 2.10).

Distribution

Asia: China, Japan, Korea. *Europe*: Russia.

Morphology

Perennial herb, 1.5–2 m tall, with creeping rhizomes, glabrescent, pubescent. Basal leaves faded before pollination. Middle leaves with petiole 2.5–3 cm; leaf blade ovate, 10–20 cm × 5–12 cm, glabrescent, abaxially gray tomentose, adaxially thinly arachnoid; pinnatipartite, segments 2 or 3 pairs, oblong or oblong-lanceolate; distal lobes larger, base attenuate, entire or occasionally few serrate. Uppermost leaves' bracts lanceolate, entire or 3-cleft. A large synflorescence, often rather narrow, panicle; ultimate branches short, without bracts. Involucre oblong, 2.5–3 mm; outer phyllaries

Fig. 2.10. *Artemisia montana* (Nakai) Pamp. (By Qwert1234, CC BY-SA 3.0, https://upload.wikimedia.org/wikipedia/commons/3/3a/Artemisia_montana_1.JPG?uselang=de.)

arachnoid tomentose. Corolla apex lacerate, marginal female florets 4–6; disk florets 8–14, bisexual. Achene obovoid, 1.5–2 mm (Bobrov *et al.*, 1934–1964).

Astragalus brachycalyx Fisch., Exsicc. (Pl. Alepp. Kurd. Moss.) 1843: no. 363 (1843)

Synonyms: *Astracantha brachycalyx* (Fisch.) Podlech, *Tragacantha brachycalyx* (Fisch.) Kuntze.

Family: *Fabaceae* (*Leguminosae*).

Arabic name: A kind of *Anzarut*.

Persian name: *Gavan gazi*.

Common name: A kind of Milkvetch.

Distribution

Asia: Iran, Iraq, Turkey.

Morphology

Lax shrub, sometimes umbraciform. Leaf rachis spiny, 25–45 mm, recurved. Leaflets elliptic, muticous, 4–6-paired, 3–10 mm; stipules lanceolate, *c.*6 mm, velutinous below, united for two-thirds their length. Flowers sessile, 2–3 per leaf axil. Inflorescence globose, 10–20-flowered. Bracts broadly ovate, white-hairy on the dorsal surface, 2–4 mm. Calyx *c.*4 mm, tomentose; tube *c.*3 mm; teeth narrowly triangular. Corolla white; 7–9 mm, platonychioid (Rechinger, 1963–2001; Davis, 1965–1988; Assadi *et al.*, 1989).

Astragalus fasciculifolius Boissier, Diagn. Pl. Orient. 2: 73 (1843)

Synonyms: *Astragalus afghanopersicus* Kitam., *A. arbusculinus* Bornm. & Gauba, *A. cornutus* Bunge, *A. fasciculifolius* subsp. *arbusculinus* (Bornm. & Gauba) Tietz, *Tragacantha fasciculifolia* (Boissier) Kuntze, *T. tebesiensis* Kuntze.

Family: *Fabaceae* (*Leguminosae*).

Arabic name: *Anzarut*.

Persian name: *Anzarut, gonjar*.

English name: Sarcocolla.

Distribution

Asia: Bahrain, Iran, Iraq, Kuwait, Oman, Qatar, Saudi Arabia, United Arab Emirates.

Morphology

Fruticose plant, shrub 30–120 cm wide, puffed up with elongated branches, most appositely gray-tonentous, armed with numerous long white thorns. Rameal leaves remotely 2–3-paired, axillary to bunch of 5–7-leaflets reduced, leaflets sub lens sparingly and appressedly puberulous or glabrate, fleshy glaucous, obovate or linear-cuneate, obtuse or notched, mucronate. Flowers of solitary bundles short; pedicellate, bracteolis very small, squamiform; calyx flowery tubular, appressed and sparingly hirsute, with teeth four times shorter than the lanceolate-subulate tube, ovate-oblong or globose, glistening in fruit, of an oblong standard, calyces of fruit smaller than those of the preceding ones; petals 1.5–3 lines long (Rechinger, 1963–2001; Assadi *et al.*, 1989).

Astragalus spp. (15 species)

1. *Astragalus gummifer* Labill., J. Phys. (Halle) 1: 46 (1790)

Synonyms: *Astracantha adanica* (Ivan.) Greuter, *A. gummifera* (Labill.) Podlech, *A. rayatensis* (Eig) Podlech, *A. tournefortii* (Boissier) Podlech, *Astragalus adanicus* Ivan., *A. adpressus* Ehrenb. ex Walp., *A. erianthus* Willd., *A. gummifer* subsp. *tournefortii* (Boissier) Ponert, *A. noemiae* Eig, *A. rayatensis* Eig, *A. tournefortii* Boissier, *Tragacantha eriantha* (Willd.) Kuntze, *T. gummifera* (Labill.) Kuntze, *T. tournefortii* (Boissier) Kuntze.

2. *Astragalus gossypinus* Fischer, Exsicc. (Pl. Alepp. Kurd. Moss.) 1843: no. 398 (1843)

Synonyms: *Astracantha gossypina* (Fisch.) Podlech, *A. xanthogossypina* (Hand.-Mazz.) Podlech, *Astragalus filagineus* Boissier, *A. gossypinoides* Hand.-Mazz. & Bornm., *A. xanthogossypinus* Hand.-Mazz., *Tragacantha filaginea* (Boissier) Kuntze, *T. gossypina* (Fisch.) Kuntze.

3. *Astragalus microcephalus* Willd., Sp. Pl., ed. 4, 3: 1332 (1802)

Synonyms: *Astracantha microcephala* (Willd.) Podlech.

4. *Astragalus verus* Olivier, Voy. Emp. Othoman 3: t. 44 (1807)

Synonyms: *Astracantha atropatana* (Bunge) Podlech, *A. chorassanica* (Bunge) Podlech, *A. consentanea* (Boriss.) Czerep., *A. densissima* (Boriss.) Czerep., *A. fallax* (Fisch.) Podlech, *A. heratensis* (Bunge) Podlech, *A. kuramensis* (Boriss.) Podlech, *A. lasiocaulos* (Bunge) Podlech, *A. longiramosa* (Boriss.) Czerep., *A. meana* (Boriss.) Podlech, *A. meraca* (Boriss.) Podlech, *A. meschhedensis* (Bunge) Podlech, *A. multifoliolata* (Boriss.) Podlech, *A. ochrobia* (Bunge) Podlech, *A. pachyacantha* (Bunge) Podlech, *A. parrowiana* (Boissier & Hausskn.) Podlech, *A. piletoclada* (Freyn & Sint.) Czerep., *A. pulvinata* (Bunge) Podlech, *A. semipellita* (Bunge) Podlech, *A. serpentinica* (Sirjev & Rech.f.) Podlech, *A. strobilifera* (Royle ex Benth.) Podlech, *A. unguiculata* (Boriss.) Czerep., *Astragalus atropatanus* Bunge, *A. brachycentros* Fisch., *A. chorassanicus* Bunge, *A. consentaneus* (Boriss.) Boriss., *A. fallax* Fisch., *A. heratensis* Bunge, *A. horridissimus* Sirjev & Bornm., *A. keyserlingii* Bunge, *A. kuramensis* (Boriss.) Czerep., *A. lasiocaulos* Bunge, *A. longemucronulatus* Sirjev & Rech.f., *A. meanus* (Boriss.) Boriss., *A. meracus* (Boriss.) Boriss., *A. meschhedensis* Bunge, *A. multifoliolatus* (Boriss.) Sirjev, *A. ochrobius* Bunge, *A. pachyacanthus* Bunge, *A. parrowianus* Boissier & Hausskn., *A. piletocladus* Freyn & Sint., *A. pulvinatus* Bunge, *A. royleanus* D. Dietr., *A. semipellitus* Bunge, *A. serpentinicus* Sirjev & Rech.f., *A. unguiculatus* (Boriss.) Boriss., *A. wrangelii* Sirjev, *Tragacantha atropatana* (Bunge) Kuntze, *T. brachycentra* (Fisch.) Kuntze, *T. chorassanica* (Bunge) Kuntze, *T. consentanea* Boriss., *T. densissima* Boriss., *T. heratensis* (Bunge) Kuntze, *T. keyserlingii* (Bunge) Kuntze, *T. kuramensis* Boriss., *T. lasiocaulos* (Bunge) Kuntze, *T. longiramosa* Boriss., *T. meana* Boriss., *T. meraca* Boriss., *T. multifoliolata* Boriss., *T. ochrobia* (Bunge) Kuntze, *T. pachyacantha* (Bunge) Kuntze, *T. piletoclada* (Freyn & Sint.) Boriss., *T. pulvinata* (Bunge) Kuntze, *T. semipellitus* (Bunge) Kuntze, *T. strobilifera* (Royle ex Benth.) Kuntze, *T. unguiculata* Boriss., *T. vera* (Olivier) Kuntze.

5. *Astragalus myriacanthus* Boissier, Diagn. Pl. Orient. 9: 91 (1849)

Synonyms: *Astracantha echidniformis* (Sirjev) Podlech, *A. eschkerensis* (Boissier & Hausskn.) Podlech, *A. lateritians* (Freyn & Bornm.) Podlech, *A. multispina* (Freyn & Bornm.) Podlech, *A. myriacantha* (Boissier) Podlech, *Astragalus echidniformis* Sirjev, *A. eschkerensis* Boissier & Hausskn., *A. janthinus* Boissier & Hausskn., *A. lateritians* Freyn & Bornm., *A. multispina* Freyn & Bornm., *A. stapfii* Sirjev, *Tragacantha eschkerensis* (Boissier & Hausskn.) Kuntze, *T. janthina* (Boissier & Hausskn.) Kuntze, *T. myriacantha* (Boissier) Kuntze.

6. *Astragalus longistylus* Bunge, Mém. Acad. Imp. Sci. Saint Pétersbourg, Sér. 7, 11(16): 81 (1868)

Synonyms: *Astracantha sofica* (Bunge) Podlech, *Astragalus soficus* Bunge, *Tragacantha longistyla* (Bunge) Kuntze, *T. sofica* (Bunge) Kuntze.

7. *Astragalus hypsogeton* Bunge, Mém. Acad. Imp. Sci. Saint Pétersbourg, Sér. 7, 11(16): 82 (1868)

Synonyms: *Astracantha hypsogeton* (Bunge) Podlech, *Tragacantha hypsogeton* (Bunge) Kuntze.

8. *Astragalus echidna* Bunge, Mém. Acad. Imp. Sci. Saint Pétersbourg, Sér. 7, 11(16): 82 (1868)

Synonyms: *Astracantha echidna* (Bunge) Podlech, *A. kuhistana* (Bunge) Podlech, *Astragalus kuhistanus* Bunge, *Tragacantha echidna* (Bunge) Kuntze, *T. kuhistana* (Bunge) Kuntze.

9. *Astragalus andalanicus* Boissier & Hausskn., P.E. Boissier, Fl. Orient. 2: 345 (1872)

Synonyms: *Astragalus globiflorus* var. *andalanicus* (Boissier & Hausskn.) C.C. Towns., *A. karadaghicus* Eig, *Tragacantha andalanica* (Boissier & Hausskn.) Kuntze.

10. *Astragalus compactus* Lam., Encycl. 1: 322 (1783)

Synonyms: *Astracantha compacta* (Lam.) Podlech, *A. insidiosa* (Borissier) Podlech, *Astragalus insidiosus* (Borissier) Borissier, *A. knappii* Bornm., *A. strictifolius* subsp. *kutepovii* (Sirjev) Ponert, *Tragacantha compacta* (Lam.) Steven, *T. insidiosa* Borissier.

11. *Astragalus globiflorus* Boissier, Diagn. Pl. Orient. 9: 81 (1849)

Synonyms: *Astracantha globiflora* (Boissier) Podlech, *Astragalus passargadensis* Maassoumi, *Tragacantha globiflora* (Boissier) Kuntze.

12. *Astragalus pycnocephalus* Fischer, Exsicc. (Pl. Alepp. Kurd. Moss.) 1843: no. 373b (1843)

Synonyms: *Astracantha alexandri* Sirjev) Podlech, *A. ferox* (Borissier) Czerep., *A. pycnocephala* (Fisch.) Podlech, *A. zahlbruckneri* (Hand.-Mazz.) Podlech, *Astragalus alexandri* Sirjev, *A. eriacmus* Boissier, *A. ferox* (Borissier) Borissier, *A. guestii* Eig, *A. karsicola* Sirjev & Bornm., *A. pycnocephalus* subsp. *munzurica* Ponert, *A. pycnocephalus* subsp. *seytunensis* (Bunge) Ponert, *A. quercetorum* Eig, *A. seytunensis* Bunge, *A. zahlbruckneri* Hand.-Mazz., *Tragacantha alexandri* (Sirjev) Soják, *T. eriacma* (Boissier) Kuntze, *T. ferox* Borissier, *T. pycnocephala* (Fisch.) Kuntze, *T. seytunensis* (Bunge) Kuntze.

13. *Astragalus albispinus* Sirjev & Bornm., Feddes Repert. Spec. Nov. Regni Veg. 52: 2 (1943)

Synonyms: *Astracantha albispina* (Sirjev & Bornm.) Podlech.

14. *Astragalus caspicus* M. Bieb., Fl. Taur. -Caucas. 2: 204 (1808)

Synonyms: *Ankylobus caspicus* (M. Bieb.) Steven, *Astracantha caspica* (M. Bieb.) Podlech, *Tragacantha caspica* (M. Bieb.) Steven.

15. *Astragalus floccosus* Boissier, Diagn. Pl. Orient. 9: 81 (1849)

Synonyms: *Astracantha floccosa* (Boissier) Podlech, *Tragacantha floccosa* (Boissier) Kuntze.

Family: *Fabaceae* (*Leguminosae*).

Arabic name: *Kathirā*.

Persian name: *Katirā*.

English name: Milkvetch.

Distribution

Asia: Iran, Iraq, Lebanon, Syria, Turkey, Caucasus, Transcaucasus, Afghanistan, Pakistan, Tajikistan, Tibet, Turkmenistan, West Himalaya.

Morphology

Spiny shrub, leaves paripinnate; rachis spiny; simple-hairy leaflets; stubs united below, free from the petioles. Flowers sessile, 2–3 per leaf axil, forming a several- to many-flowered inflorescence, ebracteolate or bracteoles minute. Calyx white simple-hairy, lobed to the base. Standard platonychioid. Legume included within the persistent corolla, unilocular, 1(2)-seeded. (Boissier, 1867–1888; Rechinger, 1963–2001; Davis, 1965–1988; Assadi *et al.*, 1989).

Bryonia alba **L., Sp. Pl.: 1012 (1753)**

Synonyms: *B. dioica* M. Bieb., *B. monoeca* E.H.L. Krause, *B. nigra* Gilib., *B. vulgaris* Gueldenst. ex Ledeb.

Family: *Cucurbitaceae*.

Arabic name: *Fāsherā*.

Persian name: *Fāsherā*.

English name: White bryony, Wild hop (Fig. 2.11).

Distribution

Asia: Kazakhstan, Kyrgyzstan, Russia, Turkey, Turkmenistan, Uzbekistan. *Europe*: Albania, Austria, Bulgaria, Denmark, Estonia, Finland, France, Germany, Greece, Hungary, Italy, Latvia, Lithuania, Norway, Poland, Romania, Sweden, Switzerland, Ukraine. *Americas*: USA.

Morphology

Stems branched, usually minutely glandular, sometimes also sparsely spiculate or hairy, rarely glabrous, up to 4 m. Leaves narrowly to broadly ovate-cordate, 4–5 cm × 10–15 cm, shallowly to usually deeply 5–7-lobed, the lobes ovate to elliptic or linear-lanceolate or triangular, triangular usually rather sharply

Fig. 2.11. *Bryonia alba* L. (By Aung commonswiki, https://upload.wikimedia.org/wikipedia/commons/5/50/Bryonia_alba1pl.jpg.)

and coarsely sinuate-toothed or lobulate, rarely subentire, acute or acuminate, sharply dentate, the central much longer than the lateral. Petiole glabrous or sparsely spiculate, sometimes sparsely minutely glandular, 1.5–7.5 cm. Male racemes 4.5–24 cm, glabrous or minutely glandular; calyx of female flowers usually as long as the corolla; sepals triangular to lanceolate, 1.5–3.5(–6.5) mm; petals 3.5–7 mm × 1.5–4 mm. Female inflorescences usually corymbose, shorter than male; sepals lanceolate, 1.1–3.5 mm; petals 1.0–4.0 mm × 0.5–1.2 mm. Stigmas glabrous, small, sometimes with a few small papillae. Fruit black, 7–8 mm. Seeds 5–6 mm × 3–3.8 mm × 2–2.2 mm (Bobrov *et al.*, 1934–1964; Tutin *et al.*, 1964; Davis, 1965–1988).

Bryonia cretica **L., Sp. Pl.: 1013 (1753)**

Synonyms: *Bryonia cretica* var. *monoica* Nábelek, *B. dioica* var. *angulosa* Mabille ex Gillies, *B. dioica* var. *elongata* Ten., *B. dioica* var. *lutea* Ser.

Family: *Cucurbitaceae*.

Arabic name: *Kermat-ol bayzā*.

Persian name: *Fāsharā*.

Common name: English mandrake, Bryony, Red bryony, White bryony, Ladies' seal.

Distribution

Africa: Algeria, Libya, Morocco, Tunisia. *Asia*: Kazakhstan, Tajikistan, Turkmenistan, Uzbekistan. *Europe*: Albania, Austria, Belgium, Bulgaria, Cyprus, Denmark, Finland, France, Germany, Great Britain, Greece, Hungary, Italy, Netherlands, Poland, Portugal, Spain, Sweden, Switzerland. *Americas*: USA.

Morphology

Stems up to 4 m, stems glabrous or coarsely spiculate, rarely glaucous, branched. Leaves ovate-cordate or broadly so, 3–10 cm × 3.5–10 cm, shallowly to deeply 5–7-lobed, lanceolate or ovate, rather obtusely or sometimes sharply sinuate-toothed or (usually 3-) lobulated, the lobes entire or with few, large, subobtuse teeth, the central usually not markedly longer than the lateral. Petiole glabrous or sparsely spiculate, sometimes glaucous, often short, 0.5–5 cm. Male racemes 4.5–13.5 cm, glabrous or almost so except for a few glands on the pedicels; calyx of female flowers usually about half as long as the corolla; stigma papillose-hairy; fruit 6–10 mm in diameter, red (Tutin *et al.*, 1964; Sell and Murrell, 2006).

Clinopodium brevifolium **(Wahlenb.) Bräuchler & Hjertson, Phytotaxa 356: 74 (2018)**

Synonyms: *Calamintha incana* (Sm.) Boissier, *Clinopodium incanum* (Sm.) Kuntze, *C. insulare* (Candargy) Govaerts, *Melissa incana* (Sm.) Benth., *Micromeria insularis* Candargy, *Satureja incana* (Sm.) Briq., *S. insularis* Greuter & Burdet, *Thymus brevifolius* Wahlenb., *Th. incanus* Sm.

Family: *Lamiaceae* (*Labiatae*).

Arabic name: *Habāghi*.

Persian name: *Reyhānak*, *Feranjemeshk*, *Vārangbou*, *Marzeh*.

Common name: Wild basil, Wild savory (Fig. 2.12).

Distribution

Africa: Libya, Iran, Iraq, Lebanon, Syria, Turkey. *Europe*: Cyprus, Greece.

Morphology

Perennial aromatic herb, pubescent to hirsute-pilose. Leaves shortly petiolate, ovate, crenate-serrate. Verticillasters dense or spreading, many-flowered, subtended by floral leaves. Bracteoles linear or subulate, as long as calyx tube or pedicels. Calyx tube curved, but not saccate, 13-veined, not accrescent, hairy, bilabiate, lower teeth always long-ciliate, usually 13-nerved. Corolla

Fig. 2.12. *Clinopodium brevifolium* (Wahlenb.). (By Radio Tonreg, CC BY-SA 2.0, https://upload.wikimedia.org/wikipedia/commons/8/8d/Clinopodium_vulgare_%288402662385%29.jpg.)

purple or pink (rarely white), tube exserted, exannulate, upper lip emarginate, lower lip 3-lobed, posterior lip usually notched. Stamens 4, lower pair longer than upper but all shorter than corolla; thecae divergent. Style branches very unequal, the lower oblong-lingulate, much larger than the upper lanceolate one. Disk usually symmetrical. Nutlets obloid, obovoid, subglobose, glabrous, often somewhat trigonal (Rechinger, 1963–2001; Davis, 1965–1988; Castroviejo *et al.*, 1980).

***Carthamus tinctorius* L., Sp. Pl.: 830 (1753)**

Synonyms: *Calcitrapa tinctoria* (L.) Röhl., *Carduus tinctorius* (L.) Falk, *Carthamus glaber* Burm.f., *C. tinctorius* var. *spinosus* Kitam., *Centaurea carthamus* E.H.L. Krause.

Family: *Asteraceae* (*Compositae*).

Arabic name: *Oʿṣfur*, *Qurṭum*.

Persian name: *Golrang*, *Kāfsheh*, *Kājirah*.

English name: Safflower, Dyer's saffron, False saffron (Fig. 2.13).

Distribution

Africa: Algeria, Egypt, Eritrea, Ethiopia, Libya, Mauritius, Morocco, Mozambique, Nicaragua, Sudan, Tunisia, Zimbabwe. *Asia*: Afghanistan, Bangladesh, Borneo, Cambodia, China, Cuba, India, Mongolia, Iran, Iraq, Lebanon, Syria, Japan, Kazakhstan, Korea, Kuwait, Laos, Myanmar, Nepal, Oman, Pakistan, Philippines, Saudi Arabia, Tajikistan, Thailand, Uzbekistan, Vietnam, Yemen. *Europe*: Austria, Baltic States, Belgium, Bulgaria, France, Germany, Great Britain, Greece, Hungary, Ireland, Italy, Netherlands, Poland, Portugal, Russia, Romania, Spain, Switzerland, Ukraine. *Americas*: Argentina, Chile, Colombia, Mexico, Paraguay, USA, Uruguay. *Oceania and Australasia*: Australia, New Zealand.

Morphology

Subglabrous annual plant 25–45 cm, cauline leaves undivided or rarely sinuate to pinnatifid, ovate to lanceolate-ovate, spinose-dentate, or entire. Stems glabrous, shining white. Cauline leaves glabrous, green, 3–9 cm × 1–2 cm, ovate to linear, sessile, margins entire to finely spiny-serrate. Capitula in loose corymbs, broadly ovoid to conical-ovoid. Outer phyllaries ovate, shorter than to slightly exceeding inner; inner arachnoid, acuminate. Flowers vermilion to orange-yellow; inner involucral bracts oblong-lanceolate, entire. Corolla yellow, orange, or reddish; filaments subglabrous. Achenes epappose, obpyramidal, more or less smooth, shiny, white. Pappus absent or rarely the inner achenes with short narrow scales (Bobrov *et al.*, 1934–1964; Rechinger, 1963–2001; Tutin *et al.*, 1964; Sell and Murrell, 2006).

Senna alexandrina* var. *alexandrina

Synonyms: *Cassia alexandrina* (Garsault) Thell., *C. acutifolia* var. *eriocarpa* Maire, *C. aethiopica*

Fig. 2.13. *Carthamus tinctorius* L. (By Pseudoanas, Public Domain, via Wikimedia Commons, https://upload.wikimedia.org/wikipedia/commons/5/56/Carthamus_tinctorius_050709b.JPG.)

Guibourt, *C. angustifolia* Vahl, *C. decipiens* Desv., *C. ehrenbergii* Bisch., *C. elongata* Lem.-Lis., *C. forsskalii* Royle, *C. italica* Lam. ex Steud., *C. lanceolata* Collad., *C. lanceolata* Forssk., *C. lanceolata* Wight & Arn., *C. lenitiva* Bisch., *C. ligustrinoides* Schrank, *C. medica* Forssk., *C. medicinalis* Bisch., *C. orientalis* Pers., *C. ovata* Mérat & Lens ex Geiger, *C. senna* L., *Senna alexandrina* Garsault, *S. angustifolia* (Vahl) Batka, *S. officinalis* Roxb.

Family: *Fabaceae* (*Leguminosae*).

Arabic name: *Sanna maki*.

Persian name: *Sannāye maki*.

Common name: Alexandrian senna (Fig. 2.14).

Distribution

Africa: Algeria, Benin, Central African Republic, Chad, Djibouti, Egypt, Eritrea, Ethiopia, Kenya, Mali, Mauritania, Niger, Nigeria, Somalia, Sudan. *Asia*: Bahrain, India, Iraq, Kuwait, Oman, Pakistan, Qatar, Saudi Arabia, Sri Lanka, Tajikistan, Transcaucasus, United Arab Emirates, Uzbekistan, Yemen. *Americas*: Brazil, Colombia, Dominican Republic, Haiti, Mexico.

Morphology

Perennial herb, up to 90 cm tall. Branches are hairless to nearly so. Stipules are lateral, about 1.5 mm long, pointed. Leaves are compound, about 45–120 mm long, 5–9 pairs of narrow lanceolate to ovate leaflets, about 12–45 mm long, 3.5–10 mm wide. Flowers are borne in racemes, up to 15 cm long, at branch ends or in leaf axils. Flowers are covered with about 7–8 mm long cup-shaped bracts; pedicel 3–4 cm long; sepals 5, nearly equal, 10–13 mm long, about 6–8 mm broad, spoon-shaped or cup-shaped, light yellow in color; petals are 5, nearly equal, 14–18 mm long, 7–10 mm wide, obovate, shortly clawed, deep yellow, veins becoming prominent after drying; stamens are 10, upper 3 reduced to staminodes, rest perfect, 2 lower largest; ovary is densely hairy, stipitate. Pods are about 4–5 cm long, about 16–25 mm broad, sparsely hairy, turning black at maturity, generally 4–10-seeded; stipe 2–3 mm (Engler and Prantl, 1897–1915).

Centaurium erythraea **Rafn, Danm. Holst. Fl. 2: 75 (1800)**

Synonyms: *Centaurium minus* Moench, *Chironia centaurium* (L.) F.W. Schmidt, *Erythraea*

Fig. 2.14. *Senna alexandrina* var. *alexandrina*. (By Lalithamba, CC BY-SA 2.0, https://upload.wikimedia.org/wikipedia/commons/6/61/Senna_alexandrina_Mill.-Cassia_angustifolia_L._%28Senna_Plant%29.jpg.)

centaurium (L.) Pers., *E. vulgaris* Gray, *Gentiana centaurium* L., *Gonipia linearis* Raf.

Family: *Gentianaceae*.

Arabic name: *Qanṭuryon ṣaghir*.

Persian name: *Qanṭuryon ṣaghir*.

English name: Common centaury (Fig. 2.15).

Distribution

Africa: Algeria, Morocco. *Asia*: Afghanistan, Caucasus, Iran, Iraq, Kazakhstan, Lebanon, Syria, Pakistan, Russia, Tajikistan, Transcaucasus, Tunisia, Turkey, Turkmenistan, Yemen. *Europe*: Albania, Austria, Baltic States, Belarus, Belgium, Bulgaria, Cyprus, Denmark, France, Germany, Great Britain, Greece, Hungary, Ireland, Italy, Netherlands, Poland, Portugal, Romania, Spain, Sweden, Switzerland, Ukraine. *Americas*: Argentina, Brazil, Canada, Costa Rica, Ecuador, Jamaica, Peru, USA. *Oceania and Australasia*: Australia, New Zealand.

Morphology

Biennial plant with a distinct basal rosette, stem usually solitary, sometimes several, usually branched in the upper part, 10–50(–80) cm, corymbosely branched above. Rosette-leaves 1–5 cm × 0.8–2 cm, obovate to elliptical, usually obtuse, 3–7-veined; basal leaves obovate to elliptical, usually much smaller and narrower, often acute, 3–7-veined, uppermost linear. Flowers sessile or shortly pedicellate, pink to pinkish-purple, rarely white, in a dense or lax corymbiform cyme. Calyx 3–7 mm, rarely nearly equaling the corolla tube, 0.5–0.75× corolla tube. Corolla pink to purple, rarely white; tube 6–11 mm, lobes 4–9 mm. (Bobrov *et al.*, 1934–1964; Rechinger, 1963–2001; Tutin *et al.*, 1964; Davis, 1965–1988; Sell and Murrell, 2006).

Fig. 2.15. *Centaurium erythraea* Rafn. (By Hans Hillewaert, CC BY-SA 3.0, https://upload.wikimedia.org/wikipedia/commons/5/5e/Centaurium_erythraea_%28plant%29.jpg.)

Ceratonia siliqua L., Sp. Pl.: 1026 (1753)

Synonyms: *Ceratonia coriacea* Salisb., *C. inermis* Stokes.

Family: *Fabaceae* (*Leguminosae*).

Arabic name: *Kharnub*.

Persian name: *Kharnub*.

English name: Locust bean, Carob.

Distribution

Africa: Algeria, Angola, Egypt, Ethiopia, Kenya, Libya, Morocco, Mozambique, Namibia, Sudan, Tanzania, Tunisia, Zambia, Zimbabwe. *Asia*: China, India, Iran, Iraq, Lebanon, Pakistan, Syria, Transcaucasus, Turkey, Yemen. *Europe*: Albania, Cyprus, France, Greece, Italy, Portugal, Spain, *Americas*: Mexico, Peru, Trinidad and Tobago, USA.

Morphology

Tall tree 15–16 m high. Leaves 8–17 cm; leaflets 2–4 pairs, shiny, obovate or suborbicular, 3.5–5.5 cm × 3–3.5 cm, leathery, glabrous, lateral veins conspicuously raised, base cuneate or broadly cuneate, margin entire, apex rounded, emarginate, or conspicuously cordate depressed. Raceme with rachis densely yellowish-brown pubescent. Flowers reddish. Legume curved, 10–25 mm × *c*.2.5 cm (Rechinger, 1963–2001; Davis, 1965–1988).

Cichorium intybus L., Sp. Pl.: 1026 (1753)

Synonyms: *Cichorium rigidum* Salisb., *C. endivia* var. *sativum* DC., *C. intybus* var. *balearicum* (Porta) Gand., *C. intybus* var. *eglandulosum* Freyn & Sint., *C. intybus* var. *indivisum* Vis., *C. intybus* var. *sylvestre* (Lam.) Vis.

Family: *Asteraceae*.

Arabic name: *Hendebā*.

Persian name: *Kāsni*.

English name: Chicory (Fig. 2.16).

Distribution

Africa: Algeria, Democratic Republic of the Congo, Egypt, Ethiopia, Guiana, Lesotho, Libya, Madagascar, Mauritania, Mauritius, Morocco, Tunisia, Zimbabwe. *Asia*: Afghanistan, Bahrain, Bangladesh, Borneo, Cambodia, Caucasus, China, India, Indonesia, Iran, Iraq, Kazakhstan, Kuwait, Kyrgyzstan, Laos, Lebanon, Mongolia, Myanmar, Oman, Pakistan, Qatar, Sakhalin, Syria, Saudi Arabia, Taiwan, Tajikistan, Thailand, Transcaucasus, Turkey, Turkmenistan, Uzbekistan, Vietnam, Yemen. *Europe*: Albania, Austria, Baltic States, Belarus, Belgium, Bulgaria, Cyprus, Denmark, Finland, France, Germany, Great Britain, Greece, Hungary, Ireland, Italy, Netherlands, Norway, Poland, Portugal, Romania, Russia, Spain, Sweden, Switzerland, Ukraine. *Americas*: Argentina, Bolivia, Brazil, Canada, Chile, Colombia, Costa Rica, Cuba, El Salvador, Guatemala, Guyana, Honduras, Mexico, Paraguay, Puerto Rico, USA, Uruguay, Venezuela. *Oceania and Australasia*: Australia, New Zealand.

Fig. 2.16. *Cichorium intybus* L. (By Christian Fischer, CC BY-SA 3.0, https://upload.wikimedia.org/wikipedia/commons/d/da/Cichorium_Intybus.jpg.)

Morphology

Herb, 40–110 cm tall, perennial, with a strong taproot. Stem usually solitary, erect; branches spreading-ascending, subglabrous. Basal leaves rosulate, obovate to oblanceolate, 15–34 cm × 2–4 cm, attenuate into a petiole-like basal portion, undivided to usually runcinately pinnatipartite, sparsely covered with long multicellular hairs, base attenuate, margin dentate; lateral lobes 3–6 pairs, triangular; terminal lobe distinctly larger than lateral ones, apex rounded to acute. Stem leaves similar to basal leaves but smaller and less divided, gradually reduced toward stem apex, base clasping, apex acute. Synflorescence of main axis and larger branches spiciform-paniculiform. Capitula axillary and terminal, solitary or in clusters of a few, sessile or on a several cm long, thick, and apically slightly inflated peduncle, with usually 15–20 florets. Involucre cylindric, 0.9–1.4 cm. Phyllaries abaxially sparsely with glandular or simple hairs, apex ± acute; outer phyllaries lanceolate, longest more than half as long as to approach inner ones in length, spreading-erect, margin ciliate; inner phyllaries linear-lanceolate. Florets blue or exceptionally pink or bluish-white. Achene brown, subcylindric to obovoid, 2–3 mm, stout, rugulose, apex truncate. Pappus (0.1–)0.2–0.3 mm (Bobrov *et al.*, 1934–1964; Rechinger, 1963–2001; Tutin *et al.*, 1964; Davis, 1965–1988; Castroviejo *et al.*, 1980; Sell and Murrell, 2006).

***Cinnamomum aromaticum* Nees, N. Wallich, Pl. Asiat. Rar. 2: 74 (1831)**

Synonyms: *Cinnamomum aromaticum* var. *longifolium* Lukman., *C. cassia* var. *medium* Lukman., *Laurus cinnamomum* Andrews, *L. malabathrum* Reinw. ex Nees, *L. zeylanica* Lam. ex Meisn.

Family: *Lauraceae*.

Arabic name: *Chinese Qurfa*.

Persian name: *Dārchin-e Chini*.

Common name: Chinese cassia, Chinese cinnamon (Fig. 2.17).

Distribution

Asia: Bangladesh, Cambodia, China, India, Indonesia, Laos, Myanmar, Taiwan, Thailand, Vietnam. *Americas*: Honduras, Trinidad and Tobago.

Morphology

Evergreen trees or shrubs, bark, branchlets, and leaves very scented; branches angular, the petioles strigose-tomentose, leaves oblong on both sides with sharp, sharp-arched veins on the underside of the triple veins, with vanishing at the tip of the leaf and narrow silky panicles. Fruit fleshy, subtended by a perianth cup; perianth lobes 6 (eFloras, 2008).

Fig. 2.17. *Cinnamomum aromaticum* Nees. (By Augustus Binu, CC BY-SA 3.0, https://upload.wikimedia.org/wikipedia/commons/e/ec/CINNAMON_BUDS.jpg.)

Cinnamomum verum J. Presl, F. Berchtold & J.S. Presl, Prir. Rostlin Aneb. Rostl. 2: 36 (1823)

Synonyms: *Camphora mauritiana* Lukman., *Camphorina cinnamomum* (L.) Farw., *Cinnamomum alexii* Kosterm., *C. aromaticum* J. Graham, *C. cinnamomum* (L.) H. Karst., *C. iners* Wight, *C. karrouwa* Lukman., *C. leptopus* A.C. Sm., *C. maheanum* Lukman., *C. mathewianum* Remya Kr., E.S.S. Kumar, Radhamany, Valsalad. & R. Jagad., *C. riedelii* Lukman., *C. zeylanicum* Blume, *C. zeylanicum* var. *antillanum* Lukman., *C. zeylanicum* var. *barthii* Lukman., *C. zeylanicum* var. *bengalense* Lukman., *C. zeylanicum* var. *biafranum* Lukman., *C. zeylanicum* var. *bonplandii* Lukman., *C. zeylanicum* var. *boutonii* Lukman.,

C. zeylanicum var. *capense* Lukman., *C. zeylanicum* var. *cayennense* Lukman., *C. zeylanicum* var. *commersonii* Lukman., *C. zeylanicum* var. *cordifolium* Blume ex Lukman., *C. zeylanicum* var. *decandollei* Lukman., *C. zeylanicum* var. *delessertii* Lukman., *C. zeylanicum* var. *ellipticum* Lukman., *C. zeylanicum* var. *erectum* Lukman., *C. zeylanicum* var. *humboldtii* Lukman., *C. zeylanicum* var. *leschenaultii* Lukman., *C. zeylanicum* var. *madrassicum* Lukman., *C. zeylanicum* var. *mauritianum* Lukman., *C. zeylanicum* var. *meissneri* Lukman., *C. zeylanicum* var. *ovatum* Lukman., *C. zeylanicum* var. *pallasii* Lukman., *C. zeylanicum* var. *pleei* Lukman., *C. zeylanicum* var. *pourretii* Lukman., *C. zeylanicum* var. *regelii* Lukman., *C. zeylanicum* var. *roxburghii* Lukman., *C. zeylanicum* var. *sieberi* Lukman., *C. zeylanicum* var. *sonneratii* Lukman., *C. zeylanicum* var. *vaillantii* Lukman., *C. zeylanicum* var. *variabile* Lukman., *C. zeylanicum* var. *wolkensteinii* Lukman., *C. zeylanicum* var. *zollingeri* Lukman., *Laurus cassia* Burm.f., *L. cinnamifera* Stokes, *L. cinnamomea* Salisb., *L. cinnamomum* L., *L. culitlaban* Buch.-Ham. ex Nees, *L. montana* Link ex Meisn., *L. rigida* Wall., *Persea cinnamomum* Spreng.

Family: *Lauraceae*.

Arabic name: *Qirfah*.

Persian name: *Dārchin*.

English name: Cinnamon (Fig. 2.18).

Fig. 2.18. *Cinnamomum verum* J. Presl. (https://commons.wikimedia.org/wiki/File:Cinnamomum_verum_002.JPG#/media/File:Cinnamomum_verum_002.JPG.)

Distribution

Africa: Comoros, Mauritius, Seychelles, Tanzania. *Asia*: Bangladesh, Borneo, Cambodia, China, India, Myanmar, Philippines, Sri Lanka, Taiwan, Vietnam. *Americas*: Argentina, Brazil, USA.

Morphology

Evergreen small trees, up to 10 m. Bark black-brown, inner bark with cinnamic aldehyde flavor. Young branchlets gray, somewhat tetragonous, white-maculate. Buds sericeous-puberulent. Leaves usually opposite; petiole *c*.2 cm, glabrous; leaf blade greenish-white abaxially, green and shiny adaxially, ovate or ovate-lanceolate, 11–16 cm × 4.5–5.5 cm, leathery or subleathery, glabrous on both surfaces, triplinerved, midrib and lateral veins elevated on both surfaces, transverse veins and veinlets reticulate, conspicuously foveolate abaxially, base acute, margin entire, apex acuminate. Panicle axillary or terminal, 10–12 cm; peduncle and rachis sericeous-puberulent. Flowers yellow, *c*.6 mm. Perianth tube obconical; perianth lobes 6, oblong, subequal, gray puberulent outside. Fertile stamens 9; filaments hairy near base, those of 3rd whorl each with 2 glands, others glandless; anthers 4-celled; cells of 1st and 2nd whorls introrse but those of 3rd whorl extrorse. Ovary ovoid, 10–15 mm, glabrous; style short; stigma discoid. Fruit ovoid, 10–15 mm, black when mature; perianth cup in fruit cupuliform, dilated, dentate, teeth truncate or acute at apex (eFloras, 2008).

***Citrullus colocynthis* (L.) Schrad., Linnaea 12: 414 (1838)**

Synonyms: *Citrullus colocynthis* subsp. *insipidus* (Pangalo) Fursa, *C. colocynthis* subsp.

stenotomus (Pangalo) Fursa, *C. colocynthoides* Pangalo, *C. pseudocolocynthis* M. Roem., *Colocynthis officinalis* Schrad., *C. vulgaris* Schrad., *Cucumis bipinnatifidus* Wight ex Naudin, *C. colocynthis* L., *C. colocynthoides* Pi. Savi, *Cucurbita colocyntha* Link.

Family: *Cucurbitaceae*.

Arabic name: *Ḥanẓal*.

Persian name: *Ḥanẓal*, *Hendewaneh abujahl*, *Kabast*.

English name: Bitter apple (Fig. 2.19).

Distribution

Africa: Algeria, Benin, Burkina Faso, Chad, Djibouti, Egypt, Eritrea, Ethiopia, Kenya, Libya, Mali, Mauritania, Morocco, Niger, Sahara, Senegal, Somalia, Sudan, Tunisia. *Asia*: Afghanistan, Bahrain, Bangladesh, India, Iran, Iraq, Kuwait, Lebanon, Myanmar, Oman, Pakistan, Qatar, Saudi Arabia, Sri Lanka, Syria, Transcaucasus, Turkey, Turkmenistan, Uzbekistan, Yemen. *Europe*: Bulgaria, Cyprus, France, Greece, Hungary, Italy, Romania, Russia, Spain. *Americas*: Argentina, Paraguay, USA.

Morphology

Trailing perennial herb; rootstock stout, somewhat woody; stems *c*.2 mm in diameter, hispid. Leaves elongate-ovate, scabrid-hairy beneath, 2.5–12 cm × 1.5–8 cm, 1- to 2-pinnatisect with 3–5 sinuate-lobed segments, lobes ± pinnately lobulate. Petals greenish-yellow, 6–10 mm × 4–6 mm. Calyx lobes subulate; corolla lobes *c*.5 mm, ovate, acute. Ovary sparsely hispid. Fruit globose-subglobose, 5–12 cm diameter, smooth, green or yellow, often striped or mottled;

Fig. 2.19. *Citrullus colocynthis* (L.). (By H. Zell, CC BY-SA 3.0, https://upload.wikimedia.org/wikipedia/commons/3/3b/Citrullus_colocynthis_004.JPG.)

pulp exceedingly bitter. Seeds 6.5–10 mm × 3.5–5 mm × 2–2.5 mm (Bobrov *et al.*, 1934–1964; Rechinger, 1963–2001; Davis, 1965–1988).

Commiphora gileadensis **(L.) C. Chr., Dansk Bot. Ark. 4(3): 18 (1922)**

Synonyms: *Balsamea gileadensis* (L.) Oken, *B. meccanensis* Gled., *B. opobalsamum* (L.) Baill., *Balsamodendrum ehrenbergianum* O. Berg, *B. gileadense* (L.) Kunth ex DC., *B. opobalsamum* (L.) Kunth ex DC., *Balsamus libanotus* Stackh., *B. meccanensis* Stackh., *B. theophrasti* Stackh., *Commiphora albiflora* Engl., *C. ancistrophora* Chiov., *C. anfractuosa* Chiov., *C. cassan* Chiov., *C. gillettii* Chiov., *C. microcarpa* Chiov., *C. opobalsamum* (L.) Engl., *C. suckertiana* Chiov., *C. velutina* Chiov., *Amyris gileadensis* L., *A. opobalsamum* L.

Family: *Burseraceae*.

Arabic name: *Balsān*.

Persian name: *Balsāne makki*.

English name: Arabian balsam.

Distribution

Africa: Djibouti, Eritrea, Ethiopia, Kenya, Somalia, Sudan. *Asia*: Oman, Saudi Arabia, Yemen.

Morphology

Shrub or tree, up to 5 m tall, unarmed, sometimes with long slender drooping branches; bark yellowish to dull orange or gray or black, smooth, sometimes peeling in small papery flakes; branchlets longitudinally ridged, glabrous to pubescent. Leaves 3–9(–11)-foliolate, glabrous to densely pubescent, sometimes with hooked hairs; petiole 2–35 mm long, sometimes narrowly winged; rhachis sometimes narrowly winged; leaflets linear to oblanceolate, obovate or suborbicular, 0.3–3.2 cm × 0.1–2(–2.5) cm, subacute to retuse at the apex, with entire margins, the lateral leaflets smaller than the terminal one. Flowers in 1- to several-flowered glabrous to pubescent cymes or clusters; peduncle *c.*0.5–5 mm long; pedicels 0.5–2.5(–6) mm long. Calyx 1–1.5 mm long, glabrous or pubescent. Petals *c.*2–5 mm long, white or cream, glabrous or pubescent outside. Fruits ellipsoid to ovoid, obtuse, and apiculate at the apex, 5–10 mm × 3–7 mm, glabrous or sparsely puberulous, purplish, marked with 4 pale lines, subsessile or usually on a distinct stalk; pericarp 4-valved; pseudaril with cup-like basal part and thinner upper part; stone ellipsoid to ovoid, flattened, 4–7.5 mm × 3–5 mm, with sterile cell much smaller than the fertile cell (Thulin, 1999).

Commiphora wightii **(Arn.) Bhandari, Bull. Bot. Surv. India 6: 327 (1965)**

Synonyms: *Balsamodendrum wightii* Arn., *Balsamea mukul* (Hook. ex Stocks) Baill., *Balsamodendrum mukul* Hook. ex Stocks, *B. roxburghii* Stocks, *Commiphora mukul* (Hook. ex Stocks) Engl., *C. roxburghii* Engl., *C. roxburghii* var. *serratifolia* Haines.

Family: *Burseraceae*.

Arabic name: *Moql el yahud*.

Persian name: *Moqle azraq*.

English name: Guggul, Mukul myrrh.

Distribution

Asia: India, Pakistan.

Morphology

Shrubby, with glandular hairs pubescent on the sides; branches scattered along the apex, prickly; leaves sessile simple or trifoliolate, terminal leaflet rhomboid-ovate, anteriorly serrate-toothed, lateral (when present) twice less serrate-incised; intermediate leaflet up to 2 cm long, 8 mm broad, lateral 6–8 mm long, 3–4 mm broad, with a few prominular lateral nerves equal to the flowers and leaves; pedicels 2–3, fascicles very short;

calyx campanulate, glandular-hairy, with triangular tubes with long teeth; petals largely linear, calyx 2.5× longer; stamens equaling half of the petals; ovary oblong-ovoid in tapering style; drupes ovoid, with acute epicarp 4-valve, 6 mm long, 5 mm gross (Nasir and Ali, 1970–1995).

Commiphora myrrha (Nees) Engl., A.L.P.P. de Candolle & A.C.P. de Candolle, Monogr. Phan. 4: 10 (1883)

Synonyms: *Balsamea myrrha* Baill., *B. myrrha* (T. Nees) Oken, *B. playfairii* Engl., *Balsamodendrum myrrha* T. Nees, *B. simplicifolium* Baker, *Commiphora coriacea* Engl., *C. cuspidata* Chiov., *C. molmol* (Engl.) Engl. ex Tschirch, *C. rivae* Engl., *C. simplicifolia* Schweinf. ex Baker.

Family: *Burseraceae*.

Arabic name: *Mor*.

Persian name: *More makki*.

English name: Myrrh, Common myrrh, African myrrh.

Distribution

Africa: Djibouti, Eritrea, Ethiopia, Kenya, Somalia. *Asia*: Oman, Saudi Arabia, Yemen.

Morphology

Spiny almost glabrous shrub or small tree, usually with a distinct though short trunk, up to 4 m tall; outer bark silvery, whitish, or bluish-gray, peeling in large or small papery flakes from the greener under-bark; exudate hardly scented, viscid, producing a hard translucent yellowish gum-resin; all branchlets spine-tipped. Leaves chartaceous, grayish-green or glaucous, very variable; petiole 1–10 mm long; a few lateral leaflets, sometimes very minute, may nearly always be found on both long- and short-shoot leaves, but most leaves may be without them or they may be up to half the size of the terminal leaflet, which maybe elliptic, spathulate, or lanceolate, attenuate, cuneate, rounded, or truncate at the base, rounded or acute apically, 6–44 mm long, 3–20 mm wide, with 3–4 rather weak main veins on each side, ultimate vein network obscure; margin entire or with up to 6 rather large broad teeth on each side. Male flowers usually precocious, 2–4 together in dichasial cymes 3–4 mm long which are often sparsely glandular; bracteoles pale brown, ovate-triangular, 0.5–0.7 mm long and wide, often lightly attached at the base and forming a fragile detachable collar; receptacle beaker-shaped, ± 2 mm deep and 2.3 mm wide; calyx tube ± 1 mm, lobes ovate-triangular. 1.5 mm long; petals oblong, tapering, pointed and recurved at the tip, 4.5 mm long, 1.5 mm wide; filaments 1.4 and 1.2 mm, anthers 1.2 and 1.0 mm long. Fruits 1–2 together on jointed stalks 2–4 mm long, flattened, ovoid, markedly beaked, ± 12.5 mm long (including a 2.5 mm beak) × 6.3 mm (1.6 × 3.5 mm); pericarp 2-valved; pseudaril variable, usually with 2 large broadly oval facial lobes and much reduced narrower sutural lobes; occasionally the 4 lobes may be subequal rather short and rounded; stone smooth or with low, gentle swellings, ovoid, cordate at the base, flattened, often with a groove down the centre of each face, ± 7 mm × 5 mm; occasionally both locules maybe fertile, a stone of this kind measured 7 mm × 5.7 mm; apical pits minute; germination shield large but often faintly demarcated (Engler and Prantl, 1924–1980).

Cordia myxa L., Sp. Pl.: 199 (1753)

Synonyms: *Bourreria glabra* (Roth) G. Don, *Cordia latifolia* Wall. ex G. Don, *C. officinalis* Lam., *C. paniculata* Roth, *C. petta-pelioporet* B. Heyne ex Roth, *C. scabrifolia* Benth. ex Griseb., *Ehretia glabra* Roth, *Gerascanthus myxus* (L.) Borhidi, *Lithocardium myxa* Kuntze, *Quarena indica* Raf., *Sebestena myxa* (L.) J. Presl, *S. officinalis* Gaertn.

Family: *Boraginaceae*.

Arabic name: *Mokhayṭ*, *Sebestan*.

Persian name: *Sepestan*.

English name: Assyrian plum, Spistan (Fig. 2.20).

Distribution

Africa: Algeria, Benin, Burkina Faso, Cameroon, Comoros, Democratic Republic of the Congo, Egypt, Ethiopia, Ghana, Guinea, Guinea-Bissau, Ivory Coast, Libya, Madagascar, Malawi, Mali, Mauritius, Mozambique, Rodrigues, Senegal, Seychelles, Tanzania, Uganda, Zimbabwe. *Asia*: Bahrain, Bangladesh, India, Indonesia, Iran, Iraq, Kuwait, Laos, Lebanon, Myanmar, Oman, Pakistan, Qatar, Saudi Arabia, Sri Lanka, Syria, Turkey, United Arab Emirates, Vietnam. *Americas*: Trinidad and Tobago, USA. *Oceania and Australasia*. Australia, Fiji.

Morphology

Glabrous or sparsely pubescent tree, to 13 m. Leaves orbicular to ovate or obovate, obtuse to mucronate, entire, lamina (2.5)4–10 cm × (2.8)3–5.8 cm, petiole (1.2)2–4.5 cm. Inflorescence thyrsoid-corymbose. *c*.25–60(–100)-flowered. Calyx ± cylindrical and 4.5–5 mm in flower, very shallowly lobed, accrescent to 9–13 mm and becoming trumpet-shaped and woody in fruit. Corolla white, tube ± equaling calyx, lobes 2.5–3.5 mm, obtuse, recurved. Stamens distinctly exserted, anthers versatile. Drupe pinkish, 12–16 mm × 8–11 mm (in sicca), broadly elliptic, mucronate (Rechinger, 1963–2001; Davis, 1965–1988; Castroviejo *et al.*, 1980).

Thymbra capitata (L.) Cav., Elench. Pl. Horti Matr.: 37 (1803)

Synonyms: *Coridothymus capitatus* (L.) Rchb.f., *Origanum capitatum* (L.) Kuntze, *Satureja capitata* L., *Thymus capitatus* (L.) Hoffmanns. & Link, *Th. creticus* Brot.

Family: *Lamiaceae* (*Labiatae*).

Arabic name: *Ḥāshā*.

English name: Mediterranean thyme, Headed savory, Conhead thyme (Fig. 2.21).

Distribution

Africa: Albania, Algeria, Egypt, Libya, Morocco, Tunisia. *Asia*: Lebanon, Syria, Turkey. *Europe*: Cyprus, France, Greece, Italy, Portugal, Spain.

Fig. 2.20. *Cordia myxa* L. (By Marco Schmidt, CC BY-SA 3.0, https://upload.wikimedia.org/wikipedia/commons/a/a7/Cordia_myxa_MS_1790.jpg.)

Fig. 2.21. *Thymbra capitata* (L.). (By Denis Barthel, CC BY-SA 3.0, https://upload.wikimedia.org/wikipedia/commons/1/1b/Thymbra_capitata_Ghajn_Tuffieha_Malta_01.jpg.)

Morphology

Very aromatic dwarf shrub, 20–50 cm high, with ascending branches up to erect, woody, clear, the young are tomentose white, which usually only have axillary leaf buds. Floral leaves imbricate, stem leaves long and falling with dryness, sessile, almost triangular, linear, pointed, 6–12 mm long and 1–1.2(–1.8) mm wide, being more or less on the margin flat, more or less glabrous, ciliated at the base, grayish green on both sides, dotted with glands. Inflorescences in dense oval-rounded whorls, narrow, spike-like. Calyx dorsally flattened, 5 mm long, upper lip tridentate, shorter than the lower 2-toothed, all teeth being ciliated; calyx tube with 20–22 nerves, unlike other *Thymus* species and flattened by the back; corolla 2-lipped, the tube straight; upper lip porrect, emarginate; lower lip 3-lobed. Stamens included, anther cells parallel, the upper lip is bifid, with 4 stamens (Engler and Prantl, 1897–1915; Davis, 1965–1988; Castroviejo *et al.*, 1980).

Corylus avellana **L., Sp. Pl.: 998 (1753)**

Synonyms: *Corylus avellana* subsp. *sylvestris* Ehrh., *C. sylvestris* Salisb.

Family: *Betulaceae*.

Arabic name: *Bondoq*, *Jilwaz*.

Persian name: *Fandogh*.

English name: Cobnut, Corkscrew hazel (Fig. 2.22).

Distribution

Asia: Caucasus, Transcaucasus, Turkey. *Europe*: Albania, Austria, Baltic States, Belarus,

Fig. 2.22. *Corylus avellana* L. (CC BY-SA 3.0, https://commons.wikimedia.org/w/index.php?curid=144925.)

Belgium, Bulgaria, Denmark, Finland, France, Germany, Great Britain, Greece, Hungary, Ireland, Italy, Netherlands, Norway, Poland, Portugal, Romania, Russia, Spain, Sweden, Switzerland, Ukraine.

Morphology

Large spreading shrub to 6 m, with brownish-gray, smooth bark. Young shoots yellowish-gray, hairy or glandular-pubescent. Leaves ovate to broadly ovate and suborbicular, acuminate at apex and cordate at base, biserrate and slightly lobulate, often shallowly lobed, cordate at base, 5–12 cm × 3–10 cm, with 6–9 pairs of veins, dark green and slightly pubescent above, pale green and pubescent beneath. Petioles 0.5–2.5 cm, glandular-hispid. Stipules oblong-ovate, obtuse, hairy. Staminate catkins to 8 cm, with broadly ovate bracts, pendent. Infructescence of 1–4 nuts; involucre mostly shorter or rarely slightly longer than nut, divided to half-way or a little more, usually irregularly dentate or deeply dissected on one side or both, stipitate-glandular below. Nut subglobose or ovoid with abruptly pointed tip, 1–2 cm × 1–1.6 cm, in clusters of 2–8 or sometimes singly, brown with a hard woody shell (Bobrov *et al.*, 1934–1964; Tutin *et al.*, 1964; Davis, 1965–1988; Sell and Murrell, 2006).

Costus spicatus **(Jacq.) Sw., Prodr. Veg. Ind. Occ.: 11 (1788)**

Synonyms: *Costus conicus* Stokes, *C. cylindricus* Jacq., *C. micranthus* Gagnep., *C. quartus* Roem. & Schult., *Alpinia spicata* Jacq., *Amomum petiolatum* Lam.

Family: *Costaceae*.

Arabic name: *Qusṭ ḥolw*.

Persian name: *Qusṭe shirin*.

English name: Crepe ginger (Fig. 2.23).

Distribution

Americas: Cuba, Dominican Republic, Haiti, Leeward Islands, Mexico Gulf, Mexico, Puerto Rico, Windward Islands.

Morphology

Perennial, herbaceous plant, often in close interactions with ants, produces a sugary

Fig. 2.23. *Costus spicatus* (Jacq.). (By Linrx, CC BY-SA 4.0, https://upload.wikimedia.org/wikipedia/commons/a/a0/Costus_spicatus_in_Singapore.jpg.)

nectar that is attractive to many different kinds of ants, generally 1.8–2.2 m tall. Leaves grow to a length of approximately 30 cm and a width of approximately 10 cm. It produces a short red cone, from which red-orange flowers emerge one at a time (Liogier, 1985–1997).

Crocus sativus L., Sp. Pl.: 36 (1753)

Synonyms: *Crocus autumnalis* Sm., *C. officinalis* (L.) Honck., *C. orsinii* Parl., *C. pendulus* Stokes, *C. sativus* var. *officinalis* L., *C. sativus* subsp. *orsinii* (Parl.) K. Richt., C. *setifolius* Stokes, *Geanthus autumnalis* Raf., *Safran officinarum* Medik.

Family: *Iridaceae*.

Arabic name: *Zaʿfarān*.

Persian name: *Zaʿfarān*.

English name: Saffron (Fig. 2.24).

Distribution

Africa: Morocco. *Asia*: Iran, Pakistan, Turkey. *Europe*: Greece, Italy, Spain.

Morphology

Corm usually symmetrical, enclosed by several tunics of variable reticulate-fibrous and color. Leaves (cataphylls) 5–11, 1.5–2 mm broad, synanthous, sheathing the aerial shoot; usually present at anthesis; prophyll present; bracteole present, much smaller than bract. Flowers deep lilac-purple with darker veins, white, or white with a purple base; throat glabrous or pubescent, white or lilac; perianth segments 3.5–5 cm × 1–2 cm, obtuse. Style branches 2.5–3.2 cm; filaments 7–11 mm, purplish, glabrous; anthers 0.5–2 cm, yellow; style dividing into 3 long red branches, each 2.5–3.2 cm (Rechinger, 1963–2001; Davis, 1965–1988; Ghahreman, 1978).

Cuscuta campestris Yunck., Mem. Torrey Bot. Club 18: 138 (1932)

Synonyms: *Cuscuta arvensis* Beyr. ex Engelm., *C. arvensis* var. *calycina* (Engelm.) Engelm., *C. gymnocarpa* subsp. *deflexa* Buia, *C. kamelorum* Pavlov, *C. pentagona* var. *calycina* Engelm., *C. pentagona* var. *subulata* Yunck., *Epithymum arvense* (Beyr. ex Engelm.)

Fig. 2.24. *Crocus sativus* L. (By Farsad Nadjafi.)

Nieuwl. & Lunell, *Grammica campestris* (Yunck.) Hadac & Chrtek.

Family: *Convolvulaceae*.

Arabic name: *Koshouth*.

Persian name: *Sess*.

Common name: Golden dodder, Field dodder, Prairie dodder, Large-seeded alfalfa dodder, Yellow dodder (Fig. 2.25).

Distribution

Africa: Botswana, Cameroon, Chad, Comoros, Egypt, Eritrea, Ethiopia, Guinea, Kenya, Lesotho, Madagascar, Mauritius, Morocco, Mozambique, Seychelles, Somalia, Sudan, Tanzania, Uganda, Zambia, Zimbabwe. *Asia*: Afghanistan, Bangladesh, Caucasus, China, India, Indonesia, Iran, Japan, Kyrgyzstan, Lebanon, Myanmar, Russia, Singapore, Sri Lanka, Syria, Tajikistan, Thailand, Transcaucasus, Turkey, Turkmenistan, Ukraine, Uzbekistan, Vietnam, Yemen. *Europe*: Albania, Austria, Belarus, Belgium, Bulgaria, Cyprus, France, Germany, Great Britain, Greece, Hungary, Italy, Netherlands, Poland, Portugal, Romania, Russia, Spain, Switzerland. *Americas*: Cuba, Ecuador, Jamaica, Mexico, Paraguay, Peru, Puerto Rico, USA, Uruguay, Venezuela. *Oceania and Australasia*: Australia, Fiji, New Zealand.

Morphology

Stems thin to medium, whitish or yellowish. Inflorescences globose, usually many-flowered and dense, sometimes few-flowered and loose, 3–12 mm diameter. Pedicels mostly shorter than flowers. Flowers 2–3 mm, white, cream, or partly reddish, smooth or glandular, 5-merous. Calyx almost as long as corolla tube, lobes ± overlapping, broadly ovate to orbicular, obtuse to acute, somewhat rigid or carinate; tube campanulate, almost as long as lobes. Corolla lobes shorter than to as long as campanulate tube, suberect then patent to recurved, triangular, acute and often inflexed at apex. Stamens generally a little shorter than corolla lobes; anthers shorter than to almost as long as subulate filaments. Scales reaching stamens, ovate or oblong-ovate, entire, abundantly fimbriate. Styles *c*.0.5× as long as capsule,

Fig. 2.25. *Cuscuta campestris* Yunck. (By Kristian Peters, CC BY-SA 3.0-migrated, https://upload.wikimedia.org/wikipedia/commons/c/cd/Cuscuta_campestris.jpeg.)

slender, suberect to divaricate, stigmas subcapitate, together almost as long as ovary. Capsule (1.5–)2–3 mm long, depressed-globose, dehiscing irregularly. Seeds 1–1.5 mm, 2–4 per capsule (Bobrov *et al.*, 1934–1964; Rechinger, 1963–2001; Tutin *et al.*, 1964; Davis, 1965–1988; Sell and Murrell, 2006).

Cuscuta epithymum (L.) L. Amoen. Acad. 4: 478 (1759)

Synonyms: *Cuscuta europaea* var. *epithymum* L., *Lepimenes epithymum* (L.) Raf.

Family: *Convolvulaceae*.

Arabic name: *Afithmun*, *Koshouth*.

Persian name: *Sess*.

English name: Clover dodder, Lesser dodder (Fig. 2.26).

Distribution

Africa: Algeria, Morocco, Libya, Tunisia. *Asia*: Afghanistan, Caucasus, Iran, Kazakhstan, Lebanon, Syria, Transcaucasus, Turkey, Turkmenistan. *Europe*: Albania, Austria, Belarus, Belgium, Bulgaria, Denmark, Estonia, Finland, France, Germany, Great Britain, Greece, Hungary, Ireland, Italy, Latvia, Lithuania, Netherlands, Norway, Poland, Portugal, Romania, Russia, Spain, Sweden, Switzerland, Ukraine. *Americas*: Argentina, Chile, Columbia, Mexico, USA. *Oceania and Australasia*: Australia, New Zealand.

Morphology

Stems thin to medium, whitish to red, rootless, parasitic plant in which leaves are reduced to minute scales, hairless, thread-like (0.25–0.40 mm in thickness) and can be yellow. Clusters of 7–25 stalkless, purplish flowers are arranged in a compact sphere. Flowers are 2.5–3.0 mm long and have 5 sepals and petals. The sepals come together to form a cup shape, each individual sepal being longer than it is wide. The corolla tube (a tubular structure formed when all petals in a flower are united) is longer than the cup-shaped sepal assemblage. Stamens (male parts) protrude from the flowers and have oval shaped anthers (pollen-bearing parts). Within flowers, the style

Fig. 2.26. *Cuscuta epithymum* (L.). (By Retama, CC BY-SA 3.0-migrated, https://upload.wikimedia.org/wikipedia/commons/2/28/Cuscuta_epithymum1.jpg.)

(female structure) extends from the top of the ovary and is usually as long as the ovary. The fruit is a spherical capsule, which opens along its circumference and is topped by the withered petals, *c.*2 mm, subglobose; each capsule contains generally 4 small seeds, *c.*1 mm, ovoid to irregularly suborbicular, scabrous (Bobrov *et al.*, 1934–1964; Rechinger, 1963–2001; Tutin *et al.*, 1964; Davis, 1965–1988; Sell and Murrell, 2006).

Cydonia oblonga Mill., Gard. Dict. ed. 8: no. 1 (1768)

Synonyms: *Crataegus serotina* Salisb., *Cydonia communis* Loisel., *C. cydonia* (L.) Pers., *C. europaea* Savi, *C. lusitanica* Mill., *C. maliformis* Mill., *C. oblonga* subsp. *integerrima* Lobachev, *C. oblonga* var. *integerrimosepala* Lobachev, *C. oblonga* var. *obpyricarpa* Lobachev, *C. oblonga* var. *obpyriformis* Lobachev, *C. oblonga* var. *orbiculatocomplanata* Lobachev, *C. oblonga* var. *ovalicarpa* Lobachev, *C. oblonga* var. *ovalis* Lobachev, *C. oblonga* var. *plano-cyclocarpa* Lobachev, *C. oblonga* var. *pomiformis* Lobachev, *C. oblonga* var. *urceolata* Lobachev, *C. silvestris* Risso, *C. sumboshia* Buch.-Ham. ex D. Don, *Pyrus cydonia* L., *P. lusitanica* (Mill.) Steud., *P. maliformis* (Mill.) Steud., *P. oblonga* (Mill.) Steud., *P. sumboshia* Spreng., *P. cydonia* Crantz.

Family: *Rosaceae*.

Arabic name: *Safarjal*.

Persian name: *Beh*.

English name: Quince (Fig. 2.27).

Distribution

Africa: Algeria, Eritrea, Ethiopia, Libya, Morocco, Tunisia. *Asia*: Caucasus, Kazakhstan,

Fig. 2.27. *Cydonia oblonga* Mill. (By Conrad Nutschan, CC BY-SA 3.0-migrated, https://upload.wikimedia.org/wikipedia/commons/0/04/Quitte_am_Baum.jpg.)

Lebanon, Iran, Iraq, Syria, Tajikistan, Transcaucasus, Turkmenistan, Turkey, Uzbekistan. *Europe*: Albania, Austria, Bulgaria, Cyprus, France, Germany, Great Britain, Greece, Hungary, Italy, Portugal, Romania, Spain, Switzerland, Ukraine. *Americas*: Bolivia, El Salvador, Guatemala, USA. *Oceania and Australasia*: Australia, New Zealand.

Morphology

Shrub or tree, 2–8 m; twigs sparsely villous, when young, becoming glabrous. Leaves ovate to oblong or occasionally suborbicular, 5–10 cm × 3–8 cm, entire, bilaterally white-tomentose at first, becoming glabrous above and densely villous beneath; petiole 1–2 cm. Pedicels short, tomentose; flowers 4–5.5 cm diameter; petals pink; sepals glandular, toothed, reflexed. Fruit pyriform or subglobose, 3–5 cm (12 cm in cultivation), tomentose, yellowish, fragrant (Rechinger, 1963–2001; Tutin *et al.*, 1964; Davis, 1965–1988; Assadi *et al.*, 1989; Sell and Murrell, 2006).

Delphinium staphisagria **L., Sp. Pl.: 531 (1753)**

Synonyms: *Staphisagria hirtella* Jord., *S. macrosperma* Spach, *S. moschata* Jord.

Family: *Ranunculaceae*.

Arabic name: *Miwizaj*.

Persian name: *Keshmeshe kauli*.

English name: Stavesacre, Larkspur, Licebane (Fig. 2.28).

Distribution

Africa: Algeria, Morocco, Tunisia. *Asia*: Lebanon, Syria, Turkey. *Europe*: Albania, Cyprus, France, Greece, Italy, Portugal, Spain.

Morphology

Biennial stout plant, softly hirsute, 30–100 cm, throughout. Leaves palmately 5- to 7-fid or -lobed, pubescent on both surfaces with mixed very short and longer hairs; segments entire or 3–5-incised, with ovate-lanceolate, or oblong, acute lobes. Bracteoles at base of pedicels. Flowers dusky blue; perianth segments 13–20 mm; lower petals obovate, glabrous, gradually narrowed into a claw; spur only one-third to one-fifth as long as sepals. Follicles ventricose, 8–11 mm wide, inflated (Davis, 1965–1988).

Ferula ammoniacum (D. Don) Spalik, M. Panahi, Piwczynski & Puchalka, Taxon 64: 780 (2015)

Synonyms: *Diserneston gummiferum* Jaub. & Spach, *Dorema ammoniacum* D. Don, *D. gummiferum* (Jaub. & Spach) K.M. Korol., *Ferula ammonifera* Oken, *Peucedanum ammoniacum* (D. Don) Baill.

Family: *Apiaceae* (*Umbelliferae*).

Arabic name: *Oshaq*.

Persian name: *Washā*.

English name: Gum ammoniacum (Fig. 2.29).

Distribution

Asia: Afghanistan, Iran, Pakistan, Turkmenistan.

Morphology

Fibrous and hairy stem 180–250 cm high, base 3–6 cm in diameter, distally paniculate-branched, first strawberry-green, finally brownish, with nodes not thickened. All basal leaves ternate, with pinnate primary segments, more rarely cut bipinnate; outermost segments large, all largely adnate to the rachis, ± connate at the base, running down the base, sometimes unevenly bipartite, 4–10 cm × 1.5–4.5 cm, ± sharp; petioles and rachis thick, finely grooved in dry groove; stem leaves generally reduced to large amplexicaules with sheaths pointed at length. Umbels short (about 10 mm) thick, pedunculate or subsessile, mostly white-floccose-lanate or less long, glabrescent or globose. The ovary is like a tinum. Fruit up to 8 mm × 5–6 mm,

Fig. 2.28. *Delphinium staphisagria* L. (By Ghislain 118 (AD) own work, CC BY-SA 3.0, https://upload.wikimedia.org/wikipedia/commons/c/c4/Delphinium_staphisagria_3.jpg.)

Fig. 2.29. *Ferula ammoniacum* (D. Don). (By Mohammad Reza Kanani.)

widely elliptic, brownish-violet, first whitish-woolly, finally glabrous; ridges filiform, prominent; valle ± distinctly united; patch 2–4-bandaged (Rechinger, 1963–2001; Ghahreman, 1978; Assadi *et al*., 1989).

Dracaena cinnabari Balf.f., Trans. & Proc. Bot. Soc. Edinburgh 30: 623 (1882)

Synonyms: *Draco cinnabari* (Balf.f.) Kuntze.

Family: *Asparagaceae*.

Arabic name: *Damm al-akhawain*.

Persian name: *Khoone siãwosh*.

English name: Dragon's blood tree (Fig. 2.30).

Distribution

Asia: Yemen.

Morphology

A tree such as palms, grows from the tip of the stem, with the leaves borne at the end, 5(–10) m tall, trunk 40 cm diameter; branches at maturity to produce an umbrella-shaped crown; thick and stout, dichotomous branching, branching tends to occur when the growth of the terminal bud is stopped, due to either flowering or traumatic events (e.g., herbivory). Leaves are dense rosettes at the end of its youngest branches, falling every 3 or 4 years before new leaves simultaneously mature, rigid, 30–60 cm × 2–3 cm. Inflorescences terminal; flowers 2–3 groups at the end of the branches, aromatic, white or green; pedicels 5 mm long; perianth cup-shaped, up to 5 mm long; stamens are slightly shorter; style thread-like ending in a capitate stigma. Fruits is small fleshy, 3 seeds, green at young and black at mature, berries exude a deep red resin, known as dragon's blood. Seeds 4–5 mm in diameter (Govaerts, 2000).

Fig. 2.30. *Dracaena cinnabari* Balf.f. (By Rod Waddington own work, CC BY-SA 2.0, https://upload.wikimedia.org/wikipedia/commons/7/79/Dragons_Blood_Tree%2C_Socotra_Island_%2810941931846%29.jpg.)

Drimia maritima **(L.) Stearn, Ann. Mus. Goulandris 4: 204 (1978)**

Synonyms: *Charybdis maritima* (L.) Speta, *Ornithogalum maritimum* (L.) Lam., *O. squilla* Ker Gawl., *Scilla maritima* L., *S. rubra* Garsault, *Squilla maritima* (L.) Steinh., *Stellaris scilla* Moench, *Urginea maritima* (L.) Baker, *U. scilla* Steinh.

Family: *Asparagaceae*.

Arabic name: *Oᶜnṣol baḥri*.

Persian name: *Oᶜnṣol*, *Piāze moush*.

English name: Maritime squill (Fig. 2.31).

Distribution

Africa: Algeria. *Europe*: Portugal, Spain.

Morphology

The bulbs (which often clump together near soil level) are up to 20 cm in diameter and can weigh up to 1 kg. Each bulb bears

Fig. 2.31. *Drimia maritima* (L.). (By Orangerind, CC BY-SA 4.0, https://upload.wikimedia.org/wikipedia/commons/2/23/Wild_Drimia_Maritima_in_Iran.jpg.)

about 10 dark green, leathery leaves, up to 100 cm long and 12 cm wide. The inflorescence is a dense, many-flowered raceme some 1–1.5 m tall. The flower stalks are 15–20 mm long. The flowers are 14–16 mm across, with 6 perianth segments, which are white with a greenish stripe. A variant form with red-tinted flowers is commonly referred to as red squill. The fruit is a capsule, 6–12 mm long, borne on an erect stalk (Chaudhary, 2001).

Ecballium elaterium **(L.) A. Rich, J.B.G. Bory de Saint-Vincent, Dict. Class. Hist. Nat. 6: 19 (1824)**

Synonyms: *Bryonia elaterium* (L.) E.H.L. Krause, *Ecballium agreste* Rchb., *Momordica elaterium* L.

Family: *Cucurbitaceae*.

Arabic name: *Qithā' el ḥimār*.

Persian name: *Khiāre khar*.

Common name: Squirting cucumber, Exploding cucumber (Fig. 2.32).

Distribution

Africa: Albania, Algeria, Libya, Morocco, Tunisia. *Asia*: Caucasus, China, Iran, Kazakhstan, Lebanon, Syria, Tajikistan, Transcaucasus, Turkey, Uzbekistan. *Europe*: Bulgaria, Cyprus, France, Great Britain, Greece, Hungary, Italy, Portugal, Russia, Spain, Ukraine. *Americas*: USA. *Oceania and Australasia*: Australia, New Zealand.

Morphology

Stems up to 60 cm. Leaves long-petiolate; lamina 4–10 cm, entire, cordate to triangular, undulate, denticulate, rather fleshy. Corolla yellowish, male flowers 18–20 mm. Fruit 4–5 cm × 2–5 cm, ovoid, green, very hispid, dehiscing explosively at the base (Bobrov *et al.*, 1934–1964; Rechinger, 1963–2001; Davis, 1965–1988; Sell and Murrell, 2006).

Phyllanthus emblica **L., Sp. Pl.: 982 (1753)**

Synonyms: *Cicca emblica* (L.) Kurz, *C. macrocarpa* Kurz, *Diasperus emblica* (L.) Kuntze, *D. pomifer* (Hook.f.) Kuntze, *Dichelactina nodicaulis* Hance, *Emblica arborea* Raf., *E. officinalis* Gaertn., *Phyllanthus glomeratus* Roxb. ex

Fig. 2.32. *Ecballium elaterium* (L.) (by Peter A. Mansfeld, CC BY-SA 3.0, https://upload.wikimedia.org/wikipedia/commons/5/5b/Ecballium_elaterium_pm2.jpg)

Benth., *P. mairei* H. Lév., *P. mimosifolius* Salisb., *P. pomifer* Hook.f., *P. taxifolius* D. Don.

Family: *Phyllanthaceae*.

Arabic name: *Amlaj*.

Persian name: *Āmoleh*.

English name: Indian gooseberry, Emblic myrobalan, Malacca tree, Amla (Fig. 2.33).

Distribution

Africa: Mauritius. *Asia*: Bangladesh, Borneo Island, Cambodia, China, India, Indonesia, Laos, Myanmar, Pakistan, Singapore, Sri Lanka, Taiwan, Thailand, Vietnam. *Americas*: Cuba, Puerto Rico, Trinidad and Tobago, Windward Islands.

Morphology

Tree, 3–8(–23) m tall, deciduous, monoecious; bark brownish; main stems terete, sparsely lenticellate, with very reduced short shoots producing groups of leafy shoots; leafy shoots angular, tawny pubescent, at start of growing season often with poorly developed leaves and densely flowered, later with fewer flowers and better-developed leaves. Leaves distichous; stipules triangular-ovate, 0.8–1.5 mm, brown, margins entire or denticulate, ciliate; petiole 0.3–0.7 mm; leaf blade oblong or linear-oblong, 8–23 cm × 1.5–6 mm, papery to leathery, paler abaxially, green adaxially, drying reddish or brownish, base shallowly cordate and slightly oblique, margin narrowly revolute, apex truncate, rounded or obtuse, mucronate or retuse at tip; lateral veins 4–7 pairs. Fascicles with many male flowers and sometimes 1 or 2 larger female flowers. Male flowers: pedicels 1–2.5 mm; sepals 6, membranous, yellow, obovate or spatulate, subequal, 1.2–2.5 mm × 0.5–1 mm, apex obtuse or rounded, margin entire or shallowly denticulate; disk glands 6, subtriangular; stamens 3; filaments coherent into column,

Fig. 2.33. *Phyllanthus emblica* L. (By L. Shymal, CC BY-SA 2.5, https://commons.wikimedia.org/wiki/File:Phyllanthus_officinalis.jpg#/media/File:Phyllanthus_officinalis.jpg.)

0.3–0.7 mm; anthers erect, oblong, 0.5–0.9 mm, longitudinally dehiscent, apex mucronate. Female flowers: pedicels *c.*0.5 mm; sepals 6, oblong or spatulate, 1.6–2.5 mm × 0.7–1.3 mm, apex obtuse or rounded, thicker, margin membranous, ± lobate; ovary ovoid, *c.*1.5 mm, 3-celled; styles 3, (1–)2.5–4 mm, connate at base, deeply bifid, lobes divided at tip. Fruit a drupe, globose, 1–1.3 cm in diameter, exocarp fleshy, pale green or yellowish white, endocarp crustaceous. Seeds reddish, 5–6 mm × 2–3 mm (Liogier, 1985–1997).

***Erysimum* × *cheiri* (L.) Crantz, Cl. Crucif. Emend.: 116 (1769)**

Synonyms: *Cheiranthus* × *cheiri* L., *Ch.* × *cheiri* f. *glaucescens* Pamp., *Ch.* × *fruticulosus* (Spreng.) L'Hér. ex DC., *Ch.* × *helveticus* Jacq., *Ch.* × *intermedius* Schleich., *Ch.* × *luteus* Dulac, *Ch.* × *muralis* Salisb., *Ch.* × *silenifolius* Willd., *Cheiri* × *montanum* Clairv., *Cheiri murale* (Lam.) Samp., *Cheiri* × *vulgare* Clairv., *Cheirinia* × *suffruticosa* Link, *Crucifera* × cheiri (L.) E.H.L. Krause, *Erysimum* × *cheiri* subsp. *inexpectans* Véla, Ouarmim & Dubset, *E.* × *helveticum* (Jacq.) DC., *E.* × *murale* Lam., *E.* × *suffruticosum* Spreng.

Family: *Brassicaceae* (*Cruciferae*).

Arabic name: *Khiri aṣfar*.

Persian name: *Shab boye khiri*.

English name: Wallflower (Fig. 2.34).

Distribution

Africa: Algeria, Morocco. *Asia*: China, Iraq, Korea, Lebanon, Syria, Turkey. *Europe*:

Fig. 2.34. *Erysimum* × *cheiri* L. (By Fabricio. Cardenas, CC BY-SA 4.0, https://upload.wikimedia.org/wikipedia/commons/1/12/20180325_-_C%C3%A9ret_-_Erysimum_cheiri.jpg.)

Albania, Austria, Belgium, Bulgaria, France, Germany, Great Britain, Greece, Hungary, Italy, Netherlands, Portugal, Romania, Spain, Switzerland. *Americas*: Canada, Ecuador, Falkland Islands, USA. *Oceania and Australasia*: Australia, New Zealand.

Morphology

Stems ascending to erect, up to 50 cm. Leaves very narrowly elliptic to lanceolate, entire or weakly dentate, canescent at least when young. Petals orange-yellow. Siliquae substrict, much flattened, canescent, somewhat strangulate between the seeds, 45–65 mm × 3–4 mm. Style 2–3 mm (Tutin *et al.*, 1964; Davis, 1965–1988; Sell and Murrell, 2006).

***Ferula assa-foetida* L., Sp. Pl.: 248 (1753)**

Synonyms: *Ferula foetida* St.-Lag., *Narthex polakii* Stapf, *N. silphium* Oerst., *Peucedanum assa-foetida* (L.) Baill., *P. hooshee* (Lindl. ex Descourt.) Baill., *Scorodosma assa-foetida* (L.) H. Karst.

Family: *Apiaceae* (*Umbelliferae*).

Arabic name: *Anjedān*, *Ḥiltit*.

Persian name: *Anghuzah*.

English name: Asafoetida (Fig. 2.35).

Distribution

Africa: Libya. *Asia*: Bangladesh, Iran, Laos, Turkmenistan, Uzbekistan.

Morphology

Perennial plant, hairy stem 100–200 cm high, in middle *c.*15 mm thick, yellowish, grooved, hairless. Basal leaves *c.*2-ternato-pinnatisect, 30–35 mm × 15–25 mm; final segments up to 100 mm × 20 mm, entire or slightly lobed. Lower stem leaves 3–4-ternate-pinnate, pubescent, up to 30 cm × 20 cm; final segments smooth, up to 40 mm × 15 mm, coarsely pinnate-lobed; upper vagina *c.*9 cm × 4 cm, sub-membranaceous, sparingly pubescent. Panicle compact, a partial inflorescence from the central fruit umbel, briefly pedunculate and from the male umbel ± very pedunculate *c.*4 composite, leaf sheath with reduced ovate,

Fig. 2.35. *Ferula assa-foetida* L. (By Valery Fassiaux, via Wikimedia commons, https://upload.wikimedia.org/wikipedia/commons/7/7e/Ferula_assa-foetida_%28Uzbekistan%29_1.jpg.)

supported by a spurious sheath for the testicles. Umbrella-bearing rays 20–25, *c.*5 cm long, glabrous; umbrella *c.*17–20-flowered. Small bracts, bracts or absent. Petals blonde, 1–5 mm long, glabrous, ovary glabrous. Mericarpia obovate or elliptic, up to 12 mm × 7 mm, glabrous (Rechinger, 1963–2001; Ghahreman, 1978; Assadi *et al.*, 1989).

Ferula gummosa Boiss., Diagn. Pl. Orient. Ser. 2, 2: 92 (1856)

Synonyms: *Ferula erubescens* Boiss., *F. galbaniflua* Boiss. & Buhse, *Peucedanum galbanifluum* (Boiss. & Buhse) Baill.

Family: *Apiaceae* (*Umbelliferae*).

Arabic name: *Qinah*.

Persian name: *Bārijah*, *Qāsni*.

English name: Galbanum (Fig. 2.36).

Distribution

Asia: Iran, Turkmenistan.

Morphology

Perennial plant, monocarpic, white. Stems 70–100 cm high, strawberry, 15 mm thick or thicker in the middle, solid, grooved. Basal leaves 30 cm × 20 cm, 4-pinna, densely pubescent; last pieces *c.*1.5 mm × 1 mm, pinnate; lobes 0.5–1 mm × 0.2 mm; upper vagina *c.*5 cm × 4 cm, leathery, glabrous. Panicle loose. Shadowy fruit rays 10–20; radius 2–4 cm long. Umbel 15–20-flowered; petals 1.5 mm long, yellowish, pubescent. Mericarpia 10–15 mm × *c.*8 mm (Rechinger, 1963–2001; Ghahreman, 1978; Assadi *et al.*, 1989).

Fig. 2.36. *Ferula gummosa* Boiss. (By Pirehelokan, CC BY-SA 4.0, https://upload.wikimedia.org/wikipedia/commons/a/ae/Goli_Lo.jpg.)

Ferula persica Willd., Sp. Pl., ed. 4, 1: 1418 (1798)

Synonyms: *Ferula puberula* Boiss. & Buhse, *Peucedanum persicum* (Willd.) Baill., *Prangos euryangiodes* Stapf & Wettst.

Family: *Apiaceae* (*Umbelliferae*).

Arabic name: *Sakbinaj*.

Persian name: *Komāye biābāni*.

English name: Persian giant fennel (Fig. 2.37).

Distribution

Asia: Iran, Transcaucasus.

Morphology

Perennial plant, polycarpical, hairy stems 30–100 cm high, thick, streaked. Basal leaves 3–4-pinnate, 30–35 cm × *c*.25 cm, above glabrous, below sparingly, with a broad-pubescent spine densely; final segments 4–20 mm × 3–10 mm, slightly to deeply lobed, margin entire or coarsely toothed; upper vagina *c*.8 cm × 2 cm, herbaceous, pubes, not inflated. Panicle loose. Umbel-bearing rays, 17–25; radius *c*.4 cm. Mericarpia 9–11 mm × 5–6 mm; wing *c*.0.5 mm wide; dorsal tubes 5–7 through the valley; joiners 16–18 (Rechinger, 1963–2001; Ghahreman, 1978; Assadi *et al*., 1989).

Fig. 2.37. *Ferula persica*. (By Mohammad Reza Kanani.)

Glycyrrhiza glabra L., Sp. Pl.: 742 (1753)

Synonyms: *Glycyrrhiza alalensis* X.Y. Li, *G. alaschanica* Grankina, *G. brachycarpa* Boiss., *G. echinata* Lepech., *G. glabra* var. *caduca* X.Y. Li, *G. glabra* subsp. *glandulifera* (Waldst. & Kit.) Ponert, *G. glabra* var. *glandulosa* X.Y. Li, *G. glabra* var. *laxifoliolata* X.Y. Li, *G. glabra* var. *pubescens* Litv., *G. glandulifera* Waldst. & Kit., *G. hirsuta* L., *G. laevis* Pall., *G. michajloviana* Grankina & E.V. Kuzmin, *G. nadezhinae* Grankina, *G. officinalis* Lepech., *G. pallida* Boiss. & Noë, *G. violacea* Boiss., *G. vulgaris* Gueldenst. ex Ledeb., *Liquiritia officinalis* Moench, *L. officinarum* Medik.

Family: *Fabaceae* (*Leguminosae*).

Arabic name: *Sus*.

Persian name: *Shirin bayan*.

English name: Liquorice, Licorice (Fig. 2.38).

Distribution

Africa: Algeria, Egypt. *Asia*: Afghanistan, Bangladesh, Caucasus, China, Iran, Iraq, Kazakhstan, Kyrgyzstan, Lebanon, Maldives, Mongolia, Pakistan, Saudi Arabia, Syria, Tajikistan, Transcaucasus, Turkey, Turkmenistan, Uzbekistan. *Europe*: Austria, Albania, Bulgaria, Cyprus, France, Greece, Hungary, Island, Italy, Portugal, Romania, Russia, Spain, Switzerland, Ukraine. *Americas*: USA. *Oceania and Australasia*: Australia.

Morphology

Perennial herb, stem 50–150 cm tall, woody at base, densely scaly glandular punctate,

Fig. 2.38. *Glycyrrhiza glabra* L. (By Farsad Nadjafi.)

white hairy. Leaves 5–14 cm, 11–17-foliolate; stipules caducous, linear, 1–2 mm; petiole densely yellow-brown glandular hairy and villous; leaflets ovate-oblong, oblong-lanceolate, or elliptic, 1.7–4 cm × 0.8–2 cm, abaxially densely yellow scaly glandular punctate and pubescent on veins, adaxially glabrescent or pilose, base rounded, apex rounded, or retuse and with mucro. Racemes much and densely flowered; rachis densely brown scaly glandular punctate, white villous and tomentose; bracts lanceolate, *c.*2 mm, membranous. Calyx campanulate, 5–7 mm, sparsely yellow glandular punctate and pubescent, 5-toothed; upper 2 teeth mostly joined. Corolla purple or light purple, 9–12 mm; standard ovate or oblong, 1–1.1 cm, base clawed, apex retuse; wings 8–9 mm; keel straight, 7–8 mm. Ovary glabrous. Legume oblong, flat, 17–35 mm × 4.5–7 mm, rarely constricted between seeds, glabrous or sparsely hairy, rarely glandular hairy. Seeds 2–8, dark green, *c.*2 mm in diameter, smooth (Bobrov *et al.*, 1934–1964; Rechinger, 1963–2001; Tutin *et al.*, 1964; Davis, 1965–1988).

***Helleborus niger* L., Sp. Pl.: 558 (1753)**

Synonyms: *Helleborus altifolius* Steud.

Family: *Ranunculaceae*.

Arabic name: *Kharbaq aswad*.

Persian name: *Kharbaqe siāh*.

English name: Christmas rose, Black hellebore (Fig. 2.39).

Distribution

Europe: Austria, France, Germany, Italy, Switzerland, Sweden. *Americas*: USA.

Fig. 2.39. *Helleborus niger* L. (By Robert.Hundsdorfer, CC BY-SA 2.0, https://upload.wikimedia.org/wikipedia/commons/f/f2/Helleborus_niger_Kaiser.jpg.)

Morphology

Perennial herb with rhizomes or erect, up to 30 cm, rather woody stems and digitate or pedate leaves, basal leaves overwintering, pedate with 7–9 segments; cauline leaves ovate, entire, pale green. Inflorescence cymose. Flowers large; perianth segments 5; honey leaves 8–12, obliquely infundibuliform, smaller than the perianth segments; involucre absent. Fruit of 3–8, several-seeded follicles. Flowers 2–3, 3–10 cm in diameter. Perianth segments white or pink-tinged, ovate. Follicles *c.*7, connate at base (Tutin *et al.*, 1964).

***Hyoscyamus niger* L., Sp. Pl.: 179 (1753)**

Synonyms: *Hyoscarpus niger* (L.) Dulac, *H. agrestis* Kit. ex Schult., *H. auriculatus* Ten., *H. bohemicus* F.W. Schmidt, *H. lethalis* Salisb., *H. niger* subsp. *agrestis* (Kit.) Hultén, *H. officinarum* Crantz, *H. pallidus* Waldst. & Kit. ex Willd., *H. persicus* Boiss. & Buhse, *H. pictus* Roth, *H. syspirensis* K. Koch, *H. verviensis* Lej., *H. vulgaris* Neck., *H. vulgaris* L.

Family: *Solanaceae*.

Arabic name: *Banj aswad*, *Sikarān*.

Persian name: *Bangdāneh*.

English name: Henbane, Black henbane, Stinking nightshade (Fig. 2.40).

Distribution

Africa: Algeria, Morocco, Tunisia. *Asia*: Afghanistan, Caucasus, China, Iran, Iraq, Japan, Kazakhstan, Korea, Kyrgyzstan, Lebanon, Mongolia, Myanmar, Nepal, Pakistan, Syria, Tajikistan, Transcaucasus, Turkey, Turkmenistan, Uzbekistan. *Europe*: Albania, Austria, Belarus, Belgium, Bulgaria, Denmark, Estonia, Finland, France, Germany, Great Britain, Greece, Hungary, Ireland, Italy, Latvia, Lithuania, Netherlands, Norway, Poland, Portugal, Romania, Russia, Spain, Sweden, Switzerland, Ukraine. *Americas*: Mexico, USA. *Oceania and Australasia*: Australia, New Zealand.

Morphology

Biennial herb, to 1 m tall, pubescent throughout with sticky glandular hairs. Roots sometimes stout, fleshy, 2–3 cm in diameter. Leaves sessile, sometimes forming a rosette;

Fig. 2.40. *Hyoscyamus niger* L. (https://commons.wikimedia.org/wiki/File:Hyoscyamus_niger_0003.JPG#/media/Datei:Hyoscyamus_niger_0003.JPG.)

blade of rosette leaves ovate-lanceolate or oblong, *c.*30 cm × 10 cm, coarsely dentate or pinnately lobed or parted, apex acute; blade of cauline leaves ovate or deltate-ovate, 4–10 cm × 2–6 cm, pubescent with sticky hairs, nearly clasping or broadly cuneate at base, lobed or entire. Flowers subsessile or on pedicels 3–5 mm. Calyx tubular-campanulate, 1–1.5 cm; lobes deltate, unequal, acute. Corolla pale yellow, usually with purple veins, campanulate, 2–3 cm. Stamens exserted. Fruiting calyx urceolate, 2–2.5 cm × 1–1.5 cm; lobes mostly erect, tipped with a sharp tooth. Capsules ovoid-rounded, *c.*1.5 cm × 1.2 cm. Seeds yellow-brown, discoid, *c.*1 mm in diameter (Bobrov *et al.*, 1934–1964; Rechinger, 1963–2001; Tutin *et al.*, 1964; Davis, 1965–1988; Sell and Murrell, 2006).

Iris florentina L., Syst. Nat. ed. 10, 2: 863 (1759)

Synonyms: *Iris albicans* Lange, *I. florentina* subsp. *albicans* (Lange) K. Richt., *I.* × *germanica* subsp. *albicans* (Lange) O. Bolòs & Vigo, *I. madonna* Sprenger, *I. majoricensis* Barceló, *I. officinalis* Salisb.

Family: *Iridaceae*.

Arabic name: *Susan abyaḍ*.

Persian name: *Zanbaqe sefid*.

English name: Florentine iris (Fig. 2.41).

Distribution

Africa: Morocco. *Asia*: Lebanon, Saudi Arabia, Syria, Turkey, Yemen. *Europe*: Bulgaria, Cyprus, France, Italy, Portugal, Romania, Spain.

Morphology

Stems 40–90 cm, stout, with branches at least 5 cm long in the upper half, the lower flowers well-exserted from the subtending bract even in bud. Leaves 30–70 cm × 20–35 mm, somewhat glaucous, more or less straight. Flowers (3–)4(–5), the lower subsessile on the main axis and enclosed by the subtending bract up to anthesis, pure white; spathes 35–55 mm, scarcely exceeding the hypanthial tube, scarious in the upper, often tinged with purple. Hypanthial tube 17–25 mm. Falls 55–90 cm × 40–60 mm, cuneate-obovate, without a distinct claw. Standards 55–90 cm × 45–60 mm, broadly elliptical, with a short, wide claw. Seeds pyriform, rugose, reddish-brown (Davis, 1965–1988; Castroviejo *et al.*, 1980).

Juniperus sabina L., Sp. Pl.: 1039 (1753)

Synonyms: *Juniperus humilis* Salisb., *J. sabina* var. *cupressifolia* Aiton, *Sabina officinalis* Garcke, *S. vulgaris* Antoine.

Family: *Cupressaceae*.

Arabic name: *Abhal*.

Persian name: *Mymarz*.

English name: Savin (Fig. 2.42).

Fig. 2.41. *Iris florentina* L. (By Alessandra. Perugi, CC BY-SA 3.0, https://upload.wikimedia.org/wikipedia/commons/0/0f/Irisflorentina_ap.jpg.)

Fig. 2.42. *Juniperus sabina* L. (By Meneerke. Bloem, CC BY-SA 3.0, https://upload.wikimedia.org/wikipedia/commons/8/8b/Juniperus_sabina_02.JPG.)

Distribution

Africa: Algeria. *Asia*: Caucasus, China, Mongolia, Kazakhstan, Korea, Kyrgyzstan, Tajikistan, Transcaucasus, Turkey. *Europe*: Albania, Austria, Bulgaria, France, Germany, Greece, Italy, Poland, Russia, Spain, Switzerland, Sweden, Ukraine.

Morphology

Shrub, rarely small tree, dioecious, rarely monoecious; bark grayish-brown; branchlets densely arranged, ascending, slender, 0.8–1 mm in diameter. Leaves both scalelike and needlelike; needlelike leaves usually present on young plants, rarely present on adult plants, decussate or in whorls of 3, closely appressed, 3–7 mm, concave adaxially, convex abaxially, apex sharply pointed; scalelike leaves decussate, rhombic or rhombic-ovate, 1–2.5 mm, abaxial gland central, prominent, elliptic. Pollen cones ellipsoid or oblong, 3–4 mm; microsporophylls 10–14, each with 2–4 pollen sacs. Seed cones light brownish-green, brown, purplish-blue, or black when ripe, often glaucous, usually irregularly globose, 5–8 mm × 5–9 mm, (1-or)2-seeded. Seeds ovoid, slightly flattened, 4–5 mm, ridged, with resin pits, apex blunt or slightly pointed (Bobrov *et al.*, 1934–1964; Tutin *et al.*, 1964; Davis, 1965–1988).

***Lawsonia inermis* L., Sp. Pl.: 340 (1753)**

Synonyms: *Alcanna spinosa* (L.) Gaertn., *A. spinosa* (L.) Gaertn., *Casearia multiflora* Spreng, *Lawsonia alba* Lam., *L. alba* var. *flavescens* Hassk., *L. alba* var. *miniata* Hassk., *L. coccinea* Sm., *L. falcifolia* Stokes, *L. inermis* var. *spinosa* (L.) Pers., *L. purpurea* Lam., *L. spinosa* L., *Rotantha combretoides* Baker.

Family: *Lythraceae*.

Arabic name: *Ḥanā*.

Persian name: *Ḥanā*.

English name: Henna, Egyptian privet (Fig. 2.43).

Distribution

Africa: Algeria, Burkina Faso, Cameroon, Chad, Comoros, Djibouti, Eritrea, Ethiopia, Ghana, Guinea, Guinea-Bissau, Kenya, Liberia, Libya, Madagascar, Mali, Mauritania, Morocco, Mozambique, Niger, Nigeria, Senegal, Sierra Leone, Somalia, Sudan, Tanzania, Togo, Tunisia, Uganda. *Asia*: Bangladesh, India, Iraq, Laos, Lebanon, Maldives, Myanmar, Oman, Pakistan, Saudi Arabia, Sri Lanka, Syria, Tajikistan, Vietnam, Yemen. *Europe*: Netherlands. *Americas*: Costa Rica, Cuba, Ecuador, Guyana, Mexico, Panamá, Suriname, Venezuela. *Oceania and Australasia*: New Guinea.

Morphology

The younger branches of the plant are often quadrangular. Leaves 12–67 mm × 5–27 mm, narrowly obovate or oblong or broadly lanceolate, often short acuminate, mucronulate, feathered, sparingly reticulate-venous, gradually tapering towards the short petiolus. Calyx 3–5 mm long. Petals slightly longer than calyx, at first white or pale sulfurous, then covered with milk. Fruit 4–6 mm × 4.5–8 mm; seeds 2–2.6 mm (Rechinger, 1963–2001).

***Liquidambar orientalis* Miller, Gard. Dict. ed. 8: no. 2 (1768)**

Synonyms: *Liquidambar imberbis* Aiton.

Family: *Altingiaceae*.

Arabic name: *Lobnā*.

Persian name: *Anbar sāel*.

English name: Oriental sweetgum, Turkish sweetgum (Fig. 2.44).

Distribution

Asia: Turkey. *Europe*: Greece (Aegean Islands).

Fig. 2.43. *Lawsonia inermis* L. (By Dinesh Valke from Thane, India – Lawsonia inermis, CC BY-SA 2.0, https://commons.wikimedia.org/w/index.php?curid=51582890.)

Fig. 2.44. *Liquidambar orientalis* Miller. (By Aroche, CC BY-SA 3.0, https://upload.wikimedia.org/wikipedia/commons/4/4c/Liquidambar_orientalis_arboretum_Breuil_2.jpg.)

Morphology

Tree up to 20 m. Leaves palmatifid, 5–10(–12) cm × 6–13 cm, divided into 5 lobes, each lobe undivided or sparsely lobulate, margin finely sinuate-dentate or -serrate, glabrous or rarely with a sparse tuft of short hairs beneath at the base of the main veins. Fruiting head 2.5–3 cm diameter, pendulous (Davis, 1965–1988).

***Malva sylvestris* L., Sp. Pl.: 689 (1753)**

Synonyms: *Malva ambigua* var. *microphylla* Rouy, *M. acutiloba* K. Koch, *M. acutiloba* Martrin-Donos, *M. acutiloba* Sennen, *M. albiflora* C. Presl, *M. albiflora* J. Presl, *M. altissima* Sennen, *M. ambigua* Guss., *M. aragonensis* Sennen, *M. ciliata* Wall., *M. elata* Pomel, *M. elata* Salisb., *M. equina* Brunfels, *M. equina* Brunfels ex Wallr., *M. erecta* C. Presl, *M. erecta* Gilib., *M. erecta* J. Presl, *M. erevaniana* Takht., *M. glabra* Desr., *M. grossheimii* Iljin, *M. gymnocarpa* Pomel, *M. hirsuta* C. Presl, *M. hirsuta* J. Presl, *M. hirsuta* Viv., *M. longelobata* Sennen, *M. longepedunculata* Sennen, *M. longilobata* Sennen, *M. lucida* Lindl., *M. major* Garsault, *M. martrinii* Rouy, *M. orientalis* Mill., *M. plebeja* Steven, *M. polymorpha* Guss., *M. pumila* Retz., *M. racemosa*

C. Presl, *M. racemosa* J. Presl, *M. recta* Opiz, *M. rotundifolia* Cav., *M. ruderalis* Salisb., *M. silvestris*, *M. sylvestris* f. *albiflora* Matvejeva, *M. sylvestris* f. *vestita* Pamp., *M. sylvestris* subsp. *ambigua* (Guss.) P. Fourn., *M. sylvestris* subsp. *ambigua* (Guss.) Thell., *M. sylvestris* subsp. *erecta* (C. Presl) Nyman, *M. sylvestris* subsp. *erecta* (J. Presl) Slavik, *M. sylvestris* subsp. *viviana* (Rouy) P. Fourn., *M. sylvestris* var. *ambigua* (Guss.) Batt., *M. sylvestris* var. *ambigua* (Guss.) Samp., *M. sylvestris* var. *angustiloba* Celak., *M. sylvestris* var. *erecta* Opiz, *M. sylvestris* var. *erecta* Wirtg., *M. sylvestris* var. *hirsuta* (J. Presl) Rouy, *M. sylvestris* var. *incanescens* Griseb., *M. sylvestris* var. *latiloba* Celak., *M. sylvestris* var. *orbicularis* Dethard., *M. sylvestris* var. *parviflora* J.J. Schmitz & Regel, *M. sylvestris* var. *polymorpha* (Guss.) Parl., *M. sylvestris* var. *procumbens* Wirtg., *M. sylvestris* var. *recta* Vollm., *M. sylvestris* var. *tomentella* C. Presl, *M. sylvestris* var. *tomentella* J. Presl, *M. tomentella* C. Presl, *M. tomentella* J. Presl, *M. veneta* hort., *M. veneta* hort. ex DC., *M. viviana* Rouy, *M. vulgaris* Gray, *M. vulgaris* Ten.

Family: *Malvaceae.*

Arabic name: *Khubbazah barriah.*

Persian name: *Panirak.*

English name: Common mallow, Cheeses, Wood mallow, Tall mallow (Fig. 2.45).

Distribution

Africa: Algeria, Angola, Egypt, Eritrea, Libya, Morocco, Tunisia, Zimbabwe. *Asia*: Afghanistan, Bangladesh, Caucasus, Iran, Iraq, Japan, Kazakhstan, Korea, Kyrgyzstan, Kyrgyzstan, Lebanon, Pakistan, Syria, Tajikistan, Transcaucasus, Turkey, Turkmenistan, Uzbekistan, Vietnam. *Europe*: Albania, Austria, Baltic States, Belarus, Belgium, Bulgaria, Canary Islands, Cyprus, Denmark, Finland, France, Germany, Great Britain, Greece (Aegean Islands), Hungary, Ireland, Italy, Netherlands, Norway, Poland, Portugal, Romania, Russia, Spain, Sweden, Switzerland, Ukraine. *Americas*: Argentina, Brazil, Canada, Chile, Ecuador, Guatemala, Honduras, Mexico, Peru, USA, Venezuela. *Oceania and Australasia*: Australia, New Zealand.

Morphology

Perennial to biennial plant, stems erect or ascending, spreading pilose with tuberculate hairs. Leaves ± orbicular, 5–7-lobed, crenate, sparsely pilose to glabrous. Flowers in fascicles in the leaf axils, and in clusters at the branch apices. Epicalyx segments ± linear to oblong or narrowly ovate. Sepals broadly triangular, 3–7 mm. Petals 18–25 mm, 4 or more times longer than the sepals, mauve to pink, emarginate or shallowly 3–5-fid. Mericarps glabrous or hairy (var. *eriocarpa* Boiss.), ± flat on the back, rugose (Bobrov *et al.*, 1934–1964; Rechinger, 1963–2001; Tutin *et al.*, 1964; Davis, 1965–1988; Sell and Murrell, 2006).

Marrubium vulgare **L., Sp. Pl.: 583 (1753)**

Synonyms: *Marrubium album* Garsault, *M. apulum* Ten., *M. ballotoides* Boiss. & Balansa, *M. germanicum* Schrank ex Steud., *M. hamatum* Kunth, *M. hyperleucum* Candargy, *M. uncinatum* Stokes, *M.* vaillantii Coss. & Germ., *M. vulgare* subsp. *apulum* (Ten.) H. Lindb., *Prasium marrubium* E.H.L. Krause.

Family: *Lamiaceae* (*Labiatae*).

Arabic name: *Farāsyun abyaḍ.*

Persian name: *Gandnāye koohi.*

English name: White horehound, Marrubium (Fig. 2.46).

Distribution

Africa: Algeria, Libya, Morocco, Tunisia. *Asia*: Afghanistan, Caucasus, Iran, Iraq, Kazakhstan, Lebanon, Pakistan, Saudi Arabia, Syria, Transcaucasus, Turkey, Turkmenistan, Uzbekistan. *Europe*: Albania, Austria, Baltic States, Belarus, Belgium, Bulgaria, Canary Islands, Cyprus, Denmark, France, Germany, Great Britain, Greece (Aegean Islands), Hungary, Ireland, Netherlands,

Fig. 2.45. *Malva sylvestris* L. (By Farsad Nadjafi.)

Poland, Portugal, Romania, Russia, Spain, Sweden, Switzerland, Ukraine. *Americas*: Argentina, Bolivia, Brazil, Canada, Chile, Ecuador, Guatemala, Mexico, Peru, USA, Venezuela.

Morphology

Perennial herb to 0.5 m tall. Leaves petiolate, broadly ovate, 20–30(–50) mm × 20–30(–50) mm, crenate, tomentose to lanate. Inflorescence lax with verticils many-flowered, dense. Calyx 10-veined; teeth 10, main 5 long, alternate with to 5 accessory teeth, 1–4 mm, subulate, hooked. Corolla white, *c.*9 mm; tube *c.*6 mm, densely pubescent outside, pilose annulate inside; upper lip as long as or slightly shorter than lower lip, straight or spreading, 2-lobed; middle lobe of lower lip reniform, undulate, 2-cleft. Nutlets triquetrous, ovoid, warty (Bobrov *et al.*, 1934–1964; Rechinger, 1963–2001; Tutin *et al.*, 1964; Davis, 1965–1988; Assadi *et al.*, 1989; Sell and Murrell, 2006).

Myrtus communis **L., Sp. Pl.: 471 (1753)**

Synonyms: *Myrtus boetica* Mill., *M. communis* f. *baetica* (L.) Cout., *M. buxifolia* Raf.,

Fig. 2.46. *Marrubium vulgare* L. (By Gidip, CC BY-SA 3.0, https://upload.wikimedia.org/wikipedia/commons/7/72/Marrubium_alysson_2.JPG.)

M. communis f. *lusitanica* (L.) Cout., *M. communis* f. *tarentinus* (L.) Cout., *M. communis* subsp. *lusitanica* (L.) Arcang., *M. communis* var. *acuminata* Rouy & E.G. Camus, *M. communis* var. *adrianii* Sennen, *M. communis* var. *augustinii* Sennen & Teodoro, *M. communis* var. *balearicus* Sennen, *M. communis* var. *baui* (Sennen & Malag.) Sennen, *M. communis* var. *baui* Sennen & Teodoro, *M. communis* var. *bercialis* Sennen, *M. communis* var. *camboanus* Sennen, *M. communis* var. *communis*, *M. communis* var. *eugeniae* Sennen, M. *communis* var. *gallachi* Sennen, *M. communis* var. *gausseni* Sennen, *M. communis* var. *gibertii* Sennen, *M. communis* var. *jonssetii* Sennen, *M. communis* var. *latifolia* Willk., *M. communis* var. *leucocarpa* DC., M. *communis* var. *macrocarpa* Sennen, *M. communis* var. *melanocarpa* DC., *M. communis* var. *microphylla* Willk. & Lange, *M. communis* var. *morawieskiana* Sennen, *M. communis* var. *petri-ludovici* Sennen & Teodoro, *M. communis* var. *rodesii* Sennen & Malag., *M. communis* var. *soleri* Sennen, *M. communis* var. *theodori* Sennen & Malag., *M. communis* var. *theodori* Sennen & Teodoro, *M. latifolia* (Willk.) Sennen, *M. sparsiflora* O. Berg, *M. tarentina* Bertol.

Family: *Myrtaceae*.

Arabic name: *Ās*.

Persian name: *Murd*.

English name: Common myrtle (Fig. 2.47).

Distribution

Africa: Algeria, Eritrea, Ethiopia, Libya, Morocco, Tunisia. *Asia*: Afghanistan, Iran, Iraq, Lebanon, Pakistan, Syria, Turkey, Yemen. *Europe*: Albania, Canary Islands, Cyprus, France, Greece (Aegean Islands), Italy, Portugal, Spain. *Americas*: Cuba, Puerto Rico, USA.

Morphology

Shrub about 1–5 m tall. Leaves ovate-lanceolate, 2–5 cm, coriaceous, glabrous, punctate-glandular, aromatic when crushed, entire; petiole *c*.2 mm. Flowers up to 3 cm diameter, white; pedicels 1.5–2.5 cm; petals orbicular, 7–15 mm. Berry subglobose to broadly ellipsoid, 7–10 mm, usually bluish-black on ripening (Rechinger, 1963–2001; Davis, 1965–1988; Ghahreman, 1978).

Nymphaea alba L., Sp. Pl.: 510 (1753)

Synonyms: *Castalia alba* (L.) Greene, *Leuconymphaea alba* (L.) Kuntze.

Family: *Nymphaeaceae*.

Arabic name: *Nilufar abyaḍ*.

Persian name: *Nilufar ābi*.

English name: White water-lily, White nenuphar (Fig. 2.48).

Distribution

Africa: Algeria, Morocco, Tunisia. *Asia*: Bangladesh, Caucasus, China, Iran, Iraq, Lebanon,

Fig. 2.47. *Myrtus communis* L. (By Giancarlo Dessi, CC BY-SA 3.0-migrated, https://upload.wikimedia.org/wikipedia/commons/1/16/Myrtus_communis1.jpg.)

Fig. 2.48. *Nymphaea alba* L. (By Peter Haas, Wikimedia Commons, https://upload.wikimedia.org/wikipedia/commons/8/81/Nymphaea_alba_flower-and-leaves-DSC_3326w.jpg.)

Myanmar, Syria, Transcaucasus, Turkey. *Europe*: Albania, Austria, Baltic States, Belarus, Belgium, Bulgaria, Denmark, Finland, France, Germany, Great Britain, Greece, Hungary, Ireland, Italy, Netherlands, Norway, Poland, Portugal, Romania, Russia, Spain, Sweden,

Switzerland, Ukraine. *Americas*: Chile. *Oceania and Australasia*: Australia, New Zealand.

Morphology

Rhizomes repent, sparsely branched, but not stoloniferous. Leaf blade suborbicular, 10–25 cm in diameter, papery, abaxially glabrous, scarcely peltate, base deeply cordate and basal lobes subparallel or spreading, margin entire. Flower floating, (7–)10–20 cm in diam. Calyx insertion on receptacle circular; sepals lanceolate, 3–5(–8) cm, obscurely veined, caducous or decaying after anthesis. Petals (1.2–)2–2.5(–3.3) cm, white, ovate-oblong, transition to stamens gradual. Filament of inner stamens ± as wide as anther; connective apically unappendaged. Carpels completely united, walls between locules of ovary single. Stigma rays usually 14–20; carpellary appendages triangular-tapered. Fruit semiglobose, 2.5–3 cm. Seeds ellipsoid, 2–3(–5) mm, smooth (Bobrov *et al.*, 1934–1964; Rechinger, 1963–2001; Tutin *et al.*, 1964; Davis, 1965–1988; Sell and Murrell, 2006).

Opopanax hispidus **(Friv.) Griseb., Spic. Fl. Rumel. 1: 378 (1843)**

Synonyms: *Ferula hispida* Friv., *F. opopanax* Spreng., *Johrenia lycaonica* Bornm., *Opopanax chironius* Guss., *O. orientalis* Boiss., *O. siculus* A. Huet ex Nyman, *O. syriacus* Boiss., *Pastinaca hispida* (Friv.) Fenzl, *P. involucrata* Stokes, *Peucedanum syriacum* (Boiss.) M. Hiroe.

Family: *Apiaceae* (*Umbelliferae*).

Arabic name: *Jāwshir*.

Persian name: *Jāvshir*.

English name: Opopanax.

Distribution

Asia: Iran, Iraq, Lebanon, Syria, Transcaucasus, Turkey. *Europe*: Albania, Bulgaria, Cyprus, Greece (Aegean Islands), Italy.

Morphology

Erect, up to 3 m tall plant; stem stiff, ± densely retrorse-hispid below, thinly hispid or glabrous above, terete, finely striate, corymbosely branched above; branches also terete and glabrous, upper commonly opposite or verticillate. Petioles and main axes of leaves furnished with flattish papillate hairs which are often hirtellous at apex; leaves diminishing upwards, finally reduced to sheaths. Leaves bipinnate, ovate in outline with short but broad and inflated sheaths, to 50 cm long and almost as wide, segments ovate to oblong-lanceolate with regularly serrate, cartilaginous borders, 1.5–10 cm × 0.7–3 cm; lower pinnae with 2 or 3 or 4 pairs of pinnules; at least the lower pinnules of upper pinnae usually ± decurrent along the main axis. Leaves with at least lower surface furnished, frequently densely, more rarely sparsely, with hispid hairs that are furcate or stellate at the tips. Peduncles 2–13 cm; umbels several, with 12–20 glabrous rays, fruiting rays 1.6–7 cm. Partial umbels 8–12-flowered, flowers ± 2.5 mm in diameter, fruiting pedicels 4–6 mm. Bracts and bracteoles 1–5, usually small and caducous. Fruit ovoid, 7–12 mm × 6–8 mm, glabrous, valleculae with 1 or 2 broad, dark vittae, those of the commissural face 6–8 and more conspicuous. Disk much flattened and only slightly convex, shallowly crenulate-edged, stylopodia very small. Styles very short, glabrous, closely deflexed on to the disk and subequaling it (Rechinger, 1963–2001; Davis, 1965–1988; Ghahreman, 1978; Assadi *et al.*, 1989).

Phoenix dactylifera **L., Sp. Pl.: 1188 (1753)**

Synonyms: *Encephalartos pungens* (L.f. ex Aiton) Lehm., *Palma dactylifera* (L.) Mill., *P. major* Garsault, *Phoenix chevalieri* D. Rivera, S. Ríos & Obón, *Ph. excelsior* Cav., *Ph. iberica* D. Rivera, S. Ríos & Obón, *Macrozamia tridentata* var. *pungens* J. Schust., *Zamia pallida* Salisb., *Z. pungens* L.f. ex Aiton.

Family: *Arecaceae* (*Palmae*).

Arabic name: *Nakhl*.

Persian name: *Nakhl*.

English name: Date palm.

Distribution

Africa: Algeria, Benin, Chad, Egypt, Gambia, Libya, Mali, Mauritania, Mauritius, Morocco, Mozambique, Senegal, Somalia, Sudan, Tunisia. *Asia*: Afghanistan, Bahrain, Bangladesh, China, India, Iran, Iraq, Oman, Pakistan, Qatar, Saudi Arabia, Turkey, Yemen. *Europe*: Canary Islands, Dominican Republic, France, Greece (Aegean Islands), Portugal, Spain. *Americas*: Bolivia, El Salvador, Leeward Islands, Mexico, Puerto Rico, Trinidad and Tobago, USA. *Oceania and Australasia*: Australia, Fiji, New Caledonia.

Morphology

Stems solitary or clustered and then with few shoots, to 30 m tall, to 50 cm in diameter, rough with persistent, diamond-shaped leaf bases. Leaves 3–5 m; sheath and petiole to 1 m; rachis 1–2 m; acanthophylls many per side of rachis; pinnae to 200 per side of rachis, linear, irregularly arranged, and spreading in different planes; middle pinnae to 40 cm × 2 cm. Male inflorescences erect, to 1 m, with many rachillae, these *c*.30 cm; female inflorescences erect, becoming pendulous, to 2 m, with to 150 rachillae, these to 40 cm. Fruits variable in shape, usually oblong, to 7 cm × 3 cm, brown or black; endosperm homogeneous (Engler and Prantl, 1897–1915; Rechinger, 1963–2001; Davis, 1965–1988).

***Pinus gerardiana* Wall. ex D. Don, A.B. Lambert, Descr. Pinus, ed. 3, 2: 144 bis (1832)**

Synonyms: *Pinus aucklandii* Lodd. ex Gordon & Glend., *P. chilghoza* Knight, *P. gerardii* J. Forbes, *P. neosa* Gouan ex W.H. Baxter.

Family: *Pinaceae*.

Arabic name: *Habb-ol sanawbar*, *Sanawbar ḏakar*.

Persian name: *Chalghozah*.

English name: Chilghoza pine (Fig. 2.49).

Distribution

Asia: Afghanistan, Pakistan, Tibet.

Morphology

Tree with white bark, fissured into irregular, thin plates; 1st-year branchlets pale green-yellow, glabrous, with projected leaf scars. Needles 3 per bundle, triangular-flabellate in cross section, 6–10 cm, stiff, vascular bundle 1, resin canals 5–7, marginal, base with sheath shed. Seed cones shortly pedunculate, almost brown at maturity, oblong or ovoid, 12–20 cm × 9–11 cm. Seed scales 4–5 cm; apophyses broad, swollen, ± recurved, obviously ridged; umbo dorsal, apex obtuse. Seeds cylindric, *c*.2.5 cm; wing rudimentary, usually adhering to adjacent scale (Rechinger, 1963–2001).

***Pinus pinea* L., Sp. Pl.: 1000 (1753)**

Synonyms: *Apinus pinea* (L.) Neck. ex Rydb., *Pinea esculenta* Opiz, *Pinus africana* Münchh., *P. aracanensis* Knight ex Gordon & Glend., *P. arctica* Carrière, *P. fastuosa* Salisb., *P. fragilis* Carrière, *P. maderiensis* Ten., *P. sativa* Garsault.

Family: *Pinaceae*.

Arabic name: *Tanub*.

Persian name: *Kāje bādāmi*.

Common name: Stone pine, Umbrella pine, Parasol pine (Fig. 2.50).

Distribution

Africa: Albania, Algeria, Libya, Morocco, Tunisia. *Asia*: Lebanon, Syria, Turkey. *Europe*: Canary Islands, Cyprus, France, Greece (Aegean Islands), Italy, Portugal, Spain. *Oceania and Australasia*: Australia.

Fig. 2.49. *Pinus gerardiana* Wall. (By GV3101992, CC BY-SA 4.0, https://upload.wikimedia.org/wikipedia/commons/c/c9/Pinus_gerardiana_India18.jpg.)

Morphology

The plant usually grows up to 20–25 m tall, either with a single trunk up to 1 m in diameter, or a trunk that forks into multiple stems not too far above the ground. The leaves (needles) have minutely toothed margins and are borne in groups of 2, held by a basal sheath. The leaves are covered with a dull grayish, waxy coating when young, but become dark green and glossy when mature. The pollen cones are yellow and are short and cylindrical in shape. The seed cones are borne singly, or sometimes in groups of 2 or 3 on short stems, and persist for several years, ripening in the 3rd year. They are 9–13 cm long and up to 13 cm wide. The egg-shaped seeds are brown, 15–20 mm long and 8–10 mm wide, with a rudimentary wing (Davis, 1965–1988).

***Pistacia atlantica* Desf., Fl. Atlant. 2: 364 (1799)**

Synonyms: *Lentiscus atlantica* Kuntze, *L. mutica* (Fisch. & C.A. Mey.) Kuntze, *Pistacia cabulica* Stocks, *P. chia* Desf., *P. choulettei* Gand., *P. mutica* Fisch. & C.A. Mey., *Terebinthus atlanticus* (Desf.) Dum. Cours.

Family: *Anacardiaceae*.

Arabic name: *Buṭm aṭlasi*.

Persian name: *Banah*.

English name: Atlas pistachio, Atlantic pistacio, Atlantic terebinth, Cyprus turpentine, Persian turpentine tree (Fig. 2.51).

Fig. 2.50. *Pinus pinea* L. (By Karora, Public Domain, https://commons.wikimedia.org/w/index.php?curid=5301388.)

Distribution

Africa: Algeria, Libya, Morocco, Sinai, Tunisia. *Asia*: Afghanistan, Caucasus, Iran, Iraq, Lebanon, Pakistan, Saudi Arabia, Syria, Tajikistan, Transcaucasus, Turkey. *Europe*: Canary Islands, Cyprus, Greece (Aegean Islands). *Americas*: USA.

Morphology

Deciduous tree up to 7 m tall, with a dense, subglobose crown. Leaves imparipinnate; leaflets 2–5-paired, ovate-oblong or lanceolate, 2.5–8 cm × 0.8–2.2 cm, obtuse, mucronate, dark green above, paler beneath, glabrous except for ciliolate margin; rachis narrowly winged. Fruits paniculate, as long as or longer than broad, 5–8 mm × 5–6 mm (Rechinger, 1963–2001; Davis, 1965–1988).

***Pistacia vera* L., Sp. Pl.: 1025 (1753)**

Synonyms: *Lentiscus vera* Kuntze, *Pistacia badghysi* K.P. Popov, *P. macrophylla* Pers., *P. narbonnensis* L., *P. nemausensis* Req. ex Planch., *P. nigricans* Crantz, *P. officinarum* Aiton, *P. reticulata* Willd., *P. trifolia* L., *P. variifolia* Salisb., *P. vera* f. *semidehiscens* K.P. Popov, *Terebinthus pistacia* Dum. Cours.

Family: *Anacardiaceae*.

Arabic name: *Fustaq*.

Persian name: *Pesteh*.

English name: Pistachio (Fig. 2.52).

Distribution

Africa: Libya, Morocco. *Asia*: Afghanistan, Kazakhstan, Kyrgyzstan, Tajikistan,

Fig. 2.51. *Pistacia atlantica* Desf. (By Eitan f, CC BY-SA 3.0, https://upload.wikimedia.org/wikipedia/commons/4/44/Pistacia_atlantica_%28Ein_Ovdat%29.JPG.)

Fig. 2.52. *Pistacia vera* L. (By Stan shebs, CC BY-SA 3.0, https://upload.wikimedia.org/wikipedia/commons/f/ff/Pistacia_vera_Kerman.jpg.)

Turkmenistan, Uzbekistan, Iran, Turkey. *Europe*: Bulgaria, France, Greece (Aegean Islands), Italy, Spain.

Morphology

Tree up to 10 m tall. Leaves deciduous, trifoliolate or imparipinnate; leaflets large, 1–2(–3)-paired, ovate or broadly lanceolate, acute or obtuse, mucronulate, 5–10(–12) cm × 3–6 cm, conspicuously reticulate-veined, terminal leaflet as large or larger than lateral ones, sparsely puberulent; rachis not winged, hairy. Fruits paniculate, ovate-oblong, 16–29 mm × 9–12 mm, seed containing large green cotyledons, edible (Rechinger, 1963–2001; Davis, 1965–1988; Ghahreman, 1978).

Plantago ovata **Forssk., Fl. Aegypt. -Arab.: 31 (1775)**

Synonyms: *Plantago albicans* Roem. & Schult., *P. argentea* Desf., *P. brunnea* E. Morris, *P. declinata* Roem. & Schult., *P. decumbens* Forssk., *P. fastigiata* E. Morris, *P. fastigiata* var. *brunnea* (E. Morris) Pilg., *P. gooddingii* A. Nelson & P.B. Kenn., *P. insularis* Eastw., *P. insularis* var. *fastigiata* (E. Morris) Jeps., *P. insularis* var. *scariosa* (E. Morris) Jeps., *P. insularis* var. *typica* Poe, *P. ispaghul* Roxb., *P. leiocephala* Wallr. ex K. Koch, *P. mauritii* Sennen, *P. microcephala* Poir., *P. minima* A.M. Cunn., *P. monspeliensis* Willd., *P. ovata* var. *fastigiata* (E. Morris) S.C. Meyers & Liston, *P. ovata* var. *insularis* (Eastw.) S.C. Meyers & Liston, *P. ovatifolia* Steud., *P. scariosa* E. Morris, *P. trichophylla* Nábelek, *P. villosa* Moench.

Family: *Plantaginaceae*.

Arabic name: *Kabāsh*.

Persian name: *Esfarzah*.

English name: Blond psyllium, Desert Indianwheat, Blond plantain, Ispagul (Fig. 2.53).

Fig. 2.53. *Plantago ovata* Forssk. (By Stan shebs, CC BY-SA 3.0, https://upload.wikimedia.org/wikipedia/commons/0/0b/Plantago_ovata_form.jpg.)

Distribution

Africa: Algeria, Egypt, Libya, Morocco, Tunisia. *Asia*: Afghanistan, Bahrain, Caucasus, China, India, Iran, Iraq, Kuwait, Lebanon, Oman, Pakistan, Saudi Arabia, Syria, Turkmenistan, Thailand, Yemen. *Europe*:

Canary Islands, Cyprus, Portugal, Spain. *Americas*: Mexico, USA.

Morphology

Annual plant, acaulous or short-stemmed, 3–10(–16) cm tall, branched at base, white villouslanate. Leaves linear to narrowly lanceolate, 3–5 cm × 0.1–0.6 cm, 3-veined, base attenuate tapering into the petiole, margins entire or remotely denticulate, apex acute, densely to sparsely white villous. Peduncle as long as leaves or longer, rarely shorter than leaves, arcuate-ascending or decumbent, rarely erect, 1–9(–12) cm, villous. Spike globose to ovoid to narrowly ovoid, 0.5–2.5(–3) cm, densely flowered; bracts glabrous, broadly ovate, 2–3 mm × 3–4 mm, obtuse, scarious except for the herbaceous midrib, lateral lamina much wider than midrib. Calyx 2.5–3 mm, glabrous, scarious except midrib; anterior sepals obovate-elliptic, posterior sepals elliptic-obovate. Corolla white, lobes broadly ovate or orbicular-cordate, 2–3 mm × 2 mm, acute, glabrous. Capsule 2-seeded. Seeds ovoid (Rechinger, 1963–2001; Ghahreman, 1978).

***Polypodium vulgare* L., Sp. Pl.: 1085 (1753)**

Synonyms: *Ctenopteris vulgaris* (L.) Newman.

Family: *Polypodiaceae*.

Arabic name: *Basfāyaj*.

Persian name: *Baspāyak*.

English name: Common polypody (Fig. 2.54).

Distribution

Africa: Lesotho, Morocco. *Asia*: Caucasus, China, Iran, Japan, Kazakhstan, Korea, Transcaucasus, Turkey. *Europe*: Albania, Austria, Baltic States, Belarus, Belgium, Bulgaria, Cyprus, Denmark, Finland, France, Germany, Great Britain, Greece, Hungary, Iceland, Ireland, Italy, Netherlands, Norway, Poland, Portugal, Romania, Russia, Spain, Sweden, Switzerland, Ukraine. *Oceania and Australasia*: Australia, New Zealand.

Fig. 2.54. *Polypodium vulgare* L. (By Andre Karwath aka Aka, CC BY-SA 2.5, https://upload.wikimedia.org/wikipedia/commons/f/f7/Polypodium_vulgare_-_close-up_%28aka%29.jpg.)

Morphology

Rhizome long creeping, 3–4 mm in diameter, densely scaly; scales pale brown, lanceolate with ovate peltate base, 4–5 mm, margin toothed, apex acuminate. Fronds remote. Stipe straw-colored, 5–10 cm, densely scaly at base, glabrescent upward. Lamina pinnatifid or pinnatisect, oblong-lanceolate in outline, 10–20 cm × 5–7 cm, herbaceous or subleathery, glabrous, apex shortly caudate. Segments 12–15 pairs, oblique, lanceolate, 3–4 cm × 0.5–0.8 cm, decurrent to adjacent lobes by very narrowly winged rachis, margin toothed, apex obtuse or acute. Veinlets hardly visible. Sori medial or slightly closer to costa (Bobrov *et al.*, 1934–1964; Rechinger, 1963–2001; Tutin *et al.*, 1964; Davis, 1965–1988; Sell and Murrell, 2006).

***Portulaca oleracea* L., Sp. Pl.: 445 (1753)**

Synonyms: *Portulaca hortensis* Rupr., *P. officinarum* Crantz.

Family: *Portulacaceae*.

Arabic name: *Rijlah*.

Persian name: *Khorfah*.

English name: Pigweed, Common purslane, Green purslane, Purslain, Purslane, Wild purslane, Pursley (Fig. 2.55).

Distribution

Africa: Algeria, Angola, Botswana, Burkina Faso, Burundi, Cameroon, Chad, Congo, Djibouti, Egypt, Eritrea, Ethiopia, Gabon, Gambia, Guinea, Kenya, Liberia, Malawi, Mali, Mauritania, Mauritius, Morocco, Mozambique, Namibia, Niger, Nigeria, Rodrigues Island, Rwanda, Senegal, Seychelles, Sierra Leone, Sudan, Tunisia, Zambia, Zimbabwe. *Asia*: Afghanistan, Bangladesh, Borneo, Cambodia, Caucasus, China, Bahrain, India, Indonesia, Iran, Iraq, Kazakhstan, Korea, Kuwait, Kyrgyzstan, Laos, Lebanon, Maldives, Myanmar, Nepal, Oman, Pakistan, Philippines, Saudi Arabia, Sri Lanka, Syria, Tajikistan, Taiwan, Transcaucasus, Turkey, Turkmenistan, Uzbekistan, Vietnam, Yemen. *Europe*: Albania, Austria, Belarus, Belgium, Bulgaria, Canary Islands, Denmark, Finland, France, Germany, Great Britain, Greece (Aegean Islands), Hungary, Italy, Netherlands, Poland, Portugal, Romania, Russia, Spain, Switzerland, Sweden, Ukraine. *Americas*: Argentina, Bahamas, Bolivia, Brazil, Canada, Cayman Islands, Chile, Colombia, Costa Rica, Cuba, Dominican Republic, Ecuador, El Salvador, Guatemala, Guyana, Haiti, Honduras, Mexico, Nicaragua, Panamá, Paraguay, Puerto Rico, Trinidad and Tobago, USA, Uruguay, Venezuela. *Oceania and Australasia*: Australia, Fiji, Nauru, New Caledonia, New Zealand, Tonga, Tuvalu.

Fig. 2.55. *Portulaca oleracea* L. (By Stefan Lefnaer, CC BY-SA 4.0, https://upload.wikimedia.org/wikipedia/commons/b/b3/Portulaca_oleracea_sl7.jpg)

Morphology

Annual herb; stems sometimes flushed red or purple, not articulated, prostrate or decumbent, less often ± erect, diffuse, much branched; leaf axils with a few inconspicuous stiff bristles. Leaves alternate or occasionally subopposite; petiole short; leaf blade flat, obovate, 10–30 mm × 5–15 mm, base cuneate, apex obtuse, rounded, truncate, or retuse. Flowers in clusters of 3–5, 0.4–0.5 cm in diameter, surrounded by involucre of 2–6 bracts. Sepals green, helmeted, *c.*4 mm, apex acute, keeled. Petals 5, yellow, obovate, 3–5 mm, slightly connate at base, apex retuse. Stamens 7–12, *c.*12 mm; anthers yellow. Ovary glabrous. Stigma 4–6-lobed. Capsule ovoid, *c.*5 mm. Seeds glossy black when mature, never iridescent, obliquely globose-reniform, 0.6–1.2 mm; testa cells stellate, usually with central peglike tubercle, sometimes without and then surface ± granular (Bobrov *et al.*, 1934–1964; Rechinger, 1963–2001; Tutin *et al.*, 1964; Davis, 1965–1988; Sell and Murrell, 2006).

***Prunus armeniaca* L., Sp. Pl.: 474 (1753)**

Synonyms: *Amygdalus armeniaca* (L.) Dumort., *Armeniaca armeniaca* (L.) Huth, *A. vulgaris* Lam., *Prunus tiliifolia* Salisb.

Family: *Rosaceae*.

Arabic name: *Meshmesh*.

Persian name: *Zardālo*.

English name: Armenian plum, Apricot (Fig. 2.56).

Distribution

Africa: Algeria, Libya, Morocco, Tunisia. *Asia*: Afghanistan, Caucasus, China, Iran, Kazakhstan, Korea, Kyrgyzstan, Lebanon, Mongolia, Pakistan, Syria, Tajikistan, Transcaucasus, Turkey, Turkmenistan, Uzbekistan. *Europe*: Albania, Bulgaria, Cyprus, France, Germany, Greece, Hungary, Italy, Portugal, Romania, Russia, Ukraine. *Americas*: Mexico, USA. *Oceania and Australasia*: Australia.

Morphology

Tree up to 15 m tall; bark gray-brown, splitting; young shoots glabrous. Leaves broadly ovate, up to 12 cm × 10 cm, finely serrate, glabrous, rarely puberulent when young. Flowers white or pink, 2–3 cm diameter, hypanthium shortly cylindrical. Drupe subglobose, 2.5–5 cm diameter, whitish, yellow, or orange, finely pubescent to ± glabrous; stone smooth (Bobrov *et al.*, 1934–1964; Rechinger, 1963–2001; Davis, 1965–1988).

Fig. 2.56. *Prunus armeniaca* L. (By Fir 0002, Wikimedia Commons, https://en.wikipedia.org/wiki/Prunus_armeniaca#/media/File:Apricots.jpg.)

Prunus amygdalus Batsch, Beytr. Entw. Gewächsreich 1: 30 (1801)

Synonyms: *Amygdalus amygdalus* (Batsch) Frye & Rigg, *A. communis* L., *Druparia amygdalus* (Batsch) Clairv., *A. amara* Duhamel, *A. amara* var. *pendula* Lavallée, *A. amara* var. *pyramidata* Lavallée, *A. amygdalina* Oken ex M. Roem., *A. cochinchinensis* Lour., *A. communis* f. *acuta* Magyar, *A. communis* f. *albaplena* Zabel, *A. communis* subvar. *amara* (Duhamel) Rouy & E.G. Camus, *A. communis* var. *amara* (Duhamel) Willd., *A. communis* f. *angustifolia* (H. Jaeger) Zabel, *A. communis* var. *angustifolia* H. Jaeger, *A. communis* f. *compacta* Zabel, *A. communis* var. *cordiformis* Risso, *A. communis* f. *cristata* Magyar, *A. communis* subvar. *dulcis* (Mill.) Rouy & E.G. Camus, *A. communis* var. *dulcis* (Mill.) Borkh., *A. communis* f. *elongata* Magyar, *A. communis* var. *elongata* Risso, *A. communis florepleno* F.J. Schultz, *A. communis foliis-variegatis* F.J. Schultz, *A. communis* var. fragilima Ficalho & Cout., *A. communis* var. *fragilis* Pers., *A. communis* var. *fragilis* Ser., *A. communis* var. *glabra* Popov, *A. communis* var. *incana* Popov, *A. communis lanceifolia* K. Koch, *A. communis* f. *lanceolata* Magyar, *A. communis* var. *macrocarpa* Ser., *A. communis* subvar. *macrocarpa* (Ser.) Rouy & E.G. Camus, *A. communis* f. *marginata* Zabel, *A. communis* var. *maxima* Risso, *A. communis* var. *media* Popov, *A. communis* f. *microcarpa* Magyar, *A. communis* var. *microcarpa* Dierb., *A. communis* var. *ossea* Gren., *A. communis* f. *ostraeiformis* Magyar, *A. communis* f. *ovata* Magyar, *A. communis* f. *pendula* (H. Jaeger) Zabel, *A. communis* var. *pendula* H. Jaeger, *A. communis* var. *pistacia* Risso, *A. communis* f. *platycarpa* Magyar, *A. communis* f. *purpurea* Zabel, *A. communis* var. *pyramidalis* Risso, *A. communis* f. *roseaplena* Zabel, *A. communis* f. *rotundata* Magyar, *A. communis* var. *rubra* Risso, *A. communis* var. *rugosa* Risso, *A. communis* f. *salicifolia* K. Koch, *A. communis* var. *sativa* L., *A. communis* var. *semifragilis* Magyar, *A. communis* var. *serotina* Risso, *A. communis* var. *sphaerocarpa* Dierb., *A. communis* var. *spontanea* Korsh., *A. communis* var. *stenocarpa* Dierb., *A. communis* var. *subrotundata* Risso, *A. communis* var.

tenuis Risso, *A. communis* var. *tristis* Risso, *A. communis* f. *variegata* Zabel, *A. decipiens* Poit. & Turpin, *A. dulcis* Mill., *A. elata* Salisb., *A. fragilis* Borkh., *A. indica* Duhamel, *A. korshinskyi* var. *bornmuelleri* Browicz, *A. sativa* Mill., *A. sinensis* Steud., *A. stocksiana* Boiss., *A. vulgaris* Weston, *A. vulgaris* plena Weston, *Persica* Mill., *Prunus amygdalus* f. *alboplena* (Zabel) Rehder, *P. amygdalus* var. *amara* (Duhamel) Focke, *P. amygdalus* f. *angustifolia* Dippel, *P. amygdalus* var. *dulcis* (Mill.) Koehne, *P. amygdalus* var. *fragilis* (Ser.) Focke, *P. amygdalus* var. *macrocarpa* Bean, *P. amygdalus* var. *ovalifolia* Franch., *P. amygdalus* f. *pendula* (H. Jaeger) Rehder, *P. amygdalus* var. *pendula* (H. Jaeger) Bean, *P. amygdalus* var. *pollardii* Anon., *P. amygdalus* var. *praecox* Bean, *P. amygdalus* f. *purpurea* (C.K. Schneid.) Rehder, *P. amygdalus* f. *roseoplena* (C.K. Schneid.) Rehder, *P. amygdalus* var. *sativa* (L.) Focke, *P. amygdalus* var. *typica* Fiori, *P. amygdalus* f. *variegata* (C.K. Schneid.) Rehder, *P. armeniaca* f. *pendula* (H. Jaeger) Rehder, *P. cochinchinensis* (Lour.) Koehne, *P. communis* (L.) Arcang., *P. communis* monstr. *albiplena* (Zabel) Asch. & Graebn., *P. communis* subvar. *amara* (Duhamel) Asch. & Graebn., *P. communis* f. *amara* (Duhamel) C.K. Schneid., *P. communis* subsp. *angustata* Arcang., *P. communis* subvar. *angustifolia* (Dippel) Asch. & Graebn., *P. communis* lusus *compacta* Zabel ex Asch. & Graebn., *P. communis* f. *dulcis* (Mill.) C.K. Schneid., *P. communis* var. *fragilis* C.K. Schneid., *P. communis* subsp. *fragilis* Arcang., *P. communis* var. *fragilis* (Ser.) C.K. Schneid., *P. communis* f. *macrocarpa* (A. Poit. & Turpin) C.K. Schneid., *P. communis* monstr. *marginata* Zabel ex Asch. & Graebn., *P. communis* f. *pendula* (H. Jaeger) C.K. Schneid., *P. communis* lusus *pendula* (H. Jaeger) Asch. & Graebn., *P. communis* lusus *purpurea* (Zabel) Asch. & Graebn., *P. communis* f. *purpurea* C.K. Schneid., *P. communis* lusus *pyramidalis* Asch. & Graebn., *P. communis* monstr. *roseiplena* (Zabel) Asch. & Graebn., *P. communis* f. *roseoplena* C.K. Schneid., *P. communis* subvar. *sativa* (L.) Asch. & Graebn., *P. communis* var. *typica* C.K. Schneid., *P. communis* f. *variegata* C.K. Schneid., *P. communis* monstr. *variegata* Zabel ex Asch. & Graebn., *P. dulcis* (Mill.) D.A. Webb, *P. dulcis* var. *amara* (Duhamel) Buchheim, *P. dulcis* var. *fragilis* (Ser.) Buchheim, *P. dulcis* var. *spontanea* (Korsh.) Buchheim, *P. intermedia* A.Sav., *P. stocksiana* (Boiss.) Brandis.

Family: *Rosaceae*.

Arabic name: *Lowz*, *Lowz marr*.

Persian name: *Bādām*, *Bādām talkh*.

Common name: Almond (Fig. 2.57).

Distribution

Africa: Algeria, Libya, Morocco, Tunisia. *Asia*: Afghanistan, Iran, Iraq, Kazakhstan, Kyrgyzstan, Lebanon, Pakistan, Syria, Tajikistan, Turkey, Turkmenistan, Uzbekistan. *Europe*: Bulgaria, Canary Islands, Cyprus, France, Great Britain, Greece (Aegean Islands), Hungary, Italy, Portugal, Spain, Switzerland. *Americas*: USA.

Morphology

Unarmed tree to 8 m; young shoots glabrous. Leaves ovate-lanceolate or elliptic, 2.5–9(–12) cm × 2–3 cm, dentate-serrate, glandular, entirely glabrous or pubescent beneath when young, leathery; petioles 10–30 mm. Flowers white or pink, up to 4 cm diameter, pedicels up to 5 mm. Drupe obliquely ellipsoid to elongate-ellipsoid, compressed, up to 50 mm × 30 mm, olive-green, velutinous; stone pitted and ± grooved, keeled (Rechinger, 1963–2001; Davis, 1965–1988; Assadi *et al.*, 1989; Sell and Murrell, 2006).

Papaver somniferum **L., Sp. Pl.: 508 (1753)**

Synonyms: *Papaver album* Mill., *P. album-nigrum* Crantz, *P. amoenum* Lindl., *P. amplexicaule* Stokes, *P. apodocarpum* Hussenot, *P. glabrum* Gilib., *P. hortense* Garsault, *P. hortense* Hussenot, *P. indehiscens* Dumort., *P. nigrum* Bubani, *P. nigrum* Garsault, *P. officinale* C.C. Gmel., *P. opiiferum* Forssk., *P. opiiferum* var. *album* DC., *P. paeoniiflorum* (Alef.) Corrêa, *P. polycephalum* Bailly, *P. polycephalum* hort., *P. polycephalum* hort. ex E. Vilm., *P. setigerum* subsp. *valdesetosum*

Fig. 2.57. *Prunus amygdalus*. (By Placid Perez Bru, CC BY-SA 3.0, https://upload.wikimedia.org/wikipedia/commons/c/c2/Ametllesjuliol.jpg.)

Maire, *P. setigerum* var. *nigrum* (Garsault) P. Fourn., *P. somniferum* subsp. *albiflorum* Wender., *P. somniferum* subsp. *album* Schübl. & G. Martens, *P. somniferum* subsp. *euroasiaticum* Vessel., *P. somniferum* subsp. *hortense* (Hussenot) Syme, *P. somniferum* subsp. *hortense* Arcang., *P. somniferum* subsp. *hortense* Huss, *P. somniferum* subsp. *nigrum* (DC.) Thell., *P. somniferum* subsp. *nigrum* (Willd.) Schübl. & G. Martens, *P. somniferum* subsp. *nigrum* Schübl. & G. Martens, *P. somniferum* subsp. *officinale* (C.C. Gmel.) Syme, *P. somniferum* subsp. *officinalis* Bonnier & Layens, *P. somniferum* subsp. *subspontaneum* Basil.,

P. somniferum subsp. *variegatum* Wender., *P. somniferum* var. *album* (Mill.) DC., *P. somniferum* var. *austroviolaceum* Vessel., *P. somniferum* var. *coerulescens* Rothm., *P. somniferum* var. *genuinovilaceum* Vessel., *P. somniferum* var. *glaucum* Rothm., *P. somniferum* var. *hortense* (Huss) Rouy & Foucaud, *P. somniferum* var. *hypoleucum* Rothm., *P. somniferum* var. *laciniatum* Wender., *P. somniferum* var. *leucum* (Rothm.) K.Hammer, *P. somniferum* var. *nigrum* DC., *P. somniferum* var. *officinale* (C.C. Gmel.) Garcke, *P. somniferum* var. *pallidum* Rothm., *P. somniferum* var. *plenum* H. Mart., *P. somniferum* var. *poecilospermum* Alef., *P. somniferum* var. *pullatum* Vessel., *P. somniferum* var. *rubellum* Vessel., *P. somniferum* var. *valdesetosum* (Maire) Maire, *P. somniferum* var. *violiflorum* Vessel., *P. stipitatum* Hussenot, *P. sylvestre* Godr.

Family: *Papaveraceae*.

Arabic name: *Ofiun*.

Persian name: *Khashkhāsh*.

English name: Opium poppy, Breadseed poppy (Fig. 2.58).

Distribution

Africa: Algeria, Angola, Ethiopia, Libya, Morocco, Tunisia. *Asia*: Afghanistan, Bangladesh, Cambodia, Caucasus, China, India, Iran, Iraq, Kazakhstan, Korea, Laos, Lebanon, Mongolia, Myanmar, Pakistan, Saudi Arabia, Syria, Tajikistan, Thailand, Transcaucasus, Turkey, Turkmenistan, Uzbekistan, Vietnam, Yemen. *Europe*: Albania, Baltic States, Belarus, Belgium, Bulgaria, Canary Islands, Cyprus, Denmark, Finland, France, Great Britain, Greece (Aegean Islands), Hungary, Ireland, Italy, Norway, Portugal, Romania, Russia, Spain, Sweden, Ukraine. *Americas*: Argentina, Canada, Chile, Ecuador, Honduras, Mexico, Peru, USA, Venezuela. *Oceania and Australasia*: Australia, New Zealand, Tasmania.

Fig. 2.58. *Papaver somniferum* L. (https://commons.wikimedia.org/wiki/File:Opium_poppy.jpg#/media/File:Opium_poppy.jpg.)

Morphology

Annual herb, 30–60(–100) cm tall (to 1.5 m in cultivation), glabrous or rarely slightly setose on plant below or on peduncle. Taproot erect, almost conical. Stems erect, glaucous, glabrous, occasionally somewhat setose. Leaves alternate; blade ovate or oblong, 7–25 cm, both surfaces glabrous, glaucous, and rather waxy, veins distinct, slightly raised, base cordate, margin irregularly undulate-serrate, apex acuminate to obtuse; leaves below shortly petiolate but above sessile and amplexicaul. Flowers solitary, deeply cup-shaped, 5–12 cm in diameter, pedicel to 25 cm, glabrous or rarely sparsely setose. Flower buds nutant at first, erect at anthesis, oval-oblong or broadly ovoid, 1.5–3.5 cm × 1–3 cm, glabrous. Sepals 2, green, broadly ovate, margin membranous. Petals 4, white, pink, red, purple, or various, often with a dark basal blotch, suborbicular or almost flabellate, 4–7 cm × 3–11 cm, margin undulate or variously lobed. Stamens many; filaments white, linear, 1–1.5 cm; anthers yellowish or cream, oblong, 3–6 mm. Ovary green, spherical, 1–2 cm in diameter, glabrous; stigmas 5–12(–18), actinomorphic, united into compressed disk, disk margin deeply divided, lobes crenulate. Capsule brown when mature, spherical or oblong-elliptic, 4–9 cm × 4–5 cm, glabrous. Seeds many, black or deep gray, adaxially alveolate (Bobrov *et al.*, 1934–1964; Rechinger, 1963–2001; Davis, 1965–1988; Assadi *et al.*, 1989; Sell and Murrell, 2006).

Punica granatum **L., Sp. Pl.: 472 (1753)**

Synonyms: *Granatum punicum* St.-Lag., *Punica florida* Salisb., *P. grandiflora* Steud., *P. nana* L., *P. spinosa* Lam., *Rhoea punica* St.-Lag.

Family: *Lythraceae*.

Arabic name: *Rommān* (*Jolnār*).

Persian name: *Anār* (*Golnār*).

English name: Pomegranate (Fig. 2.59).

Fig. 2.59. *Punica granatum* L. (By Sanjay Acharya, CC BY-SA 4.0, https://upload.wikimedia.org/wikipedia/commons/f/fa/Pomegranate_Fruit_Setting.jpg.)

Distribution

Africa: Algeria, Cameroon, Chad, Comoros, Eritrea, Ethiopia, Libya, Morocco, Niger, Somalia, Tunisia. *Asia*: Afghanistan, Bangladesh, Caucasus, China, India, Iran, Iraq, Korea, Laos, Oman, Pakistan, Sri Lanka, Tajikistan, Transcaucasus, Turkey, Turkmenistan, Uzbekistan, Vietnam, Yemen. *Europe*: Bulgaria, Canary Islands, Greece, France, Canary Islands, Italy, Portugal, Romania, Spain, Sweden, Switzerland. *Americas*: Bahamas, Cuba, Dominican Republic, Ecuador, Jamaica, Leeward Islands, Mexico, Panamá, Puerto Rico, Trinidad and Tobago, USA, Venezuela. *Oceania and Australasia*: New Caledonia, Norfolk Island.

Morphology

Shrub or small tree, 2–3 m tall, glabrous. Branches and branchlets 4-angled, becoming terete with age, often terminating as indurate spines. Petiole 2–10 mm; leaf blade adaxially shiny, lanceolate, elliptic-oblanceolate, or oblong, 2–9 cm × 1–2 cm, base attenuate, apex obtuse or mucronate. Floral tube red-orange or pale yellow, campanulate-urceolate, 2–3 cm × 1–1.5 cm; sepals 5–9, erect, deltate. Petals 5–9, bright red-orange [or white], obovate, 1.5–3 cm × 1–2 cm, apex rounded or obtuse. Stamens numerous, included to exserted. Ovary 8–13-loculed, in 2 or 3 superposed layers, lower locules with axile placentation, upper ones with apparent parietal placentation. Fruit globose, leathery berries, variable in color, red to yellow-green or red-brown, 5–12 cm in diameter, crowned by persistent sepals, irregularly dehiscent. Seeds obpyramidal within juicy sarcotestal layer, ruby-red, pink, or yellowish-white (Rechinger, 1963–2001; Davis, 1965–1988).

Quercus infectoria G. Olivier, Voy. Emp. Othoman 1: 252 (1801)

Synonyms: *Quercus infectoria* subsp. *euinfectoria* A. Camus, *Q. lusitanica* subsp. *infectoria* (G. Olivier) Mouill.

Family: *Fagaceae*.

Arabic name: *Balut ṣabghi*.

Persian name: *Dār māzu*.

English name: Aleppo oak, Gall oak (Fig. 2.60).

Distribution

Asia: Iran, Iraq, Lebanon, Syria, Transcaucasus, Turkey. *Europe*: Cyprus, Greece.

Morphology

Semi-evergreen small tree or shrub, 1–4(–10) m; young shoots finely pubescent, ± glabrescent, reddish- or yellowish-brown; buds reddish-brown, *c*.3 mm, ciliate. Leaves often overwintering till new leaves appear, very variable in size and color, 40–70(–100) mm × 10–45 mm, coriaceous, ovate to narrowly oblong, rounded or cuneate at base, margins often undulate with 4–8 crenate to serrate lobes, or entire; primary veins 6–11; indumentum of few floccose stellate-dendroid hairs beneath and sometimes above, usually glabrous later; petiole 1–15(–25) mm (± absent on summer shoots). Peduncle absent or to *c*.10 mm. Cupules solitary or in pairs, hemispherical or cyathiform, 10–18 mm diameter; scales strongly adpressed, grayish pubescent; acorn up to two-thirds exserted (Rechinger, 1963–2001; Davis, 1965–1988; Ghahreman, 1978; Assadi *et al*., 1989).

Rheum officinale Baill., Adansonia 10: 246 (1872)

Synonyms: *Rheum baillonii* F. Heim.

Family: *Polygonaceae*.

Arabic name: *Rāvand kaffi*.

Persian name: *Rivand chini*.

English name: Chinese rhubarb (Fig. 2.61).

Distribution

Asia: China, Laos, Vietnam. *Europe*: Great Britain.

Morphology

Large herb, 1.5–2 m tall with stout rhizomes and roots. Stem stout, hollow, finely sulcate, with white hairs, especially above and at nodes. Petiole of basal leaf terete, about as long as blade or slightly shorter, pubescent; leaf blade orbicular, rarely broadly ovate, large, 30–50 cm in diameter, or longer than wide, abaxially pubescent, adaxially glabrous, rarely pubescent along veins, basal veins 5–7, base subcordate; palmatilobate, apex subacute. Stem leaves smaller upward; ocrea large, to 15 cm, broad, outside with dense hairs. Panicles large; branches spreading. Pedicel 3–3.5 mm, slender, jointed below middle. Flowers 4- or 5-fascicled. Tepals 6, green to yellow-white, elliptic or narrowly elliptic, 2–2.5 mm × 1.2–1.5 mm. Stamens shorter than perianth. Style deflexed; stigma inflated. Fruit oblong-ellipsoid, 8–10 mm × 7–9 mm; wings *c*.3 mm wide, with longitudinal veins near margin. Seeds broadly ovoid (Sell and Murrell, 2006).

Rhus coriaria L., Sp. Pl.: 265 (1753)

Synonyms: *Rhus amoena* Salisb., *Rh. coriaria* var. *zebaria* Shahbaz, *Rh. heterophylla* C.C. Gmel., *Rh. sumac* O. Targ. Tozz., *Rh. variifolia* DC., *Toxicodendron coriaria* (L.) Kuntze.

Fig. 2.60. *Quercus infectoria*. (By Gideon Pisanty, CC BY 3.0, https://upload.wikimedia.org/wikipedia/commons/f/f4/Quercus_boissieri_galls_2.JPG.)

Family: *Anacardiaceae*.

Arabic name: *Somāq*.

Persian name: *Somāq*.

English name: Tanner's sumach (Fig. 2.62).

Distribution

Africa: Algeria, Morocco. *Asia*: Afghanistan, Caucasus, Iran, Iraq, Lebanon, Syria, Turkey, Turkmenistan, Tajikistan, Transcaucasus, Uzbekistan. *Europe*: Albania, Bulgaria, Canary Islands, Cyprus, France, Greece (Aegean Islands), Italy, Portugal, Spain, Ukraine.

Fig. 2.61. *Rheum officinale* Baill. (By Kurt Stueber, CC BY-SA 3.0, https://en.wikipedia.org/wiki/Rheum_officinale#/media/File:Rheum_officinale1.jpg.)

Morphology

Shrub, up to 3.5 m. Young branches densely brownish-tomentose. Leaves with 9–15, broadly lanceolate to elliptic, serrate to deeply serrate leaflets, sparsely pilose above and beneath, with small setae along the veins beneath. Petals greenish-white, 3–4.5 mm. Drupe globose, villose, reddish (Rechinger, 1963–2001; Davis, 1965–1988; Ghahreman, 1978; Assadi *et al.*, 1989).

Ruta graveolens **L., Sp. Pl.: 383 (1753)**

Synonyms: *Ruta altera* Mill., *R. ciliata* Mill., *R. crithmifolia* Moric. ex DC., *R. divaricata* Ten., *R. diversifolia* Wender., *R. holopetala* Kitt., *R. hortensis* Mill., *R. intermedia* Steud., *R. macrophylla* Sol., *R. officinalis* Pall., *R. subtripinnata* Miq.

Family: *Rutaceae*.

Arabic name: *Soḏāb bostāni*.

Fig. 2.62. *Rhus coriaria* L. (By Dedda 71, CC BY-SA 3.0, https://upload.wikimedia.org/wikipedia/commons/c/c0/Sommacco.jpg.)

Persian name: *Sodāb.*

English name: Rue, Common rue (Fig. 2.63).

Distribution

Africa: Algeria. *Asia*: Turkey, Vietnam. *Europe*: Albania, Austria, Canary Islands, Belgium, Bulgaria, France, Germany, Great Britain, Greece (Aegean Islands), Hungary, Italy, Romania, Spain, Switzerland, Ukraine. *Americas*: Canada, Columbia, Ecuador, USA.

Morphology

Strongly odoriferous evergreen herb or small shrub, stems 14–45 cm, rigid, densely foliose. Lower leaves more or less long-petiolate, the uppermost subsessile; ultimate segments 2–9 mm wide, lanceolate to narrowly oblong or obovate. Inflorescence rather lax; pedicels as long as or longer than the capsule; bracts lanceolate, leaflike. Sepals lanceolate, acute. Petals oblong-ovate, denticulate, undulate. Capsule glabrous; segments somewhat narrowed above to obtuse apex (Tutin *et al.*, 1964; Davis, 1965–1988; Sell and Murrell, 2006).

Styrax officinalis L., Sp. Pl.: 444 (1753)

Synonyms: *Styrax cotinifolius* Salisb., *S. officinarum* Crantz.

Family: *Styracaceae.*

Arabic name: *Mayaʿ̣h yābesah.*

Persian name: *Mayaʿ̣h khoshk.*

English name: Storax (Fig. 2.64).

Fig. 2.63. *Ruta graveolens* L. (By Pelenuska, CC BY-SA 4.0, https://upload.wikimedia.org/wikipedia/commons/6/66/Die_Weinraute%2C_lat._Ruta_graveolens%2C_Pflanze_mit_den_gelben_Bl%C3%BCten.jpg.)

Distribution

Asia: Lebanon, Syria, Turkey. *Europe*: Albania, Cyprus, France, Greece (Aegean Islands), Italy.

Morphology

Much branched, round-headed deciduous shrub (rarely tree) up to 6 m tall. Young shoots and buds stellate-pubescent. Leaves broadly elliptic to ovate, entire, 3.7–6.5 cm × 4.5–9.5 cm, with 5–7 pairs of lateral veins, bright green and sparingly stellate-pubescent above (later glabrous), densely stellate-tomentose beneath, obtuse. Flowers in clusters of 3–6, fragrant. Calyx densely stellate-tomentose. Corolla lobes 5–8, oblong-lanceolate, white, *c.*0.6 cm × 1.5 cm. Stamens 12, as long as petals. Fruit globose, tough, 1.2–1.4 cm diameter densely stellate-tomentose, dehiscent above (Boissier, 1867–1888; Davis, 1965–1988).

Syzygium aromaticum **(L.) Merr. & L.M. Perry, Mem. Amer. Acad. Arts, n.s., 18: 196 (1939)**

Synonyms: *Caryophyllus* aromaticus L., *C. hortensis* Noronha, *C. silvestris* Teijsm. ex Hassk., *Eugenia aromatica* (L.) Baill., *E. caryophyllata* Thunb., *E. caryophyllus* (Spreng.) Bullock & S.G. Harrison, *Jambosa caryophyllus* (Thunb.) Nied., *Myrtus caryophyllus* Spreng.

Family: *Myrtaceae*.

Arabic name: *Qaranfol*.

Persian name: *Mikhak*.

English name: Clove (Fig. 2.65).

Fig. 2.64. *Styrax officinalis* L. (By Eitan f, CC BY 3.0, https://upload.wikimedia.org/wikipedia/commons/e/ee/Styrax_officinalis_flowering.jpg.)

Distribution

Africa: Comoros, Madagascar, Seychelles, Tanzania. *Asia*: Borneo, India, Indonesia. *Americas*: Trinidad and Tobago.

Morphology

A bushy, evergreen tree with a medium-sized crown, growing 8–20 m tall. Several parts of the tree are aromatic, including the leaves and bark, but it is most valued for the aromatic flower buds, which are usually harvested by hand. Leaves are simple, bright green, and glossy, lower surface is covered in aromatic oil glands, elliptical leaves with petioles (leaf stalks) grow up to 13 cm long in opposite pairs along numerous short branches. Flowers developing in small clusters; buds are initially pale with a glossy, fleshy appearance, turning green and then bright red as they mature; consist of a long, narrow ovary, approximately 1.5–2.0 cm in length, with 4 small triangular sepals projecting outwards at one end. Sepals surround a small ball of 4 overlapping petals that protect the developing flower parts inside. Fruit matures approximately 9 months after flowering (Soepadmo *et al.*, 2011).

***Tamarix gallica* L., Sp. Pl.: 270 (1753)**

Synonyms: *Tamariscus gallicus* (L.) All., *T. narbonensis* Garsault, *T. pentander* Lam., *Tamarix aegyptiaca* Bertol., *T. anglica* Webb, *T. brachylepis* Sennen, *T. ehrenbergii* C. Presl ex Bunge, *T. esperanzana* Pau & Villar, *T. gallica* f. *divaricata* Ehrenb., *T. gallica* f. *effusa* Ehrenb., *T. gallica* var. *heterophylla* Ehrenb., *T. gallica* var. *nannifera* Ehrenb., *T. gallica* var. *narbonensis* Ehrenb., *T. gallica* var. *subtilis* Ehrenb., *T. libanotica* K. Koch, *T. matritensis* Pau & Villar, *T. narbonensis* Garsault, *T. riojana* Sennen & Elías, *T. sireti* Sennen.

Family: *Tamaricaceae*.

Arabic name: *Gaz angabin*.

Persian name: *Ṭarfā*.

English name: French tamarisk (Fig. 2.66).

Fig. 2.65. *Syzygium aromaticum* L. (By Chen Hualin, CC BY-SA 4.0, https://upload.wikimedia.org/wikipedia/commons/3/35/The_flowers_of_clove_tree_in_Pemba_island.JPG.)

Fig. 2.66. *Tamarix gallica* L. (By Teun Spaans, CC BY-SA 3.0, https://commons.wikimedia.org/w/index.php?curid=167002.)

Distribution

Africa: Morocco. *Asia*: Bangladesh, Myanmar. *Europe*: Great Britain, France, Hungary, Italy, Spain. *Americas*: Argentina, Brazil, Dominican Republic, Mexico, Puerto Rico, Trinidad and Tobago, USA.

Morphology

Shrub or bushy tree with dark brown to dark purple bark, entirely glabrous. Leaves 1–3 mm; racemes rather lax, 15–45 mm × 3–5 mm, dense, with papillose axis; bracts ± erose-denticulate, not exceeding the calyx, entire. Pedicel equaling calyx. Flowers pink; sepals 0.75–1.25 mm, entire; petals 1.5–2 mm, elliptical to elliptic-obovate; disk less fleshy; filaments expanded at the base and confluent with the lobes of the disk but disk much smaller (Sell and Murrell, 2006).

Terminalia bellirica (Gaertn.) Roxb., Pl. Coromandel 2: t. 198 (1805)

Synonyms: *Buceras bellirica* (Gaertn.) Lyons, *Myrobalanus bellirica* Gaertn., *M. punctata* (Roth) Kuntze, *Terminalia attenuata* Edgew., *T. belirica* Wall., *T. biticaria* Roxb., *T. eglandulosa* Roxb. ex C.B. Clarke, *T. gella* Dalzell, *T. laurinoides* Teijsm. & Binn., *T. moluccana* Roxb., *T. myrobalana* B. Heyne ex Roth, *T. punctata* Roth.

Family: *Combretaceae*.

Arabic name: *Balilaj*.

Persian name: *Balileh*.

English name: Baheda, Bahera, Beleric myrobalan (Fig. 2.67).

Distribution

Africa: Mauritius. *Asia*: Bangladesh, Borneo, Cambodia, China, India, Indonesia, Laos, Malaysia, Myanmar, Nepal, Pakistan, Sri Lanka, Thailand, Vietnam. *Americas*: Trinidad and Tobago.

Fig. 2.67. *Terminalia bellirica*. (By Mark Marathon, CC BY-SA 3.0, https://commons.wikimedia.org/w/index.php?curid=23476344.)

Morphology

Trees deciduous, to 35 m tall; trunk to 1 m diameter at breast height, with large buttresses. Bark gray, longitudinally ridged. Branchlets with conspicuous, spirally ascending leaf scars. Leaves spiraled, crowded into pseudowhorls at apices of branchlets; petiole 3–9 cm, glabrous but ferruginous tomentose when young, especially at base, with 2 glands above middle; leaf blade glossy, obovate, 18–26 cm × 6–12 cm, both surfaces glabrous except ferruginous tomentose when young, base obtuse-rounded or attenuate, apex obtuse or mucronate; lateral veins in 5–8 pairs. Inflorescences axillary, simple spikes, 5–18 cm, often grouped at branchlet apex and forming a panicle; axis densely ferruginous tomentose. Calyx tube distally shallowly cupular, 4–5 mm, abaxially tomentose, adaxially long villous; lobes 5. Stamens 10, exserted, 4–5 mm. Fruit shortly stipitate, subglobose to broadly ellipsoid, or ovoid, weakly to strongly 5-ridged, 2–3 cm × 1.8–2.5 cm, densely and finely velutinous or sericeous; stipe *c*.2 mm (eFloras, 2008).

Terminalia chebula Retz., Observ. Bot. 5: 31 (1788)

Synonyms: *Buceras chebula* (Retz.) Lyons, *Combretum argyrophyllum* K. Schum., *Myrobalanifera citrina* Houtt., *Myrobalanus chebula* (Retz.) Gaertn., *M. gangetica* (Roxb.) Kostel., *M. tomentella* Kuntze, *Terminalia acuta* Walp., *T. argyrophylla* King & Prain, *T. aruta* Buch.-Ham. ex G. Don, *T. gangetica* Roxb., *T. glandulipetiolata* De Wild., *T. reticulata* B. Heyne ex Roth, *T. tomentella* Kurz.

Family: *Combretaceae.*

Arabic name: *Ihlilaj kābuli.*

Persian name: *Halileh siāh.*

Common name: Myrobalan, Black myrobalan, Chebulic myrobalan (Fig. 2.68).

Distribution

Africa: Congo. *Asia*: Bangladesh, Cambodia, China, India, Laos, Malaysia, Myanmar, Nepal, Pakistan, Sri Lanka, Taiwan, Thailand, Vietnam.

Morphology

Trees to 30 m tall; trunk to 1 m diameter. Bark grayish-black to gray, coarsely split, and thick. Branchlets conspicuously white or yellowish long lenticellate, glabrous, or tomentose or appressed villous at least when young, hairs tawny, rarely silvery. Leaves alternate or subopposite, spaced along branchlets; petiole 1–3 cm, moderately stout, with 2(–4) glands 1–5 mm below apex; leaf blade elliptic, 7–18 cm × 4.5–10 cm, both surfaces glabrous, or appressed (and rarely silvery) villous at least when young, base obtuse-rounded or cuneate, oblique, apex mucronate; lateral veins in 6–12 pairs. Inflorescences axillary or terminal, simple spikes, 5–10 cm, numerous flowered, sometimes grouped at branchlet apex and forming a panicle; axis glabrous or sparsely hairy, with denser hairs near base of flowers. Flowers slightly fragrant, bisexual. Calyx tube distally cupular, 2.5–3.5 mm, abaxially glabrous, adaxially tawny tomentose; lobes 5, apex mucronate to aristate. Stamens 10, exserted, 3–4 mm. Fruit not stipitate, blackish-brown when ripe, ovoid or broadly so, ellipsoid, or cylindric-ovoid, obtusely

Fig. 2.68. *Terminalia chebula* Retz. (By Delonix, CC BY-SA 3.0, https://en.wikipedia.org/wiki/Terminalia_chebula#/media/File:Terminalia_chebula_-_flowers.JPG.)

5-ridged, 2–4.5 cm × 1.2–2.5 cm, rigid, becoming deeply wrinkled when dry, glabrous (eFloras, 2008).

Rosa × damascena Mill., Gard. Dict. ed. 8: Rosa no. 15 (1768)

Synonyms: *Rosa belgica* Mill., *R. calendarum* Münchh., *R. calendarum* Münchh. ex Borkh., *R. centifolia* var. *bifera* Poir., *R. gallica* var. *damascena* Voss, *R. polyanthos* R, *R.* × *bifera* (Poir.) Pers., *R.* × *damascena* var. *trigintipetala* Dieck, *R.* × *damascena* var. *variegata* Keller. This species is the hybrid between *R. gallica* L. and *Rosa moschata* Herrm. It was originated in Iran and then has been cultivated in European countries.

Family: *Rosaceae*.

Arabic name: *Ward juri*.

Persian name: *Gole mohamadi*.

English name: Damask rose (Fig. 2.69).

Distribution

Asia: Caucasus, Iran, Iraq, Transcaucasus, Turkey, Turkmenistan. *Europe*: Bulgaria.

Morphology

Perennial bushy shrub, deciduous, up to 1.5 m, erect, densely armed with stout prickles. Prickles small, 0.3–0.6 cm, straight to hooked, interspersed with glandular acicles or bristles, with a broad dilated base, ± uniform. Leaflets 3 or 5, ovate to suborbicular to elliptic, 2–5 cm × 1.5–2.5 cm, obtuse to subacute with rounded base, coriaceous, uni- or biserrate, teeth numerous, 10–17 on each side, rather short, blunt, glabrous and dull, bluish-green above, glabrescent to pubescent and pale gray-green beneath, nerves

Fig. 2.69. *Rosa × damascena*. (By Farsad Nadjafi.)

elevated and distinct beneath; rachis prickly, glabrous to pubescent; stipules rather broad, glandular-serrate, auricles short, ± straight. Flowers solitary or 2–4(–10) together, bracteate, strongly scented. Pedicels 1.5–5 cm, densely glandular-setose. Outer sepals pinnatifid or lobed, lobes glandular-marginate, broadly lanceolate, 1.5–2.5 cm, apiculate or with a dilated tip, reflexed after anthesis and soon deciduous. Petals 1.5–3.5 cm × 1.5–3 cm, emarginate, thick in texture, deep pink to light red. Styles lanate, stigma head compact, semiglobose, disk conoidal, broad with a wide orifice. Hypanthia globose to ovoid, up to 1–1.3 cm long, densely glandular-setose, bright red (Rechinger, 1963–2001; Davis, 1965–1988).

***Teucrium polium* L., Sp. Pl.: 566 (1753)**

Synonyms: *Chamaedrys polium* (L.) Raf.

Family: *Lamiaceae* (*Labiatae*).

Arabic name: *Misk el jin*.

Persian name: *Kalpureh*.

English name: Germander, Cat thyme, Felty germander, Poley germander (Fig. 2.70).

Distribution

Africa: Algeria, Morocco, Tunisia. *Asia*: Iran, Iraq, Turkey. *Europe*: France, Spain.

Morphology

Dwarf shrub. Stems 6–45 cm, covered with white, greenish, or golden branched hairs, often densely so. Leaves 7–27 mm, narrowly oblong to narrowly obovate, flat or with revolute margins, with 2–5(9) crenations, often in fascicles. Flowers in a simple or compound head. Bracts leaflike to entire. Calyx (2.5–)3–5 mm, densely and evenly hairy. Corolla white or red; proximal lobes rounded or triangular (Rechinger, 1963–2001; Davis, 1965–1988).

Fig. 2.70. *Teucrium polium* L. (By Petr1888, CC BY-SA 3.0, https://upload.wikimedia.org/wikipedia/commons/5/52/Plants_in_Gorges_de_Ardeche1.JPG.)

***Thymbra capitata* (L.) Cav., Elench. Pl. Horti Matr.: 37 (1803)**

Synonyms: *Coridothymus capitatus* (L.) Rchb.f., *Origanum capitatum* (L.) Kuntze, *Satureja capitata* L., *Thymus capitatus* (L.) Hoffmanns. & Link, *Th. creticus* Brot.

Family: *Lamiaceae* (*Labiatae*).

Arabic name: *Zoufā*.

Persian name: A kind of *Zoufā*, a kind of *Marzanjoush*.

Common name: Conehead thyme, Spanish oregano (Fig. 2.71).

Distribution

Africa: Algeria, Egypt, Libya, Morocco, Tunisia. *Asia*: Lebanon, Syria, Turkey. *Europe*: Albania, Cyprus, France, Greece (Aegean Islands), Italy, Portugal, Spain.

Morphology

Dwarf shrub 20–50(–150) cm, with ascending to erect woody branches bearing axillary leaf clusters (often the only leaves during the dry season). Leaves of long shoots 1.1–2 mm × 6–10 mm, sessile, linear, acute, subglabrous, sparsely ciliate at base; lateral veins not visible. Inflorescence oblong-conical; bracts *c.* 6 mm × 2 mm, imbricate, ovate to lanceolate, greenish, ciliate; bracteoles *c.*6 mm, similar to the leaves. Calyx *c.*5 mm, the upper lip shorter than the lower, all the teeth ciliate. Corolla up to 10 mm, purplish-pink, the upper lip bifid (Engler and Prantl, 1897–1915; Davis, 1965–1988).

***Trigonella foenum-graceum* L., Sp. Pl.: 777 (1753)**

Synonyms: *Buceras foenum-graecum* (L.) All., *B. odoratissima* Medik., *Foenum-graecum officinale* Moench, *Foenum-graecum sativum* Medik., *Folliculigera graveolens* Pasq., *Medicago*

Fig. 2.71. *Thymbra capitata* L. (By Lapplaender, CC BY-SA 3.0, https://upload.wikimedia.org/wikipedia/commons/a/a3/Thymus_capitatus_in_Crete.jpg.)

foenugraeca (L.) E.H.L. Krause, *Telis foenum-graecum* (L.) Kuntze, *Trigonella ensifera* Trautv., *T. graeca* St.-Lag., *T. haussknechtii* (Širj.) Vassilcz., *T. jemenensis* (Serp.) Sinskaya, *T. rhodantha* (Alef.) Vassilcz., *T. tibetana* (Alef.) Vassilcz., *Xiphostylis erectus* Gasp.

Family: *Fabaceae* (*Leguminosae*).

Arabic name: *Ḥolbah*.

Persian name: *Shanbalileh*.

English name: Fenugreek (Fig. 2.72).

Distribution

Africa: Algeria, Egypt, Ethiopia, Kenya, Libya, Mali, Morocco, Mozambique, Sudan, Tanzania, Tunisia, Zimbabwe. *Asia*: Afghanistan, Austria, Bangladesh, Caucasus, China, India, Iran, Iraq, Kuwait, Kyrgyzstan, Lebanon, Mongolia, Myanmar, Nepal, Oman, Pakistan, Saudi Arabia, Syria, Tajikistan, Transcaucasus, Turkey, Turkmenistan, Uzbekistan, Yemen. *Europe*: Albania, Baltic States, Belgium, Bulgaria, Cyprus, France, Germany, Greece (Aegean Islands), Hungary, Italy, Portugal, Romania, Russia, Spain, Switzerland, Ukraine. *Oceania and Australasia*: Fiji.

Fig. 2.72. *Trigonella foenum-graceum* L. (By Yak, CC BY-SA 3.0, https://de.wikipedia.org/wiki/Bockshornklee#/media/Datei:Junger_Bockshornklee.jpg.)

Morphology

Annual herb, 30–80 cm. Taproot to 80 cm, fibrous roots lateral. Stem erect, terete, branched. Leaves 3-foliolate; stipules entire, membranous, base adnate to petiole, apex acute; petiole spreading, 6–15 mm; leaflets narrowly obovate, ovate, to oblong-elliptic, terminal equal to laterals, 15–40 mm × 4–15 mm, sparsely pubescent or glabrescent abaxially, glabrous adaxially, lateral veins 5 or 6 pairs, base cuneate, margin dentate in upper half. Flowers sessile, axillary, solitary or in pairs. Calyx 7–8 mm, villous. Corolla creamy or pale yellow, base violet, 13–18 mm. Ovary puberulent; ovules numerous. Legume long conical, 7–12 cm × 0.4–0.5 cm, straight or slightly curved, glabrescent, veins longitudinally reticulate, apex with long beak to 2 cm, distal suture thickened. Seeds 10–20, brown, oblong-ovoid, 3–5 mm × 2–3 mm, surface uneven (Bobrov *et al.*, 1934–1964; Rechinger, 1963–2001; Tutin *et al.*, 1964; Davis, 1965–1988).

Urtica urens L., Sp. Pl.: 984 (1753)

Synonyms: *Urtica atlantica* Blume, *U. cubensis* Klotzsch ex Herder, *U. minor* Garsault, *U. minor* Lam., *U. monoica* Gilib., *U. ovalifolia* Stokes, *U. parvula* Blume, *U. quadristipulata* Dulac, *U. trianae* Rusby.

Family: *Urticaceae*.

Arabic name: *Qoraṣ ḥareq*.

Persian name: *Gazanahe sag*.

English name: Annual nettle, Nettle, Dog nettle (Fig. 2.73).

Distribution

Africa: Algeria, Egypt, Eritrea, Ethiopia, Kenya, Lesotho, Libya, Morocco, Namibia,

Fig. 2.73. *Urtica urens* L. (By H. Zell, CC BY-SA 3.0, https://upload.wikimedia.org/wikipedia/commons/6/61/Urtica_urens_002.JPG.)

Somalia, Sudan, Tanzania, Tunisia. *Asia*: China, Caucasus, Bahrain, Kuwait, Iran, Iraq, Kazakhstan, Kyrgyzstan, Lebanon, Oman, Qatar, Saudi Arabia, Syria, Transcaucasus, Turkey, Yemen. *Europe*: Albania, Austria, Baltic States, Belarus, Belgium, Bulgaria, Canary Islands, Great Britain, Cyprus, Denmark (Greenland), Finland, France, Germany, Greece (Aegean Islands), Hungary, Ireland, Italy, Netherlands, Norway, Poland, Portugal, Romania, Russia, Spain, Sweden, Switzerland, Ukraine. *Americas*: Argentina, Bolivia, Brazil, Canada, Chile, Colombia, Cuba, Ecuador, Guatemala, Mexico, Peru, USA. *Oceania and Australasia*: Australia, New Zealand.

Morphology

Annual herb. Stems branched, 10–60 cm tall, sparsely puberulent and somewhat densely armed with stinging hairs; lower internodes 4–7 cm, upper internodes 1–3 cm. Stipules free, narrowly triangular, 1–2.5 mm, ciliate; petiole 1–2.5 cm, puberulent, sparsely armed with stinging hairs; leaf blade broadly elliptic, sometimes ovate or obovate, 1.2–6 cm × 0.6–3 cm, 5-veined, often subglabrous except for sparse stinging hairs on both surfaces, base broadly cuneate or rounded, margin 6–11-dentate, apex obtuse-rounded; cystoliths punctiform, distinct adaxially. Inflorescences containing proximal female flowers and distal male flowers, spicate, 0.5–2.5 cm. Male flowers short pedicellate, in bud *c.*1.2 mm; perianth lobes connate half of length, puberulent. Female flowers: perianth lobes connate at base, dorsal-ventral lobes ovate, equaling achene, often with 1 stinging hair on dorsal rib, sparsely setulose along margin, lateral lobes ovate, *c.*5× as small as others. Achene brownish-gray, ovoid, compressed, 0.8 mm, verrucose, invested by persistent perianth lobes (Bobrov *et al.*, 1934–1964; Rechinger, 1963–2001; Tutin *et al.*, 1964; Davis, 1965–1988; Sell and Murrell, 2006).

***Valeriana officinalis* L., Sp. Pl.: 31 (1753)**

Synonyms: *Astrephia chinensis* Dufr., *Valeriana alternifolia* Bunge, *V. alternifolia* f. *angustifolia* (Kom.) Kitag., *V. alternifolia* f. *verticillata* (Kom.) S.X. Li, *V. alternifolia* subsp. *pseudocoreana* Vorosch., *V. alternifolia* subsp. *stubendorfiana* (Sumn.) Vorosch., *V. alternifolia* subsp. *stubendorfii* (Kreyer ex Kom.) Vorosch., *V. alternifolia* var. *angustifolia* (Kom.) S.H. Li, *V. alternifolia* var. *stolonifera* A.I. Baranov & Skvortsov, *V. altissima* J.C. Mikan, *V. altissima* J.C. Mikan ex Besser, *V. angustifolia* Tausch ex Host, *V. baltica* Pleijel, *V. chinensis* Kreyer, *V. chinensis* Kreyer ex Kom., *V. coreana* Briq., *V. coreana* subsp. *badzhalensis* Vorosch., *V. coreana* subsp. *leiocarpa* (Kitag.) Vorosch., *V. coreana* subsp. *pseudoumbrosa* (Vorosch.) Vorosch., *V. coreana* subsp. *sachalinensis* (Hara) Vorosch., *V. divaricata* Huter, *V. exaltata* J.C. Mikan, *V. exaltata* J.C. Mikan ex Pohl, *V. fauriei* Briq., *V. fauriei* f. *coreana* (Briq.) H. Hara, *V. fauriei* subsp. *badzhalensis* (Vorosch.) Vorosch., *V. fauriei* subsp. *sachalinensis* (Hara) Vorosch., *V. fauriei* var. *dasycarpa* H. Hara, *V. fauriei* var. *fauriei*, *V. fauriei* var. *leiocarpa* (Kitag.) Kitag., *V. jacutica* Sumn., *V. leiocarpa* Kitag., *V. lucida*

Hort. Par., *V. lucida* Hort. Par. ex DC., *V. major* Pall., *V. multiceps* Wallr., *V. multiceps* var. *dentata* Wallr., *V. multiceps* var. *incisa* Wallr., *V. nipponica* Nakai, *V. nipponica* Nakai ex Kitag., *V. officinalis* f. *pilosa* Podp., *V. officinalis* f. *pubescens* Podp., *V. officinalis* subsp. *baltica* (Pleijel) Á. Löve & D. Löve, *V. officinalis* subsp. *communis* Schübl. & G. Martens, *V. officinalis* subsp. *dunensis* P.D. Sell, *V. officinalis* subsp. *exaltata* (J.C. Mikan ex Pohl) Wirtg., *V. officinalis* subsp. *exaltata* (J.C. Mikan) Soó, *V. officinalis* subsp. *latifolia* (Vahl) Schübl. & G. Martens, *V. officinalis* subsp. *sarkayi* (Soó) Á. Löve & D. Löve, *V. officinalis* subsp. *simplicifolia* Ledeb., *V. officinalis* subsp. *vulgaris* Celak., *V. officinalis* var. *alternifolia* Ledeb., *V. officinalis* var. *altissima* (J.C. Mikan ex Besser) W.D.J. Koch, *V. officinalis* var. *angustifolia* Hayne, *V. officinalis* var. *angustifolia* Miq., *V. officinalis* var. *exaltata* (J.C. Mikan ex Pohl) Kostel., *V. officinalis* var. *latifolia* Briq., *V. officinalis* var. *latifolia* Miq., *V. officinalis* var. *latifolia* Vahl, *V. officinalis* var. *major* Wimm. & Grab., *V. officinalis* var. *media* W.D.J. Koch, *V. officinalis* var. *officinalis*, *V. officinalis* var. *palustris* Wibel, *V. officinalis* var. *sarkanyi* Soó, *V. officinalis* var. *sarkayi* Soó, *V. officinalis* var. *simplicifolia* Ledeb., *V. officinalis* var. *vulgaris* Kitt., *V. palustris* Kreyer, *V. palustris* var. *exaltata* (J.C. Mikan ex Pohl) Kreyer, *V. palustris* var. *integella* Kreyer, *V. palustris* var. *vulgaris* Kreyer, *V. pinnata* Gilib., *V. provisa* Kreyer, *V. pseudofficinalis* C.Y. Cheng & H.B. Chen, *V. pseudoumbrosa* Vorosch., *V. pulchra* Nakai, *V. sambucifolia* Eichw., *V. sambucifolia* subsp. *sachalinensis* Hara, *V. sambucifolia* var. *dasycarpa* (Hara) H. Hara, *V. sambucifolia* var. *fauriei* (Briq.) H. Hara, *V. sambucifolia* var. *sachalinensis* Hara, *V. sinensis* J.F. Gmel., *V. stubendorfiana* Sumnev., *V. stubendorfii* Kreyer, *V. stubendorfii* Kreyer ex Kom., *V. stubendorfii* f. angustifolia Kom., *V. stubendorfii* f. *verticillata* Kom., *V. stubendorfii* subsp. *angustifolia* Kom., *V. stubendorfii* subsp. *verticillata* Kom., *V. sylvestris* Garsault, *V. sylvestris* Groschke, *V. tergemina* Kreyer, *V. tianschanica* Kreyer, *V. tianschanica* Kreyer ex Hand.-Mazz., *V. vulgaris* Bubani, *Veronica officinalis* var. altissima W.D.J. Koch.

Family: *Caprifoliaceae*.

Arabic name: *Nārdin makhzani*.

Persian name: *Sonboloṭib*.

Common name: Valerian (Fig. 2.74).

Distribution

Asia: Caucasus, Iran, Transcaucasus, Turkey. *Europe*: Albania, Austria, Baltic States, Belarus, Bulgaria, Denmark, Estonia, Finland, France, Germany, Greece, Hungary, Italy, Latvia, Lithuania, Poland, Romania, Russia, Sweden, Switzerland, Ukraine. *Americas*: USA.

Morphology

Plants up to 150 cm tall, infrequently stoloniferous. Rhizomes reduced; roots fascicled, clavate, 1.5–2.5 mm in diameter. Stems hirsute or hispidulous, particularly at nodes, rarely glabrous. Basal and lowermost cauline leaves usually wilted at anthesis; cauline leaves petiolate to sessile; petiole 0–8 cm; blade ovate to broadly ovate, 5–15 cm

Fig. 2.74. *Valeriana officinalis* L. (By H. Zell, CC BY-SA 3.0, https://upload.wikimedia.org/wikipedia/commons/8/82/Valeriana_officinalis_001.JPG.)

× 3–8 cm, pinnatisect; segments 7–15, lanceolate or linear, glabrous or ± hispidulous, base decurrent, margin entire or coarsely serrate; terminal and lateral segments similar in shape and size. Inflorescence corymbiform, large and compact even in fruit, or with 2 distinct, secondary lateral inflorescences; bracts leaflike below, small and linear-lanceolate otherwise; bracteoles elliptic-oblong, oblanceolate, or linear-lanceolate, margin scarious-ciliate, apex awned. Corolla purplish-red, pink, or white, funnelform, 4–6 mm; lobes elliptic. Stamens and style strongly exserted. Achenes narrowly ovoid, 4–5 mm, both sides glabrous or pubescent (Boissier, 1867–1888; Bobrov *et al.*, 1934–1964; Rechinger, 1963–2001; Tutin *et al.*, 1964; Davis, 1965–1988).

Veratrum album **L., Sp. Pl.: 1044 (1753)**

Synonyms: *Helleborus albus* (L.) Gueldenst., *Melanthium album* (L.) Thunb., *M. bracteolare* Desr., *M. virens* Thunb., *Veratrum album albicans* Gaudin, *V. album* f. *glabrescens* (Zapal.) Soó, *V. album* subsp. *misae* (Sirj.) Tzvelev, *V. album* f. *podolicum* (Zapal.) Soó, *V. bosniacum* Beck, *V. croaticum* (Beck) O. Loes., *V. flavum* (Griseb.) O. Loes., *V. lobelianum* f. *oppositifolium* Cheshm., *V. misae* (Sirj.) O. Loes., *V. parviflorum* Bong., *V. polygamum* Gilib., *V. viride* Röhl.

Family: *Melanthiaceae*.

Arabic name: *Khrbaq abyaḍ*.

Persian name: *Khrbaqe sefid*.

Common name: False helleborine, White hellebore, European white hellebore, White veratrum (Fig. 2.75).

Fig. 2.75. *Veratrum album* L. (By Tigerente, CC BY-SA 3.0, https://upload.wikimedia.org/wikipedia/commons/8/80/Veratrum_album_Dachstein.jpg.)

Distribution

Asia: Transcaucasus, Turkey. *Europe*: Albania, Austria, Baltic States, Belarus, Finland, France, Germany, Hungary, Italy, Norway, Poland, Portugal, Romania, Russia, Spain, Switzerland, Ukraine.

Morphology

Stems 50–175 cm. Cauline leaves glabrous or subglabrous above, shortly pubescent beneath especially on the veins, the lower 10–25 cm × 6–15 cm, elliptical to broadly elliptical, becoming shorter and narrower above. Inflorescence a much-branched panicle. Pedicels 2–3 mm. Flowers 15–25 mm in diameter. Perianth segments 7–15 mm × 3–6 mm, elliptical to lanceolate, white outside and greenish inside to greenish-white or yellowish on both surfaces. Capsule sparsely pubescent (Bobrov *et al.*, 1934–1964; Tutin *et al.*, 1964; Davis, 1965–1988).

Viola odorata **L., Sp. Pl.: 934 (1753)**

Synonyms: *Viola cyclophylla* Gand., *V. gonzaloi* Sennen, *V. gonzaloi* var. *semiglabra* Sennen, *V. hirta* var. *odorata* (L.) Fiori, *V. hirta* subsp. *odorata* (L.) Fiori, *V. martia* Gilib., *V. martia* Hegetschw., *V. martii* subsp. *odorata* (L.) Schimp. & Spenn., *V. odorata* f. *odorata*,

V. odorata subsp. *rierolae* Sennen, *V. odorata* subsp. *sarotina* P. Fourn., *V. parmensis* Bailly, *V. spectabilis* K. Richt., *V. tenerrima* Halácsy & Braun, *V. vilmoriniana* Delacour & Mottet, *V. odorata* proles *brarcinonensis* Sennen, *V. odorata* subsp. *texidoris* Sennen, *V. taynensis* Elisafenko.

Family: *Violaceae*.

Arabic name: *Albanafsaji*.

Persian name: *Banafsheh*.

English name: Sweet violet (Fig. 2.76).

Distribution

Africa: Algeria, Morocco, Tunisia. *Asia*: Caucasus, China, India, Iran, Iraq, Japan, Kazakhstan, Lebanon, Myanmar, Pakistan, Syria, Tajikistan, Transcaucasus, Turkey, Vietnam. *Europe*: Albania, Austria, Baltic States, Belarus, Belgium, Bulgaria, Canary Islands, Cyprus, Denmark, Finland, France, Germany, Great Britain, Greece (Aegean Islands), Hungary, Ireland, Italy, Netherlands, Poland, Portugal, Romania, Russia, Spain, Sweden, Switzerland, Ukraine. *Americas*: Argentina, Canada, Chile, Colombia, Costa Rica, Cuba, Dominican Republic, Ecuador, El Salvador, Guatemala, Leeward Islands, Mexico, Panamá, Peru, Puerto Rico, Trinidad and Tobago, USA, Venezuela.

Fig. 2.76. *Viola odorata* L. (By Fritz Geller-Grimm, CC BY-SA 2.5, https://upload.wikimedia.org/wikipedia/commons/7/7a/Viola_odorata_fg01.JPG.)

Morphology

Perennial plant 5–15 cm, with leaf rosette and long, procumbent, rooting stolons. Leaves orbicular-reniform, deeply cordate, widest at about the middle; petiole long. Stipules ovate, glabrous, or sparsely cibate, shortly glandular-fimbriate. Bracts at or above middle of peduncle. Flowers *c.*1–5 cm, dark violet or white, fragrant. Sepals ovate, obtuse. Spur *c.*6 mm, exceeding calycine appendages. Stigmatic beak vertical, its length equaling diameter of style. Capsule pubescent (Bobrov *et al.*, 1934–1964; Rechinger, 1963–2001; Tutin *et al.*, 1964; Davis, 1965–1988; Sell and Murrell, 2006).

***Vitex agnus-castus* L., Sp. Pl.: 638 (1753)**

Synonyms: *Agnus-castus alba* Carrière, *Agnus-castus macrostachya* Carrière, *Agnus-castus robusta* (Lebas) Carrière, *Agnus-castus vulgaris* Carrière, *Vitex agnus* Stokes, *V. agnus-castus* f. *caerulea* (Rehder) Moldenke, *V. agnus-castus* subsp. *haussknechtii* Bornm., *V. agnus-castus* f. *laciniosa* (Ces.) Moldenke, *V. haussknechtii* (Bornm.) Prain, *V. hybrida* Moldenke, *V. integra* Medik., *V. latifolia* Mill., *V. lupinifolia* Salisb., *V. pseudonegundo* (Hausskn.) Hand.-Mazz., *V. robusta* Lebas, *V. verticillata* Lam.

Family: *Lamiaceae* (*Labiatae*).

Arabic name: *Fanjankosh*.

Persian name: *Panjangosht*.

English name: Chaste tree, Vitex, Abraham's balm, Lilac chastetree, Monk's pepper (Fig. 2.77).

Distribution

Africa: Algeria, Egypt, Libya, Morocco, Tunisia. *Asia*: Afghanistan, Bahrain, Caucasus, Iran, Iraq, Lebanon, Oman, Pakistan, Syria, Tajikistan, Transcaucasus, Turkmenistan, Turkey, Yemen. *Europe*: Albania, Bulgaria, Cyprus, France, Greece (Aegean Islands), Italy, Spain, Ukraine. *Americas*: Brazil, Cuba, Dominican Republic, Honduras, Mexico, Peru, Puerto Rico, USA.

Fig. 2.77. *Vitex agnus-castus* L. (By Krzsysztof Ziarnek, Kenraiz, CC BY-SA 4.0, https://upload.wikimedia.org/wikipedia/commons/9/94/Vitex_agnus-castus_kz2.jpg.)

Morphology

Fragrant plant, 1–6 m tall. Young twigs with 4 obtuse angles, gray-puberulent. Leaves petiolate, with 5–7 stipitate leaflets; leaflets 1.5–10 cm, linear-lanceolate, acuminate, entire, whitish-tomentose beneath, glabrous above. Calyx and outside of corolla tomentose. Corolla 8–10 mm, blue or pink. Drupe globose, reddish-black, slightly longer than calyx (Engler and Prantl, 1897–1915; Rechinger, 1963–2001; Davis, 1965–1988).

Ziziphus jujuba **Miller, Gard. Dict. ed. 8: no. 1 (1768)**

Synonyms: *Girtanneria jujuba* (Mill.) Neck., *Jububa mediterranea* Bubani, *Mansana arborea* J.F. Gmel., *Paliurus lucidus* Carrière, *Rhamnus circumcissa* Russell ex Wall., *R. lucida* Salisb., *R. mauritiana* Soy.-Will., *R. soporifera* Lour., *R. vulgaris* Pers., *R. zizyphus* L., *Ziziphus acidojujuba* C.Y. Cheng & M.J. Liu, *Z. acidojujuba* f. *granulata* C.Y. Cheng & M.J. Liu, *Z. acidojujuba* f. *infecunda* C.Y. Cheng & M.J. Liu, *Z. acidojujuba* f. *trachysperma* C.Y. Cheng & M.J. Liu, *Z. chinensis* Spreng., *Z. flexuosa* Wall., *Z. jujuba* f. *allochroa* C.Y. Cheng & M.J. Liu, *Z. jujuba* f. *apyrena* C.Y. Cheng & M.J. Liu, *Z. jujuba* f. *carnosicalycis* C.Y. Cheng & M.J. Liu, *Z. jujuba* f. *heteroformis* C.Y. Cheng & M.J. Liu, *Z. jujuba* f. *tortuosa* C.Y. Cheng & M.J. Liu, *Z. melanogona* Bojer, *Z. natsme* Siebold, *Z. nitida* Roxb., *Z. officinarum* Medik., *Z. sativa* Gaertn., *Z. sinensis* Lam., *Z. soporifera* (Lour.) Duhamel, *Z. sororia* Schult., *Z. spinosa* (Bunge) Hu ex F.H. Chen, *Z. trinervia* Roth, *Z. vulgaris* Lam., *Z. vulgaris* var. *spinosa* Bunge, *Z. zizyphus* (L.) H. Karst., *Zizyphon jujubum* (Mill.) St.-Lag.

Family: *Rhamnaceae*.

Arabic name: *Oʿnāb*.

Persian name: *Oʿnāb*.

English name: Jujube, Jujuba, Chinese date (Fig. 2.78).

Distribution

Africa: Algeria, Burkina Faso, Cameroon, Tunisia, Morocco, Laos, Libya. *Asia*: Afghanistan, China, Cyprus, India, Iran, Iraq, Japan, Kazakhstan, Korea, Kyrgyzstan, Lebanon, Maldives, Mongolia, Pakistan, Syria, Tajikistan, Turkey, Turkmenistan, Uzbekistan. *Europe*: Bulgaria, France, Greece, Italy, Romania, Spain. *Americas*: Cuba, Dominican Republic, Honduras, Jamaica, USA, Venezuela.

Morphology

Small tree, rarely shrub, deciduous, to 10 m tall, spinose or unarmed. Bark brown or gray-brown, with long reduced branches, without buds; branchlets (new branches) purple-red or gray-brown, flexuose, smooth, with 2 stipular spines or not; long spines erect, to 3 cm, stout; short spines recurved, developed from old branches; annual branchlets pendulous, green, solitary or 2–7-fascicled on short shoots. Stipular spines slender, caducous; petiole 1–6 mm, or to 1 cm on long shoots, glabrous or sparsely puberulent; leaf blade

Fig. 2.78. *Ziziphus jujuba* Miller. (By Philmarin, CC BY-SA 4.0, https://commons.wikimedia.org/w/index.php?curid=69284063.)

abaxially pale green, adaxially dark green, ovate, ovate-elliptic, or elliptic-oblong, 3–7 cm × 1.5–4 cm, papery, abaxially ± puberulent on major veins or glabrous, adaxially glabrous, 3-veined from base, base slightly asymmetric, subrounded, margin crenate-serrate, apex obtuse or rounded, rarely acute, mucronulate. Flowers yellow-green, bisexual, 5-merous, glabrous, solitary or 2–8 crowded in axillary cymes, shortly pedunculate. Pedicel 2–3 mm. Sepals ovate-triangular, adaxially distinctly keeled. Petals obovate, about as long as stamens, clawed at base. Disk orbicular, thick, fleshy, 5-lobed. Ovary basally slightly immersed in disk; style 2-cleft to half. Drupe red at maturity, turning red-purple, oblong or narrowly ovoid, 2–3.5 cm, (0.5–)1.5–2 cm in diameter; mesocarp fleshy, thick, sweet- or sour-tasting; stone acute or obtuse at both ends, 2-loculed, 1- or 2-seeded; fruiting pedicel 2–5 mm or longer. Seeds compressed-orbicular, *c.*1 cm × 0.8 cm (Rechinger, 1963–2001; Davis, 1965–1988).

Cultivation Methods of Some Important Medicinal and Aromatic Plants of Iran

Saffron (*Crocus sativus*)

Crocus sativus L. is a perennial stemless herb, commonly known as saffron. The dried stigmas constitute the saffron of commerce. As cultivation of this species requires intensive labor and time, and the stigmas constitute a small part of the plant, it is the most expensive spice in the world. Cultivation of this expensive traditional spice is distributed from the Mediterranean region to Europe as well as to Asia. Iran is the world's principal producer and exporter of saffron (85% of the worldwide saffron production). Due to very specific climate requirements, saffron grows in only a few countries. The main producer countries are Iran with 67,297 ha and a production amount of 239 t in 2010 and then India with 3200 ha cultivation area and a production amount of 12.5 t. The main producers in Europe are Greece with a cultivation area of 1,750 ha and total production of 7 t (in 2010) and Spain with 165 ha with a total production of 3.2 t. Between countries, Spain had the highest average yield per hectare because of good agricultural practices while the average yield in Iran was 3.5 kg/ha (UNIDO, 2014). More than 96% of saffron in Iran is being produced in Razavi Khorasan and South Khorasan provinces. Saffron is valued for its golden coloring, flavoring, and aromatizing, but recently it is also being considered by the pharmaceutical industry for its antidepressant and anti-Alzheimer activities (UNIDO, 2014).

Although about 150 volatile and nonvolatile compounds have been detected in saffron stigmas, only less than 50 constituents have been identified so far (Winterhalter and Straubinger, 2000). Terpenes, terpene alcohols, and their esters represent the main components in volatile oil, among which safranal is the major component. Various crocins, crocetin, picrocrocin, and flavonoids like quercetin and kaempferol are the main nonvolatile compounds (Liakopoulou-Kyriakides and Kyriakidis, 2002). The characteristic color of saffron is due to the presence of crocins, which are glucosyl esters of crocetin. The bitter taste of the spice is related to picrocrocin, a glycoside of safranal (Kanakis *et al.*, 2004).

Saffron is an autumnal flowering geophite characterized by a long summer rest in which the plant survives periods of drought by means of corms. Its biological cycle starts with its above-ground vegetative growth at the first autumn rains with the emission of leaves and flowers almost immediately and ends with the production of replacement corms in about 220 days. Unlike many other species of the *Crocus* genus that flower in the winter–spring period, in saffron flowering can occur from mid-October to the end of November, essentially depending on the climatic conditions. It generally starts from 60 to 90 days after planting, mainly depending on sowing time, but snowy and cold periods may retard flowering. Saffron prefers direct sun exposure, even though in India it is cultivated together with almond trees. According to Fernández (2004) and Mollafilabi (2004), the best climatic conditions for high yields are rainfall in the autumn, warm summers, and mild winters. Saffron grows on a wide range of soils. Skrubis (1990) reported that the best yields are achieved on well-drained clay-calcareous and deep soil. Fernández (2004) suggests that clay is a good soil for saffron.

Saffron is cultivated in a wide range of environments with mild to dry climates. This plant is propagated by corm and the best cultivation time is in summer (end of July) when the plant is in dormancy period (Gresta *et al.*, 2008a). Before planting, the soil should be completely cleared of weeds, ploughed at a depth of 25–30 cm, and left to rest from a few weeks to the entire winter. The application of about 20–30 t of organic manure per hectare is the most common fertilization practice all over the world (Koockeki, 2004). The best corm size is 8–10 g and the best plant density is 19–20 t of corms per hectare (Koocheki *et al.*, 2011). Koocheki *et al.* (2011) reported that best planting depth is 15 cm. They mentioned that the best planting method is row cropping with between- and inter-row distances of 15 and 20 cm, respectively (Koocheki *et al.*, 2009). Perennial crop techniques have highly variable durations from place to place: from 3–4 years in Spain, 6–8 years in India and Greece, and up to 12 years in Morocco (Ait-Oubahou and El-Otmani, 1999). With the aging of the saffron field, generally after 4–5 years, spice production declines because of increasing competition for water and nutrients, fungal infection due to overcrowding (Sampathu *et al.*, 1984), and the reduced size and reproduction capability of corms. In a 10-year experiment, Grilli Caiola (2005) observed that corms left in the soil without management techniques continue producing daughter corms for up to 3–5 years but afterwards they degenerate and are no longer able to reproduce vegetatively. Water requirements of saffron are low and can be satisfied by the scarce rainfall when cultivated in semiarid conditions. Researches in Iran (Koocheki, 2004) showed that this plant needs up to 3000 m^3 of irrigation per year. The most crucial moment for irrigation is after summer to awaken the corms but this coincides with autumn rains, so, excepting a severe drought season, this may be considered unnecessary. Weeds are probably the main problem for saffron, since it is unable to compete, above all because of its very low height. Nevertheless, due to the short time between the corm planting and flowering, severe weed problems start the following spring in perennial cultivation. From flowering, weeds can be left to grow until May and then cut without damaging the crop when the saffron leaves wither. Weeds are controlled mostly by hand. Gathering saffron flowers requires care and intensive manual labor: the flowers grow only a few centimeters above ground and, depending on vegetative activity, might be surrounded by several leaves that must not be damaged otherwise daughter corms will not be produced. Flowers are harvested manually. About 350–450 labor-hours are needed to harvest 1 kg of the spice (Gresta *et al.*, 2008b) (Fig. 2.79).

The saffron flower is highly ephemeral; given its very short life, it should be picked the same day of flowering and placed in baskets. The best practice is to pick the flower early in the morning each day, when the

Fig. 2.79. Harvesting of saffron in Torbat Heydarye, Iran. (By Farsad Nadjafi.)

corolla is still closed, thereby preventing the stigmas from losing color and quality, avoiding any sudden deterioration by wind or rain (Zanzucchi, 1987; Tammaro, 1990) and allowing a ready separation into their constituent parts. After harvest, stigmas must be separated from the petals and stamen as soon as possible by opening the corolla and cutting the stigmas with the fingers below the branching where the style changes color (from red to yellow). Drying and storage methods are very important because a poor undertaking of this procedure can completely compromise qualitative features of saffron (Carmona *et al.*, 2005). According to the ISO norm, the moisture content may range between 10 and 12% w/w (ISO-3632-2:2003, 2003). The best drying temperature is between 35 and 40°C (Gresta *et al.*, 2008b). Generally, 1 ha of saffron may produce 10–15 kg of dried stigmas, but it can range widely, depending on the above mentioned factors, from 2 to 30 kg. In Iran, the average yield of saffron is around 5.4 kg/ha (Behzad *et al.*, 1992). In New Zealand, McGimpsey *et al.* (1997) reached yields of 24 kg/ha for dried stigmas. The worst enemies of saffron are rodents and fungi (Tammaro, 1990; Goliaris, 1999). Moles, rats, and rabbits can easily damage corms or eat leaves. Fungal attacks are mostly promoted by humid conditions. To avoid fungal infection, the best practices are crop rotation along with the removal and burning of infected plants.

Cumin (*Cuminum cyminum* L.)

Cumin (*Cuminum cyminum* L.) is an important medicinal plant in Iran that has a long history of cultivation in the country. Cumin seeds are known for their spicy aroma and medicinal use (Hassanzadehdelouei *et al.*, 2013). This plant is cultivated in vast area in Iran, mostly in dryland agricultural systems. The area under cultivation and production of this plant in Iran have grown significantly in recent years. Today, the cultivation area is near 18,000 ha, with Razavi Khorasan Province (Fig. 2.80) accounting for more than 12,000 ha (Sefidkon, 2021).

Fig. 2.80. Organic cumin farm in Iran. (By Farsad Nadjafi.)

This species is a small bushy plant belonging to the *Apiaceae* family with a height of 20–30 cm and a greenish color at early stage and a grayish color at maturity (Fig. 2.81). The fruits are light brown or grayish in color and ovate or fusiform in shape (Hussein and Batra, 1998). The main producers of this plant are India, Iran, Syria, Egypt, and

Fig. 2.81. Cumin (*Cuminum cyminum* L.) plant. (By Farsad Nadjafi.)

Turkey; but it is also cultivated in Morocco, South Russia, Japan, Indonesia, Algeria, Pakistan, Saudi Arabia, Cyprus, Lebanon, Malta, Spain and on a small scale in Central America (Tuncturk and Tuncturk, 2006). Cumin fruits contain fatty oil (approximately 10% w/w), essential oil (1–5% w/w), protein, cellulose, sugar, and minerals (Li and Jiang, 2004). Volatile oil contains monoterpene hydrocarbons (α-pinene, β-pinene, ρ-cymene, and γ-terpinene) and various oxygenated monoterpenes (1,8-cineole, cuminaldehyde, cuminyl alcohol, and safranal) (Gachkar *et al.*, 2007; Rebey *et al.*, 2012). The main compound of cumin oil is cumin aldehyde (*p*-isopropyl benzaldehyde) with its content varying, depending on the individual extraction method (Behera *et al.*, 2004) and geographic origin, between less than 20% w/w and more than 40% w/w (Lawrence, 1992).

Cumin needs a Mediterranean climate with long and sunny days. The optimum growth temperature ranges between 25 and 30°C. The plants require a moderately cool and dry climate for good growth and production. High humidity, cloudy weather, too much dew, and unseasonal rain after flowering, on the other hand, are detrimental to yield (Khosh-Khui and Bonyanpour, 2010).

Seeds are used for cultivation. The plant can be cultivated in temperate regions in autumn and in cold regions in spring. November is a good time for autumn sowing, while spring planting is done in March. Autumn planting usually results to higher seed yield and essential oil content. The superiority of autumn planting was attributed to a better establishment of plants by using autumn precipitations and also lower extent of diseases (Rezvani Moghaddam *et al.*, 2014). Between 12 and 15 kg of high-quality seeds are required per hectare. Seed depth depending on soil texture is between 1.5 and 2 cm, and seeds are directly planted in the ground. Seeds should be soaked in water for 24–36 h to enhance the growing power (Faravani, 2004). Alam (2007) suggested that the use of bioinoculant *Azospirillum* or *Azotobacter* and seed priming for 8 h followed by shade drying before sowing improves germination; whereas Lal *et al.* (2014) suggested that seed treatment with Bavistin (2.5 g/kg) and *Trichoderma viride* (4 g/kg) and drying in shade in an airtight container for some hours before sowing improves germination percentage. It was also found that application of seed treatment reduces the chances of disease attacks. The control of wilt and cumin blight disease can be done by the seed treatment with *Trichoderma asperellum* and Bavistin (Trivedi *et al.*, 2019). The soil condition also plays a significant role in productivity of the crop. The most suitable soil for cumin cultivation is sandy to loamy soil with proper drainage, adequate aeration, and high oxygen availability. The preferred pH range of soil is 6.8 to 8.3 (Lal, 2018). Cumin seedlings are highly susceptible to salinity. The land required for cumin should be leveled and well pulverized. The land should be well prepared for better germination of seeds and growth of plants. A total of three or four ploughings is required. The first ploughing should be done by a soil-turning plough followed by two or three ploughing with a harrow to bring the soil to a fine tilth. At the time of sowing there should be good moisture in the soil for better germination of the seeds (Lal *et al.*, 2014).

The best cultivation method of cumin is the line sowing method in shallow furrows at a spacing of 20 to 25 cm (Kumar *et al.*, 2021).

After sowing, the first operation is to provide moisture to the soil for proper germination of seeds. The water requirement is 335 mm, throughout the crop cycle (Khajehpour, 1986). Irrigation should begin immediately after sowing and continue at 8- to 10-day intervals. The first irrigation should not be too heavy; otherwise, it would result in uneven distribution of germination (plants). The emergence of seedlings starts after the second irrigation, which is given at 10 days after the first irrigation, although, if the temperature at daytime is high or there is a dry spell, an additional irrigation may be required after 4 to 5 days for complete germination. Later, based on the individual weather conditions and soil type, the frequency of irrigation should be kept between 20 and 30 days. It was found that irrigation plays a crucial role at the time of flowering to seed-filling stage (Karimi, 1989). Hence, proper irrigation is to be given during this time period. The final irrigation should be slightly heavy at the time of seed formation to supply moisture at maturity stage.

The irrigation management is important for proper and uniform germination. Irrigation management using sprinklers, specifically in undulating areas of western Rajasthan (India), has played a significant role in increasing yield and quality (Ravindran *et al.*, 2006; Sundria *et al.*, 2014). Introducing modern methods of irrigation such as micro irrigation, drip systems, and sprinkler systems improves the water use efficiency and saves water for healthy crop production.

The optimal nutrient requirement should be met by applying the appropriate amount of fertilizers. During land preparation, application of farmyard manure at 1520 t/ha should be given as basal dose. Thereafter, the recommended fertilizer dose per hectare is 30 kg of nitrogen (N) and 20 kg of phosphorus (P) (P_2O_5) (Ehteramian, 2003). N can be applied in single dose at 30 days after sowing (DAS) or in two equivalent doses at 30 and 60 DAS (Hornok, 1992). It was found that the split applications of fertilizers as per recommendations enhanced the nutrient use efficiency with increased yield level (Ravindran *et al.*, 2006).

After, irrigation and fertilization, weeding plays an important role in increasing the productivity. Therefore, proper weeding should be done at right time. The first weeding and hoeing should be performed 30 to 40 DAS or when the plants have reached a height of 4–5 cm. If weeds reappear, another one or two hoeings and weedings will aid in better crop growth (Kumar *et al.*, 2021).

Cumin is sensitive to diseases like wilt, powdery mildew, and blight. Fusarium wilt is a seed- or soil-borne disease; it needs specific soil temperatures for epidemics to develop. It can reduce yield by up to 80% (Dar *et al.*, 2019). On the other hand, powdery mildew disease can cause severe yield losses in early development stage due to poor seed formation and, at a later stage, it can cause discolored tiny seeds. The blight disease appears as brownish spots on leaves and stems. Generally, the disease is more susceptible to cloudy weather and after flowering stage. A seed treatment by Captan at 2 g/kg of seed can be effective in controlling these diseases. Spraying the plants twice with zineb at 300 g/100 l of water is also effective in controlling this disease.

Along with disease, insect attack also hampers the productivity of cumin crop. Aphid (*Myzus persicae*) is an important insect, which attacks mostly at flowering stage. This insect feeds on plant sap from tender parts and flowers. The plant becomes yellowish in color and formation of seed is reduced, the quality of the produce is also degraded as well. Aphid can be controlled by spraying Rogor at 1 ml/l of water. Other major pests include mites, which regularly invade crops and feed mostly on young leaves; the infestation is more intense on young inflorescences. Disease and pest management by an integrated approach using both bioagents and pesticides is effective in management of wilt, blight, powdery mildew, and aphids that cause the most devastating biotic stresses in cumin crop (Israel and Lodha, 2004; Khare *et al.*, 2014). For the future, new varieties with multiple resistances to various diseases and pests such as wilt, blight, powdery mildew, and aphids need to be developed.

The crop is harvested when it turns yellow, leaves fall down, and seeds turn light grayish-brown. The cumin crop matures in 90–140 days (Balandari, 1994). The plants are harvested with hand tools such as sickles. However, attempts have been made in recent years to modify available harvesters for this purpose (Rahimi, 1993). After harvesting, threshing is done by threshers or manually. After harvesting and threshing, cumin seeds are dried in the sun up to a moisture level of 8.5–9% w/w. Seed yields range from 800 to 1000 kg/ha, depending on the individual variety, agro-management, and ecological conditions (Kumar *et al.*, 2021).

Damask rose (*Rosa* × *damascena* Mill.)

The genus *Rosa*, belonging to the *Rosaceae* family, includes more than 200 species and 18,000 cultivars (Gudin, 2000). One of the most important *Rosa* species is damask rose (*Rosa* × *damascena* Mill.), a deciduous shrub growing to 2.2 m tall that has a barbed stem, pinnate leaves with five leaflets, and fragrant pink to light red flowers. This shrub is probably a hybrid from *R. gallica* L. and *R. moschata* Herrm. Damask rose is one of the most important medicinal, aromatic, and ornamental plants, cultivated not only for its characteristic fragrance but also for its medicinal properties in many areas of the world such as Bulgaria, Turkey, India, and Iran (Tabaei-Aghdaei *et al.*, 2007; Yousefi *et al.*, 2009). Damask rose is used for the production of the hydrodistilled essential rose oil, rose water, rose concrete, and absolute, which are extremely valuable and important raw materials for the perfume, cosmetics, and flavor industries (Ayci *et al.*, 2005). The main producers of rose oil are Bulgaria, Turkey, Iran, and India (Rusanov *et al.*, 2009). In addition to its use in the flavor and fragrance industries, several other important pharmaceutical properties of damask rose oil have been reported such as anti-HIV (Mahmood *et al.*, 1996), antibacterial (Basim and Basim, 2003), antioxidant (Achuthan *et al.*, 2003; Ozkan *et al.*, 2004), anticancer (Ren *et al.*, 2003), and laxative (Gholamhoseinian *et al.*, 2010) activities. In Iran, different genotypes of this plant occur in nature (Tabaei Aghdaei *et al.*, 2003) and are widely distributed at an altutude of 300 to 3500 m above sea level (Tabaei Aghdaei *et al.*, 2005). The various genotypes differ greatly in terms of flower yield (Tabaei Aghdaei and Rezaei, 2002), essential oil content and composition (Tabaei Aghdaei *et al.*, 2003), propagation and rooting (Tabaei Aghdaei and Rezaei, 2000), morphology, phenology, and resistance to abiotic and biotic stresses (Tabaei Aghdaei and Babaei, 2001).

Iran, along with Turkey and Bulgaria, are largest producers of damask roses in the world. The area under cultivation was 18,000 ha in 2016. More than 85% of the cultivation area is located in Fars, Kerman, Isfahan, and East Azerbaijan provinces (Salehi Sardoei, 2022). The livelihood of many farmers in these regions is entirely related to cultivation and processing of this crop (Fig. 2.82). Iran has been mentioned as one of the countries of origin, and cultivation as well as consumption of *R.* × *damascena* has a long history in the country (Rusanov *et al.*, 2009). A considerable part of *R.* × *damascena* cultivation and suitable areas for the development of cultivation are located in semiarid and cool areas of Iran, including the provinces of Kurdistan, East Azerbaijan, and Zanjan.

Due to the work-intensive production of rose oil and the low oil content of rose flowers, rose oil is considered one of the most expensive essential oils; nevertheless, global demand for high-quality rose oil is expected to increase in the near future (Probir and Rakesh, 2013). More than 400 volatile compounds have been identified in the floral scent of various rose cultivars until now, which can be classified into five major chemical groups: hydrocarbons, alcohols, esters, aromatic ethers, and some others (including aldehydes, rose oxides, and norisoprenes) (Lavid *et al.*, 2002). Citronellol, nonadecane, geraniol, ethanol, heneicosane, nerol, and 1-nonadecene are reported as the major constituents of *R.* × *damascena* essential oil (Ozel *et al.*, 2006; Verma *et al.*, 2011). In

Fig. 2.82. Damask rose (*Rosa* × *damascena*), dryland farming system in Layzangan, Fars Province, Iran. (By Farsad Nadjafi.)

addition, literature indicates that environmental factors such as annual rainfall, temperature, humidity, light, soil, pruning, nutrient supply, harvesting time (Prakasa Rao *et al.*, 1995; Sangwan *et al.*, 2001; Mirali *et al.*, 2012; Probir and Rakesh, 2013), years, and cultivars (Bayrak and Akgül, 1994) have significant effects on rose oil quality. Furthermore, flower ontogeny and growing site (Verma *et al.*, 2011), flower stages, flower parts, harvesting date and fermentation period (Weiss, 1997; Baydar and Baydar, 2005; Verma *et al.*, 2011), extraction techniques (Ozel *et al.*, 2006), storing and drying treatments (Koksal *et al.*, 2015), and pressure and temperature during distillation (Kiran *et al.*, 2002) can affect the oil content and composition. For example, Kiran *et al.* (2002) reported that the monoterpene hydrocarbons and esters were found to be in traces, and stearoptene content was found to be higher in the oil distilled under high pressure, as compared to rose oil produced under atmospheric pressure. Verma *et al.* (2011) showed that oil content and composition can be different among damask rose cultivars. Yousefi and Jaimand (2018) reported that the average essential oil yield of the studied damask rose landraces in Iran is about 0.01% w/w and attributed this finding in particular to the drier environmental conditions. The individual extraction method also affects oil yield and oil components. Lawrence (1991), for example, stated found that the yield was about ten times higher with solvent extraction than with steam distillation. The main objective of damask rose cultivation in many countries such as Turkey, Bulgaria, India, and France is to extract the essential oil from the flower (Babu *et al.*, 2002); in this context, the breeding of new rose varieties with higher oil yield and optimal composition is of particular importance. In general, the main components of rose oil are the most important parameters that determine the quality of the product; therefore, the attention toward development of essential oil quantity and quality can be considered an important objective for improving essential oil yield of this species. The percentage of major components is one of the important parameters that determines the quality of rose oil (Boelens and Boelens, 1997). Good-quality rose oil should have a higher amount of monoterpene alcohols and a lower proportion of alkanes (Baser, 1992).

Nowadays, use of rose oil in pharmaceutical industries is steadily increasing. The content and relative compositions of the major components are the key parameters to determining the quality of rose oil (Nikolov *et al.*, 1977), which is chiefly governed by the genotype, ecological and agronomical factors. Nevertheless, *R.* × *damascena* is adapted to a wide range of environmental conditions.

In the main damask rose-growing region (Kazanlik) of Bulgaria, an average spring temperature of 5–15°C is considered as optimum (Oyen and Dung, 1999). If low night-time temperature of 10–12°C prevails during flowering stage, the oil synthesis reduces considerably. However, a low temperature during two weeks before flowering enhances the quality and quantity of *R.* × *damascena* flowers. The ideal temperature for growing of most of the roses is 19–27°C. Weiss (1997) found that a longer period in spring with an average temperature of 5–15°C ensures a large number of flowers per plant with a high oil content.

Low light intensity with low temperature and 88% relative humidity can increase the *R.* × *damascena* flower yield and its quantity and quality of essential oil (Misra *et al.*, 2002). During spring and early summer, 70% humidity is optimum for good harvest of damask rose. Roses do well in slightly acidic to slightly alkaline soils with a pH between 6.0 and 7.5, whereas an acidic soil inhibits plant growth and reduces flower yield, probably due to an imbalance of micronutrients. Brichet (2003) reported that rose plant is sensitive to saline and alkaline soil. Although *R.* × *damascena* does not like waterlogging, porous soils are generally unsuitable in low-rainfall areas. Poorly drained soils may lead to root diseases and nutrient deficiencies (Karlik *et al.*, 2003).

Rose plants can be propagated by either the sexual method or the vegetative method. Seed is generally used for propagation of the species, new cultivars, and for production of rootstocks (Hornok, 1992). The perfumery and cosmetics industries are considerate to procure rose oil of consistent quality. Vegetative propagation, therefore, is the most vital and sole method to produce true-to-type planting material. That is why *R.* × *damascena* is mostly vegetatively propagated, to produce consistent-quality essential oil. Generally, the damask rose is propagated by stem cutting, budding, and grafting (Hudson *et al.*, 2002). Stem cutting is the easy method for rose multiplication (Anderson and Woods, 1999). The establishment and growth rate of the cutting depend upon many factors like season of cutting, age and portion of the branch, growth media, moisture, and nutrient status (Khan *et al.*, 2006). About 20–25 cm long and 0.75–1.25 cm thick stem cuttings are prepared from 1-year-old stems, which begin to lignify. In India, the ideal time of deriving the cutting is November–December. Root initiation starts after 3 months of planting of cutting, and the sprouted cuttings are ready for transplant into the main field in about 9–12 months. The prevailing environmental conditions like light, temperature, and relative humidity play a vital role for rooting and subsequent growth and development of cuttings. Bredmose *et al.* (2004) reported that a temperature of 24.6°C is optimal to enhance the root growth of cuttings.

Damask rose is an exhaustive crop and requires abundant plant nutrients. In Bulgaria, it was estimated that 64 kg of N, 8.7 kg of P, and 36 kg of potassium (K) are annually applied per hectare on rose fields and the harvested flowers contain about 3% w/w N, 0.34% w/w P, and 1% w/w K (Koseva, 1978). Today, it is well known that quantity of flowers and oil yield of roses are closely correlated with the NPK levels in the leaf at bud development stage (Orlova, 1984). A good crop requires 200 kg N/ha, which is applied in two equal doses particularly in saline alkaline soil (Srivastava, 1975). Generally,

N is applied in two equal doses, half at the time of pruning and the other half 15 days later.

In Kashmir Valley, in India, application of 200 kg N, 100 kg P_2O_5, and 50 kg K_2O per hectare and year has been found to provide high flower production in damask rose (Jhon *et al.*, 1992). Jhon *et al.* (1992) also found maximum shoot length with N at 200 kg/ha over 100 kg/ha. Increase in the length of shoots due to higher dose of N and P application has also been reported in *R.* × *damascena* var. *trigentipetala* (Tajuddin *et al.*, 1986). Khoshgoftarmanesh *et al.* (2008) reported that micronutrient treatments reduced mildew infection in rose plants. Various effects of micronutrients on diseases and pests have also been reported (Marschner, 1995). Zinc plays an important role in enhancing tolerance of rose cultivars to the biotic stress of mildew infection (Khoshgoftarmanesh *et al.*, 2008).

Adequate water supply is essential throughout the vegetative and flowering periods of *R.* × *damascena* cultivation. In total, 10–12 irrigations are required for *R.* × *damascena* in a year. During the rainy season, it may not be necessary to irrigate the plants. However, frequency of irrigation during summer may be once weekly, and in winter it may be once fortnightly. Irrigation just after pruning is essential. Drip irrigation is most effective. Sprinkler irrigation is also efficient in terms of water use efficiency, but it encourages the infection by leaf diseases (Porter *et al.*, 1987).

Pruning seems to affect both morphological as well as yield parameters in *R.* × *damascena* (Paul *et al.*, 1995). Annual pruning is essential to attain proper shape of bushes and promote branches. In Bulgaria, light to moderate pruning increased the flower yield and flower oil content, whereas heavy pruning reduced both (Astadzhov *et al.*, 1986). Nevertheless, Hassanein (2010) reported that light and medium pruning were better than the heavy pruning in terms of petal weight and oil content, respectively. Although pruning time is mainly governed by the climate of the place, pruning of *R.* × *damascena* is usually done from October to January in different regions of India. The pruned plants take 70–90 days to flower after pruning. Time of flowering can be adjusted through date of pruning. However, Saffari *et al.* (2004) reported that the time of pruning did not affect the petal oil content. Pruning at the beginning of autumn improved flower quality and essential oil content (Hassanein, 2010). Damask roses are generally pruned annually. Astadzhov *et al.* (1986) reported that pruning once every 3 years is acceptable in terms of cost and flower yield. This cultural practice is useful in removing photosynthetic old parts and regenerates new healthy photosynthetic parts, subsequently increasing flower production.

R. × *damascena* is a deciduous shrub bearing a fewer number of leaves and branches during the initial 2 or 3 years. Thus, it faces severe weed infestation throughout the year. *R.* × *damascena* is more vulnerable to competition by weeds during the initial period of establishment of rooted stem cuttings in the field.

As mentioned before, the common way of propagation of damask rose is stem cutting. This method is very slow and time-consuming, and usually associated with various problems, such as limitation of stock plant and prolonged nursery stage (Skirvin *et al.*, 1990). In this context, tissue culture is employed for achieving faster rates of multiplication (Pati *et al.*, 2001; Kapchina-Toteva *et al.*, 2002). The tissue culture techniques for rapid multiplication have often resulted in superior-quality planting material (Billir, 1999). Different features of tissue culture propagation systems for *R.* × *damascena* have been described by several researchers (Jabbarzadeh and Khosh-Khui, 2005; Nikbakht *et al.*, 2005).

Flowering is the economic part of *R.* × *damascena*. The harvest period is short and depends on the prevailing weather conditions. In a cool and cloudy spring, it may continue for about 1 month, while in hot seasons it may last only for 16–20 days (Perry, 1925). The flowering period depends directly on the climatic conditions, but is usually completed after 20–25 days. In India, the main flowering season in the northern plains takes place during March and April, while in the hills it starts a little later in April and May. The ideal stage of harvesting is blossoming stage. Lawrence (1991) reports that the oil content in the flower is higher when the petals start to swell and two or three petals begin to open.

The flower harvesting should be started before dawn and should be stopped when the sun dries the dew on the flower. The oil content is significantly lower if the flowers are picked after 9 a.m. (Igolen, 1966). High temperature during harvest reduces the oil content of flowers (Baydar and Baydar, 2005) due to evaporation of essential oil from the trichomes of the petals (Weiss, 1997). The rose petals generally contain very little essential oil compared to other essential oil crops. One kilogram of rose oil can be obtained from about 3000 kg of rose petals (Baser, 1992), so oil content is only about 0.03% w/w. In the beginning stage of harvest, 4000 kg of flowers are required to obtain 1 kg of oil (Garnero and Buil, 1976). Oil composition varies to a large extent over the flowering stages, which is due to the different photosynthesic rates of various parts of the plants (Weiss, 1997). Mihailova *et al.* (1997) have also reported that the oil composition varies according to the flowering stages, flower parts, and the harvesting periods. The stage at which flowers are picked is not only an important factor for the oil yield but also for the quality of the oil. The results of some studies (Misra *et al.*, 2002; Baydar and Baydar, 2005) show that environmental conditions affect the quantity and quality of flowers and essential oil. Variations in yield and quality of essential oil may be due to different harvesting and distillation techniques (Bayrak and Akgül, 1994). It was also reported that the effect of storage temperature between 0 and 3°C on oil content was not significant, whereas the effect of storage duration was significant (Kazaz *et al.*, 2009). Kazaz *et al.* (2009) reported that the highest essential oil content (0.043% w/w) was obtained from petals distilled immediately after harvest, while the lowest oil content (0.022% w/w) was obtained from petals stored for 28 days. Although flower yield depends on genetic and climatic factors, agronomic management is the key factor in determining flower yield. There is a direct correlation between the level of crop management and yield.

The average annual flower yield in Bulgaria, Iran, and India is 2000–3000 kg/ha; in Turkey, 2000–2500 kg/ha; and in Russia, 1500–2000 kg/ha. However, the yields are not directly comparable as different cultivars are involved. The improved cultivars of damask rose developed at the Institute of Himalayan Bioresource Technology (IHBT) in Palampur, India, provide a flower yield of 4000–5000 kg/ha (Kaul *et al.*, 1999). Generally, flower yield follows a linear pattern up to 5 or 7 years from the date of transplantation, although the economic life of an *R.* × *damascena* plantation is 10–12 years.

Collection of Some Important Wild Medicinal Plants of Iran

Tragacanth (*Astragalus gummifer*)

Gum tragacanth is the second most important commercial gum worldwide, produced by several shrubby plants of the genus *Astragalus* that are distributed from Pakistan to Greece, particulary in Iran and Turkey (Anderson, 1989). The genus *Astragalus*, belonging to the *Papilionaceae* family, comprises about 2500 species worldwide, distributed mainly in the Near East and the Middle East but also common in southern Europe (Wahhabi, 2005). *Astragalus gummifer* is considered to be the main tragacanth-yielding species. It is a small branching thorny shrub with a height of 30–60 cm. Some species such as *Astragalus echidnaeformis* Sirjaev, *Astragalus gossypinus* Fisch., *Astragalus kurdicus* Boiss., and *Astragalus versus* are usually mentioned as a source of tragacanth. The better gum-yielding species are small, low, bushy perennials, frequently with a cushion-like form.

Among them, *A. gossypinus* (Fig. 2.83), which produces a high-quality gum, is a low, spreading bush with a height of 8–12 cm (Yazdanshenas *et al.*, 2014). The exudate is produced spontaneously on the bark of the shrub, but the yield is often increased by cutting into the bark. Gum tragacanth is obtained by tapping the branches and tap roots. The gum dries as it exudes and is

Fig. 2.83. *Astragalus gossypinus* Fisch., Fars Province, Iran. (By Mohammad Reza Kanani.)

quickly collected. The word *tragacanth* comes from the Greek meaning "goat's horn", which probably describes the appearance and texture of the crude gum. Tragacanth contains 20 to 30% v/v of a water-soluble fraction called tragacanthin (consisting of tragacanthic acid and arabinogalactan). It also contains 60 to 70% v/v of a water-insoluble fraction called bassorin. Tragacanthic acid is composed of d-galacturonic acid, d-xylose, l-fructose, d-galactose, and other sugars. Tragacanthin is composed of uronic acid and arabinose and dissolves in water to form a viscous colloidal solution, while bassorin swells to form a thick gel (Leung, 1980).

Tragacanth gum is a viscous, odorless, tasteless, water-soluble mixture of polysaccharides obtained from the sap extracted from the root of the plant and dried. It absorbs water to become a gel, which can be stirred into a paste. The most important applications of tragacanth gum today are in the food and pharmaceutical industries. It is also used in leather processing as an edge smoother and polishing agent and occasionally serves as a stiffening agent in textiles. Chemically, it is a complex mixture of acidic polysaccharides, mainly present in the form of calcium, magnesium, and potassium salts.

It is widely used in salad dressings and sauces, savory spreads, milk shakes, ice creams, and confectionery and bakery products. It is one of the oldest gums in the world and its use dates back 5000 years (Anderson and Bridgeman,1985). In Traditional Persian medicine, traganth is used to improve the human immune system and to alleviate side effects of other pharmaceuticals (Ameli, 1963).

Tragacanth swells rapidly in water to form highly viscous colloidal sols or semi-gels that act as protective colloids and stabilizing agents. The high viscosity of tragacanth solutions results from the molecular characteristics of the gum, which depend on the quality and physical form of the gum and how it is absorbed in water. The most important applications of tragacanth today are now in foods and pharmaceuticals. Its use for other industrial purposes has declined over the years as cheaper alternatives to tragacanth have been developed. Tragacanth has long been an important gum for pharmaceutical purposes and was used as a binder, suspender, or emulsifier in tablets, ointments, lubricating jellies, and oral suspensions, and particularly in dermatologic creams and lotions. It is also used in toothpastes, hair lotions, and other personal care products.

The gum is exported from the country of origin in ribbon or flake form, and has a rather horny texture. Robbins (1988) estimates the world market for tragacanth to be no more than 500 t/year; almost half of this is estimated to be consumed in Western Europe. Iran and Turkey have been the only significant producers of tragacanth for some years, with Iran being the principal source. Tragacanth is bought from origin as ribbons or flakes; loss of viscosity of gum that has been powdered and stored for long periods means powdered tragacanth is always produced in the importing country. Iranian tragacanth, which is generally regarded as superior to Turkish, is sold in about 12 different grades: five ribbons (Ribbon no. 1, Ribbon no. 2, etc.) and the remainder flake. Ribbon no. 1 is the top grade, being the palest and cleanest. Ribbon grades are usually used for pharmaceutical purposes; flake is used for food applications. The lower flake grades are appreciably darker and contain some foreign matter (Coppen, 1995). Tragacanth is sold and specified by viscosity. There is an FAO specification for food-grade tragacanth that includes limits for arsenic, lead, and heavy metals, as well as some other parameters. Tragacanth is also specified in many pharmacopoeias for pharmaceutical use, including the *British Pharmacopoeia* (Coppen, 1995).

The better gum-yielding species are small, low, bushy perennials, frequently with a cushion-like shape. However, they have relatively large tap roots that are the primary source of the gum. Gum-producing *Astragalus* species are distributed in the Iran-Turan mountains with poor flora and soil condition at altitudes of 1200 to 3100 m (Samsam Shariat, 1986). Also, habitats at altitudes of 3760–4000 m have been reported (Zarinkamar, 1997). Annual precipitation of 200–400 mm and thermal range of −10 to 36°C are the main charachteristics of their habitats in Iran (Zarinkamar, 1997).

The presently practiced harvest methods for tragacanth in Iran are incompatible with the ecological and botanical requirements of this species and cause severe damage to the plant vegetation. In some areas plants are harvested in the first year of vegetation. Harvesting in the first year shortens the total period of sprouting and decreases the quality of the gum. In warm climates, the plants are harvested in May and in cold climates in June (Safar *et al.*, 1993). It is therefore recommended to consider a 2- to 4-year fallow period for plant recovey (Safar *et al.*, 1993). Research studies have shown that the best harvesting method involves use of plants with a canopy diameter of more than 25 cm, a shallow incision, and avoiding cross-section incisions (Khosrogerdi, 1999).

The most striking feature of the gum-producing *Astragalus* is a central gum cylinder in the tap root, which is contained by the woody cylinder and may be as much as half the total diameter of the root. The gum is contained in the cylinder at high pressure and, when cut, exudes rapidly and hardens into the characteristic ribbons of tragacanth.

Some gum is collected from spontaneous exudation but most is obtained by tapping. The process of tapping entails clearing away the earth surrounding the tap root and making one or two cuts into the upper part of the root. The cuts are usually made longitudinally or cross-angled to the root, 2–5 cm long. Sometimes the branches are also cut but this usually yields an inferior gum. After a period of time that varies according to local custom or circumstances, but may last a few days or a week or longer, the tapper returns to the plants that he has cut to collect the gum. Further collections may be made thereafter but the quality of the gum soon deteriorates to a point when it is not worthwhile to continue. Flakes of gum, rather than ribbons, are usually produced later in the season. Tapping is carried out in the dry summer months and continues until the autumn rains (Coppen, 1995). Several factors are effective on the amount of gum production such as climate and especially precipitation, soil factors, type of species, age of plants, and harvesting methods (Gentry, 1957; Hagiwara *et al.*, 1992; Fahimi Poor *et al.*, 2010). Based on experimental field work, Gentry (1957) estimated the average yield of gum from *Astragalus microcephalus* to be 15 g per 100-day tapping season. Yazdanshenas *et al.* (2014) reported a gum yield of 1.19 to 2.42 g per plant in *A. gossypinus*.

Asafoetida (*Ferula assa-foetida* L.)

Asafoetida (*Ferula assa-foetida* L.) is the plant used for production of dried latex (gum oleoresin) exuded from the rhizome or the perrenial fleshy tap root. The plant grows 1 to 1.5 m tall, has deeply dissected leaves, and inconspicuous yellow flowers borne in compound umbels. The bark is wrinkled (Golmohammadi, 2013). As its name suggests, asafoetida has a fetid smell, but in cooked dishes, when diluted, it provides a smooth flavor reminiscent of leeks. Since it is very strong in taste and odour, asafoetida is often blended with diluents such as starch and flour and sold in a compounded form. An essential oil can be distilled from the oleoresin and finds minor use for flavoring purposes.

The genus *Ferula* (*Apiaceae*) consists of about 170 species worldwide, with 30 species found in Iran, of which 15 are endemic for the country (Mozaffarian, 2007). Its root is erect, succulent, and nearly thick. Its leaves are gray with short hairs. It is a monocarp plant with schizocarp fruits, and it grows up to 1000 m above sea level (Krebs, 1989). Phytochemical analyses of *F. assa-foetida* confirmed the presence of 62% w/w of resin, 25% w/w of gum, 3–7% w/w of essential oil, 1.28% w/w of free sulfuric acid, and small amounts of vanillin. Also, some novel compounds such as sesquiterpene coumarin derivatives and sesquiterpene chromone derivatives have been isolated from the roots of *F. assa-foetida*. Their structures were established by comprehensive spectroscopic analyses (Bahrami *et al.*, 2013; Meng *et al.*, 2013).

The raw material "asafoetida" is the oleoresin exudate obtained from certain *Ferula* species, particularly *F. assa-foetida*, which occur in Afghanistan, Turkey, Iran, and surrounding areas. The product is one of the few examples (another one is tragacanth gum) of an exudate that is obtained by "tapping" the roots of a shrubby plant (Fig. 2.84). The intense, characteristic smell of the plant is mainly due to various sulfur compounds. Extracts of asafoetida *hing* are used in spice blends and as a flavoring for meat sauces, pickles, curries, and other food products. The exudates from this plant are locally known as *anghuzeh*, *heng*, and *buganeh* and are traditionally applied for treatment of a vast range of diseases: urinary, gastrointestinal, respiratory infections, and epilepsy, but also as an aphrodisiac and an emmenagogue. It is furthermore used to treat snake and insect bites as well as worm infections (Hadidi *et al.*, 2003; Shweta *et al.*, 2011).

F. assa-foetida is one of the most important plants in arid regions of South Khorasan Province. Local beneficiaries in South Khorasan believe that they make a profit of US$1300–1500 annually for each family from harvesting and gum production. That is mainly the reason why this province ranks third in gum production in Iran (Golmohammadi, 2013). Unfortunately, asafoetida habitats are decreasing every year, due to conventional and nonscientific harvesting methods. Conventional harvesting involves some steps such as identification in the regions, marking, twisting, cutting, and collecting the gum (Shinwari and Gilani, 2003; Golmohammadi, 2013). India is the main importer of asafoetida and the country imports 600 to 700 t annually, mostly from Iran and Afganistan (Coppen, 1995).

A study of the effect of different pruning methods on gum yield and survival of *F. assa-foetida* showed that the amount of sap

Fig. 2.84. *Ferula assa-foetida*, Fars Provinz, Iran. (By Mohammad Reza Kanani.)

increased from 19.6 to 62.9 g per plant by different pruning methods and times of cutting. Two-way cutting of the crown was recommended as the best cutting method (Omidbeygi and Pirmoradei, 2006). Shad (1995) reported that four cuts during the harvesting season and a 6-day interval between cutting and collection is the best collection method, with 5.28 g per plant gum yield. Salar *et al.* (2006) reported that the best harvesting age for this plant is when the plant is 5 years old or older.

Askarzadeh (2000) observed a nonsignificant difference between the different cutting methods in gum yield. Results (Gholami and Faravani, 2014) showed that both harvesting methods and cutting times have a significant effect on gum yield. The highest and the lowest yields of gum were observed in the concave method and superficial method, respectively. Thirteen cutting times resulted to the highest yield of 20.93 g per plant. The lowest resin yield was found in the superficial cutting method (Gholami and Faravani, 2014). The interaction between harvesting methods and times of cutting was also significant. The highest gum yield was obtained by using the concave harvesting method with 10 cutting times. The lowest yield was achieved by using the superficial method with 5 cutting times (Gholami and Faravani, 2014). Furthermore, 25.2 g of gum per plant was collected when using the concave method with 10 cutting times, which is a considerable amount. The highest regeneration (90–100%) of the plants in the second year was observed in the superficial method with 5 cuts, but in the conventional cutting method with 13 cuts, only 10% of the plants survived. The lowest regeneration (0.10%) was observed in the conventional cutting method with 13 cuts (Gholami and Faravani, 2014). The results show that the conventional method that is mostly applied by local people provides a high gum yield but destroys the crown buds of the plant so that it cannot survive in the next year. The best harvesting method that provides a considerable yield and also gives the plants the chance to regenerate and survive is the concave harvesting method with 10 cuts. According to the results (Gholami and Faravani, 2014), regeneration rate and gum production, the treatment of 10 times cutting by the concave method could be the best recommendation for a sustainable production, with corresponding social, environmental, and global benefits.

Galbanum (*Ferula gummosa* Boiss.)

Galbanum gum is one of the most important rangeland products of Iran. The use of this gum in Iran is limited to traditional medicine and the main part of the harvested material is exported to Europe, especially Germany and France (Islami, 1994). This plant is used in traditional medicine of Iran as a stimulant, expectorant, and antispasmodic (Salar, 1997). Galbanum is an oleoresin exudate obtained from *Ferula gummosa*. It is obtained from the cut stem. Extracts of the oleoresin and the distilled essential oil contain a number of sulfurous compounds and are used to a limited extent as perfume fixatives. The main constituents are "resin" (50–70% w/w), "gum" (*c.*20% w/w), and essential oil (5–20% w/w) (McAndrew and Michalkiewicz, 1988) (Fig. 2.85).

The genus *Ferula* (*Apiaceae*) in Iran consists of 30 perennial species, which are generally distributed in mountain and desert regions. These species are classified as monocarpic plants (Molamohammadi and Masood Reza, 1989). Only three species including *F. gummosa* Boiss., *Ferula schair* Borze, and *Ferula rabriahus* Boiss. have the potential of exudate production (Batuli, 1994). *F. gummosa* is the main source of galbanum gum. This species is a perennial plant with a smooth stem and shiny, ovate, wedge-shaped, acute leaflets with finely serrated toothed edges. The flower umbels are sparse. The best tears are palish externally, about the size of a hazelnut, and when broken open are composed of clear white tears. The common kind is an agglutinated mass, showing reddish and white tears (Mozaffarian 2013; Zargari 1991).

The main habitats of *F. gummosa* are located in Pakistan, Turkmenistan, and the vast area northeast of Iran at altitudes of 2000–4000 m with an average annual precipitation of 250–400 mm (Batuli, 1994). The habitats in Iran are located mainly in the provinces of Markazi, Isfahan, Zanjan, Tehran, Semnan, Khorasan, Golestan, and Mazandaran. It is estimated that 70,000 ha are under exploitation (Ahmadi, 1991).

The yield of galbanum depends on several factors such as plant age and vigour, harvesting methods, ecological conditions, frequency and depth of cuts, spacing between cuts, and intensity of utilization (Ahmadi, 1991). Salar (1997) reported that the best time to harvest is when the plants

Fig. 2.85. *Ferula gummosa*, Khorasan Province, Iran. (By Mohammad Reza Kanani.)

are 5 years old. There are several traditional harvesting methods but some of them are not sustainable. Harvesting the whole tuber, cutting vertically and deeply, and cutting off the flowering stem are methods that give a higher yield but seriously damage the plant and affect its recovery. Research studies (Salar, 1997) have shown that three cuts in each harvest season is a suitable method and gives a yield of 25–30 g per plant. Batuli (1994) recommended a 3-year fallow for plant recovery after harvest.

References

Achuthan, C.R., Babu, B.H. and Padikkala, J. (2003) Antioxidant and hepatoprotective effects of *Rosa damascena*. *Pharmaceutical Biology* 41, 357–361.

Ahmadi, A. (1991) *Study of the Propagation and Cultivation of Galbanum (Ferula gummosa)*. Jihade Daneshgahi Research Institute Publishers, Iran (in Farsi).

Ait-Oubahou, A. and El-Otmani, M. (1999) Saffron cultivation in Morocco. In: Negbi, M. (ed.) *Saffron: Crocus sativus L.* Harwood Academic Publishers, Amsterdam, pp. 87–94.

Alam, A. (2007) Possibilities of mechanization in seed spices. In: Malhotra, S.K. and Vashishtha, B.B. (eds) *Production, Development, Quality and Export of Seed Spices (Proceedings of National Seminar on "Production, Development, Quality and Export of Seed Spices – Issues and Strategies", February 2–3, 2007, NRCSS, Ajmer)*. National Research Centre on Seed Spices Ajmer, Rajasthan, India, pp. 1–11.

Ameli, A. (1963) Physiochemical characteristics of gum tragacanth. PhD thesis, Tehran University, Tehran (in Farsi).

Anderson, D.M.W. (1989) *NFT Gums: Ancient and Modern Commercial Products. NFT Highlights*, 89-01. Nitrogen Fixing Tree Association, Hawaii.

Anderson, D.M.W. and Bridgeman, M.E. (1985) The composition of poteinaceous polysaccharides exuded by *Astragalus microcephalus*, *A. gummifer* and *A. kurdikus*. The sources of Turkish gum tragacanth. *Phytochemistry* 24(10), 2301–2304.

Anderson, R.G. and Woods, T.A. (1999) An economic evaluation of single stem cut rose production. *Acta Horticulturae* 48, 629–634.

Askarzadeh, M.A. (2000) *Investigation on Destruction Effects of Harvesting Methods on Survival and Regeneration of Ferula gummosa in Khorasan Province*. Final Reserach Report. Khorasan Agriculture and Natural Resources Reserach Center, Mashhad, Iran.

Assadi, M., Ramak Maassoumi, A.A., Mozaffarian, V. and Khatamsaz, M. (1989) *Flora of Iran*. Ministry of Agriculture, Islamic Republic of Iran, Tehran.

Astadzhov, N., Kamburova, K. and Peshov, I. (1986) Changes in flower and essential oil productivity of the Kazanlik rose caused by single pruning of shoots. *Rastenievădni Nauki* 23, 63–68.

Ayci, F., Aydinli, M., Bozdemir, O.A. and Tutas, M. (2005) Gas chromatographic investigation of rose concrete, absolute and solid residue. *Flavour and Fragrance Journal* 20, 481–486.

Babu, K.G.D., Bikran, S. and Joshi, V.P. (2002) Essential oil composition of damask rose (*Rosa damascena* Mill.) distilled under different pressures and temperatures. *Flavour and Fragrance Journal* 17, 136–140.

Bahrami, G., Soltani, R., Sajjadi, S.-E., Kanani, M.-R., Naderi, R. *et al.* (2013) Essential oil composition of *Ferula assa-foetida* L. fruits from Western Iran. *Journal of Reports in Pharmaceutical Sciences* 2, 1–8.

Balandari, A. (1994) *Botanical Characteristics of Local Population of Cumin in Iran*. Scientific Report. Iranian Scientific and Industrial Research Organization, Khorasan Center, Mashhad, Iran.

Baser, K.H.C. (1992) Turkish rose oil. *Perfumer & Flavorist* 17, 45–52.

Basim, E. and Basim, H. (2003) Antibacterial activity of *Rosa damascena* essential oil. *Fitoterapia* 74, 394–396.

Batuli, H. (1994) The effects of harvesting methods on production and viability of galbanum (*Ferula gummosa*). In: *Proceedings of the First National Conference on Rangelands and Rangeland Management in Iran*. Faculty of Natural Resources, Isfahan University, Isfahan, Iran, pp. 72–73 (in Farsi).

Baydar, H. and Baydar, N.G. (2005) The effects of harvest date, fermentation duration and Tween 20 treatment on essential oil content and composition of industrial oil rose (*Rosa damascena* Mill.). *Industrial Crops and Products* 21(2), 251–255.

Bayrak, A. and Akgül, A. (1994) Volatile oil composition of Turkish rose (*Rosa damascena*). *Journal of the Science of Food and Agriculture* 64, 441–448.

Behera, S., Nagarajan, S., Jagan, M.R. and Rao, L. (2004) Microwave heating and conventional roasting of cumin seeds (*Cuminum cyminum* L.) and effect on chemical composition of volatiles. *Food Chemistry* 87(1), 25–29.

Behzad, S., Razavi, M. and Mahajer, M. (1992) The effect of mineral nutrients (N, P, K) on saffron production. *Acta Horticulturae* 306, 426–430.

Billir, M. (1999) Use of *in vitro* culture for propagation. *Tarim ve Koyisleri Bakanligi Dergisi* 128, 32–35.

Bobrov, E.G., Shishkin, B.K., Korovin, E.P., Komarov V.L., Tzvelev N.N. *et al.* (1934–1964) *Flora of the USSR*. Akademiya Nauk SSSR, Moskva/Leningrad, 1952. Translated from Russian Published for the Smithsonian Institution, USA and the National Science Foundation, Washington, DC, by the Israel Program for Scientific Translations.

Boelens, M.H. and Boelens, H. (1997) Differences in chemical and sensory properties of orange flower and rose oils obtained from hydro distillation and from supercritical CO_2 extraction. *Perfumer & Flavorist* 22, 31–35.
Boissier, E. (1867–1888) *Flora orientalis: sive, Enumeratio plantarum in Oriente a Graecia et Aegypto ad Indiae fines hucusque observatarum*. H. Georg, Genevae et Basileae.
Bredmose, N., Kristiansen, K. and Nielson, B. (2004) Propagation temperature, PPFD, auxin treatment, cutting size and cutting position affect root formation, axillary bud growth and shoot development in miniature rose (*Rosa hybrid* L.) plants and alter homogeneity. *The Journal of Horticultural Science and Biotechnology* 79(3), 458–465.
Brichet, H. (2003) Distribution and ecology: Continental Asia and Japan. In: Roberts, A.V., Debener, T. and Gudin, S. (eds) *Encyclopedia of Rose Science*. Elsevier, Amsterdam, pp. 204–215.
Carmona, M., Zalacain, A., Pardo, J.E., Alvarruiz, A and Alonso, G.L. (2005) Influences of different drying and aging conditions on saffron constituents. *Journal of Agricultural and Food Chemistry* 53, 3974–3979.
Castroviejo, S., Laínz, M., López González, G., Montserrat, P., Muñoz Garmendia, F. *et al.* (1980) *Flora iberica; Plantas vasculares de la Península Ibéricae Islas Baleares*. Real Jardin Botanica, Consejo Superior de Investigaciones Científicas (CSIC), Madrid.
Chaudhary, S.A. (2001) *Flora of the Kingdom of Saudi Arabia*. Ministry of Agriculture & Water, Riyadh.
Coppen, J.J.W. (1995) *Gums, Resins and Latexes of Plant Origin*. Non-Wood Forest Products No. 6. Food and Agriculture Organization of the United Nations, Rome.
Dar, E.A., Mehdi, M., Ahmad, M., Bhat, F.N., Hussain, N. *et al.* (2019) The flavour of Indian cuisines – history, cultivation and uses. *Chemical Science Review Letters* 8(29), 129–135.
Davis, P.H. (1965–1988) *Flora of Turkey and the East Aegean Islands*. Edinburgh University Press, Edinburgh.
Ehteramian, K. (2003) The effects of different levels of nitrogen fertilizer and plant dating on black cumin (*Cuminum carvi* L.) in Kooshkak region in the Fars province. MSc thesis, Shiraz Universuty, Shiraz, Iran.
eFloras (2008) eFloras.org. Missouri Botanical Garden, St. Louis, Missouri & Harvard University Herbaria, Cambridge, MA. Available at: http://www.efloras.org (accessed 2 January 2023).
Engler, A. and Prantl, K. (1897–1915) *Die natürlichen Pflanzenfamilien nebst ihren Gattungen und wichtigeren Arten*. W. Engelmann, Leipzig, Germany.
Engler, A. and Prantl, K. (1924–1980) *Die Natürlichen Pflanzenfamilien nebst ihren Gattungen und wichtigeren Arten*, 2nd edn. W. Engelmann, Leipzig, Germany.
Fahimi Poor, A., Chahooky Zare, M.A. and Tavili, A. (2010) Study of relation some of range plants to environmental factors (case study: Rangelands of Central Taleghan). *Journal of Range Management* 4(1), 32–23 (in Persian).
Faravani, M. (2004) Survey on treatment of seed amount, transplanting time of black zeera from nursery to field with best cultivation practices. *Iranian Journal of Pharmaceutical Research* 2, 65.
Fernández, J.A. (2004) Biology, biotechnology and biomedicine of saffron. *Recent Research Developments in Plant Science* 2, 127–159.
Gachkar, L., Yadegari, D., Rezaei, M.B., Taghizadeh, M., Astaneh, S.A. *et al.* (2007) Chemical and biological characteristics of *Cuminum cyminum* and *Rosmarinus officinalis* essential oils. *Food Chemistry* 102, 898–904.
Garnero, J. and Buil, P. (1976) Evolution of the composition of the rose essential oils and concrete during the production campaign. *Riv. Ital Ess Prof Peante Offic Aromic Sap. Aerosol* 58, 537–540.
Gentry, H.S. (1957) Gum tragacanth in Iran. *Economic Botany* 11(1), 40–63.
Ghahreman, A. (1978) *Flore de Iran en couleur naturella*. Ministry of Agriculture, Islamic Republic of Iran, Tehran.
Ghahreman, A. and Okhovvat A.R. (2008) *Matching the Old Medicinal Plant Names With Scientific Terminology*. University of Tehran Press, Tehran.
Gholamhoseinian, A., Shahouzehi, B. and Sharififar, F. (2010) Inhibitory effect of some plant extract on pancreatic lipase. *International Journal of Pharmacology* 6, 18–24.
Gholami, B.A. and Faravani, M. (2014) Effects of different cutting methods and times of cutting on growth performance and gum resin production of *Ferula assa foetida*. *Journal of Agricultural Sciences* 59(1), 35–44.
Goliaris, A.H. (1999) Saffron cultivation in Greece. In: Negbi, M. (ed.) *Saffron: Crocus sativus L.* Harwood Academic Publishers, Amsterdam, pp. 73–85.

Golmohammadi, F. (2013) Medical plant of *Ferula assa-foetida* and its cultivating, main characteristics and economical importance in South Khorasan province – East of Iran. *Technical Journal of Engineering and Applied Sciences* 3, 2334–2346.
Govaerts, R. (2000) *World Checklist of Seed Plants Database* in Access D: 1-30141.
Gresta, F., Lombardo, G.M., Ruberto, G. and Siracusa, L. (2008a) Effect of mother corm dimension and sowing time on stigmas yields, daughter corms and qualitative aspects of saffron (*Crocus sativus* L.) in Mediterranean environment. *Journal of the Science of Food and Agriculture* 88, 1144–1150.
Gresta, F., Lombardo, G.M., Siracusa, L. and Ruberto, G. (2008b) Saffron, an alternative crop for sustainable agricultural systems. A review. *Agronomy for Sustainable Development* 28, 95–112.
Grilli Caiola, M. (2005) Embryo origin and development in *Crocus sativus* L. (*Iridaceae*). *Plant Biosystems* 139, 335–343.
Gudin, S. (2000) Rose, genetics and breeding. *Plant Breeding Reviews* 17, 159–189.
Hadidi, K., Aburjai, T. and Battah, A. (2003) A comparative study of *Ferula hermonis* root extracts and sildenafil on copulatory behavior of male rats. *Fitoterapia* 74, 242–246.
Hagiwara, A., Boonyaphiphat, P., Kawabe, M., Naito, H., Shirai, T. *et al.* (1992) Lack of carcinogenicity of tragacanth gum in B6C3F_1 mice. *Food and Chemical Toxicology* 30(8), 673–679.
Hassanein, A.M.A. (2010) Improved quality and quantity of winter flowering in rose (*Rosa* spp.) by controlling the timing and type of pruning applied in autumn. *World Journal of Agricultural Sciences* 6, 260–267.
Hassanzadehdelouei, M., Vazin, F. and Nadaf, J. (2013) Effect of salt stress in different stages of growth on qualitative and quantitative characteristics of cumin (*Cuminum cyminum* L.). *Cercetări Agronomiceîn Moldova* 46, 89–97.
Hornok, L. (1992) *Cultivation and Processing of Medicinal Plants*. Academic Kiado, Budapest, pp. 26–77.
Hudson, T.H., Dale, E.K. Jr, Davies, F.T. and Geneve, R.L. (2002) *Plant Propagation Principles and Practices*, 6th edn. Prentice Hall of India Private Limited, New Delhi.
Hussein, M. and Batra, A. (1998) *In vitro* embryogenesis of cumin hypocotyl segments. *Advances in Plant Sciences* 11, 125–128.
Igolen, G. (1966) The Turkish rose. *Soaps, Perfumery and Cosmetics* 39, 461–471.
Islami, M.B. (1994) Utilization of non-wood forest products in Iran. *Journal of Forest and Rangeland* 25, 22–30 (in Farsi).
ISO-3632-2:2003 (2003) *Part 1: Specification, Part 2: Test Methods*. International Organization for Standardization, Geneva, Switzerland.
Israel, S. and Lodha, S. (2004) Factors influencing population dynamics of *Fusarium oxysporum* f. sp. *Cumini* in the presence and absence of cumin crop in arid soils. *Phytopathologia Mediterranea* 43, 3–13.
Jabbarzadeh, Z. and Khosh-Khui, M. (2005) Factors affecting tissue culture of Damask rose (*Rosa damascena* Mill.). *Scientia Horticulturae* 105, 475–482.
Jhon, A.Q., Siddiqui, M.A.A. and Pal, T.M. (1992) Effect of different fertilizer levels on the growth and flower production of *Rosa damascena* Mill. in Kashmir. *Indian Perfumer* 36, 196–202.
Kanakis, C.D., Daferera, D.J., Tarantilis, P.A. and Polission, M.G. (2004) Qualitative determination of volatile compounds and qualitative qualitative evaluation of safranal and 4-hydroxy-2,6,6-trimethyl-l-cylohene-l-carboxaldehyde (HTTC) in Greek saffron. *Journal of Agricultural and Food Chemistry* 54(14), 4515–4521.
Kapchina-Toteva, V., Somleva, M. and Van Telgen, H.J. (2002) Anticytokinin effect on apical dominance release in *in vitro* cultured *Rosa hybrida* L. *Biologia Plantarum* 45, 183–188.
Karimi, P. (1989) Chemical analysis of essential oil of *Umbelifereae* family. DPharm thesis, Tabriz University, Tabriz, Iran.
Karlik, J.F., Becker, J.O., Pemberton, H.B. and Schuch, U.K. (2003) Production and marketing: field rose production. In: Roberts, A.V., Debener, T. and Gudin, S. (eds) *Encyclopedia of Rose Science*. Elsevier, Amsterdam, pp. 580–587.
Kaul, V.K., Singh, V. and Singh, B. (1999) Damask rose and marigold: prospective industrial crops. *Journal of Medicinal and Aromatic Plant Sciences* 21, 32.
Kazaz, S., Erbas, S. and Baydar, H. (2009) The effects of storage temperature and duration on essential oil content and composition oil rose (*Rosa damascena* Mill.). *Turkish Journal of Field Crops* 14, 89–96.
Khajehpour, M.R. (1986) *Principle of Crop Production*. Jehad-e-Daneshgahi of Isfahan University of Technology, Isfahan, Iran.
Khan, M.S., Khan, R.U. and Waseem, K. (2006) Effect of some auxins on growth of *Rosa damascena* cuttings in different growing media. *Journal of Agriculture & Social Sciences* 2, 13–16.

Khare, M.N., Tiwari, S.P. and Sharma, Y.K. (2014) Disease problems in the cultivation of I. Cumin (*Cuminum cyminum* L.) II. Caraway (*Carum carvi* L.) and their management leading to the production of high-quality pathogen free seed. *International Journal of Seed Spices* 4(1), 1–8.

Khosh-Khui, M. and Bonyanpour, A.R. (2010) Effects of some variables on seed germination and seedling growth of cumin (*Cuminum cyminum* L.). *International Journal of Agricultural Research* 5, 365–369.

Khoshgoftarmanesh, A.H., Khademi, H. and Hosseini, F. (2008) Influence of additional micronutrient supply on growth, nutritional status and flower quality of three rose cultivars in a soilless culture. *Journal of Plant Nutrition* 31, 1543–1554.

Khosrogerdi, A.H. (1999) Report of tragacanth utilization project in kooh Mish rangelands of Sabzevar. *Journal of Forest and Rangeland* 33, 22–31 (in Farsi).

Kiran, G.D., Singh, B.B., Joshi, V.P. and Singh, V. (2002) Essential oil composition of damask rose (*Rosa damascena* Mill.) distilled under different pressures and temperatures. *Flavour and Fragrance Journal* 17, 136–140.

Koksal, N., Aslancan, H. Sadighazadi, S. and Kafkas, E. (2015) Chemical investigation on *Rose damascena* Mill. volatiles; effects of storage and drying conditions. *Acta Scientiarum Polonorum* 14(1), 105–114.

Koocheki, A. (2004) Indigenous knowledge in agriculture with particular reference to saffron production in Iran. *Acta Horticulturae* 650, 175–182.

Koocheki, A., Nassiri Mahalati, M., Alizadeh, A. and Ganjali, A. (2009) Modeling the effects of climate change on flowering behavior of saffron. *Iranian Journal of Field Crops Research* 7(2), 583–590 (in Persian with English summary).

Koocheki, A., Siahgmargoie, A., Azizi, G. and Jahani, M. (2011) Effect of high corm density and depth of planting on agronomic criteria and corm behavior of saffron. *Agroecology* 3(1), 36–49 (in Persian with English summary).

Koseva, D. (1978) Effects of fertilizers on the utilisation and removal of NPK by Kazanlik rose. *Rastenievădni Nauki* 15, 107–118.

Krebs, C. (1989) *Ecological Methodology*. Harper & Row, New York.

Kumar, M., Kumar Sahoo, P., Kumar Kushwasha, D., Nath Saxena, S., Mani, I. *et al.* (2021) Cumin cultivation: present status and future prospects. *The Pharma Innovation Journal* 10(8), 1121–1131.

Lal, G. (2018) Scenario, importance and prospects of seed spices: a review. *Current Investigation in Agriculture Current Research* 8, 49–62.

Lal, G., Saran, P.L., Devi, G. and Deepak, P.R. (2014) Production technology of cumin (*Cuminum cyminum* L.). In: Choudhary, A.K., Rana, K.S., Dass, A. and Srivastav, M. (eds) *Advances in Vegetable Agronomy*, Vol. 1. Post Graduate School, Indian Agricultural Research Institute and Indian Council of Agricultural Research, Department of Agricultural Research and Education Ministry of Agriculture, Government of India, New Delhi, pp. 223–231.

Lavid, N., Wang, J., Shalit, M., Guterman, I., Bar, E. *et al.* (2002) *O*-Methyl-transferases involved in the biosynthesis of volatile phenolic derivatives in rose petals. *Plant Physiology* 129, 1899–1907.

Lawrence, B.M. (1991) Progress in essential oil and extracts. *Perfumer & Flavorist* 16, 43–77.

Lawrence, B.M. (1992) Progress in essential oils. *Perfumer & Flavorist* 17, 42–44.

Leung, A. (1980) *Encyclopedia of Common Natural Ingredients Used in Food, Drugs and Cosmetics*. Wiley, New York.

Li, R. and Jiang, Z.T. (2004) Chemical composition of the essential oil of *Cuminum cyminum* L. from China. *Flavour and Fragrance Journal* 19, 311–313.

Liakopoulou-Kyriakides, M. and Kyriakidis, D.A. (2002) *Crocus sativus* – biological active constitutents. In: ur Rahman, A. (ed.) *Studies in Natural Products Chemistry*, Vol. 26, Part G. Elsevier, Amsterdam, pp. 293–312.

Liogier, A.H. (1985–1997) *Descriptive Flora of Puerto Rico and Adjacent Islands: Spermatophyta – Dicotyledoneae. Acanthaceae to Compositae*. La Editorial, Universidad de Puerto Rico, San Juan.

Mahmood, N.S., Piacente, C., Pizza, A., Bueke, A. Khan, I. *et al.* (1996) The anti-HIV activity and mechanisms of action of pure compounds isolated from *Rosa damascena*. *Biochemical and Biophysical Research Communications* 229, 73–79.

Marschner, H. (1995) *Mineral Nutrition of Higher Plants*. Academic Press, New York.

McAndrew, B.A. and Michalkiewicz, D.M. (1988) Analysis of galbanum oils. In: *Flavors and Fragrances: A World Perspective. Proceedings of 10th International Congress of Essential Oils, Fragrances and Flavors, Washington, DC, 16–20 November 1986*. Elsevier, Amsterdam, pp. 573–585.

McGimpsey, J.A., Douglas, M.H. and Wallace, A.R. (1997) Evaluation of saffron (*Crocus sativus* L.) production in New Zealand. *New Zealand Journal of Crop and Horticultural Science* 25(2), 159–168.

Meng, H., Li, G., Huang, J., Zhang, K., Wangand, H. *et al.* (2013) Sesquiterpene coumarin and sesquiterpene chromone derivatives from *Ferula ferulaeoides* (Steud.) Korov. *Fitoterapia* 86, 70–77.
Mihailova, J., Atanasouva, R. and Balinova-Tsvetkova, A. (1997) Direct gas chromatographs of the essential oil in separate flower parts of the flower of Kazanlik rose. In: *Proceedings of the 7th International Congress on Essential Oils, Kyoto, Japan*. Maruzen, Tokyo, p. 219.
Mirali, M., Aziz, R. and Nabulsi, I. (2012) Genetic characterization of *Rosa damascena* species growing in different regions of Syria and its relationship to the quality of the essential oils. *International Journal of Medicinal and Aromatic Plants* 2(1), 41–52.
Misra, A., Sharma, S., Singh, A. and Patra, N.K. (2002) Influence of topographical and edaphic factors on rose. II. Flowering quality and quantity. *Communications in Soil Science and Plant Analysis* 33, 2771–2780.
Molamohammadi, M. and Masood Reza, A. (1989) *Galbanum (Ferula gummosa) in Iran*. Bulletin 56. Research Institute of Fotests and Rangelands, Tehran (in Farsi).
Mollafilabi, A. (2004) Experimental findings of production and echo physiological aspects of saffron (*Crocus sativus* L.). *Acta Horticulturae* 650, 195–200.
Mozaffarian, V. (2007) *Flora of Iran, Umbelliferae*. Ministry of Agriculture, Islamic Republic of Iran, Tehran.
Mozaffarian, V. (2008) *A Dictionary of Iranian Plant Names*. Farhang Moaser, Tehran.
Mozaffarian, V. (2013) *Indentification of Medicinal and Aromatic Plants of Iran*. Farhang Moaser, Tehran.
Nasir, E. and Ali, S.I. (1970–1995) *Flora of West Pakistan*. Fakhri Printing Press, Karachi, Pakistan.
Nikbakht, A., Kafi, M., Mirmasomi, M. and Babalar, M. (2005) Micropropagation of Damask Rose (*Rosa damascena* Mill.) *cvs.* Azaran and Ghamsar. *International Journal of Agriculture and Biology* 7, 535–538.
Nikolov, N., Tsoutsoulova, A. and Nenov, N. (1977) Bulgarian rose oil and other essential oil. *MBI* 2, 46–58.
Omidbeygi, R. and Pirmoradei, M. (2006) A study of the effect of root diameter and incision time ongum yield in medicinal-rangeland asafoetida (*Ferula assa-foetida* L.) plant. *Iranian Journal of Natural Resources* 59, 269–261.
Orlova, L.M. (1984) Principles of essential-oil rose nutrition determined by leaf analysis. *Referativnyi Zhurnal* 5, 569.
Oyen, L.P.A. and Dung, N.X. (1999) *PROSEA: Plant resources of South-East Asia 19, Essential Oil Plants*. Prosea Foundation, Bogor, Indonesia.
Ozel, M.Z., Göğüş, F. and Lewis, A.C. (2006) Comparison of direct thermal desorption with water distillation and superheated water extraction for the analysis of volatile components of *Rosa damascena* Mill. using GCxGC-TOF/MS. *Analytica Chimica Acta* 566(2), 172–177.
Ozkan, G., Sagdic, O., Baydarand, N.G. and Baydar, H. (2004) Antioxidant and anti-bacterial activities of *Rosa damascena* flower extracts. *Food Science and Technology International* 10, 277–281.
Pati, P.K., Sharma, M. and Ahuja, P.S. (2001) Micropropagation, protoplast culture and its implications in the improvement of scented rose. *Acta Horticulturae* 547, 147–158.
Paul, T.M., Siddique, M.A.A. and John, A.Q. (1995) Effect of severity and time of pruning on growth and flower production of *Rosa damascena* Mill. An important aromatic plant. *Advances in Plant Science* 8, 28–32.
Perry, E.J. (1925) *Perry's Cyclopaedia of Perfumery*, Vol. 2. J. & A. Churchill, London, pp. 630–664.
Porter, D.M., Wright, F.S. and Powell, N.L. (1987) Effects of sprinkler irrigation on peanut disease in Virginia. *Plant Disease* 71, 512–515.
Post, G.E. (1896) *Flora of Syria, Palestine, and Sinai: from the Taurus to Ras Muhammas and from the Mediterranean Sea to the Syrian desert*. Syrian Protestant College, Beirut.
Prakasa Rao, E.V.S., Rao, R.S.G. and Ramesh, S. (1995) Seasonal variation in oil content and its composition in two chemotypes of scented geranim (*Pelaronium* spp.). *Journal of Essential Oil Research* 7, 159–163.
Probir K.P. and Rakesh, D.S. (2013) Understanding crop-ecology and agronomy of *Rosa damascena* Mill. for higher productivity. *Australian Journal of Crop Science* 7(2), 196–205.
Rahimi, M. (1993) *Chemical Control of Weeds in Cumin*. Scientific Report. Iranian Scientific and Industrial Research Organization, Khorasan Center, Mashhad, Iran.
Ravindran, P.N., Babu, N.K., Shiva, K.N. and Kallupurackal, J.A. (2006) *Advances in Spices Research: History of Achievements of Spices Research in India Since Independence*. Agrobios (India), Jodhpur, India, p. 994.
Rebey, I.B., Bourgou, S., Debez, I.B.S., Jabri Karoui, I., Hamrouni Sellami, I. *et al.* (2012) Effects of extraction solvents and provenances on phenolic contents and antioxidant activities of cumin (*Cuminum cyminum* L.) seeds. *Food and Bioprocess Technology* 5, 2827–2836.
Rechinger, K.H. (1963–2001) *Flora Iranica*. Akademische Druck- und Verlagsanstalt, Graz, Austria.

Ren, W., Qiao, Z., Wang, H., Zhu, L. and Zhang, L. (2003) Flavonoids: promising anticancer agents. *Medicinal Research Reviews* 23(4), 519–534.
Rezvani Moghaddam, P., Moradi, R. and Mansoori, H. (2014) Influence of planting date, intercropping and plant growthpromoting rhizobacteria on cumin (*Cuminum cyminum* L.) with particular respect to disease infestation in Iran. *Journal of Applied Research on Medicinal and Aromatic Plants* 1, 134–143.
Robbins, S.R.J. (1988) Gum tragacanth. In: *A Review of Recent Trends in Selected Markets for Water-Soluble Gums*. ODNRI Bulletin No. 2. Overseas Development Natural Resources Institute [now Natural Resources Institute, Chatham], London, pp. 52–60.
Rusanov, K., Kovacheva, N., Stefanova, K., Atanassov, A. and Atanassov, I. (2009) *Rosa damascene* – genetic resources and capacity building for molecular breeding. *Biotechnology & Biotechnological Equipment* 23(4), 1436–1439.
Safar, M.T., Rezvani, M. and Movahed, R. (1993) *Utilization Project of Tragacanth and Conservation of Its Habitat in Koun-Najaf Abad Rangelands*. Report. Research Institute of Forest and Rangelands, Isfahan Natural Resource Organization, Isfahan, Iran (in Farsi).
Saffari, V.R., Khalighi, A., Lesani, H., Babalar, M. and Obermaier, J.F. (2004) Effects of different plant growth regulators and time of pruning on yield components of *Rosa damascena* Mill. *International Journal of Agriculture and Biology* 6, 1040–1104.
Salar, N. (1997) *Study of Cultivation and Propagation Methods of Ferula gummosa*. Report. Research Institute of Forest and Rangelands, Semnan, Iran (in Farsi).
Salar, N.A., Ez Aldin, H. and Taherian, K. (2006) A survey on cultivation and propagation methods of *Ferula gummosa* in Semnan province. *Journal of Pajouhesh and Sazandegi* 53, 90–97.
Salehi Sardoei, A. (2022) Iranian medicinal plants: from economically to ethnomedicine studies. *International Journal of Advanced Biological and Biomedical Research* 10(2), 98–116.
Sampathu, S.R., Shivashankar, S. and Lewis, Y.S. (1984) Saffron (*Crocus sativus* L.): cultivation, processing, chemistry and standardization. *Critical Reviews in Food Science & Nutrition* 20, 123–157.
Samsam Shariat, H.M. (1986) *Medicinal and Natural Pharmaceuticals*. Rouzbahan Publishing, Tehran (in Farsi).
Sangwan, N.S., Farooqi, A.H.A, Shabih, F. and Sangwan, R.S. (2001) Regulation of essential oil production in plants. *Plant Growth Regulation* 34, 3–21.
Sefidkon, F. (2021) National approach to make a transformation in the economic value of medicinal plants. *Iranian Nature Journal* 6(1), 135–143.
Sell, P. and Murrell, G. (2006) *Flora of Great Britain and Ireland*. Cambridge University Press, Cambridge.
Shad, G. (1995) Avetochology of *Ferula assa foetida* and an investigation on its harvesting methodsin Mohammad abad region of Chelpo, Kashmar, Iran. MSc thesis, University of Gorgan, Gorgan, Iran.
Shinwari, Z.K. and Gilani, S.S. (2003) Sustainable harvest of medicinal plants at Bulashbar Nullah, Astore (Northern Pakistan). *Journal of Ethnopharmacology* 84, 289–298.
Shweta, G., Chetna, R., Jinkal, S., Nancy, S. and Hitesh, J. (2011) Herbal plants used as contraceptives. *International Journal of Current Pharmaceutical Review and Research* 2(1), 47–53.
Skirvin, R.M., Chu, M.C. and Young, H.J. (1990) Rose. In: Ammirato, P.V., Sharp, W.R. and Evans, D.A. (eds) *Handbook of Plant Cell Culture*, Vol. 5: *Ornamental Species*. McGraw-Hill, New York, pp. 716–743.
Skrubis, B. (1990) The cultivation in Greece of *Crocus sativus* L. In: Tammaro, F. and Marra, L. (eds) *Proceedings of the International Conference on Saffron (Crocus sativus L.)*. Universita Degil Studi L'Aquila and Accademia Italiana della Cucina, L'Aquila, Italy, pp. 171–182.
Soepadmo, E., Swa, L.G., Chung, R.C.K. and Kiew, R. (2011) *Tree Flora of Sabah and Sarawak*. Forest Research Institute and Sarawak Forestry Department, Malaysia.
Srivastava, H.P. (1975) Effect of some higher doses of nitrogen on the flower yield of *Rosa damascena* mill. *Indian Perfumer*18, 21–23.
Sundria, M.M., Mehriya, M.L., Rathore, B.S. and Choudhary, B.R. (2014) Cumin (*Cuminum cyminum* L.) *Sustainable Production Technology in Rajasthan*. Agricultural University, Jodhpur, Rajasthan, India.
Tabaei Aghdaei, S.R. and Babaei, M. (2001) Study of genotype differences for drought responces at early groth stages in *Rosa damacena* Mill. *Iranian Journal of Rangelands and Forests Plant Breeding and Genetic Research* 8, 113–126.
Tabaei Aghdaei, S.R. and Rezaei, M.B. (2000) Investigation of propagation and rooting ability in cuttings of *Rosa damascene*. *Iranian Journal of Rangelands and Forests Plant Breeding and Genetic Research* 1, 76–94.
Tabaei Aghdaei, S.R. and Rezaei, M.B. (2002) Study of flower yield variation in *Rosa damascena* L. genotypes of Kashan. *Iranian Journal of Rangelands and Forests Plant Breeding and Genetic Research* 9, 99–111.

Tabaei Aghdaei, S.R., Rezaei, M.B. and Jaimand, K. (2003) Evaluation of genetic variation in floral parts and essential oils concentration of *Rosa damascena* genotypes collected from Kashan/Iran. *Iranian Journal of Rangelands and Forests Plant Breeding and Genetic Research* 11, 219–234.

Tabaei Aghdaei, S.R., Rezaei, M.B. and Jaimand, K. (2005) Study of genetic variation in essential oils yield of *Rosa damascena* Mill. genotypes from central regions of Iran. *Iranian Journal of Medicinal and Aromatic Plants* 21, 35–49.

Tabaei-Aghdaei S.R., Babaei, A., Khosh-Khui, M., Jaimand, K., Rezaee, M.B. *et al.* (2007) Morphological and oil content variations amongst damask rose (*Rosa damascena* Mill.) landraces from different regions of Iran. *Scientia Horticulturae* 113(1), 44–48.

Tajuddin, S.M.L., Singh, A.K., Shawl, A.S. and Saproo, M.L. (1986) Introduction and cultivation of Bulgarian rose as a commercial crop in India. In: Srivastava, H.C., Vatsya, B. and Menon, K.K.G. (eds) *Plantation Crops – Opportunities and Constraints*, Vol. I. Oxford and IBHP Pub. Co., New Delhi, pp. 349–359.

Tammaro, F. (1990) *Crocus sativus* L. – cv. Piano di Navelli (L'Aquila saffron): environment, cultivation, morphometric characteristics, active principles, uses. In: Tammaro, F. and Marra, L. (eds) *Proceedingsof the International Conference on Saffron (Crocus sativus L.)*. Universita Degil Studi L'Aquila and Accademia Italiana della Cucina, L'Aquila, Italy, pp. 47–57.

Thulin, M. (ed.) (1999) *Flora of Somalia*. The Royal Botanic Gardens, Kew, London.

Tristram, H.B. (1884) *The Survey of Western Palestine: The Fauna and Flora of Palestine*. Palestine Exploration Fund, London.

Trivedi, V., Sasidharan, N. and Patel, D.A. (2019) Effect of pre-sowing seed treatment on seed emergence and yield parameters in cumin (*Cuminum cyminum* L.). *International Journal of Seed Spices* 9, 73–76.

Tuncturk, R. and Tuncturk, M. (2006) Effects of different phosphorus levels on the yield and quality components of cumin (*Cuminum cyminum* L.). *Research Journal of Agriculture and Biological Sciences* 2, 336–340.

Tutin, T.G., Heywood, V.H., Burges N.A., Valentine D.H., Walters S.M. *et al.* (1964) *Flora Europaea*. Cambridge University Press, Cambridge.

UNIDO (2014) *Saffron Industry Value Chain Development in Iran: Diagnostic Study Report*. Agriculture Bank of Iran and United Nations Industrial Development Organization, Vienna.

Verma, R.S., Padalia, R.C., Chauhan, A., Singh, A., and Yadav, A.K. (2011) Volatile constituents of essential oil and rose water of damask rose (*Rosa damascena* Mill.) cultivars from North Indian hills. *Natural Product Research* 25(17), 1577–1584.

Wahhabi, M.R. (2005) Determination indicators of habitat for harvesting gum on two species of Astragalus on Isfahan Province. PhD thesis, Tehran University, Tehran (in Persian).

Weiss, E.A. (1997) *Essential Oil Crops*. CAB International, Wallingford, U.K.

Winterhalter, P. and Straubinger, M. (2000) Saffron – renewed interest in an ancient spice. *Food Reviews International* 16, 39–59.

Yazdanshenas, H., Jafari, M., Azarnivand, H. and Arzani, H. (2014) Determination of the tragacanth gum production potential of white *Astragalus (Astragalus gossypinus)* in western rangeland of Isfahan province. Presented at *Anthropogenic Changes of Environment and Landscape in Arid and Semi Arid Regions Conference, Karaj, Iran, 8–9 October 2014*.

Yousefi, B. and Jaimand, K. (2018) Chemical variation in the essential oil of Iranian *Rosa damascena* landraces under semi-arid and cool conditions. *International Journal of Horticultural Science and Technology* 5(1), 81–92.

Yousefi, B., Tabaei-Aghdaei, S.R., Darvish, F. and Assareh, M.H. (2009) Flower yield performance and stability of various *Rosa damascena* Mill. landraces under different ecological conditions. *Scientia Horticulturae* 121, 333–339.

Zanzucchi, C. (1987) Research carried out by the Consorzio Comunale Parmense on Saffron (*Crocus sativus* L.). In: *Proceedings of the Conference "Coltivazione delle Piante Officinali", Trento, Italy, 9–10 October 1986*. Siège, Villazzano, Italy, pp. 347–395.

Zargari, A. (1991) *Medicinal Plants*, Vol. 2. University of Tehran Press, Tehran (in Farsi).

Zarinkamar, F. (1997) *Anatomy–Ecology Studies of Astragalus Species (Producers of Gum Tragagantin) in Iran*. Bulletin 156. Research Institute of Forests and Rangelands, Tehran (in Farsi).

3 Relationships Between Traditional Persian/Iranian Medicine and Other Traditional Medical Systems

Hartwig Schulz*
Consulting and Project Management for Medicinal and Aromatic Plants, Stahnsdorf, Germany

Abstract

This chapter gives a detailed overview of the history of traditional Arab and Persian medical systems and their interactions with Ayurveda in India, Traditional Chinese Medicine (TCM), and Western traditional medicine (the so-called "monastic medicine") that interacted with or emerged from Islamic medicine in the Middle Ages. The first records of medicinal plants date back more than 5000 years to the time of the Sumerians in southern Mesopotamia, who prepared lists with various plant species. Furthermore, ancient Chinese and Egyptian papyrus writings dating back to 3000 BC also reveal the medicinal use of plants at that time. The "Ebers Papyrus" from ancient Egypt, for example, provides extensive information on more than 850 herbal medicines. Manuscripts of Traditional Persian/Iranian Medicine (TPM/TIM), written by Persian scholars especially in the Middle Ages, not only represent a treasure of traditional medical and pharmaceutical knowledge, but also provide a collection of ingenious studies regarding the use of medicinal plants. In present days, herbal medicines are experiencing a renaissance and are extensively used for various therapeutic purposes. In this context, this chapter contains numerous references to various ethnobotanical studies and also gives a general overview of the intercultural exchange of traditional medical systems and the resulting gain in experiential knowledge about the individual medicinal plants. Today, this traditional medical knowledge represents an important source of inspiration for the development of new efficient phytopharmaceutical drugs and new therapeutic strategies. Although traditional medicine continues to play an important role in Iran today, much of this traditional knowledge has not been adequately addressed by scientific research. Therefore, it seems advisable to verify the effectiveness of traditional treatments using the new scientific tools available today, especially in cases where there is little or no supporting data in the literature.

Introduction

Since ancient times, plants have been one of the first and most available resources usable for treating illnesses. There has always been a close relationship between man and plants, and the medicinal effects of plants and their uses have been known more or less by everybody (UNESCO, 1996). In 2002, the World Health Organization (WHO) provided the following detailed description of the term "Traditional Medicine": "Traditional medicine (TM) is a comprehensive term used to refer both to TM systems such as Traditional

*Email: hs.consulting.map@t-online.de

DOI: 10.1079/9781800621671.0003

Chinese Medicine, Indian Ayurveda, Islamic Traditional Medicine, and to various forms of indigenous medicine" (WHO, 2002) (Fig. 3.1).

Traditional medicine includes medication therapies if they involve the application of herbal medicines, animal parts, and/or minerals; and non-medication therapies if they are carried out primarily without the use of medication, as in the case of acupuncture, manual therapies, and spiritual therapies (WHO, 1978, 2002).

In general, traditional medicine offers the enormous advantage that it is considered part of the culture and therefore cultural problems that could affect the application of the medicine can usually be avoided. Moreover, it can be used in conjunction with conventional medicine and in this way often increase the healing success. The development and promotion of Traditional Medicine could also be seen as respecting and honoring the respective culture and heritage of people around the world. Thus, the extensive medical/pharmaceutical knowledge collected in ancient Persia is still successfully used today for the diagnosis, prevention, and therapy of various diseases. In this context, Traditional Persian/Iranian Medicine (TPM/TIM) is based exclusively on practical experience and observations that have been passed down from generation to generation.

Due to negligence, TPM/TIM has had a weak presence in the world for a long time. However, in recent years, a variety of research activities and faculty centers have created awareness and a suitable platform to introduce and promote Persian medicine to the world.

Today, numerous scientific articles are published in international journals dealing with evidence-based complementary and alternative medicine, as well as basic concepts and historical aspects of TPM/TIM (Moeini and Gorji, 2016; Naghizadeh *et al.*, 2020).

Pre-Islamic Period (c.8000 BC–6th Century AD)

The roots of TPM/TIM go back more than 8000 years and today it is considered a branch, if not the root, of the so-called "Arabic Unani Medicine or Islamic Traditional Medicine (ITM)" (Mosaddegh and Naghibi, 2002). The earliest records describing the use of medicinal plants come from the Assyrians-Babylonians and the Egyptians, which developed in parallel. The clay tablets of the Sumerians already list hundreds of different medicinal plants such as myrrh (*Commiphora myrrha*) and opium (*Papaver somniferum*) (Sumner, 2000). The individual medical use of many plants is also described in more detail as early as 3000 BC in the papyrus writings that have been handed down

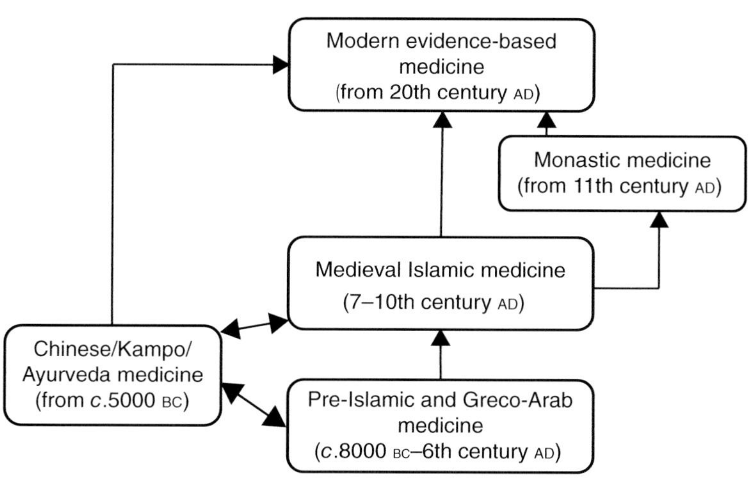

Fig. 3.1. Development of medical systems from antiquity to present days.

from China and Egypt. One of the oldest and most important collections of these documents represent the book *Papyrus Ebers* (*c*.1550 BC), which contains information how to use more than 850 herbal plants such as garlic (*Allium sativum*), juniper (*Juniperus communis*), cannabis (*Cannabis sativa*), caraway (*Carum carvi*), coriander (*Coriandrum sativum*), fennel (*Foeniculum vulgare*), anise (*Pimpinella anisum*), dill (*Anethum graveolens*), peppermint (*Mentha* × *piperita*), and mandrake (*Mandragora officinarum*) (Sumner, 2000; Bryan, 2021) in order to treat various diseases (Fig. 3.2).

Sumac (*Rhus coriaria*) was also extensively used in folk medicine during ancient times. Even today, this species is still collected and processed in mountain regions of Iran. The red fruits of sumac have a characteristic sour-salty taste, which makes them also a popular spice in the food industry in addition to their use as a herbal drug (Abdul-Jalil, 2020).

The English medical historian Cyril Elgood even assumes that TPM/TIM was already more advanced than the Assyrian one at that time and it can certainly also be assumed that several elements of this medical system were adopted by the Greeks (Elgood, 1979, 1992). Taking into account the definitions given by the WHO, Persian medicine

Fig. 3.2. *Days of the Papyrus Ebers* (1500 BC). The painting shows a head pharmacist dictating to a scribe and supervising the compounding of medicines in a temple drug room. (Artist: Robert A. Thom (1915–1979). Source: National Library of Medicine, Rockville Pike, Bethesda. Public Domain, https://collections.nlm.nih.gov/catalog/nlm:nlmuid-101651536-img.)

can be divided into a written part and another one consisting of orally transmitted, verbal traditional medicine (Naseri and Ardakani, 2004). The ancient manuscripts of Egyptian medicine from the middle of the 2nd millennium BC initially refer mainly to various symptoms and diagnoses, and, in this context, describe different treatment approaches in which mineral, herbal, and animal substances were used. Mesopotamian medicine, which is also documented by cuneiform texts on clay tablets, developed in parallel with Egyptian medicine. Surgical procedures as well as the treatment of bone fractures have also been proven to have been performed during this time. The Achaemenian kings of the Persian Empire in the 6th and 5th centuries BC also employed physicians from Egypt at their courts.

It must be emphasized, however, that the Greek medicine of the 5th century BC has some parallels to Mesopotamian medicine in terms of content, which can be explained by the cultural and economic exchange between the Orient and the Mediterranean region that already existed at that time. It can therefore be assumed that the medicine attributed to the Greek philosopher "Hippocrates of the island of Kos" was already influenced by elements of Mesopotamian culture. Evidence shows that even before the establishment of the famous medical schools of ancient Greece (Hippocrates, Homer, Asclepius) in the 5th to 6th century BC, at Cnidos in Asia Minor and on the nearby Aegean Island of Kos, different medical practices were already commonly applied in Mesopotamia, India, and Iran (Najmabadi, 2001). However, it is also undisputed that the texts in Hippocrates' handbook *De medicina* (Diller, 2021), translated and written down by Aulus Cornelius Celsus at that time (Spencer, 1989), set a new standard for medical research and education until modern times. It was only with the emergence of scientific medicine in the 19th century that Hippocratic medicine fell more and more into the background. However, the Hippocratic Oath survived into the 20th century and was only replaced in 1948 by the World Medical Association with the Geneva Declaration, to which many doctors worldwide still feel committed today. In connection with traditional horticulture and the history of medicine in the 8th century BC, the Chaldean Marduk-apla-iddina II, who came from the Bit Yakin tribe, is also worth mentioning. At that time, the plants of the royal medicinal herb garden were listed on a clay tablet. From these cuneiform records it could later be learned, for example, that purslane (*Portulaca oleracea*) was already important as a remedy in ancient times (Jastrow, 1914). Marduk-apla-iddina II, who took advantage of the political unrest in the Assyrian Empire to conquer Babylonia in 721 BC, was first mentioned historically in 729 BC.

In ancient Persia, a vast area with great differences in climate and vegetation, the inhabitants were familiar with various medicinal plants from very early on. Several medicinal herbs such as basil (*Basilicum tenuiflorum*), chicory (*Cichorium intybus*), violet (*Viola* spp.), and peppermint (*M.* × *piperita*) are already mentioned in the *Avesta*, the holy book of the Zoroastrian religion. Although the *Avesta* is one of the oldest religious documents of mankind, it only became known in Europe through the French Orientalist Abraham Anquetil-Dupperon, who traveled to the East Indies in AD 1755 to obtain a copy of the *Avesta* from the Parsi priests there. In addition, the names of 30 sacred plants are mentioned in the *Bundahishn* (Azizi, 2008; Azizi *et al*., 2008). In ancient (pre-Islamic) Iran, many medicinal plants such as barberry (*Berberis integerrima*), saffron (*Crocus sativus*), violet (*Viola odorata*), white water-lily (*Nymphaea alba*), narcissus (*Narcissus tazetta*), myrtle (*Myrtus communis*), damask rose (*Rosa* × *damascena*), citron (*Citrus medica*), pepper (*Piper nigrum*), agarwood (*Aquilaria agallocha*), haoma (*Ephedra* spp.), pomegranate (*Punica granatum*), hemp (*Cannabis sativa*), olive (*Olea europaea*), Syrian rue (*Peganum harmala*), black mustard (*Brassica nigra*), asafoetida (*Ferula assa-foetida*), gum tragacanth plant (*Astragalus tragacantha*), true indigo (*Indigofera tinctoria*), cumin (*Cuminum cyminum*), frankincense (*Boswellia sacra*), cinnamon (*Cinnamomum verum*), and labdanum (*Cistus ladanifer*) were used; and the aromatherapy with aromatic drugs such as camphor (*Cinnamomum camphora*), sandalwood (*Santalum album*), and Arabian balsam (*Commiphora gileadensis*) was also practiced. Some minerals such as alum ($K_2SO_4{\cdot}Al_2(SO_4)_3{\cdot}24H_2O$), sulfur, lapis (lapis lazuli), and nushadir salt (ammonium chloride) are among the drugs that were used in ancient Iranian medicine (Emami *et al*., 2017).

As part of the massive military expansion, the Islamized Arabs conquered Egypt, North Africa, Palestine, Syria, Mesopotamia, and Persia by the middle of the 7th century AD. At the same time, at the beginning of the 9th century AD, some philosophical writings (written for example by Aristotle and Plato) as well as various scientific and medical Greek literature were translated into Arabic on a large scale. This allowed the cultural policy of the Abbasids to follow the example of the pre-Islamic Persian dynasty of the Sasanids (AD 226–642). Greek manuscripts had been very popular at that time and occasionally even reached Baghdad as a gift from a Byzantine emperor. The translation of the mentioned books is due to the valuable and continuous efforts of some famous translators such as John Grammaticus, Aaron Alexandrian, Sergius of Rechina, Issa ibn Sahar Boxt, Sabet ibn Qorrah, Qusta ibn Luqa of Heliopolis, Sahl ibn Shapur, Shapur ibn Sahl, Ibn Serapion, al-Kendy, and members of some families such as the Boxtishu (the servant of Christ), Massarjuyah, Hunayn, Karkhy, Massuyah, and Sinan families (Tayarani-Najaran *et al*., 2014). Islamic polymaths and physicians such as Rhazes (AD 850–923) and Avicenna (AD 980–1037) adopted the writings of Hippocratic–Galian medicine that had become known to them in this way and systematically developed them further. Later, in the High Middle Ages, their works were translated into Latin.

Medicinal herbs are among the most important natural resources that may have a direct positive impact on people's health. Even today, many of the medicines we use every day are derived directly or indirectly from medical herbs. The great importance of plant medicine was already recognized in antiquity; for example, the Egyptians had used the healing power of certain herbs since the 16th century BC. But the Israelites, too, used medicinal plants since ancient times and in India the most important and authentic classical work of Ayurveda (*Charaka-Samhita*) is originally written in Sanskrit. China's oldest pharmacopoeia is attributed to one of its emperors named "Shinon" who lived around 2800 BC. From the 8th to 10th centuries AD, the famous Iranian scientists Avicenna and Rhazes (Razi) were leaders in this field of knowledge (Mah Van, 2002).

The culmination of Avicenna's philosophical work is seen in his writing *Kitāb al-Shifā* (*Book of Healing*), in which he dealt intensively with the teachings of Aristotle. In addition, he summarized the state of knowledge of Islamic medicine at that time in his *al-Qānūn fī al-Ṭibb* (*Canon of Medicine*). In this context it was particularly important for him to dedicate a concrete and independent place to medicine within the canon of the other sciences.

In Western countries, advances in this science date back to the Early Middle Ages and are still pursued today by scholars such as Constantine, Epitome Matthiol, and Dragendorff, and others (Mozaffarian, 1996; Zargari, 2014). In Iran, the most comprehensive study in this case was conducted by Ali Zargari and the result was published in five volumes (Jouri and Mahdavi, 2010). With regard to the different climatic conditions and soil quality, Iran offers optimal conditions for a high biodiversity of numerous different medicinal plant species, which often occur only in this country and can be considered endemic.

With an extent of 1,648,000 km^2, Iran has a widely varied climate; it is a habitat for very different plant species. Nearly 190 families of vascular plants are found in this country, containing 8500–9500 species, the endemic plants being relatively numerous. More details can be taken from the book *The Endemic Phanerogams of the Flora of Iran*, where 2324 species are mentioned, all of which belonging to Iranian Phanerogams (Emami and Aghazari, 2011).

In the Department of Natural Resources of Fars Province alone, more than 1600 medicinal plant species are found, of which more than 100 species are cultivated on an area of 600 ha. With an area of 133,000 km^2, this province in Iran covers about 8.1% of the total area of the country.

The long-standing use of medicinal plants provides a valuable database regarding the safety and efficacy of individual plant species. Traditional Persian Medicine (TPM) has a solid theoretical basis, which is also supported by numerous evidences of clinical efficacy. Accordingly, the chance of being able to identify individual bioactive phytopharmaceuticals is comparatively high. In traditional medicine, there is a particular focus on medicines with cardiovascular activity, which were already listed by Avicenna (Abū ʿAlī al-Ḥusayn ibn ʿAbd Allāh ibn Al-Hasan ibn Ali ibn Sīnā), a famous Persian physician of the 11th century AD, in his book entitled *The Treatise on Cardiac Drugs* dedicated to this special field. Here, Avicenna mentioned a total of 83 cardioactive drugs obtained from plant, animal, and mineral sources and described their therapeutic effects and assumed mechanisms in the treatment of various cardiovascular diseases (Javadi and Emami, 2015). In recent years, most of these cardioactive plants proposed by Avicenna have been the subject of several pharmacologic and clinical studies (Park *et al.*, 2011; Seo *et al.*, 2020). Today, it is generally accepted that loss of the sense of smell leads to a noticeable impairment of the general quality of life. However, only a small proportion of patients with olfactory loss respond to the currently available medications. Therefore, there are various efforts to use the long-standing medical knowledge of TPM in this context as well. Such therapies were already developed in Avicenna's time and it can be assumed that the olfactory changes or complete loss of the sense of smell observed in infections with COVID-19 viruses can be adequately treated in the near future by means of TPM/TIM (Sinaei *et al.*, 2017). Furthermore, in several cases the very high cost of synthetic pharmaceutical products, as well as the need for a sustainable economy, is prompting the industry to pay more attention to the use of medicinal plants today (Noorhosseini *et al.*, 2011).

Based on plant species recommended by Avicenna, for example, a link could be established between TPM/TIM and relevant modern pharmacologic data to support the development of new efficient and safe cardiovascularly active drugs (Sobhani *et al.*, 2017). Interestingly, some of these plant species showed significant cardiovascular activities via mechanisms such as calcium antagonism, angiotensin-converting enzyme inhibition, antiarrhythmic, anticoagulant and fibrinolytic activities, and reduction of cardiac fibrosis. In this context, phenolic compounds such as luteolin, rosmarinic acid, salvianolic acid A, cardamonin, cinnamic acid, punicalagin, punicalin, strictinin A, granatin B, ellagitannins, and anthocyanins were particularly responsible for cardioactive activity and are therefore considered to be promising candidates for the development of new cardiovascular drugs (Sobhani *et al.*, 2017).

Today, however, more and more the preservation of this knowledge can no longer be guaranteed, as the transmission between generations no longer works as well as it did in the past (Anyinam, 1995). Yet the loss of this traditional knowledge in a rapidly changing culture is as irreversible as the disappearance of plant species themselves (Joshi and Joshi, 2000). Considerable efforts must therefore be made to document the medicinal use of the related plant species before these traditional medicinal practices are irreversibly lost (Joshi and Joshi, 2000).

Special attention is therefore paid to support the research of medicinal plants in Iran (Jivad *et al.*, 2016; Parsaei *et al.*, 2016; Rouhi-Boroujeni *et al.*, 2016a). Phytotherapy has a long history in this country, so this topic has been extensively described in ancient Iranian pharmacopoeias, such as the works of Avicenna. In addition, numerous studies have been published showing successful approaches to treat various diseases by applying Traditional Iranian Medicine (TIM) (Ibrar *et al.*, 2007; Ahmadipour *et al.*, 2016; Rouhi-Boroujeni *et al.*, 2016b).

In Iran particularly a high diversity of species is found in the *Lamiaceae* (*Labiatae*) family and many of these species are also used in traditional folk medicine. Unfortunately, very often only few clear references concerning the ethnobotany and ethnopharmacology of the *Labiatae* species are available, and most publications and historic documents are only available in Persian or Arabic (Naghibi *et al*., 2010; Amiri *et al*., 2014).

Medieval Islamic Period (7th–10th Century AD)

TIM is a holistic and comprehensive medical school whose origins date back over 12 centuries. In this traditional medicine system, cancer was already a well-known disease with many treatment options. Especially, Razi (Rhazes; AD 865–925), Ahwazi (Hali Abbas; AD 930–994), Avicenna (AD 980–1037), Jorjāni (AD 1042–1136), and Aqili Khorasani (18th century AD) were among the most important doctors of their time who discussed various forms of cancer and their treatment in their books. They used many different types of plants in the treatment of the various tumours. In this context, *Zakhireh Khārazmshāhi* (*The Treasure of Khārazmshāh*) by Seyyed Esma'il Jorjāni (AH 433–531/AD 1041–1136) is known as the most important Persian medical book of TIM (Tayarani-Najaran *et al*., 2014).

Today, more than 80% of the world's population apply traditional medicine, especially medicinal plants, to treat the most common diseases and ailments. Especially in developing countries and rural areas, the use of medicinal plants represents an important economical alternative within individual health systems (Alexiades and Sheldon, 1996; Robineau and Soejarto, 1996). In addition, traditional medicine provides a valuable source of knowledge for research and development of new pharmaceutical products. Therefore, in recent years the use of ethnobotanical information has become particularly important in medicinal plant research (Heinrich, 2000). Knowledge about the use and properties of medicinal plants was often acquired empirically and passed on from generation to generation (Lev and Amar, 2000). From the 9th century AD onward, under the guidance of the Persians and Arabs, who at this time were considerably more advanced than the inhabitants of medieval Europe, medical schools (so-called "Bimaristans") emerged in the medieval Islamic world. As admirers of Greco-Roman culture, the Arabs translated numerous medical texts into Arabic for further studies (Castleman, 2001). As early as AD 600, the Nestorians (a Christian sect) founded translator centers in Baghdad, Damascus, Antioch, Basra, and Cairo in the Eastern Roman Empire, Syria, and Persia. The most important achievement of these translators today is generally considered to be the translation of Galen's very extensive writings into Arabic. On the basis of Galen's theory of the four humors, numerous adaptations and independent works were written later, such as Avicenna's *Canon medicinae* (*Canon of Medicine*), which had a decisive influence on the development of medicine in the Western world for centuries (Figs 3.3 and 3.4).

In the dissemination of medical knowledge, however, "cultural interfaces" also played a very important role again, and again at a later time. In this context, particular reference should be made to the famous medieval medical school of Salerno (*Civitas Hippocratica*) and the scientific languages used there: Latin, Greek, Hebrew, and Arabic. Mention should be also made here of the translator "Constantine of Africa" (AD 1010–1087), to whom a collection of textbooks is attributed that later became known as *Articella* describing the central subject areas of medieval medicine. In addition, the collection of recipes *Antidotarium Nicolai* is considered a very important work that also originated in Salerno and on which medicinal therapy was oriented until modern times (Sudhoff, 1930). The Iberian Peninsula also represented another important interface between "Orient" and "Occident" in medieval times, as the cultures of Islam and Christianity met there in close proximity. Of particular importance for the development of European medicine and science was the school of translators in Toledo founded by Gerhard of Cremona (AD 1134–1187), which translated a large number of Arabic writings into Latin.

In addition, the Arabs had access to various plant raw materials due to their

Fig. 3.3. *Canon of Medicine*, first page of the introduction to the first book (Arabic manuscript, 1597). (https://commons.wikimedia.org/wiki/File:Avicenna_canon_1597.jpg#/media/File:Avicenna_canon_1597.jpg.)

Fig. 3.4. Miniature of Ibn Sinā (Avicenna). (By Michel Bakni, CC BY-SA 4.0, https://commons.wikimedia.org/w/index.php?curid=99757546.)

intensive trade, especially with China and India. In this way, herbal books, medical textbooks, and translations of the classics of antiquity also reached them from both East and West. It must be emphasized, however, that Muslim botanists and physicians not only translated the traditional knowledge of materia medica, but considerably expanded it. For example, al-Dinawari described more than 637 herbal drugs in the 9th century and Ibn al-Bayṭār described more than 1400 different plant species in the 13th century, of which more than 300 were his own discoveries.

In medieval Islamic literature, in which the individual plant species are closely associated with medicine and agronomy, there are already numerous references to the use of medicinal herbs. Furthermore, these plant species are described in various philosophical, magical, and geographical contexts (de Vos, 2013). Especially three medieval pharmaceutical treatises, written by Yūhannā ibn Māsawayh, had an immense influence on the development of pharmacy in early modern Europe at that time. Although these works had achieved great importance, they are relatively unknown in the history of pharmacy and medicine, and the exact identity of the author cannot be precisely proven even today, so that in principle it is also conceivable that the manuscripts were written under a pseudonym. The only thing that seems certain is that these treatises can be clearly assigned to the state of knowledge of the medieval Islamic world at that time. In any case, the writings contained numerous innovations that formed the basis for the theory and practice of Western pharmacy for centuries and probably also exerted a considerable influence on medical science in Europe (Troupeau, 1995; Cambra, 2001; Cárcer, 2001).

It has been reported that several plant species were already used for therapeutic purposes in the ancient Arab world as well as in other contemporary and earlier cultures. By the 10th century, the Oriental Islamic culture was able to develop a solid knowledge of the effects of medicinal and spice plants. In particular, on the basis of the two writings, *Kitāb Jawāhir al-tayb al-Mofradah* (*Essences of Simple Perfumes*, written by Ibn Māsawayh) and *Kitāb al-Adwiyah al-Mofradah* (*Simple Drugs*, written by Ibn Wāfid), this knowledge spread in Andalusia (now Spain) in the 11th century, which is considered the so-called "Golden Age" of Andalusian pharmacology (Cárcer, 2001). Since the Muslim botanist al-Asma'i (AD 740–828) wrote his famous work *Kitāb al-Nabāt wa-'l-Shajar* (*Book of Plants and Trees*), which became the standard work on Arabic plant names and botanical terms for later generations of pharmacologists, some efforts were made to clarify or correct the botanical terms he used. Unfortunately, different botanical systematics were in use at that time, ranging from simple alphabetical sorting of individual plants to subdivisions according to their respective practical medicinal uses. However, subdivisions were also found in "trees, flowers and garden vegetables", with the group of trees (including shrubs) and plants being further subdivided according to whether or not the fruits or other plant parts could be used as food (Bolens, 1977).

Some early Islamic works, such as those of al-Shaybāni, Ibn al-Arabi, al-Bahili, and Ibn as-Sikkit, have unfortunately been lost, but their works were extensively cited in later books by Abu Hanifah al-Dinawari, Ibn Sidah, and other authors. As a result of the further spread of Islam and the intensive exchange of information with Middle Eastern, Indian, and North African sources, a rich botanical literature thus developed in which Muslim authors attempted to record the individual significance of medicinal plants and stepwise develop a botanical systematics from this knowledge. In this context, the plant terminology of the older Arabic nomenclature was also gradually expanded (Bolens, 1977). In the treatises of Jabir ibn Hayyān, the herbs needed for the production of medicinal plant extracts were described in more detail, while Abu Ubaid al-Bakri focused more on the plant and tree species growing in Andalusia (Nasr, 2003). Finally, in his book entitled *Umdat al-Ṭabib fi Mʿarifat al-Nabāt lia-kolle Labib* (*An Important Botanical Book for Physicians*, published by Mohammad Arabi al-Khattabi in 1995), Abul Khair al-Ishbili attempted to obtain a more appropriate classification of medicinal plants according to their individual genera, species, and subspecies (Samso *et al.*, 1996). Muslim botanists at that time already had an extensive knowledge regarding the cultivation of different plant species and had also gained initial insights into how different types of fruit (e.g., different species of the genus *Prunus*) could be produced by grafting on the same tree. In addition, they were already able to carry out first breeding experiments with wild rose species (e.g., *Rosa* × *damascena*) and almond trees (*Prunus dulcis*) occurring in Persia, in order to develop interesting subspecies with new quality characteristics in this way (Durant, 1950). The royal botanical gardens of that time presented an enormous variety of native and exotic plants, which were cultivated there for their colorfulness, richness of form, delightful fragrance, and, last but not least, for their culinary and medicinal benefits (Scott, 1904). The Andalusian philosopher and botanist Ibn Bajjah also went into more detail about the physiology of the various plant species in his book *Kitāb al-Nabāt* (*Book of Plants*). For the first time, he also described the individual influence of different local environmental conditions (desert areas, mountains, river banks, sea coast, and different types of soil) (Samso *et al.*, 1996).

Without any doubt, one of the most famous Muslim herbalists was Abu Dāwud Sulaymān ben Hassan, who is also mentioned in literature as "Ibn Juljul al-Andalusī". He was born in Córdoba in AD 943 and was widely respected at a very early stage for his extensive knowledge and his special ability to determine the individual geographical origin of plants (Amar *et al.*, 2014). Beside botany, Ibn Juljul also studied medicine together with a group of Hellenists chaired by Ḥasdāy ibn Shaprūṭ, a Jewish physician and vizier of Caliph Abd al-Raḥmān III, and later became the personal physician of Caliph Hishām II (AD 976–1009) (Calvo, 2008). He is also known as a famous author of the two books *Ṭabaqāt al-Aṭṭibbā'wa 'l-ḥukamā'* (*Generations of Physicians and Wise Men*), which is the oldest existing summary of the history of medicine in Arabic, and *Isḥāq ibn Ḥunayn's Tarikh al-Aṭibbā'* (*History of Physicians by Isḥāq ibn Ḥunayn, A History of Medicine with 57 Biographies from Nine Generations*). Thirty-one of them concern Asian authors, while the others refer to African and Andalusian scholars. For his pharmacopoeia, Ibn Juljul used several Eastern (Hippocrates, Galen, Dioscorides) and Western sources (Orosius, Isidore) and also described the chronological limits of Latin influence on medicine in Andalusia. This history contains brief sections on each period, beginning with the legendary fathers of medicine, but is probably of greatest interest because of Ibn Juljul's account of his predecessors and contemporaries in Andalusia. In this context, he refers in particular to the intensive contacts that existed between the Eastern Caliphate and Andalusia, as well as to students who undertook extended journeys in search of knowledge and education (Amar *et al.*, 2014).

Building on his studies of Dioscorides, Ibn Juljul wrote a supplement he called *maqālah*, containing 60 articles that focused specifically on "those drugs that Dioscorides had not mentioned" (Johnstone, 1999) (Fig. 3.5).

This work largely follows the familiar scheme of earlier authors in giving references to the appearance, action, and nature of each plant. However, in addition, their "temperament (mizādj)", their "power", and their "effect on a particular spirit or organ" are also described in detail (Ghāfiqī *et al.*, 1932–1940). Ibn Juljul also mentions the respective origin of the plants, as far as they were known to him: 28 came from India, two from Yemen, two others from Egypt, and one from the island of Ceylon. Only a few plant samples were collected in Andalusia (Johnstone, 1999). For the samples from Andalusia, Ibn Juljul sometimes added the name of that person who had first used the plant or told him something about it. For example, there is a detailed account of the collectors of monk's pepper (*Vitex agnus-castus*). But he also documents some personal descriptions of the collected Syrian rhubarb (*Rheum ribes*): "I inquired about 'ribas' from some people in the Far East, who told me that it is a sour-tasting vegetable with fleshy leaves resembling those of succulents." Another trader who was familiar with the use of "ribas" informed him "that the plant had extremely sour roots" (Savage-Smith, 2012). It is particularly noteworthy that Ibn Juljul respected the ancient knowledge of the Greeks, but did not simply translate their books; rather, he added valuable contributions to the practical knowledge and individual use of medicinal plants, and for this reason in particular was highly respected by contemporary and later scholars and physicians (Johnstone, 1999).

Several Muslim travelers also made important contributions to a more comprehensive and at the same time systematic development of botanical knowledge in Spain by providing their individual knowledge of

Fig. 3.5. Physician preparing an elixir (folio from a manuscript of *De materia medica* written by Dioscorides (*c.*AD 40–90)). (By Marie-Lan Nguyen (2011), Public Domain, https://commons.wikimedia.org/w/index.php?curid=17769084.)

nature and especially of the origins of various plant species. For example, the physician al-Ghāfiqi from Córdoba (12th century AD) collected numerous plants on his travels through Spain and North Africa and named them in Arabic, Latin, and Berber. Abu Zakariyā ibn al-ʿAwwām, another botanist and farmer who lived at the same time, wrote a treatise on agriculture (*Kitāb al-Filāha*) that is the first outstanding medieval work on this subject. He consulted both Greek and Arab specialists, but drew most of his knowledge from his own studies of agriculture in Spain itself. He described a total of 585 different plant species and also went into more detail on the cultivation of about 50 types of fruit (Hitti, 1937) (Fig. 3.6).

The Moroccan traveler and botanist Ibn Baṭṭuṭah (14th century AD) described in his travelogues (in Arabic: *Rihlah*) the various fruit varieties he discovered in Isphahan (apricots, quinces, grapes, and watermelons) and, in addition, the fruit trees he got to know on his travels through India, such as mango and sweet orange. In Malabar he discovered cinnamon (*C. verum*) and in the

Fig. 3.6. A scene of agricultural work with a man digging herbaceous plants with a spade, among cultivated trees, from a medieval Arabic manuscript from Al-Andalus (Islamic Spain). (http://www.muslimheritage.com/article/agriculture-muslim-civilisation-green-revolution-pre-modern-times, Public Domain, https://commons.wikimedia.org/w/index.php?curid=57929949.)

Maldives and the island of Java he could see other exotic species such as oil palm (*Elaeis guineensis*), lemon (*Citrus × limon*), camphor (*C. camphora*), storax (*Styrax officinalis*), and clove (*Syzygium aromaticum*) trees (Defremery and Sanguinetti, 1929). In the second half of the Abbasid Caliphate and during the Western Caliphate, Muslims also contributed in a special way to increasing the knowledge of organic and inorganic remedies as well as to improve the preparation of pharmaceutical drugs. During this period, there was also a greater separation between the medical and pharmaceutical professions for the first time, which was mainly due to the described advances in materia medica and in the knowledge of how to produce phytomedicines. Thus, the large hospitals in Iraq already had a pharmacist and a well-equipped pharmacy where the medicinal products were compiled and the doctors' prescriptions prepared and filled according to individual instructions. Medicines and spices were also stored in these pharmacies for the first time (Elgood, 1951). Among the Baghdadi scholars, the Christian physician Māsawaih al-Mārdini from Mārdin in Mesopotamia (10th century AD) deserves special mention. He lived first in Baghdad and then later in Cairo under the rule of Caliph al-Ḥākim (Wiet *et al.*, 1975). He achieved fame in later monastic medicine through the pharmacopoeia he wrote, which dealt in particular with more complex mixtures of medicines as well as specific medicines for certain diseases (Wiet *et al.*, 1975).

Abu al-Abbas al-Ashbili (Abul-Abbas of Seville) known as Ibn Rumiyyah was the first scientist to succeed in extending botanical knowledge, which until then had focused mainly on the agricultural production of food plants, to medicinal plants and their practical use in pharmacy (Scott, 1904). However, there was still no clear dividing line between food and medicines, as many plant products were clearly assigned to both categories (Johnstone, 1999).

Islamic botany also gradually developed an advanced methodology and new approach, opening the way to the current state of science. This can be seen, for example, in the treatise of Ali ben Abbās al-Madjusi known as Haly Abbas (10th century AD), who proposed different dosages for the treatment of individual diseases and the use of differently composed medicines in separate chapters, thus providing a very complete and well-organized account of medicine of that time (Levey, 1997).

Very often, Muslim scholars were even far ahead of their Greek predecessors in the use of medicinal plants, so that the Muslim list of medicines contained a large number of remedies unknown to the Greeks (Wiet *et al.*, 1975). Thus, Ibn Juljul was already aware of the fact that medicine and botany had evolved since the days of Dioscorides and that the newly identified herbal raw materials that were increasingly used to make medicines came mainly from the East or were collected in Andalusia (Johnstone, 1999). In general, Muslims gave Arabic names to the new plant species and medicines, and many of these names are still used today (al-Faruqi and al-Faruqi, 1986). For example, the work of Ibn al-Awwām lists 600 medicinal plants, while Ibn al-Bayṭār's book even describes more than 300 previously unclassified or unknown plants (Scott, 1904).

In medical history, Ābu Hanīfah Ahmad ibn Dāwūd Dīnawarī (AD 828–896), who is considered the founder of Arab botany, can certainly be called one of the earliest Muslim botanists (Ulmann, 1972). He was a polymath with profound knowledge of astronomy, agriculture, botany, and metallurgy, as well as geography, mathematics, and history. Born in Dinawar (a city in western Iran), he studied astronomy, mathematics, and mechanics in Isfahan and philology and poetry in Kufah and Basrah. His work, largely confined to the flora of Arabia (Elgood, 1951), is perhaps the most comprehensive and methodologically best work on herbs of his time. Al-Dinawari's work was long considered lost, but was made available to a wider public in 1908 in a dissertation from Breslau (Silberberg, 1908). This dissertation contains the descriptions of about 400 plant species from al-Dinawari's book, although these records only describe a part of his original work. The information from al-Dinawari is based on older written sources, on oral information from Bedouins, and occasionally on several personal observations. His book, entitled *Kitāb an-Abāt* (*Book of Plants*), consists of two parts, one of which is an alphabetical index of the names of herbal medicines (and thus the first alphabetically arranged technical dictionary). The second part contains monographs on plant species that were used for specific practical purposes at that time (Hamidullah, 1973).

The work entitled *Firdaws al-Hikmah* (*Paradise of Wisdom*), which was written by al-Tabari before the time of al-Dinawari, includes botany as well as other scientific disciplines such as climatology, astronomy, and philosophy, but nevertheless refers primarily to medicinal plants. Al-Kindī (AD 800–866), on the other hand, is considered an innovator in compiling a herbal manual that would teach useful botanical pharmacology as well as toxicology and medicines derived from minerals (Scarborough, 1980). His *Aqrabādin* (*Medical Formulary*) contains 222 recipes with at least 319 raw materials, corresponding to Muslim pharmacology of the 9th century AD.

At early Islamic scholars such as Jabir Ibn Hayyān, Hunayn Ibn Isḥāq, and others, the Greek influence is still clearly visible. For example, Ibn Waḥshiyah (AD 860–935) describes very well the state of knowledge at that time on the use of medicinal plants and the corresponding pharmacology in his herbology entitled *Kitāb al-Sumum wa-'l-Teryāqāt* (*Book of Poisons and Antidotes*) (Scarborough, 1980). In another work, he describes the *Al-Filaḥah al-Nabatiyah* (*Nabataean Agriculture*) as the first comprehensive encyclopedia on plants and their individual agricultural cultivation (Fahd, 1993/1995, 1996).

Al-Dinawari also devotes a chapter to the classification of plants ("Tajnis al-Nabāt"), whereby his descriptions clearly exceed the scope of a usual botanical dictionary, as he also tried to cover all aspects of Bedouin life (Bauer, 1988). In his treatise on soil, al-Dinawari describes a variety of different soil types, explaining, among other things, which soils are particularly suitable for cultivation in each case. He also goes into more detail about the development of the individual plants (ontogenesis), the individual growth periods, and the production of flowers and fruits. In addition, he also describes individual cultivated plant species such as cereals, grapes, and date palms (Fahd, 1996; Johnstone, 1999).

Compared to Ibn Juljul, the herbalist Ibn Samjun, who lived and worked in Córdoba (10th century AD), has been somewhat neglected. However, several manuscripts of his work, in which he wrote about Islamic pharmacology, have been preserved in Oxford and London (Kahle, 1952). His exact description of the remedies in alphabetical order later became the model for Ibn al-Bayṭār, who wrote the most comprehensive pharmacologic book in this form (Lewin, 1953). The botanical writings of Ibn al-Bayṭār (AD 1197–1248), written in Arabic, are generally considered a high point of ancient medicine. Ibn al-Bayṭār was born in Malaga, studied in Seville, where he also collected plants together with his teachers and started to journey through Spain and North Africa. Later, he lived as a leading herbalist in Cairo and from there undertook extensive travels through Syria and Asia Minor. Ibn al-Bayṭār was a disciple of Ibn Rumiyah, Abu al-Abbas al-Nabāti (the botanist), and was also strongly influenced by al-Ghāfiqi's (12th century AD) *Kitāb al-Adwiyat al-Mofradah* (*The Book of Simple Drugs*). Of his outstanding works, one referred to "*Materia medica*", and another one to "simple remedies", by which are meant medicinal preparations containing only one single ingredient. Ibn al-Bayṭār wrote his work *al-Mughni fi al-Adwiyah* (*The Sufficient*) in Arabic. In the work, which is divided into 20 chapters, he describes in particular the methods for curing various diseases of the head and ears and for making cosmetics and simple antidotes.

Ibn al-Bayṭār best known work, however, is *Al-Kitāb 'l-Jām'i fi 'l-Aghdiya wa-'l-Adwiyah al-Mofradah* (*The Comprehensive Book of Foods and Simple Remedies*), the most comprehensive encyclopedic work on simple drugs. It was the greatest medieval treatise on this subject and a fundamental work on botany, describing 3000 samples, all listed in alphabetical order (Tayarani-Najaran *et al.*, 2014).

Ibn Al-Wāfid (AD 997–1075) from Toledo was also esteemed as a very successful scientist of his time. It is reported that he spent about 20 years collecting, arranging, and checking all the names of medicines, their individual properties, and potency. In doing so, he favored dietary measures, and when medicines were needed, he preferred the simple ones before recommending compounded medicines, and when using compounded medicines, he also preferred the less complex ones (Meyer, 1854). His main work on simple medicines, *Kitāb al-Adwiya al-Mufrada*, has been partially preserved in a Latin translation (*Liber de medicamentis simplicibus*) (Wustenfeld, 1840) (Fig. 3.7).

Fig. 3.7. Farmer harvesting melons. (A plate from the Vienna *Tacuinum Sanitatis* (Codex Vindobonensis Series nova 2644 of the Austrian National Library). Scan from: Hans Biedermann – *Medicina Magica, Metaphysische Heilmethoden in spätantiken und mittelalterlichen Handschriften*, Akademische Druck- und Verlagsanstalt Graz, 1978 2. Aufl. S. 53, ISBN 3-201-01077-4. Public domain, https://commons.wikimedia.org/w/index.php?curid=6099643.)

Al-Ghafiqi (first half of the 12th century AD), who also came from the surrounding area of Córdoba, integrated into his work, which was called *Kitāb al-Jāmʿi fi 'l-Adwiyah al-Mofradah* (*Book of Simple Medicines*), mainly new plant species that had not been mentioned until then in the classical Greek sources or in comparable botanical manuals from the Muslims (Scarborough, 1980). Finally, two other botanists from the Islamic West should be mentioned here: firstly, Ibn Sirābiyun, who, however, cannot be exactly assigned in terms of time due to the lack of historical sources. His work entitled *Selecta artis medico* is mainly based on the writings of Dioscorides and Galen. Secondly, the Muslim botanist Rashid al-Din Ibn al-Ṣuri, who is described as a contemporary of Ibn al-Bayṭār, studied medicine in Damascus and then worked in a hospital in Jerusalem. Al-Ṣuri wrote a treatise on simple medicine, *al-Adwiyah al-Mofradah*, in which he mainly refuted the statements of his colleague Tāj al-Din al-Bulghāri, who had published his ideas in a similar book. Ibn al-Ṣuri was particularly distinguished by his extensive travels, especially to the mountains of Lebanon, to discover and collect medicinal and aromatic plants. During his trips he was accompanied by an artist whose task was to depict the plants in their various stages of growth as faithfully as possible and in color (Sarton, 1927–1948). His work was the first example of an Arabic book to be illustrated in color, but unfortunately this work is now considered lost (Scott, 1904).

Traditional Persian culture, with its highly developed medicine and the extensive knowledge of numerous medicinal plants, was exposed for centuries to the influences of the Greeks and Romans in particular. The use of plants as medicines or foodstuff dates back several millennia. Nowadays, various strategies are being used to find new and effective drugs from natural sources. To this end, investigating folk and traditional medicine systems, such as TPM, is a rational strategy owing to their antiquity and long-term history of human use. In TPM, considerable attention has been paid to the qualities of individual herbs. The general and specific properties of each herb such as its nature, natural habitat, energy pattern, effects, indications, duration of action, relationships to the body organs, toxicity and contraindications, dosage, types of preparations, and antidotes are discussed.

Images of medical plants can already be found on the cuneiform writing tablets of the ancient Babylonian culture, and it is transmitted that the Egyptians were already aware of the antibacterial effects related to the impact of garlic (*Allium sativum*) and onion (*Allium cepa*). The medical knowledge of these advanced civilizations, which originated primarily in the Orient, was taken up by the Greeks in ancient times and therefore they can be regarded as the founders of Western medicine. In this context the physician Hippocrates of Kos (460–370 BC), who was one of the most outstanding ancient scholars in medicine and philosophy, must be mentioned. In his written records he described not only numerous herbal preparations for the treatment of various diseases but also developed a systematic categorization of diseases known at that time. The Hippocratic school of medicine revolutionized ancient Greek medicine, establishing it as a discipline distinct from other fields with which it had traditionally been associated and thus establishing medicine as an independent profession (Fig. 3.8).

The first detailed book on medicinal plants was written around AD 60 by the Greek physician Pedanius Dioscorides (AD 40–90), who practiced in Rome. He summarized the basics of herbal medicine for the first time in his book *De materia medica*, which quickly became a standard work of European medicine with its description of around 600 medicinal plants. *De materia medica* was very popular until the 17th century, especially because there was an original illustration of every medicinal plant and the individual plant names were provided not only in Latin but also in Greek and Egyptian.

After the fall of the Roman Empire, the reference to the ancient Greek scriptures gradually disappeared. Even from Hippocrates (around 400 BC) and his important pupil Galen of Pergamon (around AD 129–216), to whom the systematization of the theory of

Fig. 3.8. Fresco in the crypt of Anagni's Cathedral (13th century AD) representing Hippocrates of Kos as teacher and Claudius Galen as student. (By Nina Aldin Thune, CC BY-SA 2.5, https://commons.wikimedia.org/w/index.php?curid=795032.)

the humors goes back, important information was lost in medieval Western Europe, except for small excerpts. The remaining writings were translated into Latin by the Roman scholar Cassiodorus (AD 485–580) after his retreat to the monastery of Monte Cassino and were thus preserved.

Almost even more important, however, was the translation of the ancient pharmacognosy "*De materia medica*" into Latin performed by Pedanios Dioscorides (1st century AD) which was later the basis to develop drug therapy in Europe.

Galen's main medical work is entitled *Methodus medendi* (*The Therapeutic Method*), where he follows the central idea that all phenomena in nature and in man fulfil a certain purpose. In this context Galen presumes that man is a body–soul entity that is mainly influenced by a dualism of mind and matter. He assumed that the balance of four elements (black bile, yellow bile, blood, and phlegm) principally influence nature (humans, animals, and plants) and that the relationship of these humors to each other is of central importance, because a disturbance of the individual balance would lead to indisposition and illness. This theory also included the idea of the Greek physician Alkmaion (around 600 BC) who assumed that health can be described as the harmony of pairs of opposites, as it is also found in the Yin–Yang principle of Traditional Chinese Medicine (TCM).

In his books entitled *De placitis Hippocratis et Platonis* (*On the Doctrines of Hippocrates and Plato*), written between 161 and 177 AD, Galen attempts to prove that the ideas of the physician Hippocrates and the philosopher Plato about the soul forces, which are supposed to control the mental and physical processes in body, coincide more or less.

The pharmacology of the Islamic, Arabic, and Occidental cultural areas was oriented toward Galen's complicated formulations until the late Middle Ages and it was only under the influence of Paracelsus' medical teachings that this situation changed. Galen's systematically developed work, which was also translated into Syriac and Arabic by Hunain ibn Ishāq (AD 808–873) in the Early Middle Ages, was of such importance for posterity in its scope and scientific level that it took 1400 years for it to be slowly overcome by more recent research.

In the 16th century AD, the so-called "Galenists" (e.g., Leonhart Fuchs) as followers of Galen's teachings and the "Arabists", who mainly referred to Avicenna (Strohmaier, 1999), were irreconcilably opposed to each other. The more medical-oriented "Galenism" was on the advance at this time, whereas "Arabism" was in retreat; likewise there were several disputes between "Galenists" and the so-called "Paracelsists" (Schmitz, 1984; Durling, 1989; Schipperges and Durling, 1989). Paracelsism was an early modern medical movement based on the theories and therapies of Paracelsus (AD 1493–1541) who was a Swiss physician, alchemist, lay theologian, and philosopher of the German Renaissance.

The great European monarch Charlemagne (AD 748–814) also made a significant contribution to the development of monastic medicine by issuing a new decree with the name *Capitulare de villis* that made it obligatory for all monasteries and towns to establish herb gardens and also prescribed in detail a kind of standard for monastery gardens. For example, a separate bed was required for each medicinal plant species and for all medicinal plants to be used, such as sage (*Salvia officinalis*), wormwood (*Artemisia absinthium*), fennel (*F. vulgare*), opium poppy (*P. somniferum*), lovage (*Levisticum officinale*), chervil (*Anthriscus cerefolium*), various types of radish (*Raphanus* spp.), peppermint (*M.* × *piperita*), and spearmint (*Mentha spicata*), he specified exactly how the individual species should be cultivated (Strank and Meurers-Balke, 2008). Since Charlemagne's empire at that time extended as far as Italy and southern France, the list of cultivated medicinal plants also contains several Mediterranean species, which could obviously be cultivated in Central Europe without any problems due to the warm period that prevailed at that time.

Future Role of Traditional Persian/Iranian Medicine

In general, Traditional Medicine has the advantage of being considered part of the respective culture, thus circumventing cultural problems that might interfere with the practice of "modern" medicine. Moreover, it can be used in conjunction with and as a support to conventional medicine. The further development and promotion of Traditional Medicine can therefore also be seen as a respect and appreciation of the individual culture and heritage of people around the world. As has been shown in this chapter, TPM/TIM contains the extensive knowledge used in Persia from ancient times to present days to diagnose, prevent, and eliminate various diseases. It is based solely on practical experiences and observations passed down from generation to generation (Rezaeizadeh *et al.*, 2009).

Most parts obtained from medicinal plants, such as leaves, roots, flowers, and seeds, are used either in the form of extracts (e.g., tea infusions from a mixture of different herbs) or as isolated, more or less purified plant constituents (Nikbakht and Kafi, 2008). The individual healing effects are due to the effectiveness of different chemical substances in more or less complex plant mixtures. The chemical active components can mainly be attributed to secondary plant metabolites such as alkaloids, amino acids, flavonoids, carotenoids, saponins, phytosterols, coumarins, and terpenoids. Even today, medicinal and aromatic plants are still of great economic importance worldwide, as they represent important, sustainable raw materials not only for the production of phytopharmaceuticals, but also in the perfume, aroma, and cosmetics industries (Najafi *et al.*, 2010).

Essential oils and solvent extracts obtained from various medicinal plants, herbs, and spices in most cases contain a complex mixture of biologically active ingredients. Even today, essential oils are widely used not only for medicinal purposes and flavoring, but also in food production due to their individual antimicrobial activity and associated preservative properties (Nejad Ebrahimi *et al.*, 2008). Compared to synthetic medicines, herbal medicines and their ingredients usually do not have adverse effects on humans (Banerjee, 2002). It is known from history that many infectious diseases have been successfully treated with herbal remedies (Nezhadali and Zarrabi Shirvan, 2010). Therefore, the use of traditional and medicinal plants in developing countries is still an important basis for maintaining health in these countries today (de Silva, 1997). Also recently, medicinal and aromatic plants have gained increasing importance in various sectors such as agroindustry, the perfume industry, the phytopharmaceutical industry, and natural cosmetics (Khorasaninejad *et al.*, 2010). Over a long period, medicinal plants have played an important role in the treatment of many diseases, especially in Asian and African countries (Fallah-Hoseini *et al.*, 2006). As ancient documents show, certain plants were already used for various medicinal purposes in ancient Persia (Nikbakht and Kafi, 2008). Today, in Iran several projects are also carried out to further explore the importance of herbal medicine, sometimes with the support of modern analytical techniques, and in this way to highlight the special therapeutic benefits of medicinal plants over some synthetic drugs (Nikbakht *et al.*, 2008). In contrast to Traditional Medicine, the current (modern) approach to evaluate the efficacy of herbal medicines is predominantly based on the aim to identify and quantify the individual plant substances occurring in medicinal herbs as precisely as possible. In this context, the individual substances related to various chemical groups are detected by means of very sensitive and reliable chromatographic and spectroscopic analytical methods.

For a long time, traditional medicine methods were marginalized compared to the steady increase in knowledge observed in modern medicine. However, in turn, the therapeutic potential of evidence-based medicine has not always been fully realized. Therefore, some studies have been initiated to quantify the descriptive classification theory of TCM and TPM/TIM (Ma *et al.*, 2010; Lu *et al.*, 2012; Rezadoost *et al.*, 2016). As in other traditional medicine systems, TPM/ITM also classifies the human body from signs and symptoms into different classes or temperaments similar to the theory of TCM (Ardekani *et al.*, 2011). The authors found a close relationship between the main chemical compounds of medicinal plants and the temperaments mentioned in traditional Iranian manuscripts. They were able to assign hot and dry temperaments to those plants containing phenolic compounds as the main constituents, with the exception of those containing tannins, which have cold and dry temperaments. Plants containing essential oils were also identified as having hot and dry temperaments, with the exception of those whose essential oils contain mainly alcoholic constituents, which have cold and dry temperaments. In the case of alkaloid-containing plants, either a cold and dry or a hot and dry temperament was found, depending on their individual alkaloid structure (Ardekani *et al.*, 2011).

Although several omics-based and also non-omics-based research studies have already been conducted to describe some of these temperaments in more detail, there is a considerable need for further research activities to clarify the outstanding questions regarding the theory of TCM and TPM/TIM in this context. A study by Rezadoost *et al.* (2016) compared the proteomic profiles of two temperaments (hot–wet and cold–dry) to further describe the molecular basis of the TPM/TIM theory for classifying the human body and possible molecular biomarkers proposed for these temperaments.

Interactions with Islamic Traditional Medicine, Traditional Chinese Medicine, Ayurveda, Kampo, and Monastic Medicine

In general, Middle Eastern medicine was strongly influenced by teachers from the Far

East, Persia, and India (Fig. 3.1). Important principles here were the influences of light, air, nutrition, exercise, rest, regulated sleep, the balance of the body humors, and a balanced mind on human well-being.

"Let your food be your medicine and your medicine be your food." This statement is attributed to the Greek physician Hippocrates of Kos (460–377 BC). However, there were also some older natural healing methods that advocated this maxim, such as TCM and Ayurveda, the traditional Indian system of medicine. Ayurveda can be traced back at least 5000 years, and here, too, nutrition and aromatic herbs play a very important role in health. Like most traditional systems of medicine, it has been developed and refined over thousands of years based on observation and experience. Ayurvedic therapies have been practiced for thousands of years, long before there were placebo-controlled experiments.

In Ayurveda, as in the other Asian traditional medical systems, foods, medicinal herbs, and other substances (including mineral and animal substances) are also considered as potential "remedies". In contrast to modern pharmacy, Ayurveda thus basically makes no distinction between food and medicine, both are considered as *dravya* ("substance"). If a "substance" (and this includes also food) is to be used for Ayurveda, it must be able to be evaluated according to the special Ayurvedic principles. In contrast to Western medicine, in Ayurveda and also in other traditional Eastern medical systems (e.g. TCM, TPM/TIM), a specific herbal mixture is developed for the individual person that best suits his or her "type". The classification according to the three *doshas* is generally reflected in the individual diet and the other Ayurvedic applications. Furthermore, food is usually cooked so that the body can better absorb the individual active plant components; raw vegetables are rarely consumed. The individual herbal raw materials such as cloves (*S. aromaticum*), turmeric (*Curcuma longa*), ginger (*Zingiber officinale*), cardamom (*Elettaria cardamomum*), coriander (*C. sativum*), cumin (*C. cyminum*), nutmeg (*Myristica fragrans*), pepper (*P. nigrum* and *Piper longum*), saffron (*C. sativus*), and cinnamon (*C. verum*) are very often attributed a spiritual effect on body and mind. In Europe, long pepper (*P. longum*), as the first introduced species of the genus *Piper*, has been more or less neglected. This pepper species originates from India and is still widely cultivated there today as a holy medicinal plant in the context of Ayurvedic medicine. In contrast to black pepper (*P. nigrum*), which is mainly used in Europe today, long pepper's aroma shows sweet, minty, and sometimes even chocolate-like aspects in addition to the pungency (mainly caused by the alkaloid piperine) that has also given this plant species the popular name "chocolate pepper" (Zaveri *et al.*, 2010).

It is not surprising that the list of medicinal and aromatic plants species used in Ayurvedic medicine corresponds very well with that mentioned by Avicenna in his book *Canon of Medicine*. In ancient times, there was already an extensive commercial trade between India, China, Arabia, and Persia. The so-called "Silk Road", which already existed during Roman times, not only supported the trade with various goods but also contributed significantly to the exchange of medical knowledge and ideas between these countries.

Even today, Ayurveda still represents an important medical system in India, Pakistan, Nepal, and Sri Lanka, which in the past has also partly influenced the medical systems in Tibet, Myanmar, and Malaysia. Another famous Eastern health system is TCM, which has also been established in Korea and Japan and spread south to Thailand, Laos, and Vietnam, where it is complemented by both Ayurveda and Unani, the ITM system used mainly in South Asian countries with significant Muslim populations. ITM has its roots in the Hippocratic theory of the humors adopted from ancient Greece. The common basic concept of all these medical systems is that there exists an underlying life energy called *prone* in Sanskrit, *chi* in China, and *ki* in Japan. In general, the aim is to combine physical, mental, and spiritual aspects and in this way to prevent diseases and strengthen the body's defences. In Ayurvedic medicine, illness is generally attributed to mental or physical imbalances. The individual therapy approach is generally based on the three Ayurvedic *doshas*, which determine the mental balance and are assigned to the five elements: space, water, earth, air, and fire. In this context, *Vāta*, consisting of air and space, is considered the life energy; *kampha*, related to earth and water, on the other hand, is responsible for the structure of the body, its growth, and flexibility. The *dosha pitta*, assigned to fire and water, is made responsible for all biochemical processes in the human body.

In a similar way, TPM/TIM too is based on the concept of the four humors: phlegm (*balgham*), blood (*dam*), yellow bile (*ṣafrā'*), and black bile (*saudā'*). This concept mainly refers to the theories developed by Rhazes and Avicenna.

Kampo medicine is the original traditional system of medicine in Japan. Here, no distinction is made between Western and Kampo medicine, and both systems are therefore subject to the same legal regulations today. Accordingly, the application for marketing authorization of ethical Kampo medicinal products follows the general notification for medicinal products. In addition, applications for marketing authorization of ethical Kampo medicinal products should follow the Kampo-specific notification "Handling of ethical Kampo medicinal products". In general, the safety, efficacy, and quality of Kampo medicinal products are comprehensively ensured by the Japanese Pharmacopoeia, Good Manufacturing Practice, Good Agricultural and Collection Practice, Certificate of Marketing Authorization, Marketing Authorization Standard, and Pharmacovigilance.

Temperament is a key concept for understanding the fundamentals of traditional medicine, on the basis of which people are classified as healthy and sick. This principle plays an important role in determining ways to maintain health and also in treating diseases.

Increasingly, conventional medicine today is also placing more emphasis on treating the patient as an individual rather than the disease in general. Accordingly, numerous studies have already been conducted to gain more knowledge regarding the relationship between temperament and proteomes in the human body (Jafari *et al.*, 2014; Shirbeigi *et al.*, 2017). Today, TPM/TIM is considered a rich and valuable source in the field of medical history to the point that in a period of history, it has been used as an important educational reference even at various universities in European countries (Aciduman *et al.*, 2009).

In Western medicine, too, "personalized medicine" has been gaining more attention again for several years. In line with the present spirit (zeitgeist) of increasing personal individualization, the challenge here is to achieve a maximum therapeutic efficacy for each patient while minimizing any possible side effects. In this context, especially high-throughput technologies and modern imaging techniques are used to gain a better understanding of the molecular processes involved in the development of various diseases.

Even before the collapse of advanced civilization in the Mediterranean region in the period from the 6th to the 8th century AD, Benedict of Nursia founded the first monastery of the Benedictine order on Monte Casino in southern Italy, which was later named after him. Benedictus founded the abbey of Montecassino in a temple of Apollo around the year AD 529. This abbey is still considered the original monastery of the Benedictine order (in Latin: *Ordo Sancti Benedicti*, OSB) which also provided the framework for monastic medicine. The monks and nuns drew on the medical knowledge of the ancient physicians and adopted the four-fluid doctrine, the so-called "humoral pathology", as a basis. The basic rule for monastic life was given by the Latin motto *ora et labora* ("pray and work"), with the care of body and soul always in the foreground. The Benedictine monks mainly healed sick people, and in this context felt responsible not only for the members of the order, but for anyone who came to them with a health problem.

This idea of placing mercy and charity above all other duties was considered revolutionary at that time, because in Greco-Roman culture weak and sick people were regarded responsible for their own fate. Apart from Benedict, his friar Cassiodorus (AD 490–580) was particularly committed to the development of monastic medicine. This work was not only focused on the study of the books handed down by Hippocrates, Dioscorides, and Galen, but also placed particular emphasis on the study of medicinal herbs and the development of new plant mixtures that could be used for various healing purposes. Apart from the ancient books cited above, the monks referred mainly to the natural history book with the Latin name *Naturalis historia* written by Plinius and thus they created a medicinal compendium that was considered a standard work at that time.

Ancient civilizations contributed much to the development of phytotherapy by identifying and describing numerous species of medicinal plants. This knowledge of herbal raw materials thus formed the basis of medieval medicine. The revival of herbal medicine can be observed since the 6th century AD, when monasteries were established where gardens were planted to grow medicinal plants. Monastic schools were also founded, and monks expanded their knowledge by studying and copying ancient books and prepared medicines of natural origin according to secret recipes, which were used to cure the sick. Although the Middle Ages is often regarded as a dark age, many achievements of phytotherapy from this period have passed into history and are still used by people today.

The monks and nuns in the monasteries copied and interpreted the writings of Greek and Roman philosophers and physicians and were thus able to store and preserve the knowledge passed down from antiquity in Europe. Monastery herb gardens typically followed the rules set down around AD 840 by Walahfrid Strabo, Benedictine monk, poet, and botanist. In his book, *Liber de cultura hortorum*, he described 24 medicinal plants, their characteristics, and individual application. The book became one of the most important botany books of the Early Middle Ages.

After the decline of the *Imperium Romanum* (Roman Empire) medical development in Europe gained new impetus with the *Lorsch Pharmacopoeia* (*Lorscher Arzneibuch*) in the middle of the 11th century AD.

In the High Middle Ages (11th and 12th centuries AD), the first important writings of monastic medicine were created, such as the translations of the book *Constantinus Africanus* in Salerno. Furthermore, the Benedictine nun Hildegard of Bingen (AD 1098–1179) described 300 plant species including their individual medicinal applications. Unfortunately, her *Book of Secrets in Nature and Creatures* (*Liber subtilitatum diversarum naturarum creaturarum*) was lost after her death. About 150 years later, several manuscripts were retrieved and attempts were made to reassemble Hildegard's original work, which was divided into two treatises, *Physica* and *Causae et curae* (*Causes and Cures*). The textbooks are mainly attributed to Hildegard, although some sections may have been influenced by later revisions.

Until the 19th century AD, the humoral pathology developed in antiquity according to Hippocrates and the numerous writings of Galen remained predominant in medicinal treatments in both the Orient and Occident. Accordingly, the balance of the four humors (eucrasia) was considered an important prerequisite for health. Hildegard of Bingen was also influenced by the ancient teachings of humoral therapies, according to which mismanagement of the humors leads to disorders of body and mind. Therefore, a central message of Hildegard of Bingen's medicine is the belief in the power of nature; a power that strives for balance.

The knowledge from the Near Eastern roots of monastic medicine and comprehensive visionary insights, shaped, among other things, the so-called "Hildegard medicine" that is still well known and appreciated in Europe of today. Hildegard of Bingen and the medicine of her time were practiced within the monastic community. Like most medieval medical cures, her practices were mainly guided by the principles described in the Bible. Today, monastic medicine is increasingly gaining influence in modern medicine and it is generally accepted that Hildegard of Bingen's work was significantly influenced by the following four works (Baader, 1979):

- *Physiologus*, a book on natural history written in Egypt in the 2nd century AD. This book of an unknown author was available in several translations, including Latin, around AD 700.
- *De materia medica* by the Greek physician Dioscorides with descriptions of about 500 medicinal plants recognized for medical treatment (Fig. 3.9).
- With regard to the cultivation of medicinal plants, she benefited predominantly from the already mentioned Benedictine monk Walahfrid Strabo, whose famous work *Hortulus* (*Garden*) has

Fig. 3.9. Page from the Vienna *Dioscorides codex* (*Codex medicus Graecus* of the Austrian National Library, 6th century), an illustrated version of *De materia medica* (1st century AD), page 29v). (Public Domain, https://commons.wikimedia.org/w/index.php?curid=43449887.)

served as an important basis for horticulture for centuries.
- The herbal medicine traditions recognized in the Middle Ages were generally known as "Macer Floridus" from the 11th century AD and coincided with the development of *Circa instans simplicibus medicinis* (*From the Simple Medicines*), which was developed by Matthew Platearius, the medieval physician of the Salerno Medical School.

After the Ottoman Empire's advance into the Iberian Peninsula, Christianity benefited from Islam's extensive knowledge of the natural sciences and medicine. In this context, the Toledo school of translators around Gerhard of Cremona (AD 1134–1187), who translated numerous Arabic writings into Latin, was of particular importance.

From around AD 1280 onward, the first German pharmacopoeia, written by Ortolf von Würzburg, a highly respected wound physician with an academic education, was used regionally until the end of the 16th century. It is characterized by a high level of expertise and can be described as a pioneering act, as it was customary until then to write medical textbooks in Latin. He addressed his textbook to fully trained practitioners, primarily wound physicians, and imparted to them profound knowledge about diagnosis, prognosis, and therapy. Ortolf drew primarily on Salerno sources for his work, which was primarily based on 12th and 13th century sources, and he also used medical literature from Parma, Toledo, and Paris. Among the most important sources is the *Compendium medicinae* (*Compendium of Medicine*) written by Gilbertus Anglicus.

In these books, mentioned above, first influences of Oriental medicine can already be found, which at that time could already look back on a longer tradition. In the 14th and 15th centuries AD, increasingly comprehensive herbal encyclopedias were published, which claimed to collect and disseminate the entire phytotherapeutic knowledge available in Europe at that time (Fuchs, 2001). These herbal records describing the use of medicinal plants thus represent in a certain way the basis of today's "European Pharmacopoeia".

Especially in the Middle Ages, Galen's teachings were favored over others and established the basis of medical knowledge at European universities until the 19th century (Schipperges and Durling, 1989). Furthermore, special attention should be also paid to the physician Ibn Buṭlan (11th century AD), who worked at the court in Baghdad and, as a Christian doctor, wrote many medical works. In his book entitled *Taqwīm aṣ-Siḥḥah* (*Tabular Overview of Health*, in Latin: *Tacuina sanitatis in medicina*), he teaches several health rules that are presented in the form of a synoptic table work. This clear reference work for questions of a healthy lifestyle at that time attracted special attention in several European cities such as Vienna, Paris, and Rome. Later a Viennese edition of the *Tacuina* was produced in the 14th century AD, probably in Lombardy (Herr, 1988).

In the middle of the 15th century AD a so-called *Paris Tacuinum sanitatis* evolved, including also several illustrations from *Circa instans* (Goehl, 2015) which were presented in the tradition of the synoptic table work, created by Ibn Butlan. These manuscripts, however, were not written for specialists but more for the upper middle classes and nobility in Europe. Accordingly, the content of this book is essentially based on ancient and medieval humoral pathology and explains which foods and human states of mind (anger, joy, etc.) represent the four elements "dry", "moist", "cold", and "warm" and for which people and in which regions they could be useful. The concept of humoral pathology assumes that the function of the human body is regulated by the four humors (blood, phlegm, yellow bile, and black bile) and only secondarily by its organs. It was assumed that humors correspond to the four elements "cold–warm" and "dry–humid". Applying a specific selection of medicines and food raw materials, the doctors tried to restore the patient's health as best as possible by establishing a balance with special regard to the human body fluids and primary elements (Table 3.1). Garlic, for example, is described in this context as "warm"

Table 3.1. Description of the so-called "inner factors" applied in diagnostics and therapy of the humoral pathology, developed by Ortolf on the basis of the four elements and several primary qualities. (According to Riha, 2017.)

Primary elements	Warm/humid	Warm/dry	Cold/humid	Cold/dry
Elements	Air	Fire	Water	Soil
Body fluids	Blood	Yellow bile	Mucus	Black bile
Temperaments	Sanguine type	Choleric type	Phlegmatic type	Melancholiac
Colors	Red	Yellow	White	Black
Taste	Sweet	Bitter	Salty	Sour
Time of day	Morning	Noon	Night	Evening
Gender		Man	Woman	
Age	Youth	Adulthood	Childhood	Senility
Organs, tissue	Heart, liver, muscles	Gall bladder	Brain, stomach/gut, kidneys/ urinary bladder, fat	Spleen, bones, tendons

and "dry". In addition, the book states that garlic "produces pungent juices and is useful against scorpion and snake bites, as well as against worms". Therefore, "garlic is especially recommended for weakened natures and in cold regions".

In the Middle Ages the manuscripts of *Circa instans* were widely circulated throughout Europe and were also translated into German and served as a model for later authors such as Adam Lonitzer (AD 1528–1568) when writing their pharmaceutical textbooks. Furthermore, it can be clearly seen that the chapter division refers to Dioscorides and the descriptions of the humoral pathology are based on Galen's theories (Goehl, 2015).

The wound physician Ortolf von Baierland (born in the first half of the 13th century) wrote a medical textbook around AD 1280 that was used regionally until the end of the 16th century.

Ortolf also focused on the model of medieval humoral pathology. He refers to this model particularly in the diagnosis and subsequent therapy. When examining his patients, Ortolf also oriented himself to certain "leading symptoms", which he assigned to the respective primary qualities (cold–warm and dry–damp, respectively) (Riha, 2017).

Philippus Aureolus Theophrastus Bombastus von Hohenheim, who called himself "Paracelsus" (AD 1493–1541), occupied a special position. Until today, the myth of Paracelsus is very widespread, especially in German-speaking countries. Especially, his well-known sentence "all things are poison and nothing is without poison; it is the dose alone that makes a thing not poison" remains unforgotten. Paracelsus created his own system of healing various diseases, which was based on alchemical medicine on the one hand and on empirical knowledge on the other. For example, he treated syphilis, which was widespread in the 16th century, with mercury and guaiac wood. During his lifetime, Paracelsus was known only to a few contemporaries, as hardly any of his writings were printed.

Herbal Medicinal Products as a Source for Modern Evidence-Based Medical Systems

Nowadays, more attention is again being paid to research related to plants used in traditional medicine, as they can be produced comparatively cheaply and usually show no undesirable side effects. Spices and herbs such as garlic, mustard, cinnamon, cumin, cloves, thyme, basil, pepper, ginger, and rosemary have been used since ancient times as food additives, flavorings, and natural preservatives. A number of spices also exhibit antimicrobial activity against numerous bacteria and fungi. Therefore, an increasing consumption of herbal medicines is observed worldwide, as they represent often a suitable alternative to the use of synthetic drugs in the treatment of a number of health problems, including heart disease, diabetes, hypertension, and even certain types of cancer.

The use of herbs to treat various diseases is almost universal, particularly in nonindustrialized societies (DaSilva *et al.*, 2002). According to estimates of the WHO, about 80% of the population in some Asian and African countries currently apply herbal medicines in order to heal certain diseases. In the USA about 25% of modern medicines available on the market are derived from herbal plants and widely used in modern medicine. About 80% of the 120 active ingredients currently isolated from higher plants show a positive correlation between their modern therapeutic and traditional use (Fabricant and Farnsworth, 2001). It is estimated that at least 7000 active medicinal ingredients in the modern pharmacopoeia are derived from plants. Significant variations in plant properties with different soil conditions have been observed for many medicinal and aromatic plants (Carrubba and Scalenghe, 2012). The growing interest in medicinal plants can be also seen, among other things, in the significant increase in plant monographs in the European Pharmacopoeia (at present clearly more than 100).

Today, in the context of clinical research, evidence-based medicine (EBM) is generally considered the standard in the Western world. However, a few phytopharmaceuticals do not correspond to these specifications, which is not surprising due to the individual underlying philosophy. However, despite the enormous scientific successes of "modern medicine", it should not be overlooked that many cultures on this earth gained extensive experience with medicinal plants over thousands of years and usually the doctors could successfully treat their patients. The isolation of individual active substances from the individual medicinal plants must also be viewed critically, since solvent extraction or steam distillation already may cause significant changes in the plant metabolite profile. The selection of the respective extraction solvent (e.g., n-hexane, ethanol, water, supercritical carbon dioxide) already generates a selection according to the different polarities. However, since the efficacy of herbal drugs often involves the complex interaction of individual constituents, the isolated extraction of individual chemical groups (e.g., terpene hydrocarbons, alkaloids, fats, sugars, amino acids) may sometimes reduce their pharmacologic efficacy.

On the other hand, however, the use of single active ingredients and active ingredient complexes isolated from the plant matrix has resulted in phytopharmaceuticals with high and reliable pharmaceutical efficacy. Undoubtedly, the very high number of plant secondary plant metabolites represents a "treasure trove" for the development of future drugs. Searching for new valuable plant substances and active ingredient complexes, the so-called "omics" techniques available today (genomics, transcriptomics, proteomics, metabolomics) and intelligent biostatistical methods in particular will play a very important role.

References

Abdul-Jalil, T.Z. (2020) *Rhus coriaria* (sumac): a magical spice. In: Akram, M. and Shabir Ahmad, R. (eds) *Herbs and Spices*. IntechOpen, London. DOI: 10.5772/intechopen.92676.

Aciduman, A., Er, U. and Belen, D. (2009) Peripheral nerve disorders and treatment strategies according to Avicenna in his medical treatise, Canon of Medicine. *Neurosurgery* 64, 172–177. DOI: 10.1227/01.NEU.0000335779.27115.D3.

Ahmadipour, S., Ahmadipour, S., Mohsenzadeh, A. and Asadi-Samani, M. (2016) The importance of some native medicinal plants of Iran effective on gastrointestinal disorders in children: a review. *Der Pharmacia Lettre* 8, 61–66.

Alexiades, N.M. and Sheldon, J.E. (eds) (1996) *Selected Guidelines for Ethnobotanical Research: A Field Manual*. New York Botanical Garden Press, New York.

Amar, Z., Lev, E. and Serri, Y. (2014) On Ibn Juljul and the meaning and importance of the list of medicinal substances not mentioned by Dioscorides. *Journal of the Malaysian Branch of the Royal Asiatic Society* 24(4), 529–555.

Amiri, M.S., Joharchi, M.R. and Yazdi, T.M.S. (2014) Ethno-medicinal plants used to cure jaundice by traditional healers of Mashhad, Iran. *Iranian Journal of Pharmaceutical Research* 13, 157–162.

Anyinam, C. (1995) Ecology and ethnomedicine: exploring links between current environmental crisis and indigenous medical practices, *Social Science & Medicine* 40(3), 321–329.

Ardekani, M.R.S., Rahimi, R., Javadi, B., Abdi, L. and Khanavi, M. (2011) Relationship between temperaments of medicinal plants and their major chemical compounds. *Journal of Traditional Chinese Medicine* 31, 27–31. DOI: 10.1016/S0254-6272(11)60006-X.

Azizi, M.H. (2008) Gondishapur school of medicine: the most important medical center in antiquity. *Archives of Iranian Medicine* 11, 116–119.

Azizi, M.H., Nayernouri, T. and Azizi, F. (2008) A brief history of the discovery of the circulation of blood in the human body. *Archives of Iranian Medicine* 11, 229–234.

Baader, G. (1979) Naturwissenschaft und Medizin im 12. Jahrhundert und Hildegard von Bingen. *Archiv für mittelrheinische Kirchengeschichte* 31, 33–54.

Banerjee, M. (2002) Public policy on Ayurveda. *Economic and Political Weekly* 37, 1136–1146.

Bauer, T. (1988) *Das Pflanzenbuch des Abu Hanifa ad-Dinawari: Inhalt, Aufbau, Quellen*. Otto Harrassowitz, Wiesbaden, Germany.

Bolens, L. (1977) L'agriculture hispano-arabe au Moyen Age. In: *Handbuch der Orientalistik, Geschichte der islamischen Länder*, Band 6. E.J. Brill, Leiden, The Netherlands/Cologne, Germany, pp. 255–275.

Bryan, C.P. (2021) *The Papyrus Ebers: Ancient Egyptian Medicine*. Translated from the German version. Martino Fine Books, Eastford, Connecticut.

Calvo, E. (2008) *Encyclopaedia of the History of Science, Technology, and Medicine in Non-Western Cultures*. Springer, Dordrecht, The Netherlands.

Cambra, L.M.A. (2001) An example of medical practice in al-Andalus: Treatise XIX of the Kitāb al-tarsīf of Abū-l-Qāsim al-Zahrāwī. *Dynamics* 21, 73–91.

Cárcer, L.F.A. (2001) Therapeutic use of aromatic substances in al-Andalus. *Dynamics* 21, 93–132.

Carrubba, A. and Scalenghe, R. (2012) Scent of Mare Nostrum – medicinal and aromatic plants in Mediterranean soils. *Journal of the Science of Food and Agriculture* 92, 1150–1170.

Castleman, M. (2001) *The New Healing Herbs: The Classic Guide to Nature's Best Medicines Featuring the Top 100 Time-Tested Herbs*. Rodale, Emmaus, Pennsylvania.

DaSilva, E.J., Baydoun, E. and Badran A. (2002) Biotechnology and the developing world. *Electronic Journal of Biotechnology* 5(1), 1–2. DOI: 10.2225/vol5-issue1-fulltext-1.

Defremery, C. and Sanguinetti, B.R. (1929) *Ibn Baṭṭuṭa: Travels in Asia and Africa*. French translation of the Arabic text edited for the first time in 1854. George Routledge & Sons Ltd, London.

de Silva, T. (1997) Industrial utilization of medicinal plants in developing countries. In: Bodeker, G., Bhat, K.K.S., Burley, J. and Vantomme, P. (eds) *Medicinal Plants for Forest Conservation and Healthcare*. Non-Wood Forest Products No. 11. Food and Agriculture Organization of the United Nations, Rome, pp. 34–44.

de Vos, P. (2013) The "Prince of Medicine": Yūhannā ibn Māsawayh and the foundations of the Western pharmaceutical tradition. *Isis* 104, 667–712. DOI: 10.1086/674940.

Diller, H. (2021) *Hippokrates/Die Heilkunst (aus dem Griechischen übersetzt)*. Reclams Universal-Bibliothek Nr. 19694. Philipp Reclam jun. Verlag GmbH, Ditzingen, Germany.

Durant, W. (1950) *The Age of Faith. A History of Medieval Civilization – Christian, Islamic, and Judaic – From Constantine to Dante: A.D. 325–1300*. Simon & Schuster, New York.

Durling, R.J. (1989) Leonhart Fuchs and his Commentaries on Galen. *Medizinhistorisches Journal* 24, 42–47.

Elgood, C. (1951) *A Medical History of Persia*. Cambridge University Press, Cambridge.

Elgood, C. (1979) *A Medical History of Persia and the Eastern Caliphate from the Earliest Times to the Year A.D.* APA-Philo Press, Amsterdam.

Elgood, C. (1992) *Iran Medical History and the Regions of Eastern Caliphate*. Translated into Persian by F. Baher. Amir Kabir Publications, Tehran, p. 22, 37.

Emami, S.A. and Aghazari, F. (2011) *Les Phanerogames Endemiques de la flore d'Iran*. Tehran University Publication, Tehran, p. VII.

Emami, S.A., Iranshahi, M. and Iranshahi, M. (2017) *Herbal Pharmacopoeia*. Behnashr and Mashhad University of Medical Sciences Publications, Mashhad, Iran, p. 28 (in Persian).

Fabricant, D.S. and Farnsworth, N.R. (2001) The value of plants used in traditional medicine for drug discovery. *Environmental Health Perspectives* 109, 69–75.
Fahd, T. (1993/1995) *L'Agriculture Nabatéene*. Translation into Arabic attributed to Abu Bakr Ahmad b. Ali al-Kasd ā ni, known as Ibn Wahshiya (4th–10th century), 2 vols, Damascus.
Fahd, T. (1996) Botany and agriculture. In: Rashed, R. (ed.) *Encyclopedia of the History of Arabic Science*, Vol. 3. Routledge, London and New York, pp. 813–852.
Fallah-Hoseini, H., Fakhrzadeh, H., Larijani, B. and Shikhsamani, A. (2006) Review of anti-diabetic medicinal plants used in traditional medicine. *Journal of Medicinal Plants* 5, 1–8.
al-Faruqi, I.R. and al-Faruqi, L.L. (1986) *The Cultural Atlas of Islam.* Prentice-Hall. New York.
Fuchs, L. (2001) *Das Kräuterbuch von 1543 (Faksimile-Druck).* Taschen GmbH, Cologne, Germany.
Ghāfiqī, M., Hebraeus, B., Meyerhof, M. and Ṣubḥī, J. (1932–1940) *The Abridged Version of "The Book of Simple Drugs" of Ahmad ibn Muhammad al-Ghāfiqī*. Govt. Press, Cairo.
Goehl, K. (2015) *Das Circa Instans. Die erste große Drogenkunde des Abendlandes*. Deutscher Wissenschafts-Verlag, Baden-Baden, Germany.
Hamidullah, M. (1973) *Le Dictionnaire botanique d'Abu hanifa al-Dinawari*. Compiled and translated according to the citations of later works, edited by the Institut Francais d'Archéologie Orientale, Cairo.
Heinrich, M. (2000) Ethnobotany and its role in drug development. *Phytotherapy Research* 14, 479–488.
Herr, M. (1988) Translation of the book: *Ibn Buṭlān, Ibn Dschezla: Schachtafelen der Gesundheyt*. Edition Straßburg (Hans Schott) 1533, with an epilogue by Marlit Leber and Elfriede Starke, Weinheim, Germany.
Hitti, P.K. (1937) *History of the Arabs*. Revised 10th edn, Palgrave Macmillan, New York, 2002.
Ibrar, M., Hussain, F. and Sultan, A. (2007) Ethnobotanical studies on plant resources of Ranyal hills, district Shangla, Pakistan. *Pakistan Journal of Botany* 39, 329–337.
Jafari, M., Rezadoost, H., Karimi, M., Mirzaie, M., Rezaie-Tavirani, M. *et al.* (2014) Proteomics and traditional medicine: new aspect in explanation of temperaments. *Forschende Komplementärmedizin* 21, 250–253. DOI: 10.1159/000366118.
Jastrow, M. (1914) Medicine of the Babylonians and Assyrians. *Proceedings of the Royal Society of Medicine* 7 (*Sect. Hist. Med.*), 109–176. Available at: https://www.ncbi.nlm.nih.gov/pmc/articles/PMC2003546/ (accessed 23 March 2023).
Javadi, B. and Emami, S.A. (2015) Avicenna's contribution to mechanisms of cardiovascular drugs. *Iranian Journal of Basic Medical Sciences* 18, 721–722.
Jivad, N., Bahmani, M. and Asadi-Samani, M. (2016) A review of the most important medicinal plants effective on wound healing on ethnobotany evidence of Iran. *Der Pharmacia Lettre* 8, 353–357.
Johnstone, P. (1999) Ibn Juljul, physician and herbalist. *Islamic Culture* LXXIII, 37–43.
Joshi, A.R. and Joshi, K. (2000) Indigenous knowledge and uses of medicinal plants by local communities of the Kali Gandaki Watershed area, Nepal. *Journal of Ethnopharmacology* 73, 175–183.
Jouri, M.H.and Mahdavi, M. (2010) *Applied Identification of Rangeland Plants*. Aeezh Press, Tehran, p. 436.
Kahle, P. (1952) Ibn Samjun's work: "Ibn Samagun und sein Drogenbuch". In: *Documenta Islamic Ibedita*. Akademie-Verlag, Berlin.
Khorasaninejad, S., Mousavi, A., Soltanloo, H., Hemmati, K. and Khalighi, A. (2010) The effect of salinity stress on growth parameters, essential oil yield and constituent of peppermint (*Mentha piperita* L.). *World Applied Sciences Journal* 11, 1403–1407.
Lev, E. and Amar, Z. (2000) Ethnopharmacological survey of traditional drugs sold in Israel at the end of 20th century. *Journal of Ethnopharmacology* 72, 191–205.
Levey, M. (1997) *Early Arabic Pharmacology: An Introduction Based on Ancient and Medieval Sources.* E.J. Brill, Leiden, The Netherlands/Cologne, Germany.
Lewin, B. (1953) *The Book of Plants of Abu Hanifa al-Dinawari*, edited from the unique manuscript in the Library of the University of Istanbul, with an introduction, notes, indices, and a vocabulary of selected words, Uppsala, Sweden/Wiesbaden, Germany.
Lu, A., Jiang, M., Zhang, C. and Chan, K. (2012) An integrative approach of linking traditional Chinese medicine pattern classification and biomedicine diagnosis. *Journal of Ethnopharmacology* 141, 549–556. DOI: 10.1016/j.jep.2011.08.045.
Ma, T., Conge, T., Zhang, H., Wang, M., Ding, W. *et al.* (2010) Bridging the gap between traditional Chinese medicine and systems biology: the connection of Cold Syndrome and NEI network. *Molecular BioSystems* 6, 613–619. DOI: 10.1039/b914024g.
Mah Van, E. (2002) *A Dictionary of Iranian Plants*, Soroush Press, Tehran.

Meyer, E.H.F. (1854) *Geschichte der Botanik*, Vol. 3. Verlag der Gebrüder Bornträger, Königsberg, Germany.
Mosaddegh, M. and Naghibi, F. (2002) *Iran's Traditional Medicine, Past and Present. Traditional Medicine and Materia Medica*, Vol. 1. TMRC, Tehran, pp. 2–20.
Moeini, R. and Gorji, N. (2016) Persian medicine in the world of research. Review of articles on Iranian traditional medicine. *Iranian Journal of Medical Sciences* 41(3 Suppl.), S7.
Mozaffarian, V. (1996) *A Dictionary of Iranian Plant Names*. Farhang Moaser Press, Tehran.
Naghibi, F., Mosaddegh, M., Motamed, M.M. and Ghorbani, A. (2010) *Labiatae* family in folk medicine in Iran: from ethnobotany to pharmacology. *Iranian Journal of Pharmaceutical Sciences* 4(2), 63–79 DOI: 10.22037/IJPR.2010.619.
Naghizadeh, A., Hamzeheian, D., Akbari, S., Mohammadi, F., Otoufat, T. *et al.* (2020) UNaProd: a universal natural product database for *materia medica* of Iranian traditional medicine. *Evidence-Based Complementary and Alternative Medicine* 13, 3690781. DOI: 10.1155/2020/3690781.
Najafi, S., Sadeghi, N.B., Deokule, S.S. and Estakhr J. (2010) Phytochemical screening of *Bidaria khandalense* (Sant.) Jagtap & N. P. Singh *Loranthus capitellatus* Wall., *Viscum articulatum* Burm.f. and *Vitex negundo* Linn. *Research Journal of Pharmaceutical, Biological and Chemical Sciences* 1(3), 388–393.
Najmabadi, M. (2001) *History of Medicine in Ancient Iran*, 2nd edn. Tehran University Press, Tehran (in Persian).
Naseri, M. and Ardakani, M.S. (2004) The School of Traditional Iranian Medicine. The definition, origin and advantages. *Iranian Journal of Pharmaceutical Sciences* 3(Suppl. 2), 17–21.
Nasr, S.H. (2003) *Science and Civilisation*, 2nd edn. The Islamic Texts Society, Cambridge.
Nejad Ebrahimi, S., Hadian, J., Mirjalili, M.H., Sonboli, A. and Yousefzadi, M. (2008) Essential oil composition and antibacterial activity of *Thymus caramanicus* at different phenological stages. *Food Chemistry* 110, 927–931.
Nezhadali, A. and Zarrabi Shirvan, B. (2010) Separation, identification and determination of volatile compounds of *Ziziphora persica* Bunge using HS-SPME/GC-MS. *International Journal of Environmental Science and Development* 1, 115–118.
Nikbakht, A. and Kafi, M. (2008) The history of traditional medicine and herbal plants in Iran. *Acta Horticulturae* 790, 255–258.
Nikbakht, A., Kafi, M. and Haghighi, M. (2008) The abilities and potentials of medicinal plants production and herbal medicine in Iran. *Acta Horticulturae* 790, 259–262.
Noorhosseini, N.S.A., Ashoori, L.D., Allahyar M.S. and Doozandeh, M.P. (2011) Socio-economic factors for adoption of medicinal plants cultivation in Eshkevarat region, north of Iran. *Journal of Medicinal Plants Research* 5, 30–38.
Park, H.M., Lee, J.H., Yaoyao, J., Jun, H.J. and Lee, S.J. (2011) Limonene, a natural cyclic terpene, is an agonistic ligand for adenosine A(2A) receptors. *Biochemical and Biophysical Research Communications* 404, 345–348.
Parsaei, P., Bahmani, M., Naghdi, N., Asadi-Samani, M., Rafieian-Kopaei, M. *et al.* (2016) Identification of medicinal plants effective on common cold: an ethnobotanical study of Shiraz, South Iran. *Der Pharmacia Lettre* 8, 90–97.
Rezadoost, H., Karimi, M. and Jafari, M. (2016) Proteomics of hot–wet and cold–dry temperaments proposed in Iranian traditional medicine: a network-based study. *Scientific Reports* 6, 30133. DOI: 10.1038/srep30133.
Rezaeizadeh, H., Alizadeh, M., Naseri, M. and Ardakani, S. (2009) The traditional Iranian medicine point of view on health and disease. *Iranian Journal of Public Health* 38(Suppl. 1), 169–172.
Riha, O. (2017) *Mittelalterliche Heilkunst. Das Arzneibuch Ortolfs von Baierland (um 1300)*. Deutscher Wissenschafts-Verlag, Baden-Baden, Germany.
Robineau, L. and Soejarto, D.D. (1996) TRAMIL: a research project on the medicinal plant resources of the Caribbean, In: Balick, M.J., Elisabetsky, E. and Laird, S.A. (eds) *Medicinal Resources of the Tropical Forest*. Columbia University Press, New York, pp. 317–325.
Rouhi-Boroujeni, H., Asadi-Samani, M. and Moradi, M.T. (2016a) A review of the medicinal plants effective on headache based on the ethnobotanical documents of Iran. *Der Pharmacia Lettre* 8, 37–42.
Rouhi-Boroujeni, H., Mosharraf, S., Gharipour, M., Asadi-Samani, M. and Rouhi-Boroujeni, H. (2016b) Anti-hyperlipidemic effects of sumac (*Rhus coriaria* L.): can sumac strengthen anti-hyperlipidemic effect of statins? *Der Pharmacia Lettre* 8, 143–147.
Samso, J., Rashed, R. and Vernet, J. (1996) Development of Arabic science in Andalusia. In: Rashed, R. (ed.) *Encyclopaedia of Arabic Science*, Vol. 1. Routledge, London/New York, pp. 243–275.

Sarton, G. (1927–1948) *Introduction to the History of Science*, 4 vols. Williams & Wilkins Co., Baltimore, Maryland/Washington, DC.
Savage-Smith, E. (2012) *A New Catalogue of Arabic Manuscripts in the Bodleian Library, University of Oxford*, Vol. I: Medicine. Oxford University Press, Oxford.
Scarborough, J. (1980) Herbals. In: Strayer, J.R. (ed.) *Dictionary of the Middle Ages*, Vol. 6. Charles Scribner's Son, New York, p. 169.
Schipperges, H. and Durling, R.J. (1989) Galen im Mittelalter. In: *Lexikon des Mittelalters*, Band 4. Artemis, Munich, Germany/Zurich, Switzerland, pp. 1082–1084.
Schmitz, R. (1984) Der Arzneimittelbegriff der Renaissance. In: Schmitz, R. and Keil, G. (eds) *Humanismus und Medizin*. Acta Humaniora, Weinheim, Germany, pp. 1–21.
Scott, S.P. (1904) *History of the Moorish Empire in Europe*, Vol. III. Facsimile, London, 2017.
Seo, S., Song, Y., Gu, S.M., Min, H.K., Hong, J.T. *et al.* (2020) D-Limonene inhibits pentylenetetrazole-induced seizure via adenosine A2A receptor modulation on GABAergic neuronal activity. *International Journal of Molecular Sciences* 21, 9277. DOI: 10.3390/ijms21239277.
Shirbeigi, L., Zarei, A., Naghizadeh, A. and Alizadeh Vaghasloo, M. (2017) The concept of temperaments in traditional Persian medicine. *Traditional and Integrative Medicine* 2(3), 143–156.
Silberberg, B. (1908) Das Pflanzenbuch des Abu Hanifa Ahmed ibn Da'ud al-Dinawari. Ein Beitrag zur Geschichte der Botanik bei den Arabern. Dissertation, Breslau, published partly in *Zeitschrift fur Assyriologie* 24(1910), 225–265.
Sinaei, F., Emami, S.A., Sahebkar, A. and Javadi, B. (2017) Olfactory loss management in view of Avicenna: focus on neuroprotective plants. *Current Pharmaceutical Design* 23, 3315–3321.
Sobhani, Z., Nami, S.R., Emami, S.A. and Javadi, B. (2017) Medicinal plants targeting cardiovascular diseases in view of Avicenna. *Current Pharmaceutical Design* 23, 2428–2443.
Spencer, W.G. (1989) *A.C. Celsus: De Medicina*. Original text translated into English, Gryphon Editions in the Classics of Medicine Library.
Strank, K.J. and Meurers-Balke, J. (2008) *Obst, Gemüse und Kräuter Karls des Großen*. Verlag Philipp von Zabern, Mainz, Germany.
Strohmaier, G. (1999) *Avicenna*. Beck, Munich, Germany.
Sudhoff, K. (1930) Konstantin der Afrikaner und die Medizinschule von Salerno. *Sudhoffs Archiv für Geschichte der Medizin* 23, 293–298.
Sumner, J. (2000) *The Natural History of Medicinal Plants*. Timber Press, Portland, Oregon.
Tayarani-Najaran, Z., Tayarani-Najaran, F. and Emami, S.A. (2014) The history of Islamic medicine at a glance. In: Watson, R.R. Preedy, V.R. and Zibadi, S. (eds) *Polyphenols in Human Health and Disease*, Vol. 1. Elsevier, Amsterdam, pp. 17–27.
Troupeau, G. (1995) The first treatise on diet: the Kitāb Khawāss al-agdiyah de Yūhannā ibn Māsawayh. *Medicina nei Secoli* 7, 121–139.
Ulmann, M. (1972) *Die Natur- und Geheimwissenschaften im Islam*. E.J. Brill, Leiden, The Netherlands.
UNESCO (1996) *Culture and Health, Orientation Texts – World Decade for Cultural Development 1988–1997*. United Nations Educational, Scientific and Cultural Organization, Paris.
WHO (1978) *The Promotion and Development of Traditional Medicine, Report of WHO Meeting [held in Geneva from 28 November to 2 December 1977]*. WHO Technical Report Series No. 622. World Health Organization, Geneva, Switzerland.
WHO (2002) *WHO Traditional Medicine Strategy 2002–2005*. World Health Organization, Geneva, Switzerland.
Wiet, G., Elisseef, V., Wolff, P. and Naudou, J. (1975) *History of Mankind*, Vol. 3: *The Great Medieval Civilisations*. Translated from the French. George Allen & Unwin/UNESCO, London/Paris.
Wustenfeld, F. (1840) *Geschichte der Arabischen Ärzte und Naturforscher*. Vandenhoeck und Ruprecht, Göttingen, Germany.
Zargari, A. (2014) *Medicinal Plants*, 5 vols. Tehran University Press, Tehran (in Persian).
Zaveri, M., Khandhar, A., Patel, S. and Patel, A. (2010) Chemistry and pharmacology of *Piper longum* L. *International Journal of Pharmaceutical Sciences Review and Research* 5, 67–76.

4 Nervous System Disorders Including Different Forms of Stresses and Alzheimer's Disease

Zahra Ayati[1,2*], Maryam Nikoosokhan[3], and Seyed Ahmad Emami[1,4]
[1]*Department of Traditional Pharmacy, School of Pharmacy, Mashhad University of Medical Sciences, Mashhad, Iran;* [2]*NICM Heath Research Institute, Western Sydney University, Westmead, Australia;* [3]*Science and Technology Park of the University of Tehran, College of Agriculture & Natural Resources, University of Tehran, Karaj, Iran;* [4]*Department of Pharmacognosy, School of Pharmacy, Mashhad University of Medical Sciences, Mashhad, Iran*

Abstract
Neuro-system conditions including neurodegenerative and neuropsychiatric disorders are a common and growing cause of mortality and morbidity worldwide. The currently available pharmaceuticals are of limited efficacy and associated with a range of adverse side effects. Traditional Persian Medicine (TPM) including Persian herbal medicine, which has been used for various neuro-system disorders over centuries, can offer alternative viable options. Moreover, Persian herbs can serve as a potential and rich source for identifying novel pharmacologic interventions. In this chapter, some of the commonly used herbal medicines for nervous system disorders mentioned in *Canon of Medicine* are discussed and the evidence of their effectiveness from preclinical and clinical studies is reviewed. Based on the current findings, these herbs have demonstrated therapeutic potential for the treatment of nervous disorders; however, more research is needed to confirm their clinical efficacy and to establish the underlying mechanisms of action.

Special Advantages of Traditional Persian Medicine in the Treatment of Nervous System Disorders

Nervous system disorders are pervasive and functionally disabling conditions that create considerable challenges for individuals and society. However, effective medicinal interventions for many nervous system disorders, such as Alzheimer's disease (AD), are generally lacking. While a range of pharmaceutical agents are currently available to alleviate these conditions, the clinical outcomes are usually limited to modest symptomatic relief and many of

*Email: z_ayati@yahoo.com.au

 Medicinal Plants Used in Traditional Persian Medicine (eds. H. Schulz, Seyed Ahmad Emami and Farsad Nadjafi)
DOI: 10.1079/9781800621671.0004

these interventions cause a range of undesirable side effects. As such, other novel treatments, as adjunctive or alternatives to existing medications, are urgently warranted for patients with neuro-system dysfunction.

Complementary and traditional medicine is a valuable source of medical knowledge that provides potentially feasible interventions which can lead to finding novel phytopharmaceuticals. Traditional Persian Medicine (TPM) has been wildly practiced in vast geographic areas including Persia and India for more than 12 centuries (Sigerist, 1987; Elgood, 2010). Combining the traditional knowledge of ancient Iran and Greece as well as archaic civilizations such as India and China, and through clinical observation and practice over many centuries, TPM has evolved into a holistic and comprehensive medical system with its own scientific theories (Gorji and Ghadiri, 2001). Herbal medicine, a key modality of TPM that uses medicinal plants in simple and compound formulations, has been used for the prevention and treatment of numerous diseases including nervous system disorders. *Canon of Medicine* is a well-known ancient literature that belongs to Avicenna, the famous philosopher and physician of the 11th century AD. This chapter will provide an overview of the scientific evidence for the commonly recommended medicinal herbs used for the management of nervous system disorders in *Canon of Medicine*. It is worth noting that TPM including *Canon of Medicine* has a holistic approach to the treatment/management of nervous system disorders and herbal therapy is often combined with lifestyle modification (such as diet therapy) and psychotherapy (Ahmadian-Attari *et al.*, 2014).

The bioenergetic basis for neurotransmission and energy consumption specific to neuronal functioning is mostly provided by deriving approximately 20% from glucose. Most communication between neurons by synapse transmission is a highly energy-demanding process. Unlike the skeletal muscles, the brain is always highly metabolically active and has minimal energy reserves, and is therefore dependent on the regular supply of energy substrates from the circulation through the blood–brain barrier. When glucose, the major energy fuel for the brain, is reduced, the brain can use alternative energy substrates such as lactate, medium-chain triglycerides, and ketone bodies. As the brain continuously consumes a considerable amount of energy, even in the absence of any particular task, it means that intrinsic electrochemical activity occurs continually, even during sleep and mental work, thus only adding a small increment (less than 5%) to the basal neural activity. Under sustained, intense mental processing, a measurable decrease in circulating levels of glucose and oxygen takes place. The brain consumes approximately 3.5 ml of oxygen per minute and per 100 g of tissue. When oxygen delivery is damaged, not only cognitive and physical performance decline, but also fatigue and tiredness are perceived. Fatigue and perceived psychological or cognitive difficulties are faced. Therefore, a selection of balanced foods and medicinal plants with hot and cold temperament based on TPM has several effects on some metabolic parameters and may play an important role in modifying brain function and/or delaying the onset of different types of dementia, particularly among those who are at high risk for the disease (Tardy *et al.*, 2020).

From the viewpoint of Avicenna, medical rationalism is an important part of medicine. From the 1990s, medical practice based on observation and experience became more important. The rationalism of Persian physicians resulted from an experimental-based educational method and clinical experiences in different historical periods passed from generation to generation. Rationalism as a complementary method can further improve the performance of physicians rather than relying solely on evidence from randomized controlled trials, meta-analyses, etc. For example, in the field of neurology, Avicenna was the first physician to differentiate between central and peripheral facial palsy by clinical examination (Ghaffari *et al.*, 2022).

The brain and nervous system disorders in the view of *Canon*

According to traditional theory, mental, psychiatric, and physical dimensions are affected by temperament as the interplay between essential body elements. A sanguine or bloody character is a person with cordiality, authoritative and dashing character, is self-confident and not depressed. A phlegmatic person has a lot of secretion in phlegm, is lazy and not active, and has slight irritability. A bilious or choleric person is associated with a fiery temper and irritability, and nervousness. A black bile character is a depressed person associated with longing depression and sadness. Furthermore, there are some distinct properties considered for brain temperament identification.

Avicenna mentioned that the brain essence is created cold to moderate. There are various theories for brain temperament creation. The first, describing the cold tendency, is the brain's willingness to endure a lot of work, which includes: (i) controlling the strong movements of body organs and many different actions and various functions in daily life that have a direct relationship with memory, thinking, and imagination; and (ii) the reactions of the five senses, which receive inspiration and orders from the brain. The soul is warm, comes from the heart to the brain, and needs to become cold. If the brain were not naturally created with a cold and moist temperament, it would suffer from excessive heat and dryness despite the huge amount of information entered into it and its processing.

The second theory says that, compared with other organs, the brain has a less cold temperament, which means it is near to a balanced temperament or a moderate temperament.

According to yet another view for brain temperament estimation, sensory actions, motor movements, thinking, and cerebration are performed in the brain, which is done with heat. According to this theory and some clinical evidence of brain temperament, the brain should be warm. It seems that nature adjuncts fat and brain tissue with each other and moves the brain away from the center of the body, placing it near the cold nature organ, like skeletal bones, to act in the direction of balancing the brain tissue (Attarfar and Ashrafzadeh, 2014).

The central nervous system (CNS), liver, and heart are mentioned as cardinal organs that need special care in both preventive and treatment approaches. The health of the CNS is affected by the healthy functions of other cardinal organs.

Most of what results in forgetfulness is due to cold and dampness, especially cold, and it may be due to brain inflammation or may be a symptom with health problems that warns of brain diseases, such as epilepsy and stroke. In addition to the principles of the traditional theory of body composition and resultant quality (temperament) therefrom, which help to know the conditions of the brain, some other indices are also considered like sensory actions, degree and quality of movement, thinking, remembrance, imagination, hallucination, the power of illusion, motor and spiritual actions, eye color and condition, and hair (considered cold to moderate), its curliness, growth rate and fall, and color (black, blonde, or red, and prone to graying). The state of sleep and wakefulness, the largeness, smallness, heaviness, and lightness of the head, smells, nasal discharge, psychic function, speech communication, and expressive style also could reveal the brain temperament and related functions. When the brain becomes cold, its optimal performance and ability are diminished. People with a cold and moist type of dementia have increased posterior nasal discharge, lacrimation, intensified deep sleep, and drowsiness. The signs of cold phlegmatic substances are perceptible cold, prolonged injury, lack of redness in the face and eyes, heavier phlegm, laziness, dullness, lethargy, and forgetfulness.

In cold and dry temperament of the brain, the most important clinical manifestations are loss of power of expression by speech, insomnia or hyposomnia, and dry nasal discharge. Some conditions are revealed during clinical examination, and they clearly show the state of the brain. Other conditions are hidden, whose action will appear in the brain later. For example, sadness and a lot of avoidance by people indicate melancholy or epilepsy, which is about to happen and will appear soon. Getting angry for no reason is followed by epilepsy, hot melancholy, or mania. Laughing without any reason is an alarm of the deterioration of the mind and the weakness of the brain. Confusion of thought may also occur in diseases of cold brain temperament. The temperament of elderly people is usually cold and dry and the temperament of the main body organs is cold. According to physiological changes during senile ages, and regarding TPM, their main organs, such as CNS, tend to be dry. However, their gastrointestinal tract tends to be cold and wet due to abnormal moisture and phlegm accumulation in it. Paying attention to mental health is necessary in older adults because they are more susceptible to stress and anxiety. The following are some indices for brain temperament identification.

Warm brain temperament is indicated by thick, dark colored, and fast-growing hair, but premature hair loss. Cold brain temperament is indicated by light colored, slow-growing, weak, and thin hair. Baldness has a direct relationship with warmness and wetness of the brain. Wet brain temperament has thin, slow-growing, light, and sparse hair. Curliness is the sign of dry hairs. Premature hair loss decreases in wet temperament.

According to Avicenna's theory, the optimal brain temperament is cold and wet. Warm and dry brain dystemperament causes insomnia. Light and short sleep is due to warm brain temperament. Severe insomnia is a sign of dry temperament. In cold brain temperament sleepiness is too much. In wet brain temperament, laziness and lethargy are excessive and deep sleep is obvious. The patients nap all the time, not having deep and pleasant sleep (Feyzabadi *et al.*, 2014).

Dementia is not a single disease but is a clinical syndrome characterized by impairment of several brain functions. Cognition skills such as thinking, memory, calculation, learning, language, and judgment change. Personality and social interactions and emotional abilities are also severely changed (Geldmacher and Whitehouse, 1996).

Alzheimer's disease or AD is the most common type of degenerative dementia and is estimated to account for 50–60% of dementia cases in persons over 65 years of age. The elderly population growing rapidly worldwide and care of the gray population causes significant psychological, social, and financial burdens. The incidence of AD is expected to accelerate over the next 30 years, especially in developing countries. There is currently no cure for most forms of dementia. Pharmacotherapy has been suggested as a useful plan for the management of AD and chronic disease, focused on symptomatic benefit and slowing disease progression (Taheri-Targhi *et al.*, 2019).

Although AD has not been defined as a distinct illness in the past, it has existed as long as the history of human beings. In 1907, Alois Alzheimer, the German physician, differentiated it from the general term "dementia" for the first time (Ahmadian-Attari *et al.*, 2020). Explanation of different types of dementia had never been linked with the concept of modern medicine before, but common symptoms such as tremors, reduced memory, and cognition destruction have always been affirmed. The most commonly recognized risk factors are midlife hypertension, low educational skills, diabetes mellitus, tobacco use, physical inactivity, obesity, unhealthy diets, and harmful use of alcohol, as well as midlife depression, social isolation, and cognitive inactivity (WHO, 2018).

Nesyan in Persian means forgetfulness and dementia is the syndrome of cognitive disorders accompanied by forgetfulness and memory problems and cognitive inability. Avicenna considered mental disorders and mental damage as *fisad-ol-zekr* (a corruption of memory), *fisad-ol-fekr* (the corruption of thinking), and *fisad-ol-takhayol* (the corruption of imagination). Generally, memory problems are categorized as the result of brain cold/moist or cold/dry dystemperament.

Fisad-ol zekr (memory corruption) is one type of memory disorder. The memory, in the posterior part of the brain, is responsible for retention, reproduction, and recognition of past experiences, which has been damaged in this case. All five senses of a person are in optimal condition, and all the things they hear and see are accurate. The objects in dreams are correct and real-world, but the problem is that everything that has been seen and heard does not remain in mind. In cold and dry brain temperament, the patient is sleepless, remembers events in the past but cannot recall recent events. If brain temperament becomes cold and wet, the person experiences moderate or excessive forgetfulness, the predominance of sleep, and can identify past events but cannot keep and recollect current events to mind. If healthy individuals suffer from forgetfulness, there is a risk for severe brain disorders like epilepsy, stroke, and meningitis.

Another type of memory disorder is *fisad-ol-fekr* (thinking corruption). In this kind of cognition disorder, an inability to distinguish between wrong and right exists, so patients of this type are incapable of distinguishing between appropriate and inappropriate behaviors. As a consequence they often fail to explain exactly what they have seen or heard.

To be distinguished from this is *fisad-ol-takhayol* (imagination corruption). Here, patients are unable to fully process the impressions they perceive, they sleep less, and forget very quickly. Avicenna believed that this could be the result of brain atrophy.

Melancholia in the View of *Canon*

Depressive and anxiety disorders ranked among the top 25 leading causes of burden worldwide in 2019. This burden of the two most disabling mental disorders was high across the entire lifespan, for both sexes, and across many locations. More interestingly, since 2019 there has been no reduction in the global prevalence or burden of either disorder (Santomauro, 2021).

Neuropsychiatric disorders are discussed precisely in the *Canon of Medicine*. Melancholia means "blackness of the bile", from (*melan-*, "black, dark, murky") and (*kholḗ*, "bile"), which comes from the ancient Greek word, *melankholía*.

Today, melancholia describes unspecified major depressive episodes (MDE), defined by the Diagnostic and Statistical Manual of Mental Disorders Text Revision Fourth Edition (DSM-IV-TR) as a mild form of depression today. Pure melancholics and atypical depressives take an intermediate position, being very similar in many characteristics. Therefore, this means that depressive disorders and melancholia have different types (Angst *et al.*, 2007).

At the first stage of melancholia, suspicion, fear with no reason, and anger are indicated. Typical clinical manifestations are guilty feelings, sadness, sleep disturbance, agitation or retardation, irritation, anorexia, and weight loss.

Avicenna wrote more than 23 books in the field of psychology and was the first physician who introduced the concept of *hallucinations* and *delusions* (Yarhosseini and Arbabi, 2019).

Even though Avicenna lived in the Middle Ages, he avoided introducing slang or seemingly scientific beliefs in explaining psychiatric issues. In the case of *delirium*, he mentioned the disease crisis in a person who wants to pick up something that does not exist, Avicenna believed this phase is dangerous and prone to the patient dying. According to traditional medicine, the melancholic disorder results from excess black bile in the brain due to moist burning in this organ. In the simple cold and wet brain distemperament, black bile humor is produced systematically and spread all over the body, it enters the brain and makes the temperament of the brain pathologic. When black bile is formed in the brain, due to the material cold and damp abnormal temperament, in addition to the signs of simple upset, there are symptoms due exclusively to the humor. In melancholic dry and cold brain detuning, there is no thirst, but dryness and a stinging sensation in the nose, and insomnia is more pronounced (Salmannegad *et al.*, 2016).

Another type of melancholia is *maraqi*, where black bile accumulates in organs like the stomach, liver, spleen, and *maraq* (in TPM, the extraperitoneal layer of the abdomen is referred to as *maraq*). By the accumulation of black bile in any of these organs that the body organ cannot expel, flatulence occurs in *maraq* (therefore, this disorder is known as *melancholia-l-maraqi*), while primary substances exist in the brain (Yousofpour, 2015).

Pathologic black bile in melancholia sometimes comes from the phlegm or blood humor that had been converted to normal black bile humor.

Mania is a complicated type of melancholic disorder due to *ehteragh* or black bile burning. In the view of Avicenna, mania manifests as anxiety, immorality, uncontrolled physical actions, and aggression, but melancholia is accompanied by fear, corruptive thinking, and problem-concentrating, and they are usually pessimistic.

Avicenna was the first physician who extensively explained the clinical description and appropriate treatment of the "love disorder" known as *adjustment disorder* today. Clinical descriptions are weak, ill, and disabled patients with prolonged fever. When the beloved one arrives, the patient suddenly returns to life. Adjustment disorders are characterized by an emotional response to a stressful event. Typically, the stressor involves financial issues, a medical illness, or a relationship problem, and the symptom complex may involve anxious or depressive affect or may present with a disturbance of conduct (Sadock *et al.*, 2017).

Depression and the heart in the view of *Canon*

According to the *Canon of Medicine*, the various mental and emotional states such as fear, sadness, and anger emanate mainly from the heart and the brain. Avicenna was also the first physician to explain the role of the heart in melancholia and to assume the origin of the spirit there. Thus he postulated the following relationships: when the heart temperament changes and becomes *sodawi* due to black bile, the temperament of the vital spirit in the heart is also changed. It reaches the brain, and the psychic soul is altered in the brain after that, making it melancholic. Abnormal humor in the brain reaches the heart through the vessels and raises an abnormal heart rate, causing sadness, wrong thinking, and depression. Therefore, Avicenna believed that it would be impossible to treat melancholia without considering participation of the heart.

In the book by Avicenna on cardiovascular drugs, *Kitab al-Adviyt-ol-Qalbiye*, which was translated into Latin under the title *Medicamenta cordialia*, drugs associated with psychiatric and cardiovascular diseases are mentioned for the first time. Although main prescriptions related are cardiologic drugs, they also have a meaningful role in the treatment of depression (Khodaei *et al.*, 2017).

Non-pharmacologic depression treatment in *Canon*

Many years ago, Avicenna recognized that lifestyle change was an important factor in reducing sources of inflammation and the risk of depression. Inflammation and free radicals are mediating pathways for both risk and neuroprogression in depression (Berk *et al.*, 2013).

Melancholic patients are mostly saddened by the events that happen to them or are afraid of something for no reason. In any case, preventing them from thinking by keeping them engaged in work is the main cure for this disease. Demanding work, staying awake, sorrow, fear, and anger prone the person to melancholia. A melancholic patient should have some entertainment, whatever it is. In all the stages necessary to create joy and happiness, other treatment plans include hearing soft and pleasant sounds, seeing beautiful sights, being in fresh and moderate air, humidifying the air in the house, wearing soft and cotton clothes, showering before meals with moderately warm water for relaxation, eating and drinking delightful foods having the wet and mild warm temperament like sweet and greasy foods, and always smelling pleasant essential oils and scents that today is known as *aromatherapy*. Such interventions amuse the patient and prevent them becoming engaged in deep thoughts, and reinforce the brain. Furthermore, sleep is one of the most evolving treatments for melancholic patients. Avicenna mentioned that pleasure results from understanding the perfections of the five senses. They are obtained from inner satisfaction, such as the feeling of sweetness, smelling pleasant for the sense of taste and smell, and feeling revenge for the power of anger. Nothing is worse than loneliness for these patients because they soon start thinking about unforeseen events characterized by hopelessness and sadness.

Pharmacologic depression treatment in *Canon*

Currently, a useful first-line therapy for the treatment of depression is pharmacotherapy or psychotherapy, as indicated in clinical guidelines. Various types of psychosocial stressors can stimulate the pro-inflammatory cytokine network and increase levels of interleukin (IL)-6 and tumor necrosis factor-α (TNF-α). Depression is usually associated with a chronic low-grade inflammatory response, activation of cell-mediated immunity and the compensatory anti-inflammatory reflex system (CIRS), which is characterized by negative immunoregulatory processes (Berk *et al.*, 2013).

When melancholy is due to extreme cold and dryness, medicines known as *mofarreh* or *cardiotonics* are helpful, depending on the individual temperament of the patient. Generally, any factor that reduces sadness, anxiety, anger, and fatigue improves mental faculties and keeps the mind bright, produces joy, and improves alertness (Kamaneh *et al.*, 2019).

Venesection is also useful when black bile spreads throughout the body, and the effect is significant when black bile "burns" (*ehteragh*). *Larj* as a strong laxative and *Itrifel* with *Cuscuta epithymum* have been extensively tested by Avicenna in various practical applications. The use of liquid dosage forms and warm tempered medicinal oils with energizing abilities such as dill and chamomile are also recommended in this context. They strengthen the head and make the patient feel good. The combination of rose water and saffron, known as "medicinal jollab", is the famous traditional syrup in TPM. Rose water and essential oil of saffron, in the form of syrup, are quickly and effectively absorbed through the mucous membrane of the tongue and may be one of the most effective treatments for depression. It also acts as a cardiotonic according to Avicenna's treatment approach for depression. This drink could also be classified as a functional drink (nutraceutical) today and marketed accordingly. Nutrition in general is known to contribute to psychopathology, especially in the two most common mental disorders (depression and anxiety), as it affects immune function and the level of systemic inflammation, which in turn predisposes to depression.

If the discharges have been blocked from menstruation or through the bowels, these discharges should flow and be expelled. Evacuation should be done once a week using softening laxatives, which are medium and mild. *Nymphaea alba*, *Cyperus rotondus*, *Terminalia chebula*, *Echium amoenum*, *Centaurea behen*, *Nepeta menthoides*, *Lavandula angustifolia*, *Citrus medica*, *Ocimum basilicum*, *Crocus sativus*, *Salix aegyptiaca*, *Melissa officinalis*, *Cinnamomum zeylanicum*, *Curcuma zedoaria*, *Cinnamomum cassia*, *Sedum spurium*, and *Rosa* × *damascena* are among the most used medicinal plants in formulations for melancholia treatment.

Herbal Medicines for Neuro-System Health

Herbal medicines have been used in TPM for thousands of years in the form of mono- or augmentation therapies to promote nervous system health and treat various nervous system disorders such as dementia, epilepsy, migraine, and Parkinson's disease. In the *Canon of Medicine*, various herbal plants such as *Piper nigrum*, *Brassica nigra*, *Zingiber officinale*, *Cinnamomum verum*, *Boswellia* sp., and *Crocus sativus* are recommended for nervous system health.[1] Tables 4.1 and Table 4.2 provide an overview of the main bioactive constituents, their pharmacological activity, and the mechanisms of action of these herbs.

Brassica nigra seed (black mustard) has been used in therapeutic preparations for centuries in different parts of the world and is recommended in TPM for the treatment of various neurological disorders (Mehri and Shojaii, 2013). *B. nigra* possesses cognitive-enhancing, anti-Parkinson, and antimigraine properties. Its key constituents include phenolic compounds, glucosinolates, flavonoids, and brassinosteroids (Kumar and Andy, 2012). The neuroprotective effects of *B. nigra* were revealed in several studies. The intraperitoneal administration of hydroalcoholic extract of *B. nigra* seed

Table 4.1. Scientific and traditional names, plant parts, and preparations of commonly used herbs in *Canon of Medicine* for treatment of neuro-system disorders.

Scientific name	English name	Arabic name	Family	Plant part	Route of administration	Main indications
Brassica nigra	Mustard	*Khardal*	*Brassicaceae*	Seeds, oil	Oral, topical (oil, *zemad*[a], *komad*[b])	Dementia, migraine, Parkinson's disease, stroke
Cinnamomum verum	Cinnamon	*Dārṣini*	*Lauraceae*	Bark	Oral (*ma'joon*[c], decoction)	Dementia, Parkinson's disease
Crocus sativus	Saffron	*Za'farān*	*Iridaceae*	Stigma	Oral (tablet and *ma'joon*), smoke, *zemad*	Paralysis, dementia, depression
Lavandula stoechas	French lavender	*Usṭukhuddūs*	*Lamiaceae*	Aerial parts	Oral (decoction), inhalation	Dementia, migraine, Bell's palsy, epilepsy, stroke, anxiety, depression
Melissa officinalis	Lemon balm	*Bādranjbūyah*	*Lamiaceae*	Leaves	Oral (infusion, *ma'joon*)	Dementia, anxiety, OCD
Phyllanthus emblica	Emblic, amla	*Āmulaj*	*Phyllanthaceae*	Fruit	Oral (*ma'joon*)	Dementia
Piper nigrum	Black pepper	*Folfol aswad*	*Piperaceae*	Fruit	Oral (*ma'joon*), topical (*zemad*), sneezing agent	Bell's palsy, paralysis, dementia, migraine
Syzygium aromaticum	Clove	*Qaranful*	*Myrtaceae*	Flower, bud	Oral (*ma'joon*), topical (*zemad*, *komad*)	Paralysis, dementia
Terminalia chebula	Black myrobalan	*Halilaj*	*Combretaceae*	Fruit	Oral (jam, sachet, and *ma'joon*)	Bell's palsy, paralysis, dementia, epilepsy, stroke
Zingiber officinale	Ginger	*Zanjibīl*	*Zingiberaceae*	Rhizome	Oral (jam and *ma'joon*), *komad*	Bell's palsy, dementia, paralysis

[a]Semisolid traditional topical dosage form.
[b]A warm mixture of plants for topical administration.
[c]Combined traditional paste.

Table 4.2. Pharmacologic actions and mechanisms of action of commonly used herbs in *Canon of Medicine* for neuroprotective protection.

Scientific name	Possible neuro-bioactive compound(s)	Pharmacologic action	Suggested mechanism(s)
Brassica nigra	Phenolic compounds (quercetin, gallic acid, ferulic acid, rutin, caffeic acid), phytic acid	Memory improvement, analgesic effect, anti-Parkinson	DPPH and OH radical scavenging, protease inhibition
Cinnamomum verum	Proanthocyanidin, cinnamaldehyde	Memory improvement, anti-Parkinson	↓ NO, inhibiting amyloid fibril formation, antioxidant, anti-cholinesterase, ↓ MAO, upregulation of Parkin and DJ-1
Crocus sativus	Crocin, safranal, crocetin	Memory improvement	↓ NO level and caspase-3 activity, ↓ oxidative stress, anti-inflammatory effect, inhibiting Aβ fibrillogenesis, inhibiting AChE
		Antidepressant, anxiolytic, anti-OCD	↑ Reuptake inhibition of monoamines (norepinephrine, dopamine, serotonin), GABA-α agonism, NMDA receptor antagonism
Lavandula stoechas	Linalool, linalyl acetate, perillyl alcohol, 1,8-cineole, aromatic phenol	Memory improvement, anticonvulsant	Antioxidant, antiapoptotic activities, ↑ VEGF expression
		Antidepressant, anxiolytic	GABAergic and cholinergic modulation, ↑ dopamine
Melissa officinalis	Flavonoids, volatile oils, triterpenes	Anti-dementia, anticonvulsant	Stimulation of acetylcholine receptors, AChE inhibitory activity, inhibition of MMP-2, anti-inflammatory, antioxidant
		Anxiolytic, antidepressant, anti-insomnia	↓ MAO, ↑ GABA
Phyllanthus emblica	Ascorbic acid, emblicanin A, emblicanin B	Anti-dementia, antiepilepsy	AChE inhibitory activity, antioxidant, anti-inflammatory, attenuating amyloid pathologies
Piper nigrum	Piperine, piperettine, piperettyline	Anti-dementia	DPPH radical-scavenging activity, Aβ aggregation inhibition, anti-cholinesterase, NMDA blocker
Syzygium aromaticum	Eugenol	Neuroprotective, memory improvement, antinociceptive	↓ NO production, ↓ IL-1β, TNF-α, and PGE_2, induces BDNF, inhibition of Aβ, AChE inhibitory
Terminalia chebula	Ascorbic acid, quercetin, tannins	Anti-dementia, anticonvulsant, anti-Parkinson	Antioxidant, anti-inflammatory, AChE inhibitory
Zingiber officinale	6-Shogaol	Anti-dementia, anti-neuroinflammation	Inhibits level of TNF-α, IL-1β, COX-2, NO, iNOS, PGE_2, and NF-κB, stimulates TrK-mediated neurites

VEGF, vascular endothelial growth factor; MMP-2, matrix metalloproteinase-2; TrK, receptor tyrosine kinase.

(75, 150, and 300 mg/kg) considerably reduced the duration and intensity of seizures in mice models of pentylenetetrazol-induced kindling (Kiasalari *et al.*, 2012; Al-Snafi, 2015). Another study evaluated the anti-Parkinson effect of orally administered *B. nigra* (400 mg/kg) in chlorpromazine-induced catalepsy in rats, an accepted Parkinson's disease model (catalepsy was measured using the actophotometer method). Pretreatment with *B. nigra* extract significantly reduced cataleptic scores, increased glutathione (GSH) level, and decreased thiobarbituric acid-reactive substances (TBARS) (an indicator of lipid peroxidation extent) (Paradkar and Sarawade, 2015). The methanol extract of *B. nigra* demonstrated considerable radical scavenging and antioxidant activities (Rajamurugan *et al.*, 2012). *B. nigra* is suggested to have considerable anti-inflammatory effects. The ethanol extract of its leaves (500 mg/kg) significantly inhibited inflammation in carrageenan-induced paw edema in rats. It also inhibited protease in a dose–response pattern (Alam *et al.*, 2011). The anti-Parkinsonism and cognitive-enhancing effects of *B. nigra* are attributed to its high antioxidant and anti-inflammatory properties; however, the exact mechanisms of action underlying these effects are yet to be studied in more detail.

In *Canon of Medicine*, *Cinnamomum verum* (cinnamon) is a brain tonic agent and is used in complex formulations for the treatment of dementia. Cinnamon includes a variety of phytochemicals such as cinnamyl acetate, cinnamaldehyde, eugenol, and cinnamyl alcohol. These chemicals may be responsible for a number of neuroprotective properties such as antioxidant and anti-neuroinflammatory effects (Essa *et al.*, 2012). More recently, cinnamon has been recommended to possess anti-Parkinson properties (Khasnavis and Pahan, 2012, 2014; Shaltiel-Karyo *et al.*, 2012). Preclinical studies suggest that the anti-Parkinson effects of cinnamon are based on enhancing sodium benzoate (NaB) levels in blood serum and the brain. The elevated level of NaB protects Parkin and DJ-1 in astrocytes by reducing the production of nitric oxide (NO) and expression of inducible nitric oxide synthase (iNOS), which are potentially toxic to neurons (Khasnavis and Pahan, 2014).

Findings of animal studies also demonstrate the potential ability of cinnamon to improve memory function (Mesripour *et al.*, 2016). Proposed mechanisms include inhibiting malondialdehyde (MDA) production (Keshvari *et al.*, 2013), radical scavenging, and inhibiting the tau aggregation and filament formation associated with AD (Peterson *et al.*, 2009). In addition, cinnamon has been shown to modulate endothelial functions and attenuate the vascular cell adhesion molecules (Peterson *et al.*, 2009). Considering the results of preclinical studies on cinnamon, together with its safety profile, randomized controlled trials (RCTs) are warranted to determine the effectiveness and safety of cinnamon in patients with Parkinson's disease and/or dementia.

Crocus sativus (saffron) stigma is a well-known herb in medieval Persia for its neuromodulatory properties; specifically as an antidepressant (Rahimi *et al.*, 2017). The neuroprotective and psychoactive properties of saffron and its major bioactive components (including crocin, safranal, and crocetin) have been demonstrated in numerous preclinical studies (Linardaki *et al.*, 2013; Soeda *et al.*, 2016). The antidepressant and anxiolytic effects of saffron were confirmed by several human clinical trials (Hausenblas *et al.*, 2013; Mazidi *et al.*, 2016). Some animal and human studies also indicate that saffron and crocin could improve obsessive-compulsive disorder (OCD) (Georgiadou *et al.*, 2012; Esalatmanesh *et al.*, 2017). The neuropsychiatric effects of saffron purportedly act by inhibiting the reuptake of some neurotransmitters such as dopamine, norepinephrine, and serotonin (Sarris *et al.*, 2013).

More recently, saffron has been suggested to possess potential anti-dementia effects, confirming its traditional claims. Saffron was shown to significantly improve learning and memory function in an animal model of dementia (Papandreou *et al.*, 2011). A 1-year single-blind clinical study in patients with mild cognitive impairment (MCI) revealed a significant improvement in Mini-Mental State

Examination (MMSE) scores with the daily administration of saffron (125 mg/day) compared to control (Tsolaki *et al.*, 2016). An RCT in AD patients revealed that a 16-week treatment with saffron significantly improved Alzheimer's Disease Assessment Scale-cognitive subscale (ADAS-cog) and Clinical Dementia Rating Scale Sum of Boxes (CDR-SB) compared to placebo (Akhondzadeh *et al.*, 2010a). The results of a 22-week double-blind RCT using saffron extract (15 mg twice daily) versus donepezil (cholinesterase inhibitor) (5 mg twice daily) in AD patients demonstrated that both treatments significantly improved cognitive function (measured by ADAS-cog and CDR-SB), but saffron appeared to possess a better safety profile (Akhondzadeh *et al.*, 2010b). In a 1-year double-blind RCT in patients with moderate to severe AD, saffron extract (30 mg daily) was compared with memantine (an *N*-methyl-D-aspartate (NMDA) receptor antagonist) (20 mg daily). The results demonstrate that saffron significantly improved cognitive function and overall functional activity comparable to that of memantine (Farokhnia *et al.*, 2014).

While the exact mechanisms underlying the anti-dementia effect of saffron are still unclear, it is thought to act through several pharmacologic pathways. Saffron and crocetin have been shown to possess anti-inflammatory effects and also to exert strong protection in repressing reactive oxygen species (ROS) production and decreasing caspase-3 activation (Poma *et al.*, 2012; Rahaiee *et al.*, 2015). Furthermore, some studies reported that crocin, a main carotenoid component of saffron, has inhibitory properties against amyloid-β (Aβ) fibrillogenesis formation (Papandreou *et al.*, 2006; Kumar *et al.*, 2012; Ghahghaei *et al.*, 2013). Additionally, saffron extract provides a moderate inhibitory activity (up to 30%) on acetylcholinesterase (AChE) and inhibits acetylcholine breakdown, which is one of the main therapeutic targets for AD. The main anti-AChE constituents of saffron are known to be crocetin and dimethylcrocetin (Geromichalos *et al.*, 2012).

Lavandula stoechas (French lavender) is native to the Mediterranean region and is one of the most studied herbs (Gorji and Ghadiri, 2002). Lavender is commonly combined with other herbs in complex formulations for the treatment of cognitive dysfunction and psychologic disorders such as anxiety and OCD (Vakili and Gorji, 2006). Recently, two of *Lavandula* species (*L. stoechas* and *L. angustifolia*) have drawn increasing interest for a number of neurodegenerative and neuropsychiatric disorders. Some experimental studies in rats showed that inhaled lavender oil has neuroprotective activity against scopolamine-induced dementia and improves spatial memory performance in different maze tasks, suggesting the potential positive effect on memory function (Hancianu *et al.*, 2012; Hritcu *et al.*, 2012). In an *in vivo* study of Aβ-induced AD model in rats, administration of aqueous extract of lavender (200 mg/kg) for 20 days significantly enhanced the learning of the maze task (Soheili *et al.*, 2014). Lavender oil has been recommended for the treatment of agitation in dementia. In a small (**n** = 15) placebo-controlled clinical study in patients with severe dementia, inhalation of a 2% lavender oil showed partial improvement in the majority of patients (Holmes *et al.*, 2002). In a short-term clinical study in healthy and young volunteers, inhalation of lavender oil significantly increased visual memory when compared to the control group (Filiptsova *et al.*, 2018).

The key bioactive components of lavender include linalool, linalyl acetate, aromatic phenol, perillyl alcohol, and 1,8-cineole. Some *in vivo* psychopharmacologic studies showed that linalool has considerable sedative effects on the CNS, including hypnotic, hypothermic, and anticonvulsant properties (Elisabetsky *et al.*, 1995). The modifying effect of linalool on the glutamatergic system is comparable with phenobarbital, a known anticonvulsant. Linalool has been shown to be able to bind to the glutamate receptors and therefore antagonize cyclic adenosine monophosphate (cAMP) regulatory mechanism (Aprotosoaie *et al.*, 2014). Also, some *in vitro* studies showed the potential antioxidant activity of the phenol fraction of lavender, remarking upon its concentration-dependent inhibition of lipid peroxidation (Denner, 2009).

Clinical studies demonstrate that lavender is potentially beneficial in some mood disorders such as depression and anxiety. In an RCT, patients undergoing diagnostic curettage were randomized to receive the inhalation of essential oil of lavender or control. The results showed a significant reduction in anxiety in the lavender-treated group compared to control (Bakhsha *et al.*, 2014). In another clinical trial, oral administration of lavender oil (once daily, 80 mg/day) reduced anxiety levels, improved anxiety associated symptoms such as somatic complaints, restlessness, and disturbed sleep, and increased general well-being. Lavender oil has the potential to modulate GABAergic and cholinergic neurotransmission. It was also found to increase dopaminergic activity by enhancing dopamine receptors (subtype D3) (Koulivand *et al.*, 2013).

Melissa officinalis (lemon balm) aerial parts are commonly used by TPM physicians for the treatment of several CNS disorders including dementia. Moreover, it is highly recommended for the treatment of mood disorders such as anxiety, depression, and OCD (Shakeri *et al.*, 2016). The neuroprotective effect of lemon balm has been evaluated in a number of preclinical and clinical studies. In an *in vitro* study, pretreatment with a methanolic extract of lemon balm revealed the significant neuroprotective effect in rat pheochromocytoma cells (PC12) against hydrogen peroxide toxicity in the 3-(4,5-dimethylthiazol-2-yl)-2,5-diphenyltetrazolium bromide (MTT) and lactate dehydrogenase (LDH) assays (López *et al.*, 2009). Lemon balm extracts contain phenolic compounds such as phenolic acids and flavonoids that are responsible for its antioxidant and free radical scavenging activities (Kamdem *et al.*, 2013; Mabrouki *et al.*, 2018).

Lemon balm has also been associated with memory-enhancing activities in different studies. Administration of lemon balm extract (200 mg/kg) significantly improved memory and learning processes in rats and ameliorated scopolamine-induced learning deficits (Soodi *et al.*, 2014). In a randomized, placebo-controlled, double-blind, crossover study, 20 healthy and young participants (between 18 and 22 years old) were randomly allocated to single doses of lemon balm dried leaf (300, 600, and 900 mg) or placebo with a washout interval of 7 days. The results demonstrated that lemon balm could modulate cognitive performance and mood in a dose-dependent manner (Kennedy *et al.*, 2002). In another double-blind clinical trial, 35 patients with mild-to-moderate AD were randomized to receive standardized (500 μg citral/ml) lemon balm extract or placebo (60 drops/day). After 16 weeks, the patients in the lemon balm group demonstrated significantly better cognitive performance than the placebo group (Akhondzadeh *et al.*, 2003).

The therapeutic effects of lemon balm on neuropsychiatric disorders have also been demonstrated in several preclinical and clinical studies. Rosmarinic acid, oleanolic acid, and ursolic acid, some of the known active constituents of lemon balm, have been shown to inhibit GABA transaminase, which is an enzyme responsible for γ-aminobutyric acid (GABA) degradation (Awad *et al.*, 2009). Moreover, lemon balm has been demonstrated to elevate GABA levels and inhibit monoamine oxidase (MAO) in mice. In an experimental study in mice, lemon balm extract (50 or 200 mg/kg) was shown to reduce corticosterone levels in the brain and serum (Yoo *et al.*, 2011). In human investigations, a double-blind RCT in 100 young females reported that a 3-month treatment with lemon balm extract (1200 mg/day, from the first day to last day of the menstrual cycle) significantly reduced stress and intensity associated with premenstrual syndrome (PMS) (Akbarzadeh *et al.*, 2015). A pilot trial on stressed participants with mild-to-moderate anxiety disorders and sleep disturbances showed a significant (15–42%) reduction in insomnia and anxiety symptoms with Cyracos® (lemon balm standardized extract) (600 mg/day) over 15 days (Cases *et al.*, 2011).

The fruit of *Phyllanthus emblica* is one of the most recommended herbs in TPM for brain health, especially for improving memory and cognition. It contains a wide range of phytochemicals including alkaloids, flavonoids, tannins, and terpenoids. *P. emblica* has shown to enhance memory function in various animal models as well as in some clinical studies. In an *in vivo* study in young and aged rats, oral administration of *P. emblica*

(50, 100, and 200 mg/kg) demonstrated significant improvements in memory scores over 15 days in a dose–response manner and reversed scopolamine- and diazepam-induced amnesia (Vasudevan and Parle, 2007). Pretreatment with *P. emblica* hydroalcoholic extract (500 and 700 mg/kg) was studied in kainic acid-induced seizures in rats. The treatment significantly increased the latency of seizure compared with vehicle, prevented the oxidative stress parameter TBARS, and ameliorated the fall in GSH. It also improved the cognitive deficit caused by kainic acid, suggesting a relationship between antioxidant activity and the memory-enhancing property of *P. emblica* (Golechha *et al.*, 2011). The aqueous extract of the fruit showed a significant antioxidant property through reducing lipid peroxidation (Naik *et al.*, 2005). Ethyl acetate extract of *P. emblica* showed significant NO radical scavenging in an *in vitro* study (Kumaran and Karunakaran, 2006). The antioxidant activity of *P. emblica* is suggested to be due to its high content of ascorbic acid, total tannins, and total phenols (Scartezzini *et al.*, 2006; Poltanov *et al.*, 2009).

A number of studies demonstrated that the anti-inflammatory activity of *P. emblica* may be related to its neuroprotective properties. In rats, oral administration of the water extract of *P. emblica* at different doses significantly inhibited carrageenan-induced paw edema in a dose-dependent manner (Jaijoy *et al.*, 2010). It has also been demonstrated that *P. emblica* extract effectively suppressed the pro-inflammatory cytokines (Wang *et al.*, 2017). In mice model of scopolamine-induced amnesia, intraperitoneal administration of hydroalcoholic extract of *P. emblica* (300, 450, and 600 mg/kg) reversed the rise of brain AChE level and improved memory dysfunction induced by scopolamine (Golechha *et al.*, 2012). Moreover, oral administration of various doses of tannoid principles of *P. emblica* showed cognitive improvement and alleviated amyloid pathologies against the aluminum chloride ($AlCl_3$)-induced model of AD in rats. This activity is suggested to be related to chelating properties of emblicanin A and emblicanin B (Thenmozhi *et al.*, 2016).

Piper nigrum fruit (black pepper) is among the most popular spices in the world and is suggested for the treatment of various CNS disorders, usually in combined formulations. It contains a variety of phytochemicals such as flavonoids, phenolics, alkaloids, amides, and steroids. The main alkaloid of *P. nigrum*, piperine (Wang *et al.*, 2017), possesses different neuropharmacologic actions including free radical scavenging, anti-neuroinflammatory, anticholinergic, and MAO inhibitor effects, and is suggested as one of the future plant-based pharmaceutical medicines for the treatment of Parkinson's disease (Sharma, 2019). Piperin in different doses significantly improved spatial memory and neurodegeneration in a rat model of AD (Chonpathompikunlert *et al.*, 2010). It was also found to have antidepressant-like activity in a corticosterone-induced depression model in mice (Damanhouri and Ahmad, 2014). Furthermore, oral administration of the methanol extract of *P. nigrum* fruits (50 and 100 mg/kg) ameliorated Aβ-induced spatial memory impairment in a rat model of AD (Hritcu *et al.*, 2014). Another study in a rat model of AD showed that *P. nigrum* could prevent neurodegeneration and alleviate the neuropsychologic symptoms that are associated with AD (Subedee *et al.*, 2015). Owing to its anti-inflammatory, antioxidant, and anti-cholinesterase activities, *P. nigrum* has been suggested to be a useful herb for the treatment of several neurodegenerative disorders (Tu *et al.*, 2016; Wang *et al.*, 2017).

Syzygium aromaticum (clove) is well-known for its neuroprotective effects both in monotherapy and combination therapy (Rahimi *et al.* 2017). Findings of animal studies suggest that *S. aromaticum*, and its bioactive constituent eugenol, may benefit several neurologic disorders. In mice, intraperitoneal administration of different clove oil doses significantly reversed scopolamine-induced memory impairment (Halder *et al.*, 2012). In another study, the dichloromethane extract of clove fruits showed AChE inhibitory activity (Adhami *et al.*, 2011), which may be responsible for its cognitive-enhancing effects. Eugenol, the main constituent of clove oil, has shown antinociceptive activity in several pain models via opioid and α-adrenergic receptors (Park, S.H. *et al.*, 2011). Eugenol has also been reported to have anticonvulsant activity (Dallmeier-Zelger *et al.*, 1983). In an experimental study, eugenol (100–300 μM) significantly reduced the oxidative neural injury (Wie *et al.*, 1997). The neuroprotective effects of eugenol can be attributed to a number of mechanisms. It has been shown to act as a pro-oxidant and antioxidant, which leads to free radical scavenging. It also exhibits anti-inflammatory properties by inhibiting the release of TNF-α, IL-1β, and prostaglandin E_2 (PGE_2) from lipopolysaccharide (LPS)-stimulated macrophages (Anuj and Sanjay, 2010). The principal mechanism responsible for the antiepileptic activity of eugenol might be its ability to modulate synaptic transmissions (Müller *et al.*, 2006).

Terminalia chebula fruit (myrobalan) is one of the most popular Persian herbs used to improve memory function. The fruit of *T. chebula* contains antioxidant ingredients, including ascorbic acid and quercetin, which are effective against oxidative stress-induced neurodegeneration (Jokar *et al.*, 2016). It also contains a high content of tannins (32–45%) including chebulic acid, chebulinic acid, punicalagin, gallic acid, ellagic acid, and tannic acid. *T. chebula* is suggested to have neuroprotective activities including anticonvulsant and anti-dementia properties. One *in vivo* study was conducted to evaluate the effect of *T. chebula* on seizures and seizure-induced cognitive impairment in rats. Hydroalcoholic extract of *T. chebula* (1000 mg/kg) showed 83.33% protection in maximal electroshock (MES) and 66.66% protection in pentylenetetrazole (PTZ)-induced seizures. It also attenuated seizure-induced cognitive impairment (Kumar *et al.*, 2018). Oral administration of *T. chebula* seed extract (100 mg/day) showed a significant neuroprotection against ischemic damage in the hippocampus of gerbils over 7 days. In addition, the level of the brain-derived neurotrophic factor (BDNF) was reported to be much higher in the *T. chebula*-treated ischemia group than the vehicle-treated control group. In mice, a 14-day oral administration of *T. chebula* hydroalcoholic extract (100 and 200 mg/kg) significantly reversed scopolamine-induced amnesia (Kim *et al.*, 2018). *T. chebula* is reported by various studies to have antioxidant activities (Park, J.H. *et al.*, 2011). Chebulic acid, one of the major constituents of the fruit, is suggested as a considerable antioxidant exhibiting free radical scavenging and ferric-reducing activities in 2,2-diphenylpicrylhydrazyl (DPPH) and ferric-reducing antioxidant power (FRAP) assays (Lee *et al.*, 2007). *T. chebula* is also known to exert anti-inflammatory effects in *in vitro* and *in vivo* studies. In a study in a model of carrageenan-induced paw edema in rats, oral administration of *T. chebula* extract (250 mg/kg) caused a significant reduction in edema (69.96%). Gallotanins of *T. chebula* showed anti-inflammatory effects by inhibiting protein expression of iNOS and cyclooxygenase-2 (COX-2) in macrophage cells (Yang *et al.*, 2014). Furthermore, *T. chebula* is suggested to have AChE inhibitory activity (Sancheti *et al.*, 2010; Sulaiman *et al.*, 2012). In an *in vitro* study, methanol extract of the fruit of *T. chebula* (5 mg/ml) significantly inhibited AChE and butyrylcholinesterase (BChE) by 89% and 95%, respectively. Moreover, gallotannins and tannic acid, present in *T. chebula*, showed inhibitory activity against AChE in some *in vitro* studies (Afshari *et al.*, 2016).

Zingiber officinale (ginger) rhizome has been used for the management of AD in many Asian countries. The principal bioactive components in ginger include gingerols, shogaols, and volatile essential oil substances such as zingiberene, bisabolene, and various monoterpenes (Essa *et al.*, 2012; Semwal *et al.*, 2015). Data from various animal and clinical studies suggest that ginger possesses neuroprotective properties (Sahardi *et al.*, 2019). In an amyloid-induced AD model in rats, ginger root extract was administered intragastrically to the animals for 35 days. The results revealed that ginger (4 g/day) significantly improved spatial memory and behavioral dysfunction. It also increased the number of neurons as well as intracellular Nissl bodies in the hippocampus (Zeng *et al.*, 2013). The results of another study in rats showed a significant effect of ginger (108 and 216 mg/kg per day) in reversing the amnesia caused by $AlCl_3$. The animals that received ginger treatments exhibited improvements in behavior, T-Maze, and Rotarod tests (Karam *et al.*, 2014). In a randomized placebo-controlled clinical trial, 60 healthy middle-aged women were randomized to receive either ginger extracts 400 or 800 mg daily or placebo over a 2-month period. The efficacy of ginger on working memory and cognitive function was evaluated using recordings of the brain's electrical activity (event-related potentials) during computerized test batteries and an auditory oddball paradigm. The results showed that ginger-treated participants had larger N100 and P300 amplitudes, and decreased P300 latencies, reflecting an improvement in attention and cognitive processes compared to those that received placebo. The ginger treatment group also exhibited enhanced memory (Saenghong *et al.*, 2012).

Ginger is also recommended to improve other neurodegenerative disorders such as Huntington's disease (Sharma *et al.*, 2012) and Parkinson's disease (Park *et al.*, 2013). Neuroprotective properties of ginger can be attributed to its several pharmacologic properties such as cholinesterase inhibition, anti-inflammatory, antioxidant, and anti-amyloidogenic activities. The AChE inhibitory activity of ginger has been evaluated in an *in vitro* study in rat brain. The results showed that ginger extract inhibits the AChE activity in a dose-dependent manner (Oboh *et al.*, 2012). Some animal studies also showed that ginger produced an inhibitory effect on cholinesterase (Aryaeian and Tavakkoli, 2015), leading to acetylcholine accumulation in synapses and improved cognitive function (Essa *et al.*, 2012). The anti-inflammatory effects of ginger are also verified in a number of *in vitro* and *in vivo* studies. Ginger inhibits the activity of nuclear factor-κB (NF-κB), iNOS, and COX-2 (Aryaeian and Tavakkoli, 2015). Some animal models of dementia demonstrate that 6-shogaol, a bioactive constituent of ginger, attenuated neuroinflammation and cognitive impairment by enhancing the production and release of nerve growth factor (NGF), which has a critical function in accelerating neurite outgrowth and improving cognitive performance (Moon *et al.*, 2014). 6-Shogaol and three other shogaols from ginger were also found to protect neural cells from Aβ attacks (Kim *et al.*, 2002). Furthermore, 6-shogaol is reported to have the potential to modulate dopaminergic neurone activities (Park *et al.*, 2013). Ginger is shown to be endowed with potent antioxidant properties in *in vitro* and *in vivo* studies (Ali *et al.*, 2008). It has a strong effect in scavenging DPPH radical and controlling the peroxidation of lipids with its high antioxidant activity (Stoilova *et al.*, 2007; Mathew and Subramanian, 2014).

Ginger is considered a safe herbal medicine and has no known acute toxicity at the usual doses consumed for dietary or medicinal purposes (Ali *et al.*, 2008).

Polyherbal formulations and synergistic effects

Although single herbs are used for the treatment of various disorders in TPM, many polyherbal formulations are also prescribed based on the belief that these combinations may produce synergistic effects, leading to better clinical outcomes. *Itriphal Ustakhuddus*, *Jawarish Jalinus*, *Maᶜjoon Falasafah*, *Maᶜjoon Baladur*, and *Jawarish Nesyian* are among the most important and complex formulations used in TPM to treat brain disorders. *Maᶜjoon Falasafah* (paste for philosophers) is one formula thought to enhance memory function, it contains 17 ingredients including *Z. officinale*, *T. chebula*, *P. emblica*, *C. zeylanicum*, and *Curcuma longa*. A number of studies were conducted to evaluate the neuroprotective effects of TMP polyherbal formulations. In a 12-week RCT, a TPM polyherbal formula, commonly known as *Dawaie Lobān* and including ginger, black pepper, sweet sedge, and incense, was evaluated for its efficacy for mild-to-moderate AD. Forty-four patients were randomized to receive *Dawaie Lobān* capsules (500 mg, three times daily) or placebo for 3 months. ADAS-cog and CDR-SB were used to evaluate the individual cognitive function. The results showed a significant improvement in ADAS-cog and CDR-SB in participants who received *DaWaie Lobān* compared to the control group. It is suggested that this formula improves cognitive performance via antioxidant, anti-inflammatory, and cholinesterase inhibition (Tajadini *et al.*, 2015).

Drugs used in the treatment of nervous system disorders

The current drugs used in AD therapy have different mechanisms of action that enhance cholinergic function, reduce free radicals, or possess anti-inflammatory activity; and a miscellaneous group of natural products exists that have multifunction activity. Donepezil, galantamine, and rivastigmine are AChE inhibitors; memantine is an NMDA receptor antagonist; statins are used in the prevention of dementia. These drugs are only symptomatic, offer temporary benefits on cognitive and behavioral functions of the disease, and have modest benefits. Thus, developing new drugs to change the progression of the disease has been a priority. Recently, new drugs have been developed that target the pathologic pathways including neurotransmitter and cell-signaling agents, neuroprotective agents, glial cell modulators, modification as anti-amyloid agents, and tau-based therapies (Vaz and Samuel Silvestre, 2020).

At an early stage of disease progression, MCI is the characteristic of the disease. The long time period exists between an asymptomatic phase when AD starts and the manifestation phase at which the first clinical signs of the disease will appear (around 15–20 years). A wide range of proteomics and metabolomics pathways involved in the disease process and development could be selected as reliable biomarkers useful for early diagnosis. They could prevent patients living in an unnecessarily extended state of suffering (François *et al.*, 2022).

Many new studies and documented ancient writings indicate herbal medicines' unique role in treating AD. In a plant, many different phytochemicals can target different biological pathways in the cell. Therefore, herbal medicines, spices, and neutraceuticals have preventive roles and could modify the disease process, but they need evidence-based recommendations. Natural products can enhance NGF biosynthesis and promote secretion, inhibiting the deterioration of nerve function.

Through the past decades, natural products have played an essential role in drug discovery and development, leading to the rational design of chemical compounds to target specific molecules. In recent years, the various "-omics" technologies have allowed scientists to detail the distinct biological effects of natural compounds on the human body, revealing multi-target action, which holds much promise for the development of new rational synergistic formulations and therapies against many devastating diseases, including dementia and cancer (Ji *et al.*, 2009).

Natural products and medicinal plants from traditional medicine have been routinely used by traditional physicians. Their toxicity and safety have been tried through long-term experiences and are relatively better understood than those of other synthetic chemical entities that have no history of human use and are entering human studies for the first time (Patwardhan and Mashelkar, 2009). However, collecting plants at random requires significantly more financial resources despite the great diversity of species (Fabricant and Farnsworth, 2001).

The multi-target-directed ligands (MTDLs) approach is characteristic of traditional prescriptions and treatment plans. Compared to the "one drug/one target" approach, it is one of the most promising alternatives designed to interact with more than one target involved in the pathogenesis of AD. Abnormal accumulations of Aβ peptides, tau proteins, sympathetic and parasympathetic disturbances in the disease process, oxidative and nitrosative stress, neuroinflammation, mitochondrial damage, increased apoptosis, alterations of calcium and other metals, as well as alterations of apolipoproteins should be considered. In addition, data from clinical trials enables us to select the right combination of biological targets and select individualized therapy for the effective treatment of the disease. It is important to note that known traditional prescriptions could reduce the time and cost of the screening process in drug discovery (González *et al.*, 2019).

Over the past 10 years, intranasal drug delivery in the field of drug development is an interesting delivery route for treating neurologic disorders because systemic approaches often fail to supply the CNS with drugs efficiently (Keller *et al.*, 2022).

Shomameh or *shomoom* (smelling) is a route of drug administration, including solid, liquid, and gaseous dosage forms, that acts rapidly for achievement of therapeutic effects, especially in the brain region. Nasal administration is considered an easy, nonaggressive, and fast drug administration route especially for CNS disorders. In *Canon of Medicine* it is used for inducing wetness and cold, as a *muhalil* (resolvent) or dissipation of a pathologic growth, and as a brain tonic. The olfactory system has a direct or indirect connection to the hypothalamus, hippocampus, and limbic system. In this way, there is no hepatic first-pass effect and due to a more targeted delivery, side effects are reduced. Active phytochemicals reach the blood by lung absorption, cross the blood–brain barrier, and enter the CNS following inhalation and dermal application. Through essential oil inhalation, brain disorders could be changed and corrected by direct activation of targeted brain pathways and can affect the modulation of the central neurotransmitter system. Natural scenes frequently have warm and cold temperaments and help modulate pathologic brain temperaments and related disorders.

Conclusion

TPM is a unique ancient medical system with more than 1000 years of history. Early evidence from preclinical and clinical studies demonstrates that several medicinal herbs used in TPM practice could produce therapeutic benefits for various nervous system disorders. Some bioactive components of the herbs may also serve as potential leads for pharmaceutical drug development as novel neuroprotective agents. Inhibition of brain AChE, antioxidant and anti-inflammatory activities are among the most important mechanisms of actions underlying the pharmacologic effects of these herbs. However, more research is needed to further evaluate the clinical effectiveness of these herbs in the treatment of nervous system disorders and their mechanisms of action.

Note

[1] For example, Zafromix™, a syrup prepared from saffron and rosewater, standardized according to the contents of crocin and picrocrocin and Rivamix™, a capsule containing *Boswellia* resin and four other medical herbs, recommended for dementia management (Patent Application no. 140050140003007256 of the Iranian intellectual property system).

References

Adhami, H.R., Farsam, H. and Krenn, L. (2011) Screening of medicinal plants from Iranian traditional medicine for acetylcholinesterase inhibition. *Phytotherapy Research* 25(8), 1148–1152. DOI: 10.1002/ptr.3409.

Afshari, A.R., Sadeghnia, H.R. and Mollazadeh, H. (2016) A review on potential mechanisms of *Terminalia chebula* in Alzheimer's disease. *Advances in Pharmacological Sciences* 2016, 8964849. DOI: 10.1155/2016/8964849.

Ahmadian-Attari, M.M., Ahmadiani, A., Kamalinejad, M., Dargahi, L., Shirzad, M. *et al.* (2014) Treatment of Alzheimer's disease in Iranian traditional medicine. *Iranian Red Crescent Medical Journal* 17(1), e18052–e18052.

Ahmadian-Attari, M.M., Eslami, S., Dargahi, L. and Noorbala, A.A. (2020) Common herbal treatments for senile dementia in ancient civilizations: Greco-Roman, Chinese, Indian, and Iranian. *Journal of Medicinal Plants* 19(73), 37–62.

Akbarzadeh, M., Dehghani, M., Moshfeghy, Z., Emamghoreishi, M., Tavakoli, P. *et al.* (2015) Effect of *Melissa officinalis* capsule on the intensity of premenstrual syndrome symptoms in high school girl students. *Nursing and Midwifery Studies* 4(2), e27001. DOI: 10.17795/nmsjournal27001.

Akhondzadeh, S., Noroozian, M., Mohammadi, M., Ohadinia, S., Jamshidi, A.H. *et al.* (2003) *Melissa officinalis* extract in the treatment of patients with mild to moderate Alzheimer's disease: a double blind, randomised, placebo controlled trial. *Journal of Neurology, Neurosurgery & Psychiatry* 74(7), 863–866. DOI: 10.1136/jnnp.74.7.863.

Akhondzadeh, S., Sabet, M.S., Harirchian, M.H., Togha, M., Cheraghmakani, H. *et al.* (2010a) Saffron in the treatment of patients with mild to moderate Alzheimer's disease: a 16-week, randomized and placebo-controlled trial. *Journal of Clinical Pharmacy and Therapeutics* 35(5), 581–588.

Akhondzadeh, S., Sabet, M.S., Harirchian, M.H., Togha, M., Cheraghmakani, H. *et al.* (2010b) A 22-week, multicenter, randomized, double-blind controlled trial of *Crocus sativus* in the treatment of mild-to-moderate Alzheimer's disease. *Psychopharmacology* 207(4), 637–643. DOI: 10.1007/s00213-009-1706-1.

Alam, M.B., Hossain, M.S. and Haque, M.E. (2011) Antioxidant and anti-inflammatory activities of the leaf extract of *Brassica nigra*. *International Journal of Pharmaceutical Sciences and Research* 2(2), 303–310. DOI: 10.13040/IJPSR.0975-8232.2(2).303-10.

Ali, B.H., Blunden, G., Tanira, M.O. and Nemmar, A. (2008) Some phytochemical, pharmacological and toxicological properties of ginger (*Zingiber officinale* Roscoe): a review of recent research. *Food and Chemical Toxicology* 46(2), 409–420. DOI: 10.1016/j.fct.2007.09.085.

Al-Snafi, A.E. (2015) Therapeutic properties of medicinal plants: A review of medicinal plants with central nervous effects (part 1). *International Journal of Pharmacology & Toxicology* 5(3), 177–192.

Angst, J., Gamma, A., Benazzi, F., Ajdacic, V. and Rossler, W. (2007) Melancholia and atypical depression in the Zurich study: epidemiology, clinical characteristics, course, comorbidity and personality. *Acta Psychiatrica Scandinavica* 115(Suppl. 433), 72–84.

Anuj, G. and Sanjay, S. (2010) Eugenol: a potential phytochemical with multifaceted therapeutic activities. *Pharmacology Online* 2, 108–120.

Aprotosoaie, A.C., Hăncianu, M., Costache, I. and Miron, A. (2014) Linalool: a review on a key odorant molecule with valuable biological properties. *Flavour and Fragrance Journal* 29(4), 193–219. DOI: 10.1002/ffj.3197.

Aryaeian, N. and Tavakkoli, H. (2015) Ginger and its effects on inflammatory diseases. *Advances in Food Technology and Nutritional Sciences* 1(4), 97–101. DOI: 10.17140/AFTNSOJ-1-117.

Attarfar, M. and Ashrafzadeh, F. (2014) Examining the temperament of the brain from the perspective of the basics of Iranian medicine. *Journal of Traditional Medicine of Islam and Iran* 6(1), 1–8.

Awad, R., Muhammad, A., Durst, T., Trudeau, V.L. and Arnason, J.T. (2009) Bioassay-guided fractionation of lemon balm (*Melissa officinalis* L.) using an *in vitro* measure of GABA transaminase activity. *Phytotherapy Research* 23(8), 1075–1081. DOI: 10.1002/ptr.2712.

Bakhsha, F., Mazandarani, M., Aryaei, M., Jafari, S.Y. and Bayate, H. (2014) Phytochemical and anti-oxidant activity of *Lavandula angustifolia* Mill. essential oil on preoperative anxiety in patients undergoing diagnostic curettage. *International Journal of Women's Health and Reproduction Sciences* 2(4), 268–271. DOI: 10.15296/ijwhr.2014.42.

Berk, M., Williams, L.J., Jacka, F.N., O'Neil, A., Pasco, J.A. *et al.* (2013) So depression is an inflammatory disease, but where does the inflammation come from? *BMC Medicine* 11, 200.

Cases, J., Ibarra, A., Feuillère, N., Roller, M. and Sukkar, S.G. (2011) Pilot trial of *Melissa officinalis* L. leaf extract in the treatment of volunteers suffering from mild-to-moderate anxiety disorders and sleep disturbances. *Mediterranean Journal of Nutrition and Metabolism* 4(3), 211–218. DOI: 10.1007/s12349-010-0045-4.

Chonpathompikunlert, P., Wattanathorn, J. and Muchimapura S. (2010) Piperine, the main alkaloid of Thai black pepper, protects against neurodegeneration and cognitive impairment in animal model of cognitive deficit like condition of Alzheimer's disease. *Food Chemical Toxicology* 48(3), 798–802. DOI: 10.1016/j.fct.2009.12.009.

Dallmeier-Zelger, K.R., Zelger, J.L. and Carlini, E.A. (1983) New anticonvulsants derived from 4-allyl-2-methoxyphenol (eugenol): comparison with common antiepileptics in mice. *Pharmacology* 27(1), 40–49. DOI: 10.1159/000137828.

Damanhouri, Z.A. and Ahmad, A. (2014) A review on therapeutic potential of *Piper nigrum* L. black pepper): the king of spices. *Medicinal & Aromatic Plants* 3, 161. DOI: 10.4172/2167-0412.1000161.

Denner, S.S. (2009) *Lavandula angustifolia* Miller: English lavender. *Holistic Nursing Practice* 23(1), 57–64. DOI: 10.1097/01.HNP.0000343210.56710.fc.

Elgood, C. (2010) *A Medical History of Persia and the Eastern Caliphate: From the Earliest Times until the Year AD 1932*. Cambridge University Press, Cambridge.

Elisabetsky, E., Marschner, J. and Souza, D.O. (1995) Effects of linalool on glutamatergic system in the rat cerebral cortex. *Neurochemical Research* 20(4), 461–465. DOI: 10.1007/BF00973103.

Esalatmanesh, S., Biuseh, M., Noorbala, A.A., Mostafavi, S.A., Rezaei, F. *et al.* (2017) Comparison of saffron and fluvoxamine in the treatment of mild to moderate obsessive-compulsive disorder: a double blind randomized clinical trial. *Iran Journal of Psychiatry* 12(3), 154–162.

Essa, M.M., Vijayan, R.K., Castellano-Gonzalez, G., Memon, M.A., Braidy, N. *et al.* (2012) Neuroprotective effect of natural products against Alzheimer's disease. *Neurochemical Research* 37(9), 1829–1842.

Fabricant, D.S. and Farnsworth, N.R. (2001) The value of plants used in traditional medicine for drug discovery. *Environmental Health Perspectives* 109(Suppl. 1), 69–75.

Farokhnia, M., Sabet S.M., Iranpour, N., Gougol, A., Yekehtaz, H. *et al.* (2014) Comparing the efficacy and safety of *Crocus sativus* L. with memantine in patients with moderate to severe Alzheimer's disease: a double-blind randomized clinical trial. *Human Psychopharmacology* 29(4) 351–359. DOI: 10.1002/hup.2412.

Feyzabadi, Z., Jafari, F., Feizabadi, P., Ashayeri. H., Esfahani, M.M. *et al.* (2014) Insomnia in Iranian traditional medicine. *Iran Red Crescent Medical Journal* 16(3), e15981.

Filiptsova, L.V., Rogozina, I.A., Timoshyna, O.I., Naboka, Ye.V., Dyomina, A.V. *et al.* (2018) The effect of the essential oils of lavender and rosemary on the human short-term memory. *Alexandria Journal of Medicine* 54(1), 41–44. DOI: 10.1016/j.ajme.2017.05.004.

François, M., Karpe, A.V., Liu, J.-W., Beale. D.J., Hor, M. *et al.* (2022) Multi-omics, an integrated approach to identify novel blood biomarkers of Alzheimer's disease. *Metabolite* 12, 949–969.

Geldmacher, D.S. and Whitehouse, P.J. (1996) Evaluation of dementia. *New England Journal of Medicine* 335, 330–336.

Georgiadou, G., Tarantilis, P. and Pitsikas, N. (2012) Effects of the active constituents of *Crocus sativus* L., crocins, in an animal model of obsessive–compulsive disorder. *Neuroscience Letters* 528(1), 27–30.

Geromichalos, G.D., Lamari, F.N., Papandreou, M.A., Trafalis, D.T., Margarity, M. *et al.* (2012) Saffron as a source of novel acetylcholinesterase inhibitors: molecular docking and in vitro enzymatic studies. *Journal of Agricultural and Food Chemistry* 60(24), 6131–6138. DOI: 10.1021/jf300589c.

Ghaffari, F., Taheri, M., Meyari, A., Karimi, Y. and Naseri, M. (2022) Avicenna and clinical experiences in Canon of Medicine. *Journal of Medicine and Life* 15(2), 168–173.

Ghahghaei, A., Bathaie, S.Z., Kheirkhah, H. and Bahraminejad, E. (2013) The protective effect of crocin on the amyloid fibril formation of Aβ42 peptide *in vitro*. *Cellular & Molecular Biology Letters* 18(3), 328–339. DOI: 10.2478/s11658-013-0092-1.

Golechha, M., Bhatia, J., Ojha, S. and Arya, D.S. (2011) Hydroalcoholic extract of *Emblica officinalis* protects against kainic acid-induced status epilepticus in rats: evidence for an antioxidant, anti-inflammatory, and neuroprotective intervention. *Pharmaceutical Biology* 49(11), 1128–1136. DOI: 10.3109/13880209.2011.571264.

Golechha, M., Bhatia, J., Arya, D.S. (2012) Studies on effects of *Emblica officinalis* (amla) on oxidative stress and cholinergic function in scopolamine induced amnesia in mice. *Journal of Environmental Biology* 33(1), 95–100.

González, J.F., Alcántara, A.R., Doadrio, A.L. and Sánchez-Montero, J.M. (2019) Developments with multi-target drugs for Alzheimer's disease: an overview of the current discovery approaches. *Expert Opinion on Drug Discovery* 14(9), 879–891.

Gorji, A. and Ghadiri, M.K. (2001) History of epilepsy in medieval Iranian medicine. *Neuroscience & Biobehavioral Reviews* 25(5), 455–461.

Gorji, A. and Ghadiri, M.K. (2002) History of headache in medieval Persian medicine. *Lancet Neurology* 1(8), 510–515. DOI: 10.1016/s1474-4422(02)00226-0.

Halder, S., Mehta, A.K., Mediratta, P.K. and Sharma, K.K. (2012) Acute effect of essential oil of *Eugenia caryophyllata* on cognition and pain in mice. *Naunyn Schmiedeberg's Archives of Pharmacology* 385(6), 587–593. DOI: 10.1007/s00210-012-0742-2.

Hancianu, M., Cioanca, O., Mihasan, M. and Hritcu, L. (2012) Neuroprotective effects of inhaled lavender oil on scopolamine-induced dementia via anti-oxidative activities in rats. *Phytomedicine* 20(5), 446-452. DOI: 10.1016/j.phymed.2012.12.005.

Hausenblas, H.A., Saha, D., Dubyak, P.J. and Anton, S.D. (2013) Saffron (*Crocus sativus* L.) and major depressive disorder: a meta-analysis of randomized clinical trials. *Journal of Integrative Medicine* 11(6), 377–383. DOI: 10.3736/jintegrmed2013056.

Holmes, C., Hopkins, V., Hensford, C., MacLaughlin, V., Wilkinson, D. *et al.* (2002) Lavender oil as a treatment for agitated behaviour in severe dementia: a placebo controlled study. *International Journal of Geriatric Psychiatry* 17(4), 305–308. DOI: 10.1002/gps.593.

Hritcu, L., Cioanca, O. and Hancianu, M. (2012) Effects of lavender oil inhalation on improving scopolamine-induced spatial memory impairment in laboratory rats. *Phytomedicine* 19(6), 529–534. DOI: 10.1016/j.phymed.2012.02.002.

Hritcu, L., Noumedem, J.A., Cioanca, O., Hancianu, M., Kuete, V. *et al.* (2014) Methanolic extract of *Piper nigrum* fruits improves memory impairment by decreasing brain oxidative stress in amyloid beta(1–42) rat model of Alzheimer's disease. *Cellular and Molecular Neurobiology* 34(3), 437–449. DOI: 10.1007/s10571-014-0028-y.

Jaijoy, K., Soonthornchareonnon, N., Panthong, A. and Sireeratawong, S. (2010) Anti-inflammatory and analgesic activities of the water extract from the fruit of *Phyllanthus emblica* Linn. *International Journal of Applied Research in Natural Products* 3(2), 28–35.

Ji, H.F., Li, X.J. and Zhang, H.Y. (2009) Natural products and drug discovery. *EMBO Reports* 10(3), 194–200.

Jokar, A., Masoomi, F., Sadeghpour, O., Nassiri-Toosi, M. and Hamedi, S. (2016) Potential therapeutic applications for *Terminalia chebula* in Iranian traditional medicine. *Journal of Traditional Chinese Medicine* 36(2), 250–254. DOI: 10.1016/s0254-6272(16)30035-8.

Kamaneh, S., Mojahedi, M., Mozafari, O., Memariani, Z. and Saravani, M. (2019) Cardiotonic medicines (Mofarrehs) and their mechanism of action in Persian Medicine. *Journal of Babol University of Medical Sciences* 21(1), 320–330.

Kamdem, J.P., Adeniran, A., Boligon, A.A., Klimaczewski, C.V., Elekofehinti, O.O. *et al.* (2013) Antioxidant activity, genotoxicity and cytotoxicity evaluation of lemon balm (*Melissa officinalis* L.) ethanolic extract: its potential role in neuroprotection. *Industrial Crops and Products* 51, 26–34.

Karam, A., Gouda, N.A., El-Fattah, A., Yassin, N.A. and El-Shenawy, S.M. (2014) Protective effect of ginger (*Zingiber officinale*) on Alzheimer's disease induced in rats. *Journal of Neuroinfectious Diseases* 5(159), 2. DOI: 10.4172/2314-7326.1000159.

Keller, L.A., Merkel, O. and Popp, A. (2022) Intranasal drug delivery: opportunities and toxicologic challenges during drug development. *Drug Delivery and Translational Research* 12, 735–757.

Kennedy, D.O., Scholey, A.B., Tildesley, N.T., Perry, E.K. and Wesnes, K.A. (2002) Modulation of mood and cognitive performance following acute administration of *Melissa officinalis* (lemon balm). *Pharmacology, Biochemistry, and Behavior* 72(4), 953–964. DOI: 10.1016/s0091-3057(02)00777-3.

Keshvari, M., Asgary, S., Jafarian-Dehkordi, A., Najafi, S. and Ghoreyshi-Yazdi, S.M. (2013) Preventive effect of cinnamon essential oil on lipid oxidation of vegetable oil. *ARYA Atherosclerosis* 9(5), 280–286.

Khasnavis, S. and Pahan, K. (2012) Sodium benzoate, a metabolite of cinnamon and a food additive, upregulates neuroprotective Parkinson disease protein DJ-1 in astrocytes and neurons. *Journal of Neuroimmune Pharmacology* 7(2), 424–435.

Khasnavis, S. and Pahan, K. (2014) Cinnamon treatment upregulates neuroprotective proteins Parkin and DJ-1 and protects dopaminergic neurons in a mouse model of Parkinson's disease. *Journal of Neuroimmune Pharmacology* 9(4), 569–581.

Khodaei, M.A., Noorbala, A.A., Parsian, Z., Targhi, S.T., Emadi, F. *et al.* (2017) Avicenna (980–1032 CE): the pioneer in treatment of depression. *Transylvanian Review* 25(17), 4377–4389.

Kiasalari, Z., Khalili, M., Rhogani, M. and Sadeghian, A. (2012) Antiepileptic and antioxidant effect of *Brassica nigra* on pentylenetetrazol-induced kindling in mice. *Iranian Journal of Pharmaceutical Research* 11(4), 1209–1217.

Kim, D.S., Kim, D.S. and Oppel, M.N. (2002) Shogaols from *Zingiber officinale* protect IMR32 human neuroblastoma and normal human umbilical vein endothelial cells from β-amyloid(25–35) insult. *Planta Medica* 68(4), 375–376. DOI: 10.1055/s-2002-26757.

Kim, M.S., Lee, D.Y., Lee, J., Kim, H.W., Sung, S.H. *et al.* (2018) *Terminalia chebula* extract prevents scopolamine-induced amnesia via cholinergic modulation and anti-oxidative effects in mice. *BMC Complementary and Alternative Medicine* 18(1), 136. DOI: 10.1186/s12906-018-2212-y.

Koulivand, P.H., Khaleghi, G.M. and Gorji, A. (2013) Lavender and the nervous system. *Evidence Based Complementary and Alternative Medicine* 2013, 681304. DOI: 10.1155/2013/681304.

Kumar, R., Arora, R., Agarwal, A. and Gupta, Y.K. (2018) Protective effect of *Terminalia chebula* against seizures, seizure-induced cognitive impairment and oxidative stress in experimental models of seizures in rats. *Journal of Ethnopharmacology* 215, 124–131. DOI: 10.1016/j.jep.2017.12.008.

Kumar, S. and Andy, A. (2012) Health promoting bioactive phytochemicals from Brassica. *International Food Research Journal* 19(1), 141–152.

Kumar, S., Okello, E.J. and Harris, J.R. (2012) Experimental inhibition of fibrillogenesis and neurotoxicity by amyloid-beta (Aβ) and other disease-related peptides/proteins by plant extracts and herbal compounds. *Subcellular Biochemistry* 65, 295–326. DOI: 10.1007/978-94-007-5416-4_13.

Kumaran, A. and Karunakaran, R.J. (2006) Nitric oxide radical scavenging active components from *Phyllanthus emblica* L. *Plant Foods for Human Nutrition* 61(1), 1–5. DOI: 10.1007/s11130-006-0001-0.

Lee, H.S., Jung, S.H., Yun, B.S. and Lee, K.W. (2007) Isolation of chebulic acid from *Terminalia chebula* Retz. and its antioxidant effect in isolated rat hepatocytes. *Archives of Toxicology* 81(3), 211–218. DOI: 10.1007/s00204-006-0139-4.

Linardaki, Z.I., Orkoula, M.G., Kokkosis, A.G., Lamari, F.N. and Margarity, M. (2013) Investigation of the neuroprotective action of saffron (*Crocus sativus* L.) in aluminum-exposed adult mice through behavioral and neurobiochemical assessment. *Food and Chemical Toxicology* 52, 163–170. DOI: 10.1016/j.fct.2012.11.016.

López, V., Martín, S., Gómez-Serranillos, M.P., Carretero, M.E., Jäger, A.K. *et al.* (2009) Neuroprotective and neurological properties of *Melissa officinalis*. *Neurochemical Research* 34(11), 1955–1961. DOI: 10.1007/s11064-009-9981-0.

Mabrouki, H., Duarte, C. and Akretche, D. (2018) Estimation of total phenolic contents and *in vitro* antioxidant and antimicrobial activities of various solvent extracts of *Melissa officinalis* L. *Arabian Journal for Science and Engineering* 43(7), 3349–3357.

Mathew, M. and Subramanian, S. (2014) *In vitro* evaluation of anti-Alzheimer effects of dry ginger (*Zingiber officinale* Roscoe) extract. *Indian Journal of Experimental Biology* 52, 606–612.

Mazidi, M., Shemshian, M., Mousavi, S.H., Norouzy, A., Kermani, T. *et al.* (2016) A double-blind, randomized and placebo-controlled trial of saffron (*Crocus sativus* L.) in the treatment of anxiety and depression. *Journal of Complementary and Integrative Medicine* 13(2), 195–199. DOI: 10.1515/jcim-2015-0043.

Mehri, A.F. and Shojaii, A. (2013) Efficacy of Iranian traditional medicine in the treatment of epilepsy. *BioMed Research International* 2013, 692751. DOI: 10.1155/2013/692751.

Mesripour, A., Moghimi, F. and Rafieian-Kopaie, M. (2016) The effect of *Cinnamomum zeylanicum* bark water extract on memory performance in alloxan-induced diabetic mice. *Research in Pharmaceutical Sciences* 11(4), 318–323.

Moon, M., Kim, H.G., Choi, J.G., Oh, H., Lee, P.K. *et al.* (2014) 6-Shogaol, an active constituent of ginger, attenuates neuroinflammation and cognitive deficits in animal models of dementia. *Biochemical and Biophysical Research Communications* 449(1), 8–13. DOI: 10.1016/j.bbrc.2014.04.121.

Müller, M., Pape, H.C., Speckmann, E.J. and Gorji, A. (2006) Effect of eugenol on spreading depression and epileptiform discharges in rat neocortical and hippocampal tissues. *Neuroscience* 140(2), 743–751. DOI: 10.1016/j.neuroscience.2006.02.036.

Naik, G.H., Priyadarsini, K.I., Bhagirathi, R.G., Mishra, B., Mishra, K.P. *et al.* (2005) *In vitro* antioxidant studies and free radical reactions of triphala, an ayurvedic formulation and its constituents. *Phytotherapy Research* 19(7), 582–586. DOI: 10.1002/ptr.1515.

Oboh, G., Ademiluyi, A.O. and A.J. Akinyemi, A.J. (2012) Inhibition of acetylcholinesterase activities and some pro-oxidant induced lipid peroxidation in rat brain by two varieties of ginger (*Zingiber officinale*). *Experimental and Toxicologic Pathology* 64(4), 315–319.

Papandreou, M.A., Kanakis, C.D., Polissiou, M.G., Efthimiopoulos, S., Cordopatis, P. *et al.* (2006) Inhibitory activity on amyloid-beta aggregation and antioxidant properties of *Crocus sativus* stigmas extract and its crocin constituents. *Journal of Agricultural and Food Chemistry* 54(23), 8762–8768. DOI: 10.1021/jf061932a.

Papandreou, M.A., Tsachaki, M., Efthimiopoulos, S., Cordopatis, P., Lamari, F.N. *et al.* (2011) Memory enhancing effects of saffron in aged mice are correlated with antioxidant protection. *Behavioural Brain Research* 219(2), 197–204. DOI: 10.1016/j.bbr.2011.01.007.

Paradkar, P.P. and Sarawade, R.D. (2015) Neuropharmacological evaluation of *Brassica nigra* seed extract in Parkinson's disease. *World Journal of Pharmaceutical Research* 4(6), 2291–2308.

Park, G., Kim, H.G., Ju, M.S., Ha, S.K., Park, Y. *et al.* (2013) 6-Shogaol, an active compound of ginger, protects dopaminergic neurons in Parkinson's disease models via anti-neuroinflammation. *Acta Pharmacologica Sinica* 34(9), 1131–1139. DOI: 10.1038/aps.2013.57.

Park, J.H., Joo, H.S., Yoo, K.Y., Shin, B.N., Kim, I.H. *et al.* (2011a) Extract from *Terminalia chebula* seeds protect against experimental ischemic neuronal damage via maintaining SODs and BDNF levels. *Neurochemical Research* 36(11), 2043–2050. DOI: 10.1007/s11064-011-0528-9.

Park, S.H., Sim, Y.B., Lee, J.K., Kim, S.M., Kang, Y.J. *et al.* (2011b) The analgesic effects and mechanisms of orally administered eugenol. *Archives of Pharmaceutical Research* 34(3), 501–507. DOI: 10.1007/s12272-011-0320-z.

Patwardhan. B. and Mashelkar, R.A. (2009) Traditional medicine approaches to drug discovery. *Drug Discovery Today* 14(15–16), 804–811.

Peterson, D.W., George, R.C., Scaramozzino, F., LaPointe, N.E., Anderson, R.A. *et al.* (2009) Cinnamon extract inhibits tau aggregation associated with Alzheimer's disease *in vitro*. *Journal of Alzheimer's Disease* 17(3) 585–597. DOI: 10.3233/JAD-2009-1083.

Poltanov, E.A., Shikov, A.N., Dorman, H.J., Pozharitskaya, O.N., Makarov, V.G. *et al.* (2009) Chemical and antioxidant evaluation of Indian gooseberry (*Emblica officinalis* Gaertn., syn. *Phyllanthus emblica* L.) supplements. *Phytotherapy Research* 23(9), 1309–1315. DOI: 10.1002/ptr.2775.

Poma, A., Fontecchio, G., Carlucci, G. and Chichiriccò, G. (2012) Anti-inflammatory properties of drugs from saffron crocus. *Anti-Inflammatory & Anti-Allergy Agents in Medicinal Chemistry* 11(1), 37–51. DOI: 10.2174/187152312803476282.

Rahaiee, S., Moini, S., Hashemi, M. and Shojaosadati, S.A. (2015) Evaluation of antioxidant activities of bioactive compounds and various extracts obtained from saffron (*Crocus sativus* L.): a review. *Journal of Food Science and Technology* 52(4), 1881–1888.

Rahimi, R., Irannejad, S. and Noroozian, M. (2017) Avicenna's pharmacological approach to memory enhancement. *Neurological Sciences* 38(7) 1147–1157.

Rajamurugan, R., Selvaganabathy, N., Kumaravel, S., Ramamurthy, C.H., Sujatha, V. *et al.* (2012) Polyphenol contents and antioxidant activity of *Brassica nigra* (L.) Koch. leaf extract. *Natural Product Research* 26(23) 2208–2210. DOI: 10.1080/14786419.2011.637215.

Sadock, B.J., Sadock, V.A. and Ruiz, P. (2017) *Kaplan & Sadok's Comprehensive Text Book of Psychiatry*, 10th edn, Vol. I/II. Wolters Kluwer, Philadelphia, Pennsylvania, pp. 5407–5409.

Saenghong, N., Wattanathorn, J., Muchimapura, S., Tongun, T., Piyavhatkul, N. *et al.* (2012) *Zingiber officinale* improves cognitive function of the middle-aged healthy women. *Evidence-Based Complementary and Alternative Medicine* 2012, 383062. DOI: 10.1155/2012/383062.

Sahardi, M., Nabilah, N.F. and Makpol, S. (2019) Ginger (*Zingiber officinale* roscoe) in the prevention of ageing and degenerative diseases: Review of current evidence. *Evidence-Based Complementary and Alternative Medicine* 2019, 5054395. DOI: 10.1155/2019/5054395.

Salmannegad, H., Mojahedi, M., Mozaffarpur, S.A. and Saghebi, R. (2016) The review of indices of mizaj-e-damagh (temperament of brain) identification in Persian medicine. *Journal of Babol University of Medical Sciences* 18(11), 71–79.

Sancheti, S., Sancheti, S., Um, B.-H. and Seo, S.-Y. (2010) 1,2,3,4,6-Penta-*O*-galloyl-β-D-glucose: a cholinesterase inhibitor from *Terminalia chebula*. *South African Journal of Botany* 76(2), 285–288. DOI: 10.1016/j.sajb.2009.11.006.

Santomauro, D.F. (2021) Global prevalence and burden of depressive and anxiety disorders in 204 countries and territories in 2020 due to the COVID-19 pandemic. *Lancet* 398, 1700–1712.

Sarris, J., McIntyre, E. and Camfield, D.A. (2013) Plant-based medicines for anxiety disorders, part 1. *CNS Drugs* 27(3), 207–219.

Scartezzini, P., Antognoni, F., Raggi, M.A., Poli, F. and Sabbioni, C. (2006) Vitamin C content and antioxidant activity of the fruit and of the Ayurvedic preparation of *Emblica officinalis* Gaertn. *Journal of Ethnopharmacology* 104(1–2), 113–118. DOI: 10.1016/j.jep.2005.08.065.

Semwal, R.B., Semwal, D.K., Combrinck, S. and Viljoen, A.M. (2015) Gingerols and shogaols: important nutraceutical principles from ginger. *Phytochemistry* 117, 554–568. DOI: 10.1016/j.phytochem.2015.07.012.

Shakeri, A., Sahebkar, A. and Javadi, B. (2016) *Melissa officinalis* L. – a review of its traditional uses, phytochemistry and pharmacology. *Journal of Ethnopharmacology* 188, 204–228. DOI: 10.1016/j.jep.2016.05.010.

Shaltiel-Karyo, R., Davidi, D., Frenkel-Pinter, M., Ovadia, M., Segal, D. *et al.* (2012) Differential inhibition of α-synuclein oligomeric and fibrillar assembly in Parkinson's disease model by cinnamon extract. *Biochimica et Biophysica Acta (BBA) – General Subjects* 1820(10), 1628–1635. DOI: 10.1016/j.bbagen.2012.04.021.

Sharma, M., Sharma, N. and Sharma, R. (2012) Neuroprotective effect of *Zingiber officinale* in 3-NP-induced Huntington disease. *IOSR Journal of Pharmacy* 2(6), 61–70.

Sharma, M.N. (2019) Prediction of prospective anti-Parkinson phytochemicals using prediction of activity spectra of substances software to justify 3R's ethics of *in vivo* evaluation. *Asian Journal of Pharmaceutics* 13(3). DOI: 10.22377/ajp.v13i3.3295.

Sigerist, H.E. (1987) *A History of Medicine: Early Greek, Hindu, and Persian Medicine*, Vol. 2. Oxford University Press, Oxford.

Soeda, S., Aritake, K., Urade, Y., Sato, H. and Shoyama, Y. (2016) Neuroprotective activities of saffron and crocin. *Advanced Neurobiology* 12, 275–292. DOI: 10.1007/978-3-319-28383-8_14.

Soheili, M., Salami, M., Haghir, A., Zali, H. and Tavirani, M.R. (2014) Aqueous extract of *Lavandula angustifolia* alter protein expression in Alzheimer rats. *Journal of Reports in Pharmaceutical Sciences* 3(1), 1–9.

Soodi, M., Naghdi, N., Hajimehdipoor, H., Choopani, S. and Sahraei, E. (2014) Memory-improving activity of *Melissa officinalis* extract in naïve and scopolamine-treated rats. *Research in Pharmaceutical Sciences* 9(2), 107–114.

Stoilova, I., Krastanov, A., Stoyanova, A., Denev, P. and Gargova, S. (2007) Antioxidant activity of a ginger extract (*Zingiber officinale*). *Food Chemistry* 102, 764–770.

Subedee, L., Suresh, R.N., Jayanthi, M.K., Kalabharathi, H.L., Satish, A.M. *et al.* (2015) Preventive role of Indian black pepper in animal models of Alzheimer's disease. *Journal of Clinical and Diagnostic Research* 9(4), FF01–FF04. DOI: 10.7860/JCDR/2015/8953.5767.

Sulaiman, C., Sadashiva, C., George, S. and Balachandran, I. (2012) Acetylcholinesterase inhibition and antioxidant activity of *Terminalia chebula*, Retz. *Journal of Tropical Medicinal Plants* 13(2), 125–127.

Taheri-Targhi, S., Gjedde, A., Araj-Khodaei, M., Rikhtegar, R., Parsian, Z. *et al.* (2019) Avicenna (980–1037 CE) and his early description and classification of dementia. *Journal of Alzheimer's Disease* 71(4), 1093–1098. DOI: 10.3233/JAD-190345.

Tajadini, H., Saifadini, R., Choopani, R., Mehrabani, M., Kamalinejad, M. *et al.* (2015) Herbal medicine Davaie Loban in mild to moderate Alzheimer's disease: a 12-week randomized double-blind placebo-controlled clinical trial. *Complementary Therapies in Medicine* 23(6), 767–772. DOI: 10.1016/j.ctim.2015.06.009.

Tardy, A.L., Pouteau, E., Marquez, D., Yilmaz, C. and Scholey, A. (2020) Vitamins and minerals for energy, fatigue and cognition: a narrative review of the biochemical and clinical evidence. *Nutrients* 12(228), 1–35.

Thenmozhi, A.J., Dhivyabharathi, M., Raja, T.R.W., Manivasagam, T. and Essa, M.M. (2016) Tannoid principles of *Emblica officinalis* renovate cognitive deficits and attenuate amyloid pathologies against aluminum chloride induced rat model of Alzheimer's disease. *Nutrutional Neuroscience* 19(6), 269–278. DOI: 10.1179/1476830515Y.0000000016.

Tsolaki, M., Karathanasi, E., Lazarou, I., Dovas, K., Verykouki, E. *et al.* (2016) Efficacy and safety of Crocus sativus L. in patients with mild cognitive impairment: one year single-blind randomized, with parallel groups, clinical trial. *Journal of Alzheimer's Disease* 54(1), 129–133. DOI: 10.3233/JAD-160304.

Tu, Y., Zhong, Y., Du, H., Luo, W., Wen, Y. *et al.* (2016) Anticholinesterases and antioxidant alkamides from *Piper nigrum* fruits. *Natural Product Research* 30(17), 1945–1949. DOI: 10.1080/14786419.2015.1089243.

Vakili, N. and Gorji, A. (2006) Psychiatry and psychology in medieval. *Journal of Clinical Psychatry* 67(12), 1862–1869. DOI: 10.4088/jcp.v67n1205.

Vasudevan, M. and Parle, M. (2007) Effect of anwala churna (*Emblica officinalis* Gaertn.): an ayurvedic preparation on memory deficit rats. *Yakugaku Zasshi* 127(10), 1701–1707. DOI: 10.1248/yakushi.127.1701.

Vaz, M. and Samuel Silvestre, S. (2020) Alzheimer's disease: recent treatment strategies. *European Journal of Pharmacology* 15(887), 173554.

Wang, C.C., Yuan, J.R., Wang, C.F., Yang, N., Chen, J. *et al.* (2017) Anti-inflammatory effects of *Phyllanthus emblica* L. on benzopyrene-induced precancerous lung lesion by regulating the IL-1β/miR-101/Lin28B signaling pathway. *Integrative Cancer Therapies* 16(4), 505–515. DOI: 10.1177/1534735416659358.

Wie, M.B., Won, M.H., Lee, K.H., Shin, J.H., Lee, J.C. *et al.* (1997) Eugenol protects neuronal cells from excitotoxic and oxidative injury in primary cortical cultures. *Neuroscience Letters* 225(2), 93–96. DOI: 10.1016/s0304-3940(97)00195-x.

WHO (2018) *Towards a Dementia Plan: A WHO Guide*. World Health Organization, Geneva, Switzerland.

Yang, M.Y., Alia, Z., Khan, I.A. and Khan, S.I. (2014) Anti-inflammatory activity of constituents isolated from *Terminalia chebula*. *Natural Product Communications* 9(7), 965–968.

Yarhosseini, A. and Arbabi, M. (2019) Avicenna's Contribution to Psychiatry. *Medical History Journal* 11(40), 107–117.

Yoo, D.Y., Choi, J.H., Kim, W., Yoo, K.Y., Lee, C.H. *et al.* (2011) Effects of *Melissa officinalis* L. (lemon balm) extract on neurogenesis associated with serum corticosterone and GABA in the mouse dentate gyrus. *Neurochemical Research* 36(2), 250–257. DOI: 10.1007/s11064-010-0312-2.

Yousofpour, M. (2015) Non-pharmacological interventions for depression in Persian medicine. *Tang* 5(4), e22.

Zeng, G.F., Zhang, Z.Y., Lu, L., Xiao, D.Q., Zong, S.H. *et al.* (2013) Protective effects of ginger root extract on Alzheimer disease-induced behavioral dysfunction in rats. *Rejuvenation Research* 16(2), 124–133. DOI: 10.1089/rej.2012.1389.

5 An Evidence-Based Review of Medicinal Plants Cited in *Canon of Medicine* for Management of Inflammatory Bowel Disease

Saeideh Momtaz[1,2,3], Roodabeh Bahramsoltani[4,5], Amir Hossein Abdolghaffari[6,3] and Roja Rahimi[4,5*]

[1]Medicinal Plants Research Center, Institute of Medicinal Plants, Karaj, Iran; [2]Toxicology and Diseases Group, Institute of Pharmaceutical Sciences, Tehran, University of Medical Sciences, Tehran, Iran; [3]GI Pharmacology Interest Group (GPIG), Universal Scientific Education and Research Network (USERN), Tehran, Iran; [4]Department of Traditional Pharmacy, School of Persian Medicine, Tehran University of Medical Sciences, Tehran, Iran; [5]PhytoPharmacology Interest Group, Universal Scientific Education and Research Network, Tehran, Iran; [6]Department of Toxicology & Pharmacology, Faculty of Pharmacy, Tehran Medical Sciences, Islamic Azad University, Tehran, Iran

Abstract

Inflammatory bowel disease (IBD) such as ulcerative colitis (UC) and Crohn's disease (CD) are chronic inflammations of the gastrointestinal (GI) tract associated with intestinal ulcers. The exact pathogenesis of the disease is not yet clear. However, several genetic and environmental factors, as well as the patient's immune response to inflammation, are assumed to be key contributors to IBD. Currently available treatments for IBD include oral and/or rectal administration of various aminosalicylates, corticosteroids, immunosuppressants, and biologic therapies such as monoclonal antibodies. However, despite these options available today, about one-third of patients cannot be cured. For this reason, therapeutic approaches of alternative, complementary, and integrative medicine are also applied today. Traditional Persian Medicine (TPM) and the *Canon of Medicine* written by Avicenna are of particular importance here, which is why this chapter briefly presents the pathology and treatments of IBD from Avicenna's point of view.

Introduction

Inflammatory bowel disease (IBD), including ulcerative colitis (UC) and Crohn's disease (CD), is chronic inflammation of the gastrointestinal (GI) tract that is accompanied by intestinal ulcers. The disease is mostly characterized with GI symptoms such as bloody stools, abdominal pain, and malabsorption, as well as extraintestinal manifestations like renal,

*Email: rojarahimi@gmail.com

(eds. H. Schulz, Seyed Ahmad Emami and Farsad Nadjafi)
DOI: 10.1079/9781800621671.0005

hepatic, pulmonary, and musculoskeletal complications, chronic fatigue, and weight loss (Larsen *et al.*, 2010). While UC mostly occurs in the rectum and colon, CD can affect any part of the GI tract. The exact pathogenesis of the disease is yet to be understood; however, several genetic and environmental factors, along with the patient's immune response to inflammation, are introduced as critical contributors in IBD. Abnormally high induction of inflammatory pathways such as nuclear factor-κB (NF-κB) and mitogen-activated protein kinases (MAPKs) results in the production of pro-inflammatory cytokines like tumor necrosis factor-α (TNF-α), different interleukins (ILs), and nitric oxide (NO), and overstimulation of toll-like receptors (TLRs), which, overall, results in a chronic inflammation (Guan, 2019).

Currently available treatments for IBD include oral and/or rectal administration of aminosalicylates (e.g., sulfasalazine, olsalazine, balsalazide), corticosteroids (hydrocortisone, budesonide), immunosuppressant agents (6-mercaptopurine, azathioprine, cyclosporine), and biologic therapies such as monoclonal antibodies blocking inflammatory cytokines like TNF-α (infliximab, adalimumab) (Pithadia and Jain, 2011). In spite of all available options, about one-third of patients do not get rid of the symptoms (Schmidt *et al.*, 2021). The low clinical response of IBD patients, along with the adverse effects of conventional therapies, prompt patients to seek alternative complementary and integrative medicine to manage their complications.

Traditional medicines of different countries have always been a source of new ideas in drug design and development based on natural compounds. Traditional Persian Medicine (TPM), as one of the most ancient doctrines of traditional medical systems, is globally well known with pioneering scientists such as Avicenna and his unique masterpiece, *Canon of Medicine*. Considering the significant role of this scientist and his manuscripts in the development of medical sciences, this chapter aims to provide a brief summary of IBD pathology and treatments in view of Avicenna as cited in *Canon of Medicine*.

Irritable Bowel Disease in View of Avicenna

In Persian medicine, information on IBD can be extracted by using keywords such as "*Zahier*", i.e., a condition of having stomach cramps, bloody diarrhea, and irritation of intestinal mucosa, as well as "*Sahj*" or "*Qorhe amā*", meaning the presence of mucosal injuries and inflammation within the intestinal wall. The first important point to know about *zahier* is to distinguish between a false colitis and real colitis. In the case of false colitis, an accumulation of feces occurs in the rectum that causes mucosal irritation and, sometimes, ulcers. Due to similar symptoms, this can be wrongly diagnosed as real colitis; however, its treatment is limited to emollients.

In real colitis (IBD), there are different reasons behind the pathogenesis of the disease. It is sometimes caused by cold dystemperament that should be treated by hot interventions. This can include topical application of a dry or wet warm compress on the anus, buttocks, and lower parts of stomach, application of hot and astringent oils like unripe olive oil, washing with warm water, taking a warm bath, and sitting on a warm place. In most cases of IBD, any form of coldness can exacerbate the symptoms and should be avoided. Additionally, intake of foods from which thick and sticky humors are produced (those kinds of humors that are difficult to be absorbed and used by body organs) usually worsens the condition.

In some cases, IBD is caused by condensation of humors and should be treated by topical medications with softening properties including dill oil, chamomile, and guggul gum, or warm olive oil. It is worth mentioning that herbal oils in TPM such as chamomile oil and dill oil are not similar to essential oils. These oils are usually prepared by soaking the dried herbal material in a fixed oil such as sesame oil, or by boiling the herbal material in water, filtration, followed by addition of sesame oil to be boiled with the herbal filtrate until the whole water is evaporated and the active compounds are trapped into the oil.

Another reason for IBD may be a hot abscess that should be treated by detoxifying the site of inflammation via excretion of excess materials (which feed the inflammation) into blood vessels or by laxatives, as well as modifying the quality and quantity of the hot humors causing the abscess. Phlebotomy and fasting (reducing food intake) in the beginning of the symptoms may also be useful. Specific types of washing solutions should be used with a mild cooling effect to act as an emollient and analgesic, as well as a tonifying agent to reduce the inflammation. Such solutions include the juice or extract of black nightshade, rice, myrtle, and rose, as well as egg white. If the patient is suffering from diarrhea, this should be first stopped, then the site should be washed with these solutions, some poultices like chamomile and dill mixed with astringent agents should be applied, and then maturant drugs should be used.

If the reason for IBD is accumulation of humors, maturant drugs should be used, followed by openers. Olive oil mixed with astringent medicines in the form of an enema, as well as oral intake of hot cow's milk can be useful in patients with IBD due to hot abscess. Topical use of mallow and fenugreek acts as maturant, dissolver, and analgesic agent. In severe cases, fried onion, guggul, some minerals, and egg yolk, mixed with rose oil and extract of black nightshade or coriander can also be administered.

Colitis is sometimes caused by a hard abscess. In such cases, a poultice made of medicines like guggul, saffron and henna, along with some minerals, birds' fat, egg yolk, and rose oil is helpful.

Another reason for colitis is accumulation of infected humors, such as infected phlegm, where honey or salted olive juice as enema would be useful.

Pathologic humors are sometimes stimulated and moved due to the use of laxative or emetic drugs; however, the residue of these humors may cause colitis symptoms. If the patient is suffering from severe diarrhea, antidiarrheal agents should be first used to manage the acute condition. If the patient can tolerate it, some mild emollient enemas or suppositories should be used to clean the bowel from remaining pathologic humors.

In the case that colitis symptoms become worsened without expelling any blood from the stool, the reason may be a hard abscess or severe cold dystemperament. In both cases, a topical compress with warm oils such as rose oil, chamomile oil, or myrtle oil and some amount of wine at the site of the anus is helpful. Sesame oil enema can also be useful.

In the case of ulcers within the intestine wall, mild emollients should be administered to prevent constipation because constipation and dry stool worsen the condition. Also, the patient should avoid taking salty, sour, sour-sweet, or spicy foods since they exacerbate the symptoms (Avicenna, AH 1426).

Cellular and Molecular Mechanisms Underlying Development of Irritable Bowel Disease

The etiology of IBD is unknown. Interplay between genetic, innate immune system microbial, and environmental factors was shown to underlie IBD initiation and progression. The etiology of IBD is partly associated with deregulation of immune response to gut microbiome dysbiosis, and dramatic changes in gut microbiota. Cross-sectional studies have indicated that the microbiomes of IBD subjects fluctuate more than those of healthy individuals (Halfvarson *et al.*, 2017; Franzosa *et al.*, 2019). Generally, intestinal health results from homeostasis between epithelial, innate, and adaptive immune cells. Various harmful stimulants can induce the innate immune system, which in turn activates tolerogenic, inflammatory, and restitutive responses. In addition to the release of the extracellular mediators, the adaptive immune cells are also activated. The colon-infiltrating immune cells, particularly macrophages, dendritic cells (DCs), and T lymphocytes, play key roles in IBD induction and progression. Pro-inflammatory M1 macrophages produce inflammatory cytokines such as TNF-α, IL-1β, IL-6, and NO, which promote intestinal inflammation. Besides, DCs enhance the production of pro-inflammatory cytokines (i.e., TNF-α, IL-1β, IL-6, and IL-12) or drive the intestinal inflammation through the activation of TLR signaling pathways. Regulatory T cells are involved in immunosuppressive actions and immune-tolerant microenvironment maintenance (Khor *et al.*, 2011; Zhang and Li, 2014; Markovic *et al.*, 2018; Samoilă *et al.*, 2020). Strong evidence confirms that inflammation and oxidative stress are interdependent and interconnected processes. Upon inflammation, some inflammatory cells release oxidative mediators such as reactive oxygen species (ROS) at the site of inflammation, triggering oxidative damage and enhancing pro-inflammatory responses (Biswas, 2016). Suppression of the antioxidant defense system and increased ROS have been observed in IBD (Patlevič *et al.*, 2016; Bourgonje *et al.*, 2020). Two other biologic pathways are also involved in UC pathogenesis: endoplasmic reticulum stress (ERS), which involves an accumulation of aberrant folded protein in the endoplasmic reticulum; and the autophagy process, which participates in elimination of protein aggregates and invading antigens (Kaser and Blumberg, 2011). Imbalanced autophagy leads to the disruption of the intestinal barrier integrity, thereby ERS is activated and production of ROS is accelerated, which together induce UC pathogenesis (Tang *et al.*, 2015). It was also shown that there is a correlation between the level of NO and IBD. Overproduction of NO via the inducible nitric oxide synthase (iNOS) enzyme can exacerbate GI inflammation. Excessive NO contributes to inflammation through nitrosation, oxidative damage, and upregulated inflammatory cytokines (Kamalian *et al.*, 2020).

Medicinal Plants Cited in *Canon of Medicine* for Irritable Bowel Disease

Here, we provide the latest scientific evidence on medicinal plants cited in *Canon of Medicine* for IBD with regard to their ability to improve GI complications, with specific emphasis on IBD (Table 5.1). Tables 5.2, 5.3, and 5.4 present *in vitro*, *in vivo*, and clinical interventions of the plant species against IBD, respectively.

Trigonella foenum-graceum L.

Fenugreek seed is rich in trigonelline and diosgenin with significant biologic activities, including anti-inflammatory effects (Bahmani *et al.*, 2016). The anti-inflammatory effects of fenugreek are attributed to downregulation of inflammatory cytokines (Kawabata *et al.*, 2011), inhibition of pro-inflammatory markers like cyclooxygenase (COX), lipoxygenase, and myeloperoxidase (MPO), and levels of nitrite and C-reactive protein (CRP). Enhancement in the activities of catalase (CAT), superoxide dismutase (SOD), glutathione peroxidase (GPx), and the levels of glutathione (GSH), and vitamin C have been linked to fenugreek antioxidant properties (Mandegary *et al.*, 2012; Sindhu *et al.*, 2012). In dextran sodium sulfate (DSS)-induced UC in rats, oral administration of *Trigonella foenum-graecum* seeds (5% and 10% w/w) reduced circulating immune cells and mast cell count. Colonic inflammation was attenuated through downregulation of inflammatory parameters including TNF-α, IL-1β, and IL-6, while increasing IL-10. In addition, fenugreek improved clinical manifestations of UC and the disease activity index (DAI). Fenugreek seeds also preserved colonic mucin, increased colonic neutral mucin, and decreased acidic mucin (Liu, 2019). In a similar study, the aqueous extract of the seeds of fenugreek at doses of 500 and 1000 mg/kg body weight (b.w.) significantly attenuated DAI and the intensity of mucosal inflammation in acetic acid-induced UC in rats, which was implicated in increase of total protein and decrease of oxidative stress-related parameters such as GSH, CAT, SOD, and malondialdehyde (MDA). Fenugreek also suppressed mRNA level of the pro-inflammatory cytokine TNF-α (Fathima *et al.*, 2021). In the study of Langmead *et al.* (2002), fenugreek was not able to scavenge superoxide radicals except at high concentrations as was tested by the luminol-enhanced chemiluminescence oxidase cell-free system; while in the phycoerythrin degradation cell-free assay, the plant exhibited a dose-dependent peroxyl-radical scavenging effect. Fenugreek (at concentration that was

Table 5.1. Medicinal plants used for management of IBD in TPM.

No.	Scientific name	Synonym	Arabic name	Common English name	Family
1	*Allium schoenoprasum* L., *Allium ampeloprasum* L.	–	*Korrāth*	Chives	*Amaryllidaceae*
2	*Althaea officinalis* L.	–	*Khaṭmi*	Marshmallow	*Malvaceae*
3	*Anethum graveolens* L.	–	*Shebet*	Dill	Apiaceae
4	*Berberis vulgaris* L., *Berberis aristata* DC.	–	*Anbarbāris*	Barberry	*Berberidaceae*
5	*Boswellia serrata* Roxb. ex Colebr.	*Boswellia carteri* Birdw.	*Kondor*	Frankincense	*Burseraceae*
6	*Bunium persicum* (Boiss.) B. Fedtsch.	–	*Kammoon*	Cumin	*Apiaceae*
7	*Commiphora mukul* (Hook. ex Stocks) Engl.	–	*Moql*	Guggul, gugul, mukul	*Burseraceae*
8	*Commiphora myrrha* (Nees) Engl.	*Commiphora molmol* (Engl.) Engl. ex Tschirch	*Morr*	Myrrh	*Burseraceae*
9	*Coriandrum sativum* L.	–	*Kozborah*	Coriander	*Apiaceae*
10	*Cydonia oblonga* Mill.	–	*Safarjal*	Quince	*Rosaceae*
11	*Lepidium angustifolia* L.	–	*Ghobayrā*	Oleaster	*Elaeagnaceae*
12	*Lepidium sativum* L.	–	*Ḥorf*, *Ḥabb-o-rreshād*	Cress	*Brassicaceae*
13	*Linum usitatissimum* L.	–	*Katān*	Flaxseed	*Linaceae*
14	*Malus* spp.	–	*Toffāḥ*	Apple	*Rosaceae*
15	*Malva sylvestris* L., *Malva neglecta* Wallr.	–	*Khobbāzi*	Mallow	*Malvaceae*
16	*Matricaria chamomilla* L.	–	*Bāboonaj*	Chamomile	*Compositae*
17	*Medicago sativa* L.	–	*Ratbeh*	Alfalfa	*Leguminosae*
18	*Myrtus communis* L.	–	*As*	Myrtle	*Myrtaceae*
19	*Pistacia lentiscus* L.	–	*Maṣṭaki*	Mastic	*Anacardiaceae*
20	*Plantago major* L., *Plantago lanceolata* L.	–	*Lesān-ol-ḥamal*	Plantain	*Plantaginaceae*
21	*Plantago ovata* Forssk.	*Plantago ispaghula* Roxb. ex Fleming, *Plantago psyllium* L.	*Bazreqatoonā*	Psyllium, Ispagula	*Plantaginaceae*
22	*Punica granatum* L.	–	*Golnār*	Pomegranate	*Lythraceae*
23	*Quercus infectoria* G. Olivier	–	*AfeṣBaluṭ*	Oak gall, Oak fruit	*Fagaceae*
24	*Rhus coriaria* L.	–	*Somāq*	Sumac	*Anacardiaceae*
25	*Rosa × damascena* Herrm.	–	*Vard*	Rose	*Rosaceae*
26	*Rumex* spp.	–	*Ḥommāḍ*	–	*Polygonaceae*
27	*Solanum americanum* Mill.	*Solanum nigrum* L.	*Enab-o-thaʿlab*	Black nightshade	*Solanaceae*
28	*Trigonella foenum-graecum* L.	–	*Ḥolbah*	Fenugreek	*Leguminosae*
29	*Viola odorata* L.	–	*Banafsaj*	Sweet violet	*Violaceae*

Table 5.2. *In vitro* studies on medicinal plants claimed to be effective in IBD in TPM.

Study	Plant	Part	Result
Langmead *et al.* (2002)	*Trigonella foenum-graecum*	–	↓ Chemiluminescence in inflamed colorectal biopsies; antioxidant activity
Hage-Sleiman *et al.* (2011)	*Althea officinalis*	Root	Anti-inflammatory, immunostimulant
Bonaterra *et al.* (2020)	*Althea officinalis*	Root	Inhibition of LPS-induced release of TNF-α and IL-6 in macrophages; antioxidative and anti-inflammatory effects
Weber *et al.* (2020)	Combination of myrrh, coffee charcoal, and chamomile	Flower	↓ Release of inflammatory mediators; preserves barrier function of endothelial cells; modulates cytokine/ chemokine signaling and inflammation
Wei *et al.* (2015)	*Cuminum cyminum*	Essential oil	Inhibition of LOX, iNOS, COX-2, IL-1, and IL-6; phosphorylation of ERK and JNK; transcriptional activation of NF-κB
Tabarsa *et al.* (2020)	*Cuminum cyminum*	Water-soluble polysaccharides	Stimulation of pro-inflammatory cytokine production; activation of NF-κB and MAPK signaling pathways
Wu *et al.* (2018)	*Malus* spp.	Proanthocyanidins	↓ Oxidative stress; ↑ SOD, HO-1, CAT, and GSH-Px mRNA expression; downregulation of inflammatory markers

likely present in the colonic lumen; 1 in 100 dilution) significantly decreased chemiluminescence in inflamed colorectal biopsies (Langmead *et al.*, 2002). In another study, co-administration of fenugreek seeds (as symbiotic) and *Lactobacillus acidophilus* (as prebiotic) ameliorated colonic inflammation in rats infected with *Escherichia coli* via improvement of intestinal mucosa adaptive immunity. Macroscopic and microscopic evaluation of duodenum revealed an improvement in intestinal lymphoid tissues, as well as increases in butyric acid production, goblet cell mucin secretion, and the immunoglobulin IgA from plasma cells (Bakheet, 2020).

Linum usitatissimum L.

Linseed or flaxseed is a rich source of useful fatty acids and unsaturated fatty acids such as *n*-6/*n*-3, α-linoleic acid, lignans, dietary and protein fiber, minerals, and vitamins such as vitamin E (Kaithwas *et al.*, 2011; Kaithwas and Majumdar, 2013; Rafieian-Kopaei *et al.*, 2017). A recent study reported that daily gavage with flaxseed oil at a dose range of 400 to 1600 mg/kg b.w. for 6 weeks improved colonic indices of colitis and alleviated DAI and colon cell damage in DSS-induced UC in rats. *Linum usitatissimum* oil consumption caused a significant reduction in colon inflammation and oxidative condition through increase in GSH, CAT, and SOD, as well as decrease in MDA and MPO, and IL-6 and monocyte chemoattractant protein-1 (MCP-1) levels. Flaxseed oil enhanced microbiota diversity and restored cecal microbiota dysbiosis in DSS-induced colitis (Zhou *et al.*, 2020). In Palla *et al.*'s study, various concentrations of the hydromethanolic extract of flaxseed were tested against acetic acid-induced colitis in mice. The plant at 300 and 500 mg/kg b.w. improved the severity of colitis, DAI, macroscopic and histologic damage scores. Flaxseed supplementation reduced the activities of MPO and MDA,

Table 5.3. *In vivo* studies on medicinal plants claimed to be effective in IBD in TPM.

Study	Model	Species	Plant	Part	Results
Liu (2019)	DSS-induced UC	Rat	*Trigonella foenum-graecum*	Seed	↓ Colon inflammation, TNF-α, IL-1β, and IL-6; ↑ IL-10; preserves colon mucin, ↑ colonic neutral mucin, ↓ acidic mucin; ↑ colon length, ↑ colon weight/length ratio
Fathima *et al*. (2021)	Acetic acid-induced UC	Rat	*Trigonella foenum-graecum*	Seed	↑ Total protein, ↓ oxidative stress-related enzymes; antioxidant and anti-inflammatory effects against IBD
Bakheet (2020)	Colonic inflammation, infection with *Escherichia coli*	Rat	*Trigonella foenum-graecum*	Seed	Improvement in intestinal lymphoid tissues; ↑ butyric acid production, goblet cell mucin secretion, IgA from plasma cells
Zhou *et al*. (2020)	DSS-induced colitis	Rat	*Linum usitatissimum*	Seed	Enhancement of microbiota diversity, restoration of cecal microbiota dysbiosis in colitis
Palla *et al*. (2016)	Acetic acid-induced colitis	BALB/c mice	*Linum usitatissimum*	Seed	↓ MPO and MDA; ↑ SOD, GPx, CAT, and total glutathione
Silva (2016)	TNBS-induced colitis	BALB/c mice	*Linum usitatissimum*	Seed	Attenuation of IBD; ↓ inflammatory mediators
Rahimi *et al*. (2013)	TNBS-induced UC	Rat	*Linum usitatissimum*	Seed	↓ Colonic histopathologic and macroscopic scores of colitis
Al-Hussaini *et al*. (2012)	Acetic acid–ethanol-induced UC	Rabbit	*Matricaria chamomilla*	Flower	↓ Serum electrolytes (Na and K); ↑ total leukocyte count, lymphocyte count, erythrocyte count, and blood hemoglobin percentage
Masoumi-Ardakani *et al*. (2014)	Acetic acid-induced colitis	Rat	*Matricaria chamomilla*	Flower	↓ Severe macroscopic and histopathologic colon injury indices
Hartmann *et al*. (2012)	Acetic-acid-induced colitis	Rat	*Boswellia serrata*	–	↑ Anal sphincter pressure; improvement of intestinal mucosal injury
Hartmann *et al*. (2014)	Acetic acid-induced acute colitis	–	*Boswellia serrata*	–	↓ LPO and expression level of NO and iNOS; improvement of colon tissue injury and anal sphincter pressure
Anthoni *et al*. (2006)	DSS-induced colitis	Mouse	*Boswellia serrata*	sAKBA	↓ DSS-induced P-selectin-mediated recruitment of inflammatory cells
Kiela *et al*. (2005)	DSS- or TNBS-induced colitis	Mouse	*Boswellia serrata*	Boswellia	↑ Histopathologic scoring criteria; exacerbates mucosal ulcerations, necrosis, and edema
Reddy *et al*. (2014)	DSS-induced colitis	Rat	*Lepidium sativum*	Seed oil	↓ Intestinal levels of MPO, leukotriene B4, MDA, NO, TNF-α, and IL-1β

Continued

Table 5.3. Continued.

Study	Model	Species	Plant	Part	Results
Hamedi *et al.* (2016)	Acetic acid-induced colitis	Rat	*Malva sylvestris*	Aerial parts	↓ Inflammation; high content of polyphenols and mucilaginous polysaccharides
Mencarelli *et al.* (2009)	T-cell-induced colitis	Mouse	*Commiphora mukul*	Gum resin	↓ Colon inflammation, mediators of adaptive immunity, DAI
Latifi *et al.* (2015)	Acetic acid-induced UC	Rat	*Rosa × damascena*	Volatile oil	↓ Histologic and macroscopic indices of colitis and MPO activity
Tafti *et al.* (2017)	Acetic acid-induced UC	Rat	*Myrtus communis*	Leaf	↓ Colonic MDA level, MPO activity. and luminol, lucigenin, NO, and peroxynitrite chemiluminescence values
Khosropour *et al.* (2019)	Acetic acid-induced UC	Rat	*Myrtus communis*	Essential oil	↓ Colonic inflammation and tissue edema, ulcer index, total colitis index, and MPO activity
Jabri *et al.* (2015)	Acetic acid-induced UC	Rat	*Myrtus communis*	Seed	↓ All histologic alterations
Joo *et al.* (2009)	DSS-induced UC	Mouse	*Solanum nigrum*	Fruit	↓ NO production, LDH release, and MDA formation
El-Meligy *et al.* (2015)	Acetic acid-induced colitis	Rat	*Solanum nigrum*	Extract	Effective against UC
Heidari *et al.* (2016)	Acetic acid-induced colitis	Rat	*Coriandrum sativum*	Essential oil, hydroalcoholic extract of the fruits	↓ Macroscopic and histopathologic scores, total colitis index, MPO activity
Jagtap *et al.* (2004)	Acetic acid-induced colitis	Mouse	Combination of bilwa (*Aegle marmelos*), dhanyak (*Coriandrum sativum*), musta (*Cyperus rotundus*), and vala (*Vetiveria zinzanioides*)	–	Effective against UC and CD
Solanki *et al.* (2011)	*N*-Ethylmaleimide-induced experimental colitis	Rat	*Quercus infectoria*	–	↓ Physical and oxidative stress parameters in colitis tissue, ↓ histologic scoring like colon mucosal damage index and DAI
Hong *et al.* (2009)	LPS-induced inflammation	Mouse	*Medicago sativa*	Sprout ethyl acetate extract	↓ TNF-α, IL-6, and IL-1β levels

Rashidian *et al*. (2021)	Acetic acid-induced colitis	Rat	*Bunium persicum*	Essential oil	↓ Macroscopic and microscopic indices of colitis and MPO and TNF-α activities; suppression of phosphorylated NF-κB protein expression
Minaiyan *et al*. (2012)	TNBS-induced UC	Rat	*Cydonia oblonga*	Juice and hydroalcoholic extract	↓ Extent of UC lesions, extravasation, and inflammation in damaged colon tissue
Rodríguez-Cabezas *et al*. (2002)	TNBS-induced colitis	Rat	*Plantago ovata*	Fiber-supplemented diet (5% w/w *P. ovata* seeds)	↓ Intestinal inflammation, TNF-α level, and NOS activity in the large intestine
Rodríguez-Cabezas *et al*. (2003)	TNBS-induced colitis	Transgenic rats	*Plantago ovata*	Fiber-supplemented diet (5% w/w *P. ovata* seeds)	Improvement of the intestinal cytoarchitecture, ↓ pro- or inflammatory markers
Larrosa *et al*. (2010)	DSS-induced colitis	Rat	*Punicum granatum*	UROA	↓ Inflammation markers; modulates the gut microbiota
Singh *et al*. (2009)	DSS-induced colonic inflammation	Mice	*Punicum granatum*	Flower and ellagic acid-rich fraction	↓ Histamine, MPO, and oxidative stress
Darji *et al*. (2010)	DNBS-induced colitis	Rat	*Punicum granatum*	Fruit	Attenuates oxidative stress parameters, i.e., MDA, NO, MPO levels, increase in SOD activity
Riaz *et al*. (2017)	TNBS model of colitis	Rat	*Punicum granatum*	Fruit	↓ Serum MPO, GSH, ALP, fibrinogen, and CRP levels
Shah *et al*. (2016)	DNBS-induced colitis		*Punicum granatum*	Juice (purified punicalagin)	↓ Disease extent and severity; represses gene expression of TNF-α, IL-18, and IL-1β through decrease of NF-κβ mRNA level
Kim *et al*. (2017)	DSS-induced colitis	Rat	*Punicum granatum*	Beverage	↓ Inflammation and ulceration in intestinal colitis
Parisio *et al*. (2020)	DNBS-induced abdominal pain	Rat	*Punicum granatum*	–	Prevention of development of abdominal pain
Kim *et al*. (2014)	DSS-induced colitis	Rat	Mango and *Punicum granatum*	Juice	Mango polyphenols inhibit the IGF-1R/AkT/mTOR pathway; pomegranate polyphenols downregulate the mTOR downstream pathway through reductions of ERK1/2
Kim *et al*. (2016)	DSS-induced colitis	Rat	Mango and *Punicum granatum*	Juice	Pomegranate polyphenols downregulate the mTOR downstream pathway through reductions in ERK1/2; decrease of p70S6K and RPS6, Rps6ka2, Map2k2, and Mapk1 mRNA

Continued

Table 5.3. Continued.

Study	Model	Species	Plant	Part	Results
Yoshioka *et al.* (2008)	DSS or oxazolone-induced IBD	Mouse	*Malus* spp.	Procyanidins	↑ Proportions of TCRγδ and TCRαβ-CD8αα T cells in intraepithelial lymphocytes; suppression of IFN-γ synthesis in stimulated intraepithelial lymphocytes
Skyberg *et al.* (2011)	DSS model of UC	Mouse	*Malus* spp.	Polyphenols extract	↓ Pro-inflammatory cytokine expression
Lee *et al.* (2011)	TNBS-induced colitis	Rat	*Malus* spp.	Flavonoid phloretin	↓ Colon inflammation and body weight loss; improvement of inflammation condition in HT-29 human colonic epithelial cells
Liu *et al.* (2021)	DSS-induced UC	Mouse	*Malus* spp.	Polyphenols extract	Restoration of bile acid metabolism disorder and gut microbiota dysbiosis; inhibition of the activation of NF-κB signaling; ↓ hyodeoxycholic acid level, ↑ abundance of *Verrucomicrobia* at phylum and *Bacteroides* and *Akkermansia* at genus level
Yeganeh *et al.* (2018)	DSS-induced UC	Mouse	*Malus* spp.	Peel polyphenols	↓ Mitochondrial dysfunction related to size, density, redox homeostasis, fatty acid β-oxidation, ATP synthesis, apoptosis, regulatory mitochondrial transcription factors
D'Argenio *et al.* (2012)	TNBS-induced colitis	Rat	*Malus* spp.	Polyphenols	↓ Transcription and protein levels of COX-2, TNF-α, calpain, tissue transglutaminase
Pastrelo *et al.* (2017)	Acetic acid-induced colitis	Rat	*Malus* spp.	–	Suppression of iNOS gene expression; ↑ CuZnSOD mRNA level
Zhou and Mineshita (2000)	TNBS-induced colitis	Rat	*Berberis vulgaris*	Fruit	↓ Histologic lesions, morphologic damage, MPO activity
Minaiyan *et al.* (2011)	Acetic acid-induced colitis	Rat	*Berberis vulgaris*	Fruit	Protection against colonic damage; ↓ ulcer index
Tanideh *et al.* (2014)	Acetic acid- induced colitis	Rat	*Berberis vulgaris*	Fruit	↓ Inflammation in colonic tissue, ↓ MDA activity
Kim *et al.* (2020)	DSS-induced colitis	Mouse	*Rumex* spp. (*R. japonicus*)	–	↓ TNF-α, IL-1β, IL-6, occludin, zonula occludens-1, p21, p53, and Bcl-2; ↑ expression of IL-10, claudin-2, and cleaved caspase-3; prevention of colitis by protecting tight junction connections in colonic tissue
Gioxari *et al.* (2011)	TNBS-induced colitis	Rat	*Pistacia lentiscus*	Fruit	Regulation of oxidant/antioxidant balance and modulation of inflammation

Naouar *et al*. (2016)	TNBS-induced colitis	Rat	*Pistacia lentiscus*	Oil	Attenuation of erosion, cryptitis, irregular crypts, and crypt loss; curative intervention improvement of ulceration, and hyperplasia, cryptitis, irregular crypts and crypt loss
Zahouani *et al*. (2020)	Acetic acid-induced UC	Rat	*Pistacia lentiscus*	Leaf	Improvement in macroscopic colonic damage; ↓ LPO and carbonylation of proteins, ↑ activities of antioxidant enzyme such as SOD, CAT, GSH, and GSTs; ↓ IL-6 level in colon and plasma
Boutemine *et al*. (2021)	DSS-induced UC	Mouse	*Pistacia lentiscus*	Leaf	↓ Clinical score of acute colitis, DAI, and histologic damages; ↓ pro-inflammatory mediator levels in plasma
Fatani *et al*. (2016)	Acetic acid-induced colitis	Rat	*Commiphora molmol*	–	↓ Pro-inflammatory cytokines, PGE_2, NO, and TBARS
Khodakarm-Tafti *et al*. (2015)	Acetic acid-induced colitis	Rat	*Elaeagnus angustifolia*	Fruit	↓ Erosion and ulceration, inflammatory cell infiltration; regeneration and improvement of mucosa

GST, glutathione *S*-transferase; TBARS, thiobarbituric acid-reactive substances.

Table 5.4. Clinical studies on medicinal plants claimed to be effective in IBD in TPM.

Study	Study design	No. of patients	Disease	Plant	Part of plant	Control group	Duration of treatment	Results
Langhorst *et al*. (2013)	Randomized, double-blind, double-dummy	96	UC	Combination of chamomile, *Commiphora molmol*, *Coffea arabica*	Flower dry extract	Mesalazine	12 months	Complete mucosal healing
Langhorst *et al*. (2014)	Randomized, double-blind, double-dummy	126	UC	Combination of myrrh, chamomile extract, coffee charcoal	Flower dry extract	Healthy control subjects	12 months	↑ $CD4^+$ T cells, $CD4^+CD25^{med}$ effector T cells, and $CD4^+CD25^{high}$ regulatory T cells
Gupta *et al*. (1997)	Randomized	–	UC in grade II and III	*Boswellia serrata*	Gum resin preparation	Sulfasalazine	6 weeks	Improvement of stool properties, histopathologic and microscopic manifestations of rectal biopsies
Madisch *et al*. (2007)	Double-blind, placebo-controlled, multicenter trial	31	Collagenous colitis	*Boswellia serrata*	Extract	Placebo	6 weeks	↑ Population of patients in clinical remission
Mohammad (2012)	Clinical study	32	IBS	*Anethum graveolens*	Leaves and stems	Mebeverine	2 weeks	Improvement in all IBS symptoms including pain severity, pain frequency, stool frequency, stool consistency, abdominal distension, incomplete evacuation, urgency, and passing of mucus

Tavakoli *et al*. (2019)	Randomized, double-blind, placebo-controlled clinical trial	20	UC	*Rosa × damascena*	Oil	Placebo	2 months	↓ Mayo Clinic (DAI) scores
Abbasi *et al*. (2018)	Randomized double-blind, placebo-controlled clinical trial	52	IBS	*Bunium persicum*	Consumption of powder	Placebo	5 weeks	No effect on symptoms and inflammatory markers of patients
Fernández-Bañares *et al*. (1999)	Open label, parallel-group, multicenter, randomized clinical trial	105	UC	*Plantago ovata*	Seed	Mesalamine	12 months	Effective to maintain remission in patients with UC
Kim *et al*. (2016)	Randomized, double-blind, placebo-controlled clinical trial	–	UC	Mango + *Punicum granatum*	Beverages	Placebo	4 weeks	↓ LCAI score
Kamali *et al*. (2017)	Randomized, double-blind, placebo-controlled clinical trial	78	UC	*Punicum granatum*	Peel extract	Placebo	4 weeks	Therapeutic response significantly higher in patients with hot temperament compared to patients with cold temperament
Baghizadeh *et al*. (2021)	Randomized, double-blind, placebo-controlled clinical trial	61	UC	*Plantago major*	Seed	Placebo	8 weeks	↓ Visible blood in stool, distension, and anal pain
Tafazoli *et al*. (2022)	Case report	1	Chronic pancolitis	*Plantago major*	–	–	2 years	Improvement in clinical examination, patient's history, colonoscopy, pathology test, and myoclonic score

Continued

Table 5.4. Continued.

Study	Study design	No. of patients	Disease	Plant	Part of plant	Control group	Duration of treatment	Results
Kaliora *et al.* (2007)	–	18	Mild to moderately active CD	*Pistacia lentiscus*	Chios mastic resin	Healthy controls treated with Chios mastic resin	4 weeks	↓ CD activity index, ↓ plasma IL-6 and CRP, ↓ total antioxidant potential
Papada *et al.* (2018)	Randomized, double-blind, placebo-controlled clinical trial	60(CD, n = 40; UC, n = 20)	Active IBD (CD and UC)	*Pistacia lentiscus*	*P. lentiscus* supplement	Placebo	3 months	↓ oxLDL, oxLDL/HDL, and oxLDL/LDL
Papada *et al.* (2019)	Randomized, double-blind, placebo-controlled clinical trial	60(CD, n = 40; UC, n = 20)	Active IBD (CD and UC)	*Pistacia lentiscus*	*P. lentiscus* supplement	Placebo	3 months	↓ Fecal lysozyme and fibrinogen, increase of Fe

oxLDL, oxidized low-density lipoprotein; LDL, HDL, high-density lipoprotein.

while inducing SOD, GPx, CAT, and total GSH. *L. usitatissimum* suppressed the immune response and neutrophil infiltration by hampering TNF-α and IFN-γ levels and increasing IL-17 levels (Palla *et al.*, 2016). The same group compared the power of the hydromethanolic extracts of flaxseed and flaxseed oil to ameliorate acetic acid-induced colitis in mice. It was shown that both flaxseed oil and its crude extract are favorable to manage IBD by means of anti-inflammatory, antibacterial, and antispasmodic activities. However, their mechanisms of actions are different. For instance, flaxseed extract was more effective in reducing mortality and improvement of colonic ulcers than flaxseed oil at 500 mg/kg; while the oil exhibited higher anti-inflammatory effect. Antispasmodic and anticholinergic activities of the crude extract (0.03 and 0.1 mg/ml) were greater than for flaxseed oil (1 and 3 mg/ml). Flaxseed oil (9 and 14 μg/ml) has shown antibacterial activity against enteropathogenic *E. coli* (EPEC), enterotoxigenic *E. coli* (ETEC), and enteroaggregative *E. coli* (EAEC); whereas the crude extract was effective against EPEC at 100 μg/ml (Palla *et al.*, 2020).

In a study by Silva, the influence of the hydrolysis with Alcalase® and simulated GI digestion on the antioxidant capacity of different flaxseed products and their protective effects on 2,4,6-trinitrobenzenesulfonic acid (TNBS)-induced colitis in mice were assessed. Flaxseed protein hydrolysates and phenolic fractions inhibited NO and TNF-α secretion in stimulated macrophage RAW cells and also reduced ROS production, weight loss, systemic indicators of colonic mucosal damage and inflammation, Th1 and Th17 responses, T-cell proliferation, and inflammatory cytokine levels (Silva, 2016). An Iranian traditional herbal formula called "Magliasa", containing 8% w/w *L. usitatissimum* seeds, remarkably attenuated colonic histopathologic and macroscopic scores of colitis in TNBS-induced UC in rats. The levels of TNF-α, IL-1β, MPO, and lipid peroxidation (LPO) were significantly lower in Magliasa-treated animals. It is noticeable that Magliasa at any dose could not increase the antioxidant capacity (Rahimi *et al.*, 2013). Together, flaxseed extract ameliorates the severity of colitis by scavenging oxygen-derived free radicals and suppression of intestinal inflammation, thereby it may be a new approach for immunotherapy in IBD.

***Althaea officinalis* L.**

Marshmallow is a rich source of acidic polysaccharides, starch, pectins, mucilage, and flavonoids with significant antioxidant and soothing properties (Capek *et al.*, 1987; Gudej, 1991; Sendker *et al.*, 2017). A recent *in vitro* study reported antioxidant and anti-inflammatory effects of the root extracts of *Althaea officinalis*. The plant was able to protect human macrophages against hydrogen peroxide (H_2O_2)-induced cytotoxicity and H_2O_2-induced ROS production. *A. officinalis* inhibited the lipopolysaccharide (LPS)-induced release of TNF-α and IL-6 in macrophages, additionally improving the migratory capacity of macrophages (Bonaterra *et al.*, 2020). Oral administration of *A. officinalis* extract, once daily for 14 days, led to a protective effect against pyloric-ligation/indomethacin-induced gastric ulceration in rats, probably via its antioxidant properties, inhibition of histamine and gastrin release, suppression of formation of pro-inflammatory cytokines like TNF-α and IL-1β, and promotion of mucin, NO, prostaglandin E_2 (PGE_2) and prostaglandin I_2 (PGI_2) contents. The expression of the protective heme oxygenase-1 (HO-1) and cystathionine β-synthase (CBS) enzymes in gastric mucosal tissue was also increased (Zaghlool *et al.*, 2019). The potential role of the aqueous extract of *A. officinalis* flower in gastric ulcer, inflammation, and platelet aggregation was the subject of another research in a rat model. *A. officinalis* at doses of 50, 100, and 250 mg/kg b.w. exhibited anti-ulcerogenic, as well as acute and chronic anti-inflammatory effects. Inhibition of platelet aggregation was observed at 500 μg/ml concentration (Hage-Sleiman *et al.*, 2011). In accordance, Zaghlool *et al.* (2015) found that administration of the aqueous extract of marshmallow flower (100 mg/kg b.w., orally for 14 days) could protect against indomethacin-induced peptic ulcer in rats presumably via increase in gastric mucosal SOD, as well as inhibition of histamine release. These findings indicate that marshmallow preparations are a proper choice for treatment of irritated mucosal tissue, and thus *A. officinalis* could be a candidate for future investigations on IBD.

Matricaria chamomilla L.

Oral administration of the ethanolic extract of chamomile flower at a dose of 500 mg/kg daily (from 48 h prior to induction of colitis) reduced serum electrolytes (sodium (Na) and potassium (K)), and enhanced total leukocyte count, lymphocyte count, erythrocyte count, blood hemoglobin percentage, and packed cell volume percentage in glacial acetic acid–ethanol-induced UC in rabbits (Al-Hussaini *et al.*, 2012). In a rat model of acetic acid-induced colitis, aqueous extract of chamomile flowers (20 and 30 mg/kg) significantly reduced severe macroscopic and histopathologic colon injury indices, which was implicated in reduction of MPO activity and MDA content of inflamed colon (Masoumi-Ardakani *et al.*, 2014).

Boswellia serrata Roxb. ex Colebr.

Oleo-gum-resin of *Boswellia* spp., known as frankincense or olibanum, contains boswellic acids as the major constituents with potent anti-inflammatory properties. *In vitro* and *in vivo* studies have shown that boswellic acids inhibit the leukotriene biosynthesis in neutrophilic granulocytes via inhibition of 5-lipoxygenase. Since the inflammatory process in IBD is associated with increased function of leukotrienes, it is thought that *Boswellia serrata* has a favorable impact on inflammatory diseases such as IBD.

In acetic acid-induced colitis in rats, oral gavage of *B. serrata* extract (34.2 mg/kg daily) for 2 days before and after the induction of colitis led to a significant increase in anal sphincter pressure, and improved the intestinal mucosal injury and histologic aspects of the colon. The protective effect of *B. serrata* was due to increases in SOD, GPx, and GSH activities, as well as reduction of LPO (Hartmann *et al.*, 2012). In another study using acetic acid-induced acute colitis, pretreatment and treatment with *B. serrata* extract (oral gavage, 34.2 mg/kg daily) for 2 days suppressed LPO and the expression level of NO and iNOS, while improving colonic tissue injury and anal sphincter pressure (Hartmann *et al.*, 2014). Intraperitoneal injection of a semisynthetic form of acetyl-11-keto-β-boswellic acid (sAKBA) (5 mg/kg/day, for 5 days), the most potent anti-inflammatory component of frankincense, decreased leukocyte and platelet infiltration into the inflamed colonic site, improving macroscopic and microscopic colitis symptoms and DAI. In addition, sAKBA reduced DSS-induced P-selectin-mediated recruitment of inflammatory cells and upregulation in mice with colitis (Anthoni *et al.*, 2006). Microarray analyses of hepatic gene expression (i.e., a large group of lipid metabolism-related genes, and detoxifying enzymes) pointed out the potential hepatotoxic effects of *Boswellia* (at high dosage) with evidence of steatosis, as confirmed *in vitro* in HepG2 cells, which should be considered in clinical use (Kiela *et al.*, 2005).

Given the positive outcomes of experimental studies on therapeutic benefits of *B. serrata* extracts or its bioactive phytochemicals on IBD, several clinical trials were performed. In 1997, Gupta *et al.* conducted a clinical trial in patients with UC in grade II and III using *B. serrata* gum resin preparation (350 mg, 3 times daily for 6 weeks). The gum resin improved stool properties, histopathologic and microscopic manifestations of rectal biopsies, as well as blood biochemicals (i.e., hemoglobin, serum iron (Fe), calcium (Ca), phosphorus (P), proteins, total leukocytes, eosinophils). While 82% of patients receiving *B. serrata* experienced remission, the remission rate in the sulfasalazine group was 75% (Gupta *et al.*, 1997). Gupta and colleagues developed another human trial in 2001. Administration of the same preparation at a daily dose of 900 mg for 6 weeks to patients with chronic colitis resulted in the same outcomes as their previous study on UC patients, although, in the *B. serrata*-treated group, the remission rate was higher than that of sulfasalazine-treated patients (Gupta *et al.*, 2001). Results of another double-blind, randomized, placebo-controlled trial revealed that oral consumption of *B. serrata* extract (400 mg, 3 times daily for 6 weeks) was promising for treatment of collagenous colitis. *B. serrata* enhanced the number of patients in clinical remission compared to the placebo group (per protocol: 63.6%, 95% confidence interval (CI) 30.8–89.1% versus 26.7%, 95% CI 7.7–55.1%, $P = 0.04$; intention-to-treat: 43.8% versus 26.7%, $P = 0.25$); whereas there was no effect on histology and quality of life (Madisch *et al.*, 2007).

Lepidium sativum L.

Garden cress seed oil is a rich source of unsaturated fatty acids such as *n*-3 fatty acids. In DSS-induced colitis in rats, addition of garden cress seed oil to the animals' diet reduced the intestinal levels of MPO, leukotriene B4, MDA, NO, TNF-α, and IL-1β. Moreover, DAI scores, colon length, macroscopic scores and morphologic damage, and the GSH content were improved in colitis subjects after receiving *L. sativum* seed oil (Reddy *et al.*, 2014). The mucilage extracted from garden cress seeds (100 mg/kg b.w.) was fed to rats with indomethacin (6 mg/kg daily)-induced colitis for 2 weeks before and 7 days after induction of colitis. Herbal treatment ameliorated IBD through mediating the erythrocyte sedimentation rate, plasma lactate dehydrogenase (LDH) activity, and reduction of intestinal TNF-α, NO, and MDA levels. Pretreatment with the mucilage increased GSH level in rats in comparison to untreated animals (Akl *et al.*, 2021). As mentioned earlier, Magliasa is a favorable herbal preparation for the treatment of colitis. It is noteworthy that the formulation of Magliasa contains 50% w/w of the intact unmilled seeds of *L. sativum* (Rahimi *et al.*, 2010). Overall, garden cress seed extracts alleviate oxidative stress, suppress inflammatory mediators, and reduce colonic damage in experimentally induced colitis models.

Anethum graveolens L.

Irritable bowel syndrome (IBS) is a long-term chronic GI condition. The effectiveness of *Anethum graveolens* (dill) in 32 patients with IBS was examined in a clinical trial against the standard control, mebeverine (135 mg, 3 times daily), for 2 weeks. In dill-treated group (1 capsule with 500 mg dill powder, three times a day), there was improvement in all IBS symptoms including pain severity, pain frequency, stool frequency, stool consistency, abdominal distension, incomplete evacuation, urgency, and passing of mucus, in both individual symptom scores and total improvement scores (Mohammad, 2012).

Malva sylvestris L.

Pretreatment with the aqueous fraction (obtained from aerial parts) and an isolated polysaccharide compound of *M. sylvestris* at a dose of 200 mg/kg daily for 5 days consecutively alleviated the inflammatory symptoms of experimental IBD in acetic acid-induced colitis in rats. The efficacy of ethanolic and *n*-hexane fractions was lower in preventing inflammation. The aqueous fraction was found more protective against UC compared to the isolated polysaccharide, suggesting that synergism of other phytochemicals in the aqueous fraction might contribute in its healing effect. The aqueous fraction of *M. sylvestris* is reported to have a high content of polyphenols and mucilaginous polysaccharides (Hamedi *et al.*, 2016).

Commiphora mukul (Hook. ex Stocks) Engl.

The gum resin of *Commiphora mukul* is known as "guggul" and has been mentioned in TPM textbooks to have anti-inflammatory function. In a mouse model of T-cell-induced colitis, it was shown that a plant sterol isolated from the gum resin of *C. mukul* tree called guggulsterone (GS) exerts immunomodulatory activities. GS (30 mg/kg, for 5 days) attenuated colon inflammation by reducing the mediators of adaptive immunity, DAI, as well as macroscopic and microscopic damage scores in both TNBS- and oxazolone-induced colitis in mice. *Cis*-GS (*E*-GS) was found more effective than *trans*-GS (*Z*-GS) in amelioration of IBD *in vivo*. Co-culture of GS and lamina propria-derived T cells obtained from mice treated with TNBS or oxazolone led to reduction of IL-2, IL-4, and interferon-γ (IFN-γ) production, as well as T-cell proliferation (Mencarelli *et al.*, 2009). Chemical analyses revealed that *Z*-GS is mainly responsible for the anti-inflammatory activity of this plant (Sosa *et al.*, 1993). Cheon and colleagues reported that GS suppressed IL-1β- or LPS-induced intercellular adhesion molecule-1 (ICAM-1) expression, NF-κB transcription activity, and IκB kinase phosphorylation in human Caco-2 cells and rat non-transformed IEC-18 cells (Cheon *et al.*, 2006). GS treatment significantly suppressed the mRNA expression of pro-inflammatory cytokines, including IL-1β, TNF-α, and iNOS, and inhibited the NF-κB activity in LPS-stimulated macrophages, a representative of its anti-inflammatory activity (Zhang *et al.*, 2016). In a recent study, following cerebral ischemic injury, administration of *Z*-GS ameliorated astrocyte-mediated neuroinflammation through downregulation of the TLR4 pathway (Liu *et al.*, 2018). Activated TLR4/NF-κB signaling pathways are involved in the pathophysiology of many chronic and inflammatory impairments such as IBD (Momtaz *et al.*, 2021).

Rosa × damascena Herrm.

Therapeutic values of *Rosa × damascena* are mainly attributed to its high content of polyphenolic components. Several animal and human studies confirmed that *R. × damascena* extracts or its volatile oil interfere in various cellular and molecular pathways and repress oxidative- or inflammation-induced damage of the body cells, tissues, or organs (Nayebi *et al.*, 2017). A hydroalcoholic extract of the plant and its volatile oil were tested in acetic acid-induced UC in rats. Oral administration of *R. × damascena* extract at all doses examined (250, 500, and 1000 mg/kg b.w, for 4 days), and the lowest doses of volatile oil given orally (100 μl/kg b.w.) or intraperitoneally (125 μl/kg b.w.), alleviated histologic and macroscopic indices of colitis, as well as MPO activity (Latifi *et al.*, 2015). In a randomized, double-blind clinical trial on 20 patients with moderate-to-severe UC, consumption of two rose oil soft capsules (1000 mg) 3 times daily for 2 months caused numerical (but not statistical) reduction of Mayo Clinic (DAI) scores in both placebo and intervention groups. Inflammatory Bowel Disease Questionnaire (IBDQ)-9 score was also elevated in both interventions before and after the treatment; however, the difference was not significant. In TPM, there are some topical remedies that are generally applied on the stomach, lower abdomen, lower back, and liver to relieve digestive discomforts such as dyspepsia, gastritis, IBD, digestive ulcers, intestinal worms, and infections. *R. × damascena* is one of the most common medicinal herbs that are used to formulate topical medications for GI diseases. It was proposed that suppression of oxidative stress, inflammatory cytokines, and signaling pathways of inflammatory responses, or improvement of analgesic factors are main mechanisms of the therapeutic effects of this plant (Tafti *et al.*, 2017).

Myrtus communis L.

Phytochemical characterization exhibited that phenolic acids, tannins, flavonoids, glycosides, and terpenes are the key components of *Myrtus communis*; whereas monoterpene and sesquiterpene derivatives are the main ingredients of the essential oil of this plant (Sisay and Gashaw, 2017). In acetic acid-induced UC in rats, treatment with *M. communis* ethanolic extract (50 and 100 mg/kg, oral gavage for 3 days) reduced colonic MDA level, MPO activity, and luminol, lucigenin, NO, and peroxynitrite chemiluminescence values. Macroscopic and microscopic signs of colitis severity were alleviated, along with a significant increase in tissue GSH level (Tafti *et al.*, 2017). In the same animal model, *M. communis* hydroalcoholic extract and its essential oil were given orally to rats, 2 h before induction of colitis and for a further 4 days. The plant extract (50, 100, and 200 mg/kg) and the essential oil (62.5 and 125 μl/kg b.w.) were statistically effective in improvement of colonic inflammation and tissue edema, ulcer index, total colitis index, and MPO activity (Khosropour *et al.*, 2019). Jabri *et al.* (2015) reported that myrtle seeds are rich in anti-inflammatory compounds such as malvidin-*O*-glucoside and delphinidin-3-*O*-glucoside. The aqueous extract of myrtle berry seeds significantly reduced all histologic alterations caused by acetic acid-induced UC in rats. The extract also inhibited ROS production by neutrophils, colonic H_2O_2, and free Fe and Ca levels during the inflammatory reaction and improved the enzymatic activity of SOD, CAT, and GPx. The extract improved the nonenzymatic antioxidants such as GSH and sulfhydryl group (–SH) in the injured colon mucosa. Inflammation retrieval and protective effect of this extract was implicated in its ROS scavenging potential or to inhibition of intracellular Ca deregulation (Jabri *et al.*, 2015). The essential oil and crude extract of the leaves of *M. communis* showed antiulcerogenic effect in an animal model of gastric ulcer, which was confirmed by its antioxidant activity *in vivo* and *in vitro*. Recovery of mucosal damage, suppression of the oxidative status associated with inflammation, and downregulation of pro-inflammatory cytokines such as IL-1β and TNF-α, as well as iNOS, ICAM-1, and cervical intraepithelial lesions 1 (CIN 1), were thought to be the main mechanisms involved in gastroprotective effect of *M. communis* leaves' essential oil and extract (Benchikh, 2018).

Solanum nigrum L.

A growing body of evidence indicates that *Solanum nigrum* (black nightshade) fruits have various valuable pharmacologic activities like antiulcer, antioxidant, and anti-inflammatory properties. In a recent study three *S. nigrum* genotypes (*S. scabrum*, *S. sacharroides*, and *S. villosum*) exhibited antiulcerogenic effects through improving both gastric mucosal texture and microscopic damage in a rat model of gastric ulcer (Mureithi, 2020). Another mechanistic study demonstrated that 14 days' supplementation with *S. nigrum* extract attenuated pyloric-ligation/indomethacin-induced gastric ulceration in rats through antioxidant action and suppression of pro-inflammatory cytokine formation like TNF-α and IL-1β, as well as promotion of mucin content (Zaghlool *et al.*, 2019).

It was shown that an isolated glycoprotein from *S. nigrum* fruits prevented NO production, LDH release, and MDA formation. This glycoprotein downregulated the activities of NF-κB (p50) and AP-1 (c-Jun) and regulated the expression of iNOS and COX-2 due to its antioxidative properties in DSS-induced UC in mice (Joo *et al.*, 2009). In another animal study in acetic acid-induced UC, *S. nigrum* at concentrations of 125, 250, and 500 mg/kg was found effective as either a prophylactic or protective administration; however, its activity was not as potent as dexamethasone at 0.1 mg/kg (El-Meligy *et al.*, 2015). Together, the potential action of different parts of this plant, especially the fruits, in both animal and clinical IBD models as well as the probable mechanisms involved remain to be elucidated.

Coriandrum sativum L.

The widespread presence of coriander in food and pharmaceutical preparations, besides its carminative, antispasmodic, anti-inflammatory, and antioxidant assets, makes *Coriandrum sativum* a proper choice for IBD management. Coriander has strong anti-inflammatory properties that are mainly correlated with suppression of pro-inflammatory cytokines and their downstream pathways such as NO, iNOS, IL-1β, IL-6, TNF-α, tumor necrosis factor receptor 1 (TNFR1), ROS, COX-2, NF-κB, and MAPK (Wu *et al.*, 2010; Al-Snafi, 2016; Prachayasittikul *et al.*, 2018; Sari *et al.*, 2021).

Heidari *et al.* (2016) provided evidence that oral intake of the essential oil and hydroalcoholic extract of the fruits of *C. sativum* possessed protective and therapeutic effects on the acetic acid-induced model of colitis. Following coriander treatment (extract at 500 and 1000 mg/kg b.w., and essential oil at 0.5 ml/kg b.w.), colon biopsies showed improved macroscopic and histopathologic scores, alleviation of total colitis index, and decreased MPO activity (Heidari *et al.*, 2016). Pretreatment with a polyherbal ayurvedic formulation containing four different plant species including *Aegle marmelos* Corrêa, *C. sativum*, *Cyperus rotundus*, and *Vetiveria zizanioides* was found to be beneficial in acetic acid-induced colitis in mice and indomethacin-induced enterocolitis in rats, which are representative of UC and CD, respectively (Jagtap *et al.*, 2004).

Viola odorata L.

In recent studies anti-inflammatory, analgesic, antioxidant, diuretic, antihypertensive, and antibacterial activities of sweet violet were reported (Feyzabadi *et al.*, 2017). Hepatoprotective activity of *Viola odorata* is the most studied property of this herb. In animal studies, *V. odorata* demonstrated protective effects against oxidative damage through suppression of ROS, LPO, and protein carbonyl, as well as increase in GSH content. Administration of *V. odorata* also regulated serum hepatic markers such as blood alanine aminotransferase (ALP), aspartate aminotransferase (AST), alkaline phosphatase (ALT), total antioxidant capacity, and total bilirubin (Qadir *et al.*, 2014; Al-Snafi, 2016; Habibi *et al.*, 2019). Anti-inflammatory and antioxidant capabilities of *V. odorata* have been documented by many other experimental models. Alipanah *et al.* (2018) investigated the cytotoxic, antioxidant, and antimetastatic properties of the hydroalcoholic extract of *V. odorata* in a breast cancer-implanted mice model. It was shown that levels of LDH, γ-glutamyl transferase (GGT), ALP, carcinoembryonic antigen, and cancer antigen 15-3 in serum, and the activities of CAT and SOD in tumor tissue, were reduced (Alipanah *et al.*, 2018). Based on our knowledge, there is no up-to-date information about apparent curative applications of *V. odorata* preparations on digestive system disorders.

Quercus infectoria G. Olivier

Most of the studies about the pharmacological application of *Quercus infectoria* galls have focused on the anti-inflammatory and antibacterial properties of this species, which is implicated by the presence of tannins and, in particular, gallic acid (Kahkeshani *et al.*, 2019; Elham *et al.*, 2020). Oral administration of the alcohol extract of *Q. infectoria* significantly inhibited carrageenan, histamine, serotonin, and PGE_2-induced paw edema, while topical application inhibited phorbol-12-myristate-13-acetate (PMA)-induced superoxide ($O_2^{\cdot -}$) production in ear inflammation. *In vitro* exposure of rat peritoneal macrophages to gall extract ameliorated LPS-stimulated PGE_2 and NO production. The extract also inhibited the iNOS and formyl-Met-Leu-Phe (FMLP)-stimulated degranulation in the neutrophils. Overall, oral or topical administration of *Q. infectoria* gall extract prevented the production of some inflammatory mediators (Kaur *et al.*, 2004).

In *N*-ethylmaleimide-induced experimental colitis in rats, oral intake of *Q. infectoria* at doses of 300 and 450 mg/kg for 18 days could significantly prevent physical and oxidative stress parameters in colon tissue through improvement of the course of experimental colitis, histological scoring like colon mucosal damage index, and DAI. Downregulation of NF-κB, NO, and improvement of SOD were described to contribute to the protective effect of *Q. infectoria* (Solanki *et al.*, 2011). Khanavi *et al.* (2014) conducted an investigation on biochemical and pathological parameters of colitis in rats treated with galls of *Q. brantii*. Both gall powder and gall hydroalcoholic extract (gavaged for 10 days at dose of 500 mg/kg) decreased TNF-α, IL-1β, LPO, and MPO activities in TNBS-induced inflammation in colon tissue (Khanavi *et al.*, 2014). During the last decade, it was shown that gallic acid and its derivatives enhance the gut microbiome activities and modulate immune responses, thereby maintaining intestinal health. Gene sequencing on mice fecal combined with metabolome analysis exhibited that gallic acid treatment could modulate the gut microbiota composition by alterations in metabolic and bacterial profiles (Yang *et al.,* 2020). Administration of gallic acid increased carbohydrate and bile acid metabolism and decreased the amino acid metabolism in DSS-induced colitis, signifying that gallic acid significantly attenuated UC by influencing the composition of the mouse gut microbiome and metabolites. Upon gallic acid treatment, the mRNA expressions of *IL-21* and *IL-23*, DAI, and the histopathologic evidence of colon injury were reduced. The compound also significantly upregulated the expression of Nrf2 and its downstream targets, including *UDP-GT* and *NQO1*, demonstrating that the protective effect of gallic acid on experimental colitis might be due to the antioxidant activity of this compound (Pandurangan *et al.*, 2015). The compound suppressed LPS-induced inflammation in transgenic mice through downregulating the NF-κB pathway and modulating the immune system responses (Hsiang *et al.*, 2013). In a TNBS-induced UC model, gallic acid increased the expressions of IL-4 and IL-10, while it downregulated IL-1, IL-6, IL-12, IL-17, IL-23, TGF-β, and TNF-α expressions. This indicates that gallic acid plays an anti-inflammatory role in UC via inhibiting the NF-κB pathway (Zhu *et al.*, 2019). Likewise, gallic acid treatment improved DAI, the histopathologic signs of epithelium injury and inflammatory cell infiltration score, and the markers of oxidative stress (i.e., LPO); whereas it decreased the inflammatory MPO activity of inflamed colon homogenate in TNBS-induced UC in rats (Khodayar *et al.*, 2018; Marinov *et al.*, 2019).

Medicago sativa L.

Phytoestrogenic (i.e., isoflavones) and antioxidant compounds (e.g., carotenoids and anthocyanins) are putative bioactive pharmacologic constituents in alfalfa. Suppression of the expression of genes encoding pro-inflammatory cytokines such as TNF-α, IL-6, and IL-1β, and downregulation of inflammation-related signaling pathways (i.e., extracellular signal-regulated kinase (ERK)/NF-kB) are supposed to be involved in anti-inflammatory action of this herb (Choi *et al.*, 2013; Chen *et al.*, 2015a,b). A recent study demonstrated that the hydroalcoholic extract of *Medicago sativa* exerted positive pathologic and behavioral impacts on brain injury caused by nicotine *in vivo*. Upon alfalfa treatment, the serum levels of antioxidant parameters including GPx and SOD increased, so the MDA concentration was reduced. Furthermore, inflammatory responses were abrogated in alfalfa-treated animals (Raeeszadeh *et al.*, 2021). In another animal model, supplementation with alfalfa sprout ethyl acetate extract lowered the serum TNF-α, IL-6, and IL-1β levels in LPS-induced inflammation in mice. The same extract significantly reduced IL-6 and IL-1β production and the NF-κB transactivation of mitogen-stimulated RAW264.7 cells (Hong *et al.*, 2009). Xie *et al.* reported that oral administration of alfalfa polysaccharides improved the function of the mouse immune system, significantly enhanced the leukocyte and lymphocyte counts, and led to improvements in spleen and thymus indices. They also reported that alfalfa polysaccharides improved the intestinal morphology in the immunological system in mice (Xie *et al.*, 2019). Thus, the plant has presented antioxidant and anti-inflammatory properties, although *in vivo* anticolitis activity still needs to be investigated.

Bunium persicum (Boiss.) B. Fedtsch., *Cuminum cyminum* L.

Despite the presence of solid data on the efficiency of herbal preparations of *Bunium persicum*, and *Cuminum cyminum* in TPM textbooks, the lack of information in this field in modern medicine is significant.

In acetic acid-induced colitis in rats, oral intake of *B. persicum* essential oil (200 and 400 mg/kg) for 5 days decreased macroscopic and microscopic indices of colitis (Rashidian *et al.*, 2021). In a double-blind, placebo-controlled study on patients with IBS, consumption of *B. persicum* powder for 5 weeks showed no significant effect on symptoms and inflammatory markers compared to the placebo group (Abbasi *et al.*, 2018).

C. cyminum contains various alkaloids and polyphenols (Rai *et al.*, 2012), as well as high levels of oxygenated monoterpenes, especially γ-terpinene, cuminaldehyde, benzenemethanol, *p*-cymene, and limonene in the essential oil (Sanei-Dehkordi *et al.*, 2016). According to the results of an animal study on rats with IBS, administration of the aqueous extract of *C. cyminum* seeds reduced movements of the GI tract and IBS, which was assumed to be associated with the presence of antioxidant substances in this plant (Teimouri *et al.*, 2019). *In vivo*, *C. cyminum* and its bioactive compounds presented strong analgesic and acute and chronic anti-inflammatory activities (Bhat *et al.*, 2014; Golabi *et al.*, 2020). *In vitro*, cumin essential oil inhibited the mRNA expressions of lipoxygenase (LOX), while blocking the transcriptional activation of NF-κB and inhibiting the phosphorylation of ERK and c-Jun N-terminal kinase (JNK) (Wei *et al.*, 2015). Recent research reported that water-soluble polysaccharides of *C. cyminum* stimulated the production of pro-inflammatory cytokines *in vitro* in macrophages and natural killer cells (Tabarsa *et al.*, 2020). Also, in ethanol-induced hepatic damage, *C. cyminum* could successfully reduce the increased activity of AST, ALT, and GGT, and inhibit LPO (Aruna *et al.*, 2005).

Allium ampeloprasum L., *Allium schoenoprasum* L.

Organosulfur and polyphenolic compounds, especially ferulic acid, are major active ingredients of *Allium* species (Putnik *et al.*, 2019). In TNBS-induced colitis in rats, ferulic acid (20 and 40 mg/kg, orally) ameliorated inflammation via inhibition of oxido-nitrosative stress, apoptosis, pro-inflammatory cytokine production, and downregulation of COX-2 synthesis (Sadar *et al.*, 2016). Administration of sodium ferulate in the rat model of acetic acid-induced colitis improved the colon mucosa damage index and the histopathologic score, probably due to the antioxidant activity, inhibition of arachidonic acid metabolism, and NF-κB expression (Dong *et al.*, 2003). Upon treatment with ferulic acid and its glycosides, particularly ferulic acid rutinoside, there was a significant increase in transforming growth factor-β1 (TGF-β1) levels in THP-1-derived DCs. Oral administration of ferulic acid rutinoside attenuated DSS-induced colitis in mice and induced the differentiation of regulatory T cells, indicating the immunomodulatory effect of this compound on intestinal inflammation (Katayama *et al.*, 2017). The seeds of *A. ampeloprasum* are a part of Magliasa herbal formulation with a positive impact on IBD (Rahimi *et al.*, 2013). In a recent study, daily treatment of onion bulb extract (*Allium cepa*) at 100–200 mg/kg reduced colonic expression and activity of pro-inflammatory chemokines and apoptotic signaling pathways and mediators in DSS-induced colitis rats (Khajah *et al.*, 2019).

Cydonia oblonga Mill.

Cydonia oblonga fruits contain various antioxidants such as caffeoylquinic acids and rutin, whereas the seeds mostly contain mucilage and organic acids (Silva *et al.*, 2005; Al-Khazraji, 2013). It was shown that intraperitoneal injection of quince juice (800 mg/kg) and quince hydroalcoholic extract (500 mg/kg) for 5 days effectively mitigated UC lesions, extravasation, and inflammation in damaged colon tissue similarly to that of the standard, dexamethasone, in a TNBS-model of UC in rats (Minaiyan *et al.*, 2012). A plethora of studies have confirmed the anti-inflammatory properties of quince pulp, peel, seed, and leaves. Furthermore, it was shown that quince presents anti-inflammatory effects through suppression of NO, LPO, IL-6, and TNF-α levels (Ahmed and Bastawy, 2014). Essafi-Benkhadir and colleagues showed that quince polyphenols extract inhibited LPS-mediated activation of three major cellular pro-inflammatory effectors: NF-κB, p38MAPK, and protein kinase B (Akt) (Essafi-Benkhadir *et al.*, 2012).

Literature data indicate that quercetin and its derivatives, as one of the constituents of quince fruit, possess remarkable protective effects against colon injury in IBD through induction of intense anti-inflammatory responses or inhibitory effects on oxidative stress pathways or mediators. They also restore the proper intestinal host–microbe relationship and immunologic hemostasis in animal models of intestinal inflammation, thereby regulating colitis-stimulated intestinal microbiota dysbiosis (Hong and Piao, 2018; Ju *et al.*, 2018; Lin *et al.*, 2019). Rutin (quercetin 3-rutinoside) was found to ameliorate inflammation in injured colon by suppressing the induction of pro-inflammatory cytokines and improving colitis histologic scores (Kwon *et al.*, 2005). In another study, rutin presented intestinal anti-inflammatory activity in chronic T-lymphocyte-dependent colitis upon quercetin release and involvement of mucosal and lymph node T cells. Rutin treatment reduced the colonic expression of pro-inflammatory genes such as interferon (*IFN*) and decreased colitis severity by modulating the colonic microenvironment, resulting in reduced colonic inflammation, promotion of colonic mucosal injury repair, and attenuation of colitis-associated microbial dysbiosis (Mascaraque *et al.*, 2014).

Plantago ovata Forssk., *Plantago psyllium* L.

Plantago ovata is a well-known plant for GI problems, i.e., constipation, diarrhea, IBS, IBD, and colon cancer (Madgulkar *et al.*, 2015). The plant seeds mainly contain hydrocolloids, acidic and neutral polysaccharides (Sarfraz *et al.*, 2017). A soluble fiber (psyllium) obtained from the husk of *P. ovata* seeds has clinical benefits in patients with IBD by remission maintenance and protection of the intestinal mucosa through reducing ulceration. This was attributed to a high content of short-chain fatty acids in the dietary fiber and consequently increased luminal production of short-chain fatty acids, which were shown to induce immunomodulatory responses, accelerate healing and regeneration processes of the intestinal epithelium, and regulate the gut microflora (Pituch-Zdanowska *et al.*, 2015). In a TNBS model of rat colitis, administration of a fiber-supplemented diet (5% w/w *P. ovata* seeds) for 2 weeks prior to colitis induction, and for 1 week after, significantly reduced the intestinal inflammation that was associated with decreased TNF-α level and NOS activity in the large intestine. A higher concentration of the short-chain fatty acid esters (mainly butyrate and propionate) was recorded in the intestinal contents in fiber-supplemented rats (Rodríguez-Cabezas *et al.*, 2002). In a transgenic rat model of intestinal inflammation, supplementation with a high-fiber diet containing 5% w/w psyllium seeds for 13 weeks improved the intestinal cytoarchitecture via downregulation of pro-inflammatory markers like NO, leukotriene B4, and TNF-α. A significantly higher production of short-chain fatty acid esters, butyrate, and propionate was also observed in psyllium-treated animals. Upon butyrate and propionate exposure, TNF-α production reduced *in vitro* (Rodríguez-Cabezas *et al.*, 2003). A clinical trial was conducted on 105 patients with UC, based on the theory that colonic fermentation of *P. ovata* seeds results in butyrate formation, which is thought to be effective in UC treatment. Oral treatment with *P. ovata* seeds (10 g, twice daily) was almost as effective as mesalamine to maintain remission in patients with UC. Thus, psyllium seems to be a promising candidate for future clinical studies in patients with IBD.

Punica granatum L.

A growing body of evidence indicates the potential therapeutic and preventive properties of dietary supplements made from pomegranate for serious chronic complications such as cancer and inflammatory diseases. Flavonoids, anthocyanins, punicic acid, and ellagitannins are the main bioactive compounds present in pomegranate. Upon intestinal metabolism, ellagitannins are hydrolyzed to ellagic acid, which is the main active compound of pomegranate (Marín *et al.*, 2013). In acute DSS-induced UC in mice, dietary supplementation with ellagic acid (2% w/w) for 7 days ameliorated the disease severity marginally by modulating inflammatory mediators. However, in a chronic UC model, consumption of ellagic acid (0.5% w/w) significantly inhibited the progression of the disease, attenuated intestinal inflammation, and decreased histologic scores. Moreover, enzymes such as COX-2 and iNOS were downregulated and the signaling pathways were blocked, proposing that ellagic acid may be favorable for UC treatment in chronic UC patients (Marín *et al.*, 2013). In 2010, Larrosa *et al.* compared the effect of pomegranate intake (250 mg/kg) and its main microbiota-derived metabolite, urolithin-A (UROA, 15 mg/kg), on colon inflammation in DSS-induced colitis in rats for 5 days. Both pomegranate and UROA decreased inflammation markers and modulated the gut microbiota. Pomegranate, but not UROA, decreased oxidative stress in plasma and colon mucosa. The authors concluded that UROA might be the most active anti-inflammatory compound derived from pomegranate ingestion in healthy subjects; whereas in colon inflammation, anti-inflammatory effects could be due to the nonmetabolized ellagitannin-related fraction (Larrosa *et al.*, 2010).

Treatment with *P. granatum* extract and its ellagic acid-rich fraction (100 and 200 mg/kg, orally) for 7 days significantly attenuated DSS-induced colonic inflammation along with reduction of histamine, MPO, and oxidative stress (Singh *et al.*, 2009). Similarly, pomegranate fruit extract improved UC severity and symptoms in dinitrobenzene sulfonic acid (DNBS)-induced colitis in rats, and attenuated oxidative stress parameters levels, while it increased SOD activity (Darji *et al.*, 2010). Pre-treatment and post-treatment of rats with pomegranate in a TNBS model of colitis decreased the serum MPO, GSH, ALP, fibrinogen, and CRP levels (Riaz *et al.*, 2017). Kamat and Singh reported that combination of *Aegle marmelos* and *P. granatum* effectively improved DAI and colitis macroscopy indices in a murine model of DSS-induced acute colitis (Kamat and Singh, 2019). Peritoneal injection of *P. granatum* juice and purified punicalagin in DNBS-induced colitis in animals significantly lowered the disease extent and severity, and repressed the gene expression of TNF-α, IL-18, and IL-1β through decrease of the NF-κB mRNA level. The results of the study suggested that *P. granatum* juice has a higher potency than purified punicalagin (Shah *et al.*, 2016). Consumption of pomegranate beverage reduced inflammation and ulceration in intestinal colitis through modulation of the miR-145/p70S6K1/HIF1α axis and downregulation of the expression of TNF-α, IL-1β, COX-2, and iNOS (at mRNA and protein levels) in DSS-induced colitis in rats and in LPS-treated CCD-18Co colon myofibroblastic cells. Moreover, the rats treated with pomegranate showed a significant regulation in their intestinal microbiome and an increase in butyrate-producing bacteria (Kim *et al.*, 2017). A dry whole pomegranate decoction (obtained from mesocarp) and its main components, polysaccharides and ellagitannins, significantly prevented the development of abdominal pain induced in rats by DNBS, although ellagitannins were found more potent than polysaccharides (Parisio *et al.*, 2020).

Administration of pomegranate beverage in rats with UC exhibited that pomegranate polyphenols downregulated the mammalian target of rapamycin (mTOR) downstream pathway through reduction of ERK1/2. *In silico* modeling indicated a high binding of docked gallic acid to the catalytic domain of insulin-like growth factor 1 receptor (IGF-1R), which may suppress the activity of the enzyme. Ellagic acid docked effectively into the catalytic domains of both IGF-1R and epidermal growth factor receptor (EGFR) (Kim *et al.*, 2016). In accordance with previous study, pomegranate juice caused reduction of colon inflammation in chronic model of colitis through reduction of p70S6K by upregulating miR-145 and downregulating the MAPK/ERK1/2 pathway (Kim *et al.*, 2014).

In a randomized, placebo-controlled clinical trial, patients with UC were randomized to receive an aqueous extract of pomegranate peel for 4 weeks as an adjuvant to standard medications. The Lichtiger Colitis Activity Index (LCAI) was measured at baseline, week 4, and week 10. The LCAI score was similarly reduced in both the pomegranate and placebo groups; however, the clinical response was higher with pomegranate compared to placebo at week 4, but not at week 10 (Kim *et al.*, 2016). In another human study including 78 individuals, the difference between therapeutic responses of hot and cold temperament patients (based on TPM) with UC to pomegranate peel extract was monitored for 4 weeks based on the LCAI scoring system. Here, therapeutic response was significantly higher in patients with hot temperament compared to patients with cold temperament in the pomegranate-treated group, indicating the significance of considering syndrome differentiation and temperaments in interpreting the effect of herbal medicine on UC (Kamali *et al.*, 2017). Conclusively, it seems that synergism between the antioxidant and anti-inflammatory actions of the bioactive constituents of pomegranate makes it beneficial as an IBD treatment (Nascimento and Araruna, 2016).

Rhus coriaria **L.**

In vitro and *in vivo* models provided strong scientific evidence on the anti-inflammatory and antioxidant effects of sumac, associated with inhibition of ROS production, NO release, and pro-inflammatory cytokines (Khalilpour *et al.*, 2018; Momeni *et al.*, 2019). Although therapeutic effects of sumac on bowel complaints have been extensively mentioned in folk medicine of various countries, only limited studies are available in this context.

A recent study reported that oral sumac improved clinical aspects of the intestinal injury in newborn rats with necrotizing enterocolitis. Treatment with sumac reduced protein degradation products in the intestinal tissues and also decreased the intestinal epithelial cell apoptosis, which was associated with suppression of caspase-3, caspase-8, and caspase-9 immunoreactivity. In ethanol-induced gastric ulcer in rats, the hydroalcoholic extract of *Rhus coriaria* at doses of 145 and 248 mg/kg (orally) significantly inhibited gastric lesions, which was confirmed by ulcer index, ulcer score, and histopathologic evaluations (Ahmad *et al.*, 2013). A later study resulted in similar outcomes in indomethacin- and water immersion-induced restraint gastric ulcer in an animal model, further confirming the antiulcer properties of this plant (Ahmad *et al.*, 2015).

Plantago major **L.,** ***Plantago lanceolata*** **L.**

Plantago major has previously presented significant anti-inflammatory effects (Hussan *et al.*, 2015; Zubair *et al.*, 2019) and is widely used to treat various GI diseases such as diarrhea, constipation, and gastric ulcers. In a randomized, double-blind clinical trial including 61 subjects with UC, after 8 weeks of intervention with *P. major* (roasted seeds, 3600 mg/day), abdominal tenderness, gastroesophageal reflux, and gastric pain were significantly less severe in the treated group compared to the control group. Visible blood in stool, distension, and anal pain were decreased in the *P. major* group, although the differences were not statistically significant between the two groups (Baghizadeh *et al.*, 2021). In a case report study involving a 35-year-old man with chronic pancolitis who received *P. major* decoction twice daily, the patient's symptoms improved within a few days after commencing treatment (Tafazoli *et al.*, 2022). Oral administration of the methanol extracts of *P. major* leaves at the dose of 400 mg/kg significantly decreased the average ulcer index and total protein in gastric juice in aspirin-induced gastric ulcer (Atta *et al.*, 2005). The polyholozidic fraction of *P. major* leaves and seeds also showed gastroprotective effects in experimental models (Hriscu *et al.*, 1990).

Malus **species**

Various species of apple represent a valuable source of quercetin glycosides, catechin, epicatechin, procyanidin, cyanidin-3-galactoside, coumaric acid, chlorogenic acid, gallic acid, and phloridzin (Boyer and Liu, 2004; Hyson, 2011; Persic *et al.*, 2017), which are implicated in reduction of oxidative stress and inflammation. Wu *et al.* (2018) found that apple proanthocyanidins attenuated oxidative stress by increase in SOD, HO-1, CAT, and GSH-Px mRNA expression, and counteracted inflammation by the downregulation of inflammatory markers *in vitro*. Orally administered apple procyanidins increased the proportions of TCRγδ and TCRαβ-CD8αα T cells in intraepithelial lymphocytes and suppressed IFN-γ synthesis in stimulated intraepithelial lymphocytes. Apple procyanidins also inhibited PMA-induced secretion of IL-8 in intestinal epithelial cells. These compounds exhibited oral preventive effect against DSS- or oxazolone-induced IBD in mice (Yoshioka *et al.*, 2008).

Polyphenol extracts from apples have shown immunomodulatory and anti-inflammatory effects in the DSS model of UC, through downregulation of pro-inflammatory cytokine expression, preliminarily mediated by T cells (Skyberg *et al.*, 2011). In another study, triterpenoids of apple peel such as ursolic acid and oleanolic acid exerted an inhibitory effect on inflammatory gene expression associated with IBD (Mueller *et al.*, 2013). In a rat model of *E. coli*-induced colon inflammation, the flavonoid phloretin significantly ameliorated colon inflammation and body weight loss, while alleviating inflammation in HT-29 human colonic epithelial cells (Lee *et al.*, 2011). Intragastric administration of apple polyphenols extract at doses of 125 or 500 mg/kg or phloretin at 100 mg/kg for 3 weeks could alleviate DSS-induced UC in mice by restoring bile acid metabolism disorder and gut microbiota dysbiosis. Both apple polyphenols extract and phloretin inhibited the activation of NF-κB signaling, decreased hyodeoxycholic acid level, and regulated intestinal microbiota. Apple polyphenols extract decreased the DAI score more profoundly than phloretin (Liu *et al.*, 2021). The preventive (200 mg/kg) and therapeutic (400 mg/kg) doses of apple peel polyphenols was shown to reduce mitochondrial dysfunction in mice with DSS-induced UC. Both treatment options improved histologic lesions, macroscopic parameters, and attenuated clinical signs, infiltration of inflammatory cells, and oxidative stress in mice with UC (Yeganeh *et al.*, 2018). Likewise, rectal administration of apple polyphenols protected rats from TNBS-induced colitis by decreasing the transcription and protein levels of COX-2, TNF-α, calpain, as well as tissue transglutaminase (D'Argenio *et al.*, 2012). Earlier, Pastrelo *et al.* reported that concentrated apple extract suppressed iNOS gene expression, while enhancing the copper and zinc superoxide dismutase (CuZnSOD) mRNA level in acetic acid-induced colitis in rats (Pastrelo *et al.*, 2017). Collectively, apple polyphenols with extensive antioxidant and anti-inflammatory actions offer a reliable source for development of natural products for IBD.

Berberis vulgaris L., *Berberis aristata* DC.

Multiple studies have demonstrated that barberry and its most well-known bioactive alkaloid, berberine, have strong anti-inflammatory activities and are capable of reducing the levels of inflammatory cytokines while upregulating the concentrations of antioxidant mediators. Therefore, they might be promising choices for treatment of inflammatory disorders such as IBD (Abd El-Wahab *et al.*, 2013; Ashrafizadeh *et al.*, 2020). Furthermore, experimental studies have shown that the anti-inflammatory effect of *Berberis vulgaris* and its main components is generally related to the shift of cell immune response to Th2, activation of T regulatory cells, inhibition of inflammatory cytokines, and stimulation of IL-4 and IL-10. Induction of apoptosis in antigen-presenting cells and other effector cells is also involved (Kalmarzi *et al.*, 2019).

Oral administration of berberine for a week at a dosage of 15 mg/kg daily reduced histologic lesions, morphologic damage, and the MPO activity in TNBS-induced colitis in rats. Addition of berberine at concentration of 10^{-5} M to the culture medium inhibited

IL-8 production of rectal mucosa (Zhou and Mineshita, 2000). The anticolitic effect of *B. vulgaris* fruit extract was compared to berberine chloride and corticosteroids in another investigation, using an animal model of acetic acid-induced colitis. Barberry fruit extract (750 and 1500 mg/kg) and berberine chloride (10 mg/kg) were effective to protect against colonic damage. Barberry fruit extract was effective in reducing ulcer index only when administered orally and for a longer period of 5 days, which might be due to its anthocyanin components (Minaiyan *et al.*, 2011). In the study of Tanideh *et al.* (2014), *B. vulgaris* extract caused a significant decrease in inflammation in acetic acid-induced colonic tissue and a significant decrease in MDA activity in both forms of intracolonic and oral administration. These data suggest a high efficacy of *B. vulgaris* and other pharmacologically active chemical groups in this plant for ameliorating IBD.

Rumex species

Rumex spp. have demonstrated anti-inflammatory activities through reduction of NO production, regulation of inflammation-related genes, and reduction of inflammatory cytokine levels (Singh *et al.*, 2013; Kim *et al.*, 2020; Eom *et al.*, 2020). *Rumex* spp. have been reported to significantly alter oxidative status by decreasing MDA and GSH levels and increasing CAT and GPx activities *in vivo* (Chelly *et al.*, 2020). Prophylactic administration of *Rumex japonicus* methanolic extract at a dose of 100 mg/kg daily for 2 weeks showed significant improvement in histopathologic changes and pro-inflammatory cytokine levels in DSS-induced colitis in mice (Kim *et al.*, 2020). Methanolic extracts of *Rumex algeriensis* and *Rumex tunetanus* showed potent antioxidant and anti-inflammatory effects in a model of acute intestinal inflammation using epithelial Caco-2 cells, as evidenced by inhibition of TNF-α-induced gene expression of IL-6 and IL-8 (Abidi *et al.*, 2020). Overall, *Rumex* spp. have the potential to be successfully applied in the prevention or treatment of colitis.

Pistacia lentiscus L.

The antibacterial, anti-inflammatory, antioxidant, antiulcer, and anticancer potentials of *Pistacia lentiscus* are well established in experimental and clinical investigations (Naouar *et al.*, 2016; Boutemine *et al.*, 2018; Pachi *et al.*, 2020). Inhibition of intestinal damage and regulation of inflammation and oxidative stress in intestinal epithelium are involved in the protective effects of this plant in IBD. Daily oral administration of *P. lentiscus* powder with a dose range of 50–200 mg/kg in TNBS-induced colitis in rats caused histologic amelioration and suppressed the MDA level, suggesting that mastic powder could possibly have a therapeutic role in CD through regulating the oxidant/antioxidant balance and modulating inflammation (Gioxari *et al.*, 2011). Naouar *et al.* (2016) investigated the preventive and curative effects of lentisk oil (cold-pressed oil from *P. lentiscus* fruit) in a TNBS model of colitis in rats. Both intervention groups had improved clinical signs of colitis including bleeding index and diarrhea. While preventive treatment attenuated erosion, cryptitis, irregular crypts and crypt loss, curative intervention improved ulceration, hyperplasia, cryptitis, and irregular crypts and crypt loss, probably via modification of arachidonic acid metabolism (Naouar *et al.*, 2016). One-week pretreatment with the aqueous leaf extract of *P. lentiscus* improved macroscopic colonic damage, reduced LPO and carbonylation of proteins, and enhanced antioxidant enzyme activities, while suppressing IL-6 level, in the colon and plasma of rats with acetic acid-induced UC. Administration of *P. lentiscus* also decreased plasma and tissue levels of H_2O_2, Fe, and free Ca. The cytoprotective effect of *P. lentiscus* was correlated with its antioxidant and anti-inflammatory activities (Boutemine *et al.*, 2021). The same extract caused a significant reduction of clinical score of acute colitis DAI and histologic damage, as well as decreases in pro-inflammatory mediator levels in plasma and culture supernatants of colonic explants and peritoneal macrophages from a DSS-induced mouse model of UC (Boutemine *et al.*, 2021).

Given the antioxidant, anti-inflammatory, and immunomodulating properties of mastic, a few clinical trials were performed in patients with active IBD. In 2007, the effectiveness of mastic administration was evaluated on the clinical course and plasma inflammatory mediators of patients with mild to moderately active CD. Ten patients and eight controls were recruited for a 4-week treatment with mastic capsules (6 caps/day, 0.37 g/cap). In the mastic-treated group, there was significant reduction of CD activity index as compared to pretreatment values. Plasma IL-6 and CRP were significantly decreased, while total antioxidant potential was significantly increased. No adverse effects were observed (Kaliora *et al.*, 2007). In a randomized, double-blind, placebo-controlled clinical trial, supplementation with natural *P. lentiscus* (2.8 g/day) for 3 months in patients with active IBD (CD, $n = 40$; UC, $n = 20$), adjunctive to stable medical treatment, resulted in favorable changes in oxidative stress biomarkers and regulation of lipid profile parameters (Papada *et al.*, 2018). In another clinical trial, it was reported that a 3-month intervention with a supplement containing the natural resinous product of *P. lentiscus* (mastiha; 2.8 g/day), adjunctive to stable pharmacologic treatments, regulated fecal biomarkers and improved quality of life in active IBD patients ($n = 60$) compared to a placebo. Upon mastic treatment, the IBDQ score was significantly improved. Also there was a significant decrease in fecal lysozyme (Papada *et al.*, 2019). The results suggest that *P. lentiscus* possesses pharmacological properties, rendering its efficacy in the treatment of IBD.

Commiphora molmol (Engl.) Engl. ex Tschirch

Natural gum myrrh, derived from various *Commiphora* spp., is well known for its anti-inflammatory and antioxidant properties. Myrrh and its bioactive constituents have been shown to modulate the functions of leukocytes and the activities of immune cells (Shalaby and Hammouda, 2014; Alsharif, 2020). Furthermore, the gum stimulates antioxidant activities and scavenges free radicals *in vitro* and *in vivo*, which was attributed to upregulation of Nrf2/ARE/HO-1 signaling (Mahmoud *et al.*, 2017, 2018).

Pretreatment with the aqueous extract of myrrh (125, 250, or 500 mg/kg daily) for 7 days diminished the levels of pro-inflammatory cytokines, PGE_2, NO, and LPO in an acetic acid-induced model of colitis. By contrast, the levels of IL-10, nonprotein sulfhydryl groups (NP-SH), total protein, and nucleic acids, and the enzymatic activities of SOD and CAT were significantly elevated (Fatani *et al.*, 2016). In human intestinal epithelial cells, addition of myrrh to the cell culture exerted barrier-stabilizing and TNF-α-antagonizing effects via inhibition of phosphatidylinositol-3 kinase (PI3K) and signal transducer and activator of transcription 6 (STAT6) signaling. Administration of myrrh inhibited the TNF-α-induced decrease in transepithelial resistance, the redistribution of the sealing tight junction protein claudin-1, the expression of claudin-2, and the change in tight junction ultrastructure. It was assumed that therapeutic application of myrrh in intestinal diseases might be associated with barrier defects and inflammation (Rosenthal *et al.*, 2017).

Elaeagnus angustifolia L.

Considering the presence of active anti-inflammatory and antioxidant phytochemicals in this plant, *Elaeagnus angustifolia* was the subject of many experiments. It has been shown that the extract of *E. angustifolia* hindered inflammation and pain induced by formalin in mice and rats (Motevalian *et al.*, 2017). Farahbakhsh *et al.* revealed that *E. angustifolia* extract lessened pain and inflammation caused by formalin through inhibition of COX-1 enzymes, as well as corticosterone release from adrenal glands in mice (Farahbakhsh *et al.*, 2011). Moreover, human clinical trials have indicated that preparations from *E. angustifolia* were able to decrease the inflammation mediators such as TNF-α and matrix metalloproteinase-1 (MMP-1), whereas increasing the anti-inflammatory cytokines such as IL-10 (Nikniaz *et al.*, 2014). A considerable proportion of *E. angustifolia* belongs to polyphenolic compounds, which are well recognized for their significant pharmacologic activities such as anti-inflammatory, wound healing, anticancer, and gastroprotective effects (Wang *et al.*, 2013; Abri and Maleki, 2016; Sajadimajd *et al.*, 2020; Sanches-Silva *et al.*, 2020; Johnson *et al.*, 2021).

Gastroprotective and antiulcer properties of this plant have been evaluated in several studies (Eliassi *et al.*, 2008). Methanolic fruit extract of *E. angustifolia* ameliorated histopathologic examination and ulcer index determination in ethanol-induced ulcer in the guinea pig, indicating a strong gastroprotective effect. In addition, leaf and flower extracts of *E. angustifolia* inhibited the activity of GI smooth muscle, which may be a reason for its advantages in the treatment of nausea, vomiting, and flatulence (Mohammed *et al.*, 2006). In acetic acid-induced colitis in rats, 2-week treatment with enema gel at 10% w/w and 20% w/w, and the fruit hydroalcoholic extract at 300 and 600 mg/kg, resulted in significant reduction of histopathologic scores such as erosion and ulceration, inflammatory cell infiltration, as well as regeneration and improvement of mucosa. It is noteworthy that the extract at a dosage of 600 mg/kg exerted the best curative effect on colon ulcers (Khodakarm-Tafti *et al.*, 2015).

Conclusion

Avicenna, as a prominent physician in TPM, has recommended several medicinal plants for the management of IBD. In this chapter, 29 of these medicinal plants were elicited from his book, *Canon of Medicine*, and the scientific evidence and possible molecular mechanisms involved in their efficacy have been investigated. The results demonstrated that these medicinal plants affect different parameters involved in the pathogenesis of IBD including pro-inflammatory cytokines and inflammatory markers, oxidative stress, endogenous antioxidant defense mechanisms, gut microbiota, as well as histopathologic features. The most investigated medicinal plants were *Malus* spp., *Boswellia serrata*, *Punica granatum*, and *Pistacia lentiscus*. *Malus* spp. may reduce pro-inflammatory cytokines and restore gut microbiota dysbiosis. *Boswellia serrata* has demonstrated its promising effects on IBD through improving intestinal mucosal injury, reducing oxidative stress, attenuation of inflammatory markers, as well as reduction of the histopathologic score of colitis. *Punica granatum* can affect different parameters involved in IBD pathogenesis including inflammatory markers, oxidative stress, antioxidant parameters, and ulcer index. *Pistacia lentiscus* may be a suitable candidate for IBD by regulation of oxidant/antioxidant balance, attenuation of inflammation and pro-inflammatory cytokines, as well as reduction of the histopathologic score of colitis.

Use of multi-ingredient drugs in various dosage forms for the management of different diseases is very common in TPM and one of the main reasons for preparation of such medicines is utilization of the synergistic effects of medicinal plants. As each medicinal herb can affect some pharmacologic mechanisms that may not be covered by other herbs, compounding these medicinal plants to produce drugs with more potent activity and better efficacy seems logical.

Most of studies conducted on medicinal plants mentioned in *Canon of Medicine* for treatment of IBD were experimental and only a few clinical trials were found. *Boswellia serrata*, *Rosa × damascena*, *Bunium persicum*, *Plantago ovata*, *Punicum granatum*, *Plantago major*, and *Pistacia lentiscus* are medicinal plants for which effectiveness in IBD has been investigated through various clinical studies. So, for achieving more conclusive results about the use of the mentioned medicinal plants for management of IBD, it is necessary to design further experimental and clinical studies to investigate the exact mechanisms of action, as well as the clinical efficacy and safety, of these natural remedies.

References

Abbasi, B., Daniali, M., Ramezani, H., Derakhshande, M. and Ghiasvand, R. (2018) The effect of *Bunium persicum* on gastrointestinal symptoms and inflammatory mediators in irritable bowel syndrome patients. *Food & Health* 1(1), 43–46.

Abd El-Wahab, A.E., Ghareeb, D.A., Sarhan, E.E.M., Abu-Serie, M.M. *et al.* (2013) *In vitro* biological assessment of *Berberis vulgaris* and its active constituent, berberine: antioxidants, anti-acetylcholinesterase, anti-diabetic and anticancer effects. *BMC Complementary and Alternative Medicine* 13(1), 218.

Abidi, J., Occhiuto, C., Cimino, F., Speciale, A., Ruberto, G. *et al.* (2020) Phytochemical and biological characterization of methanolic extracts from *Rumex algeriensis* and *Rumex tunetanus*. *Chemistry & Biodiversity* 17(8), e2000345.

Abri, A. and Maleki, M. (2016) Isolation and identification of gallic acid from the *Elaeagnus angustifolia* leaves and determination of total phenolic, flavonoids contents and investigation of antioxidant activity. *Quarterly Journal of Iranian Chemical Communication* 4, 146–154.

Ahmad, H., Wadud, A., Jahan, N., Khazir, M. and Sofi, G. (2013) Evaluation of anti-ulcer activity of hydro alcoholic extract of Post Sumaq (*Rhus coriaria* Linn.) in ethanol induced gastric ulcer in experimental rats. *International Research Journal of Medical Sciences* 1(10), 7–12.

Ahmad, H., Wadud, A., Jahan, N. and Hasan, I. (2015) Anti-ulcer activity of *Rhus coriaria* in indomethacin and water immersion restraint induced gastric ulcer in experimental rats. *International Journal of Medicine and Medical Sciences* 7(3), 61–66.

Ahmed, M.M. and Bastawy, S. (2014) Evaluation of anti-inflammatory properties and possible mechanism of action of Egyptian quince (*Cydonia oblonga*) leaf. *Egyptian Journal of Biochemistry and Molecular Biology* 32, 190–205.

Akl, E.M., Taha, F.S., Mohamed, S.S. and Mohamed, R.S. (2021) Characterization of garden cress mucilage and its prophylactic effect against indomethacin-induced enter-colitis in rats. *Biointerferance Research in Applied Chemistry* 11, 13911–13923.

Al-Hussaini, J.A.-A.S., Latif, A.D. and Abd, A.H. (2012) A comparative study of the effect of Asacol and the ethanolic extracts of *Matricaria chamomilla* and *Terminalia chebula* on induced ulcerative colitis in rabbits. *Kufa Journal for Veterinary Medical Sciences* 3(1), 36–54.

Alipanah, H., Bigdeli, M.R. and Esmaeili, M.A. (2018) Inhibitory effect of *Viola odorata* extract on tumor growth and metastasis in 4T1 breast cancer model. *Iranian Journal of Pharmaceutical Research* 17(1), 276–291.

Al-Khazraji, S.K. (2013) Phytochemical screening and antibacterial activity of the crude extract of *Cydonia oblonga* seeds. *Global Advanced Research Journal of Microbiology* 2(8), 137–140.

Alsharif, K. (2020) Potential anti-inflammatory properties effect of myrrh. *Letters in Applied NanoBioScience* 9, 1687–1694.

Al-Snafi, A.E. (2016) A review on chemical constituents and pharmacological activities of *Coriandrum sativum*. *IOSR Journal of Pharmacy* 6(7), 17–42.

Anthoni, C., Laukoetter, M.G., Rijcken, E., Vowinkel, T., Mennigen, R. *et al.* (2006) Mechanisms underlying the anti-inflammatory actions of boswellic acid derivatives in experimental colitis. *American Journal of Physiology: Gastrointestinal and Liver Physiology* 290(6), G1131–G1137.

Aruna, K., Rukkumani, R., Varma, P.S. and Menon, V.P. (2005) Therapeutic role of *Cuminum cyminum* on ethanol and thermally oxidized sunflower oil induced toxicity. *Phytotherapy Research* 19(5), 416–421.

Ashrafizadeh, M., Najafi, M., Mohammadinejad, R., Farkhondeh, T. and Samarghandian, S. (2020) Berberine administration in treatment of colitis: a review. *Current Drug Targets* 21(13), 1385–1393.

Atta, A., Nasr, S.M. and Mouneir, S.M. (2005) Antiulcerogenic effect of some plants extracts. *Natural Product Radiance* 4(4), 258–263.

Avicenna, H. (AH 1426) *The Canon of Medicine (Qanoon-fil-Tib)*. Dar-ol-Ehya al-Torath al-Arabi, Beirut, Lebanon.

Baghizadeh, A., Davati, A., Heidarloo, A.J., Emadi, F. and Aliasl, J. (2021) Efficacy of *Plantago major* seed in management of ulcerative colitis symptoms: a randomized, placebo controlled, clinical trial. *Complementary Therapies in Clinical Practice* 44, 101444.

Bahmani, M., Shirzad, H., Mirhosseini, M., Mesripour, A. and Rafieian-Kopaei, M. (2016) A review on ethnobotanical and therapeutic uses of fenugreek (*Trigonella foenum-graceum* L.). *Journal of Evidence-Based Complementary & Alternative Medicine* 21(1), 53–62.

Bakheet, N. (2020) Intestinal anti-inflammatory improvement with fenugreek seeds as a prebiotic and synbiotic with *Lactobacillus acidophilus* in rats experimentally infected with *Escherichia coli*. *The Iraqi Journal of Veterinary Medicine* 44(2), 1–14.
Benchikh, F. (2018) *Pharmacological effects of Myrtus communis L.* on the gastrointestinal tract of rats and mice. PhD thesis, Université Ferhat Abbas Sétif 1, El-Bez, Algeria.
Bhat, S.P., Rizvi, W. and Kumar, A. (2014) Effect of *Cuminum cyminum* L. seed extracts on pain and inflammation. *Journal of Natural Remedies* 14(2), 186–192.
Biswas, S.K. (2016) Does the interdependence between oxidative stress and inflammation explain the antioxidant paradox? *Oxidative Medicine and Cellular Longevity* 2016, 5698931.
Bonaterra, G.A., Bronischewski, K., Hunold, P., Schwarzbach, H., Heinrich, E.-U. *et al.* (2020) Anti-inflammatory and anti-oxidative effects of Phytohustil® and root extract of *Althaea officinalis* L. on macrophages *in vitro*. *Frontiers in Pharmacology* 11, 290.
Bourgonje, A.R., Feelisch, M., Faber, K.N., Pasch, A., Dijkstra, G. *et al.* (2020) Oxidative stress and redox-modulating therapeutics in inflammatory bowel disease. *Trends in Molecular Medicine* 26, 1034–1046.
Boutemine, I.-M., Amri, M., Amir, Z.-C., Fitting, C., Mecherara-Idjeri, S. *et al.* (2018) Gastro-protective, therapeutic and anti-inflammatory activities of *Pistacia lentiscus* L. fatty oil against ethanol-induced gastric ulcers in rats. *Journal of Ethnopharmacology* 224, 273–282.
Boutemine, I.-M., Amri, M., Dorgham, K., Amir, Z.-C., Benazzouz, S. *et al.* (2021) Beneficial role of *Pistacia lentiscus* aqueous extract in experimental colitis: anti-inflammatory and potential therapeutic effects. *Inflammopharmacology* 29(4), 1225–1239.
Boyer, J. and Liu, R.H. (2004) Apple phytochemicals and their health benefits. *Nutrition Journal* 3(1), 5.
Capek, P., Rosík, J. and Kardošova, A. (1987) Polysaccharides from the roots of the marsh mallow (*Althaea officinalis* L., var. Rhobusta): structural features of an acidic polysaccharide. *Carbohydrate Research* 164, 443–452.
Chelly, M., Chelly, S., Ben Salah, H., Athmouni, K., Bitto, A. *et al.* (2020) Characterization, antioxidant and protective effects of edible *Rumex roseus* on erythrocyte oxidative damage induced by methomyl. *Journal of Food Measurement and Characterization* 14, 229–243.
Chen, L., Liu, J., Zhang, Y., Dai, B., An, Y. *et al.* (2015a) Structural, thermal, and anti-inflammatory properties of a novel pectic polysaccharide from alfalfa (*Medicago sativa* L.) stem. *Journal of Agricultural and Food Chemistry* 63(12), 3219–3228.
Chen, L., Liu, J., Zhang, Y., Niu, Y., Dai, B. *et al.* (2015b) A novel alkaline hemicellulosic heteroxylan isolated from alfalfa (*Medicago sativa* L.) stem and its thermal and anti-inflammatory properties. *Journal of Agricultural and Food Chemistry* 63(11), 2970–2978.
Cheon, J.H., Kim, J.S., Kim, J.M., Kim, N., Jung, H.C. *et al.* (2006) Plant sterol guggulsterone inhibits nuclear factor-κB signaling in intestinal epithelial cells by blocking IκB kinase and ameliorates acute murine colitis. *Inflammatory Bowel Diseases* 12(12), 1152–1161.
Choi, K.-C., Hwang, J.-M., Bang, S.-J., Kim, B.-T., Kim, D.-H. *et al.* (2013) Chloroform extract of alfalfa (*Medicago sativa*) inhibits lipopolysaccharide-induced inflammation by downregulating ERK/NF-κB signaling and cytokine production. *Journal of Medicinal Food* 16(5), 410–420.
D'Argenio, G., Mazzone, G., Tuccillo, C., Ribecco, M.T., Graziani, G. *et al.* (2012) Apple polyphenols extract (APE) improves colon damage in a rat model of colitis. *Digestive and Liver Disease* 44(7), 555–562.
Darji, V., Bariya, A., Deshpande, S. and Patel, D. (2010) Effect of *Punica granatum* fruits in inflammatory bowel disease. *Journal of Pharmacy Research* 3(12), 2850–2852.
Dong, W.-G., Liu, S.-P., Yu, B.-P., Wu, D.-F., Luo, H.-S. *et al.* (2003) Ameliorative effects of sodium ferulate on experimental colitis and their mechanisms in rats. *World Journal of Gastroenterology* 9(11), 2533–2538.
El-Meligy, R.M., Awaad, A.S., Soliman, G.A., Bacha, A.B., Alafeefy, A.M. *et al.* (2015) Prophylactic and curative anti-ulcerative colitis activity and the possible mechanisms of action of some desert plants. *Journal of Enzyme Inhibition and Medicinal Chemistry* 30(2), 250–258.
Elham, A., Arken, M., Kalimanjan, G., Arkin, A. and Iminjan, M. (2020) A review of the phytochemical, pharmacological, pharmacokinetic, and toxicological evaluation of *Quercus infectoria* galls. *Journal of Ethnopharmacology* 273, 113592.
Eliassi, A., Mandipour, M. and Kamalinejad, M. (2008) Intragastric effect of *Elaeagnus angustifolia* L. fruit on gastric acid secretion in a rat pylorus-ligated model. *Journal of Medicinal Plants* 7(27), 82–91.
Eom, T., Kim, E. and Kim, J.-S. (2020) *In vitro* antioxidant, antiinflammation, and anticancer activities and anthraquinone content from *Rumex crispus* root extract and fractions. *Antioxidants* 9(8), 726.
Essafi-Benkhadir, K., Refai, A., Riahi, I., Fattouch, S., Karoui, H. *et al.* (2012) Quince (*Cydonia oblonga* Miller) peel polyphenols modulate LPS-induced inflammation in human THP-1-derived macrophages

through NF-κB, p38MAPK and Akt inhibition. *Biochemical and Biophysical Research Communications* 418(1), 180–185.

Farahbakhsh, S., Arbabian, S., Emami, F., Rastegar Moghadam, B., Ghoshooni, H. *et al.* (2011) Inhibition of cyclooxygenase type 1 and 2 enzyme by aqueous extract of *Elaeagnus angustifolia* in mice. *Basic and Clinical Neuroscience* 2(2), 31–37.

Fatani, A.J., Alrojayee, F.S., Parmar, M.Y., Abuohashish, H.M., Ahmed, M.M. *et al.* (2016) Myrrh attenuates oxidative and inflammatory processes in acetic acid-induced ulcerative colitis. *Experimental and Therapeutic Medicine* 12(2), 730–738.

Fathima, A., Gangachannaiah, S., Bose, U., Chakraborty, R., Se, P.K. *et al.* (2021) Effect of aqueous extract of *Trigonella foenum-graecum* L. seeds on acetic acid-induced ulcerative colitis in rats. *Research Square*, 13 May 2021 (preprint), DOI: 10.21203/rs.3.rs-490956/v1. Available at: https://www.researchsquare.com/article/rs-490956/v1 (accessed 26 March 2023).

Fernández-Bañares, F., Hinojosa, J., Sánchez-Lombraña, J.L., Navarro, E., Martínez-Salmerón, J.F. *et al.* (1999) Randomized clinical trial of *Plantago ovata* seeds (dietary fiber) as compared with mesalamine in maintaining remission in ulcerative colitis. *The American Journal of Gastroenterology* 94(2), 427–433.

Feyzabadi, Z., Ghorbani, F., Vazani, Y. and Zarshenas, M.M. (2017) A critical review on phytochemistry, pharmacology of *Viola odorata* L. and related multipotential products in traditional Persian medicine. *Phytotherapy Research* 31(11), 1669–1675.

Franzosa, E.A., Sirota-Madi, A., Avila-Pacheco, J., Fornelos, N., Haiser, H.J. *et al.* (2019) Gut microbiome structure and metabolic activity in inflammatory bowel disease. *Nature Microbiology* 4(2), 293–305.

Gioxari, A., Kaliora, A.C., Papalois, A., Agrogiannis, G., Triantafillidis, J.K. *et al.* (2011) *Pistacia lentiscus* resin regulates intestinal damage and inflammation in trinitrobenzene sulfonic acid-induced colitis. *Journal of Medicinal Food* 14(11), 1403–1411.

Golabi, S., Mohammadi, A. and Chamkouri, N. (2020) Investigation of anti-inflammatory effect of aqueous extract of *Cuminum cyminum* L. by formalin inflammatory model in male rats. *Journal of Medicinal Plants* 18(72), 236–246.

Guan, Q. (2019) A comprehensive review and update on the pathogenesis of inflammatory bowel disease. *Journal of Immunology Research* 2019, 7247238.

Gudej, J. (1991) Flavonoids, phenolic acids and coumarins from the roots of *Althaea officinalis*. *Planta Medica* 57(3), 284–285.

Gupta, I., Parihar, A., Malhotra, P., Singh, G., Lüdtke, R. *et al.* (1997) Effects of *Boswellia serrata* gum resin in patients with ulcerative colitis. *European Journal of Medical Research* 2(1), 37–43.

Gupta, I., Parihar, A., Malhotra, P., Gupta, S., Lüdtke, R. *et al.* (2001) Effects of gum resin of *Boswellia serrata* in patients with chronic colitis. *Planta Medica* 67(5), 391–395.

Habibi, E., Arab-Nozari, M., Elahi, P., Ghasemi, M. and Shaki, F. (2019) Modulatory effects of *Viola odorata* flower and leaf extracts upon oxidative stress-related damage in an experimental model of ethanol-induced hepatotoxicity. *Applied Physiology, Nutrition, and Metabolism* 44(5), 521–527.

Hage-Sleiman, R., Mroueh, M. and Daher, C.F. (2011) Pharmacological evaluation of aqueous extract of *Althaea officinalis* flower grown in Lebanon. *Pharmaceutical Biology* 49(3), 327–333.

Halfvarson, J., Brislawn, C.J., Lamendella, R., Vázquez-Baeza, Y., Walters, W.A. *et al.* (2017) Dynamics of the human gut microbiome in inflammatory bowel disease. *Nature Microbiology* 2(5), 17004.

Hamedi, A., Rezaei, H., Azarpira, N., Jafarpour, M. and Ahmadi, F. (2016) Effects of *Malva sylvestris* and its isolated polysaccharide on experimental ulcerative colitis in rats. *Journal of Evidence-Based Complementary & Alternative Medicine* 21(1), 14–22.

Hartmann, R.M., Martins, M.I.M., Tieppo, J., Fillmann, H.S. and Marroni, N.P. (2012) Effect of *Boswellia serrata* on antioxidant status in an experimental model of colitis rats induced by acetic acid. *Digestive Diseases and Sciences* 57(8), 2038–2044.

Hartmann, R.M., Fillmann, H.S., Morgan Martins, M.I., Meurer, L. and Marroni, N.P. (2014) *Boswellia serrata* has beneficial anti-inflammatory and antioxidant properties in a model of experimental colitis. *Phytotherapy Research* 28(9), 1392–1398.

Heidari, B., Sajjadi, S.E. and Minaiyan, M. (2016) Effect of *Coriandrum sativum* hydroalcoholic extract and its essential oil on acetic acid-induced acute colitis in rats. *Avicenna Journal of Phytomedicine* 6(2), 205–214.

Hong, Y.-H., Chao, W.-W., Chen, M.-L. and Lin, B.-F. (2009) Ethyl acetate extracts of alfalfa (*Medicago sativa* L.) sprouts inhibit lipopolysaccharide-induced inflammation *in vitro* and *in vivo*. *Journal of Biomedical Science* 16(1), 1–12.

Hong, Z. and Piao, M. (2018) Effect of quercetin monoglycosides on oxidative stress and gut microbiota diversity in mice with dextran sodium sulphate-induced colitis. *BioMed Research International* 2018, 8343052.

Hriscu, A., Stănescu, U., Ionescu, A. and Verbuţă, A. (1990) A pharmacodynamic investigation of the effect of polyholozidic substances extracted from *Plantago* sp. on the digestive tract. *Revista Medico-Chirurgicala a Societatii de Medici si Naturalisti din Iasi* 94(1), 165–170.

Hsiang, C.-Y., Hseu, Y.-C., Chang, Y.-C., Kumar, K.S., Ho, T.-Y. *et al.* (2013) *Toona sinensis* and its major bioactive compound gallic acid inhibit LPS-induced inflammation in nuclear factor-κB transgenic mice as evaluated by *in vivo* bioluminescence imaging. *Food Chemistry* 136(2), 426–434.

Hussan, F., Mansor, A.S., Hassan, S.N., Kamaruddin, T.N.E., Tasnim, T.N. *et al.* (2015) Anti-inflammatory property of *Plantago major* leaf extract reduces the inflammatory reaction in experimental acetaminophen-induced liver injury. *Evidence-Based Complementary and Alternative Medicine* 2015, 347861.

Hyson, D.A. (2011) A comprehensive review of apples and apple components and their relationship to human health. *Advances in Nutrition* 2(5), 408–420.

Jabri, M.-A., Rtibi, K., Tounsi, H., Hosni, K., Souli, A. *et al.* (2015) Myrtle berry seed aqueous extract inhibits human neutrophil myeloperoxidase *in vitro* and attenuates acetic acid-induced ulcerative colitis in rats. *RSC Advances* 5(80), 64865–64877.

Jagtap, A., Shirke, S. and Phadke, A. (2004) Effect of polyherbal formulation on experimental models of inflammatory bowel diseases. *Journal of Ethnopharmacology* 90(2–3), 195–204.

Johnson, J.B., Broszczak, D.A., Mani, J.S., Anesi, J. and Naiker, M. (2021) A cut above the rest: oxidative stress in chronic wounds and the potential role of polyphenols as therapeutics. *Journal of Pharmacy and Pharmacology* 74, 485–502.

Joo, H.Y., Lim, K. and Lim, K.T. (2009) Phytoglycoprotein (150 kDa) isolated from *Solanum nigrum* Linne has a preventive effect on dextran sodium sulfate-induced colitis in A/J mouse. *Journal of Applied Toxicology* 29(3), 207–213.

Ju, S., Ge, Y., Li, P., Tian, X., Wang, H. *et al.* (2018) Dietary quercetin ameliorates experimental colitis in mouse by remodeling the function of colonic macrophages via a heme oxygenase-1-dependent pathway. *Cell Cycle* 17(1), 53–63.

Kahkeshani, N., Farzaei, F., Fotouhi, M., Alavi, S.S., Bahramsoltani, R. *et al.* (2019) Pharmacological effects of gallic acid in health and diseases: a mechanistic review. *Iranian Journal of Basic Medical Sciences* 22(3), 225.

Kaithwas, G. and Majumdar, D.K. (2013) Effect of *Linum usitatissimum* (flaxseed/linseed) fixed oil against distinct phases of inflammation. *International Scholarly Research Notices* 22(3), 225–237.

Kaithwas, G., Mukherjee, A., Chaurasia, A. and Majumdar, D.K. (2011) Antiinflammatory, analgesic and antipyretic activities of *Linum usitatissimum* L. (flaxseed/linseed) fixed oil. *Indian Journal of Experimental Biology* 49, 932–938.

Kaliora, A.C., Stathopoulou, M.G., Triantafillidis, J.K., Dedoussis, G.V.Z. and Andrikopoulos, N.K. (2007) Chios mastic treatment of patients with active Crohn's disease. *World Journal of Gastroenterology* 13(5), 748–753.

Kalmarzi, R.N., Naleini, S.N., Ashtary-Larky, D., Peluso, I., Jouybari, L. *et al.* (2019) Anti-inflammatory and immunomodulatory effects of barberry (*Berberis vulgaris*) and its main compounds. *Oxidative Medicine and Cellular Longevity* 2019, 6183965.

Kamali, M.A., Khodadoost, M., Heydari, M., Tavakoli, H., Hajimehdipoor, H. *et al.* (2017) The role of syndrome differentiation in the clinical efficacy of *Punica granatum* on patients with ulcerative colitis. *Iranian Red Crescent Medical Journal* 19(3), e40810.

Kamalian, A., Sohrabi Asl, M., Dolatshahi, M., Afshari, K., Shamshiri, S. *et al.* (2020) Interventions of natural and synthetic agents in inflammatory bowel disease, modulation of nitric oxide pathways. *World Journal of Gastroenterology* 26(24), 3365–3400.

Kamat, S.K. and Singh, K.N.M. (2019) Evaluation of the effect of *Aegle marmelos* and *Punica granatum* in a murine model of dextran sulfate sodium-induced acute colitis. *National Journal of Physiology, Pharmacy and Pharmacology* 9(4), 312–319.

Kaser, A. and Blumberg, R.S. (2011) Autophagy, microbial sensing, endoplasmic reticulum stress, and epithelial function in inflammatory bowel disease. *Gastroenterology* 140(6), 1738–1747.e2.

Katayama, S., Ohno, F., Mitani, T., Akiyama, H. and Nakamura, S. (2017) Rutinosylated ferulic acid attenuates food allergic response and colitis by upregulating regulatory T cells in mouse models. *Journal of Agricultural and Food Chemistry* 65(49), 10730–10737.

Kaur, G., Hamid, H., Ali, A., Alam, M.S. and Athar, M. (2004) Antiinflammatory evaluation of alcoholic extract of galls of *Quercus infectoria*. *Journal of Ethnopharmacology* 90(2–3), 285–292.

Kawabata, T., Cui, M.-Y., Hasegawa, T., Takano, F. and Ohta, T. (2011) Anti-inflammatory and anti-melanogenic steroidal saponin glycosides from fenugreek (*Trigonella foenum-graecum* L.) seeds. *Planta Medica* 77(7), 705–710.

Khajah, M.A., Orabi, K.Y., Hawai, S., Sary, H.G. and El-Hashim, A.Z. (2019) Onion bulb extract reduces colitis severity in mice via modulation of colonic inflammatory pathways and the apoptotic machinery. *Journal of Ethnopharmacology* 241, 112008.

Khalilpour, S., Behnammanesh, G., Suede, F., Ezzat, M.O., Muniandy, J. *et al.* (2018) Neuroprotective and anti-inflammatory effects of *Rhus coriaria* extract in a mouse model of ischemic optic neuropathy. *Biomedicines* 6(2), 48.

Khanavi, M., Sabbagh-Bani-Azad, M., Abdolghaffari, A.H., Vazirian, M., Isazadeh, I. *et al.* (2014) On the benefit of galls of *Quercus brantii* Lindl. in murine colitis: the role of free gallic acid. *Archives of Medical Science* 10(6), 1225–1234.

Khodakarm-Tafti, A., Mehrabani, D., Homafar, L. and Farjanikish, G. (2015) Healing effects of *Elaeagnus angustifolia* extract in experimentally induced ulcerative colitis in rats. *Journal of Pharmacology and Toxicology* 10(1), 29–35.

Khodayar, B., Farzaei, M.H., Abdolghaffari, A.H., Bahramsoltani, R., Baeeri, M. *et al.* (2018) The protective effect of the gallic acid against TNBS-induced ulcerative colitis in rats: role of inflammatory parameters. *Journal of Iranian Medical Council* 1(1), 34–42.

Khor, B., Gardet, A. and Xavier, R.J. (2011) Genetics and pathogenesis of inflammatory bowel disease. *Nature* 474(7351), 307–317.

Khosropour, P., Sajjadi, S.-E., Talebi, A. and Minaiyan, M. (2019) Anti-inflammatory effect of *Myrtus communis* hydroalcoholic extract and essential oil on acetic acid-induced colitis in rats. *Journal of Reports in Pharmaceutical Sciences* 8(2), 204–210.

Kiela, P.R., Midura, A.J., Kuscuoglu, N., Jolad, S.D., Sólyom, A.M. *et al.* (2005) Effects of *Boswellia serrata* in mouse models of chemically induced colitis. *American Journal of Physiology: Gastrointestinal and Liver Physiology* 288(4), G798–G808.

Kim, H., Banerjee, N., Ivanov, I., Talcott, S. and Mertens-Talcott, S. (2014) Comparison of anti-inflammatory mechanisms of mango (*Mangifera indica* L.) and pomegranate (*Punica granatum* L.) in DSS-induced colitis in rats. *The FASEB Journal* 28(S1), 372.8.

Kim, H., Banerjee, N., Ivanov, I., Pfent, C.M., Prudhomme, K.R. *et al.* (2016) Comparison of anti-inflammatory mechanisms of mango (*Mangifera indica* L.) and pomegranate (*Punica granatum* L.) in a preclinical model of colitis. *Molecular Nutrition & Food Research* 60(9), 1912–1923.

Kim, H., Banerjee, N., Sirven, M.A., Minamoto, Y., Markel, M.E. *et al.* (2017) Pomegranate polyphenolics reduce inflammation and ulceration in intestinal colitis-involvement of the miR-145/p70S6K1/HIF1α axis *in vivo* and *in vitro*. *The Journal of Nutritional Biochemistry* 43, 107–115.

Kim, H.-Y., Jeon, H., Bae, C.H., Lee, Y., Kim, H. *et al.* (2020) *Rumex japonicus* Houtt. alleviates dextran sulfate sodium-induced colitis by protecting tight junctions in mice. *Integrative Medicine Research* 9(2), 100398.

Kwon, K.H., Murakami, A., Tanaka, T. and Ohigashi, H. (2005) Dietary rutin, but not its aglycone quercetin, ameliorates dextran sulfate sodium-induced experimental colitis in mice: attenuation of pro-inflammatory gene expression. *Biochemical Pharmacology* 69(3), 395–406.

Langhorst, J., Varnhagen, I., Schneider, S.B., Albrecht, U., Rueffer, A. *et al.* (2013) Randomised clinical trial: a herbal preparation of myrrh, chamomile and coffee charcoal compared with mesalazine in maintaining remission in ulcerative colitis – a double-blind, double-dummy study. *Alimentary Pharmacology & Therapeutics* 38(5), 490–500.

Langhorst, J., Frede, A., Knott, M., Pastille, E., Buer, J. *et al.* (2014) Distinct kinetics in the frequency of peripheral $CD4^+$ T cells in patients with ulcerative colitis experiencing a flare during treatment with mesalazine or with a herbal preparation of myrrh, chamomile, and coffee charcoal. *PLoS One* 9(8), e104257.

Langmead, L., Dawson, C., Hawkins, C., Banna, N., Loo, S. *et al.* (2002) Antioxidant effects of herbal therapies used by patients with inflammatory bowel disease: an *in vitro* study. *Alimentary Pharmacology & Therapeutics* 16(2), 197–205.

Larrosa, M., González-Sarrías, A., Yáñez-Gascón, M.J., Selma, M.V., Azorín-Ortuño, M. *et al.* (2010) Anti-inflammatory properties of a pomegranate extract and its metabolite urolithin-A in a colitis rat model and the effect of colon inflammation on phenolic metabolism. *The Journal of Nutritional Biochemistry* 21(8), 717–725.

Larsen, S., Bendtzen, K. and Nielsen, O.H. (2010) Extraintestinal manifestations of inflammatory bowel disease: epidemiology, diagnosis, and management. *Annals of Medicine* 42(2), 97–114.

Latifi, G., Ghannadi, A. and Minaiyan, M. (2015) Anti-inflammatory effect of volatile oil and hydroalcoholic extract of *Rosa damascena* Mill. on acetic acid-induced colitis in rats. *Research in Pharmaceutical Sciences* 10(6), 514.

Lee, J.-H., Regmi, S.C., Kim, J.-A., Cho, M.H., Yun, H. *et al.* (2011) Apple flavonoid phloretin inhibits *Escherichia coli* O157: H7 biofilm formation and ameliorates colon inflammation in rats. *Infection and Immunity* 79(12), 4819–4827.

Lin, R., Piao, M. and Song, Y. (2019) Dietary quercetin increases colonic microbial diversity and attenuates colitis severity in *Citrobacter rodentium*-infected mice. *Frontiers in Microbiology* 10, 1092.

Liu, F., Wang, X., Li, D., Cui, Y. and Li, X. (2021) Apple polyphenols extract alleviated dextran sulfate sodium-induced ulcerative colitis in C57BL/6 male mice by restoring bile acid metabolism disorder and gut microbiota dysbiosis. *Phytotherapy Research* 35(3), 1468–1485.

Liu, M. (2019) Dietary fenugreek attenuates dextran sodium sulfate-induced ulcerative colitis: role of inflammation. *Japanese Journal of Gastroenterology and Hepatology* 1(1), 1–9.

Liu, T., Liu, M., Zhang, T., Liu, W., Xu, H. *et al.* (2018) *Z*-Guggulsterone attenuates astrocytes-mediated neuroinflammation after ischemia by inhibiting toll-like receptor 4 pathway. *Journal of Neurochemistry* 147(6), 803–815.

Madgulkar, A.R., Rao, M.R. and Warrier, D. (2015) Characterization of psyllium (*Plantago ovata*) polysaccharide and its uses. In: Ramawat, K. and Mérillon, J.M. (eds) *Polysaccharides*. Springer, Cham, Switzerland, pp. 871–890.

Madisch, A., Miehlke, S., Eichele, O., Mrwa, J., Bethke, B. *et al.* (2007) *Boswellia serrata* extract for the treatment of collagenous colitis. A double-blind, randomized, placebo-controlled, multicenter trial. *International Journal of Colorectal Disease* 22(12), 1445–1451.

Mahmoud, A.M., Zaki, A.R., Hassan, M.E. and Mostafa-Hedeab, G. (2017) *Commiphora molmol* resin attenuates diethylnitrosamine/phenobarbital-induced hepatocarcinogenesis by modulating oxidative stress, inflammation, angiogenesis and Nrf2/ARE/HO-1 signaling. *Chemico-Biological Interactions* 270, 41–50.

Mahmoud, A.M., Germoush, M.O., Al-Anazi, K.M., Mahmoud, A.H., Farah, M.A. *et al.* (2018) *Commiphora molmol* protects against methotrexate-induced nephrotoxicity by up-regulating Nrf2/ARE/HO-1 signaling. *Biomedicine & Pharmacotherapy* 106, 499–509.

Mandegary, A., Pournamdari, M., Sharififar, F., Pournourmohammadi, S., Fardiar, R. *et al.* (2012) Alkaloid and flavonoid rich fractions of fenugreek seeds (*Trigonella foenum-graecum* L.) with antinociceptive and anti-inflammatory effects. *Food and Chemical Toxicology* 50(7), 2503–2507.

Marín, M., María Giner, R., Ríos, J.-L. and Carmen Recio, M. (2013) Intestinal anti-inflammatory activity of ellagic acid in the acute and chronic dextrane sulfate sodium models of mice colitis. *Journal of Ethnopharmacology* 150(3), 925–934.

Marinov, V., Tzaneva, M., Zhelyazkova-Savova, M., Gancheva, S. and Valcheva-Kuzmanova, S. (2019) Effects of gallic acid in a rat model of inflammatory bowel disease induced by trinitrobenzenesulfonic acid. *Bulgarian Chemical Communications* 51, 22–28.

Markovic, B.S., Kanjevac, T., Harrell, C.R., Gazdic, M., Fellabaum, C. *et al.* (2018) Molecular and cellular mechanisms involved in mesenchymal stem cell-based therapy of inflammatory bowel diseases. *Stem Cell Reviews and Reports* 14(2), 153–165.

Mascaraque, C., Aranda, C., Ocón, B., Monte, M.J., Suárez, M.D. *et al.* (2014) Rutin has intestinal antiinflammatory effects in the $CD4^+$ $CD62L^+$ T cell transfer model of colitis. *Pharmacological Research* 90, 48–57.

Masoumi-Ardakani, Y., Abbasnejad, M., Esmaeilpour, K., Tahami, A.N. and Ahmadi, M. (2014) Significant reduction in colonic damage by *Chamomilla recutita* L. aqueous extract in acetic acid-induced colitis in rats. *Journal of Chemical and Pharmaceutical Research* 6(12), 437–445.

Mencarelli, A., Renga, B., Palladino, G., Distrutti, E. and Fiorucci, S. (2009) The plant sterol guggulsterone attenuates inflammation and immune dysfunction in murine models of inflammatory bowel disease. *Biochemical Pharmacology* 78(9), 1214–1223.

Minaiyan, M., Ghannadi, A., Mahzouni, P. and Jaffari-Shirazi, E. (2011) Comparative study of *Berberis vulgaris* fruit extract and berberine chloride effects on acetic acid-induced colitis in rats. *Iranian Journal of Pharmaceutical Research* 10(1), 97–104.

Minaiyan, M., Ghannadi, A., Etemad, M. and Mahzouni, P. (2012) A study of the effects of *Cydonia oblonga* Miller (quince) on TNBS-induced ulcerative colitis in rats. *Research in Pharmaceutical Sciences* 7(2), 103–110.

Mohammad, I.H. (2012) Use of *Anethum graveolens* in the management of patients with irritable bowel syndrome. *Mustansiriya Medical Journal* 11(1), 94–98.

Mohammed, F.I., Al-Essa, M.K., Shafagoj, Y.A. and Afifi, F.U. (2006) Investigation of the direct effects of the alcoholic extract of *Elaeagnus angustifolia* L. (Elaeagnaceae) on dispersed intestinal smooth muscle cells of guinea pig. *Scientia Pharmaceutica* 74(1), 21–30.

Momeni, A., Maghsoodi, H., Rezapour, S., Shiravand, M. and Mardani, M. (2019) Reduction of expression of IL-18, IL-1β genes in the articular joint by sumac fruit extract (*Rhus coriaria* L.). *Molecular Genetics & Genomic Medicine* 7(6), e664.

Momtaz, S., Navabakhsh, M., Bakouee, N., Dehnamaki, M., Rahimifard, M. *et al.* (2021) Cinnamaldehyde targets TLR-4 and inflammatory mediators in acetic-acid induced ulcerative colitis model. *Biologia* 76(6), 1817–1827.

Motevalian, M., Shiri, M., Shiri, S., Shiri, Z. and Shiri, H. (2017) Anti-inflammatory activity of *Elaeagnus angustifolia* fruit extract on rat paw edema. *Journal of Basic and Clinical Physiology and Pharmacology* 28(4), 377–381.

Mueller, D., Triebel, S., Rudakovski, O. and Richling, E. (2013) Influence of triterpenoids present in apple peel on inflammatory gene expression associated with inflammatory bowel disease (IBD). *Food Chemistry* 139(1), 339–346.

Mureithi, P.M. (2020) *Morphologic and morphometric study on anti-ulcerogenic effects of selected African nightshades (Solanum nigrum L.) genotypes in the rat*. MSc thesis, University of Nairobi, Nairobi.

Naouar, M.S., Mekki, L.Z., Charfi, L., Boubaker, J. and Filali, A. (2016) Preventive and curative effect of *Pistacia lentiscus* oil in experimental colitis. *Biomedicine & Pharmacotherapy* 83, 577–583.

Nascimento, S. and Araruna, C.D.S. (2016) Phytochemicals of *Punica granatum* L. in inflammatory bowel diseases: an update. *MOL2NET* 2(14), 1–x.

Nayebi, N., Khalili, N., Kamalinejad, M. and Emtiazy, M. (2017) A systematic review of the efficacy and safety of *Rosa damascena* Mill. with an overview on its phytopharmacological properties. *Complementary Therapies in Medicine* 34, 129–140.

Nikniaz, Z., Ostadrahimi, A., Mahdavi, R., Ebrahimi, A.A. and Nikniaz, L. (2014) Effects of *Elaeagnus angustifolia* L. supplementation on serum levels of inflammatory cytokines and matrix metalloproteinases in females with knee osteoarthritis. *Complementary Therapies in Medicine* 22(5), 864–869.

Pachi, V.K., Mikropoulou, E.V., Gkiouvetidis, P., Siafakas, K., Argyropoulou, A. *et al.* (2020) Traditional uses, phytochemistry and pharmacology of Chios mastic gum (*Pistacia lentiscus* var. Chia, Anacardiaceae): a review. *Journal of Ethnopharmacology* 254, 112485.

Palla, A.H., Iqbal, N.T., Minhas, K. and Gilani, A.-H. (2016) Flaxseed extract exhibits mucosal protective effect in acetic acid induced colitis in mice by modulating cytokines, antioxidant and antiinflammatory mechanisms. *International Immunopharmacology* 38, 153–166.

Palla, A.H., Gilani, A.-U.-H., Bashir, S. and Ur Rehman, N. (2020) Multiple mechanisms of flaxseed: effectiveness in inflammatory bowel disease. *Evidence-Based Complementary and Alternative Medicine* 2020, 7974835.

Pandurangan, A.K., Mohebali, N., Norhaizan, M.E. and Looi, C.Y. (2015) Gallic acid attenuates dextran sulfate sodium-induced experimental colitis in BALB/c mice. *Drug Design, Development and Therapy* 9, 3923–3934.

Papada, E., Forbes, A., Amerikanou, C., Torović, L., Kalogeropoulos, N. *et al.* (2018) Antioxidative efficacy of a *Pistacia lentiscus* supplement and its effect on the plasma amino acid profile in inflammatory bowel disease: a randomised, double-blind, placebo-controlled trial. *Nutrients* 10(11), 1779.

Papada, E., Gioxari, A., Amerikanou, C., Forbes, A., Tzavara, C. *et al.* (2019) Regulation of faecal biomarkers in inflammatory bowel disease patients treated with oral mastiha (*Pistacia lentiscus*) supplement: a double-blind and placebo-controlled randomised trial. *Phytotherapy Research* 33(2), 360–369.

Parisio, C., Lucarini, E., Micheli, L., Toti, A., Khatib, M. *et al.* (2020) Pomegranate mesocarp against colitis-induced visceral pain in rats: effects of a decoction and its fractions. *International Journal of Molecular Sciences* 21(12), 4304.

Pastrelo, M.M., Dias Ribeiro, C.C., Duarte, J.W., Bioago Gollücke, A.P., Artigiani-Neto, R. *et al.* (2017) Effect of concentrated apple extract on experimental colitis induced by acetic acid. *International Journal of Molecular and Cellular Medicine* 6(1), 38–49.

Patlevic, P., Vašková, J., Švorc, P., Vaško, L. and Švorc, P. (2016) Reactive oxygen species and antioxidant defense in human gastrointestinal diseases. *Integrative Medicine Research* 5(4), 250–258.

Persic, M., Mikulic-Petkovsek, M., Slatnar, A. and Veberic, R. (2017) Chemical composition of apple fruit, juice and pomace and the correlation between phenolic content, enzymatic activity and browning. *LWT-Food Science and Technology* 82, 23–31.

Pithadia, A.B. and Jain, S. (2011) Treatment of inflammatory bowel disease (IBD). *Pharmacological Reports* 63(3), 629–642.

Pituch-Zdanowska, A., Banaszkiewicz, A. and Albrecht, P. (2015) The role of dietary fibre in inflammatory bowel disease. *Przeglad Gastroenterologiczny* 10(3), 135–141.

Prachayasittikul, V., Prachayasittikul, S., Ruchirawat, S. and Prachayasittikul, V. (2018) Coriander (*Coriandrum sativum*): a promising functional food toward the well-being. *Food Research International* 105, 305–323.

Putnik, P., Gabrić, D., Roohinejad, S., Barba, F.J., Granato, D. *et al.* (2019) An overview of organosulfur compounds from *Allium* spp.: from processing and preservation to evaluation of their bioavailability, antimicrobial, and anti-inflammatory properties. *Food Chemistry* 276, 680–691.

Qadir, M.I., Ali, M., Saleem, M. and Hanif, M. (2014) Hepatoprotective activity of aqueous methanolic extract of *Viola odorata* against paracetamol-induced liver injury in mice. *Bangladesh Journal of Pharmacology* 9(2), 198–202.

Raeeszadeh, M., Mortazavi, P. and Atashin-Sadafi, R. (2021) The antioxidant, anti-inflammatory, pathological, and behavioural effects of *Medicago sativa* L. (alfalfa) extract on brain injury caused by nicotine in male rats. *Evidence-Based Complementary and Alternative Medicine* 2021, 6694629.

Rafieian-Kopaei, M., Shakiba, A., Sedighi, M. and Bahmani, M. (2017) The analgesic and anti-inflammatory activity of *Linum usitatissimum* in Balb/c mice. *Journal of Evidence-Based Complementary & Alternative Medicine* 22(4), 892–896.

Rahimi, R., Shams-Ardekani, M.R. and Abdollahi, M. (2010) A review of the efficacy of traditional Iranian medicine for inflammatory bowel disease. *World Journal of Gastroenterology* 16(36), 4504–4514.

Rahimi, R., Baghaei, A., Baeeri, M., Amin, G., Shams-Ardekani, M.R. *et al.* (2013) Promising effect of Magliasa, a traditional Iranian formula, on experimental colitis on the basis of biochemical and cellular findings. *World Journal of Gastroenterology* 19(12), 1901–1911.

Rai, N., Yadav, S., Verma, A., Tiwari, L. and Sharma, R.K. (2012) A monographic profile on quality specifications for a herbal drug and spice of commerce – *Cuminum cyminum* L. *International Journal of Advanced Herbal Science and Technology* 1(1), 1–12.

Rashidian, A., Akbarzadeh, D., Asgarpanah, J. and Dehpour, A. (2021) *Bunium persicum* essential oil reduced acetic acid-induced rat colitis through suppression of NF-κB pathway. *Avicenna Journal of Phytomedicine* 11(5), 505–514.

Reddy, K.V.K., Maheswaraiah, A. and Naidu, K.A. (2014) Rice bran oil and *n*-3 fatty acid-rich garden cress (*Lepidium sativum*) seed oil attenuate murine model of ulcerative colitis. *International Journal of Colorectal Disease* 29(2), 267–269.

Riaz, A., Khan, R.A., Afroz, S. and Mallick, N. (2017) Prophylactic and therapeutic effect of *Punica granatum* in trinitrobenzene sulfonic acid induced inflammation in rats. *Pakistan Journal of Pharmaceutical Sciences* 30(1), 155–162.

Rodríguez-Cabezas, M.E., Galvez, J., Lorente, M.D., Concha, A., Camuesco, D. *et al.* (2002) Dietary fiber down-regulates colonic tumor necrosis factor α and nitric oxide production in trinitrobenzenesulfonic acid-induced colitic rats. *The Journal of Nutrition* 132(11), 3263–3271.

Rodríguez-Cabezas, M., Galvez, J., Camuesco, D., Lorente, M., Concha, A. *et al.* (2003) Intestinal anti-inflammatory activity of dietary fiber (*Plantago ovata* seeds) in HLA-B27 transgenic rats. *Clinical Nutrition* 22(5), 463–471.

Rosenthal, R., Luettig, J., Hering, N., Krug, S., Albrecht, U. *et al.* (2017) Myrrh exerts barrier-stabilising and -protective effects in HT-29/B6 and Caco-2 intestinal epithelial cells. *International Journal of Colorectal Disease* 32(5), 623–634.

Sadar, S.S., Vyawahare, N.S. and Bodhankar, S.L. (2016) Ferulic acid ameliorates TNBS-induced ulcerative colitis through modulation of cytokines, oxidative stress, iNOs, COX-2, and apoptosis in laboratory rats. *EXCLI Journal* 15, 482–499.

Sajadimajd, S., Bahramsoltani, R., Iranpanah, A., Patra, J.K., Das, G. *et al.* (2020) Advances on natural polyphenols as anticancer agents for skin cancer. *Pharmacological Research* 151, 104584.

Samoilă, I., Dinescu, S. and Costache, M. (2020) Interplay between cellular and molecular mechanisms underlying inflammatory bowel diseases development – a focus on ulcerative colitis. *Cells* 9(7), 1647.

Sanches-Silva, A., Testai, L., Nabavi, S.F., Battino, M., Devi, K.P. *et al.* (2020) Therapeutic potential of polyphenols in cardiovascular diseases: regulation of mTOR signaling pathway. *Pharmacological Research* 152, 104626.

Sanei-Dehkordi, A., Vatandoost, H., Abaei, M.R., Davari, B. and Sedaghat, M.M. (2016) Chemical composition and larvicidal activity of *Bunium persicum* essential oil against two important mosquitoes vectors. *Journal of Essential Oil Bearing Plants* 19(2), 349–357.

Sarfraz, R.M., Khan, H., Maheen, S., Afzal, S., Akram, M.R., *et al.* (2017) *Plantago ovata*: a comprehensive review on cultivation, biochemical, pharmaceutical and pharmacological aspects. *Acta Poloniae Pharmaceutica* 74(3), 739–746.

Sari, D.P., Bellatasie, R. and Ifora, I. (2021) Anti-inflammatory properties of *Coriandrum sativum* L.: a review. *Research Journal of Pharmacy and Medical Sciences* 4(2), 34–38.

Schmidt, C., Grunert, P.C. and Stallmach, A. (2021) An update for pharmacologists on new treatment options for inflammatory bowel disease: the clinicians' perspective. *Frontiers in Pharmacology* 12, 491.

Sendker, J., Böker, I., Lengers, I., Brandt, S., Jose, J. *et al.* (2017) Phytochemical characterization of low molecular weight constituents from marshmallow roots (*Althaea officinalis*) and inhibiting effects of the aqueous extract on human hyaluronidase-1. *Journal of Natural Products* 80(2), 290–297.

Shah, T.A., Parikh, M., Patel, K.V., Patel, K.G., Joshi, C.G. *et al.* (2016) Evaluation of the effect of *Punica granatum* juice and punicalagin on NFκB modulation in inflammatory bowel disease. *Molecular and Cellular Biochemistry* 419(1), 65–74.

Shalaby, M.A. and Hammouda, A.A.-E. (2014) Analgesic, anti-inflammatory and anti-hyperlipidemic activities of *Commiphora molmol* extract (myrrh). *Journal of Intercultural Ethnopharmacology* 3(2), 56–62.

Silva, B.M., Andrade, P.B., Ferreres, F., Seabra, R.M., Beatriz, M. *et al.* (2005) Composition of quince (*Cydonia oblonga* Miller) seeds: phenolics, organic acids and free amino acids. *Natural Product Research* 19(3), 275–281.

Silva, F.G.D. (2016) Antioxidant capacity of protein hydrolysates and phenolic compounds of flaxseed (*Linum usitatissimum* L.) and its modulatory effects on experimental colitis. PhD thesis, State University of Campinas, Campinas, Brazil.

Sindhu, G., Ratheesh, M., Shyni, G.L., Nambisan, B. and Helen, A. (2012) Anti-inflammatory and antioxidative effects of mucilage of *Trigonella foenum graecum* (fenugreek) on adjuvant induced arthritic rats. *International Immunopharmacology* 12(1), 205–211.

Singh, K., Jaggi, A.S. and Singh, N. (2009) Exploring the ameliorative potential of *Punica granatum* in dextran sulfate sodium induced ulcerative colitis in mice. *Phytotherapy Research* 23(11), 1565–1574.

Singh, S., Kaur, R. and Sharma, S. (2013) Antinociceptive, antiinflammatory and antipyretic activities of *Rumex hastatus* D. Don stem and roots. *Der Pharmacia Sinica* 4(3), 95–102.

Sisay, M. and Gashaw, T. (2017) Ethnobotanical, ethnopharmacological, and phytochemical studies of *Myrtus communis* Linn: a popular herb in Unani system of medicine. *Journal of Evidence-Based Complementary & Alternative Medicine* 22(4), 1035–1043.

Skyberg, J.A., Robison, A., Golden, S., Rollins, M.F., Callis, G. *et al.* (2011) Apple polyphenols require T cells to ameliorate dextran sulfate sodium-induced colitis and dampen proinflammatory cytokine expression. *Journal of Leukocyte Biology* 90(6), 1043–1054.

Solanki, R.P., Gandhi, T.R. and Patel, K.V. (2011) Evaluation of the effect of *Quercus infectoria* Olivier (Fagaceae) in experimentally induced inflammatory bowel disease in rats. *Research Journal of Pharmacy and Technology* 4(5), 782–786.

Sosa, S., Tubaro, A., Della Loggia, R. and Bombardelli, E. (1993) Anti-inflammatory activity of *Commiphora mukul* extracts. *Pharmacological Research* 27, 89–90.

Tabarsa, M., You, S., Yelithao, K., Palanisamy, S., Prabhu, N.M. *et al.* (2020) Isolation, structural elucidation and immuno-stimulatory properties of polysaccharides from *Cuminum cyminum*. *Carbohydrate Polymers* 230, 115636.

Tafazoli, V., Taherifard, E., Nimrouzi, M. and Pasalar, M. (2022) Therapeutic effect of *Plantago major* on active severe pancolitis: a case report. *Advances in Integrative Medicine* 9, 90–93.

Tafti, L.D., Shariatpanahi, S.M., Damghani, M.M. and Javadi, B. (2017) Traditional Persian topical medications for gastrointestinal diseases. *Iranian Journal of Basic Medical Sciences* 20(3), 222–241.

Tang, Y., Li, J., Li, F., Hu, C.-A.A., Liao, P. *et al.* (2015) Autophagy protects intestinal epithelial cells against deoxynivalenol toxicity by alleviating oxidative stress via IKK signaling pathway. *Free Radical Biology and Medicine* 89, 944–951.

Tanideh, N., Afaridi, E., Mehrabani, D., Azarpira, N., Hosseinzadeh, M. *et al.* (2014) The healing effect of *Berberis vulgaris* in acetic acid-induced ulcerative colitis in rat. *Middle-East Journal of Scientific Research* 21(8), 1288–94.

Tavakoli, A., Shirzad, M., Taghavi, A., Fattahi, M., Ahmadian-Attari, M. *et al.* (2019) Efficacy of rose oil soft capsules on clinical outcomes in ulcerative colitis: A pilot randomized, double-blinded, placebo-controlled clinical trial. *Galen Medical Journal* 8, e1307.

Teimouri, H., Rezaei, M., Abbaszadeh, S. and Azadpour, M. (2019) The effects of aqueous extracts of *Cuminum cyminum* L. and *Pimpinella anisum* L. seeds on the improvement of irritable bowel syndrome compared with loperamide in rats. *International Journal of Biology and Chemistry* 12(1), 41–47.

Wang, Y., Guo, T., Li, J.Y., Zhou, S.Z., Zhao, P. *et al.* (2013) Four flavonoid glycosides from the pulps of *Elaeagnus angustifolia* and their antioxidant activities. *Advanced Materials Research* 756–759, 16–20.

Weber, L., Kuck, K., Jürgenliemk, G., Heilmann, J., Lipowicz, B. *et al.* (2020) Anti-inflammatory and barrier-stabilising effects of myrrh, coffee charcoal and chamomile flower extract in a co-culture cell model of the intestinal mucosa. *Biomolecules* 10(7), 1033.

Wei, J., Zhang, X., Bi, Y., Miao, R., Zhang, Z. *et al.* (2015) Anti-inflammatory effects of cumin essential oil by blocking JNK, ERK, and NF-κB signaling pathways in LPS-stimulated RAW 264.7 cells. *Evidence-Based Complementary and Alternative Medicine* 2015, 474509.

Wu, H., Luo, T., Li, Y., Gao, Z., Zhang, K. *et al.* (2018) Granny Smith apple procyanidin extract upregulates tight junction protein expression and modulates oxidative stress and inflammation in lipopolysaccharide-induced Caco-2 cells. *Food & Function* 9(6), 3321–3329.

Wu, T.T., Tsai, C.W., Yao, H.T., Lii, C.K., Chen, H.W. *et al.* (2010) Suppressive effects of extracts from the aerial part of *Coriandrum sativum* L. on LPS-induced inflammatory responses in murine RAW 264.7 macrophages. *Journal of the Science of Food and Agriculture* 90(11), 1846–1854.

Xie, Y., Wang, L., Sun, H., Wang, Y., Yang, Z. *et al.* (2019) Immunomodulatory, antioxidant and intestinal morphology-regulating activities of alfalfa polysaccharides in mice. *International Journal of Biological Macromolecules* 133, 1107–1114.

Yang, K., Zhang, L., Liao, P., Xiao, Z., Zhang, F., Sindaye, D., *et al.* (2020) Impact of gallic acid on gut health: Focus on the gut microbiome, immune response, and mechanisms of action. *Frontiers in Immunology* 16(11), 580208.

Yeganeh, P.R., Leahy, J., Spahis, S., Patey, N., Desjardins, Y. *et al.* (2018) Apple peel polyphenols reduce mitochondrial dysfunction in mice with DSS-induced ulcerative colitis. *The Journal of Nutritional Biochemistry* 57, 56–66.

Yoshioka, Y., Akiyama, H., Nakano, M., Shoji, T., Kanda, T. *et al.* (2008) Orally administered apple procyanidins protect against experimental inflammatory bowel disease in mice. *International Immunopharmacology* 8(13–14), 1802–1807.

Zaghlool, S.S., Shehata, B.A., Abo-Seif, A.A. and Abd El-Latif, H.A. (2015) Protective effects of ginger and marshmallow extracts on indomethacin-induced peptic ulcer in rats. *Journal of Natural Science, Biology, and Medicine* 6(2), 421–428.

Zaghlool, S.S., Abo-Seif, A.A., Rabeh, M.A., Abdelmohsen, U.R. and Messiha, B.A. (2019) Gastro-protective and anti-oxidant potential of *Althaea officinalis* and *Solanum nigrum* on pyloric ligation/indomethacin-induced ulceration in rats. *Antioxidants* 8(11), 512.

Zahouani, Y., Rhouma, K.B., Kacem, K., Sebai, H. and Sakly, M. (2020) Aqueous leaf extract of *Pistacia lentiscus* improves acute acetic acid-induced colitis in rats by reducing inflammation and oxidative stress. *Journal of Medicinal Food* 24(7), 697–708.

Zhang, J.-H., Shangguan, Z.-S., Chen, C., Zhang, H.-J. and Lin, Y. (2016) Anti-inflammatory effects of guggulsterone on murine macrophage by inhibiting LPS-induced inflammatory cytokines in NF-κB signaling pathway. *Drug Design, Development and Therapy* 10, 1829–1835.

Zhang, Y.-Z. and Li, Y.-Y. (2014) Inflammatory bowel disease: pathogenesis. *World Journal of Gastroenterology* 20(1), 91–99.

Zhou, H. and Mineshita, S. (2000) The effect of berberine chloride on experimental colitis in rats *in vivo* and *in vitro*. *Journal of Pharmacology and Experimental Therapeutics* 294(3), 822–829.

Zhou, Q., Ma, L., Zhao, W., Zhao, W., Han, X. *et al.* (2020) Flaxseed oil alleviates dextran sulphate sodium-induced ulcerative colitis in rats. *Journal of Functional Foods* 64, 103602.

Zhu, L., Gu, P. and Shen, H. (2019) Gallic acid improved inflammation via NF-κB pathway in TNBS-induced ulcerative colitis. *International Immunopharmacology* 67, 129–137.

Zubair, M., Widén, C., Renvert, S. and Rumpunen, K. (2019) Water and ethanol extracts of *Plantago major* leaves show anti-inflammatory activity on oral epithelial cells. *Journal of Traditional and Complementary Medicine* 9(3), 169–171.

6 Herbal Medicine for Cardiovascular Disease in View of Avicenna

Zahra Taghipour[1]* and Mamak Hashemi Habib Abadi[2]
[1]Department of Traditional Pharmacy, School of Persian Medicine, Tehran University of Medical Sciences, Tehran, Iran; [2]Department of Persian Medicine, School of Medicine, Hamadan University of Medical Sciences, Hamadan, Iran

Abstract
Cardiovascular disease (CVD) is a spectrum of diseases affecting the heart and blood vessels and is the first cause of death worldwide. The epidemiology of CVD demonstrates a transition, driven by industrialization, urbanization, and associated lifestyle changes, and is taking place in every part of the world among all races, ethnic groups, and cultures. In the present stage, coronary heart disease (CHD), stroke, congestive heart failure, and peripheral vascular disease are the most common CVDs. However, CVDs have historically occurred in the form of rheumatic heart disease and cardiomyopathies due to infection and malnutrition. Avicenna made a great contribution to the field of cardiology and many of the current well-known concepts can be traced back to his views. He was the first scientist who established cardiology in the early Middle Ages, thus initiating the enormous progress of cardiological sciences up to the present day. Besides nutritional recommendations and specific manipulations (e.g., venesection, *hijmat*, laxation) to prevent and manage heart diseases, a lot of cardioactive drugs of herbal, animal, and mineral origins are explained by Avicenna to treat different heart diseases. In addition, he specifically mentioned 83 simple cardioactive drugs of which 50 are medicinal herbs. In this chapter the medicinal plants used by Avicenna to treat CVDs and new related research studies are presented.

Introduction

Cardiovascular disease (CVD) is a spectrum of diseases involving the heart and blood vessels and is the first cause of death globally (Mendis *et al.*, 2015). In 2015, CVDs accounted for ~17.9 million deaths worldwide (32% of total). The epidemiology of CVD demonstrates a transition, driven by industrialization, urbanization, and associated lifestyle changes, and is taking place in every part of the world among all races, ethnic groups, and cultures. In the present stage, coronary heart disease (CHD), stroke, congestive heart failure, and peripheral vascular disease are the most common CVDs. However, CVDs have historically occurred in the form of rheumatic heart disease and cardiomyopathies due to infection and malnutrition (Kasper *et al.*, 2018).

Avicenna made a great contribution to the field of cardiology and many of the current well-known concepts can be traced back to his views. He was the first scientist who

*Email: z_taghipour@ymail.com

 Medicinal Plants Used in Traditional Persian Medicine (eds. H. Schulz, Seyed Ahmad Emami and Farsad Nadjafi)
DOI: 10.1079/9781800621671.0006

founded cardiology in the Early Middle Ages, thus initiating the enormous progress of cardiological sciences up to the present day (Zarshenas and Zargaran, 2015). He discussed various heart problems in the third book of his comprehensive textbook entitled *The Canon of Medicine* (*Al-Qanun fi'l-Tibb*) (Ibn Sinā, 2002; Sobhani *et al.*, 2017) as well as in another specific book named *The Treatise on Cardiac Drugs* (*Al-Adwiyah al-Qalbiyah*) (Ibn Sinā, 2009). According to the era when Avicenna wrote his books, CHD was not the most common CVD, so as we see, he reported heart diseases similar to myocarditis and some disorders of rhythm and myocardial infarction.

Besides nutritional recommendations and specific manipulations (e.g., venesection, *hijmat*, laxation) to prevent and manage heart diseases, a lot of cardioactive drugs of herbal, animal, and mineral origins are explained by Avicenna to treat different heart diseases. In addition, he specifically mentioned 83 simple cardioactive drugs of which 50 are medicinal herbs (Sobhani *et al.*, 2017).

Cardiovascular Diseases and Treatments in View of Avicenna

According to the concepts of Traditional Persian Medicine (TPM), the heart is the origin of the "innate body heat" and so the factors affecting "innate body heat" may lead to cardiac weakness and diseases. Excess in exercise and sexual activity, emotional stress, and insomnia are the most important mentioned factors.

Heart dystemperaments

In accordance with humoral medicine, treatments for cardiac disorders that can lead to various clinical symptoms are divided into the following four types:

- warm cardiac dystemperament is treated with specific fruit sauces like apple and quince and topical cooling liquid and semiliquid drugs on the chest;
- cold cardiac dystemperament is treated with warm-temperament foods, moderate exercise, wine, oral complex drugs (*maʿajoon*), and topical warming liquid and semiliquid drugs on the chest;
- dry cardiac dystemperament is treated by drinking cool water, decreasing activity, taking a bath afterward eating, therapeutic sitz baths, and moisturizing foods and drugs; and
- wet cardiac dystemperament is treated with frequent exercises, taking a bath before eating, and administering laxative and diuretics.

It seems that heart dystemperaments are some kinds of myocarditis from the conventional medicine point of view.

Heart warm inflammation (*waram ḥār*)

This can be a fatal disease that manifests with severe palpitation (*khafaqān*) plus symptoms of heat and inflammation in the heart. In most cases venesection is necessary and lifesaving, then local administration of herbal lotions and continued drinking of cool liquid drugs. This condition resembles acute infectious myocarditis.

Syncope (*al-ghashi*)

Syncope may be caused by suppression of body motor functions due to heart weakness. Weakness in other headship organs like the brain and liver, or in related organs such as the stomach, can also lead to heart weakness and finally cause fainting (*al-ghashi*). Multiple factors are listed as its etiology such as obstruction, inflammations, enormous discharges, severe pains, and emotional disasters.

Sudden force decrease (*soqūt-al-qowah*)

This condition is similar to *al-ghashi* but different in etiology and intensity. In this condition patients do not drop and can move even with excessive effort. Hence *soqūt-al-qowah* may lead to *al-ghashi* (syncope) if it intensifies. The cause is excessive dense humors.

Palpitation (*khafaqān*)

Palpitation (*khafaqān*) is considered fasciculation (abnormal movement) triggered in the heart. Its etiology is any factor that disturbs the heart or pericardium, either a somatic factor or a psychological one, such as severe anxiety. It is an irregular pulse, characterized by changeable intensity, frequency, and speed.

Myocardial infarction (*mawt e faj'ah*)

The increase in blood viscosity and appearance of clots may lead to arterial obstruction in the heart and cause myocardial infarction (*mawt e faj'ah*). Regarding the basics of TPM, Avicenna explained that yellow bile is needed to restore permeability in the capillaries and arteries. In the case of a decrease in yellow bile, he emphasized, blood viscosity could lead to arterial obstruction and myocardial infarction.

Brief Review on Some Types of Cardiovascular Diseases and Their Treatments in Modern Medicine

Arrhythmia

An arrhythmia is any rhythm that is not a normal sinus rhythm with normal atrioventricular (AV) conduction, of which palpitation is its most common presentation. This disorder can occur as bradycardias, ventricular tachycardia, atrial fibrillation, AV block, etc. Palpitation, which is defined as an unpleasant awareness of the forceful, rapid, or irregular beating of the heart, is one of the disorders induced by arrhythmia. Abnormal impulse formation can arise from abnormal automaticity or triggered activity originating from the sinoatrial (SA) node or other sites. Hypoxia, ischemia, or excess catecholamine activity could lead to abnormal automaticity. Ion channels and currents, enhanced automaticity, and triggered activity are proposed mechanism for arrhythmia (Lévy and Olshansky, 2021).

Treatment

Some rhythm disturbances, such as rapid rates in atrial fibrillation, or sustained ventricular tachycardia, require urgent attention. Others, such as ectopic atrial and ventricular beats, may cause concern but do not need immediate management. Depending on the individual patient's situation, antiarrhythmic drugs, pacemakers, implantable cardioverter-defibrillators (ICDs), and biventricular devices may be prescribed (Lévy and Olshansky, 2021).

Shock

Shock is defined as a state of cellular and tissue hypoxia that can be due to intracardiac causes of cardiac pump failure that result in reduced cardiac output (CO). Myocardial infarction, atrial and ventricular tachyarrhythmias, and bradyarrhythmias as well as mechanical disturbance can lead to shock (Gaieski and Mikkelsen, 2021).

Treatment

Ventilatory support followed by hemodynamic support with/without norepinephrine, dopamine, and nitroglycerin, etc. can be prescribed depending on the patient's condition (Gaieski and Mikkelsen, 2021).

Scientifc Evaluation of Medicinal Plants Mentioned for the Management of Cardiovascular Diseases in Avicenna's *Canon*

The medicinal herbs mentioned in the *Canon of Medicine* for the management of CVDs and some scientific evidence supporting their efficacy are described individually in this section. In addition, the traditional application of some plants is also reported. All collected results are presented in Tables 6.1, 6.2, 6.3, and 6.4.

Table 6.1. Medicinal plants mentioned in *Canon of Medicine* for treatment of CVD.

No.	Scientific name(s)	Synonym(s)	Arabic name(s)	Common English name	Family
1	*Anchusa azurea* Mill.	*Echium orientale* C.A. Mey.	*Gāvzabān*	Bugloss	*Boraginaceae*
	Echium amoenum Fisch. & C.A. Mey.	*Anchusa italica* Retz.		Sea bugloss	
2	*Doronicum grandiflorum* Lam.	*Doronicum scorpioides* Lam.	*Dorunaj*	–	*Asteraceae*
3	*Zingiber zerumbet* (L.) Sm.	*Amomum zerumbet* L.	*Zurunbād*	Bitter ginger	*Zingiberaceae*
4	*Crocus sativus* L.	*Crocus officinalis* (L.) Honck.	*Zaʿ afarān*	Saffron	*Iridaceae*
5	*Curcuma zedoaria* (Christm.) Roscoe	*Costus nigricans* Blanco	*Jadwār*	White turmeric	*Zingiberaceae*
6	*Centaurea behen* L.	*Centaurea alata* Lam.	*Bahman abyaḍ*	White behen	*Asteraceae*
7	*Limonium vulgare* Mill.	*Statice limonium* L.	*Bahman aḥmar*	Sea lavender	*Plumbaginaceae*
8	*Syzygium aromaticum* (L.) Merr. & L.M. Perry	*Caryophyllus aromaticus* L.	*Qaranful*	Clove	*Myrtaceae*
9	*Melissa officinalis* L.	*Melissa hirsuta* Hornem.	*Bādranjbuyah, Toranjān*	Lemon balm	*Lamiaceae* (*Labiatae*)
10	*Ocimum basilicum* L.	*Ocimum album* L.	*Bādruj*	Basil	*Lamiaceae* (*Labiatae*)
11	*Tanacetum balsamita* L.	*Balsamita major* Desf.	*Shahesfaram*	Balsam weed	*Asteraceae*
12	*Amomum subulatum* Roxb.	*Cardamomum subulatum* (Roxb.) Kuntze	*Qāqolah, Hil aswad*	Black cardamom	*Zingiberaceae*
13	*Piper cubeba* L.	*Cubeba officinalis* Rafin.	*Kabābah*	Cubeb	*Piperaceae*
14	*Clinopodium acinos* (L.) Kuntze	*Calamintha acinos* (L.) Clairv.	*Falanjmeshk*	Basil thyme	*Lamiaceae* (*Labiatae*)
15	*Citrus medica* L.	*Citrus × aurantium* var. *medica* (L.) Wight & Arn.	*Otroj*	Citron	*Rutaceae*
16	*Cinnamomum citriodorum* Thwaites	*Camphora citriodora* (Thwaites) Lukman.	*Sadaj hendi*	Malabar cinnamon	*Lauraceae*
17	*Inula helenium* L.	*Aster helenium* Scop.	*Rāsan*	Elecampane	*Asteraceae*
18	*Cinnamomum camphora* (L.) J. Presl.	*Camphora camphora* (L.) H. Karst.	*Kāfur*	Camphor	*Lauraceae*
19	*Pterocarpus santalinus* L.	*Lingoum santalinum* (L.f.) Kuntze	*Ṣandal aḥmar*	Red sandalwood	*Fabaceae* (*Leguminosae*)
20	*Santalum album* L.	*Sirium myrtifolium* L.	*Ṣandal abyaḍ*	White sandalwood	*Fabaceae* (*Leguminosae*)
21	*Rosa × damascena* Mill.	*Rosa belgica* Mill.	*Ward juri*	Damask rose	*Rosaceae*
22	*Bambusa bambos* (L.) Voss	*Bambusa arundinacea* (Retz.) Willde.	*Kheizarān, Ṭabāshir*	Giant thorny bamboo	*Poaceae* (*Gramineae*)
23	*Malus domestica* Borkh.	*Pyrus malus* L.	*Toffāḥ*	Apple	*Rosaceae*
24	*Coriandrum sativum* L.	*Coriandrum majus* Gouan	*Kozbarah*	Coriander	*Apiaceae* (*Umbelliferae*)
25	*Raphanus sativus* L.	*Raphanus candidus* Vorosch.	*Fojl*	Radish	*Brassicaceae* (*Cruciferae*)
26	*Citrullus colocynthis* (L.) Schrad.	*Cucumis colocynthis* L.	*Ḥ anẓal*	Bitter apple	*Cucurbitaceae*

Continued

Table 6.1. Continued.

No.	Scientific name(s)	Synonym(s)	Arabic name(s)	Common English name	Family
27	*Laricifomes officinalis* (Vill.) Kotl. & Pouzar	*Fomitopsis officinalis* (Vill.) Fr. Bond. & Sing.	*Ghārighun*	Ghariqun	*Fomitopsidaceae*
28	*Cuscuta epithymum* Murray	*Cuscuta acutiflora* Rota	*Aftimoon*	Dodder	*Convolvulaceae*
29	*Cydonia oblonga* Mill.	–	*Safarjal*	Quince	*Rosaceae*
30	*Pyrus communis* L.	–	*Komatharā*	Common pear	*Rosaceae*
31	*Origanum majorana* L.	*Majorana hortensis* Moench	*Marzanjosh*	Sweet marjoram	*Lamiaceae* (*Labiatae*)
32	*Moringa oleifera* Lam.	*Hyperanthera moringa* (L.) Vahl .	*Bān*	Ben	*Moringaceae*
33	*Hordeum vulgare* L.	*Hordeum sativum* Pers.	*Shaʿir*	Common barley	*Poaceae* (*Gramineae*)
34	*Punica granatum* L.	*Punica nana* L.	*Rommān* (*jolnār*)	Pomegranate	*Lythraceae*
35	*Amaranthus blitum* L.	*Amaranthus alius* K. Krause	*Baqlah yamāniah*	Blite	*Amaranthaceae*
36	*Salix* sp.	–	*Khalāf*	Willow	*Salicaceae*
37	*Citrus limon* (L.) Burm.f.	–	*Laymun*	Lemon	*Rutaceae*
38	*Rheum palmatum* L.	*Rheum potaninii* Losinsk.	*Rāwand*	Rhubarb	*Polygonaceae*
39	*Lens culinaris* Medik.	*Ervum lens* Wall.	*Aʿdas*	Lentil	*Fabaceae* (*Leguminosae*)
40	*Aquilaria agallocha* Roxb.	–	*Uʿd*	Eaglewood	*Thymelaeaceae*
41	*Alhagi maurorum* Medik.	*Alhagi camelorum* Fisch.	*Taranjebin*, *Aʿqul*	Manna of hedysarum	*Fabaceae* (*Leguminosae*)
42	*Lactuca sativa* L.	*Lactuca palmata* Willd.	*Khas*	Lettuce	*Asteraceae*
43	*Cichorium intybus* L.	*Cichorium commune* Pall.	*Hindiba'*	Chicory	*Asteraceae*
	Cichorium endivia L.	*Cichorium ambiguum* Schult.		Endive	
44	*Portulaca oleracea* L.	*Portulaca fosbergii* Poelln.	*Baqlat al ḥamqā*	Common purslane	*Portulacaceae*
45	*Atropa belladonna* L.	*Atropa lethalis* Salisb.	*Lifah*	Belladonna	*Solanaceae*
46	*Papaver somniferum* L.	–	*Al'afiun*	Opium poppy	*Papaveraceae*
47	*Operculina turpethum* (L.) Silva Manso	*Ipomoea turpethum* (L.) R. Br.	*Torbod*	Turbith	*Convolvulaceae*
48	*Commiphora mukul* (Hook. ex Stocks) Engl.	*Balsamodendrum mukul* Hook. ex Stocks	*Muqil*	Guggul	*Burseraceae*
49	*Terminalia chebula* Willd. ex Flem.	*Terminalia gangetica* Roxb.	*Ihlilaj kābuli*	Chebulic myrobalan	*Combretaceae*
50	*Cicer arietinum* L.	*Cicer album* Hort.	*Ḥ immaṣ*	Chickpea	*Fabaceae* (*Leguminosae*)
51	*Cucumis sativus* L.	*Cucumis esculentus* Salisb.	*Khiār*	Cucumber	*Cucurbitaceae*
52	*Boswellia sacra* Flueck.	*Boswellia carteri* Birdw.	*Kondor*	Frankincense	*Burseraceae*
53	*Myrtus communis* L.	*Myrtus acuta* Mill.	*Ās*	Myrtle	*Myrtaceae*

54	*Paeonia officinalis* L.	*Moutan officinalis* (L.) Lindl. & Paxton	*Fāwāniā*	Common peony	*Paeoniaceae*
55	*Hyssopus officinalis* L.	*Thymus hyssopus* E.H.L. Krause	*Zufā*	Hyssop	*Lamiaceae (Labiatae)*
56	*Nardostachys jatamansi* (D. Don) DC.	*Nardostachys grandiflora* DC.	*Nāridin, Sonbol Hindi*	Nard	*Caprifoliaceae*
57	*Sinapis alba* L.	*Brassica hirta* Moench	*Khardal abyaḍ*	White mustard	*Brassicaceae*
58	*Brassica nigra* (L.) K. Koch	*Sinapis nigra* L.	*Khardal aswad*	Black mustard	*Brassicaceae*
59	*Anacyclus pyrethrum* (L.) Lag.	*Anthemis pyrethrum* L.	*Aʿqraqarhā*	Pellitory	*Asteraceae*
60	*Pistacia lentiscus* L.	*Lentiscus vulgaris* Fourr.	*Mistakiu*	Mastic	*Anacardiaceae*
61	*Lilium candidum* L.	*Lilium album* Houtt.	*Susan āzād*	Madonna lily	*Liliaceae*
62	*Commiphora myrrha* (Nees) Engl.	*Balsamodendrum myrrha* T. Nees	*Mor*	Myrrh	*Burseraceae*
	Commiphora habessinica (O. Berg) Engl.	*Commiphora abyssinica* (Engl.) Engl.			
	Commiphora schimperi (O. Bergman) Engl.	*Balsamea schimperi* Engl.			
63	*Convolvulus scammonia* L.	–	*Saqmuniā*	Scammony	*Convolvulaceae*
64	*Olea europaea* subsp. *cuspidata* (Wall. & G. Don) Cif.	*Olea africana* Mill.	*Zaytun*	*Olive*	*Oleaceae*
	Olea europaea L.	*Olea pallida* Salisb.			
65	*Prunus amygdalus* Stokes	–	*Lawz*	Almond	*Rosaceae*

Table 6.2. Recent *in vivo* studies on medicinal plants mentioned in *Canon of Medicine* for treatment of CVD.

Plant	Part/extract	Active constituent	Method	Animal	Effect	Reference
Anchusa azurea Mill.	Whole herb/60% aqueous ethanol	Total flavonoids	Chronic MI injury	Mouse	Protective effect against chronic MI injury	Wang *et al*. (2020)
	Whole plant/ aqueous extract	–	Arteriovenous shunt thrombosis model	Rat	Anticoagulant	Umar *et al*. (2015)
Zingiber zerumbet (L.) Sm.	Rhizome/aqueous extract	–	Normoglycemic and STZ-induced hyperglycemic	Rat	Antihyperglycemic	Husen *et al*. (2004)
Crocus sativus L.	Stigmas/aqueous extract	–	ISO-induced myocardial damage	Rat	Antioxidant; cardioprotective	Sachdeva *et al*. (2012)
	Stigmas/aqueous extract	–	ISO-induced MI	Rat	Antioxidant; cardioprotective	Mehdizadeh *et al*. (2013)
	–	Crocetin	Norepinephrine-induced cardiac hypertrophy	Rat	Antioxidant; cardioprotective	Alijaniha *et al*. (2015)
	Stigmas/aqueous extract	Safranal and crocin	Normotensive and desoxycorticosterone acetate-induced hypertensive	Rat	Hypotensive	Imenshahidi *et al*. (2010)
	–	Crocin	Diet-induced hyperlipidemia	Rat	Hypolipidemic	Sheng *et al*. (2006)
	–	Crocetin	HLD-induced hypercholesterolemia and aorta atherosclerosis	Rabbit	Anti-atherosclerosis	Zheng *et al*. (2005)
	Stigmas/aqueous extracts	–	Heart reperfusion-induced lethal ventricular arrhythmia	Rat	Antiarrhythmic	Joukar *et al*. (2013)
	–	Crocin	Normal and voluntary exercised	Rat	Cardioprotective effect; ↑ heart angiogenesis; ↑ Akt and ERK1/2, miR-126, miR-210 expression and CD31	Ghorbanzadeh *et al*. (2017)
	–	Crocin	$CaCl_2$-induced arrhythmia	Rat	Antioxidant; antiarrhythmic	Radmanesh *et al*. (2017)
	–	Crocin	STZ-induced diabetic cardiomyopathy	Rat	Cardiotonic	Feidantsis *et al*. (2018)
	–	Crocin	MI-induced injury	Rat	Cardiotonic; antioxidant	Wang *et al*. (2018)

	Stigmas/ hydroalcoholic extract	–	AngII-induced hypertension	Rat	Hypertension preventive induced by RAS (↓ MAP, SBP, and HR)	Plangar *et al*. (2019)
	–	Crocin	Chlorpyrifos-induced cardiotoxicity	Rat	Cardioprotective	Khalaf and El-Mansy (2019)
	Stamen/ hydroalcoholic extract	–	AngII, L-NAME, and BRS-induced hypertension	Rat	Hypotensive effects	Mohebbati *et al*. (2021)
	–	Crocetin	Cardiomyocytes induced by L-type Ca^{2+} current, contractility, and Ca^{2+} transients	Rat	Calcium channel antagonist	Zhao *et al*. (2020)
	–	Crocin	Silk ligatures-induced periodontitis	Rat	Antioxidant	Kocaman *et al*. (2021)
Curcuma zedoaria (Christm.) Roscoe	Rhizome/aqueous and methanol extract	–	Spontaneously hypertensive	Rat	Antioxidant; vasodilator	Goto *et al*. (2005)
Melissa officinalis L.	Leaves/aqueous extract	–	Lipogenic diet-induced hyperlipidemia	Rat	Antihyperlipidemic	Bolkent *et al*. (2005)
	Aerial parts/ aqueous extract	–	Ligation-induced heart ischemia/ reperfusion	Rat	Anti-ventricular arrhythmia	Joukar *et al*. (2014)
	Leaves/ethanolic extract	–	HFD-induced insulin resistance and dyslipidemia	Mouse	Antihyperglycemic; antihyperlipidemic	Weidner *et al*. (2014)
	Leaves/ethanolic extract	–	Ischemia-induced arrhythmia and heart injury	Rat	Antioxidant; cardioprotective	Sedighi *et al*. (2019)
	Leaves/methanolic extract	–	Methimazole-induced hypothyroidism	Rat	↑ Post-ischemic dysfunction by ↑ mPTP and Ca^{2+}	Ragone *et al*. (2019)
Ocimum basilicum L.	Aerial parts/ aqueous extract	–	Triton WR-1339-induced hyperlipidemia	Rat	Hypolipidemic	Amrani *et al*. (2006)
	Leaves and stalks	–	Turpentine-induced inflammation	Rat	Anti-inflammatory	Originale *et al*. (2007)
	Leaves/ethanolic extract	–	ISO-induced MI	Rat	Cardioprotective	Fathiazad *et al*. (2012)
	Leaves/aqueous extract	–	Phenylhydrazine-induced hematotoxicity	Rat	Anti-anemia and antioxidant	Zangeneh *et al*. (2019)

Continued

Table 6.2. Continued.

Plant	Part/extract	Active constituent	Method	Animal	Effect	Reference
Citrus medica L.	Seeds/petroleum ether extract	–	STZ-induced diabetic	Rat	Antidiabetic; hypocholesterolemic; hypolipidemic	Sah *et al*. (2011)
	Fruits/96% ethanolic extract	–	ISO-induced cardiomyopathy	Rat	Cardioprotective	Al-Yahya *et al*. (2013)
Pterocarpus santalinus L.f.	Bark/ethanolic extract	–	STZ-induced diabetic	Rat	Antidiabetic	Kondeti *et al*. (2010)
	Wood/methanolic extract	–	Carrageenan-induced paw edema	Rat	Anti-inflammatory; antioxidant; COX inhibition	Kumar (2011)
Bambusa bambos (L.) Voss	Leaves/methanolic extract	–	Poloxamer-407-induced hyperlipidemia	Rat	Antihyperlipidemic	Kaikini *et al*. (2015)
Malus domestica Borkh.	Fruits	–	Cholesterol diet	Rat	Antioxidant, antihyperlipidemic	Aprikian *et al*. (2001)
	Fruits	–	Cholesterol-enriched diet	Rat	Antihyperlipidemic	Serra *et al*. (2012)
	Vinegars	–	HFD	Male Wistar rat	Anti-inflammatory; anti-adiposity	Bounihi *et al*. (2017)
Coriandrum sativum L.	Fruits/70% aqueous methanolic extract	–	Voltage-dependent calcium channels blockage	Rat	Diuretic; hypotensive	Jabeen *et al*. (2009)
	Seeds/aqueous extract	–	Hypercaloric diet-induced obesity–hyperglycemia–hyperlipidemia	Rat	Cardiovascular protective; hypoglycemic	Aissaoui *et al*. (2011)
	Seed oil	–	Carrageenan-induced paw edema	Rat	Anti-inflammatory	Ibrahim *et al*. (2017)
	Seeds/aqueous extract	–	ISO-induced heart failure	Rat	Cardioprotective; antioxidant; improvement of lipid profile and endothelial dysfunction	Siddiqi *et al*. (2018)

	Seeds/aqueous extract	–	ISO-induced heart failure	Rat	Cardioprotective; improvement of lipid profile; ↓ LPO and endothelin receptors expression	Dhyani *et al*. (2018)
Pyrus communis L.	Fruits/ethyl acetate and ethanolic extract	–	Dexamethasone-induced diabetes	Rat	Antidiabetic; antihyperlipidemia	Velmurugan and Bhargava (2013)
Punica granatum L.	Flowers/methanolic extract	–	Hyperglycemia and hyperlipidemia	ZDF rat	Diminishes cardiac fibrosis	Huang *et al*. (2005)
	Seeds juice	–	ISO-induced MI	Rat	Cardioprotective	Mohan *et al*. (2010)
	Whole fruit/aqueous extract	–	Doxorubicin-induced toxicity	Rat	Cardioprotective	Hassanpour Fard *et al*. (2011)
	Fruit/hydroalcoholic extract	–	Spontaneously hypertensive	Female rat	Hypolipidemic	Delgado *et al*. (2017)
	Pomegranate juice	–	Lead-induced oxidative stress	Rat	Antioxidant	Aksu *et al*. (2017)
	Vinegars	–	HFD	Male Wistar rat	Anti-inflammatory; anti-adiposity	Bounihi *et al*. (2017)
	Pomegranate juice	–	Deoxycorticosterone acetate–salt model of hypertension	Rat	Antihypertensive	Chaudhari Pankaj *et al*. (2018)
	Leaves/ethanolic extract	–	HFD-induced obesity	Rat	Antioxidant	Rao and Krishnamurthy (2019)
	Peel/methanolic extract–	–	STZ-induced diabetic	Rat	Cardiotonic; antioxidative; anti-inflammatory; antiapoptotic	El-Mansi and Al-Kahtani (2019)
	Seeds powder	–	–	Rat	Antioxidant	Singh and Gujar (2019)
	Seed/oil	Sterol compounds	Diet-induced hypercholesterolemia	Rabbit	Improved lipid profile	Bayati and Asadi-Gharneh (2020)
	Peels/ethyl acetate extract	–	Fludrocortisones-induced hypertension	Rat	Antihypertensive	Monir *et al*. (2020)

Continued

Table 6.2. Continued.

Plant	Part/extract	Active constituent	Method	Animal	Effect	Reference
Terminalia chebula Willd. ex Flem.	Fruits/ethanol extract	–	ISO-induced myocardial injury	Rat	Cardioprotective	Suchalatha and Devi (2004)
	Fruits/ethanol extract	–	ISO-induced myocardial oxidative stress	Rat	Antioxidant	Suchalatha and Devi (2005)
	Seeds/chloroform extract	–	STZ-induced diabetic	Rat	Antidiabetic	Rao and Nammi (2006)
	Fruits/aqueous extract	–	STZ-induced diabetic	Rat	Antidiabetic	Murali *et al.* (2007)
	Mature fruits	–	Atherogenic diet-induced hyperlipidemia	Rat	Hypolipidemic	Maruthappan and Shree (2010)
Paeonia officinalis L.	Root/alcoholic extract	–	–	Rat	Antihypertensive; digitalis-like action	Wenzel and Haskell (1952)
Myrtus communis L.	Fruits/ hydroalcoholic extract	–	Normoglycemic and STZ-induced diabetic	Rat	Hypoglycemic; hypolipidemic; antioxidant	Tas *et al.* (2018)
	Leaves/methanolic extract	–	Hypoxia-induced lethality	Mouse	Antihypoxic	Kaveh *et al.* (2019)
Syzygium aromaticum (L.) Merr. & L.M. Perry	–	–	Ozone-induced oxidative stress	Mouse	Antioxidant	Feng *et al.* (2001)
	–	–	–	Rat	Antioxidant	Bafna and Balaraman (2005)
	–	–	ISO-induced MI	Rat	Cardioprotective; antioxidant	Bafna and Balaraman (2005)
	Flower buds/oil	Oleanolic acid	NaCl-induced hypertension	Rat	Antihypertensive; antioxidant	Madlala *et al.* (2015)
	–/oil	–	Carrageenan-induced paw edema	Rat	Anti-inflammatory	Ibrahim *et al.* (2017)
Piper cubeba L.	Fruits/80% methanolic extract	–	–	Rat	Antioxidant; ↓ MDA; ↑ HDL	Choi and Hwang (2005)

	Seeds/96% ethanolic extract	Cubebin	Phenylephrine-induced contraction	Rat	Vasorelaxation	Carvalho *et al*. (2013)
Raphanus sativus L.	Seeds/aqueous extract	–	Saline-induced hypertension	Rat	Antihypertensive	Ghayur and Gilani (2006)
	–	–	–	Rat	Antihypertensive	Hindo *et al*. (2007)
	Leaves/70% ethanolic extract	–	ISO-induced myocardial injury	Rat	Cardiotoxic	Shah *et al*. (2014)
	Root	–	Fattening diet-induced fatiness	Rabbit	Antiglycemic; antioxidant	Corbi *et al*. (2018)
	Fresh juice	–	Sucrose-induced postprandial glycemia	Rat	Antihypertensive	Srujana *et al*. (2019)
Citrullus colocynthis (L.) Schrad.	Pulp/70% ethanolic extract	–	Atherogenic diet + cholesterol feeding-induced hyperlipidemia	Rabbit	Hypolipidemic	Daradka *et al*. (2007)
	Pulp and seeds/70% methanolic extract	–	Cholesterol feeding-induced hyperlipidemia	Rabbit	Antihyperlipidemic	Zamani *et al*. (2007)
	Roots/aqueous extract	–	Alloxan monohydrate-induced diabetic	Rat	Antidiabetic	Agarwal *et al*. (2012)
	Leaves/hydroalcoholic extract	–	STZ-induced diabetic	Rat	Antihyperglycemic; antihyperlipidemic	Ebrahimi *et al*. (2016)
	Seeds	Total alkaloids	HFD-induced metabolic disorder	Rat	Improved metabolite disorder	Birem *et al*. (2017)
	Pulpy flesh with seeds/70% ethanol extract	–	STZ-induced diabetic	Rat	Hypoglycemic; ↓ TG and TC	Ghauri *et al*. (2020)
Cydonia oblonga Mill.	Leaves/ethanolic extracts	–	Normal and STZ-induced diabetic	Rat	Hypoglycemic; antioxidant	Aslan *et al*. (2010)
	Leaves/60% ethanolic extract	–	HFD-induced hyperlipidemia	Rat	Hypolipidemic; antioxidant; ↑ lipoprotein metabolism; ↓ lipogenesis	Abliz *et al*. (2014)
	Leaves/ethanolic extracts	–	Kidney one-clip Goldblatt model	Rat	Antihypertensive	Zhou *et al*. (2014c)

Continued

Table 6.2. Continued.

Plant	Part/extract	Active constituent	Method	Animal	Effect	Reference
	Fruit and leaf/ aqueous or ethanolic extracts	–	Kidney one-clip Goldblatt model	Rat	Antihypertensive	Zhou *et al.* (2014b)
	Leaves/aqueous extracts	–	–	Mouse and rat	Improvement of hemostasis and thrombosis	Zhou *et al.* (2014a)
	–	Total flavonoids	Spontaneously hypertensive model	Rat	Inhibition of myocardial hypertrophy; inhibition of RAAS	Zhou *et al.* (2015)
	Fruit/aqueous extracts	–	Doxorubicin-induced cardiotoxicity	Rat	Cardioprotective; ↓ cardiac mitochondrial dysfunction	Gholami *et al.* (2017)
Origanum majorana L.	Leaves/aqueous extract	–	ISO-induced MI	Rat	Cardioprotective	Ramadan *et al.* (2013)
	–	–	ISO-induced cardiotoxicity	Rat	Antioxidant; anti-inflammatory	Rababa'h and Alzoubi (2021)
Moringa oleifera Lam.	Ethanolic extract	Niazinin A, niazinin B, niazimicin	Atropine-induced hypertension	Rat	Cardiodepressant	Gilani *et al.* (1994)
	Fruits	–	Diet-induced hypercholesterolemia	Rabbit	Hypolipidemic	Mehta *et al.* (2003)
	Seed powder	–	Arsenic-induced oxidative stress	Rat	Antioxidant	Gupta *et al.* (2005)
	Leaves/aqueous extract	–	Diet-induced hypercholesterolemia	Rabbit	Hypolipidemic; anti-atherosclerotic; cardioprotective	Chumark *et al.* (2008)
	Leaves/ hydroalcohol extract	–	ISO-induced MI	Rat	Cardioprotective	Nandave *et al.* (2009)
	Leaves	–	Arsenic-induced oxidative stress	Rat	Antioxidant; ↓ LPO and ↑ SOD	Latha *et al.* (2010)

Leaves/hydroalcoholic extract	–	Diet-induced hyperlipidemia	Rat	Hypolipidemic; anti-inflammatory; antioxidant; anticoagulant; platelet antiaggregatory; ↓ TG, TC, and LDL; ↑ HDL	Rajanandh *et al.* (2012)
Leaves	N,α-L-rhamnopyranosyl vincosamide	ISO-induced cardiotoxicity	Rat	Cardioprotective	Panda *et al.* (2013)
Aerial parts/alcoholic extract	–	High-cholesterol diet-induced obesity	Rat	Anti-obesity	Ahmed *et al.* (2014)
Leaves	–	Arsenic-induced toxicity	Rat	Hypocholesterolemic	Sheikh *et al.* (2014)
Leaves/aqueous extract	–	γ-Radiation-induced oxidative stress	Rat	Antioxidant	Mansour *et al.* (2014)
Leaves/aqueous extract	–	High-salt diet-induced blood parameters change	Rat	Antihypertension	Ofem *et al.* (2015)
Leaves/butanolic extract	–	ISO-induced cardiac necrosis	Rat	Antioxidant; protection of myocardium	Panda (2015)
Seeds	–	Spontaneous hypertensive	Rat	Cardioprotective; ↓ HR	Randriamboavonjy *et al.* (2016)
Leaves/methanolic extract	–	Atherogenic diet-induced dyslipidemia	Rat	Hypolipidemic; ↓ TG, TC, and LDL; ↑ HDL	Saleem *et al.* (2016)
Leaves	–	HFD-induced glucose intolerance	Rat	Anti-inflammatory; antioxidant	Joung *et al.* (2017)
Seed	–	Spontaneously hypertensive	Rat	Antihypertensive; anti-inflammatory; antioxidant	Randriamboavonjy *et al.* (2017)
–	–	ISO-induced MI	Rat	Cardioprotective	Ishwarya *et al.* (2017)
Seed	–	Diet-induced hypercholesterolemia	Rat	Improved heart enzyme	El Rabey *et al.* (2017)
Leaves/aqueous extract	–	α-Cypermethrin-induced oxidative stress	Rat	Antioxidant	Mansour *et al.* (2018)
Leaves	–	STZ-induced diabetic	Rat	Antioxidant; ↓ TG, TC, LDL, and BS; ↑ HDL	Onyenibe *et al.* (2018)
Leaves/methanolic extract	–	Diet- and carbimazole-induced hyperlipidemia	Rat	Antihyperlipidemic; cardioprotective	Bitrus *et al.* (2018)

Continued

Table 6.2. Continued.

Plant	Part/extract	Active constituent	Method	Animal	Effect	Reference
	–	–	Aluminum phosphide-induced acute cardiac toxicity	Rat	Cardioprotective; antioxidant	Gouda *et al.* (2018)
	Leaves/methanolic extract	–	STZ-induced diabetic	Rat	Antioxidant	Aju *et al.* (2019)
	Seed powder	–	–	Rat	Vascular protective	Randriamboavonjy *et al.* (2019)
	Leaves//ethanolic extract	–	Potassium dichromate-induced toxicity and oxidative stress	Rat	Antioxidant	Kazeem *et al.* (2019)
	Leaves/aqueous extract	–	L-NAME-induced hypertension	Rat	Antihypertensive; antioxidant; ↓ vascular dysfunction	Aekthammarat *et al.* (2019)
	Leaves/methanolic extract	–	HFD-induced dyslipidemia and vascular endothelium dysfunction	Rat	Improvement in lipid profile and vascular endothelium protection	Madkhali *et al.* (2019)
	Seed oil	–	Dichlorvos-induced cardiac injury	Rat	Antioxidant	Saka *et al.* (2020)
	Leaves/methanolic extract	–	HFD-induced obesity	Rat	Anti-obesity; antioxidant	Mabrouki *et al.* (2020)
	Seed powder	–	MI-induced myocardial damage	Mouse	Anti-apoptosis; antioxidant; cardioprotective	Li *et al.* (2020)
	Leaves/ethanolic extract	–	Tilmicosin-induced cardiac injury	Rat	Antioxidant; cardioprotective	Khalil *et al.* (2020)
	Seed	Water-soluble lectin	HFD- and STZ-induced diabetic	Mouse	Hypoglycemic	Vera-Nuñez *et al.* (2021)
Hordeum vulgare L.	Barley-based diet	–	–	Chicken	Hypocholesterolemic	Qureshi *et al.* (1980)
	Barley diet	–	–	Rat	Hypotensive; ↓ SBP; ↓ TG, TC, and LDL	Li *et al.* (2004)
	Barley bran	–	Diet-induced hypocholesterolemia	Rat	Hypolipidemic; ↓ lipids, liver enzymes, LDH, and CK-MB	Abulnaja and El Rabey (2015)
Salix sp.	Flowers	Total flavones	MI-induced reperfusion injury	Rat	Cardioprotective	Fang *et al.* (2016)

	Leaves/70% ethanolic extract	–	HFD-induced hypercholesterolemia	Rabbit	Antioxidant	Amel Zabihi *et al.* (2018)
Citrus limon (L.) Burm.f.	Fruit juice	–	HFD-induced dyslipidemia	Rat	ACE inhibition; hypocholesterolemic	Oboh *et al.* (2015)
Rheum palmatum L.	–	Emodin	HFD-induced obesity	Mouse	Anti-obesity via regulating SREBP pathway	Li *et al.* (2016)
	–	Chrysophanol	Doxorubicin-induced cardiotoxicity	Rat	Cardioprotective by suppressing cellular PARylation	Lu *et al.* (2019)
Lens culinaris Medik.	–	Protein extraction process	–	Rat	Hypotriglyceridemic; ↓ VLDL and adipose-tissue LPL	Boualga *et al.* (2009)
Lactuca sativa L.	Red-pigmented leafy lettuce	–	High-fat high-cholesterol diet	Mouse	Antioxidant; cardioprotective; improvement of plasma cholesterol and lipid levels; prevention of LPO	Lee *et al.* (2009)
Cichorium intybus L.	Seed	Caffeoylquinic acid	High-fructose diet-induced metabolic disorder	Rat	Improved glycemia; ↓ atherogenic index; antioxidant	Jurgoński *et al.* (2012)
	Seeds/aqueous methanolic extracts	–	NaCl-induced hypertension	Rat	Hypotensive via inhibition of Ca^{2+} influx and via α-adrenergic receptor	Malik *et al.* (2017)
	Seed	–	STZ-induced diabetic	Rat	Cardioprotective; antioxidant; anti-inflammatory	Sharma *et al.* (2019)
	Hydroalcoholic extract	–	Model of ischemic reperfusion	Rat	Cardioprotective; antioxidant; ↑ chemokine receptor type 4	Sedighi *et al.* (2021b)
	Leaves/ethanolic extract	–	–	Rat	Antihypertension; antioxidant	Sedighi *et al.* (2021a)
Portulaca oleracea L.	Leaves	–	Vitamin A-deficient	Rat	Antioxidant	Arruda *et al.* (2004)
	Seeds	–	High-cholesterol diet-induced hyperchlosterolemia	Rat	Anti-atherogenic; hypolipidemic	Barakat and Mahmoud (2011)

Continued

Table 6.2. Continued.

Plant	Part/extract	Active constituent	Method	Animal	Effect	Reference
	Leaves/ethanolic extract	–	Dexamethasone-induced hyperlipidemia	Rat	Antihyperlipidemic	Pragda *et al*. (2012)
	Stem and leaves/ aqueous extract	–	Radiation-induced damage	Rat	Antioxidant	El-Azime *et al*. (2014)
	Seeds/ hydroalcoholic extract	–	Levothyroxine-induced thyrotoxicosis	Rat	Cardioprotective; ↓ LVSP and HR	Pakdel *et al*. (2018)
	Seeds/ hydroalcoholic extract	–	Levothyroxine-induced hyperthyroidism	Rat	Cardioprotective	Khodadadi *et al*. (2018)
	Seeds/ hydroalcoholic extract	–	Levothyroxine-induced oxidative stress and hemodynamic changes	Rat	Antioxidant	Pakdel *et al*. (2019)
	Leaves/ethanolic extract	–	Triton-X-100-induced hyperlipidemia	Rat	Hypolipidemic	Tarish *et al*. (2019)
	Leaves and stems/80% ethanolic extract	–	–	Rat	Antihyperglycemic via improving insulin resistance through the PI3K/Akt and AMPK pathways	Lee *et al*. (2020)
Operculina turpethum (L.) Silva Manso	Roots/ethanolic extract	–	N-Nitrosodimethylamine-induced toxicity	Mouse	↑ HDL; ↓LDL, TC, and TG	Sharma and Singh (2012)
Commiphora mukul (Hook. ex Stocks) Engl.	Latex	–	ISO-induced MI	Rat	Cardioprotective; prevention of lactate accumulation	Seth *et al*. (1998)
	70% methanolic extracts	–	Carrageenan-induced paw edema	Rat	Anti-inflammatory	Mary *et al*. (2003)
	–	–	Doxorubicin-induced toxicity	Rat	Cardioprotective	Jagetia *et al*. (2005)
	Whole plant/ hydroalcoholic extract	–	ISO-induced MI	Rat	Cardioprotective; antioxidant; maintaining integrity of myocardial cell membrane	Ojha *et al*. (2008)
	–	–	HFD-induced hyperlipidemia	Rat	Hypolipidemic	Gajbhiye *et al*. (2008)

	Hydroalcoholic extract	–	ISO-induced cardiotoxicity	Rat	Cardioprotective; antioxidant	Ojha *et al*. (2011)
	Gum resin/ ethanolic extract	–	STZ-induced diabetic	Rat	Antioxidant	Ramesh *et al*. (2012)
	Gum resin/ ethanolic extract	–	STZ-induced diabetic	Rat	Antidiabetic; ↑ insulin secretion; improvement of glucose and lipid metabolisms	Ramesh *et al*. (2013)
Cicer arietinum L.	–	Protein extraction process	–	Rat	Hypotriglyceridemic; ↓ VLDL and adipose-tissue LPL	Boualga *et al*. (2009)
	Seedcoat fiber	–	N-Nitrosodiethylamine-induced toxicity	Rat	Antioxidant	Mittal *et al*. (2009)
	Seeds and sprouts	–	Starch-induced postprandial glycemic spikes	Rat	Antioxidant; hypoglycemic; ↓ postprandial glycemic excursion and 2 h postprandial glycemic burden	Tiwari *et al*. (2013)
	Leaves/aqueous extract	–	Alloxan-induced diabetic	Rat	Cardioprotective	Das *et al*. (2017)
Nardostachys jatamansi (D. Don) DC.	Whole plant/ ethanolic extract	–	Triton-induced hyperlipidemic	Rat	Hypolipidemic effects; ↑ HDL/TC ratio	Dixit *et al*. (1988)
	Rhizomes	–	Doxorubicin-induced cardiac damage	Rat	Antioxidant	Subashini *et al*. (2006)
	Rhizomes/ ethanolic extract	–	Doxorubicin-induced myocardial injury	Rat	Hypolipidemic; cardioprotective	Subashini *et al*. (2006)
	Rhizomes/ aqueous methanolic extraction	–	Doxorubicin-induced oxidative stress	Rat	Cardioprotective; antioxidant; anti-inflammatory	Singh *et al*. (2020)
Pistacia lentiscus L.	Mastic gum	–	STZ-induced diabetic	Mouse	Hypocholesterolemic; hypoglycemic	Georgiadis *et al*. (2014)
	Chios mastic gum	–	Diet-induced hypercholesterolemia	Rabbit	Anti-atheromatic; hypolipidemic	Andreadou *et al*. (2016)
Commiphora myrrha (Nees) Engl.	Volatile essential oil	–	Ischemia-induced myocardial injury	Rat	Inhibition of cardiomyocytes apoptosis	Han *et al*. (2015)

Continued

Table 6.2. Continued.

Plant	Part/extract	Active constituent	Method	Animal	Effect	Reference
Olea europaea L.	Leaf/aqueous ethanolic extract	–	Spontaneously hypertensive	Rat	Antihypertensive	De A. Ribeiro *et al.* (1986)
	Leaf/ethanolic extract	–	High-carbohydrate-, high-fat-diet induced metabolic disorder	Rat	Antioxidant; anti-inflammatory; cardioprotective	Poudyal *et al.* (2010)
	Aerial parts (vegetative)/ methanol extract	–	Doxorubicin-induced toxicity	Rat	Cardioprotective	Ashour *et al.* (2012)
	Leaf/ethanolic extract	–	Thromboplastin-induced thrombosis	Rabbit	Antithrombotic; prolongation of PT	Dub and Dugani (2013)
	Leaves/70% ethanolic extract	–	Age-induced oxidative stress	Rat	Antioxidant	Çoban *et al.* (2014)
	Fruit pulp/80% ethanol extract	–	HFD-induced hypercholesterolemia	Rat	Hypocholesterolemic	Kim *et al.* (2014)
	Leaf/70% ethanol extract	–	Doxorubicin-induced cardiac toxicity	Rat	Antioxidant; cardioprotective	Kumral *et al.* (2015)
	Leaf/ hydroalcoholic extract	–	L-NAME-induced hypertension	Rat	Antihypertensive	Abdel-Rahman *et al.* (2017)
	–	Oleuropein	Porcine cardiac myosin-induced autoimmune myocarditis	Rat	Cardioprotective; anti-inflammatory	Zhang *et al.* (2017)
	Leaf/80% ethanol extract	–	Spontaneously hypertensive	Rat	Antioxidative; cardioprotective	Ivanov *et al.* (2018)
	Leaf/70% ethanol extract	–	STZ-induced diabetes mellitus	Rat	Antioxidant; antihypercholesterolemia	Gürbüz *et al.* (2020)
	Oil	–	Doxorubicin-induced cardiomyopathy	Rat	Cardioprotective; antihypertensive	AlMalki and Shahid (2020)

ACE, angiotensin-1-converting enzyme; AMPK, AMP-activated protein kinase; AngII, angiotensin II; BRS, baroreflex sensitivity; BS, blood sugar; CK-MB; creatine kinase MB; COX, cyclooxygenase; dichlorvos, 2,3-dichlorovinyl dimethyl phosphate; HDL, high-density lipoprotein cholesterol; HFD, high-fat diet; HLD, high-lipid diet; HR, heart rate; ISO, isoproterenol; LDH, lactate dehydrogenase; LDL, low-density lipoprotein cholesterol; LPL, lipoprotein lipase; LPO, lipid peroxidation; LVSP, left ventricular systolic pressure; MAP, mean aterial blood pressure; MDA, malondialdehyde; MI, myocardial infarction; miR, microRNA; mPTP, mitochondrial permeability transition pore; L-NAME, NG-nitro-L-arginine methyl ester; PI3K, phosphatidylinositol-3 kinase; PT, prothrombin time; RAS, renin–angiotensin system; SBP, systolic blood pressure; SOD, superoxide dismutase; SREBP, sterol regulatory element binding protein; STZ, streptozotocin; TC, total cholesterol; TG, triglycerides; VLDL, very low-density lipoprotein cholesterol.

Table 6.3. Recent *in vitro* studies on medicinal plants mentioned in *Canon of Medicine* for treatment of CVD.

Plant	Part/extract	Active constituent	Assay	Effect	Reference
Anchusa azurea Mill.	Flowers/aqueous and methanolic extracts	Total phenolic content	TEAC assay	Antioxidant	Alali *et al*. (2007)
	Whole plant/aqueous extract	–	Radioimmunoassay	Anticoagulant	Umar *et al*. (2015)
Zingiber zerumbet (L.) Sm.	Rhizome/methanol extracts	–	Receptor-binding inhibitory assay (bioassay)	Antiplatelet aggregation	Jantan *et al*. (2005)
	Rhizome	–	TAC, TPP, and TAT assays	Antioxidant	Lako *et al*. (2007)
Crocus sativus L.	Stigma/aqueous ethanol extract	–	–	Calcium antagonist	Boskabady *et al*. (2008)
	Stigma/aqueous extract	–	–	Aortic relaxation via nitric oxide synthase pathway	Razavi *et al*. (2018)
	Petals/methanolic extract	Kaempferol and crocin	DCFH-DA and MTT assays	Spasmolytic; antioxidant (↓ ROS)	Zeka *et al*. (2020)
	–	Crocin, crocetin, picrocrocin, and safranal	Perforated patch recording technique	↑ Risk of arrhythmia by inhibiting hERG K^+ currents	Jin *et al*. (2020)
Curcuma zedoaria (Christm.) Roscoe	Rhizome/methanolic extract	–	Nitrite assay	Anti-inflammatory	Jang *et al*. (2004)
	–	Curcumin and curcumol	MTS and caspase-3 apoptosis assays	Anti-inflammatory	Li *et al*. (2019)
Melissa officinalis L.	Aerial parts/methanolic extract	–	TBARS and DPPH assays	Antioxidant	Pereira *et al*. (2009)
Centaurea behen L.	Roots/ethanolic extract	–	DPPH assay	Antioxidant	Chougule *et al*. (2012)
	Aerial parts/methanolic extract	Luteoline	DPPH assay	Antioxidant	Esmaeili *et al*. (2013)
Ocimum basilicum L.	Aerial parts/ethyl acetate, n-butanol, and aqueous extracts	–	DPPH and TBA assays	Antioxidant	Kaurinovic *et al*. (2011)
	Leaves/methanolic extract	Total phenolics and flavonoid	ABTS assay	Antioxidant	Omoba *et al*. (2019)
	Seed/methanol extract	–	DPPH assay	Antioxidant; anti-inflammatory; antidiabetic	Shaikh *et al*. (2020)
Amomum subulatum Roxb.	Fruits/70% aqueous acetone extract	–	DPPH assay	Antioxidant	Kikuzaki *et al*. (2001)

Continued

Table 6.3. Continued.

Plant	Part/extract	Active constituent	Assay	Effect	Reference
	Seeds	Cardamonin	–	Hypertensive effect; inhibition of rho-kinase and MEK activity	Je and Jeong (2016)
Citrus medica L.	Fruits/96% ethanolic extract	–	DPPH assay	Antioxidant	Al-Yahya *et al.* (2013)
Cinnamomum camphora (L.) J. Presl.	Leaves/80% methanol extract	–	MTT assay	Anti-inflammatory; antioxidant; ↓ PGE_2, NO	Lee, J.H. *et al.* (2006)
Santalum album L.	Wood/methanolic extract	–	DPPH assay	Antioxidant	Owen and Johns (2002)
	Aqueous extract	–	Lipid uptake and cholesterol assay	Anti-atherosclerotic; ↓ CD36 and PPAR-γ	Park *et al.* (2018)
Rosa × damascena Mill.	Flower methanolic extract	–	Phosphormolybdenum complex methods	Antioxidant	Özkan *et al.* (2004)
	Flowers/aqueous ethanolic extract	–	Heart rate and contractility determination	Inotropic and chronotropic	Boskabady *et al.* (2011)
	Flowers/aqueous ethanolic extract	–	Heart rate and contractility determination	Inotropic and chronotropic	Boskabady *et al.* (2013)
Malus domestica Borkh.	Peels	Flavonoids	Biochemical assay of ACE inhibition and HUVEC model	ACE inhibitor	Balasuriya and Rupasinghe (2012)
	Flesh and peel/methanolic extract	Arabinofuranoside and 2-oxopomolic acid	sEH assay	Inhibiting sEH	Cho *et al.* (2020)
Coriandrum sativum L.	Fruits/70% aqueous methanolic extract	–	Bioassay	Cholinergic, Ca^{2+} antagonist	Jabeen *et al.* (2009)
	Leaves/70% ethanolic extract	–	DPPH assay	Antioxidant	Harsha and Anilakumar (2014)
	Seeds/methanolic extract	–	DPPH and DNA protection assays	Antioxidants	Afsheen *et al.* (2018)
Pyrus communis L.	Fruits	Polyphenols	TBARS and TQ assays	Antioxidant	Ergamo *et al.* (2002)
Punica granatum L.	Fruit/aqueous, ethanolic, and aqueous ethanolic extracts	–	DPPH and OH assays	Antioxidant	Sharma *et al.* (2018)

	Pomegranate juice	–	Ransod assay	Cardioprotective; ↓ CK-MB, LDH, and troponin-I	Kazemirad and Kazerani (2018)
	Seeds	–	DPPH assay	Antioxidant	Singh and Gujar (2019)
	Seeds/oil/peel/70% ethanolic extract	–	Flow cytometry, MTS and LDH assays	Antioxidant	Dițescu *et al*. (2019)
	Peel/acetone extract	–	NO assay	Antioxidant; α-amylase inhibitory; ↓ GSH depletion	Chukwuma *et al*. (2020)
	Flowers/ethyl acetate extract	–	DPPH and FRAP assays	Antioxidant; anti-inflammatory	Mehrabani *et al*. (2020)
	Pomegranate juice	–	–	Cardioprotective; antiarrhythmic	Rahimi and Kazerani (2021)
Aquilaria agallocha Roxb.	Heartwood/methanolic extract	–	–	Cardiotonic	Mahmood *et al*. (2013)
Terminalia chebula Willd. ex Flem.	Fruits/aqueous, ethanolic, aqueous ethanolic extracts	–	OH and DPPH assays	Antioxidant	Sharma *et al*. (2018)
Lilium candidum L.	Bulb/aqueous extracts	Polysaccharides	MTT assay	Antidiabetic	Li *et al*. (2012)
Syzygium aromaticum (L.) Merr. & L.M. Perry	Leaves/hexane, ethyl acetate, and methanol extracts	–	DPPH assay	Antioxidant	Ramadhania *et al*. (2017)
	Flower buds/methanolic extracts	–	DPPH assay	Cardioprotective; antioxidant	Owen and Johns (2002)
	–	Eugenyl acetate	–	Cardiodepressant	Marques *et al*. (2020)
Piper cubeba L.	Fruits/aqueous extract	–	DPPH assay and RBC method	Antioxidant; anti-inflammatory	Karthiga *et al*. (2016)
Raphanus sativus L.	Seeds/aqueous extract	–	–	Cardiovascular inhibitory; cholinergic effect	Ghayur and Gilani (2006)
Citrullus colocynthis (L.) Schrad.	Pulp/10% ethanol extract	Coloside A	–	Cardiodepressant	The *et al*. (1967)
	Pulpy flesh with seeds/70% ethanolic extract	–	DPPH assay	Antioxidant; inhibition of α-glucosidase	Ghauri *et al*. (2020)
Origanum majorana L.	Aerial sections/methanolic extract	–	Platelet adhesion assay	Anti-thrombolytic; ↓ platelet adhesion, aggregation, and protein secretion induced by thrombin	Yazdanparast and Shahriyary (2008)
	Leaves/ethanolic extract	–	cGMP assay	Vasorelaxation	Badran *et al*. (2019)

Continued

Table 6.3. Continued.

Plant	Part/extract	Active constituent	Assay	Effect	Reference
Moringa oleifera Lam.	Bark/aqueous extract	–	–	Dose-dependent hypotensive	Limaye *et al*. (1995)
	Leaves/aqueous extract	Alkaloids	–	Cardiodepressant	Dangi *et al*. (2002)
	Leaves/aqueous extract	–	DPPH assay	Antioxidant	Chumark *et al*. (2008)
	Pod coat and pod seed/	–	NO assay	Antioxidant	Singh *et al*. (2012)
	Leaves	N,α-L-rhamnopyranosyl vincosamide	–	Antioxidant	Panda *et al*. (2013)
	Flowers/ethanolic extract	–	DPPH assay	Antioxidant	Alhakmani *et al*. (2013)
	Leaves/80% ethanolic extract	Total flavonoids	DCFH-DA assay	Antioxidant; maintaining lipid homeostasis by suppression of gene expression	Sangkitikomol *et al*. (2014)
	Flowers/methanolic extract	–	DPPH assay	Antioxidant	Bhuvaneswari *et al*. (2014)
	Leaves	–	AutoZyme creatinine assay kit	↓ Serum creatinine	Sahoo *et al*. (2014)
	Leaves	N,α-L-rhamnopyranosyl vincosamide	MDA, GSH, and SOD assays	Antioxidant	Cheraghi *et al*. (2017)
	Leaves/aqueous extract	–	DPPH assay	Antioxidant; ACE inhibitory	Oboh *et al*. (2018)
	Leaves/methanol extract	–	DPPH assay	Antioxidant; anti-inflammatory; antidiabetic	Shaikh *et al*. (2020)
	Seeds	Niazirin	FRAP and DPPH assays	Antioxidant; attenuating high glucose-induced oxidative stress by regulating PKCζ/Nox4/ROS signaling pathway	Wang *et al*. (2021)
Hordeum vulgare L.	Bark/extraction with water, methanol, n-hexane, and chloroform	–	GPx and SOD activity	Anti-inflammatory; cardioprotective	Gul *et al*. (2014)
Salix sp.	Leaves/ethanolic extracts	–	DPPH assay	Antioxidant	Lee, M.-H. *et al*. (2006)

	Leaves/70% ethanolic extract	–	DPPH assay	Antioxidant	Amel Zabihi *et al.* (2018)
Citrus limon (L.) Burm.f.	Leaves/methanolic extracts	–	DPPH assay	Antioxidant	Prakash Udaya *et al.* (2013)
	Fruit juices	–	DPPH assay	ACE inhibition; antioxidant	Oboh *et al.* (2015)
	Fruit juices	–	DPPH assay	Antioxidant	Oikeh *et al.* (2016)
	Fruit juices	–	ROS assay	Antioxidant	Dahima *et al.* (2016)
	Peel/methanolic extract	–	ACE inhibition assay	ACE inhibition	Ademosun *et al.* (2019)
	Fruits/olive oil extract	–	DPPH assay	Antioxidant	Flori *et al.* (2020)
Rheum palmatum L.	–	Emodin	–	Anti-inflammatory via deactivation of JNK signaling pathway through downregulation of miR-223	Yang *et al.* (2019)
	–	Aloe emodin	–	Antihypertensive via promoting ubiquitination-mediated NLRP3 inflammasome inactivation	Zhang *et al.* (2020)
Lens culinaris Medik.	Aerial part/methanolic and aqueous extracts	–	DPPH assay	Antioxidant	Spanou *et al.* (2007)
	Seeds/ethanolic extract	–	DPPH and β-carotene bleaching assays	Antioxidant	Talukdar (2012)
	Seed/hydromethanolic extract	–	–	Hypotensive via blocking Ca^{2+} channels	Khan *et al.* (2014)
Cichorium intybus L.	Aerial parts/aqueous extract	–	DPPH assay	Antioxidant	Alvatore *et al.* (2005)
	Leaves/methanolic extract	–	DPPH and SOD assays	Antioxidant	Malik *et al.* (2014)
	Leaves/70% methanolic extract	–	FRAP and DPPH assays	Antioxidant	Epure *et al.* (2020)
Portulaca oleracea L.	Aerial part/fresh juice	–	Isolated frog heart	Cardiotonic	Bidkar *et al.* (2011)
	Aerial parts/organic extract	–	Pancreatic lipase inhibition assay	Anti-lipase	Jaradat *et al.* (2017)
	Germplasm/aqueous extract	–	FRAP assay	Antioxidant	Melilli *et al.* (2020)
Commiphora mukul (Hook. ex Stocks) Engl.	70% methanolic extract	–	Superoxide assay	Anti-atherogenic effect	Mary *et al.* (2003)

Continued

Table 6.3. Continued.

Plant	Part/extract	Active constituent	Assay	Effect	Reference
Cicer arietinum L.	Seeds/methanolic extract	–	COX-2 assay	Antithrombotic; anti-inflammatory	Zia-Ul-Haq *et al.* (2012)
Cucumis sativus L.	Aerial parts/aqueous extract	–	Radioimmunoprecipitation assay	Antioxidant; anti-inflammatory	Trejo-Moreno *et al.* (2018)
	Fruit/aqueous ethanolic extract	–	Capillary-like tube formation assay	Anti-inflammatory	Bernardini *et al.* (2018)
Boswellia sacra Flueck.	Oleogum resin/methanol extract	Olibanum	NO and ABTS assays	Cardioprotective; antioxidant	Zaki *et al.* (2014)
Nardostachys jatamansi (D. Don) DC.	Ethanolic extract	–	Bioassay	Cardiodepressant	Sajid *et al.* (1996)
	Root and rhizome/aqueous extract	–	ROS assay	Antioxidant	Maiwulanjiang *et al.* (2013)
	Root and rhizome/volatile oil	–	MTT assay	Antioxidant	Maiwulanjiang *et al.* (2014)
	Root and rhizome/volatile oil	–	NO assay	Vascular dilation	Maiwulanjiang *et al.* (2015)
Commiphora myrrha (Nees) Engl.	–	–	ABTS and TBARS assays	Antioxidant	Huerta *et al.* (2010)
Olea europaea L.	Leaves/hexane, ethyl acetate, methanol, and 80% aqueous methanolic extracts	Oleanolic acid, ursolic acid, methyl maslinate, and uvaol	–	Vasodepressor; cardiotonic; anti-dysrhythmic	Somova *et al.* (2004)
	Leaves	–	–	Blocking calcium channels	Scheffler *et al.* (2008)
	–-	Hydroxytyrosol	NO assay	Apoptosis of vascular smooth muscle cells via NO production and PP2A activation	Zrelli *et al.* (2011a)
	–	Hydroxytyrosol	ROS assay	Antioxidant via phosphorylation of AMPK	Zrelli *et al.* (2011b)
	Aerial parts (vegetative)/methanolic extract	–	DPPH assay	Antioxidant	Ashour *et al.* (2012)

	–	Oleacein and oleuropein	SA-β-gal assay	Protecting endothelial progenitor cells via induction of Nrf2 and increasing HO-1 expression	Parzonko *et al*. (2013)
	Leaves/seeds oil/70% ethanolic extract	–	DPPH assay	Antioxidant	Abd El-Hady and El-Shahid (2014)
	Powdered leaves	–	Biological assay	Vasorelaxant; cardiodeppresant	Micucci *et al*. (2016)
	Juice	–	LDH assay	Antioxidant; anti-inflammatory	Recinella *et al*. (2019)
	Leaves/70% ethanolic extract	–	MTT and dichlorofluorescein assays	Antioxidant	Torul *et al*. (2020)
	–	Oleuropein	–	Antihypertensive	Ilic *et al*. (2021)
Prunus amygdalus Stokes	Dried pericarp/aqueous extract	–	α-Amylase inhibition assay	Antidiabetic	Jyoti (2020)

ABTS, 2,2′-azino-bis(3-ethylbenzothiazoline-6-sulfonic acid); ACE, angiotensin-1-converting enzyme; AMPK, AMP-activated protein kinase; CK-MB, creatine kinase MB; COX-2, cyclooxygenase-2; DCFH-DA, 2′,7′-dichlorodihydrofluorescein diacetate; DPPH, 1,1-diphenyl-2-picrylhydrazyl; FRAP, ferric-reducing antioxidant power; GPx, glutathione peroxidase; GSH, glutathione; hERG, human ether-a-go-go-related gene; HO-1, heme oxygenase-1; HUVEC, human umbilical vein endothelial cell; JNK, c-Jun N-terminal kinase; LDH, lactate dehydrogenase; MDA, malondialdehyde; MEK, mitogen-activated protein kinase; miR, microRNA; MTS, (3-(4,5-dimethylthiazol-2-yl)-5-(3-carboxymethoxyphenyl)-2-(4-sulfophenyl)-2H-tetrazolium); MTT, 3-(4,5-dimethylthiazol-2-yl)-2,5-diphenyl-tetrazolium bromide; NLRP3, NOD-like receptor family pyrin domain containing-3; NO, nitric oxide; Nox 4, NADPH oxidase 4; Nrf2, NF-E2-related factor 2; OH, hydroxyl; PGE_2, prostaglandin E_2; PKCζ, protein kinase Cζ; PP2A, protein phosphatase 2A; PPAR-γ, peroxisome proliferator-activated receptor γ; RBC, red blood cell; ROS, reactive oxygen species; SA-β-gal, senescence-associated β-galactosidase; sEH, soluble epoxide hydrolase; SOD, superoxide dismutase; TAC, total antioxidant capacity; TAT, total anthocyanin; TBA, thiobarbituric acid; TBARS, thiobarburic acid-reactive substances; TPP, total polyphenol; TEAC, Trolox-equivalent antioxidant capacity; TQ, tocopherolquinone.

Table 6.4. Clinical studies on medicinal plants mentioned in *Canon of Medicine* for treatment of CVD.

Plant	Treatment group	Control group	Study design	No. of patients	Treatment duration	Result	Reference
Cinnamomum camphora (L.) J. Presl.	*C. camphora* forest	–	Randomized clinical trial	31	–	Anti-inflammatory; $\downarrow$ DBP, hs-CRP	Wu *et al*. (2020)
Punica granatum L.	Seed/oil	Placebo	Double-blind, placebo-controlled trial	51	4 weeks	Antihyperlipidemia	Mirmiran *et al*. (2010)
	Pomegranate juice	–	Before-and-after single-arm study	13	SD	Hypotensive	Asgary *et al*. (2013)
	Pomegranate juice	Water	Single-blind clinical trial	21	2 weeks	Hypotensive; $\downarrow$ VCAM-1; $\uparrow$ E-selectin	Asgary *et al*. (2014)
	Peel of pomegranate in combination with other herbs	Placebo	Randomized, double-blind, placebo-controlled study	24	22 days	Antioxidant; $\uparrow$ physical fitness and stamina	Gopi *et al*. (2017)
	Whole pomegranate capsule	Placebo capsules	Double-blind, placebo-controlled, randomized controlled trial	55	8 weeks	Cardiopreventive; hypotensive ($\downarrow$ DBP)	Stockton *et al*. (2017)
	Pomegranate juice	–	Randomized clinical trial	100	5 days	Cardioprotective; $\downarrow$ intensity, occurrence, and duration of angina pectoris; $\downarrow$ troponin and MDA	Razani *et al*. (2017)
Crocus sativus L.	*C. sativus* L.	Placebo	Randomized, double-blind, placebo-controlled trial	44	12 weeks	$\downarrow$ TC, LDL, TG, FBS, and hs-CRP; $\uparrow$ HDL and VEGF; anti-inflammatory	Kermani *et al*. (2017a)
	Crocin tablets	Placebo	Randomized, double-blind, placebo-controlled trial	48	6 weeks	$\downarrow$ Chol; no effect on metabolic syndrome (BP, TG, HDL, LDL, FBS, and anthropometric measures)	Kermani *et al*. (2017b)
	Crocin tablets	Placebo tablet	Randomized, double-blind, placebo-controlled trial	44	8 weeks	No effect on lipid profile	Javandoost *et al*. (2017)

	Saffron capsule	Placebo capsule	Randomized, double-blind, placebo-controlled trial	63	6 weeks	↓ miR-21; no effect on anthropometric indices and biochemical parameters	Khatir *et al*. (2018)
	Saffron tablets	Placebo tablet	Randomized, double-blind, placebo-controlled trial	80	12 weeks	↓ Waist circumference and MDA	Ebrahimi *et al*. (2019)
	Saffron stigma in combination with training	Placebo in combination with training	Randomized, double-blind, placebo-controlled trial	23	5 days/week for 2 weeks	↓ LDH; no effect on CK	Ghanbari-Niaki *et al*. (2019)
	Saffron capsule	Placebo capsule	Randomized, double-blind, placebo-controlled trial	12	SD	Improvement in WAnT	Abassi *et al*. (2020)
	Crocin	Placebo	Randomized, double-blind, placebo-controlled trial	84	8 weeks	Improvement in CAD by ↑ expression of SIRT1 and AMPK and ↓ expression of LOX1 and NF-κB	Abedimanesh *et al*. (2020)
Melissa officinalis L.	Lyophilized aqueous extract of *M. officinalis* leaves	Placebo	Randomized, double-blind, placebo-controlled trial	55	14 days	Improvement of palpitation	Alijaniha *et al*. (2015)
	Hydroalcoholic extract	Placebo	Randomized, double-blind, placebo-controlled trial	70	12 weeks	Improvement of ApoA-I, ApoB/ApoA-I and lipid ratios; cardioprotective	Asadi *et al*. (2018)
	Dried aerial parts capsules	Placebo capsule	Randomized, double-blind, placebo-controlled trial	80	2 months	↑ EF, maximum workload, and NO; ↓ LDH and BP	Javid *et al*. (2018)
	Hydroalcoholic extract	Placebo	Randomized, double-blind, placebo-controlled trial	62	12 weeks	Improvement in lipid profile and glycemic control; anti-inflammatory	Asadi *et al*. (2019)
Terminalia chebula Willd. ex Flem.	Aqueous extract of *T. chebula*	Placebo	Randomized double-blind, placebo-controlled	60	12 weeks	Improved endothelial function; ↑ NO; anti-inflammatory; ↓ MDA; ↓ risk of CVD and hyperglycemia; modulation of blood lipid and HbA1c levels	Pingali *et al*. (2020)

Continued

Table 6.4. Continued.

Plant	Treatment group	Control group	Study design	No. of patients	Treatment duration	Result	Reference
	Aqueous extract of *T. chebula* in combination with Guggulu	Placebo capsules	Placebo-controlled, double-blind, randomized trial	90	3 months	No effect on LDL level, BMI, and waist	Donato *et al.* (2021)
Syzygium aromaticum (L.) Merr. & L.M. Perry	Polyphenolic clove extract	–	Open-label pilot study	13	30 days	Maintained healthy glucose metabolism; inhibition α-amylase and α-glucosidase activities	Mohan *et al.* (2019)
Moringa oleifera Lam.	Malunggay (*M. oleifera*) leaves capsule supplements	–	Prospective cohort study	56	12 weeks	Anti-inflammatory; hypoglycemic	Mozo and Caole-Ang (2015)
	Natural sports supplement	Placebo	Randomized double-blind, placebo-controlled	24	24 h	↑ NO	Jacob *et al.* (2018)
Hordeum vulgare L.	β-Glucan-enriched barley	–	Randomized controlled crossover study	18	4 weeks	No clinically significant improvement in CVD risk factors	Keogh *et al.* (2003)
	Diets with brown rice/ whole wheat, barley	–	Controlled clinical trial	21	5 weeks	↓ Cardiovascular risk	Hallfrisch *et al.* (2003)
	Consumption of barley	–	Controlled clinical trial	25	5 weeks	↓ LDL	Behall *et al.* (2004a)
	Diets containing barley	–	Controlled clinical trial	18	5 weeks	↓ Cardiovascular risk	Behall *et al.* (2004b)
	Higher consumption of barley	–	Case–control study	95	10 years	Preventionn of metabolic syndrome, diabetes, and CVD	Valachovičová *et al.* (2006)
Salix sp.	Salicis cortex extract	Placebo	Randomized double-blind, placebo-controlled	51	4 weeks	Platelet deaggregation	Krivoy *et al.* (2001)

Citrus limon (L.) Burm.f.	Lemon detox diet	Placebo	Before-and-after clinical intervention trial	84	7 days	↓ Body fat and insulin resistance through caloric restriction; ↓ hs-CRP without hematologic changes	Kim *et al*. (2015)
	CitraVes™	–	Pilot open-label study	20	12 weeks	↓ Waist circumference and LDL	Raimondo *et al*. (2021)
Cichorium intybus L.	Chicory coffee	–	Clinical study	27	1 week	Antithrombotic; anti-inflammatory	Schumacher *et al*. (2011)
Portulaca oleracea L.	*P. oleracea* seeds capsule	Placebo capsules	Triple-blinded, randomized, placebo-controlled clinical trial	94	1 month	↓ LDL, TG	Sabzghabaee *et al*. (2014)
	Purslane extract capsule	Placebo capsules	Randomized, placebo-controlled clinical trial	63	12 weeks	↓ SBP, HbA1c	Wainstein *et al*. (2016)
Atropa belladonna L.	*A. belladonna* tincture	Placebo	Single-blind placebo-controlled study	8	SD	Parasympathetic	Bettermann *et al*. (2001)
Commiphora mukul (Hook. ex Stocks) Engl.	Organic extract of Guggulu	Placebo	Clinical trial	406	6 months	Cardioprotective; regulation of abnormal lipid metabolism; modifying BP	Agrawal *et al*. (2000)
	Policol®	–	Observational prospective study	–	3 months	↓ TC, LDL; ↑ HDL	Llabrés *et al*. (2017)
Cicer arietinum L.	Diet including canned drained chickpeas	Wheat diet	Randomized, crossover interventions	27	5 weeks	↓ TC, LDL	Pittaway *et al*. (2007)
Cucumis sativus L.	Diet containing cucumber	–	Clinical study	12	2 weeks	Anti-inflammatory; antioxidant; ↓ F_2-isoprostanes, PGE_2, and MCP-1	Sánchez-Moreno *et al*. (2006)
Nardostachys jatamansi (D. Don) DC.	–	–	Open-label, prospective study	–	–	Anti-inflammatory; antioxidant	Rajyalakshmi *et al*. (2018)
	N. jatamansi capsule	Placebo	Single-blind, randomized, placebo-controlled study	40	4 weeks	Hypotensive; ↓ SBP and DBP	Bhat and Malik (2020)

Continued

Table 6.4. Continued.

Plant	Treatment group	Control group	Study design	No. of patients	Treatment duration	Result	Reference
Pistacia l entiscus L.	Chios mastic powder	Placebo	Clinical study	133	12–18 months	Cardioprotective; ↓ TC, LDL, TC/HDL	Triantafyllou *et al*. (2007)
	Mastic gum	Placebo	Prospective, randomized, placebo-controlled, pilot study	156	8 weeks	Hypocholesterolemic; hypoglycemic	Kartalis *et al*. (2016)
Olea europaea L.	Olive leaf extract	–	Randomized single-blinded study	11	7 days	Inhibition of platelet activation	Singh *et al*. (2008)
	Olive leaf extract tablet	No medication	Open, controlled, parallel-group, co-twin study	20	8 weeks	Antihypertensive; hypocholesterolemic	Perrinjaquet-Moccetti *et al*. (2008)
	Olive leaf extract capsules	Placebo	Randomized, placebo-controlled, crossover trial	46	12 weeks	Improvement of insulin sensitivity and pancreatic β-cell secretory capacity	de Bock *et al*. (2013)
	Olive leaf extract capsule	Matched control	Randomized, double-blind, placebo-controlled, crossover trial	18	SD	Cardioprotective; modulates vascular function and IL-8 production	Lockyer *et al*. (2015)
	Tablet olive oil by-product called pâté	Placebo tablet	Crossover trial with two intervention periods	19	2 months	Improvement of lipid profile, oxidative stress, and MCP-1 levels	Dinu *et al*. (2021)
	95% ethanolic extract from leaves	Captopril	Phase-II, randomized, double-blind, captopril-controlled clinical trial	134	8 weeks	Antihypertensive	Elkafrawy *et al*. (2020)

AMPK, AMP-activated protein kinase; ApoA-I, apolipoprotein A-I, ApoB; apolipoprotein B; BMI, body mass index; BP, blood pressure; CAD, coronary artery disease; Chol, cholesterol; CK, creatine kinase; DBP, diastolic blood pressure; EF, ejection fraction; FBS, fasting blood sugar; HbA1c, glycated hemoglobin; HDL, high-density lipoprotein cholesterol; hs-CRP, high-sensitivity C-reactive protein; IL-8, interleukin-8; LDH, lactate dehydrogenase; LDL, low-density lipoprotein cholesterol; LOX1, lectin-like oxidized LDL receptor 1; MCP-1, monocyte chemoattractant protein-1; MDA, malondialdehyde; miR, microRNA; NF-κB, nuclear factor-κB; NO, nitric oxide; SD, single dose; SIRT1, Sirtuin 1; TC, total cholesterol; TG, triglycerides; VCAM-1, vascular endothelial adhesion molecule 1; VEGF, vascular endothelial growth factor; WAnT, Wingate anaerobic test.

Crocus sativus L.

The aqueous extract of the stigmas of *Crocus sativus* L. (i.e., saffron) has shown antioxidant and cardioprotective effects in various animal studies, such as hypotensive, antiarrhythmic, hypolipidemic, and anti-atherosclerotic. Crocin, one of its valuable constituents, has demonstrated cardioprotective effect via increasing heart angiogenesis through a mechanism involved in elevating protein kinase B (Akt) and extracellular signal-regulated protein kinase 1/2 (ERK1/2), microRNA (miR)-126, miR-210 expression and CD31. In addition, it has shown antagonistic effects on calcium (Ca) channels and aortic relaxation via the nitric oxide synthase pathway (Ghorbanzadeh *et al.*, 2017).

Melissa officinalis L.

Various extraction methods of leaves and aerial parts of *Melissa officinalis* L. (lemon balm) have demonstrated cardioprotective effects such as anti-ventricular arrhythmia, antihyperglycemic, antihyperlipidemic, and antioxidant. It also has improved post-ischemic dysfunction by elevating mitochondrial permeability transition pore (mPTP) and Ca^{2+}. Patients receiving lemon balm in clinical trials have significantly improved palpitation and ejection fraction as well as lipid and glycemic profiles (Ragone *et al.*, 2019).

Punica granatum L.

Parts of *Punica granatum* L. (pomegranate) plants have been studied in CVD, showing improvement in cardiac fibrosis and cardiotonic as well as antiarrhythmic effects (Huang *et al.*, 2005; Rao and Krishnamurthy, 2019). Fifty patients with unstable angina were shown to have significant reduction in the intensity, occurrence, and duration of angina pectoris by receiving pomegranate juice (Razani *et al.*, 2017).

Cydonia oblonga Mill.

Cydonia oblonga Mill. (quince) improved hemostasis and thrombosis in addition to exhibiting other cardioprotective effects such as hypotensive and improvement of metabolic parameters (Zhou *et al.*, 2014a). Furthermore, inhibition of myocardial hypertrophy and the renin–angiotensin–aldosterone system (RAAS) has been shown in animal studies (Zhou *et al.*, 2015).

Moringa oleifera Lam.

The cardiprotective effects of *Moringa oleifera* Lam. have been demonstrated in many studies, and hypolipidemic, anti-inflammatory, antioxidant, anticoagulant, and platelet antiaggregatory effects have been obtained (Rajanandh *et al.*, 2012).

References

Abassi, W., Ouerghi, N., Jebabli, N., Feki, M. and Bouassida, A. (2020) Effect of a single dose of saffron (*Crocus sativus*) on anaerobic power in young active males. *Science & Sports* 35(3), 176–179. DOI: 10.1016/j.scispo.2019.06.008.

Abd El-Hady, F.K. and El-Shahid, Z.A. (2014) Influence of some culinary plants on the suppression of human low density lipoprotein (LDL) peroxidation (*in-vitro*). *International Journal of Pharmaceutical Sciences Review and Research* 26(1), 209–215. Available at: https://www.scopus.com/inward/record.uri?eid=2-s2.0-84900471183&partnerID=40&md5=cc34fb1696cf55452514694b15cb67ca (accessed 20 January 2023).

Abdel-Rahman, R.F., Hessin, A.F., Abdelbaset, M., Ogaly, H.A., Abd-Elsalam, R.M. *et al.* (2017) Antihypertensive effects of roselle–olive combination in L-NAME-induced hypertensive rats. *Oxidative Medicine and Cellular Longevity* 2017, 9460653. DOI: 10.1155/2017/9460653.

Abedimanesh, N., Motlagh, B., Abedimanesh, S., Bathaie, S.Z., Separham, A. *et al.* (2020) Effects of crocin and saffron aqueous extract on gene expression of *SIRT1*, *AMPK*, *LOX1*, *NF-κB*, and *MCP-1* in patients with coronary artery disease: a randomized placebo-controlled clinical trial. *Phytotherapy Research* 34(5), 1114–1122. DOI: 10.1002/ptr.6580.

Abliz, A., Aji, Q., Abdusalam, E., Sun, X., Abdurahman, A. *et al.* (2014) Effect of *Cydonia oblonga* Mill. leaf extract on serum lipids and liver function in a rat model of hyperlipidaemia. *Journal of Ethnopharmacology* 151(2), 970–974. DOI: 10.1016/j.jep.2013.12.010.

Abulnaja, K.O. and El Rabey, H.A. (2015) The efficiency of barley (*Hordeum vulgare*) bran in ameliorating blood and treating fatty heart and liver of male rats. *Evidence-Based Complementary and Alternative Medicine* 2015, 740716. DOI: 10.1155/2015/740716.

Ademosun, A.O., Adebayo, A.A. and Oboh, G. (2019) Modulatory effect of some *Citrus* (*Citrus limon*, *Citrus reticulata*, *Citrus maxima*) peels on monoamine oxidase, phosphodiesterase-5 and angiotensin-1 converting enzyme activities in rat heart homogenate. *Journal of Complementary & Integrative Medicine* 18, 20180067. DOI: 10.1515/jcim-2018-0067.

Aekthammarat, D., Pannangpetch, P. and Tangsucharit, P. (2019) *Moringa oleifera* leaf extract lowers high blood pressure by alleviating vascular dysfunction and decreasing oxidative stress in L-NAME hypertensive rats. *Phytomedicine* 54, 9–16. DOI: 10.1016/j.phymed.2018.10.023.

Afsheen, N., Jahan, N., Ijaz, M., Manzoor, A., Khan, K.M. *et al.* (2018) Cardioprotective and metabolomic profiling of selected medicinal plants against oxidative stress. *Oxidative Medicine and Cellular Longevity* 2018, 9819360. DOI: 10.1155/2018/9819360.

Agrawal, A., Instt, P., Medical, O., Dubey, G.P., Agrawal, A. *et al.* (2000) Individuals at risk of coronary heart disease (CHD), its prevention and management by an indigenous compound. *Ancient Science of Life* 20(1–2), 48–57.

Agarwal, V., Sharma, A.K., Upadhyay, A., Singh, G. and Gupta, R. (2012) Hypoglycemic effects of *Citrullus colocynthis* roots. *Acta Poloniae Pharmaceutica* 69(1), 75–79.

Ahmed, H.H., Metwally, F.M., Rashad, H., Zaazaa, A.M., Ezzat, S.M. *et al.* (2014) *Moringa oleifera* offers a multi-mechanistic approach for management of obesity in rats. *International Journal of Pharmaceutical Sciences Review and Research* 29(2), 98–106. Available at: https://www.scopus.com/inward/record.uri?eid=2-s2.0-84976294036&partnerID=40&md5=33ebc738bdb14157db2af5c163ad2a13 (accessed 20 January 2023).

Aissaoui, A., Zizi, S., Israili, Z.H. and Lyoussi, B. (2011) Hypoglycemic and hypolipidemic effects of *Coriandrum sativum* L. in Meriones shawi rats. *Journal of Ethnopharmacology* 137(1), 652–661.

Aju, B.Y., Rajalakshmi, R. and Mini, S. (2019) Protective role of *Moringa oleifera* leaf extract on cardiac antioxidant status and lipid peroxidation in streptozotocin induced diabetic rats. *Heliyon* 5(12), e02935. DOI: 10.1016/j.heliyon.2019.e02935.

Aksu, D.S., Sağlam, Y.S., Yildirim, S. and Aksu, T. (2017) Effect of pomegranate (*Punica granatum* L.) juice on kidney, liver, heart and testis histopathological changes, and the tissues lipid peroxidation and antioxidant status in lead acetate-treated rats. *Cellular and Molecular Biology (Noisy-Le-Grand, France)* 63(10), 33–42. DOI: 10.14715/cmb/2017.63.10.5.

Alali, F.Q., Tawaha, K., El-Elimat, T., Syouf, M., El-Fayad, M. *et al.* (2007) Antioxidant activity and total phenolic content of aqueous and methanolic extracts of Jordanian plants: an ICBG project. *Natural Product Research* 21(12), 1121–1131.

Alhakmani, F., Kumar, S. and Khan, S.A. (2013) Estimation of total phenolic content, *in-vitro* antioxidant and anti-inflammatory activity of flowers of *Moringa oleifera*. *Asian Pacific Journal of Tropical Biomedicine* 3(8), 623–627. DOI: 10.1016/S2221-1691(13)60126-4.

Alijaniha, F., Naseri, M., Afsharypuor, S., Fallahi, F., Noorbala, A. *et al.* (2015) Heart palpitation relief with *Melissa officinalis* leaf extract: double blind, randomized, placebo controlled trial of efficacy and safety. *Journal of Ethnopharmacology* 164, 378–384. DOI: 10.1016/j.jep.2015.02.007.

AlMalki, W.H. and Shahid, I. (2020) Characterization of antihypertensive and cardioprotective effects of extra virgin olive oil against doxorubicin induced cardiomyopathy in rats. *Journal of Pharmacy and Pharmacognosy Research* 8(4), 316–326. Available at: https://www.scopus.com/inward/record.uri?eid=2-s2.0-85082751767&partnerID=40&md5=e849df22f60d72f222af9d149e7b68f6 (accessed 20 January 2023).

Alvatore, S., Pellegrini, N., Brenna, O.V., Del Rio, D., Frasca, G. *et al.* (2005) Antioxidant characterization of some Sicilian edible wild greens. *Journal of Agricultural and Food Chemistry* 53(24), 9465–9471. DOI: 10.1021/jf051806r.

Al-Yahya, M.A., Mothana, R.A., Al-Said, M.S., El-Tahir, K.E., Al-Sohaibani, M. *et al.* (2013) *Citrus medica* "Otroj": attenuates oxidative stress and cardiac dysrhythmia in isoproterenol-induced cardiomyopathy in rats. *Nutrients* 5(11), 4269–4283.

Amel Zabihi, N., Mahmoudabady, M., Soukhtanloo, M., Hayatdavoudi, P., Beheshti, F. *et al.* (2018) *Salix alba* attenuated oxidative stress in the heart and kidney of hypercholesterolemic rabbits. *Avicenna Journal of Phytomedicine* 8(1), 63–72. DOI: 10.22038/ajp.2017.24804.1901.

Amrani, S., Harnafi, H., Bouanani, N.E.H., Aziz, M., Caid, H.S. *et al.* (2006) Hypolipidaemic activity of aqueous *Ocimum basilicum* extract in acute hyperlipidaemia induced by triton WR-1339 in rats and its antioxidant property. *Phytotherapy Research* 20(12), 1040–1045.

Andreadou, I., Mitakou, S., Paraschos, S., Efentakis, P., Magiatis, P. *et al.* (2016) *Pistacia lentiscus* L. reduces the infarct size in normal fed anesthetized rabbits and possess antiatheromatic and hypolipidemic activity in cholesterol fed rabbits. *Phytomedicine* 23(11), 1220–1226. DOI: 10.1016/j.phymed.2016.06.002.

Aprikian, O., Levrat-Verny, M.-A., Besson, C., Busserolles, J., Rémésy, C. *et al.* (2001) Apple favourably affects parameters of cholesterol metabolism and of anti-oxidative protection in cholesterol-fed rats. *Food Chemistry* 75(4), 445–452. DOI: 10.1016/S0308-8146(01)00235-7.

Arruda, S.F., Siqueira, E.M.A. and Souza, E.M.T. (2004) Malanga (*Xanthosoma sagittifolium*) and purslane (*Portulaca oleracea*) leaves reduce oxidative stress in vitamin A-deficient rats. *Annals of Nutrition & Metabolism* 48(4), 288–295. DOI: 10.1159/000081075.

Asadi, A., Shidfar, F., Safari, M., Malek, M., Hosseini, A.F. *et al.* (2018) Safety and efficacy of *Melissa officinalis* (lemon balm) on ApoA-I, Apo B, lipid ratio and ICAM-1 in type 2 diabetes patients: a randomized, double-blinded clinical trial. *Complementary Therapies in Medicine* 40, 83–88. DOI: 10.1016/j.ctim.2018.07.015.

Asadi, A., Shidfar, F., Safari, M., Hosseini, A.F., Fallah Huseini, H. *et al.* (2019) Efficacy of *Melissa officinalis* L. (lemon balm) extract on glycemic control and cardiovascular risk factors in individuals with type 2 diabetes: a randomized, double-blind, clinical trial. *Phytotherapy Research*, 33(3), 651–659. DOI: 10.1002/ptr.6254.

Asgary, S., Keshvari, M., Sahebkar, A., Hashemi, M. and Rafieian-Kopaei, M. (2013) Clinical investigation of the acute effects of pomegranate juice on blood pressure and endothelial function in hypertensive individuals. *ARYA Atherosclerosis* 9(6), 326.

Asgary, S., Sahebkar, A., Afshani, M.R., Keshvari, M., Haghjooyjavanmard, S. *et al.* (2014) Clinical evaluation of blood pressure lowering, endothelial function improving, hypolipidemic and anti-inflammatory effects of pomegranate juice in hypertensive subjects. *Phytotherapy Research* 28(2), 193–199. DOI: 10.1002/ptr.4977.

Ashour, O.M., Abdel-Naim, A.B., Abdallah, H.M., Nagy, A.A., Mohamadin, A.M. *et al.* (2012) Evaluation of the potential cardioprotective activity of some Saudi plants against doxorubicin toxicity. *Zeitschrift für Naturforschung. C, Journal of Biosciences* 67(5–6), 297–307. DOI: 10.1515/znc-2012-5-609.

Aslan, M., Orhan, N., Orhan, D.D. and Ergun, F. (2010) Hypoglycemic activity and antioxidant potential of some medicinal plants traditionally used in Turkey for diabetes. *Journal of Ethnopharmacology* 128(2), 384–389. DOI: 10.1016/j.jep.2010.01.040.

Badran, A., Baydoun, E., Samaha, A., Pintus, G., Mesmar, J. *et al.* (2019) Marjoram relaxes rat thoracic aorta via a PI3-K/eNOS/cGMP pathway. *Biomolecules* 9(6), 227. DOI: 10.3390/biom9060227.

Bafna, P.A. and Balaraman, R. (2005) Antioxidant activity of DHC-1, an herbal formulation, in experimentally-induced cardiac and renal damage. *Phytotherapy Research* 19(3), 216–221. DOI: 10.1002/ptr.1659.

Balasuriya, N. and Rupasinghe, H.P.V. (2012) Antihypertensive properties of flavonoid-rich apple peel extract. *Food Chemistry* 135(4), 2320–2325.

Barakat, L.A.A. and Mahmoud, R.H. (2011) The antiatherogenic, renal protective and immunomodulatory effects of purslane, pumpkin and flax seeds on hypercholesterolemic rats. *North American Journal of Medical Sciences* 3(9), 351–357. DOI: 10.4297/najms.2011.3351.

Bayati, R. and Asadi-Gharneh, H.A. (2020) Study of steroidal compounds from peel and seed of some pomegranate cultivars (*Punica granatum* L.) and investigating the effect of pomegranate seed oil on blood lipid levels in hypercholesterolemic rabbits. *Journal of Medicinal Plants* 19(73), 152–162. DOI: 10.29252/jmp.1.73.152.

Behall, K.M., Scholfield, D.J. and Hallfrisch, J. (2004a) Diets containing barley significantly reduce lipids in mildly hypercholesterolemic men and women. *American Journal of Clinical Nutrition* 80(5), 1185–1193. DOI: 10.1093/ajcn/80.5.1185.

Behall, K.M., Scholfield, D.J. and Hallfrisch, J. (2004b) Lipids significantly reduced by diets containing barley in moderately hypercholesterolemic men. *Journal of the American College of Nutrition* 23(1), 55–62. DOI: 10.1080/07315724.2004.10719343.

Bernardini, C., Zannoni, A., Bertocchi, M., Tubon, I., Fernandez, M. *et al.* (2018) Water/ethanol extract of *Cucumis sativus* L. fruit attenuates lipopolysaccharide-induced inflammatory response in endothelial cells. *BMC Complementary and Alternative Medicine* 18(1), 194. DOI: 10.1186/s12906-018-2254-1.

Bettermann, H., Cysarz, D., Portsteffen, A. and Kümmell, H.C. (2001) Bimodal dose-dependent effect on autonomic, cardiac control after oral administration of *Atropa belladonna*. *Autonomic Neuroscience: Basic & Clinical* 90(1–2), 132–137. DOI: 10.1016/S1566-0702(01)00279-X.

Bhat, M.D.A. and Malik, S.A. (2020) Efficacy of *Nardostachys jatamansi* (D. Don) DC. in essential hypertension: a randomized controlled study. *Complementary Therapies in Medicine* 53, 102532. DOI: 10.1016/j.ctim.2020.102532.

Bhuvaneswari, S., Deepa, S., Sripriya, N., Prameela, L. and Udaya Prakash, N.K. (2014) Antioxidant activity and phytochemistry of various flowers from Tamil Nadu, India. *International Journal of Research in Pharmaceutical Sciences* 5(1), 40–45. Available at: https://www.scopus.com/inward/record.uri?eid=2-s2.0-84899805504&partnerID=40&md5=17dcd0473dcfca9175fa9e198bcb5177 (accessed 20 January 2023).

Bidkar, J.S., Ghanwat, D.D., Bhujbal, M.D. and Dama, G.Y. (2011) Cardiotonic activity of portulaca oleracea on isolated frog heart. *International Journal of Pharmacy and Pharmaceutical Sciences* 3(Suppl. 5), 118–120. Available at: https://www.scopus.com/inward/record.uri?eid=2-s2.0-84858306632&partnerID=40&md5=9f0b66cc12378594843d53a39f83548a (accessed 20 January 2023).

Birem, Z., Tabani, K., Lahfa, F., Djaziri, R., Hadjbekkouche, F. *et al.* (2017) Effects of colocynth alkaloids and glycosides on Wistar rats fed high-fat diet. A biochemical and morphological study. *Folia Histochemica et Cytobiologica* 55(2), 74–85. DOI: 10.5603/FHC.a2017.0011.

Bitrus, N.E., Kingsley, U.I. and Chukwuma, O.O. (2018) The role of crude methanol leaf extract of moringa oleifera in protection against hyperlipidaemia and cardiomyopathy in albino rat fed a high cholesterol diet and carbimazole. *PharmacologyOnLine* 3, 261–271. Available at: https://www.scopus.com/inward/record.uri?eid=2-s2.0-85059639204&partnerID=40&md5=a272a5a2b580fe4756e46aa917a6aa91 (accessed 20 January 2023).

Bolkent, S., Yanardag, R., Karabulut-Bulan, O. and Yesilyaprak, B. (2005) Protective role of *Melissa officinalis* L. extract on liver of hyperlipidemic rats: a morphological and biochemical study. *Journal of Ethnopharmacology* 99(3), 391–398. DOI: 10.1016/j.jep.2005.02.038.

Boskabady, M.H., Shafei, M.N., Shakiba, A. and Sefidi, H.S. (2008) Effect of aqueous-ethanol extract from *Crocus sativus* (saffron) on guinea-pig isolated heart. *Phytotherapy Research* 22(3), 330–334. DOI: 10.1002/ptr.2317.

Boskabady, M.H., Vatanprast, A., Parsaee, H. and Boskabady, M. (2013) Possible mechanism of inotropic and chronotropic effects of *Rosa damascena* on isolated guinea pig heart. *DARU Journal of Pharmaceutical Sciences* 21(1), 38. DOI: 10.1186/2008-2231-21-38.

Boskabady, M.H., Vatanprast, A., Parsee, H. and Ghasemzadeh, M. (2011) Effect of aqueous-ethanolic extract from *Rosa damascena* on guinea pig isolated heart. *Iranian Journal of Basic Medical Sciences* 14(2), 116–121. DOI: 10.22038/ijbms.2011.4974.

Boualga, A., Prost, J., Taleb-Senouci, D., Krouf, D., Kharoubi, O. *et al.* (2009) Purified chickpea or lentil proteins impair VLDL metabolism and lipoprotein lipase activity in epididymal fat, but not in muscle, compared to casein, in growing rats. *European Journal of Nutrition* 48(3), 162–169. DOI: 10.1007/s00394-009-0777-4.

Bounihi, A., Bitam, A., Bouazza, A., Yargui, L. and Koceir, E.A. (2017) Fruit vinegars attenuate cardiac injury via anti-inflammatory and anti-adiposity actions in high-fat diet-induced obese rats. *Pharmaceutical Biology* 55(1), 43–52. DOI: 10.1080/13880209.2016.1226369.

Carvalho, M.T., Rezende, K.C., Evora, P.R., Bastos, J.K., Cunha, W.R. *et al.* (2013) The lignan (–)-cubebin inhibits vascular contraction and induces relaxation via nitric oxide activation in isolated rat aorta. *Phytotherapy Research* 27(12), 1784–1789. DOI: 10.1002/ptr.4932.

Chaudhari Pankaj, M., Baviskar Dheeraj, T. and Mahajan Sachin, N. (2018) Antihypertensive effect of the punica *Granatum* juice in deoxycorticosterone acetate–salt model of hypertension in rats. *Asian Journal of Pharmaceutical and Clinical Research* 11(6), 498–504. DOI: 10.22159/ajpcr.2018.v11i6.24950.

Cheraghi, M., Namdari, M., Daraee, H. and Negahdari, B. (2017) Cardioprotective effect of magnetic hydrogel nanocomposite loaded *N*,α-L-rhamnopyranosyl vincosamide isolated from *Moringa oleifera* leaves against doxorubicin-induced cardiac toxicity in rats: *in vitro* and *in vivo* studies. *Journal of Microencapsulation* 34(4), 335–341. DOI: 10.1080/02652048.2017.1311955.

Cho, I.S., Kim, J.H., Lin, Y., Su, X.D., Kang, J.S. *et al.* (2020) Inhibitory activity of quercetin 3-*O*-arabinofuranoside and 2-oxopomolic acid derived from *Malus domestica* on soluble epoxide hydrolase. *Molecules (Basel, Switzerland)* 25(18), 4352. DOI: 10.3390/molecules25184352.

Choi, E.-M. and Hwang, J.-K. (2005) Effect of some medicinal plants on plasma antioxidant system and lipid levels in rats. *Phytotherapy Research* 19(5), 382–386. DOI: 10.1002/ptr.1464.

Chougule, P., Pawar, R., Limaye, D., Joshi, Y.M. and Kadam, V. (2012) *In-vitro* antioxidant activity of ethanolic extract of *Centaurea behen*. *Journal of Applied Pharmaceutical Science* 2(4), 106–110.

Chukwuma, C.I., Mashele, S.S. and Akuru, E.A. (2020) Evaluation of the *in vitro* α-amylase inhibitory, antiglycation, and antioxidant properties of *Punica granatum* L. (pomegranate) fruit peel acetone extract and its effect on glucose uptake and oxidative stress in hepatocytes. *Journal of Food Biochemistry* 44(5), e13175. DOI: 10.1111/jfbc.13175.

Chumark, P., Khunawat, P., Sanvarinda, Y., Phornchirasilp, S., Morales, N.P. *et al.* (2008) The *in vitro* and *ex vivo* antioxidant properties, hypolipidaemic and antiatherosclerotic activities of water extract of *Moringa oleifera* Lam. leaves. *Journal of Ethnopharmacology* 116(3), 439–446. DOI: 10.1016/j.jep.2007.12.010.

Çoban, J., Öztezcan, S., Doğru-Abbasoğlu, S., Bingül, I., Yeşil-Mizrak, K. *et al.* (2014) Olive leaf extract decreases age-induced oxidative stress in major organs of aged rats. *Geriatrics & Gerontology International* 14(4), 996–1002. DOI: 10.1111/ggi.12192.

Corbi, G., Conti, V., Komici, K., Manzo, V., Filippelli, A. *et al.* (2018) Phenolic plant extracts induce Sirt1 activity and increase antioxidant levels in the rabbit's heart and liver. *Oxidative Medicine and Cellular Longevity* 2018, 2731289. DOI: 10.1155/2018/2731289.

Dahima, R., Mishra, S. and Rathore, D. (2016) *In-vitro* antioxidant and reducing potential activity of extracts of *Citrus limon* and *Solanum lycopersicum*. *Der Pharmacia Lettre* 8(10), 258–269. Available at: https://www.scopus.com/inward/record.uri?eid=2-s2.0-84978234780&partnerID=40&md5=8634bbfeb0c112e1dbaafe95ebb75214 (accessed 20 January 2023).

Dangi, S.Y., Jolly, C.I. and Narayanan, S. (2002) Antihypertensive activity of the total alkaloids from the leaves of *Moringa oleifera*. *Pharmaceutical Biology* 40(2), 144–148. DOI: 10.1076/phbi.40.2.144.5847.

Daradka, H., Almasad, M.M., Qazan, W.S., El-Banna, N.M. and Samara, O.H. (2007) Hypolipidaemic effects of *Citrullus colocynthis* L. in rabbits. *Pakistan Journal of Biological Sciences* 10(16), 2768–2771. DOI: 10.3923/pjbs.2007.2768.2771.

Das, K.K., Chadchan, K.S., Reddy, R.C., Biradar, M.S., Kanthe, P.S. *et al.* (2017) Effects of some indigenous plants of North Karnataka (India) on cardiovascular and glucose regulatory systems in alloxan-induced diabetic rats. *Cardiovascular & Hematological Agents in Medicinal Chemistry* 15(1), 49–61. DOI: 10.2174/1871525715666170712121347.

De A. Ribeiro, R., De Melo, M.M.R.F., De Barros, F., Gomes, C. and Trolin, G. (1986) Acute antihypertensive effect in conscious rats produced by some medicinal plants used in the state of São Paulo. *Journal of Ethnopharmacology* 15(3), 261–269. DOI: 10.1016/0378-8741(86)90164-9.

de Bock, M., Derraik, J.G.B., Brennan, C.M., Biggs, J.B., Morgan, P.E. *et al.* (2013) Olive (*Olea europaea* L.) leaf polyphenols improve insulin sensitivity in middle-aged overweight men: a randomized, placebo-controlled, crossover trial. *PLoS One* 8(3), e57622. DOI: 10.1371/journal.pone.0057622.

Delgado, N.T., Rouver, W.D., Freitas-Lima, L.C., de Paula, T.D., Duarte, A. *et al.* (2017) Pomegranate extract enhances endothelium-dependent coronary relaxation in isolated perfused hearts from spontaneously hypertensive ovariectomized rats. *Frontiers in Pharmacology* 7, 522. DOI: 10.3389/fphar.2016.00522.

Dhyani, N., Parveen, A., Siddiqi, A. and Hussain, M.E. (2018) Cardioprotective efficacy of *Coriandrum sativum* (L.) seed extract in heart failure rats through modulation of endothelin receptors and antioxidant potential. *Journal of Dietary Supplements* 17(1), 13–26. DOI: 10.1080/19390211.2018.1481483.

Dinu, M., Pagliai, G., Scavone, F., Bellumori, M., Nediani, C. *et al.* (2021) Effects of an olive by-product called pâté on cardiovascular risk factors. *Journal of the American College of Nutrition* 40(7), 617–623. DOI: 10.1080/07315724.2020.1813060.

Dițescu, C., Băbeanu, N., Olariu, L., Dumitriu, B., Papacocea, R. *et al.* (2019) Antioxidant effects on vascular endothelial cells induced by active components of *Punica granatum* and *Amaranthus* sp. extracts. *Farmacia* 67(4), 722–728. DOI: 10.31925/farmacia.2019.4.22.

Dixit, V.P., Jain, P. and Joshi, S.C. (1988) Hypolipidaemic effects of *Curcuma longa* L. and *Nardostachys jatamansi* DC. in triton-induced hyperlipidaemic rats. *Indian Journal of Physiology and Pharmacology* 32(4), 299–304.

Donato, F., Raffetti, E., Toninelli, G., Festa, A., Scarcella, C. *et al.* (2021) Guggulu and Triphala for the treatment of hypercholesterolaemia: a placebo-controlled, double-blind, randomised trial. *Complementary Medicine Research* 28(3), 216–225. DOI: 10.1159/000510985.

Dub, A.M. and Dugani, A.M. (2013) Antithrombotic effect of repeated doses of the ethanolic extract of local olive (*Olea europaea* L.) leaves in rabbits. *The Libyan Journal of Medicine* 8(1), 20947. DOI: 10.3402/ljm.v8i0.20947.

Ebrahimi, E., Mohammadzadeh, G., Mansouri, E. and Aberomand, M. (2016) Effects of hydro-alcoholic leaf extract of *Citrullus colocynthis* on biochemical factors and histopathological changes in streptozotocin-induced diabetic rats. *Jundishapur Journal of Natural Pharmaceutical Products* 11(3), e33214. DOI: 10.17795/jjnpp-33214.

Ebrahimi, F., Sahebkar, A., Aryaeian, N., Pahlavani, N., Fallah, S. *et al.* (2019) Effects of saffron supplementation on inflammation and metabolic responses in type 2 diabetic patients: a randomized, double-blind, placebo-controlled trial. *Diabetes, Metabolic Syndrome and Obesity: Targets and Therapy* 12, 2107–2115. DOI: 10.2147/DMSO.S216666.

El-Azime, A.S.H.A., Hussein, E.M., Ashry, O.M., Abd El-Azime, A.S.H., Hussein, E.M. *et al.* (2014) Synergestic effect of aqueous purslane (*Portulaca oleracea* L.) extract and fish oil on radiation-induced damage in rats. *International Journal of Radiation Biology* 90(12), 1184–1190. DOI: 10.3109/09553002.2014.926040.

Elkafrawy, N., Younes, K., Naguib, A., Badr, H., Kamal Zewain, S. *et al.* (2020) Antihypertensive efficacy and safety of a standardized herbal medicinal product of *Hibiscus sabdariffa* and *Olea europaea* extracts (NW Roselle): a phase-II, randomized, double-blind, captopril-controlled clinical trial. *Phytotherapy Research* 34(12), 3379–3387. DOI: 10.1002/ptr.6792.

El-Mansi, A.A. and Al-Kahtani, M.A. (2019) Calcitriol and *Punica granatum* extract concomitantly attenuate cardiomyopathy of diabetic mother rats and their neonates via activation of Raf/MEK/ERK signalling and mitigation of apoptotic pathways. *Folia Biologica* 65(2), 70–87.

El Rabey, H.A., Khan, J.A., Almutairi, F.M. and Elbakry, M.A. (2017) The low dose of drumsticks (*Moringa oleifera* L.) seed powder ameliorates blood cholesterol in hypercholesterolemic male rat. *Indian Journal of Biochemistry and Biophysics* 54(6), 306–313. Available at: https://www.scopus.com/inward/record.uri?eid=2-s2.0-85040243289&partnerID=40&md5=65feb0e55af72d195d5cb10a1a891007 (accessed 20 January 2023).

Epure, A., Alina, E.P., Vlase, L., Benedec, D., Hanganu, D. *et al.* (2020) Phytochemical profile, antioxidant, cardioprotective and nephroprotective activity of romanian chicory extract. *Plants (Basel, Switzerland)* 10(1), 64. DOI: 10.3390/plants10010064.

Ergamo, P.A.B., Appelloni, M.A.C., Carbonaro, M., Mattera, M., Nicoli, S. *et al.* (2002) Modulation of antioxidant compounds in organic vs conventional fruit (peach, *Prunus persica* L., and pear, *Pyrus communis* L.). *Journal of Agricultural and Food Chemistry* 50(19), 5458–5462. DOI: 10.1021/jf0202584.

Esmaeili, A., Mousavi, Z., Shokrollahi, M. and Shafaghat, A. (2013) Antioxidant activity and isolation of luteoline from *Centaurea behen* L. grown in Iran. *Journal of Chemistry* 2013, 620305. DOI: 10.1155/2013/620305.

Fang, H., Zhang, Y., Li, J., Su, S.-S. and Chen, W. (2016) Protective effect of total flavones from *Salix caprea* flowers against myocardial ischemia-reperfusion injury in rats. *Chinese Journal of New Drugs* 25(8), 949–953, 960. Available at: https://www.scopus.com/inward/record.uri?eid=2-s2.0-84971264088&partnerID=40&md5=6e7247ae6eaa06b42761b2530a6a55c6 (accessed 20 January 2023).

Fathiazad, F., Matlobi, A., Khorrami, A., Hamedeyazdan, S., Soraya, H. *et al.* (2012) Phytochemical screening and evaluation of cardioprotective activity of ethanolic extract of *Ocimum basilicum* L. (basil) against isoproterenol induced myocardial infarction in rats. *DARU Journal of Pharmaceutical Sciences* 20(1), 87. DOI: 10.1186/2008-2231-20-87.

Feidantsis, K., Mellidis, K., Galatou, E., Sinakos, Z. and Lazou, A. (2018) Treatment with crocin improves cardiac dysfunction by normalizing autophagy and inhibiting apoptosis in STZ-induced diabetic cardiomyopathy. *Nutrition, Metabolism, and Cardiovascular Diseases* 28(9), 952–961. DOI: 10.1016/j.numecd.2018.06.005.

Feng, R., He, W. and Ochi, H. (2001) A new murine oxidative stress model associated with senescence. *Mechanisms of Ageing and Development* 122(6), 547–559. DOI: 10.1016/S0047-6374(01)00232-9.

Flori, L., Macaluso, M., Taglieri, I., Sanmartin, C., Sgherri, C. *et al.* (2020) Development of fortified citrus olive oils: from their production to their nutraceutical properties on the cardiovascular system. *Nutrients* 12(6), 1557. DOI: 10.3390/nu12061557.

Gaieski, D.F. and Mikkelsen, M.E. (2021) Definition, classification, etiology, and pathophysiology of shock in adults. *UpToDate®*, Wolters Kluwer. Available at:. https://www.uptodate.com/contents/definition-classification-etiology-and-pathophysiology-of-shock-in-adults?search=Definition,%20classification,%20etiology,%20and%20pathophysiology%20of%20shock%20in%20adults&source=search_result&selectedTitle=1~150&usage_type=default&display_rank=1 (accessed 30 March 2023).

Gajbhiye, S.S., Naik, S., Sharma, G., Kohli, K., Kundnani, P.K.M. *et al.* (2008) Hypolipidemic activity evaluation of DRF/AY/4013, an herbal formulation in experimentally induced hyperlipidemic rats. *Journal of Complementary and Integrative Medicine* 5(1), 7–11. DOI: 10.2202/1553-3840.1148.

Georgiadis, I., Karatzas, T., Korou, L., Agrogiannis, G., Vlachos, I.S. *et al.* (2014) Evaluation of Chios mastic gum on lipid and glucose metabolism in diabetic mice. *Journal of Medicinal Food* 17(3), 393–399. DOI: 10.1089/jmf.2013.0069.

Ghanbari-Niaki, A., Saeidi, A., Kolahdouzi, S., Aliakbari-Baydokhty, M., Ardeshiri, S. *et al.* (2019) Effect of *Crocus sativus* Linnaeus (saffron) supplementations combined with circuit resistance training on total creatine kinase, lactate dehydrogenase and creatine kinase MB levels in young untrained men. *Science & Sports* 34(1), e53–e58. DOI: 10.1016/j.scispo.2018.10.004.

Ghauri, A.O., Ahmad, S. and Rehman, T. (2020) *In vitro* and *in vivo* anti-diabetic activity of *Citrullus colocynthis* pulpy flesh with seeds hydro-ethanolic extract. *Journal of Complementary & Integrative Medicine* 17(2), 1–9. DOI: 10.1515/jcim-2018-0228.

Ghayur, M.N. and Gilani, A.H. (2006) Radish seed extract mediates its cardiovascular inhibitory effects via muscarinic receptor activation. *Fundamental & Clinical Pharmacology* 20(1), 57–63. DOI: 10.1111/j.1472-8206.2005.00382.x.

Gholami, S., Hosseini, M.-J., Jafari, L., Omidvar, F., Kamalinejad, M. *et al.* (2017) Mitochondria as a target for the cardioprotective effects of *Cydonia oblonga* Mill. and *Ficus carica* L. in doxorubicin-induced cardiotoxicity. *Drug Research* 67(6), 358–365. DOI: 10.1055/s-0043-101824.

Ghorbanzadeh, V., Mohammadi, M., Dariushnejad, H., Abhari, A., Chodari, L. *et al.* (2017) Cardioprotective effect of crocin combined with voluntary exercise in rat: role of miR-126 and miR-210 in heart angiogenesis. *Arquivos Brasileiros de Cardiologia* 109(1), 54–62. DOI: 10.5935/abc.20170087.

Gilani, A.H., Aftab, K., Suria, A., Siddiqui, S., Salem, R. *et al.* (1994) Pharmacological studies on hypotensive and spasmolytic activities of pure compounds from *Moringa oleifera*. *Phytotherapy Research* 8(2), 87–91. DOI: 10.1002/ptr.2650080207.

Gopi, S., Jacob, J., Varma, K., Amalraj, A., Sreeraj, T.R. *et al.* (2017) Natural sports supplement formulation for physical endurance: a randomized, double-blind, placebo-controlled study. *Sport Sciences for Health* 13(1), 183–194. DOI: 10.1007/s11332-017-0352-y.

Goto, H., Sasaki, Y., Fushimi, H., Shibahara, N., Shimada, Y. *et al.* (2005) Effect of *Curcuma* herbs on vasomotion and hemorheology in spontaneously hypertensive rat. *The American Journal of Chinese Medicine* 33(3), 449–457. DOI: 10.1142/S0192415X05003053.

Gouda, A.S., El-Nabarawy, N.A. and Ibrahim, S.F. (2018) *Moringa oleifera* extract (Lam) attenuates aluminium phosphide-induced acute cardiac toxicity in rats. *Toxicology Reports* 5, 209–212. DOI: 10.1016/j.toxrep.2018.01.001.

Gul, S., Ahmed, S., Kifli, N., Uddin, Q.T., Batool Tahir, N. *et al.* (2014) Multiple pathways are responsible for anti-inflammatory and cardiovascular activities of *Hordeum vulgare* L. *Journal of Translational Medicine* 12, 316. DOI: 10.1186/s12967-014-0316-9.

Gupta, R., Kannan, G.M., Sharma, M. and S Flora, S.J. (2005) Therapeutic effects of *Moringa oleifera* on arsenic-induced toxicity in rats. *Environmental Toxicology and Pharmacology* 20(3), 456–464. DOI: 10.1016/j.etap.2005.05.005.

Gürbüz, M., Öğüt, S. and Ogut, S. (2020) Antidiabetic effect of olive leaf extract on streptozotocin-induced diabetes mellitus in experimental animals. *Nutricion Hospitalaria* 37(5), 1012–1021. DOI: 10.20960/nh.03051.

Hallfrisch, J., Daniel J. and Scholfield, D.J. (2003) Blood pressure reduced by whole grain diet containing barley or whole wheat and brown rice in moderately hypercholesterolemic men. *Nutrition Research* 23(12), 1631–1642. DOI: 10.1016/j.nutres.2003.08.014.

Han, X.-D., Zhou, Z.-W., Yang, W., Ye, H.-C., Xu, Y.-Z. *et al.* (2015) A computational and functional study elicits the ameliorating effect of the Chinese herbal formula Huo Luo Xiao Ling Dan on experimental ischemia-induced myocardial injury in rats via inhibition of apoptosis. *Drug Design, Development and Therapy* 9, 1063–1102. DOI: 10.2147/DDDT.S76336.

Harsha, S.N. and Anilakumar, K.R. (2014) *In vitro* free radical scavenging and DNA damage protective property of *Coriandrum sativum* L. leaves extract. *Journal of Food Science and Technology* 51(8), 1533–1539. DOI: 10.1007/s13197-012-0648-5.

Hassanpour Fard, M., Ghule, A.E., Bodhankar, S.L. and Dikshit, M. (2011) Cardioprotective effect of whole fruit extract of pomegranate on doxorubicin-induced toxicity in rat. *Pharmaceutical Biology* 49(4), 377–382. DOI: 10.3109/13880209.2010.517758.

Hindo, M.S., Asai, T.K., Be, A.A., Ondo, Y.K., Shindo, M. *et al.* (2007) Effects of dietary administration of plant-derived anthocyanin-rich colors to spontaneously hypertensive rats. *Journal of Nutritional Science and Vitaminology* 53(1), 90–93. DOI: 10.3177/jnsv.53.90.

Huang, T.H.W., Yang, Q., Harada, M., Li, G.Q., Yamahara, J. *et al.* (2005) Pomegranate flower extract diminishes cardiac fibrosis in Zucker diabetic fatty rats: modulation of cardiac endothelin-1 and nuclear factor-kappaB pathways. *Journal of Cardiovascular Pharmacology* 46(6), 856–862. DOI: 10.1097/01.fjc.0000190489.85058.7e.

Huerta, V., Mihalik, K., Beckett, K., Maitin, V. and Vattem, D.A. (2010) Anti-diabetic and anti-energy harvesting properties of common traditional herbs, spices and medicinal plants from India. *Journal of Natural Remedies* 10(2), 123–135. Available at: https://www.scopus.com/inward/record.uri?eid=2-s2.0-78650187593&partnerID=40&md5=e4c3c60210d811d1f2e112e96a51c13d (accessed 20 January 2023).

Husen, R., Pihie, A.H.L. and Nallappan, M. (2004) Screening for antihyperglycaemic activity in several local herbs of Malaysia. *Journal of Ethnopharmacology* 95(2–3), 205–208. DOI: 10.1016/j.jep.2004.07.004.

Ibn Sinā, H.A. (2002) *Al-Qanun fi'l-Tibb (The Canon of Medicine)*. ar al-Fekr, Beirut.

Ibn Sinā, H.A. (2009) *The Book on Drugs for Cardiovascular Diseases*, tr. to Persina by H. Razavi Borghei. Nashre Ney, Tehran.

Ibrahim, F.M., Attia, H.N., Maklad, Y.A.A., Ahmed, K.A. and Ramadan, M.F. (2017) Biochemical characterization, anti-inflammatory properties and ulcerogenic traits of some cold-pressed oils in experimental animals. *Pharmaceutical Biology* 55(1), 740–748. DOI: 10.1080/13880209.2016.1275705.

Ilic, S., Stojiljkovic, N., Stojanovic, N., Stoiljkovic, M., Mitic, K. *et al.* (2021) Effects of oleuropein on rat's atria and thoracic aorta: a study of antihypertensive mechanisms. *Canadian Journal of Physiology and Pharmacology* 99(1), 110–114. DOI: 10.1139/cjpp-2020-0363.

Imenshahidi, M., Hosseinzadeh, H. and Javadpour, Y. (2010) Hypotensive effect of aqueous saffron extract (*Crocus sativus* L.) and its constituents, safranal and crocin, in normotensive and hypertensive rats. *Phytotherapy Research* 24(7), 990–994. DOI: 10.1002/ptr.3044.

Ishwarya, M., Narendhirakannan, R.T., Sharma, S. and Emilin Renitta, R. (2017) Effect of cardioprotective polyherbal formulation on isoproterenol induced myocardial infarction in experimental animals. *Indian Journal of Experimental Biology* 55(6), 357–364. Available at: https://www.scopus.com/inward/record.uri?eid=2-s2.0-85020422379&partnerID=40&md5=5429133f8014e1a05c9021facf134c01 (accessed 20 January 2023).

Ivanov, M., Vajic, U., Mihailovic-Stanojevic, N., Miloradovic, Z., Jovovic, D. *et al.* (2018) Highly potent antioxidant *Olea europaea* L. leaf extract affects carotid and renal haemodynamics in experimental hypertension: the role of oleuropein. *EXCLI Journal* 17, 29–44. DOI: 10.17179/excli2017-1002.

Jabeen, Q., Bashir, S., Lyoussi, B. and Gilani, A.H. (2009) Coriander fruit exhibits gut modulatory, blood pressure lowering and diuretic activities. *Journal of Ethnopharmacology* 122(1), 123–130. DOI: 10.1016/j.jep.2008.12.016.

Jacob, J., Gopi, S. and Divya, C. (2018) A randomized single dose parallel study on enhancement of nitric oxide in serum and saliva with the use of natural sports supplement in healthy adults. *Journal of Dietary Supplements* 15(2), 161–172. DOI: 10.1080/19390211.2017.1331944.

Jagetia, G.C., Reddy, T.K., Malagi, K.J., Nayak, B.S., Naidu, M.B.R. *et al.* (2005) Antarth, a polyherbal preparation protects against the doxorubicin-induced toxicity without compromising its antineoplastic activity. *Phytotherapy Research* 19(9), 772–778. DOI: 10.1002/ptr.1713.

Jang, M.K., Lee, H.J., Kim, J.S. and Ryu, J.-H. (2004) A curcuminoid and two sesquiterpenoids from *Curcuma zedoaria* as inhibitors of nitric oxide synthesis in activated macrophages. *Archives of Pharmacal Research* 27(12), 1220–1225. DOI: 10.1007/BF02975885.

Jantan, I., Rafi, I.A.A. and Jalil, J. (2005) Platelet-activating factor (PAF) receptor-binding antagonist activity of Malaysian medicinal plants. *Phytomedicine* 12(1–2), 88–92. DOI: 10.1016/j.phymed.2003.06.006.

Jaradat, N., Zaid, A.N. and Zaghal, E.Z. (2017) Anti-lipase activity for *Portulaca oleracea*, *Urtica urens*, *Brassica napus* and *Lathyrus hierosolymitanus* wild plants from Palestine. *Marmara Pharmaceutical Journal* 21(4), 828–836. DOI: 10.12991/mpj.2017.9.

Javandoost, A., Afshari, A., Nikbakht-Jam, I., Khademi, M., Eslami, S. *et al.* (2017) Effect of crocin, a carotenoid from saffron, on plasma cholesteryl ester transfer protein and lipid profile in subjects with metabolic syndrome: a double blind randomized clinical trial. *ARYA Atherosclerosis* 13(5), 245–252. Available at: https://www.scopus.com/inward/record.uri?eid=2-s2.0-85038438343&partnerID=40&md5=348c4466406e8650c4e4fa980f230fb0 (accessed 20 January 2023).

Javid, A.Z., Haybar, H., Dehghan, P., Haghighizadeh, M.H., Mohaghegh, S.M. *et al.* (2018) The effects of *Melissa officinalis* on echocardiography, exercise test, serum biomarkers, and blood pressure in patients with chronic stable angina. *Journal of Herbal Medicine* 11, 24–29. DOI: 10.1016/j.hermed.2017.10.002.

Je, H.D. and Jeong, J.H. (2016) Cardamonin inhibits agonist-induced vascular contractility via rho-kinase and MEK inhibition. *The Korean Journal of Physiology & Pharmacology* 20(1), 69–74. DOI: 10.4196/kjpp.2016.20.1.69.

Jin, W., Xue, Y., Xue, Y., Liang, Y., Zhang, Y. *et al.* (2020) Inhibitory effects of four active components in saffron on human ether-a-go-go-related gene (hERG) K^+ currents. *General Physiology and Biophysics* 39(5), 491–498. DOI: 10.4149/gpb_2020025.

Joukar, S., Ghasemipour-Afshar, E., Sheibani, M., Naghsh, N. and Bashiri, A. (2013) Protective effects of saffron (*Crocus sativus*) against lethal ventricular arrhythmias induced by heart reperfusion in rat: a potential anti-arrhythmic agent. *Pharmaceutical Biology* 51(7), 836–843.

Joukar, S., Zarisfi, Z., Sepehri, G., Bashiri, A. and Paper, O. (2014) Efficacy of *Melissa officinalis* in suppressing ventricular arrhythmias following ischemia-reperfusion of the heart: a comparison with amiodarone. *Medical Principles and Practice* 23(4), 340–345. DOI: 10.1159/000363452.

Joung, H., Kim, B., Park, H., Lee, K., Kim, H.-H.H. *et al.* (2017) Fermented *Moringa oleifera* decreases hepatic adiposity and ameliorates glucose intolerance in high-fat diet-induced obese mice. *Journal of Medicinal Food* 20(5), 439–447. DOI: 10.1089/jmf.2016.3860.

Jurgoński, A., Juśkiewicz, J., Zduńczyk, Z. and Król, B. (2012) Caffeoylquinic acid-rich extract from chicory seeds improves glycemia, atherogenic index, and antioxidant status in rats. *Nutrition* (*Burbank, Los Angeles County, Calif.*) 28(3), 300–306. DOI: 10.1016/j.nut.2011.06.010.

Jyoti (2020) Comparative *in vitro* antidiabetic activity of *Prunus amygdalus* (Bail.) and *Hibiscus rosa-sinensis* (Linn.). *Indian Drugs* 57(4), 29–31. Available at: https://www.scopus.com/inward/record.uri?eid=2-s2.0-85089914623&partnerID=40&md5=92b7448c423ab522f985d4fddd2b9a80 (accessed 20 January 2023).

Kaikini, A., Dhande, S. and Kadam, V. (2015) Antihyperlipidemic activity of methanolic extract of leaves of *Bambusa bambos* druce against Poloxamer-407 induced hyperlipidemia in rats. *International Journal of Pharmacy and Pharmaceutical Sciences* 7(2), 393–398. Available at: https://www.scopus.com/inward/record.uri?eid=2-s2.0-84961293666&partnerID=40&md5=ada3b3fdaeea24e96cc23ee1dea412d7 (accessed 20 January 2023).

Kartalis, A., Didagelos, M., Georgiadis, I., Benetos, G., Smyrnioudis, N. *et al.* (2016) Effects of Chios mastic gum on cholesterol and glucose levels of healthy volunteers: A prospective, randomized, placebo-controlled, pilot study (CHIOS-MASTIHA). *European Journal of Preventive Cardiology* 23(7), 722–729. DOI: 10.1177/2047487315603186.

Karthiga, T., Venkatalakshmi, P., Vadivel, V. and Brindha, P. (2016) *In vitro* anti-obesity, antioxidant and anti-inflammatory studies on the selected medicinal plants. *International Journal of Toxicological and Pharmacological Research* 8(5), 332–340. Available at: https://www.scopus.com/inward/record.uri?eid=2-s2.0-84994134116&partnerID=40&md5=afa6c071cb0e9d40d435587ccc6788aa (accessed 20 January 2023).

Kasper, D.L., Fauci, A.S., Hauser, S.L., Longo, D.L., Jameson, J.L. *et al.* (2018) *Harrison's Principles of Internal Medicine*, 20 th edn. McGraw Hill Professional, New York.

Kaurinovic, B., Popovic, M., Vlaisavljevic, S., Trivic, S. and Extracts, O.L. (2011) Antioxidant capacity of *Ocimum basilicum* L. and *Origanum vulgare* L. extracts. *Molecules (Basel, Switzerland)* 16(9), 7401–7414. DOI: 10.3390/molecules16097401.

Kaveh, K., Mohamadyan, M. and Ebrahimzadeh, M.A. (2019) Antihypoxic activities of *Sambucus ebulus* leaf and fruit and *Myrtus communis* leaf in mice. *Journal of Mazandaran University of Medical Sciences* 29(176), 61–73. Available at: https://www.scopus.com/inward/record.uri?eid=2-s2.0-85073288089&partnerID=40&md5=4d1a3a3edbbd2596ca70f72d603d03ff (accessed 20 January 2023).

Kazeem, A.A., Olabode, O.O., Afusat, J.J. and Olaitan, D.O. (2019) Modification of hexavalent chromate hepatotoxicity by ethanol extract of *Moringa oleifera* in Wistar rats. *Pharmacognosy Journal* 11(4), 764–770. DOI: 10.5530/pj.2019.11.121.

Kazemirad, H. and Kazerani, H.R. (2018) Nitric oxide plays a pivotal role in cardioprotection induced by pomegranate juice against myocardial ischemia and reperfusion. *Phytotherapy Research* 32(10), 2069–2077. DOI: 10.1002/ptr.6150.

Keogh, G.F., Cooper, G.J.S., Mulvey, T.B., McArdle, B.H., Coles, G.D. *et al.* (2003) Randomized controlled crossover study of the effect of a highly β-glucan-enriched barley on cardiovascular disease risk factors in mildly hypercholesterolemic men. *American Journal of Clinical Nutrition* 78(4), 711–718. DOI: 10.1093/ajcn/78.4.711.

Kermani, T, Zebarjadi, M., Mehrad-Majd, H., Mirhafez, S.R., Shemshian, M. *et al.* (2017a) Anti-Inflammatory effect of *Crocus sativus* on serum cytokine levels in subjects with metabolic syndrome: a randomized, double-blind, placebo-controlled trial. *Current Clinical Pharmacology* 12(2), 122–126. DOI: 10.2174/1574884712666170622082737.

Kermani, T., Kazemi, T., Molki, S., Ilkhani, K., Sharifzadeh, G. *et al.* (2017b) The efficacy of crocin of saffron (*Crocus sativus* L.) on the components of metabolic syndrome: a randomized controlled clinical trial. *Journal of Research in Pharmacy Practice* 6(4), 228–232. DOI: 10.4103/jrpp.JRPP_17_26.

Khalaf, H.A. and El-Mansy, A.A.E.-R. (2019) The possible alleviating effect of saffron on chlorpyrifos experimentally induced cardiotoxicity: histological, immunohistochemical and biochemical study. *Acta Histochemica* 121(4), 472–483. DOI: 10.1016/j.acthis.2019.03.003.

Khalil, S.R., Abdel-Motal, S.M., Abd-Elsalam, M., Abd El-Hameed, N.E. and Awad, A. (2020) Restoring strategy of ethanolic extract of *Moringa oleifera* leaves against tilmicosin-induced cardiac injury in rats: targeting cell apoptosis-mediated pathways. *Gene* 730, 144272. DOI: 10.1016/j.gene.2019.144272.

Khan, M., Khan, A.-U., Najeeb-ur-Rehman and Gilani, A.-H. (2014) Cardiovascular inhibitory properties of *Lens culinaris*. *Bangladesh Journal of Pharmacology* 9(3), 303–306. DOI: 10.3329/bjp.v9i3.19090.

Khatir, S.A., Bayatian, A., Barzegari, A., Roshanravan, N., Safaiyan, A. *et al.* (2018) Saffron (*Crocus sativus* L.) supplements modulate circulating microRNA (miR-21) in atherosclerosis patients; a randomized, double-blind, placebo-controlled trial. *Iranian Red Crescent Medical Journal* 20(10), e80260. DOI: 10.5812/ircmj.80260.

Khodadadi, H., Pakdel, R., Khazaei, M., Niazmand, S., Bavarsad, K. *et al.* (2018) A comparison of the effects of *Portulaca oleracea* seeds hydro-alcoholic extract and vitamin C on biochemical, hemodynamic and functional parameters in cardiac tissue of rats with subclinical hyperthyroidism. *Avicenna Journal of Phytomedicine* 8(2), 161–169.

Kikuzaki, H., Kawai, Y., Nakatani, N. and Vitaminol, J.N.S. (2001) 1,1-Diphenyl-2-picrylhydrazyl radical-scavenging active compounds from greater cardamom (*Amomum subulatum* Roxb.). *Journal of Nutritional Science and Vitaminology* 47(2), 167–171.

Kim, M.J., Hwang, J.H., Ko, H.J., Na, H.B. and Kim, J.H. (2015) Lemon detox diet reduced body fat, insulin resistance, and serum hs-CRP level without hematological changes in overweight Korean women. *Nutrition Research* 35(5), 409–420. DOI: 10.1016/j.nutres.2015.04.001.

Kim, M.-S., Koppula, S., Sung, S.-J., Lee, S.-R., Park, Y.-D. *et al.* (2014) *Olea europaea* L. (Oleaceae) fruit pulp exhibits hypocholesterolemic and hepatoprotective effects via regulation of peroxisome proliferation-activated receptor alpha in high-fat diet-fed rats. *Tropical Journal of Pharmaceutical Research* 13(1), 31–39. DOI: 10.4314/tjpr.v13i1.5.

Kocaman, G., Altinoz, E., Erdemli, M.E., Gul, M., Erdemli, Z. *et al.* (2021) Crocin attenuates oxidative and inflammatory stress-related periodontitis in cardiac tissues in rats. *Advances in Clinical and Experimental Medicine: Official Organ Wroclaw Medical University* 30(5), 517–524. DOI: 10.17219/acem/133753.

Kondeti, V.K., Badri, K.R., Maddirala, D.R., Thur, S.K.M., Fatima, S.S. *et al.* (2010) Effect of *Pterocarpus santalinus* bark, on blood glucose, serum lipids, plasma insulin and hepatic carbohydrate metabolic enzymes in streptozotocin-induced diabetic rats. *Food and Chemical Toxicology* 48(5), 1281–1287. DOI: 10.1016/j.fct.2010.02.023.

Krivoy, N., Pavlotzky, E., Chrubasik, S., Eisenberg, E. and Brook, G. (2001) Effect of salicis cortex extract on human platelet aggregation. *Planta Medica* 67(3), 209–212. DOI: 10.1055/s-2001-12000.

Kumar, D. (2011) Anti-inflammatory, analgesic, and antioxidant activities of methanolic wood extract of *Pterocarpus santalinus* L. *Journal of Pharmacology & Pharmacotherapeutics* 2(3), 200. DOI: 10.4103/0976-500X.83293.

Kumral, A., Giriş, M., Soluk-tekkeşin, M., Olgaç, V., Doğru-Abbasoğlu, S. *et al.* (2015) Effect of olive leaf extract treatment on doxorubicin-induced cardiac, hepatic and renal toxicity in rats. *Pathophysiology* 22(2), 117–123. DOI: 10.1016/j.pathophys.2015.04.002.

Lako, J., Trenerry, V.C., Wahlqvist, M., Wattanapenpaiboon, N., Sotheeswaran, S. *et al.* (2007) Phytochemical flavonols, carotenoids and the antioxidant properties of a wide selection of Fijian fruit, vegetables and other readily available foods. *Food Chemistry* 101(4), 1727–1741. DOI: 10.1016/j.foodchem.2006.01.031.

Latha, R., Nagashree, R., Shylaja, M. and Karthikeyan, V. (2010) Effect of administration of *Moringa oleifera* leaves in the remediation of arsenic-induced oxidative stress in rats. *Biomedicine* 30(3), 278–281. Available at: https://www.scopus.com/inward/record.uri?eid=2-s2.0-78649475838&partnerID=40&md5=7350ca915bba486630e06be27c73d923 (accessed 20 January 2023).

Lee, J.H., Hyun, E.-A., Yoon, W.J., Kim, B.H., Rhee, M.H. *et al.* (2006a) *In vitro* anti-inflammatory and antioxidative effects of *Cinnamomum camphora* extracts. *Journal of Ethnopharmacology* 103(2), 208–216. DOI: 10.1016/j.jep.2005.08.009.

Lee, J.H., Felipe, P., Yang, Y.H., Kim, M.Y., Kwon, O.Y. *et al.* (2009) Effects of dietary supplementation with red-pigmented leafy lettuce (*Lactuca sativa*) on lipid profiles and antioxidant status in C57BL/6J mice fed a high-fat high-cholesterol diet. *The British Journal of Nutrition* 101(8), 1246–1254. DOI: 10.1017/S0007114508073650.

Lee, J.H., Park, J.E. and Han, J.S. (2020) *Portulaca oleracea* L. extract reduces hyperglycemia via PI3k/Akt and AMPK pathways in the skeletal muscles of C57BL/Ksj-db/db mice. *Journal of Ethnopharmacology* 260, 112973. DOI: 10.1016/j.jep.2020.112973.

Lee, M.-H., Jiang, C.-B., Juan, S.-H., Lin, R.-D. and Hou, W.-C. (2006b) Antioxidant and heme oxygenase-1 (HO-1)-induced effects of selected Taiwanese plants. *Fitoterapia* 77(2), 109–115. DOI: 10.1016/j.fitote.2005.11.012.

Lévy, S. and Olshansky, B. (2021) Arrhythmia management for the primary care clinician. *UpToDate®*, Wolters Kluwer. Available at: https://www.uptodate.com/contents/arrhythmia-management-for-the-primary-care-clinician?search=Arrhythmia management for the primary care clinician&source=search_result&selectedTitle=1~150&usage_type=default&display_rank=1 (accessed 20 January 2023).

Li, J., Wang, J., Kaneko, T., Qin, L.-Q. and Sato, A. (2004) Effects of fiber intake on the blood pressure, lipids, and heart rate in Goto Kakizaki rats. *Nutrition* 20(11–12), 1003–1007. DOI: 10.1016/j.nut.2004.08.010.

Li, J., Ding, L., Song, B., Xiao, X., Qi, M. *et al.* (2016) Emodin improves lipid and glucose metabolism in high fat diet-induced obese mice through regulating SREBP pathway. *European Journal of Pharmacology* 770, 99–109. DOI: 10.1016/j.ejphar.2015.11.045.

Li, N., Liu, T.-H., Yu, J.-Z., Li, C.-X., Liu, Y. *et al.* (2019) Curcumin and curcumol inhibit NF-κB and TGF-β (1)/Smads signaling pathways in CSE-treated RAW246.7 cells. *Evidence-Based Complementary and Alternative Medicine* 2019, 3035125. DOI: 10.1155/2019/3035125.

Li, Y., Chang, J., Liu, C., Xiong, X., Gong, Y. *et al.* (2012) Lily polysaccharide prevents alloxan-induced HIT-T15 cell damage by reducing ROS generation. *2012 International Conference on Biomedical Engineering and Biotechnology*, Macau, Macao, 2012, 364–367. DOI: 10.1109/iCBEB.2012.264.

Li, Y.-J., Ji, Q.-Q., Wang, Z., Shen, L.-H. and He, B. (2020) *Moringa oleifera* seeds mitigate myocardial injury and prevent ventricular failure induced by myocardial infarction. *American Journal of Translational Research* 12(8), 4511–4521.

Limaye, D.A., Nimbkar, A.Y., Jain, R. and Ahmad, M. (1995) Cardiovascular effects of the aqueous extract of *Moringa pterygosperma*. *Phytotherapy Research* 9(1), 37–40. DOI: 10.1002/ptr.2650090109.

Llabrés, J.M., Urcola, I., Boada, J.M., Araujo, J.C. and Durán, J. (2017) Effect of Policol® One in the reduction of midhigh cholesterol levels. *Revista de Fitoterapia* 17(1), 63–70. Available at : https://www.scopus.com/inward/record.uri?eid=2-s2.0-85040081177&partnerID=40&md5=b7762634ce4addab0cf89d7eb0bdbfb3 (accessed 20 January 2023).

Lockyer, S., Corona, G., Yaqoob, P., Spencer, J.P.E. and Rowland, I. (2015) Secoiridoids delivered as olive leaf extract induce acute improvements in human vascular function and reduction of an inflammatory cytokine: a randomised, double-blind, placebo-controlled, cross-over trial. *The British Journal of Nutrition* 114(1), 75–83. DOI: 10.1017/S0007114515001269.

Lu, J., Li, J., Hu, Y., Guo, Z., Sun, D. *et al.* (2019) Chrysophanol protects against doxorubicin-induced cardiotoxicity by suppressing cellular PARylation. *Acta Pharmaceutica Sinica. B* 9(4), 782–793. DOI: 10.1016/j.apsb.2018.10.008.

Mabrouki, L., Rjeibi, I., Taleb, J. and Zourgui, L. (2020) Cardiac ameliorative effect of moringa oleifera leaf extract in high-fat diet-induced obesity in rat model. *BioMed Research International* 2020, 6583603. DOI: 10.1155/2020/6583603.

Madkhali, H.A., Alharthy, K.M., Asiri, M.A., Ganaie, M.A., Ansari, M.N. *et al.* (2019) *Moringa oleifera* Lam. (family Moringaceae) leaf extract attenuates high-fat diet-induced dyslipidemia and vascular endothelium dysfunction in Wistar albino rats. *Tropical Journal of Pharmaceutical Research* 18(12), 2597–2604. DOI: 10.4314/tjpr.v18i12.20.

Madlala, H.P., Van Heerden, F.R., Mubagwa, K. and Musabayane, C.T. (2015) Changes in renal function and oxidative status associated with the hypotensive effects of oleanolic acid and related synthetic derivatives in experimental animals. *PLoS One* 10(6), e0128192. DOI: 10.1371/journal.pone.0128192.

Mahmood, Q., Ahmad, M., Akhtar, M.S. (2013) Inotropic and chronotropic effects of methanol extract of *Aquilaria agallocha* on rabbit heart. *Bangladesh Journal of Pharmacology* 8(3), 311–316. DOI: 10.3329/bjp.v8i3.15080.

Maiwulanjiang, M., Zhu, K.Y., Chen, J., Miernisha, A., Xu, S.L. *et al.* (2013) Song Bu Li decoction, a traditional Uyghur medicine, protects cell death by regulation of oxidative stress and differentiation in cultured PC12 cells. *Evidence-Based Complementary and Alternative Medicine* 2013, 687958. DOI: 10.1155/2013/687958.

Maiwulanjiang, M., Chen, J., Xin, G., Gong, A.G., Miernisha, A. *et al.* (2014) The volatile oil of Nardostachyos Radix et Rhizoma inhibits the oxidative stress-induced cell injury via reactive oxygen species scavenging and Akt activation in H9c2 cardiomyocyte. *Journal of Ethnopharmacology* 153(2), 491–498. DOI: 10.1016/j.jep.2014.03.010.

Maiwulanjiang, M., Bi, C.W., Lee, P.S., Xin, G., Miernisha, A. *et al.* (2015) The volatile oil of Nardostachyos Radix et Rhizoma induces endothelial nitric oxide synthase activity in HUVEC cells. *PLoS One* 10(2), e0116761. DOI: 10.1371/journal.pone.0116761.

Malik, A., Mehmood, M.H., Channa, H., Akhtar, M.S. and Gilani, A. (2017) Pharmacological basis for the medicinal use of polyherbal formulation and its ingredients in cardiovascular disorders using rodents. *BMC Complementary and Alternative Medicine* 17(1), 142. DOI: 10.1186/s12906-017-1644-0.

Malik, B., Pirzadah, T.B., Tahir, I. and Rehman, R.U. (2014) Chemo-profiling, antioxidant potential and ionomic analysis of *Cichorium intybus* L. *Pharmacognosy Journal* 9(6), 917–928. DOI: 10.5530/pj.2017.6.144.

Mansour, H.H., Ismael, N.E.R. and Hafez, H.F. (2014) Modulatory effect of *Moringa oleifera* against gamma-radiation-induced oxidative stress in rats. *Biomedicine & Aging Pathology* 4(3), 265–272. DOI: 10.1016/j.biomag.2014.04.006.

Mansour, S.A., Mohamed, R.I., Ali, A.R. and Farrag, A.H. (2018) The protective effect of moringa tea against cypermethrin-induced hepatorenal dysfunction, oxidative stress, and histopathological alterations in female rats. *Asian Journal of Pharmaceutical and Clinical Research* 11(10), 111–117. DOI: 10.22159/ajpcr.2018.v11i10.24993.

Marques, L.P., Beserra, S.S., Roman-Campos, D. and Gondim, A.N.S. (2020) Cardiodepressive effect of eugenyl acetate in rodent heart. *Arquivos Brasileiros de Cardiologia* 115(5), 967–970. DOI: 10.36660/abc.20190823.

Maruthappan, V. and Shree, K.S. (2010) Hypolipidemic activity of Haritaki (*Terminalia chebula*) in atherogenic diet induced hyperlipidemic rats. *Journal of Advanced Pharmaceutical Technology & Research* 1(2), 229–235.

Mary, N.K., Babu, B.H. and Padikkala, J. (2003) Antiatherogenic effect of Caps HT2, a herbal Ayurvedic medicine formulation. *Phytomedicine* 10(6–7), 474–482. DOI: 10.1078/094471103322331412.

Mehdizadeh, R., Parizadeh, M., Khooei, A.-R., Mehri, S. and Hosseinzadeh, H. (2013) Cardioprotective effect of saffron extract and safranal in isoproterenol-induced myocardial infarction in Wistar rats. *Iranian Journal of Basic Medical Sciences* 16(1), 56–63.

Mehrabani, M., Raeiszadeh, M., Najafipour, H., Tarzi, M.E., Amirkhosravi, A. *et al.* (2020) Evaluation of the cytotoxicity, antibacterial, antioxidant, and anti-inflammatory effects of different extracts of *Punica granatum* var. *pleniflora*. *Journal of Kerman University of Medical Sciences* 27(5), 414–425. DOI: 10.22062/jkmu.2020.91474.

Mehta, K., Balaraman, R., Amin, A.H., Bafna, P.A., Gulati, O.D. *et al.* (2003) Effect of fruits of *Moringa oleifera* on the lipid profile of normal and hypercholesterolaemic rabbits. *Journal of Ethnopharmacology* 86(2–3), 191–195. DOI: 10.1016/S0378-8741(03)00075-8.

Melilli, M.G., Di Stefano, V., Sciacca, F., Pagliaro, A., Bognanni, R. *et al.* (2020) Improvement of fatty acid profile in durum wheat breads supplemented with *Portulaca oleracea* L. quality traits of purslane-fortified bread. *Foods (Basel, Switzerland)* 9(6), 764. DOI: 10.3390/foods9060764.

Mendis, S., Davis, S. and Norrving, B. (2015) Organizational update: the World Health Organization global status report on noncommunicable diseases 2014; one more landmark step in the combat against stroke and vascular disease. *Stroke* 46(5), e121–e122. DOI: 10.1161/STROKEAHA.115.008097.

Micucci, M., Angeletti, A., Cont, M., Corazza, I., Aldini, R. *et al.* (2016) *Hibiscus sabdariffa* L. flowers and *Olea europea* L. leaves extract-based formulation for hypertension care: *in vitro* efficacy and toxicological profile. *Journal of Medicinal Food* 19(5), 504–512. DOI: 10.1089/jmf.2015.0072.

Mirmiran, P., Fazeli, M.R., Asghari, G., Shafiee, A. and Azizi, F. (2010) Effect of pomegranate seed oil on hyperlipidaemic subjects: a double-blind placebo-controlled clinical trial. *The British Journal of Nutrition* 104(3), 402–406. DOI: 10.1017/S0007114510000504.

Mittal, G., Vadhera, S., Brar, A.P.S. and Soni, G. (2009) Protective role of chickpea seed coat fibre on *N*-nitrosodiethylamine-induced toxicity in hypercholesterolemic rats. *Experimental and Toxicologic Pathology* 61(4), 363–370. DOI: 10.1016/j.etp.2008.07.006.

Mohan, M., Patankar, P., Ghadi, P. and Kasture, S. (2010) Cardioprotective potential of *Punica granatum* extract in isoproterenol-induced myocardial infarction in Wistar rats. *Journal of Pharmacology & Pharmacotherapeutics* 1(1), 32–37. DOI: 10.4103/0976-500X.64533.

Mohan, R., Jose, S., Mulakkal, J., Karpinsky-Semper, D., Swick, A.G. and Krishnakumar, I.M. (2019) Water-soluble polyphenol-rich clove extract lowers pre- and post-prandial blood glucose levels in healthy and prediabetic volunteers: an open label pilot study. *BMC Complementary Medicine and Therapies* 19(1), 99. DOI: 10.1186/s12906-019-2507-7.

Mohebbati, R., Kamkar-Del, Y. and Shafei, M.N. (2021) Effect of standardized hydroalcoholic extract of saffron stamen on high blood pressure and baroreflex sensitivity in anesthetized rats. *Endocrine, Metabolic & Immune Disorders Drug Targets* 21(8), 1476–1484. DOI: 10.2174/1871530320666201029142513.

Monir, R., Abd Elkarim, A.S., Shalaby, M.F., Zaki, A.A. and Shabana, S. (2020) Anti-hypertensive activity of *Punica granatum* peels ethyl acetate extract on fludrocortisones induced hypertension in wistar rats. *Pharmacognosy Journal* 12(5), 1135–1142. DOI: 10.5530/PJ.2020.12.160.

Mozo, R.N. and Caole-Ang, I. (2015) The effects of malunggay (*Moringa oleifera*) leaves capsule supplements on high specificity C-reactive protein and hemoglobin A1c levels of diabetic patients in hospital in Manila medical center: a prospective cohort study. *Phillippine Journal of Internal Medicine* 53(4). Available at: https://www.scopus.com/inward/record.uri?eid=2-s2.0-84952685279&partnerID=40&md5=2d05185103b2d43b6738e38c6528c4fb (accessed 20 January 2023).

Murali, Y.K., Anand, P., Tandon, V., Singh, R., Chandra, R. *et al.* (2007) Long-term effects of *Terminalia chebula* Retz. on hyperglycemia and associated hyperlipidemia, tissue glycogen content and *in vitro* release of insulin in streptozotocin induced diabetic rats. *Experimental and Clinical Endocrinology & Diabetes* 115(10), 641–646. DOI: 10.1055/s-2007-982500.

Nandave, M., Ojha, S.K., Joshi, S., Kumari, S. and Arya, D.S. (2009) *Moringa oleifera* leaf extract prevents isoproterenol-induced myocardial damage in rats: evidence for an antioxidant, antiperoxidative, and cardioprotective intervention. *Journal of Medicinal Food* 12(1), 47–55. DOI: 10.1089/jmf.2007.0563.

Oboh, G., Bello, F.O., Ademosun, A.O., Akinyemi, A.J. and Adewuni, T.M. (2015) Antioxidant, hypolipidemic, and anti-angiotensin-1-converting enzyme properties of lemon (*Citrus limon*) and lime (*Citrus aurantifolia*) juices. *Comparative Clinical Pathology* 24(6), 1395–1406. DOI: 10.1007/s00580-015-2088-x.

Oboh, G., Ademosun, A.O., Oyetomi, O.J. and Adewuni, T.M. (2018) Influence of moringa (*Moringa oleifera*) leaf extracts on the antioxidant and angiotensin-1 converting enzyme inhibitory properties of lisinopril. *Oriental Pharmacy and Experimental Medicine* 18(4), 317–324. DOI: 10.1007/s13596-018-0317-y.

Ofem, O.E., Ani, E.J., Archibong, A.N. and Ufford, J.M. (2015) Variations in blood parameters of high salt loaded rats following administration of *Moringa oleifera* leaf extract. *Trends in Medical Research* 10(4), 97–105. DOI: 10.3923/tmr.2015.97.105.

Oikeh, E.I., Omoregie, E.S., Oviasogie, F.E. and Oriakhi, K. (2016) Phytochemical, antimicrobial, and antioxidant activities of different citrus juice concentrates. *Food Science & Nutrition* 4(1), 103–109. DOI: 10.1002/fsn3.268.

Ojha, S., Bhatia, J., Arora, S., Golechha, M., Kumari, S. *et al.* (2011) Cardioprotective effects of *Commiphora mukul* against isoprenaline-induced cardiotoxicity: a biochemical and histopathological evaluation. *Journal of Environmental Biology* 32(6), 731–738.

Ojha, S.K., Nandave, M., Arora, S., Mehra, R.D., Joshi, S. *et al.* (2008) Effect of *Commiphora mukul* extract on cardiac dysfunction and ventricular function in isoproterenol-induced myocardial infarction. *Indian Journal of Experimental Biology* 46(9), 646–652.

Omoba, O.S., Olagunju, A.I., Salawu, S.O. and Boligon, A.A. (2019) HPLC-DAD phenolic profiling and *in vitro* antioxidant activities of three prominent Nigerian spices. *Preventive Nutrition and Food Science* 24(2), 179–186. DOI: 10.3746/pnf.2019.24.2.179.

Onyenibe, N.S., Isreal, F. and Babatunji, O. (2018) *Moringa oleifera* leaves modulate blood glucose concentration, lipid profile and carbohydrate metabolizing enzymes in streptozotocin diabetic rats. *World Heart Journal* 10(3), 213–222. Available at: https://www.scopus.com/inward/record.uri?eid=2-s2.0-85064073800&partnerID=40&md5=27bd099f0bd0c20b1e8ccd32d9d6ad23 (accessed 20 January 2023).

Originale, A., Benedec, D., Pârvu, A.E., Oniga, I., Toiu, A. *et al.* (2007) Effects of *Ocimum basilicum* L. extract on experimental acute inflammation. *Revista medico-chirurgicala a Societatii de Medici si Naturalisti din Iasi* 111(4), 1065–1069.

Owen, P.L. and Johns, T. (2002) Antioxidants in medicines and spices as cardioprotective agents in Tibetan highlanders. *Pharmaceutical Biology* 40(5), 346–357. DOI: 10.1076/phbi.40.5.346.8461.

Özkan, G., Sağdiç, O., Baydar, N.G. and Baydar, H. (2004) Antioxidant and antibacterial activities of *Rosa damascena* flower extracts. *Food Science and Technology International* 10(4), 277–281. DOI: 10.1177/1082013204045882.

Pakdel, R., Niazmand, S., Mouhebati, M., Vahedi, M.M., Aghaee, A. *et al.* (2018) A comparison between the effects of *Portulaca oleracea* seeds extract and valsartan on echocardiographic and hemodynamic parameters in rats with levothyroxine-induced thyrotoxicosis. *Avicenna Journal of Phytomedicine* 8(3), 276–285.

Pakdel, R., Vatanchian, M., Niazmand, S., Beheshti, F., Rahimi, M. *et al.* (2019) Comparing the effects of *Portulaca oleracea* seed hydro-alcoholic extract, valsartan, and vitamin E on hemodynamic changes, oxidative stress parameters and cardiac hypertrophy in thyrotoxic rats. *Drug and Chemical Toxicology* 45(1), 14–21. DOI: 10.1080/01480545.2019.1651330.

Panda, S. (2015) Butanolic fraction of *Moringa oleifera* L. (moringaceae) attenuates isoprotrenol–induced cardiac necrosis and oxidative stress in rats: an EPR study. *EXCLI Journal* 14, 64–74. DOI: 10.17179/excli2014-431.

Panda, S., Kar, A., Sharma, P. and Sharma, A. (2013) Cardioprotective potential of *N*,α-L-rhamnopyranosyl vincosamide, an indole alkaloid, isolated from the leaves of *Moringa oleifera* in isoproterenol induced cardiotoxic rats: *in vivo* and *in vitro* studies. *Bioorganic & Medicinal Chemistry Letters* 23(4), 959–962. DOI: 10.1016/j.bmcl.2012.12.060.

Park, K.-S., Ahn, S.H., Lee, K.P., Park, S.-Y., Cheon, J.H. *et al.* (2018) The natural compound dansameum reduces foam cell formation by downregulating CD36 and peroxisome proliferator-activated receptor-gamma; expression. *Pharmacognosy Magazine* 13(Suppl. 4), S868–S874. DOI: 10.4103/pm.pm_112_17.

Parzonko, A., Czerwińska, M.E., Kiss, A.K. and Naruszewicz, M. (2013) Oleuropein and oleacein may restore biological functions of endothelial progenitor cells impaired by angiotensin II via activation of Nrf2/heme oxygenase-1 pathway. *Phytomedicine* 20(12), 1088–1094. DOI: 10.1016/j.phymed.2013.05.002.

Pereira, R.P., Fachinetto, R., de Souza Prestes, A., Puntel, R.L., Santos da Silva, G.N. *et al.* (2009) Antioxidant effects of different extracts from *Melissa officinalis*, *Matricaria recutita* and *Cymbopogon citratus*. *Neurochemical Research* 34(5), 973–983. DOI: 10.1007/s11064-008-9861-z.

Perrinjaquet-Moccetti, T., Busjahn, A., Schmidlin, C., Schmidt, A., Bradl, B. *et al.* (2008) Food supplementation with an olive (*Olea europaea* L.) leaf extract reduces blood pressure in borderline hypertensive monozygotic twins. *Phytotherapy Research* 22(9), 1239–1242. DOI: 10.1002/ptr.2455.

Pingali, U., Sukumaran, D. and Nutalapati, C. (2020) Effect of an aqueous extract of *Terminalia chebula* on endothelial dysfunction, systemic inflammation, and lipid profile in type 2 diabetes mellitus: a randomized double-blind, placebo-controlled clinical study. *Phytotherapy Research* 34(12), 3226–3235. DOI: 10.1002/ptr.6771.

Pittaway, J.K., Ahuja, K.D.K., Robertson, I.K. and Ball, M.J. (2007) Effects of a controlled diet supplemented with chickpeas on serum lipids, glucose tolerance, satiety and bowel function. *Journal of the American College of Nutrition* 26(4), 334–340. DOI: 10.1080/07315724.2007.10719620.

Plangar, A.F., Anaeigoudari, A., KhajaviRad, A. and Shafei, M.N. (2019) Beneficial cardiovascular effects of hydroalcoholic extract from *Crocus sativus* in hypertension induced by angiotensin II. *Journal of Pharmacopuncture* 22(2), 95–101. DOI: 10.3831/KPI.2019.22.012.

Poudyal, H., Campbell, F. and Brown, L. (2010) Olive leaf extract attenuates cardiac, hepatic, and metabolic changes in high carbohydrate-, high fat-fed rats. *The Journal of Nutrition* 140(5), 946–953. DOI: 10.3945/jn.109.117812.

Pragda, S.S., Kuppast, I.J., Mankani, K.L. and Ramesh, L. (2012) Evaluation of antihyperlipidemic activity of leaves of *Portulaca oleracea* Linn against dexamethasone induced hyperlipidemia in rats. *International Journal of Pharmacy and Pharmaceutical Sciences* 4(Suppl. 4), 279–283. Available at: https://www.scopus.com/inward/record.uri?eid=2-s2.0-84866330309&partnerID=40&md5=3c85036fac7a32d783f184bfcf48a613 (accessed 20 January 2023).

Prakash Udaya, N.K., Bhuvaneswari, S., Balamurugan, A., Karthik Ashwin, N., Deepa, S. *et al.* (2013) Studies on bio activity and phytochemistry of leaves of common trees. *International Journal of Research in Pharmaceutical Sciences* 4(3), 476–481. Available at: https://www.scopus.com/inward/record.uri?eid=2-s2.0-84897545953&partnerID=40&md5=3bcb6ae5917555633029008cc14285db (accessed 20 January 2023).

Qureshi, A.A., Burger, W.C., Prentice, N., Bird, H.R. and Sunde, M.L. (1980) Regulation of lipid metabolism in chicken liver by dietary cereals. *The Journal of Nutrition* 110(3), 388–393. DOI: 10.1093/jn/110.3.388.

Rababa'h, A.M. and Alzoubi, M.A. (2021) *Origanum majorana* L. extract protects against isoproterenol-induced cardiotoxicity in rats. *Cardiovascular Toxicology* 21(7), 543–552. DOI: 10.1007/s12012-021-09645-2.

Radmanesh, E., Dianat, M. and Radan, M. (2017) The effect of crocin on electrocardiogram parameters, oxidative stress, antioxidant enzymes, and its comparison with amiodarone in $CaCl_2$ induced arrhythmia. *Latin American Journal of Pharmacy* 36(8), 1606–1612. Available at: https://www.scopus.com/inward/record.uri?eid=2-s2.0-85027239704&partnerID=40&md5=235c820773c12b31046af727e7e434f7 (accessed 20 January 2023).

Ragone, M.I., Bayley, M., Lazarte, M.L., Clavellino, L., Prieto, J.J. *et al.* (2019) Oral administration of *Melissa officinalis* or *Bahuinia forficata* infusions reduce the postischemic recovery of hearts from hypothyroid rats. *PharmacologyOnLine* 3, 1–14. Available at: https://www.scopus.com/inward/record.uri?eid=2-s2.0-85077722198&partnerID=40&md5=6f2537fed7aece4b009327857d070c39 (accessed 20 January 2023).

Rahimi, K. and Kazerani, H.R. (2021) Antiarrhythmic effects of pomegranate (*Punica granatum*) juice on isolated rat hearts following ischemia and reperfusion. *Pharmaceutical Chemistry Journal* 55(1), 81–85. DOI: 10.1007/s11094-021-02376-2.

Raimondo, S., Nikolic, D., Conigliaro, A., Giavaresi, G., Lo Sasso, B. *et al.* (2021) Preliminary results of CitraVes™ effects on low density lipoprotein cholesterol and waist circumference in healthy subjects after 12 weeks: a pilot open-label study. *Metabolites* 11(5), 276. DOI: 10.3390/metabo11050276.

Rajanandh, M.G., Satishkumar, M.N., Elango, K. and Suresh, B. (2012) *Moringa oleifera* L. A herbal medicine for hyperlipidemia: a pre-clinical report. *Asian Pacific Journal of Tropical Disease* 2(Suppl. 2), S790–S795. DOI: 10.1016/S2222-1808(12)60266-7.

Rajyalakshmi, I., Rao, A. V, Potti, R.B. and Murty, K.V.G.S. (2018) Effects of *Nardostachys jatamansi* extract on inflammatory and oxidative stress markers in hypertensive, hyperglycaemic patients: an open-label, prospective study. *International Journal of Research in Pharmaceutical Sciences* 9(4), 1096–1098. DOI: 10.26452/ijrps.v9i4.1637.

Ramadan, G., El-Beih, N.M., Arafa, N.M. and Zahra, M.M. (2013) Preventive effects of Egyptian sweet marjoram (*Origanum majorana* L.) leaves on haematological changes and cardiotoxicity in isoproterenol-treated albino rats. *Cardiovascular Toxicology* 13(2), 100–109. DOI: 10.1007/s12012-012-9189-4.

Ramadhania, Z.M., Insanu, M., Gunarti, N.S., Wirasutisna, K.R., Sukrasno, S. *et al.* (2017) Antioxidant activity from ten species of Myrtaceae. *Asian Journal of Pharmaceutical and Clinical Research* 10(14), 5. DOI: 10.22159/ajpcr.2017.v10s2.19470.

Ramesh, B., Karuna, R., Sreenivasa, R.S., Haritha, K., Sai, M.D. *et al.* (2012) Effect of *Commiphora mukul* gum resin on hepatic marker enzymes, lipid peroxidation and antioxidants status in pancreas and heart of streptozotocin induced diabetic rats. *Asian Pacific Journal of Tropical Biomedicine* 2(11), 895–900. DOI: 10.1016/S2221-1691(12)60249-4.

Ramesh, B., Karuna, R., Sreenivasa Reddy, S., Sudhakara, G. and Saralakumari, D. (2013) Ethanolic extract of *Commiphora mukul* gum resin attenuates streptozotocin-induced alterations in carbohydrate and lipid metabolism in rats. *EXCLI Journal* 12, 556–568.

Randriamboavonjy, J.I., Loirand, G., Vaillant, N., Lauzier, B., Derbré, S. *et al.* (2016) Cardiac protective effects of *Moringa oleifera* seeds in spontaneous hypertensive rats. *American Journal of Hypertension* 29(7), 873–881. DOI: 10.1093/ajh/hpw001.

Randriamboavonjy, J.I., Rio, M., Pacaud, P., Loirand, G. and Tesse, A. (2017) *Moringa oleifera* seeds attenuate vascular oxidative and nitrosative stresses in spontaneously hypertensive rats. *Oxidative Medicine and Cellular Longevity* 2017, 4129459. DOI: 10.1155/2017/4129459.

Randriamboavonjy, J.I., Heurtebise, S., Pacaud, P., Loirand, G. and Tesse, A. (2019) *Moringa oleifera* seeds improve aging-related endothelial dysfunction in Wistar rats. *Oxidative Medicine and Cellular Longevity* 2019, 2567198. DOI: 10.1155/2019/2567198.

Rao, N.K. and Nammi, S. (2006) Antidiabetic and renoprotective effects of the chloroform extract of *Terminalia chebula* Retz. seeds in streptozotocin-induced diabetic rats. *BMC Complementary and Alternative Medicine* 6(1), 17. DOI: 10.1186/1472-6882-6-17.

Rao, S.P. and Krishnamurthy, V. (2019) Ethanol extract of *Punica granatum* L. leaf in combination with gallic acid ameliorates liver dysfunction in high-fat diet-induced obese rats. *Indian Journal of Pharmaceutical Sciences* 81(4), 673–680. Available at: https://www.scopus.com/inward/record.uri?eid=2-s2.0-85072332335&partnerID=40&md5=4dc6fd5e96bc432f8f47bf86ce410f99 (accessed 20 January 2023).

Razani, Z., Dastani, M. and Kazerani, H.R. (2017) Cardioprotective effects of pomegranate (*Punica granatum*) juice in patients with ischemic heart disease. *Phytotherapy Research* 31(11), 1731–1738. DOI: 10.1002/ptr.5901.

Razavi, B.M., Alyasin, A., Hosseinzadeh, H. and Imenshahidi, M. (2018) Saffron induced relaxation in isolated rat aorta via endothelium dependent and independent mechanisms. *Iranian Journal of Pharmaceutical Research* 17(3), 1018–1025. Available at: https://www.scopus.com/inward/record.uri?eid=2-s2.0-85049839033&partnerID=40&md5=d727590b64cebd4601ef6dd0d70a6c0b (accessed 20 January 2023).

Recinella, L., Chiavaroli, A., Orlando, G., Menghini, L., Ferrante, C. *et al.* (2019) Protective effects induced by two polyphenolic liquid complexes from olive (*Olea europaea*, mainly cultivar Coratina) pressing

juice in rat isolated tissues challenged with LPS. *Molecules (Basel, Switzerland)* 24(16), 3002. DOI: 10.3390/molecules24163002.
Sabzghabaee, A.M., Kelishadi, R., Jelokhanian, H., Asgary, S., Ghannadi, A. *et al.* (2014) Clinical effects of *Portulaca oleracea* seeds on dyslipidemia in obese adolescents: a triple-blinded randomized controlled trial. *Medical Archives (Sarajevo, Bosnia and Herzegovina)* 68(3), 195–199. DOI: 10.5455/medarh.2014.68.195-199.
Sachdeva, J., Tanwar, V., Golechha, M., Siddiqui, K.M., Nag, T.C. *et al.* (2012) *Crocus sativus* L. (saffron) attenuates isoproterenol-induced myocardial injury via preserving cardiac functions and strengthening antioxidant defense system. *Experimental and Toxicologic Pathology* 64(6), 557–564. DOI: 10.1016/j.etp.2010.11.013.
Sah, A.N., Joshi, A., Juyal, V. and Kumar, T. (2011) Antidiabetic and hypolipidemic activity of *Citrus medica* Linn. seed extract in streptozotocin induced diabetic rats. *Pharmacognosy Journal* 3(23), 80–84. DOI: 10.5530/pj.2011.23.12.
Sahoo, S., Raghavendra, K.M. and Biswas, S. (2014) Identification of a proteinaceous component in the leaf of *Moringa oleifera* Lam. with effects on high serum creatinine. *Indian Journal of Pharmaceutical Sciences* 76(1), 78–81.
Sajid, T.M., Rashid, S., Ahmad, M. and Khan, U. (1996) Estimation of cardiac depressant activity of ten medicinal plant extracts from Pakistan. *Phytotherapy Research* 10(2), 178–180. DOI: 10.1002/(SICI)1099-1573(199603)10:2<178::AID-PTR802>3.0.CO;2-T.
Saka, W.A., Ayoade, T.E., Akhigbe, T.M. and Akhigbe, R.E. (2020) *Moringa oleifera* seed oil partially abrogates 2,3-dichlorovinyl dimethyl phosphate (Dichlorvos)-induced cardiac injury in rats: evidence for the role of oxidative stress. *Journal of Basic and Clinical Physiology and Pharmacology* 32(3), 237–246. DOI: 10.1515/jbcpp-2019-0313.
Saleem, H., Al-Dujaily, A.N.G. and Al-Murshidi, M.H.H. (2016) Comparison study between effect of methanolic extract of *Moringa oleifera* and exogenous ghrelin on lipid profile in atherogenic rats. *Research Journal of Pharmaceutical, Biological and Chemical Sciences* 7(3), 2210–2221. Available at: https://www.scopus.com/inward/record.uri?eid=2-s2.0-84973332946&partnerID=40&md5=71ed8b9c2e6070d967204f91db648695 (accessed 20 January 2023).
Sánchez-Moreno, C., Cano, M.P., De Ancos, B., Plaza, L., Olmedilla, B. *et al.* (2006) Mediterranean vegetable soup consumption increases plasma vitamin C and decreases F_2-isoprostanes, prostaglandin E_2 and monocyte chemotactic protein-1 in healthy humans. *Journal of Nutritional Biochemistry* 17(3), 183–189. DOI: 10.1016/j.jnutbio.2005.07.001.
Sangkitikomol, W., Rocejanasaroj, A. and Tencomnao, T. (2014) Effect of *Moringa oleifera* on advanced glycation end-product formation and lipid metabolism gene expression in HepG2 cells. *Genetics and Molecular Research* 13(1), 723–735. DOI: 10.4238/2014.January.29.3.
Scheffler, A., Rauwald, H.W., Kampa, B., Mann, U., Mohr, F.W. *et al.* (2008) *Olea europaea* leaf extract exerts L-type Ca^{2+} channel antagonistic effects. *Journal of Ethnopharmacology* 120(2), 233–240. DOI: 10.1016/j.jep.2008.08.018.
Schumacher, E., Vigh, É., Molnár, V., Kenyeres, P., Fehér, G. *et al.* (2011) Thrombosis preventive potential of chicory coffee consumption: a clinical study. *Phytotherapy Research* 25(5), 744–748. DOI: 10.1002/ptr.3481.
Sedighi, M., Faghihi, M., Rafieian-Kopaei, M., Rasoulian, B. and Nazari, A. (2019) Cardioprotective effect of ethanolic leaf extract of *Melissa officinalis* L. against regional ischemia-induced arrhythmia and heart injury after five days of reperfusion in rats. *Iranian Journal of Pharmaceutical Research* 18(3), 1530–1542. DOI: 10.22037/ijpr.2019.1100761.
Sedighi, M., Cheraghi, M., Faghihi, M., Rahimi-Madiseh, M., Kiani, A.A. *et al.* (2021a) Hypotensive effect of *Cichorium intybus* extract in rats. *Journal of HerbMed Pharmacology* 10(2), 257–261. DOI: 10.34172/jhp.2021.29.
Sedighi, M., Ghorbanzadeh, V., Abaszadeh, S., Karimi, A., Cheraghi, M. *et al.* (2021b) Up-regulation of chemokine receptor type 4 expression in the ischemic reperfused heart by alcoholic extract of *Cichorium intybus* rescue the heart from ischemia injury in male rat. *The Journal of Pharmacy and Pharmacology* 73(10), 1351–1360. DOI: 10.1093/jpp/rgab076.
Serra, A.T., Rocha, J., Sepodes, B., Matias, A.A., Feliciano, R.P. *et al.* (2012) Evaluation of cardiovascular protective effect of different apple varieties – correlation of response with composition. *Food Chemistry* 135(4), 2378–2386.
Seth, S.D., Maulik, M., Katiyar, C.K. and Maulik, S.K. (1998) Role of Lipistat in protection against isoproterenol induced myocardial necrosis in rats: a biochemical and histopathological study. *Indian Journal of Physiology and Pharmacology* 42(1), 101–106.

Shah, M., Shahu, G., Tamrakar, A.K., Malshetthy, S., Janadri, S. *et al.* (2014) Cardiotoxic activity of leaves extract of *Raphanus sativus* Linn. in adult male albino rats. *International Journal of Pharmaceutical and Phytopharmacological Research* 4(1), 66–69. Available at: https://www.scopus.com/inward/record.uri?eid=2-s2.0-85042950012&partnerID=40&md5=a9cd35522596b6061e01354c4f1dc195 (accessed 20 January 2023).

Shaikh, S., Shaikh, S., Sayyed, F., Khan, P., Shaikh, S. *et al.* (2020) Traditional plants: *in vitro* analysis of medicinal properties. *International Journal of Pharmaceutical Research* 12, 2449–2458. DOI: 10.31838/ijpr/2020.SP1.350.

Sharma, G., Sharma, V. and Mishra, T. (2018) Enhancement of *in vitro* antioxidant potential of *Terminalia chebula* by various fruit extracts and optimization of concentration by response surface methodology. *Asian Journal of Pharmaceutical and Clinical Research* 11(7), 228–233. DOI: 10.22159/ajpcr.2018.v11i7.25786.

Sharma, M., Afaque, A., Dwivedi, S., Jairajpuri, Z.S., Shamsi, Y. *et al.* (2019) *Cichorium intybus* attenuates streptozotocin induced diabetic cardiomyopathy via inhibition of oxidative stress and inflammatory response in rats. *Interdisciplinary Toxicology* 12(3), 111–119. DOI: 10.2478/intox-2019-0013.

Sharma, V. and Singh, M. (2012) Alterations induced by *N*-nitrosodimethylamine and ethanolic root extract of *Operculina turpethum* in serum lipid profile of male albino mice. *Asian Journal of Pharmaceutical and Clinical Research* 5(3), 69–73. Available at: https://www.scopus.com/inward/record.uri?eid=2-s2.0-84875412428&partnerID=40&md5=0e60761996e3a7e6fd43ca3fb8895a72 (accessed 20 January 2023).

Sheikh, A., Yeasmin, F., Agarwal, S., Rahman, M., Islam, K. *et al.* (2014) Protective effects of *Moringa oleifera* Lam. leaves against arsenic-induced toxicity in mice. *Asian Pacific Journal of Tropical Biomedicine* 4(Suppl. 1), 353–358. DOI: 10.12980/APJTB.4.201414B44.

Sheng, L., Qian, Z., Zheng, S. and Xi, L. (2006) Mechanism of hypolipidemic effect of crocin in rats: crocin inhibits pancreatic lipase. *European Journal of Pharmacology* 543(1–3), 116–122. DOI: 10.1016/j.ejphar.2006.05.038.

Siddiqi, A., Parveen, A., Dhyani, N., Hussain, M.E., Fahim, M. *et al.* (2018) Effects of *Coriandrum sativum* extract and simvastatin in isopreterenol induced heart failure in rats. *Serbian Journal of Experimental and Clinical Research* 19(2), 119–129. DOI: 10.1515/SJECR-2017-0080.

Singh, D., Rana, R. and Mishra, M. (2012) *In vitro* free radical scavenging assay of pod coat and pod seed of *Moringa oleifera* by spectrophotometric method. *International Journal of Pharmaceutical Sciences Review and Research* 17(1), 94–96. Available at: https://www.scopus.com/inward/record.uri?eid=2-s2.0-84870379948&partnerID=40&md5=d7ff720aff6d1f0a9de261665a792416 (accessed 20 January 2023).

Singh, I., Mok, M., Christensen, A.-M., Turner, A.H. and Hawley, J.A. (2008) The effects of polyphenols in olive leaves on platelet function. *Nutrition, Metabolism, and Cardiovascular Diseases* 18(2), 127–132. DOI: 10.1016/j.numecd.2006.09.001.

Singh, M., Khan, M.A., Ahmad, J., Fahmy, U.A., Kotta, S. *et al.* (2020) Effect of *Nardostachys jatamansi* DC. on apoptosis, inflammation and oxidative stress induced by doxorubicin in Wistar rats. *Plants (Basel, Switzerland)* 9(11), 579. DOI: 10.3390/plants9111579.

Singh, M.C. and Gujar, K.N. (2019) Preparation and evaluation of nutraceutical product mixture of seeds of *Cucumis melo*, *Punica granatum*, *Linum usitatissimum*, for antioxidant, prebiotic and nutraceutical potential. *Pharmacognosy Journal* 11(2), 383–387. DOI: 10.5530/pj.2019.11.59.

Sobhani, Z., Nami, S.R., Emami, S.A., Sahebkar, A. and Javadi, B. (2017) Medicinal plants targeting cardiovascular diseases in view of Avicenna. *Current Pharmaceutical Design* 23(17), 2428–2443. DOI: 10.2174/1381612823666170215104101.

Somova, L.I., Shode, F.O. and Mipando, M. (2004) Cardiotonic and antidysrhythmic effects of oleanolic and ursolic acids, methyl maslinate and uvaol. *Phytomedicine* 11(2–3), 121–129. DOI: 10.1078/0944-7113-00329.

Spanou, C., Stagos, D., Tousias, L., Angelis, A., Aligiannis, N. *et al.* (2007) Assessment of antioxidant activity of extracts from unique Greek varieties of Fabaceae (Leguminosae) plants using *in vitro* assays. *Anticancer Research* 27(5A), 3403–3410. Availabe at: https://www.scopus.com/inward/record.uri?eid=2-s2.0-35348981250&partnerID=40&md5=ad57e88eda732b2945bc9cf50224db65 (accessed 20 January 2023).

Srujana, M.S., Thota, R.S.P., Anand, A., Anusha, K., Zehra, A. *et al.* (2019) *Raphanus sativus* (Linn.) fresh juice priming moderates sucrose-induced postprandial glycemia as well as postprandial glycemic excursion in rats. *Indian Journal of Traditional Knowledge* 18(2), 339–345.

Stockton, A., Farhat, G., McDougall, G.J. and Al-Dujaili, E.A.S.S. (2017) Effect of pomegranate extract on blood pressure and anthropometry in adults: a double-blind placebo-controlled randomised clinical trial. *Journal of Nutritional Science* 6(11), e39. DOI: 10.1017/jns.2017.36.

Subashini, R., Yogeeta, S., Gnanapragasam, A. and Devaki, T. (2006) Protective effect of *Nardostachys jatamansi* on oxidative injury and cellular abnormalities during doxorubicin-induced cardiac damage in rats. *The Journal of Pharmacy and Pharmacology* 58(2), 257–262. DOI: 10.1211/jpp.58.2.0014.

Suchalatha, S. and Devi, C.S. (2004) Protective effect of *Terminalia chebula* against experimental myocardial injury induced by isoproterenol. *Indian Journal of Experimental Biology* 42(2), 174–178.

Suchalatha, S. and Devi, C.S.S. (2005) Antioxidant activity of ethanolic extract of *Terminalia chebula* fruit against isoproterenol-induced oxidative stress in rats. *Indian Journal of Biochemistry and Biophysics* 42(4), 246–249.

Talukdar, D. (2012) Total flavonoids, phenolics, tannins and antioxidant activity in seeds of lentil and grass pea. *International Journal of Phytomedicine* 4(4), 537–542. Available at: https://www.scopus.com/inward/record.uri?eid=2-s2.0-84885986513&partnerID=40&md5=696944bc3a4b9d0dee98833530adafa6 (accessed 20 January 2023).

Tarish, R.J., Ghazi, A.M. and Abd-Alhassen, J.K. (2019) Hypolipidemic effect of *Portulaca oleracea* and salvia officinalis comparing to atorvastatin in rats. *International Journal of Pharmaceutical Quality Assurance* 10(4), 613–618. DOI: 10.25258/ijpqa.10.4.9.

Tas, S., Tas, B., Bassalat, N. and Jaradat, N. (2018) *In-vivo*, hypoglycemic, hypolipidemic and oxidative stress inhibitory activities of *Myrtus communis* L. fruits hydroalcoholic extract in normoglycemic and streptozotocin-induced diabetic rats. *Biomedical Research (India)* 29(13), 2727–2734. DOI: 10.4066/biomedicalresearch.29-18-708.

The, N., Banerjee, S.P. and Dandiya, P.C. (1967) Smooth muscle and cardiovascular pharmacology of α-elaterin-2-D-glucopyranoside glycoside of *Citrullus colocynthis*. *Journal of Pharmaceutical Sciences* 56(12), 1665–1667. DOI: 10.1002/jps.2600561229.

Tiwari, A.K., Sahana, C., Zehra, A., Madhusudana, K., Kumar, D.A. *et al.* (2013) Mitigation of starch-induced postprandial glycemic spikes in rats by antioxidants-rich extract of *Cicer arietinum* Linn. seeds and sprouts. *Journal of Pharmacy & Bioallied Sciences* 5(4), 270–276. DOI: 10.4103/0975-7406.120077.

Torul, H., Küçükboyaci, N., Tamer, U. and Karasu, Ç. (2020) Evaluation of phenolic compounds and protective effects of olive (*Olea europaea* L.) leaf extracts on endothelial cells against hydrogen peroxide-induced toxicity. *Journal of Research in Pharmacy* 24(4), 497–507. DOI: 10.35333/jrp.2020.198.

Trejo-Moreno, C., Méndez-Martínez, M., Zamilpa, A., Jiménez-Ferrer, E., Perez-garcia, M.D. *et al.* (2018) *Cucumis sativus* aaqueous fraction inhibits angiotensin II-induced inflammation and oxidative stress *in vitro*. *Nutrients* 10(3), 276. DOI: 10.3390/nu10030276.

Triantafyllou, A., Chaviaras, N., Sergentanis, T.N., Protopapa, E. and Tsaknis, J. (2007) Chios mastic gum modulates serum biochemical parameters in a human population. *Journal of Ethnopharmacology* 111(1), 43–49. DOI: 10.1016/j.jep.2006.10.031.

Umar, A., Yimin, W., Tohti, I., Upur, H., Berké, B. *et al.* (2015) Effect of traditional Uyghur medicine abnormal Savda Munziq extract on rabbit platelet aggregation *in vitro* and rat arteriovenous shunt thrombosis *in vivo*. *Journal of Ethnopharmacology* 159, 184–188. DOI: 10.1016/j.jep.2014.11.006.

Valachovičová, M., Krajčovičová-Kudláčková, M., Blažíček, P. and Babinská, K. (2006) No evidence of insulin resistance in normal weight vegetarians: a case control study. *European Journal of Nutrition* 45(1), 52–54. DOI: 10.1007/s00394-005-0563-x.

Velmurugan, C. and Bhargava, A. (2013) Anti-diabetic and hypolipidemic activity of fruits of *Pyrus communis* L. in hyperglycemic rats. *Asian Journal of Pharmaceutical and Clinical Research* 6(Suppl. 5), 108–111.

Vera-Nuñez, N., Guirao, A.R.Y., Silva, J.D.F.D.A., Ramos, I.P., Torres, M.K.S. *et al.* (2021) Water-soluble lectin (WSMoL) from *Moringa oleifera* seeds treatment recovers glycemic levels and improves left ventricular ejection fraction on type-2 diabetes mice model. *Anais da Academia Brasileira de Ciencias* 93(3), e20201596. DOI: 10.1590/0001-3765202120201596.

Wainstein, J., Landau, Z., Dayan, Y.B., Jakubowicz, D., Grothe, T. *et al.* (2016) Purslane extract and glucose homeostasis in adults with type 2 diabetes: a double-blind, placebo-controlled clinical trial of efficacy and safety. *Journal of Medicinal Food* 19(2), 133–140. DOI: 10.1089/jmf.2015.0090.

Wang, F., Bao, Y., Shen, X., Zengin, G., Lyu, Y. *et al.* (2021) Niazirin from *Moringa oleifera* Lam. attenuates high glucose-induced oxidative stress through PKCζ/Nox4 pathway. *Phytomedicine* 86, 153066. DOI: 10.1016/j.phymed.2019.153066.

Wang, S., Zhao, Y., Song, J., Wang, R., Gao, L. *et al.* (2020) Total flavonoids from *Anchusa italica* Retz. improve cardiac function and attenuate cardiac remodeling post myocardial infarction in mice. *Journal of Ethnopharmacology* 257, 112887. DOI: 10.1016/j.jep.2020.112887.

Wang, Y., Wang, Q., Yu, W. and Du, H. (2018) Crocin attenuates oxidative stress and myocardial infarction injury in rats. *International Heart Journal* 59(2), 387–393. DOI: 10.1536/ihj.17-114.

Weidner, C., Wowro, S.J., Freiwald, A., Kodelja, V., Abdel-Aziz, H. *et al.* (2014) Lemon balm extract causes potent antihyperglycemic and antihyperlipidemic effects in insulin-resistant obese mice. *Molecular Nutrition & Food Research* 58(4), 903–907. DOI: 10.1002/mnfr.201300477.

Wenzel, D.G. and Haskell, A.R. (1952) Pharmacological actions of *Paeonia officinalis*. *Journal of the American Pharmaceutical Association* 41(3), 162–164. DOI: 10.1002/jps.3030410317.

Wu, Q., Ye, B., Lv, X., Mao, G., Wang, S. *et al.* (2020) Adjunctive therapeutic effects of *Cinnamomum camphora* forest environment on elderly patients with hypertension. *International Journal of Gerontology* 14(4), 327–331. DOI: 10.6890/IJGE.202011_14(4).0014.

Yang, Y., Jiang, Z. and Zhuge, D. (2019) Emodin attenuates lipopolysaccharide-induced injury via down-regulation of miR-223 in H9c2 cells. *International Heart Journal* 60(2), 436–443. DOI: 10.1536/ihj.18-048.

Yazdanparast, R. and Shahriyary, L. (2008) Comparative effects of *Artemisia dracunculus*, *Satureja hortensis* and *Origanum majorana* on inhibition of blood platelet adhesion, aggregation and secretion. *Vascular Pharmacology* 48(1), 32–37. DOI: 10.1016/j.vph.2007.11.003.

Zaki, A.A., Hashish, N.E., Amer, M.A. and Lahloub, M.-F. (2014) Cardioprotective and antioxidant effects of oleogum resin "Olibanum" from *Boswellia carteri* Birdw. (Bursearceae). *Chinese Journal of Natural Medicines* 12(5), 345–350. DOI: 10.1016/S1875-5364(14)60042-X.

Zamani, M., Rahimi, A.O., Mahdavi, R., Nikbakhsh, M., Jabbari, M.V. *et al.* (2007) Assessment of anti-hyperlipidemic effect of *Citrullus colocynthis*. *Revista Brasileira de Farmacognosia* 17(4), 492–496. DOI: 10.1590/S0102-695X2007000400003.

Zangeneh, M.M., Zangeneh, A., Salmani, S., Jamshidpour, R. and Kosari, F. (2019) Protection of phenylhydrazine-induced hematotoxicity by aqueous extract of *Ocimum basilicum* in Wistar male rats. *Comparative Clinical Pathology* 28(2), 331–338. DOI: 10.1007/s00580-018-2845-8.

Zarshenas, M.M. and Zargaran, A. (2015) A review on the Avicenna's contribution to the field of cardiology. *International Journal of Cardiology* 182, 237–241. DOI: 10.1016/j.ijcard.2014.12.145.

Zeka, K., Marrazzo, P., Micucci, M., Ruparelia, K.C., Arroo, R.R.J. *et al.* (2020) Activity of antioxidants from *Crocus sativus* L. petals: potential preventive effects towards cardiovascular system. *Antioxidants (Basel, Switzerland)* 9(11), 1102. DOI: 10.3390/antiox9111102.

Zhang, J., Yang, Z., Fang, K., Shi, Z., Ren, D. *et al.* (2017) Oleuropein prevents the development of experimental autoimmune myocarditis in rats. *International Immunopharmacology* 48(208), 187–195. DOI: 10.1016/j.intimp.2017.05.013.

Zhang, Y., Song, Z., Huang, S., Zhu, L., Liu, T. *et al.* (2020) Aloe emodin relieves Ang II-induced endothelial junction dysfunction via promoting ubiquitination mediated NLRP3 inflammasome inactivation. *Journal of Leukocyte Biology* 108(6), 1735–1746. DOI: 10.1002/JLB.3MA0520-582R.

Zhao, Z., Zheng, B., Li, J., Wei, Z., Chu, S. *et al.* (2020) Influence of crocetin, a natural carotenoid dicarboxylic acid in saffron, on L-type Ca^{2+} current, intracellular Ca^{2+} handling and contraction of isolated rat cardiomyocytes. *Biological & Pharmaceutical Bulletin* 43(9), 1367–1374. DOI: 10.1248/bpb.b20-00298.

Zheng, S., Qian, Z., Tang, F. and Sheng, L. (2005) Suppression of vascular cell adhesion molecule-1 expression by crocetin contributes to attenuation of atherosclerosis in hypercholesterolemic rabbits. *Biochemical Pharmacology* 70(8), 1192–1199. DOI: 10.1016/j.bcp.2005.07.034.

Zhou, W., Abdurahman, A., Umar, A., Iskander, G., Abdusalam, E. *et al.* (2014a) Effects of *Cydonia oblonga* Miller extracts on blood hemostasis, coagulation and fibrinolysis in mice, and experimental thrombosis in rats. *Journal of Ethnopharmacology* 154(1), 163–169. DOI: 10.1016/j.jep.2014.03.056.

Zhou, W., Abdusalam, E., Abliz, P., Reyim, N., Tian, S. *et al.* (2014b) Effect of *Cydonia oblonga* Mill. fruit and leaf extracts on blood pressure and blood rheology in renal hypertensive rats. *Journal of Ethnopharmacology* 152(3), 464–469. DOI: 10.1016/j.jep.2014.01.018.

Zhou, W.T., Abdurahman, A., Abdusalam, E., Yiming, W., Abliz, P. *et al.* (2014c) Effect of *Cydonia oblonga* Mill. leaf extracts or captopril on blood pressure and related biomarkers in renal hypertensive rats. *Journal of Ethnopharmacology* 153(3), 635–640. DOI: 10.1016/j.jep.2014.03.014.

Zhou, W.-T., Wang, X.-F., Yiming, W., Abdurahman, A., Ma, H. *et al.* (2015) Effects of total flavonoids of *Cydonia oblonga* Mill. on myocardial hypertrophy in spontaneously hypertensive rats and mechanisms. *Chinese Pharmacological Bulletin* 31(11), 1540–1546. DOI: 10.3969/j.issn.1001-1978.2015.11.013.

Zia-Ul-Haq, M., Khan, B.A., Landa, P., Kutil, Z., Ahmed, S. *et al.* (2012) Platelet aggregation and anti-inflammatory effects of garden pea, Desi chickpea and Kabuli chickpea. *Acta Poloniae Pharmaceutica Drug Research* 69(4), 707–711. Available at: https://www.scopus.com/inward/record.uri?eid=2-s2.0-84864063933&partnerID=40&md5=670cc170cae571a0f23a727060058c22 (accessed 20 January 2023).

Zrelli, H., Matsuka, M., Araki, M., Zarrouk, M., Miyazaki, H. *et al.* (2011a) Hydroxytyrosol induces vascular smooth muscle cells apoptosis through NO production and PP2A activation with subsequent inactivation of Akt. *Planta Medica* 77(15), 1680–1686. DOI: 10.1055/s-0030-1271073.

Zrelli, H., Matsuoka, M., Kitazaki, S., Zarrouk, M. and Miyazaki, H. (2011b) Hydroxytyrosol reduces intracellular reactive oxygen species levels in vascular endothelial cells by upregulating catalase expression through the AMPK–FOXO3a pathway. *European Journal of Pharmacology* 660(2–3), 275–282. DOI: 10.1016/j.ejphar.2011.03.045.

7 An Evidence-Based Review of Medicinal Plants Prescribed by Avicenna for Management of Diabetes

Zahra Boghrati[1], Seyed Ahmad Emami[1,2] and Maryam Akaberi[2*]
[1]*Department of Traditional Pharmacy, School of Pharmacy, Mashhad University of Medical Sciences, Mashhad, Iran;* [2]*Department of Pharmacognosy, School of Pharmacy, Mashhad University of Medical Sciences, Mashhad, Iran*

Abstract

Natural products have always been a source of drugs for human beings and animals. Drug discovery is a process of finding active biologic compounds from natural resources. This natural products-based drug discovery (NPD) is a complicated process that usually needs a long time and large expenses. A very beneficial strategy in NPD is using the knowledge of different systems of traditional medicine. Traditional Iranian Medicine (TIM) is a traditional system of medicine that flourished during the Islamic golden age. During this period, many scientists and physicians emerged, including Avicenna (AD 980–1037), one of the most important Persian physicians. In his book *Canon of Medicine*, he discussed the pathology and treatments for various diseases. In this chapter, we aim to have a glance at his point of view on the pathology and treatments for diabetes management, particularly those based on herbs that are discussed in the 19th volume of his abovementioned book. In addition, we provide some evidence from *in vitro*, *in vivo*, and clinical studies for the medicinal plants prescribed by Avicenna for this purpose.

Introduction

Avicenna's *Canon of Medicine*, which became the standard medical textbook for 600 years, is divided into five volumes and in the third volume, the author discussed the pathology and treatment of various diseases such as diabetes along with other health problems including bladder and urine problems, and lower urinary tract disorders like bladder stones, cystitis, urinary incontinence, hematuria, and overactive bladder. In this chapter, we review Avicenna's point of view on the diagnosis and treatment of diabetes mellitus.

Diabetes in Avicenna's Point of View

In the *Canon of Medicine* and other Arabic medical textbooks, diabetes is referred to as

*Email: akaberim@mums.ac.ir

(eds. H. Schulz, Seyed Ahmad Emami and Farsad Nadjafi)
DOI: 10.1079/9781800621671.0007

ḍyābiṭes, *dūlāb*, and *al-bawl as-sokkary*, which shows that traditional Islamic physicians were aware of diabetic conditions pretty well. They found that a diabetic person would get thirsty, but this thirst is not quenched by drinking water, since water is excreted directly through the kidneys without reaching the targeted cells.

In his scientific texts, Avicenna mentioned different pathologies for diabetes. He believed that it could be due to kidney failure and urinary tract crevice dilation, caused by a coldness overcoming the whole body or the liver. The other reason could be an intense absorption power caused by an unnatural heat in the body that forces the kidney to absorb and excrete too much from the liver, so that the liver absorbs the water from the intestine, which leads to thirst and drinking even more water resulting in a vicious circle. As a result, diabetes is considered a detrimental disease draining body fluids that leads to organ atrophy and losing weight.

Treatments Prescribed by Avicenna

Avicenna prescribed various therapeutic strategies for treating diabetes including oral and topical administration of medicines, conducting enemas and phlebotomy (*faṣd*), taking *abzan* (sitz bath: a sitz bath is a warm, shallow bath that cleanses the perineum, which is the space between the rectum and the vulva or scrotum. A sitz bath can also provide relief from pain or itching in the genital area), and dietary modifications. Avicenna believed that diabetes is mostly due to excessive heat in the body, therefore the treatment was usually based on cooling and moistening the body by using various nondiuretic vegetables, fruits, and extracts like lettuce (*Lactuca sativa*) and opium poppy (*Papaver somniferum*) as well as living in cool and humid climates or using cool *abzan*. Interestingly, Avicenna prescribed his patients to smell some aromatic plants like the bark of the camphor tree (*Cinnamomum camphora*) and water-lily (*Nymphaea alba/Nymphaea lotus*) for the management of diabetes. Administration of water-rich foods like celery and melons, chewing gums, and sucking sour foods in order to reduce water intake were among other helpful strategies for treating diabetes.

In order to prevent the quick secretion of the consumed water from the kidneys, Avicenna prescribed that patients either induce vomiting and excessive sweating or benumb the renal area. He forbade diabetic patients from performing heavy physical activities and consuming diuretics; instead, he encouraged them to do moderate enemas to increase the bowel movements. In addition, *faṣd* (phlebotomy) was an essential strategy in the early stages of the disease. He also prescribed some juices for diabetic patients such as chilled sour Doogh (a traditional Iranian beverage from yogurt; the thicker the better, especially the one produced from sheep's milk), roasted pumpkin juice, cucumber juice with psyllium, sour pomegranate juice, mulberry juice, plum juice, rose water, and the extracted water from Doogh, etc. These drinks were administered instead of drinking water to help control the thirst.

Medicinal Plants Prescribed by Avicenna for Diabetes Management

Oral formulations

Cucurbita pepo and *Cucurbita maxima*

Cucurbita pepo and *Cucurbita maxima*, commonly known as courgette (zucchini) and pumpkin, are among the most nutritious and health-improving vegetables. Pumpkin has traditionally been used in China, Mexico, North America, India, and Iran as an antidiabetic agent (Lal *et al.*, 2011; Mahmoodpoor *et al.*, 2018; Thanh *et al.*, 2021). Several studies including *in vitro*, *in vivo*, and clinical trials have been carried out to investigate the potential of these plants as antidiabetic and hypoglycemic medicinal herbs. The constituents of the fruit pulp and seeds are considered to be responsible for this property (Lal *et al.*, 2011; Kushawaha *et al.*, 2016; Mahmoodpoor *et al.*, 2018). A clinical trial conducted on diabetic critically ill patients who were admitted to the intensive care unit (ICU) demonstrated the efficacy of *C. maxima* pulp in decreasing blood glucose level. In that study, the powder of *C. maxima* (5 g) was administrated twice daily for 3 days. During the time period, the average glucose level and insulin dose were decreased significantly (Lal *et al.*, 2011). In an *in vitro* study, the aqueous extract of *C. maxima* seeds presented inhibitory action on carbohydrate-hydrolyzing enzymes like α-amylase and α-glucosidase (Kushawaha *et al.*, 2016). Furthermore, pumpkin seeds demonstrated hypoglycemic and hypolipidemic effects in a rat model (Farid *et al.*, 2015). In an *in vitro* study, the hypoglycemic activity of fruit juice and hydroalcoholic extract of *C. maxima* was evaluated. Oral dose of both fruit juice and extract (200 mg/kg) for 28 days could decrease hyperglycemia in streptozotocin-induced diabetic rats, significantly. The results showed that the extract was more potent than the control group (Lal *et al.*, 2011).

Several studies have investigated the active antidiabetic constituents of pumpkin. The phyto-elemental analysis of the extract of *C. maxima* by laser-induced breakdown spectroscopy (LIBS) showed the presence of magnesium, calcium, potassium, and sodium, which could be responsible for the α-glucosidase and α-amylase enzyme inhibitory activity (Kushawaha *et al.*, 2016). Studies show that other constituents of *Cucurbita* species including flavonoids, saponins, and tocopherol isomers might be responsible for the hypoglycemic properties. It has also been observed that oxidative stress plays a major role in the development of insulin resistance and diabetes (Akhigbe and Ajayi, 2021). Flavonoids, saponins, and tocopherols present in *Cucurbita* species could alleviate diabetes through their antioxidant activities and reducing oxidative stress (Mahmoodpoor *et al.*, 2018; Dotto and Chacha, 2020). Additionally, pumpkin polysaccharides such as pectins can reduce insulin and postprandial glucose levels (Ullah *et al.*, 2019). In a study, the antidiabetic effect of a polysaccharide (PP-PE) isolated from *C. pepo* aqueous extract was evaluated. Administration of PP-PE (100 mg/kg) for 7 days could decrease blood glucose level in alloxan-induced diabetic mice. This hypoglycemic property was related to the α-glucosidase and α-amylase inhibition activity of the compound (Thanh *et al.*, 2021). Moreover, fatty acids like oleic and linoleic acids could be involved in antidiabetic property through their modulatory effects on insulin secretion (Farid *et al.*, 2015). Another compound present in pumpkin that might be responsible for its antidiabetic activity is trigonelline. This alkaloid, occurring in many plant species such as *Trigonella foenum-graecum*, *Raphanus sativus* cv. Sakurajima Daikon, *Strophanthus* species, and *Dichapetalum cymosum*, has beneficial effect for diabetes through increasing insulin sensitivity index and insulin content, decreasing blood glucose and lipid levels, decreasing lipid peroxidation, and upregulating antioxidant enzyme activity (Adams *et al.*, 2014).

Cucumis sativus

In an animal model of diabetes, the methanol extract of *Cucumis sativus* (cucumber) could reduce blood glucose in streptozotocin-induced diabetic rats, significantly. The hypoglycemic effect was attributed to enhancing insulin expression (Atta *et al.*, 2020). Moreover, the ethanol extract of *C. sativus* could reduce blood glucose levels (67%) at 12 h after a single intraperitoneal injection in alloxan-induced diabetic rats (Sahu and Sahu, 2015). In addition, Heidari *et al.* (2016) have investigated the protective mechanisms of *C. sativus* in diabetes-related models of oxidative stress and carbonyl stress. They found that an aqueous extract of *C. sativus* fruit (40 μg/ml) could prevent all cytotoxicity markers in both the carbonyl stress and oxidative models, such as formation of reactive oxygen species (ROS), cell lysis, depletion of glutathione, membrane lipid peroxidation, lysosomal labialization, mitochondrial membrane potential decline, and proteolysis (Heidari *et al.*, 2016).

Plantago indica

Plantago indica (synonym: *Plantago psyllium*) (psyllium) mucilage is a type of dietary water-soluble fiber. Consuming dietary fibers is proven to reduce development of type 2 diabetes as well as attenuate fasting blood glucose levels and glycosylated hemoglobin percentage in type 2 diabetic patients (McRae, 2018). The hypolipidemic and hypoglycemic effect of *P. indica* as an adjunct to dietary therapy in type 2 diabetes was determined in a double-blind placebo-controlled study. The study including 125 patients showed a significant reduction in fasting plasma glucose and lipid levels (Rodríguez-Morán *et al.*, 1998). Another trial investigated the efficacy of acarbose (200 mg) and *P. psyllium* (15 g) together with the test meal in regulating the glycemic index in type 2 diabetic patients. It was observed that this combination could improve the glycemic index of carbohydrate foods (Naveen and Urooj, 2021). Because of the high amount of fiber, the glucose-lowering mechanism of psyllium is suggested to be related to an induced delay in gastric emptying, a delay in intestinal absorption of glucose, and disturbance in carbohydrate digestion and absorption (Hashem Dabaghian *et al.*, 2012; Ota and Ulrih, 2017).

Rosa × damascena

In research conducted by Safari and co-workers, a methanolic extract of *Rosa × damascena* (damask rose) showed high antioxidant and glycation inhibitory activities (Safari *et al.*, 2018). In a similar study, both methanolic and aqueous extracts presented high inhibitory activity against α-glucosidase by more than 75% (Gholamhoseinian *et al.*, 2008). Another study, the antihyperglycemic effect of *R. × damascena* aqueous extract was attributed to upregulating expression of peroxisome proliferator-activated receptor-γ (PPAR-γ) protein. The findings indicated that the aqueous extract could increase insulin sensitivity in insulin-resistant rats (Mohammadi *et al.*, 2017). In a triple-blind, placebo- and acarbose-controlled randomized trial in type 2 diabetic patients, *R. × damascena* methanolic extracts could decrease postprandial and fasting plasma glucose levels significantly. Besides, no adverse effect was observed in comparison with the acarbose or placebo group (Sanjari *et al.*, 2019).

Acacia arabica

Studies show that *Acacia arabica* (babul) extract has a stimulating effect for stimulating insulin release (Shamim *et al.*, 2017; Hamiduddin *et al.*, 2018). The oral consumption of powdered seeds of *A. arabica* showed a significant hypoglycemic effect in normal rabbits but not in alloxan-induced diabetic ones (Wadood *et al.*, 1989). Moreover, the chloroform extract of *A. arabica* bark exhibited a significant dose-dependent decrease in serum glucose and insulin resistance in streptozotocin-induced diabetic rats (Hegazy *et al.*, 2013).

Punica granatum

Several researches have been conducted to evaluate the efficacy of different parts and extracts of *Punica granatum* (pomegranate) in glucose metabolism (Akaberi *et al.*, 2021). The flower extract showed hypoglycemic effects in non-fasting diabetic rats, but not in healthy or fasting ones. Moreover, the extract could increase the sensitivity of insulin receptors and reduce serum glucose in type 2 diabetic patients by upregulating PPAR-γ expression and inhibiting α-glucosidase, respectively (Ge *et al.*, 2021). Various *in vivo* investigations and trials revealed that pomegranate peel extract could reduce hyperglycemia and increase insulin secretion (Doostkam *et al.*, 2020). On the other hand, fresh pomegranate juice appears to induce no changes in fasting glucose and insulin in type 2 diabetic subjects (Kim *et al.*, 2016). Similarly, a systematic review and meta-analysis assessing the effect of pomegranate supplementation on metabolic status in type 2 diabetic patients demonstrated no significant effect on metabolic parameters (Jandari *et al.*, 2020). Obviously, more well-designed investigations are needed to evaluate pomegranate efficacy in diabetes management.

Lactuca sativa

Administration of *Lactuca sativa* leaves (lettuce) (ground powder and methanol extract) could decrease hyperglycemia and enhance insulin sensitivity in mice models (Cheng *et al.*, 2014; Han *et al.*, 2018). The antidiabetic activity of lettuce may be attributed to its polyphenol contents, specifically chlorogenic acid and quercetin (Cheng *et al.*, 2014), as well as the carotenoid lactucaxanthin, which is an α-amylase and α-glucosidase inhibitor (Gopal *et al.*, 2017). Recently, a polyherbal formulation named *Qurṣ Ṭabasheer* traditionally used in TIM has been investigated for its pharmacologic efficiency against diabetes. The formulation consists of five medicinal plants. Among them are *R.* × *damascena* flower, *P. granatum* flower, and *L. sativa* seeds. Thus, it could be shown that administration of *Qurṣ Ṭabasheer* for a period of 28 days produced antihyperglycemic and antihyperlipidemic activity in streptozotocin-induced diabetic rats (Ahmed *et al.*, 2013).

Papaver somniferum

The poppy plant has a history of use as antidiabetic agent in Pakistan. However, there is a huge controversy about the efficacy of *Papaver somniferum* on metabolic disorders like diabetes (Hedayati-Moghadam *et al.*, 2022).

Topical formulations

Interestingly, Avicenna prescribed some medicinal plants topically for the management of diabetes. Tendrils and unripe fruit of *Vitis vinifera* (grapes), buds of *Cydonia oblonga* (quince), buds of *Malus domestica* (apple), buds of *Crataegus azarolus* (azarole), *Rheum ribes* (Syrian rhubarb), *Polygonum aviculare* (knotgrass), *Boswellia sacra* (frankinscense), *Tragopogon pratensis* (meadow goat's beard), *Cistus ladaniferus/Cistus criticus* (rockrose), *Tumor plantarum* (oak apple or oak gall), and juice of *Myrtus communis* (myrtle) are some of these plants. The assumed mechanism for these plants is through transdermal delivery of their active ingredients.

Conclusion

Our literature review through *Canon of Medicine* showed that Avicenna knew diabetes very well and he applied several strategies for its control. Besides other treatments, he prescribed herbs both orally and topically for this purpose. *Cucurbita pepo/Cucurbita maxima*, *Cucumis sativus*, *Plantago psyllium*, *Rosa* × *damascena*, *Acacia arabica*, *Punica granatum*, *Lactuca sativa*, and *Papaver somniferum* are among the most important medicinal plants used by Avicenna for the management of diabetes. The literature review in scientific databases revealed that there are plenty of *in vitro*, *in vivo*, and in some cases clinical trials confirming the antidiabetic activities of these plants. Although the beneficial effects of some of these plants are well studied, many more complementary studies are necessary for evaluating their efficacy for diabetes management.

References

Adams, G.G., Imran, S., Wang, S., Mohammad, A., Kok, M.S. *et al.* (2014) The hypoglycemic effect of pumpkin seeds, trigonelline (TRG), nicotinic acid (NA), and D-chiro-inositol (DCI) in controlling glycemic levels in diabetes mellitus. *Critical Reviews in Food Science and Nutrition* 54, 1322–1329.

Ahmed, D., Sharma, M., Mukerjee, A., Ramteke, P.W. and Kumar, V. (2013) Improved glycemic control, pancreas protective and hepatoprotective effect by traditional poly-herbal formulation "Qurs Tabasheer" in streptozotocin induced diabetic rats. *BMC Complementary and Alternative Medicine* 13, 10.

Akaberi, M., Boghrati, Z., Sahebkar, A. and Emami, S.A. (2021) Therapeutic potential of pomegranate in metabolic disorders. *Advances in Experimental Medicine and Biology* 1328, 421–440.

Akhigbe, R. and Ajayi, A. (2021) The impact of reactive oxygen species in the development of cardiometabolic disorders: a review. *Lipids in Health and Disease* 20(1), 23.

Atta, A., Saad, S., Atta, S., Mouneir, S., Nasr, S. *et al.* (2020) *Cucumis sativus* and *Cucurbita maxima* extract attenuate diabetes-induced hepatic and pancreatic injury in a rat model. *Journal of Physiology and Pharmacology* 71(4), 507–518.

Cheng, D.M., Pogrebnyak, N., Kuhn, P., Krueger, C.G., Johnson, W.D. *et al.* (2014) Development and phytochemical characterization of high polyphenol red lettuce with anti-diabetic properties. *PLoS One* 9, e91571.

Doostkam, A., Bassiri-Jahromi, S. and Iravani, K. (2020) *Punica granatum* with multiple effects in chronic diseases. *International Journal of Fruit Science* 20, 471–494.

Dotto, J.M. and Chacha, J.S. (2020) The potential of pumpkin seeds as a functional food ingredient: a review. *Scientific African* 10, e00575.

Farid, H.E., El-Sayed, S.M. and Abozid, M.M. (2015) Pumpkin and sunflower seeds attenuate hyperglycemia and protect liver in alloxan-induced diabetic rats. *Research Journal of Pharmaceutical Biological and Chemical Sciences* 6, 1269–1279.

Ge, S., Duo, L., Wang, J., Yang, J., Li, Z. *et al.* (2021) A unique understanding of traditional medicine of pomegranate, *Punica granatum* L. and its current research status. *Journal of Ethnopharmacology* 271, 113877.

Gholamhoseinian, A., Fallah, H., Sharifi-Far, F. and Mirtajaddini, M. (2008) The inhibitory effect of some Iranian plants extracts on the alpha glucosidase. *Iranian Journal of Basic Medical Sciences* 11(1), 1–9.

Gopal, S.S., Lakshmi, M.J., Sharavana, G., Sathaiah, G., Sreerama, Y.N. *et al.* (2017) Lactucaxanthin – a potential anti-diabetic carotenoid from lettuce (*Lactuca sativa*) inhibits α-amylase and α-glucosidase activity *in vitro* and in diabetic rats. *Food & Function* 8, 1124–1131.

Hamiduddin, M., Ali, W., Jahangeer, G. and Al, A. (2018) Unani formulations for management of diabetes: an overview. *International Journal of Green Pharmacy* 12(4), S769–S783.

Han, Y., Zhao, C., He, X., Sheng, Y., Ma, T. *et al.* (2018) Purple lettuce (*Lactuca sativa* L.) attenuates metabolic disorders in diet induced obesity. *Journal of Functional Foods* 45, 462–470.

Hashem Dabaghian, F., Kamalinejad, M., Shojaii, A., Abdollahi Fard, M. and Ghushegir, S.A. (2012) Review of antidiabetic plants in Iranian traditional medicine and their efficacy. *Journal of Medicinal Plants* 11(Suppl. 8), 1–11.

Hedayati-Moghadam, M., Moezi, S.A., Kazemi, T., Sami, A., Akram, M. *et al.* (2022) The effects of *Papaver somniferum* (opium poppy) on health, its controversies and consensus evidence. *Toxin Reviews* 41(3), 1030–1043.

Hegazy, G.A., Alnoury, A.M. and Gad, H.G. (2013) The role of *Acacia arabica* extract as an antidiabetic, antihyperlipidemic, and antioxidant in streptozotocin-induced diabetic rats. *Saudi Medical Journal* 34, 727–733.

Heidari, H., Kamalinejad, M., Noubarani, M., Rahmati, M., Jafarian, I. *et al.* (2016) Protective mechanisms of *Cucumis sativus* in diabetes-related models of oxidative stress and carbonyl stress. *BioImpacts* 6, 33–39.

Jandari, S., Hatami, E., Ziaei, R., Ghavami, A. and Yamchi, A.M. (2020) The effect of pomegranate (*Punica granatum*) supplementation on metabolic status in patients with type 2 diabetes: a systematic review and meta-analysis. *Complementary Therapies in Medicine* 52, 102478.

Kim, Y., Keogh, J.B. and Clifton, P.M. (2016) Polyphenols and glycemic control. *Nutrients* 8, 17.

Kushawaha, D., Yadav, M., Chatterji, S., Srivastava, A. and Watal, G. (2016) α-Amylase and α-glucosidase inhibitory activity assessment of *Cucurbita maxima* seeds – a LIBS based study. *International Journal of Phytomedicine* 8, 312–318.

Lal, V., Gupta, P., Pandey, A. and Tripathi, P. (2011) Effect of hydro-alcoholic extract of *Cucurbita maxima*, fruit juice and glibenclamide on blood glucose in diabetic rats. *American Journal of Pharmacology and Toxicology* 6, 84–87.

Mahmoodpoor, A., Medghalchi, M., Nazemiyeh, H., Asgharian, P., Shadvar, K. *et al.* (2018) Effect of *Cucurbita maxima* on control of blood glucose in diabetic critically ill patients. *Advanced Pharmaceutical Bulletin* 8, 347.

McRae, M.P. (2018) Dietary fiber intake and type 2 diabetes mellitus: an umbrella review of meta-analyses. *Journal of Chiropractic Medicine* 17, 44–53.

Mohammadi, A., Fallah, H. and Gholamhosseinian, A. (2017) Antihyperglycemic effect of *Rosa damascena* is mediated by PPARγ gene expression in animal model of insulin resistance. *Iranian Journal of Pharmaceutical Research* 16, 1080.

Naveen, Y. and Urooj, A. (2021) A review on medicinal plants evaluated for anti-diabetic potential in clinical trials: present status and future perspective. *Journal of Herbal Medicine* 28, 100436.

Ota, A. and Ulrih, N.P. (2017) An overview of herbal products and secondary metabolites used for management of type two diabetes. *Frontiers in Pharmacology* 8, 436.
Rodríguez-Morán, M., Guerrero-Romero, F. and Lazcano-Burciaga, G. (1998) Lipid- and glucose-lowering efficacy of *Plantago psyllium* in type II diabetes. *Journal of Diabetes and Its Complications* 12, 273–278.
Safari, M.R., Azizi, O., Heidary, S.S., Kheiripour, N. and Ravan, A.P. (2018) Antiglycation and antioxidant activity of four Iranian medical plant extracts. *Journal of Pharmacopuncture* 21, 82.
Sahu, T. and Sahu, J. (2015) *Cucumis sativus* (cucumber): a review on its pharmacological activity. *Journal of Applied Pharmaceutical Research* 3, 04–09.
Sanjari, M., Gholamhoseinian Najar, A. and Ghaseminejad Tafreshi, A. (2019) The safety and efficacy of *Rosa damascena* extract in patients with type II diabetes: preliminary report of a triple blind randomized acarbose controlled clinical trial. *Journal of Kerman University of Medical Sciences* 26, 22–35.
Shamim, A., Siddiquil, H.H., Mahmood, T., Bagga, P. and Kumar, R. (2017) A comprehenisive study on literature evidence, clinical studies and practices of herbal drugs for diabetic neuropathy and cardiomyopathy. *Asian Journal of Pharmaceutical and Clinical Research* 10, 30–37.
Thanh, T.T.T., Quach, T.T.M., Yuguchi, Y., Nguyen, N.T., Van Ngo, Q. *et al.* (2021) Molecular structure and anti-diabetic activity of a polysaccharide extracted from pumpkin *Cucurbita pepo*. *Journal of Molecular Structure* 1239, 130507.
Ullah, S., Khalil, A.A., Shaukat, F. and Song, Y. (2019) Sources, extraction and biomedical properties of polysaccharides. *Foods (Basel, Switzerland)* 8, 304.
Wadood, A., Wadood, N. and Shah, S. (1989) Effects of *Acacia arabica* and *Caralluma edulis* on blood glucose levels of normal and alloxan diabetic rabbits. *Journal of the Pakistan Medical Association* 39, 208–212.

8 An Evidence-Based Review of Medicinal Plants for Cough Management Cited by Avicenna

Motahare Nayebzadeh[1], Fatemeh Etemadpour[1], and Parmis Badr[2,3*]

[1]Department of Phytopharmaceuticals (Traditional Pharmacy), Shiraz University of Medical Sciences, Shiraz, Iran; [2]Phytopharmaceutical Technology and Traditional Medicine Incubator, School of Pharmacy, Shiraz University of Medical Sciences, Shiraz, Iran; [3]Pharmaceutical Sciences Research Center, Shiraz University of Medical Sciences, Shiraz, Iran

Abstract

Cough is a defensive response to clear the airways from excessive secretions or irritating foreign objects. Some patients suffer from excessive or troublesome coughs that need to be treated repeatedly. Cough may also be due to a respiratory infection such as pneumonia, flu, or tuberculosis. Acute bronchitis, asthma, lung cancer, and smoking are other reasons for coughs. In *Canon of Medicine*, cough (*soal*) is defined as a corporeal effort or movement to repulse an annoyance from the lungs or related organs. This action contains the two phases of thoracic expansion and thoracic contraction, leading to movement of the diaphragm. Sometimes, other organs like the esophagus, stomach, liver, or the whole body (e.g., in fevers) are involved. Avicenna has specifically focused on the causation for each disease. Like other diseases, cough can be categorized according to four temperaments. Each type has a distinct protocol for treatment; therefore, physicians have to consider this point for choosing the most practical prescriptions for each patient. The principles and traditional methods of treating this type of disease, especially the medicinal plants used, and the associated new research findings are discussed in this chapter.

Introduction

Cough is a defensive response to clear the airways from excessive secretions or irritant foreign objects. Some patients experience overactive or troublesome coughs that need repetitive treatments (McGovern *et al.*, 2018). Cough can also be due to a respiratory tract infection such as pneumonia, flu, or tuberculosis. Acute bronchitis, asthma, lung cancer, and smoking are other reasons for coughs (Chung and Pavord, 2008; Sultana *et al.*, 2016). In *Canon of Medicine*, cough (*soal*) is defined as a corporeal effort or movement to repulse an annoyance from the lungs or related organs. This action contains the two phases of thoracic expansion and thoracic contraction, leading to movement of the diaphragm. Sometimes, other organs like the esophagus, stomach, liver, or

*Email: badrp@sums.ac.ir

DOI: 10.1079/9781800621671.0008

the whole body (e.g., in fevers) are involved. Avicenna has specifically focused on the causation for each disease. Based on his viewpoint, three types of causes underlie the cough (Avicenna, 1993):

- *Bādieh* causes: Meaning external reasons, for example experiencing cold weather; drinking cold water; breathing smoky or dusty air; eating foods with sour, caustic, astringent, or acrid taste; and suffering tracheal obstruction while eating or drinking.
- *Wāseleh* causes: Meaning internal intermediate-free reasons, including warmth, coldness, wetness, and dryness according to traditional medicine contexts, or exceeding in one humor including bile, sanguine, phlegm, and black bile in the respiratory system.
- *Sabeghe* causes: Meaning internal reasons acting via intermediates, such as fullness (*emtel' a*).

Like other diseases, cough can be categorized according to four temperaments. Each type has a distinct protocol for treatment; therefore, physicians have to consider this point for choosing the most practical prescriptions for each patient:

- Cold (*bāred*) cough: Cold weather intensifies it, while warm weather makes cold cough lighter. The patients more or less complain about cold catarrh.
- Warm (*ḥār*) cough: It is accompanied by inflammation, thirst, redness of face, and large pulses.
- Wet (*raṭb*) cough: Old people with wet temperaments experience this kind of cough. They often suffer from croaky voice and snoring.
- Dry (*yābes*) cough: Movement and hunger worsen it. Motionlessness, bathing, and drinking liquids have positive effects on this cough type.

Medicinal Plants Cited in *Canon of Medicine* for Cough Management

In this chapter, antitussive medicinal plants recommended in *Canon of Medicine* are reviewed, and supportive relevant evidence for their efficacy is presented. Tables 8.1, 8.2, 8.3, and 8.4 present the collected results.

Acacia senegal (L.) Willd.

Gum acacia (gum arabic) or senegal gum is extracted from the stems and branches of *Acacia senegal*. This gum consists of glycoproteins and branched complex polysaccharides rich in galactose and arabinose. It is known to be a demulcent and is also used in medications for throat inflammation and sore throat (Kariduraganavar *et al.*, 2014; Lopez-Torrez *et al.*, 2015). It is reported to be used traditionally as an internal and external anti-inflammatory agent (Ali *et al.*, 2009). Gum arabic is widely used in the pharmaceutical and food industries as a stabilizer, emulsifier, food additive, etc. (Dave and Gor, 2018). Other uses of gum arabic include treatment of kidney diseases, asthma, oral cavity lesions, diarrhea, diabetes, and dry coughs in amoebic dysentery. Antipyretic, astringent, emollient, analgesic, antimicrobial, and liver tonic properties are also known. The antioxidant and protective activity against toxicities of the heart, kidney, and liver in rats was also claimed (Singh *et al.*, 2015). In an *in vitro* study, bark extracts of *A. senegal* in different solvents showed activities against selected Gram-negative and Gram-positive multidrug-resistant (MDR) organisms (Lawrence *et al.*, 2015).

Adiantum capillus-veneris L.

Adiantum capillus-veneris plant contains several compounds including triterpenoids, flavonoids, phenyl propanoids, carotenoids, mucilages, tannins, saponins, oleananes, glycolic acid, lactic acid, coumarins, phytosterols, and a bitter substance named capillarin. Filicane and adiantone plant substances from the triterpenoid group include isoadiantone, 3-methoxy-4-hydroxyfilicane, isoadiantone-B, and 3,4-dihydroxyfilicane in the hexane fraction, and capillirone and cotillion in the ethanolic extract of *A. capillus-veneris* fronds. Additionally, the most

Table 8.1. Antitussive medicinal plants cited in *Canon of Medicine*.

No.	Scientific name	Synonym	Arabic name	Persian name	Common name	Family
1	*Acacia senegal* (L.) Willd.	*Acacia senegal* var. *fistula* (Schweinf.) Oliv. *Acacia senegal* subsp. *modesta* (Wall.) Roberty	*Aqāqyā*	*Samgh-i 'raby*	Gum arabic	*Fabaceae*
2	*Adiantum capillus-veneris* L.	*Adianthum truncatum* Burm.f.	*Sh'r al-'arz*	*Pari syāvoshān*	Maidenhair	*Pteridaceae*
3	*Althaea officinalis* L.	*Althaea officinalis* subsp. *micrantha* Dostál *Althaea officinalis* var. *pseudoarmeniaca* Polg.	*Khaṭmy*	*Khatmy*	Marsh mallow	*Malvaceae*
4	*Commiphora gileadensis* (L.) C. Chr.	*Commiphora opobalsamum* L.	*Balasān*	*Balasān*	Opobalsam	*Burseraceae*
5	*Commiphora myrrha* (Nees) Engl.	*Commiphora myrrha* var. *molmol* Engl.	*Mur makky*	*Mur*	True myrrh	*Burseraceae*
6	*Cordia myxa* L.	*Cordia myxa* var. *domestica* (Roth) C.B. Clarke *Cordia myxa* var. *ixiocarpa* (F. Muell.) Domin	*Dabq*	*Sipistān*	Assyrian plum	*Boraginaceae*
7	*Cucurbita pepo* L.	*Cucurbita pepo* subsp. *ovifera* (L.) D.S. Decker *Cucurbita pepo* var. *texana* (Scheele) D.S. Decker	*Qarʿ*	*Kadū*	Pumpkin	*Cucurbitaceae*
8	*Cydonia oblonga* Mill.	*Cydonia oblonga* f. *lusitanica* (Mill.) Rehder *Cydonia oblonga* var. *maliformis* (Mill.) Rehder *Cydonia oblonga* f. *marmorata* (Dippel) C.K. Schneid. *Cydonia oblonga* f. *pyramidalis* (Dippel) Rehder *Cydonia oblonga* f. *pyriformis* (Dierb.) Rehder	*Safarjal*	*Bih*	Quince	*Rosaceae*
9	*Drimia maritima* (L.) Stearn	*Drimia maritima* subsp. *maura* (Maire) Förther & Podlech	*Isqyl*	*Pyāz-i dashty*	Squill	*Asparagaceae*
10	*Ferula gummosa* Boiss.	–	*Bārzad*	*Bārzad*	Galbanum	*Apiaceae*
11	*Ferula persica* Willd.	*Ferula puberula* Boiss. & Buhse, *Ferula szowitziana* var. *kandavanensis* Bornm. & Gouba	*Sakbīnaj*	*Sakbīnih*	Sagapenum	*Apiaceae*
12	*Glycyrrhiza glabra* L.	*Glycyrrhiza glabra* var. *glabra*, *Glycyrrhiza glabra* var. *glandulifera* (Waldst. & Kit.) Regel & Herder	*Sūs*	*Mihak*	Licorice	*Fabaceae*
13	*Hyssopus officinalis* L.	*Hyssopus officinalis* subsp. *aristatus* (Godr.) Nyman *Hyssopus officinalis* subsp. *canescens* (DC.) Nyman *Hyssopus officinalis* subsp. *montanus* (Jord. & Fourr.) Briq.	*Zūfā*	*Zūfā*	Hyssop	*Lamiaceae*

14	*Linum usitatissimum* L.	*Linum usitatissimum* var. *stenophyllum* (Boiss.) Rech.f.	*Katān*	*Katān*	Flax, linseed	*Linaceae*
15	*Liquidambar orientalis* Mill.	*Liquidambar imberbis* Aiton	*Mī'ih sā'ilah*	*Mī'ih*	Oriental sweetgum	*Altingiaceae*
16	*Malva sylvestris* L.	–	*Khubbāzy*	*Panirak*	Common mallow	*Malvaceae*
17	*Myrtus communis* L.	*Myrtus communis* Blanco	*ās*	*Mūrd*	Myrtle	*Myrtaceae*
18	*Nymphaea alba* L.	*Nymphaea alba* var. *alba*	*Nylūfar*	*Nylūfar*	White water rose, white nenuphar, European white water-lily	*Nymphaeaceae*
19	*Opopanax chironius* (L.) W.D.J. Koch	*Dorema chironia* (L.) M. Hiroe *Laserpitium chironium* L. *Maspeton chironium* (L.) Raf. *Opopanax chironium* (L.) W.D.J. Koch	*Jāwshīr*	*Jāwshīr*	–	*Apiaceae*
20	*Papaver somniferum* L.	*Papaver somniferum* subsp. *setigerum* (DC.) Arcang.	*Khashkhāsh al-aswad*	*Kvknār*	Poppy	*Papaveraceae*
21	*Pinus gerardiana* Wall. ex D. Don	*Pinus gerardii* J. Forbes	*Jalghūzih*	*Chalghūzih*	Chilgoza	*Pinaceae*
22	*Pistacia terebinthus* L.	*Pistacia terebinthina* St.-Lag. *Terebinthus communis* Dum. Cours. *Terebinthus vulgaris* Dum. Cours.	*Habbat al-khazrā'*	*Banih*	Terebinth	*Anacardiaceae*
23	*Pistacia vera* L.	–	*Fustuq*	*Pistah*	Pistachio	*Anacardiaceae*
24	*Plantago ovata Forssk.*	*Plantago ovata* var. *fastigiata* (E. Morris) S.C. Meyers & Liston	*Bazr qatūnā'*	*Isfarzih*	Psyllium	*Plantaginaceae*
25	*Prunus dulcis* (Mill.) D.A. Webb	–	*Al-lawz al-hulw*	*Bādām-i shīrīn*	Sweet almond	*Rosaceae*
26	*Viola odorata* L.	*Viola odorata* f. *odorata*	*Banafsaj*	*Banafshih*	Sweet violet	*Violaceae*

Table 8.2. *In vitro studies* on antitussive medicinal plants cited in *Canon of Medicine*.

Plant	Part	Result	Reference
Acacia senegal (L.) Willd.	Bark extract	Acetone extract was found to be effective against all organisms such as *Escherichia coli*, *Stretptococcus pyogenes*, *Bacillus cereus*, *Vibrio cholerae*, *Staphylococcus aureus*, *Bacillus subtilis*, *Pseudomonas aeruginosa*, *Shigella dysenteriae*, *Clostridiium perfringens*, *Salmonella typhi*, and *Listeria monocytogenes*, except *Campylobacter jejuni*, followed by methanol, chloroform, ethyl acetate, and hexane extracts, while petroleum ether extract was found least effective	Lawrence *et al.* (2015)
Adiantum capillus-veneris L.	Leaf	Methanol fraction showed maximum ZI against MDR bacterial strains	Hussain *et al.* (2014)
	Leaf	Ethanolic extract can normalize p38 and NF-κB activations in mice as well as increase spleen index by LPS induction	Yuan *et al.* (2013)
Althaea officinalis L.	Seed	The 200 mg/ml concentration of essential oil and methanol extract is more effective against respiratory pathogens	Gautam *et al.* (2015)
Commiphora gileadensis (L.) C. Chr.	Aerial parts	Chloroform and petroleum ether fractions inhibited COX-2 enzyme activity at IC_{50} of 10 and 5 mg/ml, respectively	Abbas *et al.* (2007)
	Aerial parts	Syringic acid exhibited weak antimalarial activity in comparison to chloroquin and artemisinin against the tested strains	Abbas *et al.* (2007)
	Aerial parts	Triterpenes and the flavonols mearnsetin and quercetin showed antioxidant activitities without exhibiting any cytotoxic effects on HL-60 cells	Abbas *et al.* (2007)
	Plant extract	Ethyl acetate fraction extract of dry plant material exhibited antibacterial activity against *S. aureus*, MRSA, and *P. aeruginosa* with IC_{50} of 75, 95, and 150 mg/ml, respectively	Abbas *et al.* (2007)
		Four compounds exhibited weak antimicrobial activity against *Candida glabrata* and *Mycobacterium intracellulare* with IC_{50} values of 10 and 15 mg/ml, respectively, in comparison to amphotericin B at 0.15 mg/ml for *C. glabrata* and ciprofloxacin at 0.3 mg/ml for *M. intracellulare*	
Commiphora myrrha (Nees) Engl.	Oil	Myrrh oil has antiviral activity against influenza A Puerto Rico 8/34/H1N1 virus	Madia *et al.* (2021)
		Myrrh oil at 25 and 50 μg/ml in mixture with vitamin E acetate showed a reduction in viral replication in a dose-dependent manner	
	Resin	Myrrhterpenes A and B have potent anti-inflammatory activity by inhibiting NO oxide production	Bao *et al.* (2021)
Cucurbita pepo L.	Leaves	Antibacterial activity against *Providencia stuartii*, *P. aeruginosa*, *Klebsiella pneumoniae*, *E. coli*, *Enterobacter aerogenes*, and *Enterobacter cloacae*	Ratnam *et al.* (2017)
	Oil	Activity against *Actinetobacter baumannii*, *Aeromonas veronii* biogroup *sobria*, *Candida albicans*, *Enterococcus faecalis*, *E. coli*, *K. pneumoniae*, *P. aeruginosa*, *Salmonella enterica* subsp. *enterica* serotype *typhimurium*, *Serratia marcescens*, and *S. aureus* at the concentration of 2.0% (v/v)	Ratnam *et al.* (2017)

	Seed	Ethanolic extract has the most antibacterial effect against *K. pneumoniae*, *E. coli*, and *E. aerogenes* in comparison with aqueous one	Alizadeh *et al.* (2013)
Cydonia oblonga Mill.	Seed	Methanolic extract of freeze-dried seeds is the most powerful antioxidant in this plant	Silva *et al.* (2004)
	Seed	Seed extract at 500 μg/ml concentration has antiproliferative activity against renal cancer cells	Al-Snafi (2016b)
Drimia maritima (L.) Stearn	Bulb	Activates caspase, also increases ROS, Bax/Bcl-2 ratio, and mRNA expression of *CHOP, TRIB3, GADD34, and ATF-4 as stress markers*	*Hamzeloo-Moghadam et al. (2018)*
Ferula gummosa Boiss.	Essential oil/ oleogum resin	Alcoholic essential oil and oleogum resin of *F. gummosa* had inhibitory effect on *Acinetobacter* strains	Afshar *et al.* (2016)
Ferula persica Willd.	Root	Umbelliprenin at concentration of 500 μg/ml has activity against *B. subtillis*, *B. cereus*, *E. coli*, *K. pneumoniae*, *S. typhi*, *S. aureus*, and *Staphylococcus epidermidis*	*Sattar and Iranshahi* (2017)
	Root	Chloroform extract containing compounds persicasulfide A and persicasulfide B showed potent antifungal activity	*Sattar and Iranshahi* (2017)
	Root	Umbelliprenin has inhibitory effect on the growth of *Leishmania major*	*Sattar and Iranshahi* (2017)
Glycyrrhiza glabra L.	Root	Bronchodilatory effect by ↑ cGMP and ↓ PDE5 activation	Liu *et al.* (2008)
Hyssopus officinalis L.	Aerial part	5–10 μl of the essential oil can inhibit *E. coli*, *S. aureus*, and *S. pyogenes*	Süleyman *et al.* (2010)
Malva sylvestris L.	Stem/seed	Antimicrobial activity against *Aggregatibacter actinomycetemcomitans*	Vahabi *et al.* (2019)
Myrtus communis L.	Fruit	↓ Expression of IL-1β, TNF-α, eNOS, and IL-6	Kutlu *et al.* (2021)
	Fruit	Antioxidant activity through scavenging free radicals and inhibiting lipid peroxidation	Serce *et al.* (2010)
Nymphaea alba L.	Flower	The extract can be significantly effective in the treatment of HCV with or without another anti-HCV drug	Koushik *et al.* (2015)
Pistacia vera L.	Nut	Pretreatment with natural raw shelled pistachios (0.01, 0.1 and 0.5 mg/ml) and roasted salted pistachio kernels exhibited a considerable protection against LPS-induced inflammation	Paterniti *et al.* (2017)
Plantago ovata Forssk.	seed	HCV drug extract has anti-inflammatory and bacterial activity against *Porphyromonas gingivalis*, *A. actinomycetemcomitans*, *Fusobacterium nucleatum, and Prevotella intermedia* pathogens	Reddy *et al.* (2018)

IC_{50}, half maximal inhibitory concentration.

Table 8.3. *In vivo* studies on antitussive medicinal plants cited in *Canon of Medicine*.

Plant	Part	Model	Species	Result	Reference
Adiantum capillus-veneris L.	Leaf	Acute asthma by histamine aerosol	Guinea pig	Anti-asthmatic effect of the ethanolic extract is more effective than that of the other plant	Kumar *et al.* (2012)
	Leaf	LPS-induced PGE_2 production	RAW264.7 mouse	Ethanolic extract can normalize p38 and NF-κB activations in mice as well as increase spleen index by LPS induction	Yuan *et al.* (2013)
	Leaf	Hypoxia condition	Rat	Ethanol extraction can modulate Bax/Bcl-2 apoptotic index ratio and remodeling of pulmonary epithelial cells	Yadegari *et al.* (2019)
	Leaf	Paw edema induced by formalin	Mouse and rat	Anti-inflammatory activity of ethyl acetate fraction by ↓ NO and TNF-α levels	Haider *et al.* (2011)
	Leaf	Colitis induction by rectal injection of acetic acid 3% w/w	Wistar rat	Aqueous and hydroalcoholic extracts have ulcer healing, anti-inflammatory, and antioxidant activities by ↓ IL-6, NO, TNF-α release, and NF-κB activation	Khoramian *et al.* (2020)
Cordia myxa L.	Fruit	Tracheal muscle	Sheep	Tracheal relaxation activity of alcoholic extract through ↑ NO synthesis by cytosolic Ca^{2+}	Al-Bayati (2008)
	Fruit	Acetic acid and formalin test	Mouse	Alcoholic extract has analgesic and anti-inflammatory activity in the formalin test	Ranjbar *et al.* (2013)
	Fruit	Colitis induced by intrarectal administration of acetic acid	Rat	Anti-inflammatory effect by modulating Zn, Mn, Cu, and Se levels	Al-Awadi *et al.* (2001)
	Fruit	DTH and MI test	BALB/c mice	Aqueous extract has an immunomodulatory effect on immune cells	Rashid Ali *et al.* (2015)
Cydonia oblonga Mill.	Seed	Tracheal tissues	Rabbit	Bronchodilatory effect by Ca^{2+} antagonism	Janbaz *et al.* (2013)
Glycyrrhiza glabra L.	Root	Tracheal rings precontracted with acetylcholine, KCl, and histamine	Guinea pig	↑ Latency time of collapse and ↓ lung overflow in TSMCs	Liu *et al.* (2008)
	Root	DTH reaction	CBA/CaLac mouse	↑ Cellular immune response	Borsuk *et al.* (2011)
	Root	Citric acid-induced cough	Guinea pig	↓ Citric acid as a major cause of cough	Saha *et al.* (2011)
	Root	Inflammation and airway hyperresponsiveness induced by intraperitoneal ovalbumin and $Al(OH)_3$ administration	BALB/c mouse	↓ IgE level, ↓ leukoctye count, ↓ total protein	Dogan *et al.* (2020)

	Root	Cough induced by SO_2 gas	Albino mouse	↓ Up to 35.62% of coughs in 60 min	Jahan and Siddiqui (2012)
	Root	Acute lung injury induced by LPS	BALB/c mouse	↓ TLR-4/NF-κB pathway activity and protects lung tissue	Lee *et al*. (2019)
Hyssopus officinalis L.	Aerial part	Asthma model by ovalbumin	BALB/c mouse	↓ Expression of MMP-9 and TIMP-1 by collagen deposition, the proliferation of smooth muscle, and mucus secretion	Ma *et al*. (2014a)
	Aerial part	Asthma model by ovalbumin	BALB/c mouse	↓ IgG level and ↑ ratio of EOS and modulates the immune system	Ma *et al*. (2014b)
Myrtus communis L.	Fruit	Hepatic ischemia-induced	Rat	↓ AST, MDA, and ALT markers, additionally ↑ MEGX concentration	Ferchichi (2012)
Nymphaea alba L.	Flower	Renal oxidative trauma by Fe-NTA	Wistar rat	↓ γ-Glutamyl transpeptidase, xanthine oxidase, lipid peroxidation, and H_2O_2 generation	Koushik *et al*. (2015)
Pistacia vera L.	Nut	Treatment with natural raw shelled pistachios	Rat	↓ CAR-induced histological paw damage, neutrophil infiltration, and nitrotyrosine formation	Paterniti *et al*. (2017)
Plantago ovata Forssk.	Seed	Intestinal inflammatory	HLA-B27 rat	↓ Leukotriene B4, NO, and TNF-α	Sarfraz *et al*. (2017)
Viola odorata L.	Flower	Lung damage by formalin	NMR rat	*V. odorata* can prevent lung injury equal to the efficacy of hydrocortisone	Koochek *et al*. (2002)
	Flower	Hepatotoxicity by ethanol	Wistar rat	Both leaf and flower extracts have significant antioxidant activity	Habibi *et al*. (2019)

CAR, carrageenan; DTH, delayed type hypersensitivity; MI, mitotic index; PGE_2, prostaglandin E_2.

Table 8.4. Clinical studies on antitussive medicinal plants cited in *Canon of Medicine*.

Study design	No. of patients	Disease	Plant(s)	Part(s)	Control group	Treatment duration	Results	Reference
Double-blind randomized, placebo-controlled clinical trial	46	Asthma	*Adiantum capillus-veneris* *Matricaria chamomilla* *Hyssopus officinalis* *Glycyrrhiza glabra* *Althaea officinalis* *Malva sylvestris* *Ziziphus jujuba*	Flower Aerial part Aerial part Root Flower Flower Fruit	Placebo	5 days	↓ Nighttime awakening and severity of cough	Javid *et al.* (2019)
Triple-blind randomized, placebo-controlled clinical trial	60	Moderate to severe persistent asthma	*Drimia maritima* (L.) Stearn	Bulb	Placebo	6 weeks	The syrup significantly improved FEV_1	Nejatbakhsh *et al.* (2017)
Randomized, single-blind, controlled, crossover feeding study	25	CHD risk factors such as inflammation and hemostasis	*Prunus dulcis* (Mill.) D.A. Webb	Nut	Cholesterol-lowering control diet without nuts	6 weeks	High-almond diet significantly decreased serum E-selectin and almond diets can reduce CRP In control and high-almond diets, tPA antigen was significantly lower	Rajaram *et al.* (2010)
Double-blind, randomized controlled trial	128	Intermittent asthma	*Viola odorata* L.	Flower	Placebo	5 days	The syrup enhanced cough suppression with short-acting β-agonist	Qasemzadeh *et al.* (2015)
Randomized single-blind, placebo-controlled trial	400	Recurrent tonsillitis and peritonsillar abscess	*Viola odorata* L.	Flower	Co-amoxiclav (625 mg) and fexofenadine (60 mg) twice daily	7 days	Decoction of *V. odorata* flower (1 mg/150 ml water) can decrease the frequency of these abscesses in one year	Ali *et al.* (2013)

CHD, coronary heart disease; tPA, tissue plasminogen activator.

important flavonoid compounds in *A. capillus-veneris* are quercetin, quercetin-3-*O*-rutinoside, and quercetin-3-*O*-glycoside in the ethyl acetate fraction (Singh *et al.*, 2008; Ibraheim *et al.*, 2011; Gaikwad-Samdani, 2012; Ishaq *et al.*, 2014; Al-Snafi, 2015; Dehdari and Hajimehdipoor, 2018; Khan *et al.*, 2018; Nazim *et al.*, 2018; Zhang *et al.*, 2019). In a study, the antibacterial effect of the methanol fraction of *A. capillus-veneris* fronds against MDR bacterial strains was investigated, and the maximum zone of inhibition (ZI) against *Klebsiella pneumoniae*, *Providencia*, *Shigella*, *Staphylococcus aureus*, *Vibrio cholerae*, *Salmonella typhi*, and *Proteus vulgaris* was determined by the disk diffusion method (Hussain *et al.*, 2014). In another study, the anti-inflammatory effect of ethanolic extract of *A. capillus-veneris in vitro* using HEK293 and HepG2 cell lines and *in vivo* was investigated experimentally on 32 mice. The results showed that this extract could normalize p38 and nuclear factor-κB (NF-κB) activation in mice as well as the increase in spleen index by lipopolysaccharide (LPS) induction (Yuan *et al.*, 2013). In an *in vivo* study that aimed to compare the anti-asthmatic effect of the ethanolic extract of *A. capillus-veneris* fronds with that of *Sargassum wightii* on 24 guinea pigs with acute asthma induced by histamine aerosol, it was found that *A. capillus-veneris* was more effective than the other plant (Kumar *et al.*, 2012). Twenty-seven healthy rats were examined in hypoxia condition to investigate the Bcl2-associated-X protein (Bax)/B-cell lymphoma 2 (Bcl-2) apoptotic index ratio and remodeling of pulmonary epithelial cells in the control group compared to the group supplemented with a *A. capillus-veneris* ethanol extract obtained using the Soxhlet method. The result showed this extract could significantly modulate the apoptotic ratio (Yadegari *et al.*, 2019). In another *in vivo* study, the anti-inflammatory activity of an alcoholic extraction of *A. capillus-veneris* was evaluated by inducing paw edema with formalin in 66 mice and nine rats. The results showed that this fraction could significantly inhibit the release of nitric oxide (NO) and tumor necrosis factor-α (TNF-α) (Haider *et al.*, 2011). Another study examined the efficacy of aqueous and hydroalcoholic extracts of *A. capillus-veneris* in the prevention of inflammation and incidence of colitis. During this examination, 60 Wistar rats were induced with colitis by rectal injection of acetic acid 3% w/w and the efficacy of two extracts of *A. capillus-veneris* was compared to that of prednisolone and mesalazine. Results showed that these extracts have beneficial, dose-dependent, ulcer healing, anti-inflammatory, and antioxidant activities by suppressing interleukin-6 (IL-6), NO, and TNF-α release as well as inhibiting NF-κB activation (Khoramian *et al.*, 2020).

Althaea officinalis L.

The secondary metabolites of *Althaea officinalis* include alkaloids, glycosides, flavonoids (like isoscutellarein, kaempferol, hypolaetin, and luteolin), coumarines (like scopoletin), phenolic acids, terpenes, tannins, and saponins (Table 8.5). Mucilage of *A. officinalis* contains rhamnogalacturonan, glucans, arabinans, and arabinoglucans (Gautam *et al.*, 2015; Mahboubi, 2020).

In an *in vitro* study, the antibacterial activity of the essential oil and the petroleum ether, acetone, methanol, and aqueous extracts of *A. officinalis* seeds was evaluated against various respiratory pathogens such as *Haemophilus influenzae, Pseudomonas aeruginosa, S. aureus, Streptococcus pneumoniae*, and *Streptococcus pyogenes* by the agar diffusion method. The results showed that both the essential oil and the seed extracts possess antimicrobial activity. In particular, the 200 mg/ml concentration of the essential oil was more effective against *S. pyogenes* and *H. influenza,* and the methanol extract showed high antimicrobial activity against *P. aeruginosa* (Gautam *et al.*, 2015).

Table 8.5. Major phytochemicals of *Althaea officinalis* seeds. (From Gautam *et al.*, 2015.)

Extract	Compounds
Methanol	Alkaloids, flavonoids, glycosides, terpenes, saponin, tannins
Petroleum ether	Alkaloids, steroids
Acetone	Alkaloids, terpenes, saponins, tannins
Aqueous	Alkaloids, flavonoids, glycosides, terpenes, saponins, tannins

Cordia myxa L.

Major compounds of the extract obtained from Assyrian plum fruit include the essential oil, flavonoids, glycosides, sterols, carbohydrates, terpenoids, saponins, alkaloids, coumarins, phenolic acids, mucilages, resins, tannins, and gums (Al-Snafi, 2016a; Shwaish and Imarah, 2017). *Cordia myxa* fruit extract contains high amounts of phenolic compounds, and the antitumor and antioxidant activities of this plant are related to the individual concentrations of these substances including gallic acid, catechin, caffeic acid, *p*-coumaric acid, rutin, quercetin, and kaempferol (El-Massry *et al.*, 2021). A study evaluated the mechanism of tracheal relaxation activity of alcoholic extracts obtained from *C. myxa* fruits in sheep tracheal muscle and showed this extract can induce the expression of NO synthesis, which is related to the cytosolic calcium ions (Ca^{2+}) originating from extracellular space (Al-Bayati, 2008).

The anti-inflammatory activity of the aqueous alcoholic extract of *C. myxa* fruit was evaluated by acetic acid and formalin assay in 72 mice. Tramadol and indomethacin administered intraperitoneally and orally, respectively, were compared with this fruit extract. The study showed that hydroalcoholic extracts of this fruit possess significant anti-inflammatory activity in both chronic and acute phases (Ranjbar *et al.*, 2013). In another *in vivo* study, colitis was induced in 48 rats by intrarectal administration of acetic acid to show the anti-inflammatory effect of the suspension of the section surrounding the seeds of *C. myxa* fruit, which have the largest amount of mucilage in the whole fruit. The results showed that this effect of *C. myxa* is related to the fixation of zinc (Zn), manganese (Mn), copper (Cu), and selenium (Se) in the destroyed tissue (Al-Awadi *et al.*, 2001).

The immunomodulatory activity of *C. myxa* fruit was evaluated in 40 BALB/c (albino, laboratory-bred) mice through comparison between immunization by hydatid cyst fluid antigens (HCFAg) and treatment by aqueous extract of the fruit in four mice groups, which revealed that this extract could stimulate the immune responses in mice and cell-mediated cases (Rashid Ali *et al.*, 2015). In another study, the anticancer and anti-inflammatory activity of hydroalcoholic extracts of *C. myxa* fruit was investigated through inhibition of cyclooxygenase-2 (COX-2) and 5-lipoxygenase (5-LOX) enzymes with the molecular docking of α-amyrins as a triterpenoid of *C. myxa* and both enzymes. These enzymes are related to occurrence of cancer, so this extract can inhibit cancer pathways in the lung, breast, and colon neoplasm (Ranjbar *et al.*, 2016).

Commiphora myrrha (Nees) Engl.

Commiphora myrrha or true myrrh from the family *Burseraceae* has long been used as a medicinal plant. It is a gum-resin that is extracted from the bark of the trees of *Commiphora* species (Shen *et al.*, 2012). Ethanol and ether extracts of *C. myrrh* have been traditionally used in Greece to treat cough, wounds, sepsis, and mouth infections. In Britain, the resin has traditionally been used to treat cough, gingivitis, tonsillitis, pharyngitis, sinusitis, ulcers, and skin inflammation (Shen *et al.*, 2012). As mentioned in Chinese culture, the extract of *C. myrrha* showed activity against cancer diseases such as lung cancer. Due to its anti-inflammatory, antioxidant, antibacterial, and expectorant properties, *C. myrrha* it is also used to cure chest infections and sore throat. Furthermore, furanosesquiterpenoids occurring in the volatile essential oil have been shown to have antiseptic and antifungal activities (Alyafei, 2020). This genus is known to have anti-inflammatory, antibacterial, antiseptic, and local anesthetic properties (Latha *et al.*, 2021). Myrrhterpenes A and B are two cadinane sesquiterpenes that are found in the resin of this species and showed potent anti-inflammatory effects by inhibiting NO production (Bao *et al.*, 2021). A mixture of *C. myrrha* oil with vitamin E acetate showed antiviral activity in one study (Madia *et al.*, 2021).

Commiphora gileadensis (L.) C. Chr.

Commiphora gileadensis is from the family *Burseraceae*. Its extract has been used traditionally in Arabic medicine to treat headache. The total extract has shown antiulcerogenic property. The aerial parts were reported to contain triterpenes such as oleanonic acid, friedelin, and canophyllal. The ethyl acetate fraction of dry plant material extraction showed antibacterial activity against *P. aeruginosa*, *S. aureus*, and methicillin-resistant *Staphylococcus aureus* (MRSA). Quercetin and mearnsetin are the flavonols found in this plant that show antioxidant activity. Syringic acid is another compound that is related to antimycobacterial and antimalarial properties (Abbas *et al.*, 2007). The essential oil has also antibacterial and antioxidant activity (Al-Massarany *et al.*, 2008).

Cydonia oblonga Mill.

Functional compounds of *Cydonia oblonga* include steroids, phenolics, terpenoids, flavonoids, tannins, organic acids, sugars, and glycosides (Table 8.6). *C. oblonga* seeds contain sterols, tannins, and triterpenes, with a high amount of phenolic compounds such as apigenin, D-(–)-quinic acid derivatives, 6-*C*-pentosyl-8-*C*-glucosyl chrysoeriol, and stellarin, while the major phenolic component of *C. oblonga* fruit is 5-*O*-caffeoylquinic acid (Silva *et al.*, 2005).

In another study, the antimicrobial activity of fruit and seed extracts of *C. oblonga* was evaluated by diffusion agar and microdilution methods. The study showed that the ethanolic extract of *C. oblonga* seeds had the most effect against *K. pneumoniae*, *Escherichia coli*, and *Enterobacter aerogenes* in comparison with aqueous ones (Alizadeh *et al.*, 2013). In a study, the antiradical and antioxidant actions of *C. oblonga* peel, pulp, jam, and seed were evaluated through the 2,2-diphenylpicrylhydrazyl (DPPH) method, showing that the methanolic extract of freeze-dried seeds is the most powerful antioxidant in this plant (Silva *et al.*, 2004). A study evaluated the antispasmodic activity of *C. oblonga* in the tracheal tissues of rabbits. Results showed that this action was related to the antagonist Ca^{2+} activity, which relaxes the trachea in bronchitis and asthma patients; however, this extract can cause mild spasm because of stimulation of the muscarine receptors (Janbaz *et al.*, 2013). An *in vitro* study evaluated the anti-inflammatory effect of the *C. oblonga* seed extract on the human cell line THP-1 that was polluted by LPS. Results showed that the extract of the seeds had an antiproliferative activity against renal cancer cells in the 500 μg/ml concentration (Al-Snafi, 2016b).

Table 8.6. Major phytochemicals of *Cydonia oblonga* seeds. (From Ashraf *et al.*, 2016.)

Compound	Components
Mucilages	L-Arabinose, aldobionic acids, D-xylose, 4-*O*-methyl glucose, 2,3,4-tri-*O*-methyl-D-xylose, D-glucose, 2,3-di-*O*-methyl-D-xylose, 3-*O*-methyl-D-xylose
Phenolics	3-, 4- and 5-*O*-Caffeoyl quinic acids, apigenin derivatives, 3,5-dicaffeoyl quinic acid, leucenin-2, 6-*C*-pentosyl-8-*C*-glucosyl chrysoeriol, stellarin-2, 6-*C*-glucosyl-8-*C*-pentosyl chrysoeriol
Flavones	Isoschaftoside, 5-*O*-caffeoylquinic acid, caffeoylquinic acids
Amino acids	L-Aspartic acid, L-asparagine, L-glutamic acid, L-glycine, L-isoleucine, L-valine, L-proline, L-alanine, L-leucine
Fatty acids	Palmitic acid, stearic acid, linoleic acid, eicosanoic acid, oleic acid

Cucurbita pepo L.

Cucurbita pepo (pumpkin) belongs to the *Cucurbitaceae* family. This plant is a good source of protein, carbohydrate, minerals, and vitamins. It has traditionally been used as an antioxidant, anti-inflammatory, and antiviral. In Chinese Medicine, the seeds are said to be good for the lung (Gutierrez, 2016). Regarding Iranian traditional use, Avicenna in the *Canon of Medicine* recommends a mixture of bread-paste and pumpkin for cough. It is mentioned that pumpkin has a role in preventing lung cancer and controlling its symptoms. Containing high levels of vitamin A, it is capable of preventing pulmonary emphysema in cigarette smokers (Batool and Dadkhah Tehrani, 2014). In another study, compounds cucurbitacin B and cucurbitacin E had antiproliferative property in lung cancer cells (A549) and lung cancer cell lines (NCI-H460) (Gutierrez, 2016). Seeds and seed oil of *C. pepo* are a rich source of polyunsaturated fatty acids, mainly palmitic and palmitoleic acid, proteins, phytosterols, tocopherols, carotenoids, and antioxidant vitamins. The phenolic compounds, β- and δ-tocopherol, and polysaccharides are related to the antioxidant activity. The extract of leaves and also seed oils have exhibited antibacterial activity (Ratnam *et al.*, 2017).

Ferula gummosa **Boiss.**

Ferula gummosa is a plant that has been used as an anti-inflammatory, antiseptic, anti-spasm, and painkiller. The oleoresin of *F. gummosa* is also used in asthma, earache, and as an expectorant in chronic bronchitis. New studies showed different properties such as spasmolytic, antinociceptive, anti-inflammatory, and antimicrobial properties. The oleoresin contains terpenes and terpenoids, arabinogalactan, and some inorganic substances (Hajifattahi *et al.*, 2016). The essential oil yield was about 11% in a study and the major compounds of that determined by one-dimensional gas chromatography–mass spectrometry (1D GC–MS) were β- and α-pinene and myrcene (Jalali *et al.*, 2012). Antibacterial activity of *F. gummosa* was examined against some *Acinetobacter* strains. The experiment showed that both the essential oil and oleogum resin had considerable activity and the alcoholic essential oil had more effect on the strains of *Acinetobacter* (Afshar *et al.*, 2016).

Ferula persica **Willd.**

Ferula persica is indigenous to Iran and belongs to the family *Apiaceae*. It has been traditionally used as antispasmodic and expectorant (Razavi and Janani, 2015) and in sinusitis (Ahvazi *et al.*, 2012). Chemical constituents of *Ferula* species are sesquiterpenes, sesquiterpene coumarins, sesquiterpene coumarin glycosides, and sulfur-containing compounds. Compounds auraptene and umbelliprenin found in *Ferula* species have both anti-inflammatory and cancer chemopreventive properties. Chloroform extract of *F. persica*'s root showed antifungal activity related to the compounds persicasulfide A and persicasulfide B. In addition, the coumarine umbelliprenin at a concentration of 500 μg/ml is responsible for antibacterial activity against *Bacillus subtilis*, *Bacillus cereus*, *E. coli*, *K. pneumoniae*, *S. typhi*, *S. aureus*, and *Staphylococcus epidermidis*. It is also capable of inhibiting the growth of *Leishmania major (Sattar and Iranshahi*, 2017*)*. Another study showed umbelliprenin to be anti-inflammatory and a lipoxygenase inhibitor, so it can be used for treating asthma and chronic obstructive pulmonary disease (COPD) (Iranshahi *et al.*, 2009).

Drimia maritima **(L.) Stearn**

Drimia maritima bulb contains several phytochemicals including phenolic compounds (like pelargonidin 3-monoglucoside and cyanidin 3-monoglucoside), phytosterols (like β- and γ-sitosterol), cardiac glycosides, and alkaloids, which significantly have several pharmacological effects (Bozorgi *et al.*, 2017). An *in vitro* study investigated the cytotoxic and anticancer activity of the methanolic extract of *D. maritima* bulb in MDA-MB-468 and MCF-7 cells. The results showed that this extract could significantly activate caspase and also increase ROS, Bax/Bcl-2 ratio, and mRNA expression of *CHOP* (C/EBP homologous protein), *TRIB3* (tribbles-related protein 3), *GADD34* (growth arrest and DNA damage-inducible 34), and *ATF-4* (activating transcription factor 4) as stress markers *(Hamzeloo-Moghadam et al.*, 2018*)*. A randomized clinical trial evaluated the anti-asthmatic activity of *D. maritima* syrup. For this purpose, 10 ml of the syrup was administrated to the 60 patients taking anti-asthmatic medicine for 6 weeks, and then symptoms and spirometry parameters were evaluated. The results showed that this syrup could significantly improve forced expiratory volume in the first second (FEV_1) (Nejatbakhsh *et al.*, 2017).

Glycyrrhiza glabra L.

The main compounds of *Glycyrrhiza glabra* root are triterpene saponins like glycyrrhizin, flavonoids like glabridin and isoliquiritigenin, isoflavonoids, chalcones, mucilage, coumarin-GU-12, phytosterols, and arabinogalactan protein (Table 8.7) (Sharma *et al*., 2017).

A study investigated the effect of isoliquiritigenin, as a flavonoid of *G. glabra* root, on tracheal smooth muscle *in vivo* and *in vitro*. To learn the mechanism of tracheal relaxation, the researchers experimented with isoliquiritigenin in guinea-pig tracheal smooth muscle cells (TSMCs) pretreated by acetylcholine, potassium chloride (KCl), and histamine. It was found that this compound could relax the trachea by increasing cyclic guanosine monophosphate (cGMP) and inhibiting cGMP-specific phosphodiesterase type 5 (PDE5); also, isoliquiritigenin significantly increased the latency time of collapse and inhibited the lung overflow induced by histamine and acetylcholine (Liu *et al*., 2008). In an *in vivo* study, the activity of *G. glabra* extract on the cellular and humoral immune activation and nonspecific resistance was evaluated in 90 CBA/CaLac mice. The results showed that *G. glabra* root extract enhanced the cellular immune response, and this efficacy was higher than for *Echinacea purpurea* (Borsuk *et al*., 2011). In another study, the antitussive effect of an aqueous extract of *G. glabra* was evaluated in 24 guinea pigs in comparison with codeine, showing that the polymeric fraction of *G. glabra* decreased the amount of citric acid in the animals, which is the major cause of cough (Saha *et al*., 2011).

Table 8.7. Major phytochemicals of *Glycyrrhiza glabra* roots. (From Liu *et al*., 2008; Ng *et al*., 2013.)

Compound	Components
Flavonoids	Glabridin, isoliquiritigenin, liquiritin, rhamnoliquirilin, liquiritigenin, prenyllicoflavone A, glucoliquiritin apioside, 1-methoxyphaseolin, shinpterocarpin, shinflavanone, licopyranocoumarin, glisoflavone, licoarylcoumarin
Phenolics	Isoangustone A, semilicoisoflavone B, licoriphenone, 1-methoxyficifolinol
Volatile components	Pentanol, tetramethyl pyrazine, hexanol, terpinen-4-ol, linalool oxide A, linalool oxide B, geraniol, α-terpineol

The anti-inflammatory effect of glabridin was evaluated in an *in vivo* study through inducing inflammation and airway hyperresponsiveness in BALB/c mice by intraperitoneal ovalbumin and aluminum hydroxide ($Al(OH)_3$) administration. The study showed that glabridin at a dose of 20–30 mg/kg significantly decreased immunoglobulin (Ig)E level, leukoctye count, and total protein, so it can be effective in asthma treatment (Dogan *et al*., 2020). The antitussive activity of the ethanol extract of G. *glabra* roots and rhizomes was investigated in 36 albino mice with cough induced by sulfur dioxide (SO_2) gas. The results showed that this extract, in comparison with codeine sulfate, could significantly inhibit up to 35.62% of coughs in 60 min (Jahan and Siddiqui, 2012). The anti-inflammatory effect of glycyrrhizin, as a saponin compound in *G. glabra*, was evaluated in 36 BALB/c mice with acute lung injury induced by LPS to assay its mechanism; it was shown that glycyrrhizin inhibited the toll-like receptor-4 (TLR-4)/NF-κB pathway and protected the lung tissue (Lee *et al*., 2019).

Hyssopus officinalis L.

Hyssopus officinalis contains several compounds such as flavonoids (like quercetin and apigenin-7-glucoside), polyphenols (like gallic acid), volatile oil, tannins, terpenoids, and murrain. The important compounds of hyssop essential oil are pinocamphone, pinocarvone, β-pinene, terpinene-4-ol, isopinocamphone, *p*-cymene, myrtenal, and carvacrole. The phenolic compounds of the alcoholic extract of hyssop comprise caffeic acid, kaempferol, apigenin, cinnamic acid, naringenin, and cateachin (Süleyman *et al*., 2010; Karegar-Borzi *et al*., 2016; Abdel-Megeed *et al*., 2020). An *in vitro* study examined the antioxidant and antimicrobial effects of hyssop essential oil. To evaluate the antimicrobial activity of this oil, the disk diffusion test was done for *E. coli*, *P. aeruginosa*, *S. aureus*, and *Staphylococcus pyogenes*, showing that 5–10 μl of the oil can inhibit the *E. coli*, *S. aureus*, and *S. pyogenes*. For assaying the antioxidant activity of this oil, in comparison with acetic acid and butylated hydroxytoluene (BHT), the DPPH method was used, showing that hyssop oil has lower antioxidant action in comparison with BHT and acetic acid (Süleyman *et al.*, 2010). In an *in vivo* study, the anti-remodeling and anti-inflammatory effects of the aqueous extract of hyssop were evaluated. Asthma was induced in 32 BALB/c mice by $Al(OH)_3$ and ovalbumin; then, the efficacy of this extract in comparison with dexamethasone was evaluated. The results showed that hyssop extract decreased the expression of matrix metalloproteinase (MMP)-9 and tissue inhibitor of metalloproteinase-1 (TIMP-1) by collagen deposition, proliferation of smooth muscle, and mucus secretion (Ma *et al.*, 2014a). Thirty-two BALB/c mice were investigated to determine the effect of aqueous extract of hyssop on asthmatic condition in comparison with dexamethasone; it was concluded that hyssop extract decreased the level of IgG and increased the ratio of eosinophils (EOS). These results showed that this extract, via modulating the immune system, had a positive effect on asthma management (Ma *et al.*, 2014b). A randomized clinical study investigated the anti-asthmatic action of a mixed herbal mixture containing *Matricaria chamomilla*, *A. capillus-veneris*, *H. officinalis*, *G. glabra*, *A. officinalis*, and *Malva sylvestris* on 46 children suffering from asthma. The patients took this herbal mixture from the first day of starting signs to five days. The results showed that this herbal mixture significantly reduced the nighttime awakening and severity of cough, whereas it did not improve tachypnea, wheezing, and respiratory distress (Javid *et al.*, 2019).

Linum usitatissimum L.

Linum usitatissimum (flax or linseed) from family *Linacaea* has been used as a natural laxative for many years. The seed, bark, leaves, flower, and oil of this plant have traditionally been used for cough and asthma in Pakistan (Alamgeer *et al.*, 2018). In Iran, it has been used traditionally to treat respiratory disorders with sticky and thick secretions. It contains minerals, vitamins, fat, protein, phenolic, trypsin inhibitor, linatine, lignans (phytoestrogens), and Se. These compounds are shown to have roles in affecting cystic fibrosis pulmonary pathogenesis (Tahermohammadi *et al.*, 2021). The seeds are a rich source of α-linolenic acid (ALA; C18 : 3*n*-3) and linoleic acid (LA; C18 : 2*n*-6), which are essential fatty acids. *n*-3 Fatty acids can help prevent some inflammatory diseases. *L. usitatissimum* is a rich source of lignan that seems to have antioxidant property depending on the presence of specific bacteria. This plant also contains soluble and insoluble fibers. Mucilage is the most soluble fiber existing in *L. usitatissimum* (Jhala and Hall, 2010).

Liquidambar orientalis Mill.

Liquidambar orientalis, known as Asian storax, is produced when the bark of sweetgum trees is damaged. It has been used for a long time to cure cough, skin problems, and wounds. Major components of *L. orientalis* are cinnamic derivatives, vanillin, and styrene. Cinnamic acid presents antibacterial and antioxidant activities. Oxygenated monoterpenes and monoterpene ketone groups also account for considerable antioxidant activity. The sap also possesses antifungal and immunosuppressant properties (Lingbeck *et al.*, 2015). The wound healing property was tested on young Yorkshire pigs with excised wounds on their back; the results showed that wound healing was related to cinnamic acid and esters such as cinnamyl cinnamate and storesinol that help the proliferation phase to be potent and increase both granulation and re-epithelialization (Ocsel *et al.*, 2012).

Malva sylvestris L.

Malva sylvestris contains various phytochemicals with several pharmacologic activities; these compounds include flavonols, tannins, mucilage, polyphenols, flavonoids, and folic acid (Irfan *et al.*, 2021). The antimicrobial activity of the hydroalcoholic extract of *M. sylvestris* was evaluated in an *in vitro* study by using disk diffusion and microdilution methods against *Aggregatibacter actinomycetemcomitans*. The result showed that *M. sylvestris*, with minimum inhibitory concentration of 156.2 μg/ml, had an antimicrobial activity against periodontal diseases (Vahabi *et al.*, 2019). An *in silico* study evaluated the potential of the methanolic extract of *M. sylvestris* in the treatment of COVID-19 by acting on the 6LU7 viral protein. On the other hand, the *in vitro* examination of the antioxidant effect of this plant was evaluated by DPPH and NO bioassay. The results showed that this plant extract was significantly effective in the treatment of SARS-CoV-2 resistance (Irfan *et al.*, 2021).

***Myrtus communis* L.**

Myrtus communis has several important secondary metabolites such as tannins, flavonoids, and essential oils. Studies showed that *M. communis* fruit contains 14 fatty acids, its major ones being oleic acid, palmitic acid, and stearic acid. Additionally, the pharmacologic effect of *M. communis* fruit essential oil is important because of the existence of 1,8-cineole, α-terpineol, and α-pinene (Serce *et al.*, 2010; Kutlu *et al.*, 2021). An *in vitro* study was conducted on the anti-inflammatory action of *M. communis* fruit through inducing endotoxemia in human umbilical vein endothelial cells (HUVECs) by LPS administration. The results showed that the 1,8-cineole and α-pinene of *M. communis* significantly reduced the expression of inflammation markers such as IL-1β, TNF-α, endothelial nitric oxide synthase (eNOS), and IL-6 (Kutlu *et al.*, 2021). The antioxidant capacity of the methanol extract of *M. communis* fruit was evaluated by DPPH and β-carotene/LA; the study showed that this extract was a good antioxidant through scavenging free radicals and inhibiting lipid peroxidation (Serce *et al.*, 2010). An *in vivo* study compared the effect of black and white *M. communis* fruit and leaves on hepatic ischemia-induced rats. For this purpose, the researchers used the leaves and fruit extract 15 min before reperfusion and then evaluated the metabolism capacity of the liver. The concentration of total phenols (TP) in these extracts was different; the TP of white fruit and black leaves (27 and 94.3 μg/ml, respectively) showed more effectiveness. The extract of white fruit showed a reduction in the markers aspartate aminotransferase (AST), malondialdehyde (MDA), and alanine aminotransferase (ALT), and an increase in the concentration of monoethylglycinexylidide (MEGX) (Ferchichi, 2012). An open-label randomized clinical trial investigated the efficacy of *M. communis* fruit syrup in 18–65-year-old outpatients with mild to moderate COVID-19; now, this trial is in phase three of evaluation (Azimi and Hasheminasab, 2020).

***Nymphaea alba* L.**

Nymphaea alba has a high content of several phytochemicals including flavonoids (quercetin), tannins, tocopheryl esters, phenolic acids, kaempferol, apigenin, and isokaempferide. Particularly, *N. alba* conains ellagic acid and gallic acid such as *p*-hydroxybenzoic, ferulic, and vanillic acids as methyl and ethyl esters of them. Among these compounds, the major compounds of *N. alba* are quercetin, apigenin, and ellagic acid (Cudalbeanu *et al.*, 2018). The antiviral activity of the methanolic and acetonic extracts of *N. alba* was evaluated *in vitro*, which was used to examine the anti-hepatitis C virus (HCV) effect of these extracts in the Huh-7 cell line. The result showed that *N. alba* flower extract was significantly effective in the treatment of HCV with or without another anti-HCV drug (Rehman *et al.*, 2018). The anticarcinogenic activity of methanolic extract of *N. alba* flowers was evaluated in Wistar rats with renal oxidative trauma induced by iron nitrilotriacetate (Fe-NTA) administration. Results showed that administration of the plant extract at 100–200 mg/kg dose significantly decreased γ-glutamyl transpeptidase, xanthine oxidase, lipid peroxidation, and hydrogen peroxide (H_2O_2) generation, which also recovered the normal level of antioxidant enzymes and glutathione-metabolizing enzymes (Khan and Sultana, 2005).

Opopanax chironius (L.) W.D.J. Koch

Opopanax chironius is a plant from the family *Apiaceae* with deobstruent, antispasmodic, antipyretic, and expectorant activity that has also been used in asthma (Maggio *et al.*, 2013; Önder *et al.*, 2020). This plant is rich in coumarins, especially furanocoumarins and simple prenylated coumarins. Phthalides are another phytochemical observed in *O. chironius*. Gum and malic acid are active components for treating asthma. Imperatorin and heraclenin are the compounds related to the anticancer activity of this plant that induces apoptosis in Jurkat leukemia cells (Önder *et al.*, 2020).

Papaver somniferum L.

Papaver somniferum, with the common name poppy, belongs to the family *Papaveraceae*. More than 40 alkaloids exist in this plant including morphine, codeine, thebaine, and noscapine as the most important ones. Avicenna (AD 980–1037) introduced poppy as the most powerful analgesic. Morphine reduces the sensation of pain and suppresses the cough reflex. Codeine is also a very important antitussive and is used in many antitussive products. Other compounds such as ethyl morphine, phal codeine, hydrocodone, and oxycodone also have antitussive properties (Yazdani *et al.*, 2003). The seeds and fruits of *P. somniferum* have been used for cough and the seeds have also been used in Unani Medicine to treat asthma as well (Masihuddin *et al.*, 2018). In Pakistan, seeds and fruit have been used as antitussives (Alamgeer *et al.*, 2018).

Pinus gerardiana Wall. ex D. Don

Pinus gerardiana, also known as chilgoza, is considered to have carminative, stimulant, and expectorant activity. In a study, the oils, proteins, and carbohydrates were found to be relatively abundant. Also, the oil yield on a dry weight basis was observed to be 49.45% (Thakur, 2015). The kernel and the oil obtained from *P. gerardiana* are a rich source of fatty acids, especially unsaturated fatty acids like LA (*n*-6) and oleic acid (*n*-9) that have many health benefits (Cai *et al.*, 2013; Sheikhahmadi *et al.*, 2020). γ-Tocopherol and α-tocopherol are the most abundant tocopherols in the oil (Sheikhahmadi *et al.*, 2020). Albumenoids, oil starch, phytosterol, polyphenols, xanthenes, carotenoids, tocopherols, catechin, etc. are also found in *P. gerardiana* nut. According to the Ayurvedic Pharmacopoeia, this plant is used in the treatment of cough and asthma (Singh *et al.*, 2021).

Pistacia terebinthus L.

Pistacia terebinthus that belongs to the family *Anacardiaceae* is commonly known as terebinth. Smoke of the resin has been used as an air purifier and antiseptic in Iran. In Greece, it has been considered an expectorant. The fruit in Turkey is known as an antitussive, stimulant, and for cold and flu treatment, while the resin has traditional uses with anti-inflammatory, antipyretic, and respiratory antiseptic effects and was considered to be good for asthma. The major compounds in the essential oil of *P. terebinthus* are hydrocarbon and oxygenated monoterpenes such as α-pinene, and the resin commonly contains triterpenes. Masticadienonic acid, masticadienolic acid, and morolic acid are compounds with anti-inflammatory effect. Flavonoids, fatty acids like oleic acid, and sterols such as β-sitosterol are also reported in the fruits of *P. terebinthus* (Bozorgi *et al.*, 2013).

Pistacia vera L.

Known as pistachio, *Pistacia vera* is a nut from the *Anacardiaceae* family. *P. vera* is an excellent source of unsaturated fatty acids, antioxidants, dietary fibers, proteins, and minerals (Kashaninejad and Tabil, 2011).

P. vera kernels contain 50 to 60% fat which mostly includes linoleic, linolenic, and oleic acids (Kashaninejad *et al.*, 2006). In Iran, it was used traditionally as a food and in Turkey, the resin was used for treating asthma. New researches show that anti-inflammatory and antimicrobial properties are related to different terpenoids. Also, antioxidant effects are associated with high levels of flavonoids and phenols (Bozorgi *et al.*, 2013). Regarding the anti-inflammatory effect, an experiment showed that a polyphenols-rich extract of natural raw shelled pistachio considerably reduced the paw edema and neutrophilic infiltration in rats. It is also believed that epicatechin and isoquercetin concentrations are responsible for antioxidant and anti-inflammatory properties. A significant reduction in TNF-α and IL-1β levels was observed following consumption of natural raw shelled and roasted salted pistachios (Paterniti *et al.*, 2017).

Plantago ovata Forssk.

Plantago ovata seeds have several compounds in the outer coat including mucilage, and compounds in the neutral and acidic polysaccharides fractions from mucilage are D-galactose, L-arabinose, L-rhamnose, D-galacturonic acid, and D-xylose. Gallic acid and rutin are the main polyphenol compounds of *P. ovata* seeds (Talukder *et al.*, 2016; Sarfraz *et al.*, 2017). The antibacterial and anti-inflammatory action of the aqueous extract of the leaves and seed of *P. ovata* was evaluated. For this purpose, the antibacterial activity of these extracts was evaluated by the tube dilution method against periodontal pathogens including *Porphyromonas gingivalis*, *A. actinomycetemcomitans*, *Fusobacterium nucleatum*, and *Prevotella intermedia*. On the other hand, the anti-inflammatory effect of *P. ovata* aqueous extract was investigated by zymogen gel electrography against MMP-2 and MMP-9. The results showed that these extracts were effective in inflammatory and bacterial infection of periodontal diseases (Reddy *et al.*, 2018). An *in vivo* study in HLA-B27 rats evaluated the anti-inflammatory effect of psyllium in the intestinal inflammatory pathway and its intermediates, which showed that this plant seed decreased the leukotriene B4, NO, and TNF-α in inflammatory processes (Sarfraz *et al.*, 2017).

Prunus dulcis (Mill.) D.A. Webb

Almond (*Prunus dulcis*) is a good source of vitamin E, magnesium, manganese, copper, phosphorus, fiber, riboflavin, as well as phenolic and polyphenolic compounds. Studies show that the kernels are rich in monounsaturated fatty acids (MUFAs), which are known to be beneficial, especially for the cardiovascular system. The oil contains high amounts of oleic acid as the major fatty acid, and other compounds such as sterols, tocopherols, and squalene (Čolić *et al.*, 2019). As to treatment, it is suggested that the paste of soaked almond with butter helps to relieve cough (Sultana *et al.*, 2016). The anti-inflammatory property of almond was investigated in a randomized, single-blind, crossover, controlled feeding study. The group receiving a high-almond diet had lower levels of E-selectin than the control and low-almond groups. Both groups with almond diets showed a decrease in the serum level of C-reactive protein (CRP). MUFAs seem to be responsible for the decrease in CRP and E-selection levels and, as a result, for the anti-inflammatory effect (Rajaram *et al.*, 2010). Another study shows the role of *P. dulcis* as the core drug in treating exogenous cough and acute bronchitis. The mechanism may involve arachidonic acid metabolism, peroxidase, estrogen metabolism, and tryptophan metabolism pathways (Sun *et al.*, 2020).

Viola odorata L.

Viola odorata contains several compounds including alkaloids (like the emetocathartic alkaloid violine), saponins, glycosides, tannins, mucilage, coumarin, methyl salicylate, monoterpenes and sesquiterpene terpenoids (like phenyl butanone, phenyl butanone, and viridiflorol), vitamin C, and flavonoids (like gallic acid and quercetin). Additionally, the essential oil of *V. odorata* includes geraniol, citronella, salicylaldehyde, and linalool (Akhbari *et al.*, 2012; Feyzabadi *et al.*, 2017; Habibi *et al.*, 2019). A study investigated the anti-inflammatory activity of *V. odorata* flower on six rats before and after inducing lung damage by formalin, in comparison to hydrocortisone. The results showed that the *V. odorata* preparation prevented lung injury equal to the efficacy of hydrocortisone (Koochek *et al.*, 2002). An *in vivo* study experimented with the antioxidant effect of the ethyl acetate flower and leaf extracts of *V. odorata* on Wistar rats with hepatotoxicity induced by ethanol. By exploring oxidative markers in the liver tissue and plasma of rats it was shown that both leaf and flower extracts had a significant antioxidant activity (Habibi *et al.*, 2019). The anti-asthmatic efficacy of *V. odorata* flower in 128 children (aged 2–12 years) was evaluated in a randomized controlled trial. The flower was formulated as a violet syrup and given to one of the groups of the study. The results showed that the anti-asthmatic action of the syrup was significantly related to the children's age. As a result, the violet syrup with a short-acting β-agonist can increase cough suppression (Qasemzadeh *et al.*, 2015). In a randomized clinical trial, the efficacy of *V. odorata* decoction in the recurrence of peritonsillar and tonsillitis abscess was evaluated in comparison to fexofenadine (60 mg) and co-amoxiclav (625 mg). In 400 patients, tonsillitis and peritonsillar abscess recurred two or three times a year. Decoction of *V. odorata* flower (1 mg/150 ml water) can decrease the frequency of these abscesses in one year (Ali *et al.*, 2013). A clinical trial study evaluated the efficacy of *V. odorata* decoction on 30 patients suffering from chronic rhinosinusitis. *V. odorata* flower (10 g) in the form of a decoction was given to the patients for 30 days. The results showed that this decoction improved the symptoms of rhinosinusitis; however, X-ray of the paranasal sinus revealed that it was not a significant alternative (Mulla *et al.*, 2019).

Conclusion

Acacia senegal mainly consists of polysaccharides and glycoproteins. Apart from being useful in industries, this gum is antibacterial, antioxidant, and protective in some toxicities. The most important components of *Adiantum capillus-veneris* leaf are mucilage, tannins, quercetin, adiantone, and capillarin that cause pharmacologic effects such as modulation of apoptotic ratio and tracheal tissue remodeling, decreased IL-6, NO, TNF-α, and NF-κB, and antibacterial effect on pulmonary pathogens. *Althaea officinalis* seed has coumarin, tannins, mucilage, and kaempferol that affect respiratory infections and other diseases. *Commiphora gileadensis* compounds also show antibacterial and antimycobacterial properties. *Commiphora myrrha* has a potent anti-inflammatory effect due to myrrhterpenes A and B, and the oil has antiviral activity. *Cordia myxa* fruit has mucilage, tannins, quercetin, gallic acid, and kaempferol, which cause several pharmacologic effects on pulmonary tissue such as tracheal relaxation, anti-inflammatory, and immunomodulatory effect by inhibition of COX-2 and 5-LOX enzymes. *Cucurbita pepo* possesses anticancer activities and is involved in the prevention and control of many lung diseases. *Cydonia oblonga* seed affects asthma symptoms and cough via antispasmodic function on the tracheal muscle, bronchial relaxation, as well as antioxidant and antibacterial activity. This seed has several compounds such as mucilage, tannins, and flavonoids. *Drimia maritima* bulb contains phenolic compounds and alkaloids that cause anti-asthmatic effects by improving FEV_1. Both oleogum resin and essential oil of *Ferula gummosa* are antibacterial. *Ferula persica* from the same genus mostly contains sesquiterpenes. It has anti-inflammatory effect related to umbelliprenin and auraptene. *Glycyrrhiza glabra* has glycyrrhizin, mucilage, coumarins, and flavonoids that cause several effects on the respiratory system such as tracheal relaxation by increasing cGMP and decreasing PDE5 activity, suppressing cough by decreasing citric acid, inhibition of lung overflow, and having immunomodulatory effect through decreasing IgE as well as an anti-inflammatory effect by inhibition of TLR-4/NF-κB. *Hyssopus officinalis* aerial part has quercetin, apigenin, and tannins, which affect cough and asthmatic symptoms by modulating the immune system, increasing IgG and ratio of EOS, enhancing mucus secretion, anti-remodeling, antibacterial, and anti-inflammatory activity. *Linum usitatissimum* is effective in respiratory disorders. The essential fatty acids that exist in its seeds are so beneficial for health and can prevent some inflammatory diseases. *Liquidambar orientalis* has a potent wound healing property, and some existing monoterpenes provide the antioxidant activity. *Malva sylvestris* stem and seed have mucilage and flavonoids that cause cough and COVID-19 management by antibacterial and antioxidant activity. *Myrtus communis* has tannins, essential oil, fatty oils, and flavonoids, which cause antioxidant and anti-inflammatory effects by decreasing IL-1β, TNF-α, eNOS, and IL-6. *Nymphaea alba* flower has quercetin, apigenin, kaempferol, tannins, and ellagic acid. It has antiviral and anticancer effects by decreasing lipid peroxidation, γ-glutamyl transpeptidase, and xanthine oxidase. *Opopanax chironius* is a rich source of coumarines and is used in asthma due to the gum and malic acid. Different alkaloids in *Papaver somniferum* are responsible for the antitussive activity through suppression of cough reflex. A variety of unsaturated fatty acids and tocopherols exist in *Pinus gerardiana* oil. *Pistacia terebinthus* consists of masticadienonic acid, masticadienolic acid, and morolic acid with anti-inflammatory effect. Epicatechin and isoquercetin in *Pistacia vera* are believed to be responsible for anti-inflammatory and antioxidant activity. *Prunus dulcis* is a rich source of MUFAs, which seem to be related to decreased CRP and E-selection levels and thus anti-inflammatory effect. *Plantago ovata* has mucilage and gallic acid components that cause antibacterial and anti-inflammatory activity by decreasing leukotriene-B4, NO, and TNF-α. *Viola odorata* has mucilage, tannins, saponin, and essential oil, which can treat the lung injury as well as hydrocortisone; also, it has anti-asthmatic, antioxidant, and anti-rhinosinusitis activity in the respiratory system.

Acknowledgments

The authors would like to thank Shiraz University of Medical Sciences, Shiraz, Iran and also Center for Development of Clinical Research of Nemazee Hospital and Dr Nasrin Shokrpour for editorial assistance.

References

Abbas, F.A., Al-Massarany, S.M., Khan, S., Al-Howiriny, T.A., Mossa, J.S. *et al.* (2007) Phytochemical and biological studies on Saudi *Commiphora opobalsamum* L. *Natural Product Research* 21, 383–391.

Abdel-Megeed, R.M., El Newary, S.A., Kadry, M.O., Ghaneh, H.Z., El-Shesheny, R.A. *et al.* (2020) *Hyssopus officinalis* exerts hypoglycemic effects on streptozotocin-induced diabetic rats via modulating GSK-3β, C-fos, NF-κB, ABCA1 and ABGA1 gene expression. *Diabetes and Metabolic Disorders* 19, 483–491.

Afshar, F.F., Saffarian, P., Hosseini, H.M., Sattarian, F., Amin, M. *et al.* (2016) Antimicrobial effects of *Ferula gummosa* Boiss. gum against extended-spectrum β-lactamase producing *Acinetobacter* clinical isolates. *Iranian Journal of Microbiology* 8, 263–273.

Ahvazi, M., Khalighi-sigaroodi, F., Charkhchiyan, M.M., Mojab, F., Mozaffarian, V.A. *et al.* (2012) Introduction of medicinal plants species with the most traditional usage in Alamut region. *Iranian Journal of Pharmaceutical Research* 11, 185–194.

Akhbari, M., Batooli, H. and Kashi, F.J. (2012) Composition of essential oil and biological activity of extracts of *Viola odorata* L. from central Iran. *Natural Product Research* 26, 802–809.

Alamgeer, Younis, W., Asif, H., Sharif, A., Riaz, H., Bukhari, I.A. and Assiri, A.M. (2018) Traditional medicinal plants used for respiratory disorders in Pakistan: a review of the ethno-medicinal and pharmacological evidence. *Chinese Medicine* 13, 48–48.

Al-Awadi, F.M., Srikumar, T.S., Anim, J.T. and Khan, I. (2001) Antiinflammatory effects of *Cordia myxa* fruit on experimentally induced colitis in rats. *Nutrition* 17, 391–396.

Al-Bayati, M. (2008) Mechanism of the tracheal smooth muscle relaxant activity of the *Cordia myxa* plant extract in sheep. *Iraqi Journal of Veterinary Sciences* 32, 214–226.
Ali, B.H., Ziada, A. and Blunden, G. (2009) Biological effects of gum arabic: a review of some recent research. *Food and Chemical Toxicology* 47, 1–8.
Ali, M., Qadir, M., Rasool, S., Abbas, G., Ahmed, S. *et al.* (2013) Effectiveness of *Viola odorata* flower decoction for tonsillitis and peritonsillar abscess. *Journal of Rashid Latif Medical College* 2, 51–53.
Alizadeh, H., Rahnema, M., Semnani, S.N. and Hajizadeh, N. (2013) Detection of compounds and antibacterial effect of quince (*Cydonia oblonga* Miller) extracts *in vitro* and *in vivo*. *Journal of Biologically Active Products from Nature* 3, 303–309.
Al-Massarany, S.M., Abbas, F.A., Demirci, B., Baser, K.H., Khan, S.I. *et al.* (2008) Chemical composition and biological evaluation of the essential oil of *Commiphora opobalsamum* L. *Journal of Herbs, Spices and Medicinal Plants* 13, 111–121.
Al-Snafi, A. (2015) The chemical constituents and pharmacological effects of *Adiantum capillus-veneris* – a review. *Asian Journal of Pharmaceutical Science & Technology* 5, 106–111.
Al-Snafi, A. (2016a) The pharmacological and therapeutic importance of *Cordia myxa* – a review. *IOSR Journal of Pharmacy* 6, 47–57.
Al-Snafi, A.E. (2016b) The medical importance of *Cydonia oblonga* – a review. *IOSR Journal of Pharmacy* 6, 87–99.
Alyafei, N. (2020) Can myrrh combat COVID-19? *Iberoamerican Journal of Medicine* 03, 223–229.
Ashraf, M.U., Muhammad, G., Hussain, M.A. and Bukhari, S.N.A. (2016). *Cydonia oblonga* M., a medicinal plant rich in phytonutrients for pharmaceuticals. *Frontiers in Pharmacology* 7, 163–163.
Avicenna (1993) *Canon of Medicine.* Jamia Hamdard, New Delhi.
Azimi, M. and Hasheminasab, F.S. (2020) Evaluating the efficacy and safety of the myrtle (*Myrtus communis*) in treatment and prognosis of patients suspected to novel coronavirus disease (COVID-19): study protocol for a randomized controlled trial. *Trials Journal* 21, 978.
Bao, S., Li, C., Li, H., Wang, J., Luan, J. *et al.* (2021) Myrrhterpenes A and B, anti-inflammatory cadinane sesquiterpenes from the resin of *Commiphora myrrha*. *Phytochemistry Letters* 44, 190–194.
Batool, H. and Dadkhah Tehrani, T. (2014) Properties of *Cucurbita pepo* L. in Islamic Persian medicine. *Health, Spirituality and Medical Ethics* 1, 24–28.
Borsuk, O.S., Masnaya, N.V., Sherstoboev, E.Y., Isaykina, N.V., Kalinkina, G.I. *et al.* (2011) Effects of drugs of plant origin on the development of the immune response. *Bulletin of Experimental Biology and Medicine* 151, 194–196.
Bozorgi, M., Memariani, Z., Mobli, M., Salehi Surmaghi, M.H., Shams-Ardekani, M.R. *et al.* (2013) Five *Pistacia* species (*P. vera*, *P. atlantica*, *P. terebinthus*, *P. khinjuk*, and *P. lentiscus*): a review of their traditional uses, phytochemistry, and pharmacology. *The Scientific World Journal* 2013, 219815–219815.
Bozorgi, M., Amin, G., Shekarchi, M. and Rahimi, R. (2017) Traditional medical uses of *Drimia* species in terms of phytochemistry, pharmacology and toxicology. *Journal of Traditional Chinese Medicine* 37, 124–139.
Cai, L., Liu, C. and Ying, T. (2013) Changes in quality of low-moisture conditioned pine nut (*Pinus gerardiana*) under near freezing temperature storage. *CyTA – Journal of Food* 11.
Chung, K.F. and Pavord, I.D. (2008) Prevalence, pathogenesis, and causes of chronic cough. *The Lancet* 371, 1364–1374.
Čolić, S., Zec, G., Natić, M. and Fotiric Aksic, M. (2019) Almond (*Prunus dulcis*) oil. In: Ramadan, M. (ed.) *Fruit Oils: Chemistry and Functionality*. Springer, Cham, Switzerland, pp. 149–180
Cudalbeanu, M., Ghinea, I.O., Furdui, B., Dah-Nouvlessounon, D., Raclea, R. *et al.* (2018) Exploring new antioxidant and mineral compounds from *Nymphaea alba* wild-grown in Danube Delta biosphere. *Molecules* 23, 1247.
Dave, P.N. and Gor, A. (2018) Natural polysaccharide-based hydrogels and nanomaterials: recent trends and their applications. In: Mustansar Hussain, C. (ed.) *Handbook of Nanomaterials for Industrial Applications.* Elsevier, Amsterdam, pp. 36–66.
Dehdari, S. and Hajimehdipoor, H. (2018) Medicinal properties of *Adiantum capillus-veneris* Linn. in traditional medicine and modern phytotherapy: a review article. *Iran Journal of Public Health* 47, 188–197.
Dogan, M.F., Parlar, A., Cam, S.A., Tosun, E.M., Uysal, F. *et al.* (2020) Glabridin attenuates airway inflammation and hyperresponsiveness in a mice model of ovalbumin-induced asthma. *Pulmonary Pharmacology and Therapeutics* 63, 101936.
El-Massry, K.F., Farouk, A., Mahmoud, K.F., El-Ghorab, A.H., Musa, A. (2021) Chemical characteristics and targeted encapsulated *Cordia myxa* fruits extracts nanoparticles for antioxidant and cytotoxicity potentials. *Saudi Journal of Biological Sciences* 28, 5349–5358.

Ferchichi, H. (2012) Effect of *Myrtus communis* L. extracts on attenuation of liver normothermic ischemia-reperfusion injury. *Journal of Transplantation Technologies and Research* S03, 001.
Feyzabadi, Z., Ghorbani, F., Vazani, Y. and Zarshenas, M.M. (2017) A critical review on phytochemistry, pharmacology of *Viola odorata* L. and related multipotential products in Traditional Persian Medicine. *Phytotherapy Research* 31, 1669–1675.
Gaikwad-Samdani, K. (2012) Evaluation of phytochemicals, antioxidant activity and elemental content of *Adiantum capillus-veneris* leaves. *Journal of Chemical and Pharmaceutical Research* 4, 365.
Gautam, S.S., Navneet, S.K. and Chauhan, R. (2015) Antimicrobial efficacy of *Althaea officinalis* Linn. seed extracts and essential oil against respiratory tract pathogens. *Journal of Applied Pharmaceutical Science* 5, 115–119.
Gutierrez, R. (2016) Review of *Cucurbita pepo* (pumpkin) its phytochemistry and pharmacology. *Medicinal Chemistry* 6, 012–021.
Habibi, E., Arab-Nozari, M., Elahi, P., Ghasemi, M. and Shaki, F. (2019) Modulatory effects of *Viola odorata* flower and leaf extracts upon oxidative stress-related damage in an experimental model of ethanol-induced hepatotoxicity. *Applied Physiology, Nutrition, and Metabolism* 44, 521–527.
Haider, S., Nazreen, S., Alam, M.M., Gupta, A., Hamid, H. *et al.* (2011) Anti-inflammatory and anti-nociceptive activities of ethanolic extract and its various fractions from *Adiantum capillus veneris* Linn. *Journal of Ethnopharmacology* 138, 741–747.
Hajifattahi, F., Moravej-Salehi, E., Taheri, M., Mahboubi, A. Kamalinejad, M. (2016) Antibacterial effect of hydroalcoholic extract of *Punica granatum* Linn. petal on common oral microorganisms. *International Journal of Biomaterials* 2016, 8098943.
Hamzeloo-Moghadam, M., Aghaei, M., Abdolmohammadi, M.H., Khalaj, A. and Fallahian, F. (2018) Cytotoxic effect of *Drimia maritima* bulb extract and induction of mitochondrial apoptotic signaling in human breast cancer cells, MCF-7 and MDA-MB-468. *OncoTargets and Therapy* 11, 7669–7677.
Hussain, M.M., Ahmad, B., Rashid, E., Hashim, S., Marwat, K.B. *et al.* (2014) *In vitro* antibacterial activity of methanol and water extracts of *Adiantum capillus-veneris* and *Tagetes patula* against multidrug resistant bacterial strains. *Pakistan Journal of Botany* 46, 363–368.
Ibraheim, Z.Z., Ahmed, A.S. and Gouda, Y.G. (2011) Phytochemical and biological studies of *Adiantum capillus-veneris* L. *Saudi Pharmaceutical Journal* 19, 65–74.
Iranshahi, M., Askari, M., Sahebkar, A. and Hadjipavlou-Litina, D. (2009) Evaluation of antioxidant, anti-inflammatory and lipoxygenase inhibitory activities of the prenylated coumarin umbelliprenin. *DARU Journal of Pharmaceutical Sciences* 17, 99–103.
Irfan, A., Imran, M., Khalid, M., Sami Ullah, M., Khalid, N. *et al.* (2021) Phenolic and flavonoid contents in *Malva sylvestris* and exploration of active drugs as antioxidant and anti-COVID19 by quantum chemical and molecular docking studies. *Journal of Saudi Chemical Society* 25, 101277.
Ishaq, M.S., Hussain, M.M., Afridi, M.S., Ali, G., Khattak, M. *et al.* (2014) *In vitro* phytochemical, antibacterial, and antifungal activities of leaf, stem, and root extracts of *Adiantum capillus-veneris*. *ScientificWorld Journal* 2014, 269793.
Jahan, Y. and Siddiqui, H. (2012) Study of antitussive potential of *Glycyrrhiza glabra* and *Adhatoda vasica* using a cough model induced by sulphur dioxide gas in mice. *International Journal of Pharmaceutical Sciences and Research* 3, 1668.
Jalali, H.T., Petronilho, S., Villaverde, J.J., Coimbra, M.A., Domingues, M.R.M. *et al.* (2012) Deeper insight into the monoterpenic composition of *Ferula gummosa* oleo-gum-resin from Iran. *Industrial Crops and Products* 36, 500–507.
Janbaz, K., Shabbir, A., Mehmood, M.H. and Gilani, A.-U. (2013) Insight into mechanism underlying the medicinal use of *Cydonia oblonga* in gut and airways disorders. *Journal of Animal and Plant Sciences* 23, 330–336.
Javid, A., Motevalli Haghi, N., Emami, S.A., Ansari, A., Zojaji, S.A. *et al.* (2019) Short-course administration of a traditional herbal mixture ameliorates asthma symptoms of the common cold in children. *Avicenna Journal of Phytomedicine* 9, 126–133.
Jhala, A.J. and Hall, L.M. (2010) Flax (*Linum usitatissimum* L.): current uses and future applications. *Australian Journal of Basic and Applied Sciences* 4, 4304–4312.
Karegar-Borzi, H., Salehi, M. and Rahimi, R. (2016) Laūq: a sustained-release dosage form for respiratory disorders in traditional Persian medicine. *Journal of Evidence-Based Complementary & Alternative Medicine* 21, 63–70.
Kariduraganavar, M.Y., Kittur, A.A. and Kamble, R.R. (2014) Polymer synthesis and processing. In: Kumbar, S.G., Laurencin, C.T. and Deng, M. (eds) *Natural and Synthetic Biomedical Polymers.* Elsevier, Oxford, pp. 1–31.

Kashaninejad, M. and Tabil, L.G. (2011) Pistachio (*Pistacia vera* L.). In: Yahia, E.M. (ed.) *Postharvest Biology and Technology of Tropical and Subtropical Fruits*, Vol. 4: *Mangosteen to White Sapote*. Woodhead Publishing, Cambridge, pp. 218–246.

Kashaninejad, M., Mortazavi, A., Safekordi, A. and Tabil, L. (2006) Some physical properties of pistachio (*Pistacia vera* L.) nut and its kernel. *Journal of Food Engineering* 72, 30–38.

Khan, M.F., Tang, H., Lyles, J.T., Pineau, R., Mashwani, Z.U. *et al.* (2018) Antibacterial properties of medicinal plants from Pakistan against multidrug-resistant escape pathogens. *Frontiers in Pharmacology* 9, 815.

Khan, N. and Sultana, S. (2005) Anticarcinogenic effect of *Nymphaea alba* against oxidative damage, hyperproliferative response and renal carcinogenesis in Wistar rats. *Molecular and Cellular Biochemistry* 271, 1–11.

Khoramian, L., Sajjadi, S.E. and Minaiyan, M. (2020) Anti-inflammatory effect of *Adiantum capillus-veneris* hydroalcoholic and aqueous extracts on acetic acid-induced colitis in rats. *Avicenna Journal of Phytomedicine* 10, 492–503.

Koochek, M., Pipelzadeh, M. and Mardani, H. (2002) The effectiveness of *Viola odorata* in the prevention and treatment of formalin-induced lung damage in the rat. *Journal of Herbs, Spices & Medicinal Plants* 10, 95–103.

Koushik, O., Himaja, V., Babu, P. and Karthikeyan, R. (2015) *In vitro* anti-bacterial activity of flowers of *Nymphea alba*. *Indian Journal of Drugs* 3(2), 47–48.

Kumar, A.A., Ashwini, R.S., Swaroop, J. and Anjana, M.A. (2012) Influence of ethanolic leaf extract of *Sargassum wightii* and *Adiantum capillus* on histamine induced asthma in animal model. *International Journal of Pharmacy and Pharmaceutical Sciences* 4, 121–123.

Kutlu, Z., Gulaboglu, M., Halıcı, Z., Cınar, İ. and Dıyarbakır, B. (2021) Biochemical research of the effects of essential oil obtained from the fruit of *Myrtus communis* L. on cell damage associated with lipopolysaccharide-induced endotoxemia in a human umbilical cord vein endothelial cells. *Biochemical Genetics* 59, 315–334.

Latha, S., Selvamani, P. and Prabha, T. (2021) Pharmacological uses of the plants belonging to the genus *Commiphora*. *Cardiovascular and Hematological Agents in Medicinal Chemistry* 19, 101–117.

Lawrence, R., Jeyakumar, E. and Gupta, A. (2015) Antibacterial activity of *Acacia arabica* (bark) extract against selected multi drug resistant pathogenic bacteria. *International Journal of Current Microbiology and Applied Sciences* 1, 213–222.

Lee, S.A., Lee, S.H., Kim, J.Y. and Lee, W.S. (2019) Effects of glycyrrhizin on lipopolysaccharide-induced acute lung injury in a mouse model. *Journal of Thoracic Disease* 11, 1287–1302.

Lingbeck, J.M., O'Bryan, C.A., Martin, E.M., Adams, J.P. and Crandall, P.G. (2015) Sweetgum: an ancient source of beneficial compounds with modern benefits. *Pharmacognosy Reviews* 9, 1–11.

Liu, B., Yang, J., Wen, Q. and Li, Y. (2008) Isoliquiritigenin, a flavonoid from licorice, relaxes guinea-pig tracheal smooth muscle *in vitro* and *in vivo*: role of cGMP/PKG pathway. *European Journal of Pharmacology* 587, 257–266.

Lopez-Torrez, L., Nigen, M., Williams, P., Doco, T. and Sanchez, C. (2015) *Acacia senegal* vs. *Acacia seyal* gums – Part 1: Composition and structure of hyperbranched plant exudates. *Food Hydrocolloids* 51, 41–53.

Ma, X., Ma, X., Ma, Z., Sun, Z., Yu, W. *et al.* (2014a) The effects of Uygur herb *Hyssopus officinalis* L. on the process of airway remodeling in asthmatic mice. *Evidence-Based Complementary and Alternative Medicine* 2014, 710870.

Ma, X., Ma, X., Ma, Z., Wang, J., Sun, Z. *et al.* (2014b) Effect of *Hyssopus officinalis* L. on inhibiting airway inflammation and immune regulation in a chronic asthmatic mouse model. *Experimental and Therapeutic Medicine* 8, 1371–1374.

Madia, V.N., De Angelis, M., De Vita, D., Messore, A., De Leo, A. *et al.* (2021) Investigation of *Commiphora myrrha* (Nees) Engl. oil and its main components for antiviral activity. *Pharmaceuticals* 14, 243.

Maggio, A., Bruno, M., Formisano, C., Rigano, D. and Senatore, F. (2013) Chemical composition of the essential oils of three species of Apiaceae growing wild in Sicily: *Bonannia graeca*, *Eryngium maritimum* and *Opopanax chironium*. *Natural Product Communications* 8, 1934578X1300800640.

Mahboubi, M. (2020) Marsh mallow (*Althaea officinalis* L.) and its potency in the treatment of cough. *Complementary Medicine Research* 27, 174–183.

Masihuddin, M., Jafri, M.A., Siddiqui, A. and Chaudhary, S. (2018) Traditional uses, phytochemistry and pharmacological activities of *Papaver somniferum* with special reference of Unani Medicine an updated review. *Journal of Drug Delivery and Therapeutics* 8, 110–114.

McGovern, A.E., Short, K.R., Moe, A.A.K. and Mazzone, S.B. (2018) Translational review: neuroimmune mechanisms in cough and emerging therapeutic targets. *The Journal of Allergy and Clinical Immunology* 142, 1392–1402.
Mulla, İ., Roqaiya, M. and Khan, M. (2019) Efficacy of *Viola odorata* flower decoction in chronic rhinosinusitis. *Medical Journal of Islamic World Academy of Sciences* 27, 77–84.
Nazim, M.D., Aslam, M. and Chaudhary, S. (2018) Hansraj (*Adiantum capillus-veneris*) – a review. *Journal of Drug Delivery and Therapeutics* 8, 105–109.
Nejatbakhsh, F., Karegar-Borzi, H., Amin, G., Eslaminejad, A., Hosseini, M. *et al.* (2017) Squill Oxymel, a traditional formulation from *Drimia Maritima* (L.) Stearn, as an add-on treatment in patients with moderate to severe persistent asthma: a pilot, triple-blind, randomized clinical trial. *Journal of Ethnopharmacology* 196, 186–192.
Ng, S.C., Lam, Y.T., Tsoi, K.K., Chan, F.K., Sung, J.J. *et al.* (2013) Systematic review: the efficacy of herbal therapy in inflammatory bowel disease. *Alimentary Pharmacology and Therapeutics* 38, 854–863.
Ocsel, H., Teke, Z., Sacar, M., Kabay, B., Duzcan, S.E. *et al.* (2012) Effects of oriental sweet gum storax on porcine wound healing. *Journal of Investigative Surgery* 25, 262–270.
Önder, A., Nahar, L., Nath, S. and Sarker, S.D. (2020) Phytochemistry, traditional uses and pharmacological properties of the genus *Opopanax* W.D.J. Koch: a mini-review. *Pharmaceutical Sciences* 26, 99–106.
Paterniti, I., Impellizzeri, D., Cordaro, M., Siracusa, R., Bisignano, C. *et al.* (2017) The anti-inflammatory and antioxidant potential of pistachios (*Pistacia vera* L.) *in vitro* and *in vivo*. *Nutrients* 9, 915.
Qasemzadeh, M.J., Sharifi, H., Hamedanian, M., Gharehbeglou, M., Heydari, M. *et al.* (2015) The effect of *Viola odorata* flower syrup on the cough of children with asthma: a double-blind, randomized controlled trial. *Journal of Evidence-Based Complementary & Alternative Medicine* 20, 287–291.
Rajaram, S., Connell, K.M. and Sabate, J. (2010) Effect of almond-enriched high-monounsaturated fat diet on selected markers of inflammation: a randomised, controlled, crossover study. *The British Journal of Nutrition* 103, 907–912.
Ranjbar, M., Varzi, H.N., Sabbagh, A., Bolooki, A. and Sazmand, A. (2013) Study on analgesic and anti-inflammatory properties of *Cordia myxa* fruit hydro-alcoholic extract. *Pakistan Journal of Biological Sciences* 16, 2066–2069.
Ranjbar, M.M., Assadolahi, V., Yazdani, M., Nikaein, D. and Rashidieh, B. (2016) Virtual dual inhibition of COX-2/5-LOX enzymes based on binding properties of alpha-amyrins, the anti-inflammatory compound as a promising anti-cancer drug. *EXCLI Journal* 15, 238–245.
Rashid Ali, W., Al-Asady, Z.T. and Ibrahim, A. (2015) Immunomodulatory of *Cordia myxa* (L.) aqueous extract fruit in immunized mice with hydatid cyst fluid. *Journal of Natural Sciences Research* 5, 75–82.
Ratnam, N., Naijibullah, M. and Ibrahim, M. (2017) A review on *Cucurbita pepo*. *International Journal of Pharmacognosy and Phytochemical Research* 9, 1190–1194.
Razavi, S.M. and Janani, M. (2015) A new ester coumarin from *Ferula persica* wild, indigenous to Iran. *Natural Product Research* 29, 717–721.
Reddy, P.R.T., Vandana, K.V. and Prakash, S. (2018) Antibacterial and anti-inflammatory properties of *Plantago ovata* Forssk. leaves and seeds against periodontal pathogens: an *in vitro* study. *Ayu* 39, 226–229.
Rehman, S., Ashfaq, U.A., Ijaz, B. and Riazuddin, S. (2018) Anti-hepatitis C virus activity and synergistic effect of *Nymphaea alba* extracts and bioactive constituents in liver infected cells. *Microbial Pathogenesis* 121, 198–209.
Saha, S., Nosál'ová, G., Ghosh, D., Flešková, D., Capek, P. *et al.* (2011) Structural features and *in vivo* antitussive activity of the water extracted polymer from *Glycyrrhiza glabra*. *International Journal of Biological Macromolecules* 48, 634–638.
Sarfraz, R.M., Khan, H., Maheen, S., Afzal, S., Akram, M.R. *et al.* (2017) *Plantago ovata*: a comprehensive review on cultivation, biochemical, pharmaceutical and pharmacological aspects. *Acta Poloniae Pharmaceutica* 74, 739–746.
Sattar, Z. and Iranshahi, M. (2017) Phytochemistry and pharmacology of *Ferula persica* Boiss.: a review. *Iranian Journal of Basic Medical Sciences* 20, 1–8.
Serce, S., Ercisli, S., Sengul, M., Gunduz, K. and Orhan, E. (2010) Antioxidant activities and fatty acid composition of wild grown myrtle (*Myrtus communis* L.) fruits. *Pharmacognosy Magazine* 6, 9–12.
Sharma, V., Katiyar, A. and Agrawal, R.C. (2017) *Glycyrrhiza glabra*: chemistry and pharmacological activity. In: Merillon, J.M. and Ramawat, K. (eds) *Sweeteners: Pharmacology, Biotechnology, and Applications*. Reference Series in Phytochemistry. Springer, Cham, Switzerland, pp. 87–100.
Sheikhahmadi, M., Sahari, M.A. and Barzegar, M. (2020) Evaluation of physicochemical and antioxidant properties of *Pinus gerardiana* nuts and oil. *Journal of Food Engineering and Technology* 9, 38–47.

Shen, T., Li, G.H., Wang, X.N. and Lou, H.X. (2012) The genus *Commiphora*: a review of its traditional uses, phytochemistry and pharmacology. *Journal of Ethnopharmacology* 142, 319–330.

Shwaish, T. and Imarah, F. (2017) Chemical composition of *Cordia myxa* fruit: phytochemical screening and identification of some bioactive compounds. *International Journal of Advanced Research* 5, 1255–1260.

Silva, B.M., Andrade, P.B., Valentão, P., Ferreres, F., Seabra, R.M. *et al.* (2004) Quince (*Cydonia oblonga* Miller) fruit (pulp, peel, and seed) and jam: antioxidant activity. *Journal of Agricultural and Food Chemistry* 52, 4705–4712.

Silva, B.M., Andrade, P.B., Ferreres, F., Seabra, R.M., Oliveira, M.B. *et al.* (2005) Composition of quince (*Cydonia oblonga* Miller) seeds: phenolics, organic acids and free amino acids. *Natural Product Research* 19, 275–281.

Singh, B.R., Dubey, S. and Siddiqui, M.Z. (2015) Antimicrobial activity of natural edible gums. *World Journal of Pharmaceutical Sciences* 3, 2217–2221.

Singh, G., Kumar, D. and Dash, A.K. (2021) *Pinus gerardiana Wallichex. D. Don. – a review. Phytomedicine Plus* 1, 100024.

Singh, M., Singh, N., Khare, P.B. and Rawat, A.K. (2008) Antimicrobial activity of some important *Adiantum* species used traditionally in indigenous systems of medicine. *Journal of Ethnopharmacology* 115, 327–329.

Süleyman, K., Nesrin, H., Veysel, T., Ersin, K. and Hakan, K. (2010) Chemical composition, antimicrobial and antioxidant activities of hyssop (*Hyssopus officinalis* L.) essential oil. *Notulae Botanicae Horti Agrobotanici Cluj-Napoca* 38, 99–103.

Sultana, S., Khan, A., Safhi, M.M. and Alhazmi, H.A. (2016) Cough suppressant herbal drugs: a review. *International Journal of Pharmaceutical Science Invention* 5, 15–28.

Sun, J.Y., Fan, Y.H. and Zhou, Q.Y. (2020) Analysis on the mechanism of core drugs in treating exogenous cough in clinical guidelines and medical records. *History and Philosophy of Medicine* 2, 110–118.

Tahermohammadi, H., Kaveh, S. and Sadr, S. (2021) Possible effect and mechanism of action of flaxseed on cystic fibrosis pulmonary disorder: a medical hypothesis. *International Journal of Pediatrics* 9, 14400–14412.

Talukder, P., Talapatra, S., Ghoshal, N. and Sen Raychaudhuri, S. (2016) Antioxidant activity and high-performance liquid chromatographic analysis of phenolic compounds during *in vitro* callus culture of *Plantago ovata* Forsk. and effect of exogenous additives on accumulation of phenolic compounds. *Journal of the Science of Food and Agriculture* 96, 232–244.

Thakur, N. (2015) Studies on quality characteristics of chilgoza nut and oil. *International Journal of Farm Sciences* 5, 78–82.

Vahabi, S., Hakemi-Vala, M. and Gholami, S. (2019) *In vitro* antibacterial effect of hydroalcoholic extract of *Lawsonia inermis*, *Malva sylvestris*, and *Boswellia serrata* on *Aggregatibacter actinomycetemcomitans*. *Advanced Biomedical Research* 8, 22.

Yadegari, M., Sellami, M., Riahy, S., Mirdar, S., Hamidian, G. *et al.* (2019) Supplementation of *Adiantum capillus-veneris* modulates alveolar apoptosis under hypoxia condition in Wistar rats exposed to exercise. *Medicina (Kaunas)* 55, 401.

Yazdani, D., Rezazadeh, S. and Shahnazi, S. (2003) Review of poppy (*Papaver somniferum* L.). *Journal of Medicinal Plants* 2, 1–12.

Yuan, Q., Zhang, X., Liu, Z., Song, S., Xue, P. *et al.* (2013) Ethanol extract of *Adiantum capillus-veneris* L. suppresses the production of inflammatory mediators by inhibiting NF-κB activation. *Journal of Ethnopharmacology* 147, 603–611.

Zhang, X., Chen, H.L., Hong, L., Xu, L.L., Gong, X.W. *et al.* (2019) Three new hopane-type triterpenoids from the aerial part of *Adiantum capillus-veneris* and their antimicrobial activities. *Fitoterapia* 133, 146–149.

9 An Evidence-Based Review of Medicinal Plants for Tuberculosis Management Cited by Avicenna

Seyedeh Mahnaz Karimi*
Department of Traditional Pharmacy, School of Traditional Medicine, Tehran University of Medical Sciences, Tehran, Iran

Abstract
Tuberculosis (TB) is a potentially serious infectious disease and one of the ten leading causes of death worldwide. Even today, available treatments are long and difficult, and TB control is complicated by problems such as inadequate global public health infrastructure, human immunodeficiency virus (HIV) epidemics, and the emergence of drug resistance. In spite of various modern treatments, some reasons like total cost, efficacy, safety, and drug resistance have caused a global willingness towards natural remedies. Therefore, the aim of this chapter is to focus on medicinal plants mentioned in Avicenna's *Canon of Medicine*, one of the primary medicinal books. Twenty-eight plants belonging to 22 botanical families have been mentioned for treatment of TB in the *Canon*. Most of them have a history of use in ethnomedicine and some are used in respiratory products nowadays. Investigating the literature has shown that anti-inflammatory and antibacterial properties are the proposed mechanisms for the treatment of some types of TB. Traditional Persian Medicine (TPM) can give new insights for the development of multiple natural treatments, the use of which in human medicine has been tested for thousands of years. By confirming their efficacy and safety, traditional herbal remedies are appropriate adjuvants and complementary to the chemicals mainly used for treating TB as an infectious agent. In 2018, approximately 1.7 billion people (23% of the world's population) were estimated to have latent TB, about 90 % of which live in developing countries.

Introduction

Tuberculosis (TB) is an infectious bacterial disease caused by Koch's bacillus *Mycobacterium tuberculosis* (Mtb) that is transmitted through the respiratory route and mainly affects the lungs and sometimes also other tissues (Bloom *et al.*, 2017). Droplets and small respiratory particles containing bacilli are scattered in space by infected people when they cough, sneeze, and talk, and can remain in space for a long time (Turner and Bothamley, 2011). Although vaccination has been performed for more than 90 years, TB is still recognized as the cause of a high number of deaths, with India having the highest number of TB deaths, followed by Indonesia, China, Philippines, Pakistan, and Nigeria. TB treatment involves multiple medications over several months that force heavy costs

*Email: karimi-sm@razi.tums.ac.ir

 Medicinal Plants Used in Traditional Persian Medicine (eds. H. Schulz, Seyed Ahmad Emami and Farsad Nadjafi)
DOI: 10.1079/9781800621671.0009

on individuals and communities, especially in cases of drug-resistant TB. However, chemotherapy for tuberculosis is an effective treatment (McKee and Atun, 2006). In 2021 an estimated 10.6 million people fell ill with TB and 1.6 million deaths from TB were reported, including 187,000 among people with human immunodeficiency virus (HIV) infection (WHO, 2022). The way to diagnose TB infection is to examine the sputum smear microscopically, and in some areas more advanced methods like rapid molecular diagnostics to directly monitor the patient's disease status following the use of drugs have made great strides in controlling TB over the past two decades. The prevalence of multi-drug-resistant (MDR) TB, especially to the two main TB drugs, isoniazid (INH) and rifampicin (RMP), is increasing worldwide and remains a public health crisis.

Another problem is the prevalence of TB in vulnerable people, such as those having HIV/acquired immune deficiency syndrome (AIDS) and TB at the same time. Although Bacille Calmette–Guérin (BCG) is the most widely used vaccine in the world, a more effective vaccine is needed in high-prevalence areas. TB control requires the following three preconditions: new diagnostic methods; more effective drug regimens; and the availability of more effective vaccines (Bloom *et al.*, 2017). Most patients' concerns about TB treatments are related to their cost, efficacy, and side effects, thus leading them to use natural products and herbal medicines that have been introduced by botanists or traditional physicians (Welz *et al.*, 2018). The use of plants with antimicrobial properties along with synthetic antimicrobial agents can increase the effectiveness of TB treatments and reduce their side effects (Iwu *et al.*, 1999). The current positive attitude of society toward the use of medicinal plants and traditional medicine, as well as the side effects of existing TB treatments and the emergence of resistant strains, led our team to review the types of medicinal plants mentioned for the treatment of TB in *Canon of Medicine*, a book by the great Iranian philosopher and physician, Avicenna (Ibn Sinā, 2015a).

In the present chapter, plants effective against TB were extracted from Avicenna's book and their scientific names were identified using reliable sources (Soltani, 2004; Ibn Sinā, 2015b).

Tuberculosis in View of Avicenna

Brief review on the respiratory system as described in *Canon of Medicine*

Avicenna has given a clear description of the respiratory system and its diseases as well as some common treatments that can be noteworthy nowadays. According to the *Canon*, the lung is a white organ made up of many interconnected circular shutters, which are intertwined by tendons and covered by a membrane. The lung is a porous organ that is responsible for inhaling, preparing, and storing the suitable air for the heart, and it consists of two parts, right (including three parts) and left (including two parts). According to Persian medicinal literature, all the materials in the world exhibit four main qualities: "warmth", "coldness", "wetness", and "dryness", and *mizāj* (temperament) is a predominant quality. According to this idea, the temperament of the lungs may be normal or out of balance. How a person is breathing and the type of voice are the clearest indications for the mood of our lungs. Avicenna studied the causes and treatment of different types of breathing: *ḍayq alnafas*, which means shortness of breath caused by inflammation in the lungs or respiratory tract; and *rabw*, which means asthma to the state where the breaths are small, short, and consecutive. Pulse status, cough, and discharge from the mouth indicate the health condition the lungs are in. In cold weather, especially in a rainy autumn following a dry summer, lung diseases occur a lot. Avicenna has divided pulmunary diseases into five categories:

- diseases with similar details;
- instrumental diseases and obstruction of the arteries and components of the pulmonary system;
- diseases in the porous mass of the lung;

- diseases of obstruction of small and large vessels; and
- participatory lung diseases with other organs.

Avicenna has given a description of pulmonary diseases and treatments of them, such as pleurisy (*d̲at al janb*), pneumonia (*d̲at* or *riah'*), pulmonary edema, pimples on the lungs, plural effusion, pus in the lungs, and lung ulcers. Ulcers may occur in different parts of the pulmonary system. Lung ulcers have the worst prognosis.

In the treatment of respiratory diseases, Avicenna's *Canon* describes single herbs or a combination of cleansing and expectorant herbs, such as flaxseed, almonds, caraway, black cumin, ginger, rhubarb, and squill, to be taken orally. In this context, especially lozenges or compresses on the chest are described in detail. In addition, non-medicinal treatments such as blowing and other special exercises are also used.

Tuberculosis

Ulcers in the lungs are called *sel* (tuberculosis). According to Persian medicinal literature in the *Canon*, lung ulcers can have a variety of causes: corrosive agent (poured from upper organs into the lungs); infectious agent from nearby organs or distant organs; infectious agent from the lungs themselves; untreated pneumonia and pleurisy; and injury. Avicenna considered lung ulcers following an infectious agent or untreated pneumonia to be tuberculosis (TB). Symptoms listed for TB in the *Canon* include:

- Cough and discharge that comes out of the mouth and has a special color and smell.
- Fever: Daily fever (*deq*) that increases especially after eating and at night (although sometimes *deq* fever occurs along with other types of fever).
- Sweating a lot.
- Other symptoms, expressed in the altered appearance of the patient: weakness and wilting, hair loss, stretched forehead skin, long neck, high thirst, low appetite, weakness of arms and legs, excretion of wounds or blood with cough, diarrhea, and shortness of breath.

The longevity of patients with tuberculosis varies depending on the type of ulcer and its cause. Children with TB are more treatable than other ages. People who are more susceptible or at risk for TB are those with curvy stature, with forward-leaning shoulders, a long neck, a prominent neck knot, a small chest, thin hair, white or yellow skin, and a stiff body. Also, people with cold and wet temperaments, people between 18 and 30 years of age, and people living in cold regions are more susceptible to infection (Ibn Sinā, 2015a).

Common treatments for tuberculosis mentioned in *Canon of Medicine*

Avicenna has stated two treatments for TB: *real treatment* when the disease is curable; and *tolerance treatment* if there is no hope of complete cure of the disease, with the goal to give the patient time and increase longevity. For the tolerance treatment, the disease should be prevented from getting worse and the wound should be dried. Lung ulcer drying and cleansing drugs such as *ma olᶜsal* (honey that is boiled slightly in water in a special way), lozenge of squill (*Urginea maritima*, called *ᶜonṣol* in Arabic), and lozenge of ervil (*Vicia ervilia*, called *kersannah* in Arabic) are recommended. Sometimes, narcotics that prevent cough are used. Milk is not recommended. Common foods such as ervil soup with chives and chickpea flour and corn flour are recommended.

In the real treatment, the lungs are cleansed and new foreign substances are prevented from falling into the lungs. Preventing the falling of foreign substances into the lungs is done by the treatment methods of bloodletting (*faṣd*) or cupping (*alḥijama*) and using herbal medicines such as poppy syrup, guggul (*Commiphora mukul*), arabic gum, dodder (*Cuscuta* sp.), and honey. Then drying, cleansing, emollient drugs and finally wound healing drugs are used. Some cleansing and emollient herbal medicines are juice of jujube (*Ziziphus jujuba*), violet, poppy, root of dog rose, flower of mallow, seeds of sebesten (*Cordia myxa*), chicory (*Cichorium intybus*), quince (*Cydonia oblonga*), and traganth gum. Some multi-herbal drying medicines are "caraway maᶜ-jun" (*maᶜjun* is an oral, semisolid traditional medicine consisting of several herbs with honey), *athānasyā maᶜjun*, lozenge of linseed, *mitrādāte maᶜjun*, and Armenian bole poultice. If the lung ulcer is old, tar with honey is recommended for a small spoon only once. For treatment of fever, sugar syrup including fennel seeds, licorice root paste, and maidenhair fern extract is very effective. Avicenna has considered the incense (administered by funnel) of natural compounds such as pepper, sweet olive leaves, birthwort (*Aristolochia* spp.), caper root bark, and poplar bark to clean and dry lung ulcers. It is stated in *Canon* that consuming a herbal mixture made from rose petals and sugar (which is mentioned in Traditional Persian Medicine (TPM) books) every day for a year is very effective for treating TB.

For people with TB, consumption of certain foods – such as poultry meat (*Francolinus* spp.), barley juice, wheat bread, rice soup, pumpkin, raisins, crabs, vegetables, lentils, cucumber, casaba melon, cabbage, and asparagus – as well as beverages such as sweet white wine, oxymel, or a drink containing sorrel is recommended. These patients should not be angry and should rest. Drinking buttermilk is highly recommended in patients with TB in accordance with the principles. The lactating animal should eat plants that are effective in healing lung ulcers like knotgrass.

Myrtle seed, violet seed, bamboo cane, traganth, burnt crabs, hemp, gum arabic, lily juice with purslane juice, and cucumber juice are useful for bloody cough. Rubbing carved red or white sandalwood in rose water with a little clay topically on the right side of the chest is useful for the treatment of uvulitis (Ibn Sinā, 2015a).

Brief Review on the Most Common Types of Tuberculosis and Their Treatments in Modern Medicine

Tuberculosis

TB is caused by *M. tuberculosis*, which is a nonmotile, aerobic, rod-shaped bacterium and is transmitted through droplet infection (Suárez *et al.*, 2019). After inhalation, the bacterium grows intercellularly and extracellularly in the well-ventilated lung, especially in the alveolar macrophages. Intracellular bacteria also survive the body's defense mechanism and form the tuberculous granulomas (Stewart *et al.*, 2003). Most patients develop a latent tuberculosis infection (LTBI) without any clinical symptoms (WHO, 2018). The risk of reactivation of the TB infection is higher in the first 2 years after the initial infection and in children and those with untreated HIV infection (WHO, 2018). Also, the risk of TB reactivation increases in conditions of immunosuppression such as diabetes mellitus and terminal kidney failure (Suárez *et al.*, 2019).

Latent tuberculosis infection

LTBI is an infection with a vital nonreplicating TB pathogen. The results of immunologic tests (interferon-γ release assays) are positive but signs cannot be detected on the radiograph. These people have no clinical symptoms and they are not contagious, but their infections can become active by weakening the cellular immunity (Suárez *et al.*, 2019).

Forms of disease

Pulmonary tuberculosis

In 2017, almost three-quarters of TB cases registered in Germany affected the lungs (Brodhun *et al.*, 2019). These people had shown symptoms such as fever, night sweats, fatigue, violent cough, and hemoptysis. In people without immunodeficiency, pulmonary TB progresses slowly, but in children and people with immunodeficiency, it progresses much faster (Suárez *et al.*, 2019).

Extrapulmonary tuberculosis

Prevalence of extrapulmonary TB is very low; however, reports from some regions have shown that it is increasing. It is more common in children and immunocompromised individuals and its symptoms vary depending on the individual organ involved (Oliveira *et al.*, 2007). In 2017, half of extrapulmonary TB cases registered in Germany involved lymph node TB (Brodhun *et al.*, 2019). Disseminated TB is an infection in which mycobacteria have spread from the lungs to two or more organ systems; it was previously seen only in children or persons with immune suppression, but has now also been identified in people without defective immune systems (Suárez *et al.*, 2019).

Treatment

The World Health Organization (WHO) has developed a new strategy to control TB called DOTS (Directly Observed Treatment, Short-course). Treating TB generally requires taking several antibiotics for a long time (Bansal *et al.*, 2018). The aim of TB treatment is to cure the disease and reduce its transmission. Anti-TB drug regimens are usually successful with efficacy of up to 95%, but this percentage varies with location. The DOTS strategy has five elements: political commitment and financial support to control TB; identifying patients; providing an effective and standard medication regimen; ensuring regular supply of medicine; and reporting of the results. The importance of using this strategy is greater in children, the elderly, and homeless persons, and for people with drug-resistant TB (Rabahi *et al.*, 2017).

Standard treatment

Conservative treatment for pulmonary TB includes 2 months of quadruple therapy with INH, RMP, pyrazinamide (PZA), and ethambutol (EMB), followed directly by a further 4 months of dual administration of RMP and INH (Bass *et al.*, 1994; Suárez *et al.*, 2019).

In extrapulmonary forms, depending on the organ involved, higher dosage and longer treatment are needed, such as TB of the central nervous system, which requires 12 months of treatment (Suárez *et al.*, 2019).

Resistance to standard drugs and preventive treatment

In some patients, resistance to one of the abovementioned TB drugs (INH, EMB, RMP, PZA, streptomycin) may occur. Multiple resistance occurs when resistance to more than one drug (including RMP or INH) is observed. This requires prolonged treatment in specialized centers where the use of fluoroquinolone is replaced accordingly (Suárez *et al.*, 2019).

Preventive treatment means treating the bacteria at rest. It is important for people with HIV or those taking immunosuppressive drugs to reduce the risk of bacterial activation (Suárez *et al.*, 2019).

Adverse effects

Side effects of applied TB drugs include skin complications such as flushing and itching (mostly with RMP and INH), gastrointestinal complications such as nausea and vomiting, visual side effects (with EMB and streptomycin), hepatotoxicity, arthralgia, and neuropathy such as tingling. Side effects are often mild, such as gastrointestinal side effects or exanthema, which are treated symptomatically, but sometimes these complications are severe (4 to 9% of patients). In severe hepatotoxicity, temporary discontinuation of treatment is required. Ophthalmologic examination at 4-week intervals is recommended for the detection of visual neuroma (less than 1% of patients with EMB). Neurologic complications such as peripheral neuropathy and psychosis are rare (less than 1%) (Gülbay *et al.*, 2006).

However, in high-risk individuals, such as in pregnancy, pyridoxine with INH is recommended for prevention (WHO, 2016). PZA can increase serum uric acid; therefore, dose adjustment is required for EMB and PZA in renal failure (Suárez *et al.*, 2019).

Scientific Evaluation of Medicinal Plants Mentioned for the Management of Tuberculosis in *Canon of Medicine*

The medicinal herbs mentioned for the management of TB in Avicenna's *Canon* and some scientific evidences substantiating their efficacy are described individually in this section. All the collected results are shown in Tables 9.1, 9.2, 9.3, and 9.4.

Aristolochia species

Aristolochia (birthwort) is a genus with more than 500 species belonging to the *Aristolochiaceae* family. *Aristolochia* spp. are subshrubs or shrubs and are widely distributed in tropical, subtropical, and temperate regions of the world. Their secondary metabolites are aristolochic acids and esters, aristolactams, aporphines, flavonoids, lignans, protoberberines, isoquinolines, benzylisoquinolines, amides, biphenyl ethers, coumarins, tetralones, terpenoids, benzenoids, and steroids (Kuo *et al.*, 2012). Aristolochic acids are the main compounds in most *Aristolochia* spp., which have nephrotoxicity and genotoxicity (Bartha *et al.*, 2019). The roots of *Aristolochia* spp. were used in traditional medicine. Avicenna has considered the incense of roots of caper, poplar, and *Aristolochia* to be effective in treating TB (Ibn Sinā, 2015a).

Several studies have shown that the extracts of *Aristolochia brevipes* Benth., *Aristolochia cymbifera* Mart., *Aristolochia indica* L., *Aristolochia mollissima* Hance, and *Aristolochia taliscana* Hook. & Arn. have significant antimicrobial properties (Kuo *et al.*, 2012). Isolated licarin (5 mg/kg) from *A. taliscana* root was evaluated for 30 and 60 days in mice infected with drug-sensitive or MDR strains and showed a significant decrease in pulmonary bacillary burdens and no abnormalities in main organs. The study proved the antitubercular and anti-inflammatory properties of licarin in an animal model (León-Díaz *et al.*, 2013). Navarro-García and co-workers showed that the dichloromethane extract obtained from the rhizomes of *A. brevipes* has strong *in vitro* antimycobacterial activity against *M. tuberculosis* H37Rv and aristolactam I was identified as the most active compound against all mycobacterial strains. In traditional medicine, *A. brevipes* was used for diarrhea, arthritis, cleansing wounds, and bloody cough (Navarro-García *et al.*, 2011). Isolated eupomatenoid-1 from *Aristolochia elegans* Mast. rhizomes has antimycobacterial activity against *M. tuberculosis* H37Rv (Jiménez-Arellanes *et al.*, 2012). Another study showed that the substances cubebin and fargesin isolated from *A. elegans* rhizome have antimycobacterial activity against *M. tuberculosis* H37Rv, four monoresistant variants, and two MDR *M. tuberculosis* clinical isolates (Oliveira *et al.*, 2007). *A. elegans* Mast. (guaco) is known for its expectorant, antitussive, antiasthmatic, analgesic, antihistamine, and detoxicant effects (Shi *et al.*, 2004). As a result, due to their antimicrobial, antitussive, and anti-inflammatory activity, *Aristolochia* spp. are considered in the treatment of TB, but the effects of nephrotoxicity should be further investigated.

Capparis spinosa L.

Capparis spinosa L. (caper), belonging to the family *Capparaceae*, is a perennial shrub that grows mainly in warm climates and grows sparingly in most parts of the Middle East as a wild plant (Ehsanifar *et al.*, 2017; Annaz *et al.*, 2022). Roots and flower contain pectin, saponin, aresin material, aminoglycosides, and kaparyroutin. It has been found that the main chemical active

Table 9.1. Medicinal plants for treatment of TB mentioned in Avicenna's *Canon of Medicine*.

No.	Scientific name	Synonym	Arabic name	Common English name	Family
1	*Acacia senegal* (L.) Britton Basionym: *Mimosa senegal* L.	*Acacia cufodontii* Chiov. *Acacia senegal* (L.) Willd.	*Ṣamgh*	Gum arabic, gum acacia,	*Fabaceae* (*Leguminosae*)
2	*Asparagus officinalis* L.	*Asparagopsis adscendens* Roxb. *Asparagus satawur* *Protasparagus adscendens* Roxb.	*Haliun*	Asparagus	Asparagaceae
3	*Adiantum capillus-veneris* L.	*Adiantum africanum* R. Br. *Adiantum capillus* Sw.	*Barshiāvoshān*	Venus's hair, maidenhair fern	*Pteridaceae*
4	*Aristolochia clematitis* L.	*Aristolochia infesta* Salisb. *Aristolochia tenuis* Houtt.	*Zarawand*	Birthwort	*Aristolochiaceae*
5	*Astragalus gummifer* Labill.	–	*Kathyr a'*	Traganth, gum dragon	*Fabaceae* (*Leguminosae*)
6	*Capparis spinosa* L.	–	*Kabar*	Caper, caperberry	*Capparaceae*
7	*Carum carvi* L.	*Bunium carvkammun* L. *Carum aromaticum* Salisb.	*Kammun*	Caraway	Apiaceae
8	*Cichorium intybus* L.	–	*Hendeba'*	Chicory	*Asteraceae* (*Compositae*)
9	*Cinnamomum camphora* L.	*Camphora camphora* L.	*Kāfur*	Camphor laurel	*Lauraceae*
10	*Commiphora mukul* (Hook. ex Stocks) Engl. Basionym: *Balsamodendrum mukul* Hook. ex Stocks	*Balsamea mukul* Baill.	*Moql*	Indian bdellium tree, guggul, mukul	*Burseraceae*
11	*Cordia myxa* L.	–	*Sebestān*	Assyrian plum, lasura, laveda	*Boraginaceae*
12	*Cuscuta epithymum* (L.) L. Basionym: *Cuscuta europaea* var. *epithymum* L.	–	*Aftimun*	Clover dodder	*Convolvulaceae*
13	*Cydonia oblonga* Mill.	*Cydonia communis* Loisel. *Cydonia cydonia* (L.) Pers.	*Safarjal*	Quince	*Rosaceae*
14	*Drimia maritima* (L.) Stearn Basionym: *Scilla maritima* L.	*Charybdis maritima* (L.) Speta *Urginea maritima* (L.) Baker	*Ashqeil*, *ʿonṣol*	Squill	*Asparagaceae*
15	*Foeniculum vulgare* Mill.	*Anethum dulce* DC. *Anethum foeniculum* L. *Foeniculum azoricum* Mill.	*Rāziānj*	Fennel	*Apiaceae*

Continued

Table 9.1. Continued.

No.	Scientific name	Synonym	Arabic name	Common English name	Family
16	*Glycyrrhiza glabra* L.	*Glycyrrhiza glabra* subsp. *glandulifera* (Waldst. & Kit.) Ponert *Glycyrrhiza glabra* var. *laxifoliolata* X.Y. Li	*Sus*	Licorice, liquorice	*Fabaceae* (*Leguminosae*)
17	*Linum usitatissimum* L.	–	*Katān*	Linseed, flaxseed	Linaceae
18	*Malva sylvestris* L.	–	*Khobbāzi*	Cheeses, high mallow, tall mallow	*Malvaceae*
19	*Myrtus communis* L.	–	*Ās*	Myrtle	*Myrtaceae*
20	*Papaver somniferum* L.	–	*Khashkhāh*	Opium poppy	*Papaveraceae*
21	*Populus alba* L.	–	*Ḥawr abyaḍ*	Silver-leaf poplar, white-leaf poplar	Salicaceae
21	*Portulaca oleracea* L.	–	*Baqlah hamqā'*	Duck-weed, little hogweed, purslane	Portulacaceae
20	*Santalum album* L.	*Sirium myrtifolium* L.	*Ṣandal*	Sandalwood	Santalaceae
22	*Bambusa bambos* (L.) Voss Basionym: *Arundo bambos* L.	–	*Ṭabāshir*	Indian thorny bamboo	Poaceae
23	*Rosa canina* L.	*Rosa acanthina* *Rosa actinodroma* *Rosa adenocalyx*	*Nisrin*	Dog rose	Rosaceae
24	*Rosa* × *damascena* Herrm.	–	*Wardat al-ḥamra'*	Damask rose, Persian rose	*Rosaceae*
25	*Rumex acetosella* L.	*Acetosa acetosella* (L.) Mill.	*Ḥommaḍ*	Red sorrel, sheep's sorrel	*Polygonaceae*
26	*Vicia ervilia* (L.) Willd. Basionym: *Ervum ervilia* L.	*Ervilia sativa* *Lens pygmaea* Grossh.	*Kersannah*	Bitter vetch, ervil	*Fabaceae* (*Leguminosae*)
27	*Viola odorata* L.	–	*Banafsaj*	Sweet violet	*Violaceae*
28	*Ziziphus jujuba* Mill.	–	*Oʿnnāb*	Jujube	*Rhamnaceae*

Table 9.2. Recent *in vivo* studies on medicinal plants for treatment of TB mentioned in Avicenna's *Canon of Medicine*.

Plant	Part/extract	Active constituent	Method	Animal	Effect	Reference
Aristolochia taliscana	Roots/hexane extract	Licarin	Inducing disease with *Mycobacterium tuberculosis* H37Rv or MDR, treating with licarin (5 mg/kg); post treatment, determining lung bacilli loads and pneumonia percentage	BALB/c mouse	Antitubercular effect against both mycobacterial strains, no major signs of damage in subacute toxicity	León-Díaz *et al.* (2013)
Carum carvi	Fruits/butanol extract	Triflorin (kaempferol-3-β-D-galactoside)	Determining bioavailability indices of the anti-TB drugs RMP, PZA, and INH	Wistar rat	Herbal bioenhancer of the anti-TB drugs	Sachin *et al.* (2009)
Cichorium intybus	Roots/aqueous extract	Inulin	INH-induced hepatotoxicity in adult male mice	BALB/c albino mouse	Hepatoprotective	Ali *et al.* (2016)
	Roots/aqueous extract	Inulin	PZA-induced hepatotoxicity in male mice	BALB/c albino mouse	Hepatoprotective	Munir *et al.* (2019)
Cuminum cyminum, *Carum carvi*	Fruits/aqueous extract	3,5-Dihydroxyflavone-7-O-β-D-galacturonide-4-O-β-D-glucopyranosid (cumin)	Determining plasma levels and bioavailability of RIF, INH, and PZA	Wistar new line rat	Bioenhancer of the anti-TB drugs	Bhusari (2010)
Glycyrrhiza glabra	Root/LDPI containing licorice extract	Glabridin	Film hydration technique and freeze-dried	Mouse	Antitubercular activity	Viswanathan *et al.* (2019)

Table 9.3. Recent *in vitro* studies on medicinal plants for treatment of TB mentioned in Avicenna's *Canon of Medicine*.

Plant	Part/extract	Active constituent	Assay	Effect	Reference
Aristolochia brevipes	Rhizome/ dichloromethane extract	Aristolactam I	Antimycobacterial activity by fluorometric microplate Alamar Blue assay, column chromatography, NMR	Antimycobacterial activity against resistant *Mycobacterium tuberculosis* (MDR-TB)	Navarro-García *et al.* (2011)
Aristolochia elegans	Rhizomes/hexane extract	Fargesin, cubebin	^{1}H-NMR, microplate Alamar Blue assay	Antimycobacterial activity	Jiménez-Arellanes *et al.* (2012)
Aristolochia triangularis Cham.	Leaf, root and stem/ aqueous and hydroalcoholic extracts	Nerolidol	Time-to-kill assay	Antimycobacterial activity	Oliveira *et al.* (2007)
Aristolochia triangularis	Rhizome and stem/ methanol extract	Nerolidol	Microplate Alamar Blue assay	Antimycobacterial activity	Pereira *et al.* (2018)
Capparis spinosa	Fruits/methanol extract	–	MIC of bacterial growth by the proportional method	Antimycobacterial activity	Ehsanifar *et al.* (2017)
Glycyrrhiza glabra	Root/LDPI containing licorice extract (LE)	Glabridin	Film hydration technique and freeze-drying	Antitubercular activity	Viswanathan *et al.* (2019)
	Root/ethanolic extract	Glabridin	TLC, NMR, MIC by the twofold serial dilution technique, standard REMA	Antimicrobial activity against both Gram-positive and Gram-negative bacteria, antitubercular activity	Gupta *et al.* (2008)
	Rhizome	Isoliquiritigenin and liquiritigenin, analog	Molecular docking and in silico ADME studies	Anti-tuberculosis activity	Gaur *et al.* (2015)
Populus alba L.	Leaves/aqueous and ethanol extract	Flavonoids, polyphenols	Disk method, TLC	Antimycobacterial effect against three mycobacteria: *Mycobacterium smegmatis*, *Mycobacterium aurum*, and *Mycobacterium bovis*	Haouat *et al.* (2013)
Myrtus communis	Leaf extract	Limonene, 1,8-cineole, and α-pinene	Proportional method	Antimicrobial activity	Zanetti *et al.* (2010)
Rosa × damascena	Flowers	Citronellol, nerol	Hydrodistillation and solvent extraction	*M. bovis inactivation*	Fatemeh *et al.* (2015)

Santalum album	Endophytic fungus	Asperterpenoid A	Single-crystal XRD using Cu Kα radiation, antimicrobial activity by conventional broth dilution assay	Anti-inflammatory activity, potent inhibitory activity against mPTPB	Yan *et al*. (2018)
	Sandalwood essential oil	Sesquiterpene alcohols α-santalol, β-santalol, epi-β-santalol, and trans-α-bergamotol with α- and β-santalol	LCMS, colorimetric microdilution broth assay	Antituberculosis activity, MtSK inhibition	Rants'o *et al*. (2017)
Viola odorata	Flower	Salicin and phenyl alanine ethyl ester	Preparative HPLC, LCMS, determining MIC	Antimicrobial activity against *M. tuberculosis* H37Rv	Hassan and Naeem (2014)

ADME, absorption, distribution, metabolism, excretion; HPLC, high-performance liquid chromatography; LCMS, liquid chromatography–mass spectrometry; MDR-TB, multidrug-resistant TB; mPTPB, *M. tuberculosis* protein tyrosine phosphatase B; MtSK, *M. tuberculosis* shikimate kinase; NMR, nuclear magnetic resonance spectroscopy; REMA, resazurin microtitre plate assay; TLC, thin layer chromatography; XRD, X-ray diffraction.

Table 9.4. Clinical studies on medicinal plants for treatment of TB mentioned in Avicenna's *Canon of Medicine*.

Plant	Treatment group	Control group	Study design	No. of patients	Treatment duration	Result	Reference
Carum carvi	A single FDC containing RMP (450 mg), INH (300 mg), and PZA (1000 mg), plus capsule of *C. carvi* extract (100 mg)	–	Prospective, two-period, open-label, crossover experiment	20	10 days	↑ Plasma levels, C_{max}, and AUC of RMP, INH, and PZA	Choudhary *et al.* (2014)
Glycyrrhiza glabra	Capsule (450 mg) of Rasayana daily along with the standard regimen of pulmonary TB	Standard regimen of pulmonary TB, which includes INH (600 mg) RMP (450 mg), PZA (150 mg), and EMB (1200 mg), 3 times weekly on alternate days	Single-blind controlled study	133	60 days	↓ Cough, fever, dyspnea, and hemoptysis; ↑ body weight	Vyas *et al.* (2012)
	Licorice, nettle, tansy, mint	Essentiale, legilon, Liv-52, hemodez	Single-blind controlled study	112		↓ Occurrence of HTR	Galitskiĭ *et al.* (1997)

AUC, area under the plasma drug concentration versus time curve; C_{max}, maximum serum concentration.

compounds are flavonoids (Heidari *et al.*, 2003). In TPM, different parts of the plant like fruits, roots, bark, leaves, buds, and stems were used. Therapeutic effects of caper in traditional medicine include especially improving liver function, anti-flatulence, prevention of atherosclerosis, and improving anemia, gout and arthritis (Rhizopoulou *et al.*, 2006). It has been shown that *C. spinosa* fruits possess considerable antimycobacterial properties in the laboratory (Ehsanifar *et al.*, 2017).

Carum carvi L.

Caraway (*Carum carvi* L.) belongs to the family *Apiaceae*. This species is cultivated all over the world and widely used in food products (Rasooli and Allameh, 2016). Caraway has many uses in traditional medicine such as for gastrointestinal problems (cramps, nervous cardiac-gastric complaints, flatulence, irritable intestinal, indigestion, low appetite, infant colic) and for improving lactation. It is a tranquilizer, diuretic, emenagogue, aphrodisiac, and astringent (Al-Snafi, 2015; Rasooli and Allameh, 2016). In caraway essential oil, more than 30 compounds have been identified of which carvone and limonene represent the main substances (Wichtmann and Stahl-Biskup, 1987; Edláková *et al.*, 2003). The essential oil obtained from caraway fruits showed promising inhibitory activity against *M. tuberculosis*, nine other pathogenic bacteria, and six phytopathogenic fungi (Sadowska and Obidoska, 1998). Another study showed that caraway active principles act as a bioenhancer and are able to modify the kinetics of drugs used in TB treatment. After administration of caraway extracts in fixed dose combination (FDC) to 20 healthy human volunteers, increased plasma levels of RMP, INH, and PZA were observed (Choudhary *et al.*, 2014). Due to caraway's pharmacologic properties such as antimicrobial, antioxidant, analgesic, bronchial relaxant effects, and enhancement of TB drug kinetics, it can be considered a useful phytopharmaceutical drug in the treatment of TB (Al-Snafi, 2015).

Cichorium intybus L.

Chicory (*Cichorium intybus* L.), a Mediterranean erect perennial plant, belongs to the family *Asteraceae*. Previous studies showed that application of chicory root extracts provides liver protective, immunomodulatory, antiglycemic, anti-inflammatory, lipid-lowering, neuroprotective, and gastrointestinal tract protective effects (Munir *et al.*, 2019). Previous studies were published proving that the extract of *C. intybus* roots has significant hepatoprotective activity in high doses (Ali *et al.*, 2016; Munir *et al.*, 2019).

Glycyrrhiza glabra L.

Licorice (*Glycyrrhiza glabra* L.) is a traditional medicinal plant, the root of which is very sweet, moist, and soothing and contains flavonoids and triterpenoids (Lee *et al.*, 2009). Some bioactive compounds identified in licorice root are glabridin and liquiritin (a glycosidic form of liquiritigenin) (Lee *et al.*, 2009; Sharma *et al.*, 2016; Roque *et al.*, 2018). Previous studies have shown that these compounds possess anti-inflammatory, antiulcer, antimicrobial, and antifungal activities (Roque *et al.*, 2018).

In Avicenna's *Canon* the decoction or extract of roots was considered useful for TB treatment. Avicenna has suggested to use fennel fruits, licorice root paste, and maidenhair fern extract for treatment of TB fever (Ibn Sinā, 2015a). It was shown that glabridin is an active compound against *M. tuberculosis* H37Ra and H37Rv strains at 29.16 μg/ml concentration (Gupta *et al.*, 2008). *In vivo* evaluation of a liposomal dry powder for inhalation (LDPI) containing licorice extract showed a significant reduction in bacterial counts in the lungs and spleen of TB-infected mice (Viswanathan *et al.*, 2019). Administration of a herbal medicine containing several herbs including licorice, called Rasayana, along with anti-Koch's treatment (AKT) showed that, due to its antioxidative, immunoprotective, and free radical scavenging properties, Rasayana therapy was able to significantly decrease TB complications like cough, fever, dyspnea, and hemoptysis (Vyas *et al.*, 2012).

The use of licorice root as a preventive hepatoprotective therapy in TB patients led to a decrease in hepatotoxic responses (HTR) and had positive effects on treatment efficacy (Galitskiĭ *et al.*, 1997). These and previous results on the anti-inflammatory and antiulcer effects, as well as immunomodulatory activity indicate that licorice is a very useful anti-TB agent (Jalilzadeh-Amin *et al.*, 2015).

Myrtus communis L.

Myrtle (*Myrtus communis* L.) belonging to the family *Myrtaceae* is an evergreen shrub of the Mediterranean area and was widely used in various traditional medical systems. The various parts of this species, especially its leaves and berries, have different properties such as antioxidant, antimicrobial, and antiseptic effects. The chemical composition of myrtle essential oil as determined by gas chromatography includes myrtenyl acetate (20.75%), 1,8-cineole (16.55%), α-pinene (15.59%), linalool (13.30%), limonene (8.94%), linalyl acetate (3.67%), geranyl acetate (2.99%), and α-terpineol (2.88%) as the major components (Alem *et al.*, 2008; Zanetti *et al.*, 2010; Ben Hsouna *et al.*, 2014). Zanetti *et al.* showed that myrtle leaves possess considerable antimycobacterial properties in the laboratory; the first promising results of this study require further detailed research activities including also reliable clinical studies (Zanetti *et al.*, 2010).

Populus alba L.

White poplar tree (*Populus alba* L.), from the *Salicaceae* family, is mostly found in riverside and humid areas. In traditional medicine, various therapeutic effects such as antifungal, antioxidant, antitumoral, antiseptic, and antiviral properties have been described (Haouat *et al.*, 2013) and also recent studies have proven its antibacterial and antifungal activity (Talib and Mahasneh, 2010; Al-Hussaini and Mahasneh, 2011). Haouat and coworkers showed that the combination of aqueous and ethanolic extracts obtained from leaves and branches has an antimycobacterial effect against mycobacterial growth by the disk diffusion method. In this context, especially flavonoids and polyphenols were found to be responsible for this biologic activity (Haouat *et al.*, 2013). Other studies with polyphenolic extracts obtained from various medicinal plants also identified inhibiting effects on *M. tuberculosis* growth *in vivo* and *in vitro* (Newton *et al.*, 2002; Anand *et al.*, 2006).

Rosa × damascena Mill.

Damask rose (Rosa × damascena Mill. L.), which is well-known as "Gole Mohammadi" in Iran, belongs to the *Rosaceae* family and is especially used worldwide in the food and perfume industries. Its flowers and petals possess pharmacologic importance relating to their antioxidative, anti-inflammatory, antibacterial, antitussive, hypnotic, antidiabetic, and antidepressant activities, which are mainly attributed to the phenolic compounds (Boskabady *et al.*, 2011). The anti-inflammatory effects result by inhibiting the mediators of acute inflammation (Maleev *et al.*, 1972) and the antioxidant effect against lipid peroxidation is due to the presence of quercetin-3-*O*-glucoside and other flavonoids (Yassa *et al.*, 2009). The antibacterial effects of rose extract and rose oil against various bacteria, *Mycobacterium smegmatis*, and *Mycobacterium bovis* have been studied due to the high phenyl ethyl alcohol content (Özkan *et al.*, 2004; Fatemeh *et al.*, 2015). In one study, the antitussive effect of ethanolic and aqueous extracts of this plant in guinea pigs was investigated and found to significantly reduced the number of coughs (Shafei *et al.*, 2003). Another study on guinea pig tracheal smooth muscle showed that ethanolic extracts and essential oil had a strong relaxant effect comparable to that of theophylline (Boskabady *et al.*, 2006) due to their tachykinin content (Advenier *et al.*, 1997).

Santalum album L.

Sandelwood (*Santalum album*) belonging to the *Santalaceae* family is an evergreen parasitic tree and is especially used in aromatherapy and cosmetics (Tu *et al.*, 2013). It has anti-inflammatory and antioxidant properties through inhibition of the oxidative enzyme 5-lipoxygenase. The mechanism of the antimicrobial effect of album oil has not been precisely elucidated, but the described effects are mainly due to impaired membrane integrity (Moy and Levenson, 2017). In a study performed in 2017 it was shown that East Indian Sandalwood oil displayed antitubercular activity against *M. tuberculosis* H37Rv at a minimum inhibitory concentration (MIC) of 0.002% v/v (Rants'o *et al.*, 2017). Another study applying the Unani formulation Qurs-e-Sartan Kafoori, which contains 19 components including Scilla Crab and some herbs like *S. album*, was able to show that this mixture is effective against *M. tuberculosis* through increasing the susceptibility to INH and RMP, impairing membrane homeostasis, and modulating the immune system (Saif *et al.*, 2021). Avicenna stated that the use of sandalwood along with some other herbs topically is useful for uvulitis (Ibn Sinā, 2015a).

Viola odorata L.

A member of the the *Violaceae* family, violet (*Viola odorata*) is native to Asia, North Africa, and Europe. The plant species is mainly used for treatment of respiratory problems in different traditional medical systems and has various biologic properties, including anti-inflammatory, analgesic, antihypertensive, weight loss and antilipid, and expectorant properties. Avicenna has recommended the use of *Viola* for coughs, especially bloody coughs in TB (Ibn Sinā, 2015a). *Viola* compounds include alkaloids, various glycosides, saponins, tannins, methyl salicylate, mucilage, vitamin C, coumarins, and flavonoids. In this context cyclotides including cycloviolacin have been found to be the most important active compounds of violet. It was also shown that salicin and phenyl alanine ethyl ester derived from *V. odorata* possess antimicobacterial activity against *M. tuberculosis* H37Rv and *Mycobacterium avium* (Hassan and Naeem, 2014). In a parallel, double-blind, randomized controlled trial in children with intermittent asthma, it was shown that the use of violet flower syrup with short-acting β-agonist effect increases the suppression of cough (Qasemzadeh *et al.*, 2015).

Concluding Remarks

Traditional medicine and its documents, such as Avicenna's *Canon of Medicine*, have always been valuable treasures for modern medicine and pharmacy. The *Canon* describes a wide range of herbs with different medicinal and therapeutic effects. In this chapter, descriptions of TB, susceptible individuals, TB treatments, and dietary recommendations were extracted from the original *Canon*. Treatments are natural formulations often including one or a mixture of various medicinal plants. Some of the 28 plants listed in the *Canon* have been studied clinically or *in vitro*. Different mechanisms of action can be considered for these medicinal plants, including antimicrobial, immune-modulating, anti-inflammatory, and anti-cough activities. Since most of the studied plants have shown inhibitory growth of *M. tuberculosis*, this effect can be considered as the main mechanism of TB treatments. This is even more valuable in case studies on drug-resistant strains. However, detailed research work investigating the antibacterial effects of many of the plants listed in *Canon* and their mechanisms and effective components has not yet been conducted.

References

Advenier, C., Lagente, V. and Boichot, E. (1997) The role of tachykinin receptor antagonists in the prevention of bronchial hyperresponsiveness, airway inflammation and cough. *European Respiratory Journal* 10(8), 1892–1906. DOI: 10.1183/09031936.97.10081892.

Alem, G., Mekonnen, Y., Tiruneh, M. and Mulu, A. (2008) *In vitro* antibacterial activity of crude preparation of myrtle (*Myrtus communis*) on common human pathogens. *Ethiopian Medical Journal* 46(1), 63–69.

Ali, A., Jehangir, A. and Ayub, F. (2016) Hepatoprotective effect of aqueous extract of *Chichorium intybus* roots on isoniazid induced hepatotoxicity. *Journal of Islamic International Medical College* 11(3), 99–102.

Al-Hussaini, R. and Mahasneh, A.M. (2011) Antibacterial and antifungal activity of ethanol extract of different parts of medicinal plants in Jordan. *Journal of Pharmaceutical Sciences* 4(1), 57–69.

Al-Snafi, A.E. (2015) The chemical constituents and pharmacological effects of *Carum carvi* – a review. *Indian Journal of Pharmaceutical Sciences* 5(2), 72–82.

Anand, P.K., Kaul, D. and Sharma, M. (2006) Green tea polyphenol inhibits *Mycobacterium tuberculosis* survival within human macrophages. *International Journal of Biochemistry & Cell Biology* 38(4), 600–609. DOI: 10.1016/j.biocel.2005.10.021.

Annaz, H., Sane, Y., Bitchagno, G.T.M., Ben Bakrim, W., Drissi, B. *et al.* (2022) Caper (*Capparis spinosa* L.): an updated review on its phytochemistry, nutritional value, traditional uses, and therapeutic potential. *Frontiers in Pharmacology* 13(1), 878749. DOI: 10.3389/fphar.2022.878749.

Bansal, R., Sharma, D. and Singh, R. (2018) Tuberculosis and its treatment: an overview. *Mini-Reviews in Medicinal Chemistry* 18(1), 58–71. DOI: 10.2174/1389557516666160823160010.

Bartha, G.S., Tóth, G., Horváth, P., Kiss, E., Papp, N. *et al.* (2019) Analysis of aristolochlic acids and evaluation of antibacterial activity of *Aristolochia clematitis* L. *Biologia Futura* 70(4), 323–329. DOI: 10.1556/019.70.2019.36.

Bass, J.B. Jr, Farer, L.S., Hopewell, P.C., O'Brien, R., Jacobs, R.F. *et al.* (1994) Treatment of tuberculosis and tuberculosis infection in adults and children. American Thoracic Society and the Centers for Disease Control and Prevention. *American Journal of Respiratory and Critical Care Medicine* 149(5), 1359–1374. DOI: 10.1164/ajrccm.149.5.8173779.

Ben Hsouna, A., Hamdi, N., Miladi, R. and Abdelkafi, S. (2014) *Myrtus communis* essential oil: chemical composition and antimicrobial activities against food spoilage pathogens. *Chemistry & Biodiversity* 11(4), 571–580. DOI: 10.1002/cbdv.201300153.

Bhusari, S. (2010) Modulation of bioavailability of selected anti tuberculosis drugs by herbal preparations of *Cuminum cyminum* and *Carum carvi*. PhD thesis, Birla Institute of Technology and Science–Pilani, Rajasthan, India.

Bloom, B.R., Atun, R., Cohen, T., Dye, C., Fraser, H. *et al.* (2017) Tuberculosis. In: Holmes, K.K., Bertozzi, S., Bloom, B.R. and Jha, P. (eds) *Major Infectious Diseases*, 3rd edn. International Bank for Reconstruction and Development/World Bank, Washington, DC, pp. 233–314.

Boskabady, M.H., Kiani, S. and Rakhshandah, H. (2006) Relaxant effects of *Rosa damascena* on guinea pig tracheal chains and its possible mechanism(s). *Journal of Ethnopharmacology* 106(3), 377–382. DOI: 10.1016/j.jep.2006.01.013.

Boskabady, M.H., Shafei, M.N., Saberi, Z. and Amini, S. (2011) Pharmacological effects of *Rosa damascena*. *Iranian Journal of Basic Medical Sciences* 14(4), 295–307.

Brodhun, B., Altmann, D. and Hauer, B. (2019) *Bericht zur Epidemiologie der Tuberkulose in Deutschland für*. Epidemiologisches Bulletin. Robert Koch-Institut, Berlin, pp. 1–99.

Choudhary, N., Khajuria, V., Gillani, Z.H., Tandon, V.R. and Arora, E. (2014) Effect of *Carum carvi*, a herbal bioenhancer on pharmacokinetics of antitubercular drugs: a study in healthy human volunteers. *Perspectives in Clinical Research* 5(2), 80–84. DOI: 10.4103/2229-3485.128027.

Edláková, J., Kocourková, B., Lojková, L. and Kubáň, V. (2003) Determination of essential oil content in caraway (*Carum carvi* L.) species by means of supercritical fluid extraction. *Plant, Soil and Environment* 49(6), 277–282.

Ehsanifar, A., Kazemipoor, N. and Fouladi, B. (2017) Anti-mycobacterial activity of *Capparis spinosa* L. extract against clinical isolates of *Mycobacterium tuberculosis*. *Asian Journal of Pharmaceutical and Clinical Research* 10(8), 181–185. DOI: 10.22159/ajpcr.2017.v10i8.17629.

Fatemeh, F., Karimi, A., Abbassion, M., Khiabani Rad, P. and Karimi Yazdi, M. (2015) Extraction of *Rosa damascena* as an antibacterial agent. *International Journal of Mycobacteriology* 4(5), 169.

Galitskiĭ, L.A., Barnaulov, O.D., Zaretskiĭ, B.V., Malkov, M.I., Konenkov, S.I. *et al.* (1997) Bliianie fitoterapii na vozmozhnost' preduprezhdeniia i ustraneniia gepatotoksicheskikh reaktsiĭ u bol'nykh tuberkulezom legkikh, nositeleĭ markerov virusa gepatita B (Effect of phytotherapy on the prevention and elimination of hepatotoxic responses in patients with pulmonary tuberculosis, carriers of hepatitis B virus markers). *Problemy Tuberkuleza* 4, 35–38.

Gaur, R., Thakur, J.P., Yadav, D.K., Kapkoti, D.S.,Verma, R.K. *et al.* (2015) Synthesis, antitubercular activity, and molecular modeling studies of analogues of isoliquiritigenin and liquiritigenin, bioactive

components from *Glycyrrhiza glabra*. *Medicinal Chemistry Research* 24, 3494–3503. DOI: 10.1007/s00044-015-1401-1.
Gülbay, B.E., Gürkan, O.U., Yildiz, O.A., Onen, Z.P., Erkekol, F.O. *et al.* (2006) Side effects due to primary antituberculosis drugs during the initial phase of therapy in 1149 hospitalized patients for tuberculosis. *Respiratory Medicine* 100(10), 1834–1842. DOI: 10.1016/j.rmed.2006.01.014.
Gupta, V.K., Fatima, A., Faridi, U., Negi, A.S., Shanker, K. *et al.* (2008) Antimicrobial potential of *Glycyrrhiza glabra* roots. *Journal of Ethnopharmacology* 116(2), 377–380. DOI: 10.1016/j.jep.2007.11.037
Haouat, A.C., Guendouzi, S., Haggoud, A., David, S., Sqalli, H. *et al.* (2013) Antimycobacterial activity of *Populus alba* leaf extracts. *Journal of Medicinal Plants Research* 7, 1015–1021.
Hassan, F. and Naeem, I. (2014) Biological activity of *Viola odorata* Linn. against *Mycobacterium tuberculosis*. *International Journal of Pharma and Bio Sciences* 5(3), 61–69.
Heidari, M., Bidoki, K. and Vafazade, J. (2003) Evaluation of the analgesic effect of methanolic extract of *Capparis spinosa* in mice. Dissertation, Kerman Medical University, Kerman, Iran, pp. 44–47 (in Persian).
Ibn Sinā, H.A. (2015a) *Al-Qanun fi'l-Tibb* (*The Canon of Medicine*), Vol. 3. Egypt University of Sciences and Technology, Cairo, pp. 416–422, 493–515 (in Arabic).
Ibn Sinā, H.A. (2015b) *Al-Qanun fi'l-Ṭibb* (*The Canon of Medicine*), ed. A. Masoudi. Almaʿee Publishing House, Tehran.
Iwu, M.W., Duncan, A.R. and Okunji, C.O. (1999) New antimicrobials of plant origin. In: Janick, J. (ed.) *Perspectives on New Crops and New Uses.* ASHS Press, Alexandria, Virginia, pp. 457–462.
Jalilzadeh-Amin, G., Najarnezhad, V., Anassori, E., Mostafavi, M. and Keshipour, H. (2015) Antiulcer properties of *Glycyrrhiza glabra* L. extract on experimental models of gastric ulcer in mice. *Iranian Journal of Pharmaceutical Research* 14(4), 1163–1170.
Jiménez-Arellanes, A., León-Díaz, R., Meckes, M., Tapia, A., Molina-Salinas, G.M. *et al.* (2012) Antiprotozoal and antimycobacterial activities of pure compounds from *Aristolochia elegans* rhizomes. *Evidence-Based Complementary and Alternative Medicine* 2012, 593403. DOI: 10.1155/2012/593403.
Kuo, P.C., Li, Y.C. and Wu, T.S. (2012) Chemical constituents and pharmacology of the *Aristolochia* (mădōu ling) species. *Journal of Traditional and Complementary Medicine* 2(4), 249–266. DOI: 10.1016/s2225-4110(16)30111-0.
Lee, J.Y., Lee, J.H., Park, J.H., Kim, S.Y., Choi, J.Y. *et al.* (2009) Liquiritigenin, a licorice flavonoid, helps mice resist disseminated candidiasis due to *Candida albicans* by Th1 immune response, whereas liquiritin, its glycoside form, does not. *International Immunopharmacology* 9(5), 632–638. DOI: 10.1016/j.intimp.2009.02.007.
León-Díaz, R., Meckes-Fischer, M., Valdovinos-Martínez, L., Campos, M.G., Hernández-Pando, R. *et al.* (2013) Antitubercular activity and the subacute toxicity of (–)-licarin A in BALB/c mice: a neolignan isolated from *Aristolochia taliscana*. *Archives of Medical Research* 44(2), 99–104. DOI: 10.1016/j.arcmed.2012.12.006.
Maleev, A., Neshtev, G., Stoianov, S. and Sheikov, N. (1972) The ulcer protective and antiinflamatory effect of Bulgarian rose oil. *Eksperimentalna Meditsina i Morfologiia* 11, 55–60.
McKee, M. and Atun, R. (2006) Beyond borders: public-health surveillance. *The Lancet* 367 (9518), 1224–1226. DOI: 10.1016/S0140-6736(06)68520-6.
Moy, R.L. and Levenson, C. (2017) Sandalwood album oil as a botanical therapeutic in dermatology. *Journal of Clinical and Aesthetic Dermatology* 10(10), 34–39.
Munir, A., Jehangir, A. and Ayub, F. (2019) Hepatoprotective effect of aqueous extract of *Cichorium intybus* (Kasni) on pyrazinamide induced hepatotoxicity in male mice. *Journal of Islamic International Medical College* 14(2), 72–75.
Navarro-García, V.M., Luna-Herrera, J., Rojas-Bribiesca, M.G., Álvarez-Fitz, P. and Ríos, M.Y. (2011) Antibacterial activity of *Aristolochia brevipes* against multidrug-resistant *Mycobacterium tuberculosis*. *Molecules (Basel, Switzerland)* 16(9), 7357–7364. DOI: 10.3390/molecules16097357.
Newton, S.M., Lau, C., Gurcha, SS., Besra, GS. and Wright, CW. (2002) The evaluation of forty-three plant species for *in vitro* antimycobacterial activities; isolation of active constituents from *Psoralea corylifolia* and *Sanguinaria canadensis*. *Journal of Ethnopharmacology* 79(1), 57–67.
Oliveira, D.G., Prince, K.A., Higuchi, C.T., Santos, A.C.B., Lopes, L.M.X. *et al.* (2007) Antimycobacterial activity of some Brazilian indigenous medicinal drinks. *Revista de Ciências Farmacêuticas Básica e Aplicada* 28(2), 165–169.
Özkan, G., Sagdiç, O. and Baydar, H. (2004) Antioxidant and antibacterial activities of *Rosa damascena* flower extracts. *Food Science and Technology International* 10(4), 277–281. DOI: 10.1177/1082013204045882.

Pereira, A., Avila, J., Do Carmo, G., Siqueira, F., Campos, M. *et al.* (2018) Chemical composition, antimicrobial and antimycobacterial activities of *Aristolochia triangularis* Cham. from Brazil. *Industrial Crops and Products* 121, 461–467. DOI: 10.1016/j.indcrop.2018.05.052.

Qasemzadeh, M.J., Sharifi, H., Hamedanian, M., Gharehbeglou, M., Heydari, M. *et al.* (2015) The effect of *Viola odorata* flower syrup on the cough of children with asthma: a double-blind, randomized controlled trial. *Journal of Evidence-Based Complementary & Alternative Medicine* 20(4), 287–291. DOI: 10.1177/2156587215584862.

Rabahi, M.F., Silva Júnior, J., Ferreira, A., Tannus-Silva, D. and Conde, M.B. (2017) Tuberculosis treatment. *Jornal Brasileiro de Pneumologia* 43(6), 472–486. DOI: 10.1590/S1806-37562016000000388.

Rants'o, A., Alturki, M., Levenson, C. and Calderón, A.I. (2017) LC-MS-based screening of East Indian Sandalwood oil for *Mycobacterium tuberculosis* shikimate kinase and *Plasmodium falciparum* thioredoxin reductase inhibitory activities. *Natural Product Communications* 12(9), 1491–1494. DOI: 10.1177/1934578X1701200928.

Rasooli, I. and Allameh, A. (2016) Caraway (*Carum carvi* L.) In: Preedy, R (ed.) *Essential Oils in Food Preservation, Flavor and Safety.* Elsevier, Amsterdam, pp. 287–293.

Rhizopoulou, S., Ioannidi, E., Alexandredes, N. and Argiropoulos, A. (2006) A study on functional and structural traits of the nocturnal flowers of *Capparis spinosa* L. *Journal of Arid Environments* 66(4), 635–647. DOI: 10.1016/j.jaridenv.2005.12.009.

Roque, L., Duarte, N., Bronze, M.R., Garcia, C., Alopaeus, J. *et al.* (2018) Development of a bioadhesive nanoformulation with *Glycyrrhiza glabra* L. extract against *Candida albicans*. *Biofouling* 34(8), 880–892. DOI: 10.1080/08927014.2018.1514391.

Sachin, B.S., Monica, P., Sharma, S.C., Satti, N.K., Tikoo, M.K. *et al.* (2009) Pharmacokinetic interaction of some antitubercular drugs with caraway: implications in the enhancement of drug bioavailability. *Human & Experimental Toxicology* 28(4), 175–184. DOI: 10.1177/0960327108097431.

Sadowska, A. and Obidoska, G. (1998) *Pharmacological Uses and Toxicology of Caraway*. In: Nemeth, E. (ed.) *Caraway. The Genus Carum*. Harwood Academic Publishers, London, pp. 165–174.

Shafei, M.N., Rakhshandahb, H. and Boskabady, M.H. (2003) Antitussive effect of *Rosa damascena* in guinea pigs. *Iranian Journal of Pharmaceutical Research* 2(4), 231–234. DOI: 10.22037/ijpr.2010.61

Saif, H., Sandeep, H., Shiv, N. and Zeeshan, F. (2021) Mechanistic insights into the antimycobacterial action of Unani formulation, Qurs Sartan Kafoori. *Journal of Traditional and Complementary Medicine* 12(2), 162–171. DOI: 10.1016/j.jtcme.2021.07.009.

Sharma, H., Yunus, G.Y., Agrawal, R., Kalra, M., Verma, S. *et al.* (2016) Antifungal efficacy of three medicinal plants *Glycyrrhiza glabra*, *Ficus religiosa*, and *Plantago major* against oral *Candida albicans*: a comparative analysis. *Indian Journal of Dental Research* 27(4), 433–436. DOI: 10.4103/0970-9290.191895.

Shi, LS., Kuo, PC., Tsai, YL., Damu, AG. and Wu, TS. (2004) The alkaloids and other constituents from the root and stem of *Aristolochia elegans*. *Bioorganic & Medicinal Chemistry* 12(2), 439–446. DOI: 10.1016/j.bmc.2003.10.031.

Soltani, A. (2004) *Encyclopedia of Traditional Medicine (Dictionary of Medicinal Plants)*. Arjmand Publisher, Tehran.

Stewart, G.R., Robertson, B.D. and Young, D.B. (2003) Tuberculosis: a problem with persistence. *Nature Reviews Microbiology* 1(2), 97–105. DOI: 10.1038/nrmicro749.

Suárez, I., Fünger, S.M., Kröger, S., Rademacher, J., Fätkenheuer, G. *et al.* (2019) The diagnosis and treatment of tuberculosis. *Deutsches Aerzteblatt International* 116(43), 729–735. DOI: 10.3238/arztebl.2019.0729.

Talib, W. and Mahasneh, A.M. (2010) Antiproliferative activity of plant extracts used against cancer in traditional medicine. *Scientia Pharmaceutica* 78, 33–45. DOI: 10.3797/scipharm.0912-11.

Tu, X., Ling, F., Huang, A., Zhang, Q. and Wang, G. (2013) Anthelmintic efficacy of *Santalum album* (Santalaceae) against monogenean infections in goldfish. *Parasitology Research* 112(8), 2839–2845. DOI: 10.1007/s00436-013-3455-7.

Turner, R.D. and Bothamley, G.H. (2011) Cough and the transmission of tuberculosis. *The Journal of Infectious Diseases* 211(9), 1367–1372. DOI: 10.1093/infdis/jiu625.

Viswanathan, V., Pharande, R., Bannalikar, A., Gupta, P., Gupta, U. and Mukne, A. (2019) Inhalable liposomes of *Glycyrrhiza glabra* extract for use in tuberculosis: formulation, *in vitro* characterization, *in vivo* lung deposition, and *in vivo* pharmacodynamic studies. *Drug Development and Industrial Pharmacy* 45(1), 11–20. DOI: 10.1080/03639045.2018.1513025.

Vyas, P., Chandola, H.M., Ghanchi, F. and Ranthem, S. (2012) Clinical evaluation of Rasayana compound as an adjuvant in the management of tuberculosis with anti-Koch's treatment. *Ayu* 33(1), 38–43. DOI: 10.4103/0974-8520.100307.

Welz, A.N., Emberger-Klein, A. and Menrad, K. (2018) Why people use herbal medicine: insights from a focus-group study in Germany. *BMC Complementary and Alternative Medicine* 18(1), 92. DOI: 10.1186/s12906-018-2160-6.

WHO (2016) *WHO Treatment Guidelines for Drug-Resistant Tuberculosis: 2016 Update. October 2016 Revision*. World Health Organization, Geneva, Switzerland. Available at: https://www.who.int/publications/i/item/9789241549639 (accessed 2 April 2023).

WHO (2018) *Latent Tuberculosis Infection: Updated and Consolidated Guidelines for Programmatic Management*. World Health Organization, Geneva, Switzerland. Available at: https://apps.who.int/iris/handle/10665/260233 (accessed 2 April 2023).

WHO (2022) Tuberculosis: Key facts. Available at: https://www.who.int/news-room/fact-sheets/detail/tuberculosis (accessed 27 October 2022).

Wichtmann, E.M. and Stahl-Biskup, E. (1987) Composition of the essential oils from caraway herb and root. *Flavour and Fragrance Journal* 2(2), 83–89. DOI: 10.1002/ffj.2730020207.

Yan, C., Liu, W., Li, J., Deng, Y., Chen, S. *et al.* (2018) Bioactive terpenoids from *Santalum album* derived endophytic fungus *Fusarium* sp. YD-2. *RSC Advances* 8(27), 14823–14828. DOI: 10.1039/c8ra02430h.

Yassa, N., Masoomi, F. and Hadjiakhoondi, A. (2009) Correspondence chemical composition and antioxidant activity of the extract and essential oil of *Rosa damascena* from Iran, population of Guilan. *DARU Journal of Pharmaceutical Sciences* 17, 175–180.

Zanetti, S., Cannas, S., Molicotti, P., Bua, A., Cubeddu, M. *et al.* (2010) Evaluation of the antimicrobial properties of the essential oil of *Myrtus communis* L. against clinical strains of *Mycobacterium* spp. *Interdisciplinary Perspectives on Infectious Diseases* 2010, 931530. DOI: 10.1155/2010/931530.

10 Application of Certain Medicinal Herbs to Cure Various Viral Diseases and Enhance the Human Immune System

Zahra Boghrati[1], Elaheh Zibaee[1], Maryam Nikoosokhan[2], Hamed Baharara[3], Sajad Azad[1], and Seyed Ahmad Emami[1,4*]

[1]Department of Traditional Pharmacy, School of Pharmacy, Mashhad University of Medical Sciences, Mashhad, Iran; [2]Science and Technology Park of University of Tehran, College of Agriculture & Natural Resources, University of Tehran, Karaj, Iran; [3]Department of Clinical Pharmacy, School of Pharmacy, Mashhad University of Medical Sciences, Mashhad, Iran; [4]Department of Pharmacognosy, School of Pharmacy, Mashhad University of Medical Sciences, Mashhad, Iran

Abstract

Virology has a long history dating back to the late 1800s. The first viruses to be discovered were the tobacco mosaic virus and the virus responsible for foot-and-mouth disease, as well as other viruses that cause various infectious diseases in plants and animals, respectively. Although infectious diseases such as measles, rabies, catarrh, warts and others have long been studied clinically, there is no section in Avicenna's *Canon of Medicine* that deals explicitly with viral infections. Notwithstanding this, this chapter summarizes several viral infections mentioned in the third and fourth volumes of the *Canon*, as well as a brief pathology and holistic approaches to the subject. The infections discussed are: nasal discharge and catarrh (in the nose section of the *Canon*), smallpox and measles (in the section on infectious fever), herpes (in the section on inflammation and pustules), rabies (in the section on bites in humans and animals), and warts (in the section on cosmetic art).

Nasal Discharge and Catarrh

Nasal discharge and catarrh in view of Avicenna

In the third volume of the *Canon*, in the section on nose illnesses, Avicenna covered *zokam* (nasal discharge) and *nazleh* (catarrh) (Ibn Sinā, 2015). He believed that the two share a common cause with the discharge originating from the brain, while in *zokam* the secretions flow down the nasal passages and in *nazleh* they descend down the back of the throat.

*Email: emamia@mums.ac.ir

 Medicinal Plants Used in Traditional Persian Medicine (eds. H. Schulz, Seyed Ahmad Emami and Farsad Nadjafi)
DOI: 10.1079/9781800621671.0010

Nasal discharges are dilute and continuous, affecting eyes and face color and impairing the sense of smell.

Catarrh can cause lesions in the mouth, throat, lungs, esophagus, and stomach. Bulimia, pharyngitis, pleuritis, pulmonary inflammation, and tuberculosis are the most common side effects.

Avicenna classified catarrh into two types: warm and cold, each with its own symptoms and treatments. Temperamental or external heat, such as the heat of the sun or toxins, or the odor of hot-temperament medications like saffron and onion, could be the cause of warm catarrh. Cold catarrh can be caused by a cold temperament or external coldness, such as being outside in cold weather, especially if the head is exposed, or after a shower or exercise.

Catarrh clinical manifestation

If warm catarrh discharges into the nasal canal, symptoms include thin, stinging secretions, redness of the face and eyes, and a rise in body temperature. In this case, inducing fever would not help the situation. If secretions discharge to the back of the throat, symptoms include thin, yellowish, and red mucus, which causes an intense burning and irritation in the throat, titillation in the nose, and nasal congestion.

If cold catarrh discharges into the nasal canal, the following symptoms may occur: thick and cold-nature secretions, nasal titillation, and stuffiness and tension in the forehead. If secretions flow to the back of the throat, the symptoms include cold and whitish mucus. In this case, fever may help to improve the state of health.

Treatments

Avicenna lists a variety of remedies for catarrh and nasal discharge, including lifestyle modification, dietary changes, oral or topical medications, and *fasd* (phlebotomy). These practices are intended to reduce the matter and cause of catarrh, reduce secretions, and prevent disease side effects such as nose sores, impaired sense of smell, throat irritation, and cough. Patients should keep their heads warm and avoid overeating, drinking cold water, and napping during the day. If they are suffering from the symptoms of a dry cough, phlebotomy in small amounts may also help. Generally, showering should be avoided when fever is present.

Phlebotomy, expelling hot humors, and employing enemas to draw materials to the lower portions of the body are beneficial for warm catarrh. The patient should keep the head cool by showering early in the morning before breakfast, applying violet (*Viola odorata*) oil to the head, limbs, and belly button, applying douches of barley (*Hordeum vulgare*), poppy (*Papaver somniferum*), violet (*V. odorata*), and chamomile (*Anthemis nobilis*) to the head, eating foods in cold and wet nature, and eating rose (*Rosa* × *damascena*) blossom honey during the day. Some of the incenses beneficial for warm catarrh are prepared from the following plant species: *Tamarix gallica*, *R* × *damascena*, *Saccharum officinarum*, *Faba vulgaris*, and *Cinnamomum camphora*.

Enemas and oral medications help clear phlegm from the head for cold catarrh. Generally, it is necessary to eat and drink less, and if cold catarrh is accompanied by throat titillation and sneezing, the patient must first warm his head with warm rags. In addition, rubbing the hands and feet with cold oils prepared from dill (*Anethum graveolens*), chamomile (*A. nobilis*), sweet marjoram (*Origanum majorana*), and stronger oils from bearded iris (*Iris germanica*), common rue (*Ruta gravolens*), moringa (*Moringa pterygosperma*), and bay tree (*Laurus nobilis*) is quite beneficial. The food should be hot and dry in nature, with a soothing effect on the throat, and it should be consumed only in modest quantities. Cooling remedies such as cold water, rose water, lentil (*Lens esculenta*), and coriander (*Coriandrum sativum*) water can be gargled to reduce secretions. Pomegranate (*Punica granatum*) juice can be also applied as a gargle or mouth rinse to treat both warm and cold catarrh.

Diffusing various incenses such as frankincense (various *Boswellia* species), sandarac tree (*Tetraclinis articulata*), agarwood (various *Aquilaria* species), and crepe ginger (*Cheilocostus speciosus*) helps to minimize discharges by penetrating astringent substances into the nostrils.

Another method of treating catarrh is to change the consistency of the secretions;

for example, linctures containing tragacanth (*Astragalus tragacantha* gum) and quince (*Cydonia oblonga*) seeds thicken the thin discharges and make them easier to cough up.

Dietary changes may also assist in reducing the symptoms of catarrh on the lungs and throat. Rubbing the chest with violet (*V. odorata*) oil, eating barley (*H. vulgare*) water with honey preserved violets, sweet pomegranate juice, soups made of starch, broad bean (*Vicia faba*) and barley flour, milk (however milk is forbidden in cases of fever!), as well as consuming soft and cool tinctures and hyssop (*Hyssopus officinalis*) beverages are all useful to treat warm catarrh. For cold catarrh, the instructions are as follows: rubbing the chest with violet and *M. pterygosperma* oils, drinking warm water, eating warm and soothing soups like water of wheat bran decoction with almond oil, honey or bread and grape juice (boiled until one-third remains), and consuming warm and soft tinctures and warm hyssop (*H. officinalis*) drinks.

Common cold in modern medicine

The respiratory epithelium is affected by some rhinoviruses. Half of all common cold diseases are caused by rhinoviruses, which belong to the RNA viruses and are commonly referred to as "cold virus". They cause infections of the upper respiratory system including symtoms like coughing, nasal discharge, and sore throat. Adenoviruses, parainfluenza viruses, influenza viruses, and coronaviruses can also cause colds with similar symptoms of illness.

Pathobiology

Very often, a nonspecific host inflammatory response is responsible for the respective disease symptoms. Usually, virus propagation is triggered by microdroplets during direct contact with an infected person or objects contaminated with various viruses.

Clinical manifestation

Nasal blockage and rhinorrhea are common clinical symptoms, as is a sore throat. A change in the color of nasal discharge is also prevalent. Nasal epithelium infection is the initial manifestation, and congestion is due to blood pooling in capacitance vessels of the nose, nasal blood flow, and increased vascular permeability. Serum leakage into nasal secretions causes rhinorrhea, and postnasal drip with pharyngeal irritation might induce cough.

Treatment

The treatment protocol is symptomatic therapy. Usually, antibacterial and specific antiviral therapy is ineffective. Symptomatic treatments are not recommended for children under the age of 4 years. Adrenergic medications for nasal congestion and antihistamines for rhinorrhea are commonly prescribed. Cough is unaffected by codeine, dextromethorphan, or guaifenesin, but it responds to an antihistamine and decongestant combination (Turner, 2012a).

Smallpox and Measles

Smallpox in view of Avicenna

Smallpox, according to Avicenna, appears when the blood "boils" due to an infection (Ibn Sinā, 2015). This "boiling" could be caused by blood purging inharmonious and inappropriate materials or compensating for its consolidation. According to *Canon*, smallpox is most common in children. However, even the elderly might become infected, particularly if the weather is hot and humid. Spring and early summer are the most common times for the disease to appear. Avicenna indicated some early signs of smallpox, such as: back pain, sleep terrors, nose itching, pins and needles sensation, redness of the face and eyes, overflowing tears, hoarseness and dyspnea, and parched mouth. These symptoms are accompanied by a persistent fever. The rashes of smallpox can affect not just the skin but also various internal organs. Itching is the initial symptom of smallpox. Following that, some miliary pinpoint rashes form, which raise and become filled with pus and finally develop, sloug, and shed after evacuation.

In general, it has been found that in light-colored smallpox, the course of the disease is milder, while in dark-colored smallpox, a more severe course of the disease is often observed. Asphyxia, pharyngitis, or severe diarrhea leading to significant weakness are the most common causes of mortality in patients.

Smallpox in modern medicine

The poxviruses include the genus *Orthopoxvirus*, the human variola viruses (smallpox), the genus *Molluscipoxvirus*, molluscum contagiosum, and some viruses that naturally infect animals but can cause incidental infection in humans (zoonoses). Poxvirus infection in humans is characterized by a cutaneous manifestation that can be localized or systemic. Naturally acquired variola virus infection induces fever and a distinctive rash (Murray *et al.*, 2016).

Pathobiology

Infection is caused by inhalation of respiratory droplets and virus implantation on the respiratory and oropharyngeal mucosa. The virus replicates when it travels to local lymph nodes. The characteristic macular, popular, and vesicular lesions are seen in the epidermis. Clinical descriptions of smallpox types are fever, rash, pustules, erythema, and skin edema. There is an asymptomatic period of 10–14 days followed by a fever rising quickly to around 38–40°C, occasionally with multiple petechiae, and a systematic rash appearing a day or two later. In 90% of cases, ordinary smallpox results in rash, fever, and viremia (Turner, 2012b).

Treatment

Supportive treatment entails maintaining hydration and electrolyte balance, skin care, and monitoring for infection consequences. While some antiviral medications may help treat smallpox, there is no smallpox treatment that has yet proven successful in persons who are sick with the disease. Smallpox is preventable by vaccination. Currently, smallpox immunizations are not recommended for the general public because smallpox has been eradicated. However, if a smallpox outbreak were to occur one day, health officials would use smallpox vaccines to control it.

There are drugs approved by the US Food and Drug Administration (FDA) for the treatment of smallpox but these medications have not been tried on people who have smallpox. However, if a smallpox outbreak occurs, their usage may be contemplated. The FDA approved Tembexa (brincidofovir) to treat smallpox on June 4, 2021 (Hutson *et al.*, 2021).

Prevention

INFECTION CONTROL PRECAUTIONS. For patients with an acute generalized vesicular or pustular rash, contact precautions should include properly fitted N95 respirators, gloves, and gowns.

VACCINATION. There are two available smallpox vaccines: a replication-deficient modified vaccinia Ankara vaccine (MVA) and a replication-competent smallpox vaccine (ACAM2000).

1. The MVA (marketed as Imvamune in Canada, Imvanex in Europe, and Jynneos in the USA) has a high safety profile, even in immunocompromised patients and those with skin diseases. This vaccine can be used for the prevention of both smallpox and monkeypox (Overton *et al.*, 2018).
2. The ACAM2000 (Smallpox [Vaccinia] Vaccine, Live), a replication-competent vaccinia virus vaccine for use in emergency and routine nonemergency settings, is approved for active immunization against smallpox disease in persons assessed to be at high risk for smallpox infection (Brett and Petersen, 2015).

Measles in view of Avicenna

Avicenna believed that measles is generally similar to yellow smallpox (Ibn Sinā, 2015). The only differences are:

1. Measles is made up of yellow bile humor.
2. The measles rashes are much smaller than smallpox and are usually flat.

3. Measles is less common in general.
4. Measles rarely affects the eyes.

Additionally, the onset signs of measles and smallpox have some distinctions:

5. Measles patients experience more nausea, depression, and hot flashes.
6. The back pain is less in measles.
7. The appearance of measles is sudden but smallpox develops gradually.

Treatment of smallpox and measles

Phlebotomy (*fasd*) is the first and best treatment for smallpox and measles (if hyperemia is present). Nasal phlebotomy protects the upper organs from smallpox and is more convenient, especially for youngsters.

DIETARY RECOMMENDATIONS. During the early stages of the disease, nutritious, anti-inflammatory, and mildly laxative meals, such as tamarind (*Tamarindus indica*), jujube, and skinless lentil pottages, should be served. It is important for patients to have regular bowel movements, so tamarind is highly recommended as a proven laxative in this context. If it is not successful on its own, manna or hedysarum, as well as soaked plums, may be added.

The physician should facilitate the exposure based on the appearance of the rashes in order to protect the inner organs from injury. The skin pores should be opened by topical use of fennel (*Foeniculum vulgare*) or celery (*Apium graveolens*) juices mixed with sugar and by smelling saffron (*Crocus sativus*). According to Avicenna, fig (*Ficus carica*) juice is also extremely effective for this purpose. Other medications include rhubarb (*Rheum ribes*) paste, vert jus (a juice obtained from unripe grapes), cold-tempered fruits, and date palm blooms (*Phoenix dactylifera*).

Furthermore, the patient should be kept warm and not exposed to cold temperatures or cold beverages.

In the late stages, again the bowel movements should be regular. However, the diet changes and whole lentils substitute the skinless ones, and sour pomegranate juice, sumac (*Rhus coriaria*), or vert jus replace tamarind.

After around 7 days, a gold needle can be used to puncture the fluid-filled lumps. Blisters can only be rubbed with salt water at this point. To improve efficiency, saffron and rose water can be added to salt water. This process helps the bumps to dry up and scab over.

Anointing the blisters should be avoided at all stages since it stops them from drying out. Topical application of barley, rice, broad bean (*V. faba*), or proso millet (*Panicum miliaceum*) flour with saffron, on the other hand, aids in the drying process.

Preventive remedies

In this section, preventive remedies suggested by Avicenna to protect vital organs such as the throat, eyes, nostrils, lungs, and intestines from smallpox harm are discussed.

PROTECTION OF EYES. Drops of fermented rice water, coriander (*C. sativum*) juice, sumac (*R. coriaria*), and camphor (*C. camphora*) should be placed in the eyes as early as possible to provide them an optimal protection. The application of pistachio oil is also described as beneficial in this context.

PROTECTION OF MOUTH AND THROAT. Sucking on pomegranate and mulberry (*Morus nigra*), chewing pomegranate seeds, and gargling mulberry paste are also described as very helpful.

PROTECTION OF NOSTRILS. Topical applications of extracts from yellow horned poppy (*Glaucium flavum*) and sandalwood (*Santalum album*), unripe grape extracts (vert jus), and vinegar, as well as smelling vinegar alone, are reported to be effective in this context according to Avicenna.

PROTECTION OF LUNGS. Also *Linctus*, a dish made from cooked lentils and poppy seeds, is described as very effective in protecting against smallpox and measles.

PROTECTION OF INTESTINES. To protect the digestive organs, the consumption of astringent foods is generally said to be helpful.

If diarrhea should occur at the end of the disease, a tablet of *Tabasheer* (obtained from the cavities of bamboo cane) is recommended as a polytherapeutic product, which should be given as a mixture with rhubarb juice and tablets of sorrel (*Rumex acetosa*).

Treatment of smallpox scars

In the section on cosmetics, Avicenna mentions more details on this topic. Just a few of the most effective treatments for smallpox scars are listed below:

- Dried *Arundo donax* roots, *V. faba* flour, *Astragalus fasciculifolius*, *Cucumis melo* seeds and dried peels, washed rice, egg white, *Iris florentina* oil, pistachio oil, and other ingredients.
- Sweet pomegranate, chickpea, soft-boiled egg yolk and chicken, partridge and meaty chicken broth are mentioned as foods that are also suitable to improve complexion.

Measles in modern medicine

The measles virus causes a highly contagious, sometimes fatal, but vaccine-preventable disease. Following exposure to the measles virus, approximately nine out of ten people who are not immune will become infected. Measles outbreaks are announced when the number of cases recorded in an area exceeds the expected number of cases, and outbreaks are more likely to spread in countries with weak health systems, low immunization rates, or coverage gaps.

Pathobiology

Respiratory droplets primarily transmit measles virus over short distances. Symptoms include fever, maculopapular rash, cough, coryza, or conjunctivitis, and people are infectious for several days before and after the onset of rash. Vaccinated individuals may experience milder or even no symptoms, The host immune response at sites of virus replication is responsible for the signs and symptoms. Measles and rubella can result in serious illness, birth abnormalities, and even death. Rubella, also known as "German measles", is a mild illness with distinctive lymphadenopathy but can have serious consequences for pregnant women and their infants (Moss, 2020).

Treatment

Because there is no specific antiviral treatment for measles, control of the disease is largely limited to prevention.

Despite the availability of safe and effective vaccination, measles is still endemic in many countries and causes considerable morbidity and mortality, particularly among children living in resource-poor environments (Hübschen *et al.*, 2022).

Fluids and antipyretics should be provided to the patient as supportive care. Bacterial infections should be treated promptly with appropriate antimicrobials. However, prophylactic antibiotics are ineffective and, accordingly, should not be used (Bleck and Rupprecht, 2000).

Herpes

Herpes in view of Avicenna

Herpes sores normally appear as one or more lumps swell and spread across the body, causing minor inflammations. The blisters are yellowish, inflamed, and spherical, resembling warts with a burning sensation like an ant sting. If left untreated, the herpes sores may become purulent or fade away (Ibn Sinā, 2015).

Treatment

The treatment should begin as soon as the outbreaks appear by purging the skin of harmful humors, as they might create recurrent ulcers. Skin erosion can occur if ulcers are left untreated.

Avicenna named a few herbal medications for the treatment of herpes. *Convolvulus scammonia* and whey are used for purging the herpes matter out. In erosive herpes, herbs that cool and moisturize the skin (such as *Lactuca sativa*, *Nymphaea lotus*, *Aeonium arboretum*, *Lemna minor*, and *Portulaca oleracea*) should be avoided. However, if these types of herbs are required, ground dry *Solanum nigrum* is preferred due to its exsiccation effect. Other recommended herbal treatments are: *Plantago major*, *Rubus fruticosus*, *L. esculenta*, barley flour, pomegranate peel, and grapevine cordons. Sick children should be exposed to heat in the bathroom, then immediately taken out and the affected areas of the skin rubbed with rose water and rose oil.

Herpes in modern medicine

Herpesviridae is a family of lipid-enveloped DNA viruses. The two herpes simplex virus (HSV) serotypes HSV-1 and HSV-2 are members of the *Alphaherpesvirinae* subfamily that grow rapidly and efficiently destroy the host cells. Lesions in the epidermis are characteristic of HSV infection and often involve mucosa, sometimes spreading to the nervous system. According to the immune response, the immune status of the host, and the site of infection, the clinical manifestations of HSV infection appear. Primary severe infections develop in persons without preexisting immunity to HSV serotypes, frequently with systemic signs and symptoms, with a higher rate of complications than recurrent episodes (Marques and Cohen, 2012).

Treatment

Many HSV infections require no specific treatment. All that is required for HSV infections is keeping the lesions clean and dry while they heal by themselves. Highly symptomatic or complicated infections need to be treated. The treatment protocol depends on the clinical manifestation, host immune status, and whether it is a primary or recurrent episode or considering suppressive therapy. Acyclovir, an acyclic guanosine analog, has a high therapeutic index because of its preferential activation in infected cells and the viral DNA polymerase inhibition. The l-valyl ester of acyclovir is valacyclovir, which achieves three- to fivefold higher bioavailability after oral administration. Treatment is effective only if used very early in the disease, especially in the prodromal or erythema lesion stages. Patients should have the medication available and be vigilant for the earliest signs and symptoms of recurrence (Marques and Cohen, 2012).

Rabies

Rabies in view of Avicenna

A painful ulcer is initially the only symptom after being bitten by a rabid dog. After a few days, other symptoms emerge, including bad thoughts and fantasies, nightmares, aggression, mental disorder, talking to oneself, fear of light, hiccup, thirst, fear of crowds, finger spasm, redness of the body (particularly the face), presentation of purulent wounds on the face, increasing pain, hoarse voice, crying, fear of water (hydrophobia), seizure, cold sweat, and death (Ibn Sinā, 2015).

When a patient refuses to drink water, there is the risk of death from dehydration. On the other hand, the patient may drink, but is unable to swallow the water and suffocates.

Avicenna considered that there was no possibility of surviving following the onset of hydrophobia. The time between the commencement of the ailment and the person's death, according to him, might range from 7 days to 6 months, with an average of 40 days. The patient's prognosis is better if the bite site bleeds profusely after being bitten by a rabid dog or if the person has hematuria after taking antidotes.

Treatment

First and foremost, Avicenna emphasizes the significance of delaying the healing of the bite wound for at least 40 days. For this, sucking, cupping, and the use of deobstruent medications are indicated.

Opopanax chironium in vinegar, onion, garlic, *Eruca sativa* boiled in water, *Ferula assa-foetida* gum, and *Beta vulgaris* are some of the plants that might assist open a bite wound.

The physician has to attempt to pull out the toxin rather than do a bodily evacuation, since the latter attracts the humors and toxin to the inside of the body. Cauterization is the best procedure for pulling the poison out through the bite site. In cases of interdiction, alternative therapies such as salt ointment and rubefacient medications like mustard poultice can act similarly. The purgative drugs should be melanagogue (expelling the black bile) like *Helleborus niger*, *Terminalia chebula*, *Cuscuta epithymum*, *Polypodium vulgare*, and others. Mild enemas, such as olive oil, beet juice, and cheese water with *C. epithymum*, should be used every day or every other day when using powerful cathartic drugs.

DIETARY RECOMMENDATIONS. Dietary modification is quite important. Meals should be moist and moderate, such as fowl broth soups, white bread soaked in cold water, and drinking the water that has extinguished a molten iron in it several times. In general, garlic and onions are quite useful in the treatment of toxins and the prevention of poison spread in the body. However, in rabies, combination drugs such as applied in cancer medicine, Galenus medicine, and some well-known antidotes (*teryagh*) are frequently recommended.

Some plant species used in single-drug therapy recommendations are: *Lycium afrum*, *F. assa-foetida*, *Artemisia absinthium*, *Teucrium polium*, *Nigella sativa* (highly effective), *Commiphora myrrha*, *Gentiana lutea*, and *Teucrium chamaedrys*.

TOPICAL MEDICATIONS. To remove the poison and maintain the wound open, topical medicines are utilized. Several doctors agreed that applying a rabid dog's liver to a bite wound is really effective. *F. assa-foetida* poultice, garlic, salty fish meat, vineyard ashes, salt and peppermint, *O. chironium*, *C. melo* leaves, and *F. vulgare* roots are also recommended.

INDUCING THE PATIENT TO DRINK WATER. As previously said, the rabies sufferer becomes extremely thirsty, but hydrophobia and swallowing difficulties prevent him from drinking water. However, the uptake of water can lead to a more convenient treatment. To attain this goal, people perform a variety of tricks and maneuvers. Some recommend using long and reedy iron or gold tubes to pour water into the patient's throat. Another approach is to use beeswax or solid honey to produce hollow pieces, then fill them with water before sealing them and feeding them to the patient.

Rabies in modern medicine

Rabies is a zoonotic, vaccine-preventable viral illness that causes acute infection of the central nervous system (CNS). Domestic canines are responsible for transmission to humans in 99% of cases through bites or scratches via contamination of mucous membranes with infected saliva. Although effective human rabies vaccines and immunoglobulins are available, they are not available to everyone in need; 80% of cases occur in rural regions, and children aged 5 to 14 years are the most common victims (Fooks and Jackson, 2000).

Diagnosis

Usually, diagnosis is based on epidemiologic and clinical findings; but without laboratory confirmation, and with only a clinical diagnosis, the disease in humans is often misdiagnosed. Every patient with neurologic signs or symptoms or unexplained encephalitis should be questioned if there is a history of animal bite exposure in a rabies-endemic area, because an acute progressive encephalitis (rabies) is virtually always fatal once clinical symptoms develop (Fooks and Jackson, 2000).

Pathobiology

After inoculation, the rabies virus may penetrate the peripheral nervous system and travel to the brain, or it may replicate in muscle tissue, remaining sequestered at the entrance site throughout incubation, before invading and replicating further in the CNS. The infection then spreads centrifugally to a variety of other organs. Fever, chills, malaise, anorexia, lethargy, insomnia, irritability, pain, or paresthesia in the limb are nonspecific symptoms; an encephalitic variant develops in about 80% of patients, causing behavioral disorders. Early in the course of illness, rabies can "mimic" many other encephalitides, such as those caused by herpesviruses and arboviruses (Singh *et al.*, 2020).

Treatment

Human rabies is best prevented by avoiding exposure to the disease. Oral vaccination of susceptible wild free-ranging animal populations with rabies vaccines offers hope for controlling the disease. By controlling rabies in dogs, the burden of human rabies is reduced. When exposure is suspected, postexposure prophylaxis should be started right away. After-bite treatment with local antirabies immunoglobulin, local wound cleansing, passive immunization with human rabies immune globulin (HRIG), and active immunization with rabies vaccination are appropriate (Singh *et al.*, 2020).

When critical care is not feasible due to medical or financial reasons, palliative care requires attention. Guidelines for palliative care in rabies are absent in endemic regions (Fooks and Jackson, 2000).

Warts

Warts in view of Avicenna

Avicenna believed that the accumulation of thick melancholic (black bile) humor in the blood is responsible for various raised bumps on the skin. As for warts, he believed that the cold and dry temperament of one organ can spread to its neighbor organs, causing the warts to spread over the skin. Treating or removing one of them aids in the removal of the others. Common warts, corns (plantar warts), and horn-shaped warts are in this category (Ibn Sinā, 2015).

Treatment

In the case of many spreading warts, phlebotomy and evacuating the black bile are required and the topical drugs should have bitterness and astringent properties. In case of mild disease, the treatment is also moderate.

Some examples of common topical treatments for warts are as follows: application of pistachio oil, wild leek juice with sumac (*R. coriaria*) and *M. pterygosperma* oil, *Capparis spinosa* leaves, cypress fruits, dried figs and green walnut shells; *Ceratonia siliqua* is beneficial for large resistant warts and *Myrtus communis* fresh leaves for both mild and severe cases. For severe warts, stronger and more effective combination medications are used that usually comprise mineral compounds.

Avicenna also instructed how to remove warts. This may be done with silver or iron tubes that have a sharp edge and are hollowed out to the extent that the wart is affixed with some hardness. Avicenna wrote:

> Wrap the tubes around the wart and immerse a bit at its base before cutting it out, or pull the wart out with a hook until you get the whole root out, then carefully cut it off with a sharp keen tool. Ghee is applied after cutting.
>
> Other approaches are to apply acute medicine on wart to loosen it. After that apply ghee and leave it until it removes. Or press a thin dimpled iron tool near the wart to make it fully visible, then cut it and treat it with acute medicine. Massage the area with soap, *Cyperus rotundus* and roses until the blood that flows from it drips and gets trapped and after that what remains will fall off.

Papillomavirus in modern medicine

The papovavirus family (*Papovaviridae*) is divided into *Papillomaviridae* and *Polyomaviridae*.

Human papillomavirus (HPV) is a group of DNA viruses that can cause both benign and malignant skin and mucous membrane lesions. Warts are a kind of benign epidermal neoplasm caused by HPVs. HPVs infect and replicate in the squamous epithelium of skin (warts) and mucous membranes (genital, oral, and conjunctival papillomas) to induce epithelial proliferation. The HPV types are very tissue specific (Murray *et al.*, 2016).

Warts come in over 100 varieties and commonly occur in children and young adults, but they may develop at any age and appear at sites of trauma, on the hands, in regions of nail biting, and on plantar surfaces.

Pathobiology

All types of HPV infection are more likely in people with weakened immune systems. Warts are normally harmless and most resolve spontaneously in weeks or months without treatment, especially in youngsters, but they can be uncomfortable, particularly on the foot. Some treatments can help get rid of warts faster, but if some viruses remain, new warts can emerge and others may last years or a lifetime.

Clinical manifestation

HPV infections can have a variety of clinical manifestations. Common warts are exophytic and hyperkeratotic. HPV strains 1, 2, 4, 27, and 57 are the most prevalent cause of papules on the hand, but they can form on any skin surface, including the vaginal area (Turner, 2012c).

Treatment

Topical salicylic acid (SA) solutions are over-the-counter medicines that are excellent at eliminating the top layer of a wart. They are well tolerated, but they may be more effective on hands than on feet. Cryotherapy, which involves freezing a wart with liquid nitrogen, is effective for common warts but was not proven to be more successful than no treatment for warts on the soles of the feet or plantar warts. There was no significant difference in effectiveness when comparing cryotherapy to placebo, and the same was true when comparing cryotherapy to SA. Cryotherapy causes blisters, scarring, inflammation, and discoloration, and usually hurts a little (Kwok *et al.*, 2012).

Liquid nitrogen and very light electrocautery are the best methods of initial therapy. Nightly application of the immunomodulatory drug imiquimod 5% cream may be effective for recalcitrant warts. There are some successful claims with apple cider vinegar: this involves soaking a small piece of cotton in vinegar and sticking it to the wart overnight. This process is repeated until the wart has disappeared.

Flat warts (verruca plana) are flat-topped papules caused mostly by HPV-3 and HPV-10. Typical sites of involvement include the mouth, forehead, backs of the hands, and shaved areas. Imiquimod 5% cream, administered every day or every other day, may be beneficial in the treatment of flat warts. Patients who are concerned about their cosmetic appearance and desire quick results can have lesions frozen using liquid nitrogen. For dramatic clearing of flat warts, treatment with 5-fluorouracil cream may be useful. A study of patients with recalcitrant facial plana (ages 5 to 35 years) reported that oral isotretinoin 0.5 mg/kg daily, prescribed for 2 months, resulted in a complete response in 73% of patients (Dinulos, 2021).

Pharmacologic Aspects

In this section, scientific evidences are provided on medicinal herbs cited in *Canon of Medicine* for various viral diseases. However, only around 30 out of more than 140 mentioned plants have been researched in detail for this purpose (Table 10.1). Tables 10.2, 10.3, and 10.4 respectively present the *in vitro*, *in vivo*, and clinical studies of these medicinal herbs against viral diseases.

Table 10.1. Medicinal plants mentioned in *Canon of Medicine* for treatment of viral diseases.

Scientific name	Family	Arabic name	English name
Acacia arabica (Lam.) Willd. Synonym: *Acacia nilotica* (L.) Willd. ex Delile	*Fabaceae* (*Leguminosae*)	*Sanṭ*	Gum arabic tree
Adiantum capillus-veneris L.	*Pteridaceae*	*Shaʿr el arḍ*, *shaʿr el kenzir*	Southern maidenhair fern
Allium sativum L.	*Amaryllidaceae*	*Thawm*	Garlic
Artemisia absinthium L.	*Asteraceae* (*Compositae*)	*Afsantin*	Common absinth
Capparis spinosa L.	*Capparaceae*	*Kabbār*	Caper bush
Chelidonium majus L.	*Papaveraceae*	*Māmirān*	Greater celandine
Cinnamomum cassia (L.) J. Presl	*Lauraceae*	*Salikhah*	Chinese cassia
Cupressus sempervirens L.	*Cupressaceae*	*Sarv*	Cypress tree
Cyperus rotundus L.	*Cyperaceae*	*Soʿd mostadir*	Nut grass, sweet cyperus
Descurainia sophia (L.) Webb ex Prantl	*Brassicaceae* (*Cruciferae*)	*Khākshir*	Flixweed
Ferula assa-foetida L.	*Apiaceae* (*Umbelliferae*)	*Anjedān*, *hiltit*	Asafoetida
Ficus carica L.	*Moraceae*	*Tin*	Common fig tree
Glycyrrhiza glabra L.	*Fabaceae* (*Leguminosae*)	*Sus*	Licorice
Lactuca sativa L.	*Asteraceae* (*Compositae*)	*Khas bostāni*	Lettuce
Moringa pterygosperma	*Morigaceae*	*Bān*	Moringa
Myrtus communis L.	*Myrtaceae*	*Ās*	Myrtle
Narcissus tazetta L.	*Amaryllidaceae*	*Aʿbhar*, *narjes*	Bunch-flowered narcissus
Nigella sativa L.	*Ranunculaceae*	*Shuniz*	Black cumin, Roman coriander
Peganum harmala L.	*Nitrariaceae*	*Ḥ armal*	Esfand, Syrian rue
Pimpinella anisum L.	*Apiaceae* (*Umbelliferae*)	*Anison*	Anise
Pistacia vera L.	*Anacardiaceae*	*Fostoq*	Pistachio
Portulaca oleracea L.	*Portulacaceae*	*Rijlah*	Common purslane
Prunus domestica L.	*Rosaceae*	*Barquq*	Plum
Prunus dulcis (Mill.) D.A. Webb	*Rosaceae*	*Lawz*	Almond
Punica granatum L.	*Lythraceae*	*Rommān* (*jolnār*)	Pomegranate
Ricinus communis L.	*Euphorbiaceae*	*Khrwaʿ*	Castor oil plant
Syzygium aromaticum (L.) Merr. & L.M. Perry Synonym: *Eugenia caryophyllus* (Spreng.) Bullock & S.G. Harrison	*Myrtaceae*	*Qaranfol*	Clove
Vitis vinifera L.	*Vitaceae*	*Eʿnab*	Grape/vine
Ziziphus jujube Mill.	*Rhamnaceae*	*Oʿnāb*	Jujube

Table 10.2. *In vitro* studies on medicinal plants mentioned in *Canon of Medicine* for treatment of viral diseases.

Plant	Part	Extract/phytochemical	Activity	Tested cell line/assay	Result	Reference
Acacia arabica Synonym: *Acacia nilotica*	Leaves	Methanol and acetone extracts	Anti-HCV	Huh-7 cell line	↓ Virus titer by 73% (methanol) and 82% (acetone); direct inactivation by acetone extract	Rehman *et al.* (2011)
Allium sativum	Bulbs	Methanol (98%) and ethanol (70%) extracts	Anti-influenza A (H1N1)	MDCK cell line	Hemagglutination inhibition: 14% for both extracts	Chavan *et al.* (2016)
Capparis spinosa	Buds	Methanol extract	Anti-HSV-2	Human PBMCs	↑ IL-12, ↑ TNF-α, ↑ IFN-γ; improves immune system	Arena *et al.* (2008)
	Seeds	A purified protein	Anti-HIV	ELISA kit	Inhibition of HIV-1 reverse transcriptase with IC_{50} of 0.23 μM	Lam and Ng (2009)
Chelidonium majus	Aerial parts	Aqueous extract	Anti-HIV	$CD4^+$ AA2 cell line and $CD4^+$ H9 cells	Prevention of HIV-1 at dose of 25 μg/ml	Gerencer *et al.* (2006)
Cinnamomum cassia	Dried twigs	Aqueous extract	Anti-HRSV	Human upper (HEp-2) and lower (A549) respiratory tract cell lines	Dose-dependent inhibition of HRSV-induced plaque formation by inhibiting viral attachment, internalization, fusion protein production, and syncytium formation	Yeh *et al.* (2013)
Cupressus sempervirens	Green cones	Hydromethanolic (80%) extract/ proanthocyanin polymer fraction	Anti-HIV, anti-HTLV-IIIB	T lymphoblasts Jurkat J6 infectable by HIV, H9 infectable by HTLV	IC_{50} = 1.5–15 mg/ml against HIV; IC_{50} = 5–25 mg/ml against HTLV	Amouroux *et al.* (1998)
Cyperus rotundus	Rhizomes	Ethanol extract	Anti-HBV	HepG2.2.15 cells	Dose-dependent inhibition of HBsAg expressions	Parvez *et al.* (2019)
Ferula assa-foetida	Gum-resin	Sesquiterpene coumarins (badrakemin acetate, kellerin, samarcandin diasteromer)	Anti-HSV-1	African green monkey kidney (Vero) cells	↓ Viral titer by kellerin at 10, 5 and 2.5 μg/ml	Ghannadi *et al.* (2014)

Continued

Table 10.2. Continued.

Plant	Part	Extract/phytochemical	Activity	Tested cell line/assay	Result	Reference
Ficus carica	Leaves and fruits	Latex	Anti-HSV-2	HEp-2 cell line	Synergist effect of F. carica and acyclovir on HSV-2	Ay and Duran (2018)
	Fruits	Latex	Anti-HPV	Cervical cancer cell lines CaSki and HeLa positive for HPV types 16 and 18	↓ Rapid growth and invasion of cervical cancer-transformed cells; ↓ expression of p16 and HPV onco-proteins E6, E7	Ghanbari *et al.* (2019)
	Fruit latex	Methanol, hexane, ethyl acetate, hexane–ethyl acetate, and chloroform extracts	Anti-HSV-1, anti-ECV-11, anti-ADV	Green monkey (Cercopithecus aethiops) kidney cells	↓ Virus multiplication by hexane and hexane–ethyl acetate (v/v) extracts at 78 μg/ml	Lazreg Aref *et al.* (2011)
	Leaves	Aqueous extract	Anti-HSV-1	Hep-2, BHK21, and PRK cells	MTC = 0.5 mg/ml; TI = 30.0	Wang *et al.* (2004)
Glycyrrhiza glabra	Roots	Aqueous extract	Anti-HSV-1	Vero cells infected by HSV-1	Antiviral activity depends on incubation period and pretreatment of cells with extract	Ghannad *et al.* (2014)
Lactuca sativa	Leaves	Aqueous extract, methanol extract	Anti-HCMV, anti-Cox-B3	HCMV strain AD-169 and Cox-B3	Antiviral activity with IC_{50} = 200 μg/ml for both extracts	Edziri *et al.* (2011)
Narcissus tazetta	Bulbs	Mannose-binding lectin	Anti-influenza A and B, anti-HRSV	Human larynx epidermoid carcinoma cell line (HEp-2) cells, MDCK cells, and RSV (Long strain)	↓ Plaque formation by RSV (IC_{50} = 2.30 μg/ml); dose-dependent anti-influenza activity (IC_{50} = 0.20–1.33 μg/ml)	Ooi *et al.* (2010)
Nigella sativa	Seeds	Herbal formulation from N. sativa seeds (Alphazam)	Anti-HCV	Subgenomic HCV replicon cells carrying luciferase gene (LucNeo#2) and the full genome (NNC#2)	Selective inhibition of HCV replication	Oyero *et al.* (2016)
Peganum harmala	Aerial parts	Ethyl acetate, chloroform, butanol, and methanol extracts	Anti-HCMV, anti-Cox-B3	CPE reduction assay on MRC-5 cells	Highest antiviral activity by methanol extract (anti-HCMV and anti-Cox-B3: IC_{50} = 25 μg/ml)	Edziri *et al.* (2010)

Pimpinella anisum	Seeds	Aqueous extract/ lignin–carbohydrate–protein complexes	Anti HSV-1 and -2, anti-HCMV, anti-measles	Vero and MRC-5 cells	Inactivation of viruses by interfering with virus adsorption; ↑ NO, IL-1β, and IL-10	Lee *et al*. (2011)
Pistacia vera	Various plant parts	n-Hexane extract	Anti-HSV and parainfluenza viruses	Vero cell line	Anti-HSV activity: extracts of the skin of processed woody shells, fresh kernels, and Turkish unripe seeds; anti-parainfluenza activity: extracts of the Turkish ripe seeds, Turkish processed seeds, Iranian processed seeds, and fresh skin of natural woody shells	Özçelik *et al*. (2005)
Portulaca oleracea	Aerial parts	Aqueous extract	Anti-influenza A	MDCK cells	Virucidal and antiviral activity at an early stage of infection	Li *et al*. (2019)
	Leaves	Methanol extract	Anti-HCV	Huh-7 cells	↓ HCV NS3 protease expression in a dose-dependent manner	Noreen *et al*. (2015)
Prunus domestica	Fruit	Ethyl acetate extract/ rutin	Anti-HCV	Human hepatocellular carcinoma cells Huh-7 and Huh-7.5	Inhibition of binding of HCV-LPs to hepatoma cells by crude extract (62%); inhibition of viral entry and infection by rutin	Bose *et al*. (2017)
Prunus dulcis	Skin	n-Hexane extract	Anti HSV-1	Vero cell line	Dose-dependent anti-HSV type 1	Bisignano *et al*. (2017)
	Skin	Mix of polyphenols	Anti HSV-1	Vero cell line	↓ Viral titer and DNA accumulation	Musarra-Pizzo *et al*. (2019)
Punica granatum	Fruits	Polyphenol extract/ punicalagin	Anti-influenza	MDCK cells	↓ Influenza virus replication	Haidari *et al*. (2009)
	Leaves	Aqueous, ethanol, and hydroalcoholic extracts	HHV-3	HEp-2 cells	Potential antiviral activity by aqueous extract; interaction of leaf phytochemicals with HHV-3 protease	Angamuthu *et al*. (2019)
Ricinus communis	Leaves	Methanol (70%) extract (petroleum ether, dichloromethane, ethyl acetate, and n-butanol fractions)	Anti-HAV, anti-Cox-B4, anti-HSV-1	Vero cells	Highly anti-replicative against HAV (especially methanol and dichloromethane fractions); moderate anti-Cox-B4; protective effect against HAV (especially butanol extract)	Elkousy *et al*. (2021)

Continued

Table 10.2. Continued.

Plant	Part	Extract/phytochemical	Activity	Tested cell line/assay	Result	Reference
Sambucus nigra	Fruits	Ethanol (80%) extract	Anti-IBV	Vero cells	Inhibition of viral replication; ↓ viral titer dose-dependently	Chen *et al*. (2014)
Syzygium aromaticum Synonym: *Eugenia caryophyllus*	Flower buds	Ethanol and methanol extracts/eugenol	Anti-HSV, anti-influenza A virus, anti-Ebola	–	Viral inhibition	Vicidomini *et al*. (2021)
	Flower buds	Essential oil/eugenol	Anti-HSV	GMK cells	Crude extract: anti-HSV; ED_{50} = 72.8 μg/ml (HSV-1) and 74.4 μg/ml (HSV-2); eugenol more potent than crude extract	Tragoolpua and Jatisatienr (2007)
Vitis vinifera	Leaves	Polyphenol extract	Anti HSV-1, anti-SARS-CoV-2	Vero, HCT-116, A549, MCF7, H9c2, HepG2, and Hacat cells	Inhibition of viral replication at early stages (concentration = 10 μg/ml)	Zannella *et al*. (2021)

ADV, adenovirus; Cox-B3, coxsackievirus B type 3; Cox-B4, coxsackievirus B type 4; CPE, cytopathic effect; ED_{50}, median effective dose; ECV-11, echovirus type 11; GMK, green monkey kidney; HAV, hepatitis A virus; HBsAg, hepatitis B surface antigen; HCMV, human cytomegalovirus; HCV, hepatitis C virus; HCV-LP, hepatitis C virus-like particle; HHV-3, human herpes virus-3; HIV, human immunodeficiency virus; HRSV; human respiratory syncytial virus; HTLV, human T-lymphotropic virus; IBV, infectious bronchitis virus; IC_{50}, half-maximal inhibitory concentration; IL, interleukin; IFN-γ, interferon-γ; MDCK, Madin–Darby canine kidney; MTC, minimum toxic concentration; NO, nitric oxide; PBMC, peripheral blood mononuclear cell; SARS-CoV-2, severe acute respiratory syndrome coronavirus 2; TI, therapeutic index; TNF-α, tumor necrosis factor-α.

Table 10.3. *In vivo* studies on medicinal plants mentioned in *Canon of Medicine* for treatment of viral diseases.

Plant	Part	Extract/ phytochemical	Activity	Study design	Result	Reference
Allium sativum	Cloves	Aqueous extract	Anti-NDV	Time-of-addition assay in embryonated chicken eggs	↓ Virus infectivity and titer (especially in pretreatment trial)	Doostmohammadian *et al.* (2020)
Chelidonium majus	Aerial parts	Aqueous extract (purified active substance)	Antiretroviral	C57B1/6 murine AIDS model	Prevention of splenomegaly and the enlargement of cervical lymph nodes	Gerencer *et al.* (2006)
Glycyrrhiza glabra	Leaves	Aqueous, methanol, and ethanol extracts	Anti-NDV	9-day-old embryonated chicken eggs	↑ Survival rate and virus control by all extracts; strong antiviral activity: methanol extract at 300 μg/ml	Ashraf *et al.* (2017)
Nigella sativa	Seeds	Seed oil	Anti-SARS-CoV-2	Murine (BALB/c) cytomegalovirus model	↓ Virus load in spleen and liver	Khazdair *et al.* (2021)

AIDS, acquired immunodeficiency syndrome; NDV, Newcastle disease virus; SARS-CoV-2, severe acute respiratory syndrome coronavirus 2.

Table 10.4. Clinical studies on medicinal plants mentioned in *Canon of Medicine* for treatment of viral diseases.

Plant	Part	Extract/ phytochemical	Activity	Study design/no. of patients	Result	Reference
Nigella sativa	Seeds	Seed oil in capsule (450 mg)	Anti-HCV	Prospective, single-armed, self-controlled pilot study/30 patients	↓ Viral load, improving oxidative stress and clinical condition	Barakat *et al*. (2013)
	Seeds	Oil	Anti-COVID-19	Prospective, two-arm, parallel-group, randomized, controlled, open-label, phase II clinical trial/173 patients	↑ Symptom recovery	Koshak *et al*. (2021)
Allium sativum	Bulbs	Ethanol extract	Anti-HPV (genital warts)	Double-blind, randomized, placebo-controlled clinical trial/35 male patients	Use of 10% garlic extract for 2 months has similar effects to cryotherapy	Mousavi *et al*. (2018)
Myrtus communis	Leaves	–	Anti-HPV	Quantitative randomized controlled clinical trial/100 patients	↓ Size and number of warts; more effective than SA	Yazdi *et al*. (2014)
Descurainia sophia	Leaves	–	Anti-HPV	Quantitative randomized controlled clinical trial/100 patients	↑ Excretion of wart-inducing materials	Yazdi *et al*. (2014)
Adianthum capillus-veneris *Glycyrrhiza glabra* *Ziziphus jujube*	Aerial parts Roots Fruits	Herbal mixture (with other medicinal herbs)	Anti-asthma symptoms in viral respiratory infections	Randomized, double-blind, placebo-controlled clinical trial/46 patients	↓ Cough and nighttime awakening	Javid *et al*. (2019)
Nigella sativa *Allium sativum*	Seed oil Bulbs	Herbal mixture (with other medicinal herbs)	Anti-rhinovirus infections	Prospective, double-blind, randomized controlled manner in an outpatient setting/62 patients	↓ Severity and duration of symptoms	Yakoot and Salem (2011)
Artemisia absinthium	Whole plant	Decoction	Anti-HBV	Prospective, single-arm, pilot clinical trial/30 patients	↓ HBV DNA, HBeAg, and HBsAg,normalizing ALT	Ansari *et al*. (2018)

ALT, alanine transaminase; HBV, hepatitis B virus; HBeAg, hepatitis B e-antigen; HBsAg, hepatitis B surface antigen; HCV, hepatitis C virus.

References

Amouroux, P., Jean, D. and Lamaison, J.L. (1998) Antiviral activity *in vitro* of *Cupressus sempervirens* on two human retroviruses HIV and HTLV. *Phytotherapy Research* 12, 367–368.

Angamuthu, D., Purushothaman, I., Kothandan, S. and Swaminathan, R. (2019) Antiviral study on *Punica granatum* L., *Momordica charantia* L., *Andrographis paniculata* Nees, and *Melia azedarach* L., to human herpes virus-3. *European Journal of Integrative Medicine* 28, 98–108.

Ansari, S., Siddiqui, M.A., Malhotra, S. and Maaz, M. (2018) Antiviral efficacy of qust (*Saussurea lappa*) and afsanteen (*Artemisia absinthium*) for chronic hepatitis B: a prospective single-arm pilot clinical trial. *Pharmacognosy Research* 10, 282–290.

Arena, A., Bisignano, G., Pavone, B., Tomaino, A., Bonina, F.P. *et al.* (2008) Antiviral and immunomodulatory effect of a lyophilized extract of *Capparis spinosa* L. buds. *Phytotherapy Research* 22, 313–317.

Ashraf, A., Ashraf, M.M., Rafiqe, A., Aslam, B., Galani, S. *et al.* (2017) *In vivo* antiviral potential of *Glycyrrhiza glabra* extract against Newcastle disease virus. *Pakistan Journal of Pharmaceutical Sciences* 30, 567–572.

Ay, E. and Duran, N. (2018) Investigation of the antiviral activity of *Ficus carica* L. latex against HSV-2. In: *International Conference on Advanced Materials and Systems (ICAMS)*. The National Research & Development Institute for Textiles and Leather-INCDTP, Bucharest, pp. 33–37.

Barakat, E.M., El Wakeel, L.M. and Hagag, R.S. (2013) Effects of *Nigella sativa* on outcome of hepatitis C in Egypt. *World Journal of Gastroenterology* 19, 2529–2536.

Bisignano, C., Mandalari, G., Smeriglio, A., Trombetta, D., Pizzo, M.M. *et al.* (2017) Almond skin extracts abrogate HSV-1 replication by blocking virus binding to the cell. *Viruses* 9, 178.

Bleck, T.P. and Rupprecht, C.E. (2000) Rabies virus. *In:* Mandell, G.L., Douglas, R.G., Bennett, J.E. and Dolin, R. (eds) *Mandell, Douglas, and Bennett's Principles and Practice of Infectious Diseases*, 5th edn. Churchill Livingstone, Philadelphia, Pennsylvania, pp. 1811–1820.

Bose, M., Kamra, M., Mullick, R., Bhattacharya, S., Das, S. *et al.* (2017) Identification of a flavonoid isolated from plum (*Prunus domestica*) as a potent inhibitor of hepatitis C virus entry. *Scientific Reports* 7, 3965.

Brett, W. and Petersen, B.W. (2015) Clinical guidance for smallpox vaccine use in a postevent vaccination program. *Morbidity and Mortality Weekly Report* 64(RR02), 1–26.

Chavan, R.D., Shinde, P., Girkar, K., Madage, R. and Chowdhary, A. (2016) Assessment of anti-influenza activity and hemagglutination inhibition of *Plumbago indica* and *Allium sativum* extracts. *Pharmacognosy Research* 8, 105–111.

Chen, C., Zuckerman, D. M., Brantley, S., Sharpe, M., Childress, K. *et al.* (2014) *Sambucus nigra* extracts inhibit infectious bronchitis virus at an early point during replication. *BMC Veterinary Research* 10, 24.

Dinulos, J.G.H. (2021) Warts, herpes simplex, and other viral infections. In: *Habif's Clinical Dermatology*, 7th edn. Elsevier Inc., Philadelphia, Pennsylvania, pp. 445–450.

Doostmohammadian, F., Shomali, T., Mosleh, N. and Mohammadi, M. (2020) *In ovo* evaluation of antiviral effects of aqueous garlic (*Allium sativum*) extract against a velogenic strain of Newcastle disease virus. *Journal of HerbMed Pharmacology* 9, 232–238.

Edziri, H., Mastouri, M., Mahjoub, M. A., Patrich, G., Matieu, M. *et al.* (2010) Antibacterial, antiviral and antioxidant activities of aerial part extracts of *Peganum harmala* L. grown in Tunisia. *Toxicological and Environmental Chemistry* 92, 1283–1292.

Edziri, H.L., Smach, M.A., Ammar, S., Mahjoub, M.A., Mighri, Z. *et al.* (2011) Antioxidant, antibacterial, and antiviral effects of *Lactuca sativa* extracts. *Industrial Crops and Products* 34, 1182–1185.

Elkousy, R.H., Said, Z.N.A., Abd El-Baseer, M.A. and Abu El Wafa, S.A. (2021) Antiviral activity of castor oil plant (*Ricinus communis*) leaf extracts. *Journal of Ethnopharmacology* 271, 113878. DOI: 10.1016/J.JEP.2021.113878.

Fooks, A.R. and Jackson, A.C. (2000) Strategies for the elimination of dog-mediated human rabies by 2030. In: Terrence, P.S., Coetzer, A. and Nel, L.H. (eds) *RABIES Scientific Basis of the Disease and Its Management*, 4th edn. Academic Press, London, pp. 682–684.

Gerencer, M., Turecek, P.L., Kistner, O., Mitterer, A., Savidis-Dacho, H. *et al.* (2006) *In vitro* and *in vivo* anti-retroviral activity of the substance purified from the aqueous extract of *Chelidonium majus* L. *Antiviral Research* 72, 153–156.

Ghanbari, A., Le Gresley, A., Naughton, D., Kuhnert, N., Sirbu, D. *et al.* (2019) Biological activities of *Ficus carica* latex for potential therapeutics in human papillomavirus (HPV) related cervical cancers. *Scientific Reports* 9, 1013.

Ghannad, M.S., Mohammadi, A., Safiallahy, S., Faradmal, J., Azizi, M. *et al.* (2014) The effect of aqueous extract of *Glycyrrhiza glabra* on herpes simplex virus 1. *Jundishapur Journal of Microbiology* 7, e11616.

Ghannadi, A., Fattahian, K., Shokoohinia, Y., Behbahani, M. and Shahnoush, A. (2014) Anti-viral evaluation of sesquiterpene coumarins from *Ferula assa-foetida* against HSV-1. *Iranian Journal of Pharmaceutical Research* 13(2), 523–530.

Haidari, M., Ali, M., Casscells, S.W. III and Madjid, M. (2009) Pomegranate (*Punica granatum*) purified polyphenol extract inhibits influenza virus and has a synergistic effect with oseltamivir. *Phytomedicine* 16, 1127–1136.

Hübschen, J.M., Gouandjika-Vasilache, I. and Dina, J. (2022) Measles. *The Lancet* 12, 399(10325), 678–690.

Hutson, C.L., Kondas, A.V., Mauldin, M.R., Doty, J.B., Grossi, I.M. *et al.* (2021) Pharmacokinetics and efficacy of a potential smallpox therapeutic, brincidofovir, in a lethal monkeypox virus animal model. *mSphere* 6(1), e00927-20.

Ibn Sinā, A.H. (2015) *Al-Qânun fi'l-Tibb (Canon of Medicine)*, ed. M.A. Rashad and M.T. Mahmoud. Misr University for Science and Technology, Cairo. Vol. 3, pp. 330–336; Vol 4, pp. 174–184, 274–276, 537–545, 613–616 (in Arabic).

Javid, A., Haghi, N.M., Emami, S.A., Ansari, A., Zojaji, S.A. *et al.* (2019) Short-course administration of a traditional herbal mixture ameliorates asthma symptoms of the common cold in children. *Avicenna Journal of Phytomedicine* 9, 126–133.

Khazdair, M.R., Ghafari, S. and Sadeghi, M. (2021) Possible therapeutic effects of *Nigella sativa* and its thymoquinone on COVID-19. *Pharmaceutical Biology* 59, 694–701.

Koshak, A.E., Koshak, E.A., Mobeireek, A.F., Badawi, M.A., Wali, S.O. *et al.* (2021) *Nigella sativa* for the treatment of COVID-19: an open-label randomized controlled clinical trial. *Complementary Therapies in Medicine* 61, 102769.

Kwok, C.S., Gibbs, S., Bennett, C., Holland, R. and Abbott, R. (2012) Topical treatments for cutaneous warts. *The Cochrane Database of Systematic Reviews* 2012(9), CD001781.

Lam, S.K. and Ng, T.B. (2009) A protein with antiproliferative, antifungal and HIV-1 reverse transcriptase inhibitory activities from caper (*Capparis spinosa*) seeds. *Phytomedicine* 16, 444–450.

Lazreg Aref, H., Gaaliche, B., Fekih, A., Mars, M., Aouni, M. *et al.* (2011) *In vitro* cytotoxic and antiviral activities of *Ficus carica* latex extracts. *Natural Product Research* 25, 310–319.

Lee, J.B., Yamagishi, C., Hayashi, K. and Hayashi, T. (2011) Antiviral and immunostimulating effects of lignin-carbohydrate-protein complexes from *Pimpinella anisum*. *Bioscience, Biotechnology, and Biochemistry* 75, 459–465.

Li, Y.H., Lai, C.Y., Su, M.C., Cheng, J.C. and Chang, Y.S. (2019) Antiviral activity of *Portulaca oleracea* L. against influenza A viruses. *Journal of Ethnopharmacology* 241, 112013.

Marques, A. R. and Cohen, J. I. (2012) In: Lowell Goldsmith *et al.* (eds) *Fitzpatrick's Dermatology in General Medicine*, 8th edn. McGraw-Hill Education/Medical, New York.

Moss, W.J. (2020) Viral infections with cutaneous lesions, measles. In: Ryan, E.T., Hill, R.D., Solomon, T., Naomi E., Aronson, N.E. *et al.* (eds) *Hunter's Tropical Medicine and Emerging Infectious Diseases*, 10th edn. Elsevier Inc., Philadelphia, Pennsylvania, pp. 267–270.

Mousavi, Z.B., Mehrabian, A., Golfakhrabadi, F. and Namjoyan, F. (2018) A clinical study of efficacy of garlic extract versus cryotherapy in the treatment of male genital wart. *Dermatologica Sinica* 36, 196–199.

Murray, P.R., Rosenthal, K.S. and Pfaller, M.A. (2016) Poxviruses. In: *Medical Microbiology*, 8th edn. Elsevier Inc., Philadelphia, Pennsylvania, p. 450.

Musarra-Pizzo, M., Ginestra, G., Smeriglio, A., Pennisi, R., Sciortino, M.T. *et al.* (2019) The antimicrobial and antiviral activity of polyphenols from almond (*Prunus dulcis* L.) skin. *Nutrients* 11, 2355.

Noreen, S., Hussain, I., Tariq, M.I., Ijaz, B., Iqbal, S. *et al.* (2015) *Portulaca oleracea* L. as a prospective candidate inhibitor of hepatitis C virus NS3 serine protease. *Viral Immunology* 28, 282–289.

Ooi, L.S., Ho, W.S., Ngai, K.L., Tian, L., Chan, P.K. *et al.* (2010) *Narcissus tazetta* lectin shows strong inhibitory effects against respiratory syncytial virus, influenza A (H1N1, H3N2, H5N1) and B viruses. *Journal of Biosciences* 35, 95–103.

Overton, E.T., Lawrence, S.J., Wagner, E., Nopora, K., Rösch, S. *et al.* (2018) Immunogenicity and safety of three consecutive production lots of the non-replicating smallpox vaccine MVA: a randomised, double blind, placebo controlled phase III trial. *PLoS One* 13(4), e0195897.

Oyero, O.G., Toyama, M., Mitsuhiro, N., Onifade, A.A., Hidaka, A. *et al.* (2016) Selective inhibition of hepatitis C virus replication by Alphazam, a *Nigella sativa* seed formulation. *African Journal of Traditional, Complementary and Alternative Medicines* 13, 144–148.

Özçelik, B., Aslan, M., Orhan, I. and Karaoglu, T. (2005) Antibacterial, antifungal, and antiviral activities of the lipophylic extracts of *Pistacia vera*. *Microbiological Research* 160, 159–164.

Parvez, M.K., Al-Dosari, M.S., Arbab, A.H. and Niyazi, S. (2019) The *in vitro* and *in vivo* anti-hepatotoxic, anti-hepatitis B virus and hepatic CYP450 modulating potential of *Cyperus rotundus*. *Saudi Pharmaceutical Journal* 27, 558–564.

Rehman, S., Ashfaq, U.A., Riaz, S., Javed, T. and Riazuddin, S. (2011) Antiviral activity of *Acacia nilotica* against hepatitis C virus in liver infected cells. *Virology Journal* 8, 220.

Singh, B., Rupprecht, C.E. and Bleck, T.P. (2020) Rabies (rhabdoviruses). In: Bennett, J.E., Dolin, R. and Blaser, M.J. (eds) *Mandell, Douglas, and Bennett's Principles and Practice of Infectious Diseases*, 9th edn. Churchill Livingstone, Philadelphia, Pennsylvania, pp. 2127–2137.e3.

Tragoolpua, Y. and Jatisatienr, A. (2007) Anti-herpes simplex virus activities of *Eugenia caryophyllus* (Spreng.) Bullock & SG Harrison and essential oil, eugenol. *Phytotherapy Research* 21, 1153–1158.

Turner, R.B. (2012a) The common cold. In: Goldman, L. and Schafer, A.I. (eds) *Cecil Medicine*, 24th edn. Elsevier Saunders, Philadelphia, Pennsylvania, p. 2089.

Turner, R.B. (2012b) The common cold. In: Goldman, L. and Schafer, A.I. (eds) *Cecil Medicine*, 24th edn. Elsevier Saunders, Philadelphia, Pennsylvania, p. 2117.

Turner, R.B. (2012c) The common cold. In: Goldman, L. and Schafer, A.I. (eds) *Cecil Medicine*, 24th edn. Elsevier Saunders, Philadelphia, Pennsylvania, p. 2121.

Vicidomini, C., Roviello, V. and Roviello, G.N. (2021) Molecular basis of the therapeutical potential of clove (*Syzygium aromaticum* L.) and clues to its anti-COVID-19 utility. *Molecules* 26, 1880.

Wang, G., Wang, H., Song, Y., Jia, C., Wang, Z. *et al.* (2004) Studies on anti-HSV effect of *Ficus carica* leaves. *Zhong yao cai = Zhongyaocai = Journal of Chinese Medicinal Materials* 27, 754–756.

Yakoot, M. and Salem, A. (2011) Efficacy and safety of a multiherbal formula with vitamin C and zinc (Immumax) in the management of the common cold. *International Journal of General Medicine* 4, 45–51.

Yazdi, E.G., Minaei, M.B., Dabaghian, F.H., Ardakani, M.E.Z., Ranjbar, A.M. *et al.* (2014) Efficacy of *Myrtus communis* L. and *Descurainia sophia* L. versus salicylic acid for wart treatment. *Iranian Red Crescent Medical Journal* 16, e16386.

Yeh, C.F., San Chang, J., Wang, K.C., Shieh, D.E. and Chiang, L.C. (2013) Water extract of *Cinnamomum cassia* Blume inhibited human respiratory syncytial virus by preventing viral attachment, internalization, and syncytium formation. *Journal of Ethnopharmacology* 147, 321–326.

Zannella, C., Giugliano, R., Chianese, A., Buonocore, C., Vitale, G.A. *et al.* (2021) Antiviral activity of *Vitis vinifera* leaf extract against SARS-CoV-2 and HSV-1. *Viruses* 13, 1263.

11 Treatment of Various Skin Diseases, Including Psoriasis, Skin Cancer, Ultraviolet Protection, Skin Dryness, and Allergic Reactions

Maryam Nikoosokhan*

Science and Technology Park of University of Tehran, College of Agriculture & Natural Resources, University of Tehran, Karaj, Iran

Abstract

The skin is the largest body organ and acts as a physical barrier established in the epidermis, it is exposed to fluctuating external factors and shows remarkable functional and structural diversity. Over the last two decades, understanding the complex phenotype and mechanisms of dermal diseases has led to a variety of new biologic targets and strategies for pharmacologic intervention and significant potential for a medical approach to the management of skin disorders. Atopic dermatitis, eczema, psoriasis, and abscesses are common acute or chronic skin diseases increasing dramatically due to industrialization and unhealthy lifestyle habits. The comparatively high failure rate in treatment of skin diseases makes alternative treatments urgently needed. Fast approaches to drug discovery through traditional medical therapies reduce process time and costs. Traditional Persian Medicne has unique therapeutic methods in the treatment of skin disorders. These traditional treatments and the latest research findings in this field are presented in this chapter.

Introduction

The skin is the largest body organ and acts as a physical barrier established in the epidermis, it is exposed to fluctuating external factors and shows remarkable functional and structural diversity. The outermost layer or epidermis has powerful metabolic and endocrine capabilities with squamous cells or keratinocytes as principal cells. The dermis is the inner layer composed of the fibrillar structural protein collagen (Arnold *et al.*, 1990). Better understanding of the complex phenotype and mechanisms of dermal disease over the past two decades offers a variety of new biologic targets and strategies for pharmacologic intervention and significant potential for a medical approach to the management of skin disorders. Atopic dermatitis, eczema, psoriasis, and abscesses are common acute or chronic skin diseases that have increased dramatically due to industrialization and unhealthy lifestyle habits. Many cellular activities of the nervous, immune, and endocrine systems carry on towards the skin (O'Sullivan *et al.*, 1988) and several metabolic events contribute to the rapid restoration of barrier function caused by adverse mental and emotional conditions (Altemus *et al.*, 2001). The high treatment failure rate in skin diseases requires urgent alternative treatment methods. Fast approaches to drug discovery through traditional medical therapies reduce process time and costs (Bieber, 2022).

*Email: maryam.nikoosokhan@gmail.com

 Medicinal Plants Used in Traditional Persian Medicine (eds. H. Schulz, Seyed Ahmad Emami and Farsad Nadjafi)

DOI: 10.1079/9781800621671.0011

Skin in View of *Canon*

Avicenna considered that skin and hair diseases are temperamental illnesses and that the skin covers the whole body, creating cohesive capillary and nervous networks and a sensitive barrier for protection of internal organs. Many sensors, early awareness of receiving a slight dystemperament, engaging negative stimuli from inside or outside the body, and expelling waste products out of its pores as an incense vapor are the duties of the skin as a large organ. Humors from the diet and environmental conditions will determine the intrinsic body temperament (*mezadj*), the transformation of ileus into blood takes place in the liver and the form of food is destroyed and becomes humor, the result of the second or liver digestion is the preliminary moisture of the four humors (Sepehrikia *et al*., 2022). The skin has a moderate temperament in the body and like the other organs, proper nutrition with appropriate quality and quantity is one of the most important factors in skin health. Avicenna's views on dermatology were transferred to Europe in the 13th century upon translating the *Canon of Medicine* into Latin, which had a considerable effect on medieval European medicine (Atarzadeh *et al*., 2016). Many traditional medical approaches in the *Canon of Medicine* can help prevent the incidence and development of illnesses and treat the cause, not just disease symptoms. Traditional formulations for skin disorders in the *Canon* were prescribed in the form of powder, solid, semisolid, paste, and liquids by local or systemic route of administration. Both the brain and the skin may be influenced by phytochemicals that can activate many biochemical and cellular processes.

Six Necessary Measures of Hygiene Principles

As Avicenna mentioned, six measures of hygiene (*Settehe–Ḍarurieah*) play a vital role in preventing skin weakness and strengthening it:

1. Healthy air that causes depletion of waste materials from the body and has a positive effect on the skin.

2. Modification of nutrition that contains variable amounts of hot-, cold-, dry-, and wet-nature foods and which, in terms of quality and quantity, has different degrees of impact on the skin.

3. Movement and rest, in that taking moderate physical exercise and adequate rest helps in better digestion and absorption.

4. Sleep and wakefulness, and average sleep time with good quality, make the body temperament in the normal state and prevent skin aging.

5. Relationships, values, life purpose, and mental condition.

6. Evacuation and retention processes are good for skin health when they are enough and typical.

Proper nutrition as the most critical point in health, humor production with appropriate quality and quantity, and lifestyle modification are the hallmarks of treatment approaches in the *Canon*. Some evidence suggests that the quality of life in patients with dermatologic conditions is the same or worse than the quality of life in patients with cancer or heart diseases (Parvizi *et al*., 2021).

Holistic Management in the Modern Era

There are two perspectives at each end of a spectrum of approaches to disease management: the holistic approach, or considering the psychological or spiritual condition of the patient and attention to the whole patient in their wider environment, at one end; and the technical or scientific issues, focusing on the hard facts and concentrating purely on diagnosis and treatment, at the other. The holistic approach is particularly pertinent to the discipline of dermatology. *Settehe–Ḍarurieah* is a complete concept of holistic management.

Decoding of Humors in Traditional Persian Medicine

Yellow bile humor is similar to fire, dry and warm, invisible, and can penetrate into every hole of the body, thus acute bacterial and/or viral infections of the skin are usually related to choleric humor. Blood is warm and wets like air, increasing the heat and wetness, thus skin inflammatory disorders and swellings can result from blood humor. Melancholic humor is cold and dry, similar to earth, hard, and tends to stasis, thus cancerous diseases, eczema, and dermatitis are related to melancholic humor. Phlegmatic humor is cold and wet, similar to water or liquid it has a constant movement, thus phlegmatic humor can be leading to diseases related to the pathology of water like obesity (Rafieian Kopaei *et al.*, 2016).

General Treatment Protocols in Traditional Medicine

Several accepted approach steps are generally considered in the treatment of skin disorders in traditional medicine. The first step is diagnosing the responsible humor; the second step is general cleansing or blood purification by different traditional processes for reducing waste toxic materials in the body, because this is one of the primary ways the body maintains homeostasis; and finally is skin healing. According to patient temperament, different dosage forms are prepared in the *Qarabadin* or pharmacopeias section of the *Canon*, known as "individualized therapy" today. Due to the multiplicity of formulas, only general treatment protocols are mentioned. The most common herbs used in various formulas in the *Qarabadin* are listed in the tables in this chapter.

Makeup and Beauty in *Canon*

Specific chapters in *Canon of Medicine* are allocated to beautification, which deals with body care. A variety of cosmetic products formulated by different carriers and plants/minerals for beauty and makeup (*zinat*), as well as face, hair, and nail hygienic/cosmetic products as preventive and healing methods are found in this preeminent textbook. Hair curling and straightening methods, depilatories, hair loss, dandruff, hair lice, alopecia, nail disorders like splitting and pitting, scent and body aroma, obesity, and emaciation, all of which affect the appearance, are also discussed in a separate part in Chapter 7 of the *Canon*'s fourth volume. Here are to be found some of the most widely used traditional dosage forms such as toner/lotion (*ghasool*), liniment (*tela*), ointments (*marham*), eyeliner (*koḥl*), and hair and nail color (*ṣabgh*). In Avicenna's *Canon*, the characterization of skin diseases is better described for some of the previously known conditions. The section on skin diseases discusses different dermatologic conditions such as psoriasis, vitiligo, acne, melasma, erysipelas, etc. Also explained are some cutaneous problems in the genitalia. The first descriptions of elephantiasis, leprosy, various hair and nail disorders, and pemphigus have been attributed to Avicenna (Bagherani and Smoller, 2015).

Inflammatory Skin Lesions, Vesicle

Inflammatory dermatoses such as eczema, psoriasis, and acne have a significant role on general physical health. However, their impact on psychological and social well-being is also considerable and needs to be understood. Five characteristic signs of inflammation are redness, heat, pain, swelling, and loss of function. However, not all signs may be observed simultaneously in all inflammatory skin lesions.

Redness (*rubor*) reflects increased blood flow in arterioles and capillaries and may also result from stasis without the increased flow of blood. Inflammation-induced plasma extravasation leads to enhanced migration of leukocytes to the site of inflammation. Heat (*calor*) is also a consequence of increased blood flow through the skin. In acute lesions, the skin temperature may increase as a direct result of an elevated local metabolic rate, pH, and activation of receptors or ion channels by inflammatory mediators or metabolites (Steinhoff, 2016).

Bosur (plural of *basarah*) is a minor skin rash, *awrām* (plural of *waram*), which can be produced by the abnormal formation and the accumulation of waste products in the subcutaneous tissue are abnormal bulgings of the external or internal organs that are equivalent to edema or swelling. Changes in the quality of humors due to mixing are the leading cause of skin inflammation, the two general types of swelling are hot and cold. The leading cause of hot *basareh* and hot *oram* is blood and yellow bile humor combination divided into "good, unfavorable blood" and "dense, watery yellow bile humor", subsequently leading to the creation of a range of different inflammatory disorders (Table 11.1). Accumulation of abnormal humors in the skin leads to signs and symptoms; for example, frequent use of hot-nature, thick, salty, spicy, and sweet foods corrupts the blood vessels, and the skin as a weak organ accepts it. According to modern medicine, skin dystemperament can be due to:

- any gene variation;
- changes in the skin's external layer like damage related to pruritis; and
- penetration of microorganisms or environmental allergens causing skin component interruption (*taffaroq-e-eteṣ āl*).

The main cause of *batharah* or abnormal swelling often occurs in the body's soft organs, although sometimes it is also found in bone, which thickens bone volume. Avicenna believed that if the skin structure as a weak organ becomes too weak or if natural body force sends too much waste material produced in the body to skin layers, or if the material quality that spills on the skin has changed, it will appear as dryness, itching, burning, scales, crust and vesicles, and swellings, and sometimes will increase excessively the size of an organ. Abnormal humor of the body accumulates beneath the epidermis. These abnormal metabolites not only act as culture media for pathogenic organisms like *jarab* or scabies but also change skin temperament and viscosity and cause skin diseases (Eghbalian *et al.*, 2017).

Treatment

General treatment protocols for *oram* and *bosur* are both systemic and local drugs. The systemic treatment is used for blood purification and normalizing the disturbed body humor, and local remedies with different dosage forms are used for reducing inflammation, healing properties, and reduction of invasive behavior action of the disease. Manual interventions such as venesection (*fasd*) and wet cupping (*hijamat*) are also applied. However, according to the severity of the disease and the patient's condition, the order of treatment may change.

***Bosur labanieh* (acne)**

Acne vulgaris is a prevalent skin condition that affects 85% of adolescents, often starts in preadolescence, persists into adulthood, and is associated with a significant negative impact on mental health. The mainstay of acne treatment has remained largely unchanged over recent years. *Bosur* means "skin lesions", "vesicle" and *labanieh* means "oily". *Namlah* means "wound" or "pustule" and is red skin lesions with severe itching or hot swelling. There are different types of acne in *Canon of Medicine* due to abnormal yellow bile humor that natural body force transmits to the skin to protect internal organs against the danger of its accumulation. Commonly it may be related to dystemperament of the liver. Characteristics of *namlah* and *bosur-e-labanieh* are similar to acne (Shirbeigi *et al.*, 2020). According to Table 11.1, there are different types of acneform lesions in the *Canon*, such as a mild form with light yellow bile humor that passes the layer of the skin and conflicts with subcutaneous tissue, called the "progressive *namlah*"; *akalla* or corrosive progressive acne; and *javresiyeh*, a moderate form with a less inflamed skin lesion.

Treatment

Nutrition modification is essential in disease management, using a wet/heat-lowering temperament diet and oral herbal medicines for blood purification and excretion of abnormal mixed humors such as phlegm and black bile combined with yellow bile or yellow bile alone. Whey powder is prescribed in the therapy of acne and scarring often associated with the

Table 11.1. Inflammatory skin lesions/vesicles mentioned in *Canon of Medicine*.

Feleghmuni	The general type of any hot swelling. It is of sanguine (blood) type that is distended, severe inflammation, large and dense, congestion of tissue with blood, involving skin surface and deep penetration with pulse and pain
Ḥomrah	Red–glossy yellow inflammation with severe burning and localized heat due to yellow bile alone or mixed with blood (sanguine). Erysipelas is the related modern terminology, which is inflammation with large red and bright patches on the face and legs with sudden blistering, fever, thirst, and restlessness
Jāvarsiah	Less inflamed, progressive *namlah* with white head and red base rashes with burning, dry, ulcerative, sometimes self-improving, sometimes creating wounds due to phlegm and black bile combined with hot yellow bile humor
Namlah	Skin lesions with burning and mild round swelling, severe pruritis, dry skin due to dilute yellow bile that can produce small wounds
Ākallah	*Akal* means corrosive organ and *ākallah* is corrosive penetrating skin ulcer, high swelling, infectious and progressive due to oxidation of yellow bile
Qorooh (plural of *qorḥah*)	Ghorhe or wound is an injury, especially one that breaks the skin, that is due to the changing integrity of oram and bosreh
Nār-e-farsi or *jamrah*	Includes a lot of diluted liquid vesicles due to yellow bile mixing into blood, with rashes at the site of lesions accompanied by erythema, severe pain, itching, burning, and shaped as a linear flame of fire. It was more common among people of Persia or Molk-e-Fars (Nasir *et al.*, 2016)
Erqe-madani	(Caused by *Filaria medinensis*, or guinea worm.) Severe itching and tingling in the organ before the rash, then blister rash, blister perforation, permanently white, or red to black exudation, which causes dracunculiasis (Maleki, 2017)
Jarab	Scabies is the related modern terminology and is defined as minor red rashes with severe itching and burning. Avicenna defined *jarab* as a contagious disease in that pruritis is an essential symptom caused by blood corruption due to an abnormal mixture of humors, including yellow bile, black bile, and salty phlegm in the blood
Salʿāh	Small subcutaneous gland of cold phlegm and is bounded by a membrane, can be moved under pressure in different directions, may be soft or rigid
Ghanghariā	Gangrene is the mortification of tissue in a living body, due to the closure of blood vessels

disease. To evacuate acne-related humor, moistening the body prevents the formation of a progressive form of acne (*alkalla*). Various studies report that morning glory (*Ipomea carnea*) has profound potential against many diseases such as cancer, diabetes, inflammation, sleep disorders, and cardiovascular disorders (Singla *et al.*, 2021). Plantago (*Plantago major*), pomegranate (*Punica granatum*) peels, and barley (*Hordeum vulgare*) are recommended as topical remedies in the form of pastes or ointments, and honey can also be used in combination with them when skin lesions do not respond to these treatments and develop into ulcers. Finally, if there is no response to these remedies, the surgical removal technique is recommended. To treat the cancer form *javarsieh*, turpeth (*Ipomea turpethum)* in Ayurveda: (*Operculina turpethum*) and dodder (*Cuscuta campestris*) are usually applied to remove mucus and black bile, then gall oak (*Quercus infectoria*) bark and tamarisk (*Tamarix* sp.) are used as an astringent topical paste.

Treatment of *feleghmuni* depends on the individual medical history, as bad mood accumulates in the body. Elimination of exotic substances that cause swelling and application of topical dosage forms such as *zomad* (paste) made from wheat flour boiled in oil and water are helpful. Removal of morbid humor that aggravates the disease process if it remains in the body is done by *ishal* (a laxative) or vein removal (*fasd*) depending on the age, season, environment, and physical condition of the patient. Topical dosage forms that gradually loosen the skin, containing cold herbs such as purslane (*Portulaca oleracea*), chicory (*Cichorium intybus*) juice, fine solanum (*Solanum nigrum*) powder, and morning glory, cool the body, slow blood circulation, and slow swelling when applied as a topical paste. The cold is brought on with a sponge dipped in water and vinegar. Repulsive herbs that divert the substance from one part of the body to another, such as a decoction of pomegranate peel with vinegar and sumac (*Rhus coriaria*), are effective; in addition, acacia (*Acacia arabica*), red horned poppy (*Glaucium corniculatum*), and saffron (*Crocus sativus*) are used when a strong effect is needed. In the final stage of treatment of swelling, medicines such as saffron help healing as a desiccant, as they can reduce exudation from blood vessels and inhibit infection. The warm and moist nature of herbs such as flaxseed (*Linum usitatissimum*), chamomile (*Matricaria chamomilla*), and dill (*Anethum graveolens*), due to their dissolving properties, facilitates the elimination of congested mucus from the skin pores when swelling becomes stiff. Thermogenesis also promotes the production of digestive enzymes that eliminate waste and toxins. Excessive use of cold-producing herbs slows down blood and energy circulation and leads to corruption of organs by restraining toxic humor. In this way, the local application of barley flour and field bindweed is beneficial. Treatment of *homrah* or erysipelas (which is generally more difficult to cure than *feleghmuni*) is done by excretion of the yellow bile by appropriate laxatives and/or phlebotomy, depending on the clinical findings. Then, the cold-inducing drugs mentioned in the treatment of *feleghmuni* are used if there is evidence of stain distribution. A liquid dosage form of spinach beet leaves boiled in astringent wine is used and poured on the body surface, also a rue (*Ruta graveolens*) extract with solid Persian rose oil in the form of an ointment can be used in this regard. For the treatment of *nar-e-farsi* or *jamrah*, vein removal is required to expel the yellow bile, and a paste of plantago or wheat with a high content of bran and oak juice with vinegar is recommended.

Skin Cancer

Skin cancer is the most commonly diagnosed cancer in the USA, but most cases are preventable (NLM, 2014). Skin cancer global rates have continued to increase and it is by far the most common type of cancer, dramatically affecting the quality of life and imposing high costs on communities (Lomas *et al.*, 2012). There are some terms in *Canon of Medicine* that refer to skin cancer; Avicenna mentioned that cancer is a type of inflammation caused by changes in the quality and quantity of main humors, including yellow bile, phlegm, black bile, and blood (Tables 11.2, 11.3, 11.4, and 11.5). Some types of swellings such as *varame solb* and *varame saratani* (different forms of cancer) are the result of abnormal black bile. During a process of excessive heat in the body, there is a burning or charring of the humor, similar to the burning of food on a stove when the evaporation of water leaves charred material behind. (Saher *et al.*, 2021). When these substances cannot be eliminated adequately and reach the skin or lymph nodes as weak organs, they spread with great penetrating power.

Treatment

Avicenna, in his *Canon*, mentioned that skin cancer treatment is difficult. According to the six hygiene principles, lifestyle modification includes using wet foods and nutrition correction to produce healthy materials, proper absorption, and reduce congested waste products and carcinogenic materials. Depleting abnormal black bile and sometimes phlebotomy and modifying liver and spleen function are used to prevent increased swelling, prevent increased inflammation and ulcer, and inhibit invasive behavior of the tumor, and represent strategic approaches for cancer treatment. The spleen is in close connection with the liver, and any spleen dysfunction leads to changes in liver function, mood, skin, and maybe cancer formation, so it is important for body empowerment

Table 11.2. Skin malignancies from the viewpoints of *Canon of Medicine* and modern medicine.

Khanāzir	Different types of *khanazir* or cold swelling type, created from thick phlegm and black bile, adhere to the surrounding tissues and cannot be moved. Swollen lymph nodes are more common in the axilla, neck, and soft parts of the body
Khilān	A type of skin color that changes due to abnormal black bile and is a nevus that may be susceptible to malignancy (Iranzadasl *et al.*, 2020)
Warame sodawi (*saraṭani*)	A cancerous inflammation, carcinoma, *varame soda*, or melancholic swelling with erythema that has a black bile origin produced from the *ehteragh* (burning out) of yellow bile, phlegm combined with yellow bile, or black bile alone. Avicenna believed that it is difficult to cure, and increases rapidly day by day with pain and pulse. It is like a solid mass with deep penetration
Warame ṣolb	This is complicated with phlegm and black bile combination, one kind is very similar to the body color and curable. One kind with black bile origin is black in color and noncurable. It is static without numbness and pain, may be similar to a solid tumor
Saqirus	A firm type of inflammation that is caused by black bile, phlegm, or both, which may transform into cancer

(Zakeri *et al.*, 2018). Agents that resolve viscous or thick humor are used for body cleansing. Systemic dosage forms are used in the form of *habb* (pill), *sharbat* (syrup), *Ayarij*, a purgative, and *Itrifal* (Triphala), which is a semisolid dosage form with the essential constituents consisting of chebulic myrobalan (*Terminalia chebula*), beleric (*Terminalia bellirica*), and amla (*Emblica officinalis*). Topical dosage forms like *dohn* (medicated oil), *tela* (liniment), and *zomad* (paste) are prepared for topical cancer treatment.

Mastic (*Pistacia lentiscus*), myrrh (*Commiphora myrrha*), dorema or gum ammonicum (*Dorema ammoniacum*) combined with animal fats, mucilages from flaxseed, fenugreek (*Trigonella foenum-graecum*), and plantago are also recommended. Fixed oils of henna (*Lawsonia inermis*), daylily (*Hemerocallis fulva*), damask rose (*Rosa × damascena*), or wallflower (*Cheiranthus cheiri*) are combined with some kinds of inorganic substances as an ointment or paste in treating the severe form of malignant cancer, the composition of which is mentioned in detail in the *Canon* of Avicenna. Sometimes also althea (*Althea officinalis*) and wild cucumber (*Ecballium elaterium*) with storax (*Styrax camporum*) are applied to the surface of carcinogenic swellings. Natural remedies and formulations in the *Canon* not only improve the quality of life for patients but also help get the desired local and systemic pharmacologic action.

Judham (leprosy)

Leprosy is a chronic, generalized, infectious disease, caused by the bacterium *Mycobacterium leprae*, that manifests as granulomatous or neurotrophic lesions in the skin, mucous membranes, nerves, and viscera, starting in the distal parts of fingers and toes. There is strong evidence for genetic determinants of susceptibility to leprosy. Peripheral sensory nerve damage is the leading cause of functional morbidity in people with leprosy. Peripheral nerve damage in leprosy can be mediated directly by *M. leprae*, as well as by the immune response to *M. leprae* (Cybele *et al.*, 2010). Early symptoms are darkening of the facial color and of the eyes, difficult and hoarse respiration, blocking of the nose, thinning and loss of hair, stinking sweat and respiration, melancholic dreams, micronecrosis and ulceration of extremities, and disruption of tissue connection in a cold region of the body such as finger joints; generalized nodules and destruction of the nasal septum with an ulcer may occur later. The etiology of *judham* is caused by a lot of black bile humor and a melancholic temperament which diffuses in the whole body. Contact with lepers, hot climate, foods that produce black bile humor (e.g., salted fish, dried and processed meats), and coarse and leftover foods that produce black bile are predisposing factors. The final clinical manifestation

Table 11.3. *In vitro* studies on medicinal plants mentioned in *Canon of Medicine* for treatment of skin disorders.

Scientific name (English name)	Arabic name	Part	Extract/ phytochemical	Activity	Tested cell lines/ assay	Result	Reference
Allium cepa L. (onion)	*Basal*	Dried skin	Quercetin-glucoside (methanol extract)	Whitening agent; antioxidant activity	B16 melanoma cell line	Anti-tyrosinase and melanin synthesis inhibition	Arung *et al.* (2011)
Aristolochia bottae Jaub. & Spach. (birthwort)	*Zaravand*	Aerial parts	Hexane extract	Tyrosinase inhibition	Mushroom tyrosinase cell lines	More than 50% tyrosinase inhibition	Hashemi and Zarei (2014)
Carthamus tinctorius (safflower)	*Ehriz*	Seeds	Arbutin/N-feruloyl serotonin	Melanogenesis inhibition	ELISA kit	Inhibitory activity on tyrosinase and melanin formation	Roh *et al.* (2004); Delshad *et al.* (2018)
Citrus sp. (orange tree)	–	Fruit peel	Naringin	Clinical use in psoriasis and other inflammation-associated diseases	Human peripheral blood mononuclear cells	↓ Production of all pro-inflammatory cytokines	Deenonpoe *et al.* (2019)
Citrullus colocynthis (bitter melon)	*Hanzal*	Pulp	Whole extract	Inhibition of the growth of cancer cells such as human larynx carcinoma	Hep2 cell line	Dose-dependent cytotoxicity effect by MTT assay	Afshari Tavakkol *et al.* (2005)
Commiphora myrrha (myrrh)	*Moghl*	Dried twigs	Aqueous extract	IL-31 and histamine supression	HMC-1 cells and WST cell viability assay	Anti-inflammatory effect by regulation of the itch-associated IL-31 cytokine release	Shin *et al.* (2019)
Convolvulus arvensis (bindweed)	*Saghmoonia*	Leaves, stems, roots	Aqueous and methanol extracts	Cytotoxic effect	Rhabdomyosarcoma (RD) cell line	Cytogenetic effect	Al-Asady *et al.* (2014)
Costus spicatus (Indian head ginger)	*Quost*	Leaves	Propylene glycol/ water extract	Skin pigmentation disorders	Mushroom tyrosinase	Extracts present poor anti-tyrosinase activity (less than 50%) compared to the reference	Arnoult *et al.* (2002)

Continued

Table 11.3. Continued.

Scientific name (English name)	Arabic name	Part	Extract/ phytochemical	Activity	Tested cell lines/ assay	Result	Reference
Cuscuta compestris (dodder)	–	Seeds	Quercetin and kampferol	Anti-inflammation	Chang Liver cell lines	Inhibition of gene expression of iNOS, COX-2, and CRP	García-Mediavilla *et al.* (2007)
Cuscuta epithymum (clover dodder)	*Shajar-ol-zabaa*	Aerial parts	Total phenols	Cell viability	MTT assay	Cytotoxicity against Hela, HT29, MDA-MB-468 tumor cells	Jafarian *et al.* (2014)
Ecballium elaterium (wild cucumber)	*Qesa-ol-hemar*	Fruit juice	Cucurbitacins	Modification of the levels of bilirubin in plasma	Bilirubin assay	Glucurono-conjugation increases the binding of bilirubin to albumin	Greige-Gerges (2007)
Emblica officinalis (amla)	*Amlaj*	Fruits	Ethanol extract	Cytoprotective activity	Determination of cell proliferation, IFN-γ and ROS production, and phagocytosis	Cytoprotective properties against oxidative injury and inhibited chromium-induced immunosuppression	Sai Ram *et al.* (2003)
Glaucium corniculatum (red horned poppy)	*Khashkhash-e-bari*	?	Extract	Neuroprotection	PC-12 cells	Apoptosis suppression activity in neurite-like cells	Kocanci and Aslim (2021)
Hemerocallis fulva (daylily)	*Sousan*	Flowers	Phenanthrenes	Antiproliferative effect	Breast carcinoma MDA-MB-231, hepatocellular carcinoma HepG2, and lung carcinoma A549 cell lines	Cytotoxic effects	Fu *et al.* (2022)
Lepidium sativum (garden cress)	*Horf*	Leaves	Phenolic acids	Anti-inflammatory and antiandrogenic activity	Human keratinocyte cells (HaCaT)	Beneficial for some noninvasive skin tumors and inflammatory disorders such as psoriasis, atopic dermatitis, and alopecia areata, melasma, and for some androgen-dependent diseases such as acne and androgenetic alopecia	Türkoğlu *et al.* (2018)

Lupinus termis (lupine)	*Tormes*	Seeds	Alkaloids, phytosterols, resin, coumarins	Antioxidant activity	DPPH method compared to ascorbic acid	Radical scavenging activity	Abdallah *et al.* (2017)
Nigella sativa (black cumin)	*Shoniz*	Seeds	Monosaccharides	Regulatory effects on immunity and anti-inflammation	Immunomodulatory effects in RAW 264.7 cells, evaluated by phagocytosis, NO and cytokine release, mRNA transcription	Inhibition of the excessive release of pro-inflammatory cytokines and excessive transcription at the mRNA level	Niu *et al.* (2021)
Piper nigrum (black pepper)	*Folfol*	Fruits	Piperine	Vitiligo treatment	Melan-a cells	Stimulates melanocyte proliferation and dendrite formation	Soumyanath *et al.* (2006)
Pistacia lentiscus (mastic tree)	*Mastaki*	Leaves and fruits	Polyphenols	Metabolic syndrome management	Pancreatic lipase, α-amylase, and α-glucosidase	Highly significant dose-related inhibition of pancreatic amylase, amyloglucosidase, and lipase	Foddai *et al.* (2015)
Portulaca oleracea (purslane)	*Baghlat-ol-hamgha*	Leaves	Linoleic acid, vitamins B, C, E	Antiaging; anti-skin wrinkling	Mushroom tyrosinase, fibroblast culture, DPPH	↑ Tyrosinase inhibitive activity compared to kojic acid and arbutin; ↑ collagen synthesis by protein synthesis	Kim *et al.* (2013)
Punica granatum (pomegranate)	*Romman*	Flower and fruit peel	Hydroalcoholic extract	Anti-leishmaniasis activity	Strain of *Leishmania major* (MRHO/75/ER)	*L. major* promastigotes were studied based on F_{90} and IC_{50} using the MTT assay	Mardani *et al.* (2020)
		Rinds	Total tannins	Anti-inflammatory effect on COX-2 expression	Porcine ears	Western blotting and immunocytochemistry for COX-2	Houston *et al.* (2017)
Raphanus sativus (common radish)	*Fojl*	Roots	Aromatic aldehyde, aromatic acid, and polyphenols	Skin whitening	Mushroom tyrosinase	Anti-tyrosinase activity	Kamkaen *et al.* (2007)

Continued

Table 11.3. Continued.

Scientific name (English name)	Arabic name	Part	Extract/ phytochemical	Activity	Tested cell lines/ assay	Result	Reference
Rosa × damascena (Persian rose)	*Vard-e-ahmar*	Waste water of distillation	Polyphenols	Anti-inflammatory activity targeting gene expression	Human non-tumorigenic HaCaT keratinocytes	↓ IL-1β gene expression level and gene expression of IL-8 was significantly downregulated	Wedler *et al.* (2016)
Rubia tinctorum (madder)	*Fovvah*	Extract	Silver nanoparticles	Inhibiting cancerous cells	ABTS and DPPH free radical scavenging assays, MTT assay on liver cancer cells (HepG2)	Antioxidant activity without toxic effects on normal HDF cells	Ghandehari *et al.* (2018)
Ruta graveolens (rue)	*Sodab*	Aerial parts and roots	Furanocridone alkaloids	Antiproliferative effects on human breast cancer cell lines	Human breast adenocarcinoma cell lines, MCF- 7, MDA-MB-361, MDA-MB-231, T-47D, MRC-5 (noncancerous fetal lung fibroblast) cells	Significant activation of caspase-9 and caspase-3	Schelz *et al.* (2016)
Styrax camporum (storax)	*Lobni, astorak*	Stems	Lignans	Cytotoxic activity	Hep-2 (larynx epidermoid carcinoma), HeLa (human cervix carcinoma), and C6 (rat glioma)	Cytotoxicity evaluation (IC_{50}) in tumor cells	Teles (2005)
Syzygium aromaticum (clove)	*Gharanfol*	Flower buds	Methnol extracts	Immunomodulation, anti-inflammation	Cytokine measurement with macrophages	Preventive and therapeutic action on IL-1β and IL-8 production	Bachiega *et al.* (2012)
Vicia faba (broad bean)	*Baghella*	Legumes	Quercetin	Skin whitening	Mushroom tyrosinase	Anti-tyrosinase activity	Kramell *et al.* (2005)

ABTS, 2,2′-azino-bis(3-ethylbenzothiazoline-6-sulfonic acid); COX-2, cyclooxygenase-2; CRP, C-reactive protein; DPPH, 2,2-diphenylpicrylhydrazyl; F_{90}, 90% fatality; IC_{50}, half maximal inhibitory concentration; IFN-γ, interferon-γ; IL, interleukin; iNOS, inducible nitric oxide synthase; MTT, 3-(4,5-dimethylthiazol-2-yl)-2,5-diphenyltetrazolium bromide; NO, nitric oxide; ROS, reactive oxygen species; WST, 4-[3-(4-iodophenyl)-2-(4-nitro-phenyl)-2H-5-tetrazolio]-1,3-benzene sulfonate.

Table 11.4. *In vivo* studies on medicinal plants mentioned in *Canon of Medicine* for treatment of skin disorders.

Scientific name (English name)	Arabic name	Part	Extract/ phytochemical	Activity	Study animal	Result	Reference
Acacia arabica (acacia)	*Aghaghia*	Gum	Gum powder gel	Wound healing	Rabbit	Reepithelialization, anti-inflammation	Abdullah *et al.* (2022)
Anethum graveolens (dill)	*Shebet*	Seeds	Carvone, essential oil	Analgesic, anti-inflammation	*Rats*	Significantly reduces paw inflammation	Naseri *et al.* (2012)
Carthamus tinctorius (safflower)	*Osfor*	Seeds	Carthamins	Whole blood viscosity	Rats	Plasma viscosity and platelet aggregation reduced	Li *et al.* (2009)
Cassia angustifolia (senna)	*Senna-e-makki*	Leaves	Ethanolic extract	Anti-inflammation	Albino mice	Significantly inhibited the edema induced by TPA, oxazolone, and arachidonic acid	Cuéllar *et al.* (2001)
Cheiranthus cheiri (wallflower)	*Kheiri*	Seeds	Extract	Chemopreventive agent and alleviation of cutaneous carcinogenesis	Mice	Recovery of antioxidant enzymes	Saleem *et al.* (2001b)
Cichorium intybus (chicory)	*Hendoba*	Roots	Aqueous, methanol, and ethanol extracts	Positive effect on lipid metabolism, anti-inflammation	Male piglets	Significant downregulation of proteins engaged in the innate immune response but also those related to blood clotting	Lepczynski *et al.* (2015)
Cicer arietinum (chickpea)	*Hemmas*	Seeds	Chloroform extract	Estrogenic effect	Rats	Increases reproductive organ weight	Wikhe *et al.* (2013)
Citrullus lanatus (watermelon)	*Batikh-e-akhzar*	Fruits	Juice	Antioxidant activity and inhibition of LPO	rats	Decrease of LPO level in the liver, kidney, and brain tissues	Altaş *et al.* (2011)
Citrus medica (citron)	*Otroj*	Seeds	Alkaloids, fat, amino acids, flavonoids	Uterine weight gain	Rats	Estrogenic activity	Patil Saraswati (2008)
Convolvulus arvensis (bindweed)	*Saghmoonia*	Whole plant	Tannins and flavonoids	Topically prevents skin carcinogenesis	Mice	Significant decline in cumulative number of papillomas	Saleem *et al.* (2015)
Crocus sativus (saffron)	*Zafran*	Stigma	Essential oil	Immunoregulation	Guinea pigs	↓ Serum levels of total NO, nitrite, and IL-4; enhancing effect on IFN-γ levels and Th1/Th2 balance	Boskabady *et al.* (2014)
Cucurbita pepo (pumpkin)	*Qar*	Seeds	Cucurbitacins	Anti-inflammatory agent, it is effective against acute and chronic skin inflammatory processes	Female mice	Inhibition of the productionof eicosanoids via modulation of phospholipase A2, COX, and LOX enzymes	Oliveira *et al.* (2013)

Continued

Table 11.4. Continued.

Scientific name (English name)	Arabic name	Part	Extract/ phytochemical	Activity	Study animal	Result	Reference
Cydonia oblonga (quince)	*Safarjal*	Seeds	Cream made of mucilage	Reduction of dermal irritation, wounds, frosting	Rabbit	Amount of growth factors involved in the wound healing process is positively altered	Hemmati *et al.* (2012)
Eruca sativa (arugula)	*Jerjir*	Extract	Silver nanoparticles	Improved testicular functions and structure and oxidative stress reduction	Rats	Decrease of testicular DNA fragmentation, improvement of normal seminiferous tubules, and marked increase in sperm number	Altwaijry *et al.* (2020)
Ferula assa-foetida (asafoetida)	*Sakbinaj*	Oleogum resin	Acetone extract	Effective chemoprevention, alleviating cutaneous carcinogenesis, recovery in the level and activity of cellular antioxidants	Mice	↓ TPA-induced cutaneous oxidative stress	Saleem *et al.* (2001a)
Foeniculum vulgare (fennel)	*Razianj*	Fruit	Methanol extract	Anti-inflammatory, anti-type IV allergic, and central analgesic activities	Mice and rats	↓ Carrageenan-induced paw and arachidonic acid-induced ear edema	Choi and Hwang (2004)
Gentiana lutea (gentian)	*Gentiana*	Roots	Secoiridoids	Stomachic and antiulcerogenic properties	Rats	Suppressive effect on ethanol-induced gastritis	Niiho *et al.* (2005)
Hordeum vulgare (barley)	*Shaeer*	Unhulled seeds	Methanolic extract	Antioxidant activity	Rats	Protection against increased LPO and maintained glutathione contents	Shah *et al.* (2009)
Lactuca sativa (lettuce)	*Khas*	Seeds	Saponins, triterpenoids, phenolics	Analgesic and anti-inflammatory activity	Rats	Carrageenan and formalin test: analgesic effect in formalin test and also a dose-dependent anti-inflammatory activity in a carrageenan model of inflammation	Sayyah *et al.* (2004)
Moringa olifera (horseradish tree)	*Ban*	Seeds	Isothiocyanates	Modulating inflammatory and antioxidant responses	Mice	Relieves symptoms in DSS-induced acute and chronic ulcerative colitis	Kim *et al.* (2017)
Nardostachys jatamansi (Indian valerian)	*Nardin*	Rhizomes	Ethnolic extract	Potent free radical scavenger, stress modulator	Rats	Lowering LPO, nitrite level, and catalase activity in the brain	Lyle (2009)
Pistacia atlantica) Atlas pistachio)	*Habat-ol-khazra*	Resin	Oil	Antioxidant, wound healing	Rats	↑ VEGF and hydroxyproline levels in wound area	Shahouzehi *et al.* (2018)

Prunus amygdalus (almond)	*Loz*	Tree gum	Polysaccharides	Anti-inflammatory and antibacterial activity for wound healing	Rats	Promotion of neo-blood vessels and collagen formation	Bouaziz *et al.* (2014)
Rheum officinale (rhubarb)	*Ravand*	Rhizome	Raspberry ketone	Depigmentation activity	Zebrafish	↓ Skin melanin content	Lin *et al.* (2011)
Rosa × damascena (Persian rose)	*Vard-e-ahmar*	Fixed oil	Geraniol	Anti-inflammatory and immunosuppressive potential	Rats	↑ Colon/spleen mass indices; ↓ Wnt/β-catenin, p38MAPK, NF-κB, PPAR-γ, and caspase-3	Soubh *et al.* (2015)
Terminalia bellirica (beleric)	*Balilaj*	Fruits	Ethanol extract	Hepatoprotection, is effective against CCl_4-induced liver and kidney damage	Female albino rats	Serum biochemical parameter against CCl_4-induced toxicity	Jadon *et al.* (2007)
Terminalia chebula (haritaki or chebulic myrobalan)	*Ehlilaj-e-kaboli*	Fruits	Polyphenols	Hepatoprotection	Mice	Protects the liver against acute and severe injury by reducing oxidative stress and the inflammatory reaction via activation of pro-inflammatory cytokines	Choi *et al.* (2015)
Trigonella foenum-graceum (fenugreek)	*Holbah*	Seeds	Ethanol extract	Hepatoprotection and reduces liver cirrhosis	Rats	Inhibiting the activity of certain enzymes, mainly xanthine oxidase, and scavenging free radicals	Zargar (2014)
Viola odorata (violet)	*Benafsej*	Aerial part	Total phenols	Hepatoprotective, and vasculoprotective potential	Rats	Limiting hepatic damage by lowering AST, ALT, and GGT; protecting against vascular damage by preventing hyperuricemia and hyperphosphatemia	Jamshed *et al.* (2019)
Ziziphus jujuba (jujube)	*Onab*	Fruits	Polysaccarides	Immunomodulation	Mice	Spleen and thymus index assays indicated the immunomodulator activity	Zou *et al.* (2018)

ALT, alanine aminotransferase; AST, aspartate aminotransferase; CCl4, carbon tetrachloride; DSS, dextran sulfate sodium; GGT, γ-glutamyl transferase; IFN-γ, interferon-γ; IL, interleukin; LOX, lipoxygenase; LPO, lipid peroxidation; NF-κB, nuclear factor-κB; NO, nitric oxide; p38MAPK, p38 mitogen-activated protein kinase; PPAR-γ, peroxisome proliferator-activated receptor-γ; TPA, 12-*O*-tetradecanoylphorbol-13-acetate; VEGF, vascular endothelial growth factor.

Table 11.5. Clinical studies on medicinal plants mentioned in *Canon of Medicine* for treatment for skin disorders.

Scientific name (English name)	Arabic name	Part	Extract/ phytochemical	Activity	Study design	Result	Reference
Astragalus sp. (astragal)	*Katira*	Gum	Tragacanth gel cushion	Prevention of pressure ulcer	Triple-blind randomized controlled trial/ ICU patients	Elimination of frictional forces between the patient's skin and bed and delaying the onset of erythema in ICU patients	Shakibamehr *et al.* (2019)
Dorema ammoniacum (dorema)	*Oshagh*	Gum	Gum extract topical formula	Positive effect of the cream formulation of *D. ammoniacum* extract on melasma	Randomized, double-blind, placebo-controlled trial/49 patients	Mean melasma severity index due to melanin hyperpigmentation was significantly decreased in the drug group	Javedan *et al.* (2021)
Fumaria parviflora (fumitory)	*Baghlat-ol-molk*	Aerial parts	4% Ethanolic extract cream	Improvement of all types of hand eczema	Randomized, double-blind, placebo-controlled trial/44 patients	Inhibition of T-lymphocyte proliferation as well as inhibition of pro-inflammatory granulocyte cytokines and modulation of monocyte cytokine production	Jowkar *et al.* (2011)
Lawsonia inermis (henna)	*Henna*	Leaves	Traditional application of henna dye	Anti-inflammatory, antipyretic, and analgesic effects	Case study of capecitabine-induced hand–foot syndrome	Clinical improvement	Yucel and Guzin (2008)
Helleborus niger (Christmas rose)	*Kharbagh*	Roots or seeds	Extract	Integrative anthroposophic therapy	Clinical case report	Long-term survival with MPM	Werthmann *et al.* (2017)

Plantago major (great plantain)	*Lesan-ol-haml*	Leaves	Gel formulation	↓ Burn, pain, and dermatitis	Randomized, double-blind, phase 2 clinical trial/80 patients	Evaluating prevention of acute radiation dermatitis grade in breast cancer patients	Tavakoli (2021)
Portulaca oleracea (purslane)	*Baghlat-ol-hamgha*	Leaves	Whitening lotion and antiaging essence	Collagen synthesis,free radical scavenging	Human volunteers (after 6 weeks, $n = 6$), application twice daily for 6 weeks	Fine wrinkles diminished	Kim *et al.* (2013)
Quercus infectoria (gall oak)	*Afs*	Nutgall	Topical formulation	Rapid wound closure	Multicenter, open label, pre- and post-test randomized controlled trial	Inhibits diabetes-induced inflammation and oxidative stress in macrophages through suppression of the Set7/NF-κB pathway	Chokpaisarn *et al.* (2020)
Viola odorata (violet)	*Benafsej*	Flowers	Fixed oil	Reduce the dryness score, severity, location, and frequency of pruritus in hemodialysis patients	Randomized controlled trial	Massage with violet oil was more effective in treatment of uremic pruritus than massage alone	Khorsand (2019)

CTCAE-4, Common Terminology Criteria for Adverse Events, v4; ICU, intensive care unit; MPM, malignant pleural mesothelioma; VAS, visual analog scale.

of leprosy, which is accompanied by necrosis, is similar to a cancerous state and chronic ulcers (Avijgan *et al.*, 2018). Avicenna relates leprosy to cancer: "cancer, is Judham of a single organ and is incurable, but '*Judham*' is a cancer of the whole body". He clearly stated that black *baras* was an early sign of *judham*. Because of the popularity of Avicenna's work in medieval Europe, the signs that he attributed to leprosy reappeared in Western records of the disease (Dols, 1979).

Treatment

Avicenna mentioned it is most important to accelerate the process of discharging the offending material and to purify the blood by the opening of large veins in order to draw the corrupt blood from the liver and spleen, after that to pay attention to diet in order to renormalize the unhealthy temperament. For treating advanced leprosy, palliative care is indicated, like moistening the drying body and strengthening the heart and other principal organs to protect the body's natural defensive mechanism by use of various traditional dosage forms for minimizing deformities. In the advancing stage of leprosy, vipers were to be utilized in food, drink, ointments, and "grand theriac" as a healing remedy that is effective in antagonizing poisons. *Theriac*, a word derived from ancient Greek, means "wild animal" but according to many changes in the original Greek formulations of theriacs by Avicenna, it is used for protection against the many compounds that may disrupt the normal functioning of the human body (Taghizadieh *et al.*, 2020).

Skin Pigmentation Disorders

Baraṣ (vitiligo)

Vitiligo is an acquired localized pigmentary skin disorder caused by the progressive loss of melanocytes from the epidermis and has a huge psychosocial impact on patients. There are depigmented macules and patches on the skin, mucous membranes, or white hair due to destroyed skin melanocytes. It is characterized by milky-white sharply demarcated macules. Autoimmune, autocytotoxic, and neural mechanisms are responsible for vitiligo in that immune cells invade and damage melanocytes. T cells and cytokines such as interleukins and interferon-γ may have a role in the vitiligo process (Haghir *et al.*, 2021). Many hypotheses have been put forward but the main etiology is a few autoimmune disorders and thyroid abnormalities that are the commonest ones. One hypothesis suggested that melanocytes destroy themselves due to a defect in a natural protective mechanism that removes toxic melanin precursors. These compounds are released at peripheral nerve endings in the skin and have a toxic effect on melanocytes (Van Geel and Speeckaert, 2016). Vitiligo is named *baras*, and imbalanced temperament plays a significant role in the etiology of the disease. Temperament imbalance due to increasing wetness and coldness and increasing the condensed and viscous white phlegm humor within the dermis causes skin surface depression, which appears as white and shiny patches. The metabolic power of the body is weakened, and under these conditions, food fails to produce normal blood transformation in the skin.

Treatment

Vitiligo treatment consists of administration of some medicinal plants to modify the quality and quantity of the humor to produce a physiologic condition and make a healthy internal environment and to help drugs have appropriate effects. At first, prior to evacuation of abnormal phlegmatic humor, an approach named *munzij* (conoctive) is applied for humor maturation and abnormal phlegm ripening, making it ready for easy elimination by changing its consistency. Sometimes the matter is too thick or too thin to be eliminated.

Then a laxative (*mushil*) such as *Ayarij* is used to remove the excess impurities by inducing diarrhea, to effectively move the waste from the vessels, membranes, and every part of the body to the gastrointestinal tract. In this context, *Ayarij-e-Faiqra* is one of the best known remedies. The main ingredients are barks of cinnamon (*Cinnamomum zeylanicum* and *Cinnamomum cassia*), rhizomes of Indian valerian (*Nardostachys jatamansi*), fruits of cubeb (*Piper cubeba*), flowers of Persian rose, mastic (an oleoresin of *Pistacia lentiscus*), and juice of various *Aloe vera* subspecies. But Avicenna recommended another version of this dosage form with *C. zeylanicum* and *C. cassia*, rhizomes of Indian valerian, *Cinnamomum opobalsamum*, *Cinnamomum citriodorum*, wild ginger (*Asarum* sp.), saffron, and water mint (*Mentha aquatica*). Topical dosage forms containing the following herbs are also used for skin coloring: asafoetida (*Ferula assa-foetida*), black pepper (*Piper nigrum*), common radish (*Raphanus sativus*), madder (*Rubia tinctorum*), Indian head ginger (*Costus spicatus*), gentian (*Gentiana lutea*), and black cumin (*Nigella sativa*) in various combinations. All cinnamon species contain cinnamaldehyde, which accounts for between 65 and 80% of the essential oil, and various studies have been conducted on extracts and isolated compounds of *Cinnamomum* species investigating antibacterial, antidiabetic, anti-inflammatory, antioxidant, antitumor, and neuroprotective properties (Sharifi-Rad *et al.*, 2021).

Skin hyperpigmentation

Melanin, hemoglobin, and carotenoids determine the standard skin color. Most pigment cells are melanocytes within the epidermis. Human pigmentation is regulated by ultraviolet (UV) light as well as endocrine, paracrine, and autocrine factors. Endocrine factors and external agents like sunlight play a significant role, which is the most common cause of facial melanosis and manifests as hyperpigmented macules on the face. The etiology and pathologic mechanism of melasma are poorly understood, but the number of melanocytes is not increased; instead they become enlarged and more dendritic, suggesting that a hypermetabolic state leads to increased melanin deposition in the epidermis and dermis (Van Geel and Speeckaert, 2016).

Kalaf, *bahaq*, and *namash* are the most used terms denoting skin hyperpigmentation. Changes in skin color may be due to homeostatic imbalances in the body, corruption of blood, or an acute mixture of temperament with blood. In people with melancholic temperament, pathogenic black bile is congested in the epidermis (as the first place for the accumulation of waste and the most extensive defense line), which is manifested as itching, black and white patches, rough eruption, and dryness. *Kalaf* is a dark or brown to black pigmentation patch of skin but not elevated and usually occurs on the face, with clinical manifestations similar to melasma. *Bahaq* was identical to *kalaf* except that *bahaq* was associated with sloughing and roughness. *Namash* is equivalent to freckles and is defined as abnormal black bile congestion in skin layers with increased concentration that causes dark color spots on the face.

Treatment

The systemic treatment should be taken seriously for a proper result. Appropriate nutritional programs based on temperament can potentiate the effectiveness of topical drugs and ease the situation, blood purification, and pathogenic temperament relief. Medicinal plants such as lupine seeds (*Lupinus termis*), horseradish tree (*Moringa olifera*), common radish, wild cucumber, pumpkin seeds (*Cucurbita pepo*), garden cress (*Lepidium sativum*), arugula (*Eruca sativa*), onion (*Allium cepa*), birthwort (*Aristolochia bottae*), and safflower (*Carthamus tinctorius*) are also administered through the skin. Today various cosmetic companies produce different topical dosage forms for hyperpigmentation and compete to win the production advantage, the products are expensive with a moderate treatment outcome (Mojtabaee *et al.*, 2016).

Dermatitis

Diseases like *hekkah* and *taghashofe-atraf* have been described to resemble the characteristics of dermatitis, which is currently accepted as eczema. Atopic dermatitis, atopic eczema, and infantile eczema sufferers are inheritors of a chronic, relapsing, inflammatory pruritic skin disease modulated by immunity and environmental factors. Redness, pruritus, pain, and burning sensation are the main similarities. The wet type or exudative eczema is accompanied by pruritus and discharge and may lead to ulcers. Itching and erythema without any discharge are characteristics of dry type or non-exudative eczema. Psychological factors have been recognized as causative agents in some skin conditions. There is evidence of the endogenous type of depression, disturbed sleep with early waking, loss of weight and appetite, and low mood associated with pruritis. It is generally recognized that pruritus with skin reaction may be due to melancholia (Eghbalian *et al.*, 2017). According to traditional medicine theory, eczema is highly associated with an imbalance in the liver temperament, as the liver is the primary source of humoral production. Corruption of blood or an acute mixture of humors with each other changes the original skin and also the whole body temperament. The abnormal humor cannot be absorbed into the target organs and enters beneath the skin by vessels, which can cause itching, burning, rash, and vesicles. Melancholic type is very similar to chronic and resistant atopic dermatitis with low pain, pruritus, and dark rash, but bloody and phlegmatic types are remarkably identical to childhood and adolescent atopic dermatitis characterized by itching, white rash, and diluted blood secretion (Edwards, 1954). Signs and symptoms of acute eczema are redness, itching, and burning due to the blood type. Blood and phlegmatic eczema are of the wet type because of the wet dystemperament of the liver; *ṣafrā* (yellow bile) and black bile or melancholic eczema are of the dry type, because of the dry dystemperament of the liver (Mansouri *et al.*, 2016).

Atopic dermatitis

Atopic dermatitis or temperamental eczema is a chronic, relapsing, inflammatory pruritic skin disease that develops mainly from childhood (2–10%) to adulthood (10–30%), it is a genetically complicated disorder accompanied by food sensitivity, asthma, and allergic rhinitis. Skin dryness and pruritus are characteristic features of atopic dermatitis (Arnold *et al.*, 1990).

Treatment

Treatment for eczema and related pruritis was based on temperament modification, and some preventive factors overlap in traditional and allopathic medicines. In Avicenna's *Canon*, lifestyle modifications are essential in correcting liver temperament for correct humor production as a first and important approach.

The diet must be adjusted to avoid mixed foods and to avoid thick, leftover, spicy, and dry/hot foods. On the other hand, generally easily digestible and moist/hot foods such as chicory, watermelon, and lettuce should be consumed, because cold foods cool down the body and slow down blood and energy circulation. Attention should also be paid to the process by which foods are prepared, as sometimes this may significantly change their texture and composition (Dashtdar *et al.*, 2022).

Daily bowel movements are necessary for the elimination of harmful substances to achieve an overall balanced state. First of all, it is better to eliminate the "bile melancholy" caused by the burning of yellow bile and/or abnormal salty or alkaline mucus that tends to heat and dryness. Recommended for this purpose are different *Terminalia* species known as myrobalans and a combination with dodder, fumitory (*Fumaria parviflora*), senna (*Cassia angustifolia*), Persian rose, and common wormwood (*Artemisia absinthium*). Stress and anxiety were found to raise the strange temperature in the body and reduce the instinctive heat, and the effect of emotional aspects was greater and faster than that of other factors (Eghbalian *et al.*, 2017). Mental relaxation and strengthening of the brain by a variety of local and systemic traditional formulations according to body temperament are mentioned in the *Canon of Medicine*. Natural mineral waters with sulfate improved skin lesions in scabies, and traditional formulations had an advantage via restoring the epidermal barrier's balance (Jedkareh *et al.*, 2016).

Qoobā (Psoriasis)

Avicenna believes that *ghoobaye moteghasher* or "flaky *qoobā*" is very similar to *sʿafah-yābes* and treatment approaches are the same. *Baraṣ aswad*, *sʿafah-yābes*, and *qoobā* have similar symptoms (Mohammadian Rasanan *et al.*, 2022). *Sheqaqe poosti* and *taqshofe atraf* are also skin scaling, or roughness, and increased thickness of the skin due to abnormal melancholic humor that has occurred in the hands, feet, face, lips, and anus, and the main reason is dry skin. *Ḥakkah* is the general term for pruritis without any skin lesion. The dryness of skin may be the result of exposure to extreme hot or cold weather and wind, and many other systemic diseases.

Avicenna has defined *ghooba* as redness, itching, desquamation, and localized scaly patches mostly on the head. Some kinds of *ghooba* have pruritis and autumn is an exacerbating factor for the disease. Psoriasis is similar to *ghooba* due to salty phlegm or blood mixed with black bile humor.

Baraṣ aswad

Baraṣ aswad is a severe itching, dry skin disorder with rough and black flakes due to accumulation of abnormal black bile humor in the skin that can spread all over the body surface.

Sʿafah

Sʿafah (favus) is a skin condition that causes a scattered rash with mild itching that can develop into an ulcerative, crusty plaque with scales. Favus is typically a childhood disease, yet adult cases are not uncommon. In tinea capitis favosa, a chronic inflammatory dermatophyte infection of the scalp, over 90% of patients are affected with *Trichophyton schoenleinii*. *Trichophyton violaceum*, *Trichophyton verrucosum*, zoophilic *Trichophyton mentagrophytes*, *Microsporum canis*, and geophilic *Microsporum gypseum* have also been recovered from favic lesions. *T. schoenleinii* fungus can be isolated from tinea capitis lesions appearing as gray patches, but can also colonize the scalp without causing any symptoms (Ilkit, 2010).

Sʿafah-yābes is a skin condition that causes dry, crusted plaque with scales. It occurs all over the skin surface but mainly on the head/face. It manifests as two types: dry and wet, due to the combination of soda with salty phlegm or the combination of burned yellow bile humor for which winter is an exacerbating factor; *shirinaj* or moist favus has plasma discharge that decreases rapidly.

Psoriasis treatment

Lifestyle modifications, known as the six essential measures or *Settehe–Ḍarurieah*, are important external factors. Expelling waste humor and balancing dry temperament is the essential treatment protocol. Black-bile-producing foods such as canned foodstuffs, frozen meat that is kept for a long time in the freezer, all precooked and processed meats, fast foods, the meat of old animals, salted or smoked fish, old cheese, eggplant, dense and leftover foods can produce excessive black bile humor in the body. Excessive activity, extreme cold, hot weather, and astringent baths should be avoided. A short-duration bath moisturizes the body, and humid air needs more consideration. Dry and polluted air is also a negative external factor, dry weather exacerbates disease symptoms, and stress or anxiety can trigger or worsen psoriasis symptoms. Healthy and adequate sleep, vomiting to eliminate abnormal bodily fluids, and methods of defecation are also important and should be considered accordingly. Internal factors such as a healthy stomach, liver, and spleen for proper digestion produce normal humor and generate usual black bile. Systemic and local herbal medicines and physical manipulations like wet cupping and/or venesection are also recommended. Wetting is stimulated by the consumption of moist food and whey powder (Walzem *et al.*, 2002). Some local treatment examples are gum arabic, almond tree (*Prunus amygdalus*) and plum tree (*Prunus domestica*) gums combined with vinegar, fixed oils of wheat, viola (*Viola odorata*) and chickpea (*Cicer arietinum*), a decoction of barley (*H. vulgare*) and citron seeds (*Citrus medica*).

A Brief Review of Some Common Modern Dermatologic Treatments

Vitiligo

There are different approaches for vitiligo treatment that aim to prevent the spread of the disease and facilitate repigmentation of affected lesions. The mainstay of treatment has been topical steroids, calcineurin inhibitors, phototherapy, cosmetic products, immunomodulators, and surgery. However, systemic treatments are increasingly being shown to have a significant impact on the course of the disease as monotherapy or adjunctive therapy. Oral mini-pulsed corticosteroid therapy, methotrexate, minocycline, ciclosporin, Janus kinase inhibitors, and certain supplements have been used in the systemic treatment of vitiligo. Selenium, zinc, polyphenols, and vitamins A, C, and E induced visible repigmentation in mouse studies (Searle *et al.*, 2021).

Atopic dermatitis

General management of atopic dermatitis in allopathic medicine involves eliminating irritant factors, controlling emotional stress, restoring the skin barrier function by using hydrophilic creams and lotions for hydration of xerotic skin, patient education, and pharmacologic treatment of skin inflammation, including oral antihistamine, topical and oral corticosteroids that can cause side effects in long-term use (Howe, 2022). Physical activity like jogging and a brisk walk is recommended, a change of environment every 2–3 months is beneficial, and meditation is helpful in most melancholic psychological problems (Rafieian Kopaei *et al.*, 2016).

Acne

Compliance to acne vulgaris treatment is poor because of numerous factors. Topical and oral antibiotic monotherapy is not recommended because of increasing worldwide bacterial resistance. There was even use of topical antibiotics as a monotherapy that have additional anti-inflammatory effects. Topical retinoids and four combination oral contraceptives (COCs) are approved by the US Food and Drug Administration (FDA). Isotretinoin, which is generally safe and well tolerated, is FDA approved for the treatment of severe recalcitrant acne vulgaris and is also recommended for moderate acne that is treatment-resistant, leads to scarring, or causes significant psychosocial distress. There is increasing evidence regarding the role of diet in acne, high-glycemic-index diets, dairy especially skim milk, and whey protein consumption should be avoided. Dietary modifications and natural treatments will likely play an increasing role in acne treatment as further evidence accrues. In general, high-quality randomized controlled trials and analytical studies are needed regarding the safety and comparative efficacy of various acne treatments (Habeshian and Cohen, 2020).

Psoriasis

At present, there is no cure and no alteration in the overall course of the disease. All treatments are suppressive. Mild psoriasis affects less than 3 to 5% of the body surface area. The treatment approach begins with evaluation for psoriatic arthritis, and treatment options include topical corticosteroids, vitamin D analogs, calcineurin inhibitors, and keratolytic and targeted phototherapy. The moderate form covers 3 to 10% of the body surface area, and severe psoriasis is typically considered with a body surface area coverage of 10% or greater. The American Academy of Dermatology–National Psoriasis Foundation guidelines recommend the consideration of prescribing biologics, oral agents, and phototherapy concurrently for patients with moderate to severe psoriasis. Biologics have higher efficacy compared to oral medications or phototherapy, and topical therapies can be used as adjunct treatments but not as monotherapy in treating moderate to severe psoriasis. Physical and mental rest help to back up the specific management of acute episodes. Anxiety and depression should be treated concomitantly (Armstrong and Read, 2020).

Melasma

Melasma, also known as chloasma, often occurs predominantly in women. There is a definite relationship between high estrogen levels and melasma. Because of sun-induced pigment darkening and the important role of UV-A radiation, advice is to use broad-spectrum sunblock, especially outdoors. First-line therapy with hydroquinone as whitening agent, tretinoin, corticosteroids, and triple combination, because of better efficacy, and second-line therapy with azelaic acid and liquiritin is recommended (Griffiths, 2006).

References

Abdallah, E.M., Qureshi, K.A. and Musa, K.H. (2017) Antimicrobial, antioxidant and phytochemical screening of lupine seeds (*Lupinus termis* Forrsk.) from Sudan. *CIBTech Journal of Microbiology* 6(1), 1–8.

Abdullah, E., Taha, S., Sulaiman, N. and Ahmed, M. (2022) Impact of *Acacia arabica* topical gel on skin wound healing: an experimental study. *Pharmacia* 69(1), 77–83.

Afshari Tavakkol, J., Rakhshandeh, H., Zamani, A., Shahri, N.-M., Ghazezadeh, L. *et al.* (2005) Cytotoxicity effects of *Citrullus colocynthis* on Hep2 and L929 cell lines. *Hakim Research Journal* 8(2), 47–54.

Al-Asady, A.A., Suker, D.H. and Hassan, K.K. (2014) Cytotoxic and cytogenetic effects of *Convolvulus arvensis* extracts on rhabdomyosarcoma (RD) tumor cell line *in vitro*. *Journal of Medicinal Plants Research* 8(15), 588–598.

Altaş, S., Kızıl, G., Kızıl, M., Ketani, A. and Haris, P.I. (2011) Protective effect of Diyarbakır watermelon juice on carbon tetrachloride-induced toxicity in rats. *Food and Chemical Toxicology* 49(9), 2433–2438.

Altemus, M., Rao, B., Dhabhar, F.S., Ding, W. and Granstein, R.D. (2001) Stress-induced changes in skin barrier function in healthy women. *The Journal of Investigative Dermatology* 117(2), 309–317.

Altwaijry, N., El-Masry, T., Badriyah Alotaibi, B., Tousson E. and Saleh, A. (2020) Therapeutic effects of rocket seeds (*Eruca sativa* L.) against testicular toxicity and oxidative stress caused by silver nanoparticles injection in rats. *Environmental Toxicology* 35(9), 952–960.

Armstrong, A.W. and Read, C. (2020) Pathophysiology, clinical presentation, and treatment of psoriasis – a review. *JAMA* 323(19), 1945–1960.

Arnold, H.L., Odom, R.B. and James, W.D. (1990) The skin: basic structure and function. In: *Andrews' Diseases of the Skin: Clinical Dermatology*, 8th edn. Saunders, Philadelphia, Pennsylvania, pp. 1–3.

Arnoult, E., Scior, T., Do, Q. and Bernard, P. (2002) Preliminary screening of some tropical plants for anti-tyrosinase activity. *Journal of Ethnopharmacology* 82(2), 155–158.

Arung, E.T., Furuta, S., Ishikawa, H., Tanaka, H. and Shimizu, K. (2011) Melanin biosynthesis inhibitory and antioxidant activities of quercetin-3′-*O*-β-D-glucoside isolated from *Allium cepa*. *Zeitschrift für Naturforschung C* 66(5–6), 209–214.

Atarzadeh, F., Heydari, M., Sanaye, M.R. and Amin, G. (2016) Avicenna, a pioneer in dermatology. *JAMA Dermatology* 152(11), 1257.

Avijgan, M., Ansari, M. and Salhzadeh F. (2018) Mucous-cutaneous infectious diseases in Persian medicine. *ARC Journal of Dermatology* 3(1), 6–9.

Bachiega, F., Barreto de Sousa, J.P., Kenupp Bastos, J. and Sforcin, J.M. (2012) Clove and eugenol in noncytotoxic concentrations exert immunomodulatory/anti-inflammatory action on cytokine production by murine macrophages. *Journal of Pharmacy and Pharmacology* 64(4), 610–616.

Bagherani, N. and Smoller, B.R. (2015) The history of dermatology and dermatopathology in different countries. *Iran Global Dermatology* 2(7), S39–S49.

Bieber, T. (2022) Atopic dermatitis: an expanding therapeutic pipeline for a complex disease. *Nature Reviews Drug Discovery* 21, 21–40.

Boskabady, M.H., Byrami, G. and Feizpour, A. (2014) The effect of safranal, a constituent of *Crocus sativus* (saffron), on tracheal responsiveness, serum levels of cytokines, total NO and nitrite in sensitized guinea pigs. *Pharmacological Reports* 66(1), 56–61.

Bouaziz, F., Ben Romdhane, M., Boisset Helbert, C., Buon, L., Bhiri, F. *et al.* (2014) Healing efficiency of oligosaccharides generated from almond gum (*Prunus amygdalus*) on dermal wounds of adult rats. *Journal of Tissue Viability* 23, 98–108.

Choi, E.Ch. and Hwang, J.K. (2004) Antiinflammatory, analgesic and antioxidant activities of the fruit of *Foeniculum vulgare*. *Fitoterapia* 75(6), 557–565.

Choi, M.K. Kim, H.-G., Han, J.-M., Lee, J.-S., Lee, J.S. *et al.* (2015) Hepatoprotective effect of *Terminalia chebula* against t-BHP-induced acute liver injury in C57/BL6 mice. *Evidence-Based Complementary and Alternative Medicine* 2015, 1–11.

Chokpaisarn, J., Chusri, S., Piyawan, S. and Voravuthikunchai, S.P. (2020) Clinical randomized trial of topical *Quercus infectoria* ethanolic extract for the treatment of chronic diabetic ulcers. *Journal of Herbal Medicine* 21, 100301.

Cuéllar, M.J., Giner, R.M. and Recioh, M.C. (2001) Topical anti-inflammatory activity of some Asian medicinal plants used in dermatological disorders. *Fitoterapia* 72(1), 221–229.

Cybele, A., Renault, C.A., Joel, D. and Ernst, J.D. (2010) *Mycobacterium leprae*. In: Mandell, G.L., Douglas, J.E. and Dolin, R. (eds) *Mandell, Douglas, and Bennett's Principles and Practice of Infectious Diseases*, 7th edn. Churchill Livingstone, Philadelphia, Pennsylvania, pp. 3168–3169.

Dashtdar, M., Dashtdar, M.R., Dashtdar, F., Dashtdar, B. and Mojab, E. (2022) The perspectives of traditional and modern medicine on the nature of hot and cold foods. *Medical Research Journal* 7(1), 82–93.

Deenonpoe, R., Prayong, P. and Thippamom, N. (2019) Anti-inflammatory effect of naringin and sericin combination on human peripheral blood mononuclear cells (hPBMCs) from patient with psoriasis. *BMC Complementary and Alternative Medicine* 19(1), 168–179.

Delshad, E., Yousefi, M., Sasannezhad, P., Rakhshandeh, H. and Ayati, Z. (2018) Medical uses of *Carthamus tinctorius* L. (safflower): a comprehensive review from traditional medicine to modern medicine. *Electronic Physician* 10(4), 6672–6681.

Dols, M.W. (1979) Leprosy in medieval arabic medicine. *Journal of the History of Medicine and Allied Sciences* XXXIV(3), 314–333.

Edwards, K. (1954) Pruritis in melancholia. *British Medical Journal* 25, 27–29.

Eghbalian, F., Atyabi, A., Esmaeili, N. and Mehrdad Karimi, M. (2017) Treatment of atopic dermatitis (Jarab) in a 38-year-old female patient with traditional medicine at Tehran University of medical sciences. *Asian Journal of Clinical Case Reports for Traditional and Alternative Medicine* 1(1), 37–42.

Foddai, M, Kasabri, V., Afifi, F.U., Azara, E., Petretto, G.L. *et al.* (2015) *In vitro* inhibitory effects of Sardinian *Pistacia lentiscus* L. and *Pistacia terebinthus* L. on metabolic enzymes: pancreatic lipase, α-amylase, and α-glucosidase. *Starch* 67(1–2), 204–212.

Fu, R., Xia Wang, X., Zhao, B., Yang, M., Gou, J. *et al.* (2022) Hemecitones A and B: two phenanthrenes with cytotoxicity from *Hemerocallis fulva*. *Natural Product Research* 36(5), 1266–1272.

García-Mediavilla, V., Crespo, I., Collado, P.S., Esteller, A., Sánchez-Campos, S. *et al.* (2007) The antiinflammatory flavones quercetin and kaempferol cause inhibition of inducible nitric oxide synthase, cyclooxygenase-2 and reactive C-protein, and down-regulation of the nuclear factor kappaB pathway in Chang Liver cells. *European Journal of Pharmacology* 557, 221–229.

Ghandehari, S., Homayouni Tabrizi, M. and Ardalan, P. (2018) Study of antioxidant properties and toxicity of silver nanoparticles synthesized by aqueous extract of *Rubia tinctorum* on liver cancer cells (HepG2) compared to normal HDF cells. *Journal of Ilam Medical Sciences* 26(2), 57–67.

Greige-Gerges, H. (2007) Cucurbitacins from *Ecballium elaterium* juice increase the binding of bilirubin and ibuprofen to albumin in human plasma. *Chemico-Biological Interactions* 169, 53–62.

Griffiths, C.E.M. (2006) Melasma. In: Lebwohl, M.G., Heymann, W.R., Berth-Jones, J. and Coulson, I. (eds) *Treatment of Skin Disease: Comprehensive Therapeutic Strategies*, 2nd edn. Mosby Elsevier, Philadelphia, Pennsylvania, pp. 394–395.

Habeshian, K.A. and Cohen, B.A. (2020) Current issues in the treatment of acne vulgaris. *Pediatrics* 145, S225–S230.

Haghir, E.M., Sadati, L.N. and Shirbeigi, L. (2021) Immunomodulatory effects of medicinal plants used for vitiligo in traditional persian medicine. *Current Drug Discovery Technologies* 18(2), 160–178.

Hashemi, F. and Zarei, M.A. (2014) Tyrosinase inhibitory activity within hexane extract of ten screened plants from Kurdistan province of Iran. *International Journal of Advanced Biological and Biomedical Research* 2(11), 2795–2799.

Hemmati, A.A., Hibatoallah Kalantari, H., Amir Jalali, A., Somie Rezai, S. and Haghighi Zadeh, H. (2012) Healing effect of quince seed mucilage on T-2 toxin-induced dermal toxicity in rabbit. *Experimental and Toxicologic Pathology* 64, 181–186.

Houston, D.M., Bugert, J., Denyer, S.P. and Heard, C.M. (2017) Anti-inflammatory activity of *Punica granatum* L. (pomegranate) rind extracts applied topically to *ex vivo* skin. *European Journal of Pharmaceutics and Biopharmaceutics* 112, 30–37.
Howe, W. (2022) Treatment of atopic dermatitis (eczema). *UpToDate.* Available at: https://www.uptodate.com/contents/treatment-of-atopic-dermatitis-eczema (accessed 15 April 2022).
Ilkit, M. (2010) Favus of the scalp: an overview and update. *Mycopathologia* 170, 143–154.
Iranzadasl, M., Parvin Mansouri, P., Kamalinejad, M. and Narjes Gorji, N. (2020) Review of skin cancers terminology, etiology and treatment from ancient Persian medicine view point. *Medicinal Plants* 19(73), 27–36.
Jadon, A., Monika Bhadauria, M. and Sangeeta Shukla, S. (2007) Protective effect of *Terminalia belerica* Roxb. and gallic acid against carbon tetrachloride induced damage in albino rats. *Journal of Ethnopharmacology* 109, 214–218.
Jafarian, A., Ghannadi, A. and Mohebi, B. (2014) Cytotoxic effects of chloroform and hydroalcoholic extracts of aerial parts of *Cuscuta chinensis* and *Cuscuta epithymum* on Hela, HT29 and MDA-MB-468 tumor cells. *Research in Pharmaceutical Sciences* 9(2), 115–122.
Jamshed, H., Siddiqi, H.S., Gilani, A.-Ul-H., Arslan, J., Qasim, M. *et al.* (2019) Studies on antioxidant, hepatoprotective, and vasculoprotective potential of *Viola odorata* and *Wrightia tinctoria*. *Phytotherapy Research* 33(9), 2310–2318.
Javedan, K., Hydarpur, F., Pour, P.M., Najafi, F., Mirzaeei, S. *et al.* (2021) The formulation and efficacy of topical *Dorema ammoniacum* in treating melasma: a randomized double-blind, placebo-controlled trial. *Journal of Complementary and Integrative Medicine* 19(3), 743–751.
Jedkareh, A., Esmaeili, S., Alembagheri, A. and Mortazavi, S.A. (2016) Comparing the etiology and treatment of skin fissure in traditional and conventional medicine; a brief review. *Research Journal of Pharmacognosy* 3(1), 49–54.
Jowkar, F., Jamshidzadeh, A., Mirzadeh Yazdi, A. and Pasalar, M. (2011) The effects of *Fumaria parviflora* L. extract on chronic hand eczema: a randomized double-blind placebo controlled clinical trial. *Iranian Red Crescent Medical Journal* 13(11), 824–828.
Kamkaen, N., Mulsri, N. and Treesak, C. (2007) Screening of some tropical vegetables for anti-tyrosinase activity. *Thai Pharmaceutical and Health Science Journal* 2(1), 15–19.
Khorsand, A. (2019) The effect of massage and topical violet oil on the severity of pruritus and dry skin in hemodialysis patients: a randomized controlled trial. *Complementary Therapies in Medicine* 45, 248–253.
Kim, I.-Y., Lee, M.-H., Shim, S.-B. and Chun, Y.-J. (2013) Skin lightening and wrinkle improving efficacy of organic *Portulaca oleracea* extract in skin care cosmetic. *International Journal of Bio-Science and Bio-Technology* 5(5), 75–84.
Kim, Y., Wu, A.G., Jaja-Chimedza, A., Graf, B.L., Waterman, C. *et al.* (2017) Isothiocyanate-enriched moringa seed extract alleviates ulcerative colitis symptoms in mice. *PLoS One* 12(9), e0184709.
Kocanci, F.G. and Aslim, B. (2021) Chemical composition and neurotherapeutic potential of *Glaucium corniculatum* extracts. *Pharmacognosy Magazine* 17, 67–75.
Kramell, R., Schmidt, J., Herrmann, G. and Schliemann, W. (2005) *N*-(Jasmonoyl) tyrosine-derived compounds from flowers of broad beans (*Vicia faba*). *Journal of Natural Products* 68(9), 1345–1349.
Lepczynski, A., Herosimczyk, A., Ozgo, M., Skomial, J., Taciak, M. *et al.* (2015) Dietary supplementation with dried chicory root triggers changes in the blood serum proteins engaged in the clotting process and the innate immune response in growing pigs. *Journal of Physiology and Pharmacology* 66(1), 47–55.
Li, H.-X., Han, S.-Y., Wang, X.-W., Ma, X., Zhang, K. *et al.* (2009) Effect of the carthamins yellow from *Carthamus tinctorius* L. on hemorheological disorders of blood stasis in rats. *Food and Chemical Toxicology* 47(8), 1797–1802.
Lin, C.-H.V., Ding, H.-Y., Kuo, S.-Y., Chin, L.-W., Wu, J.-Y. *et al.* (2011) Evaluation of *in vitro* and *in vivo* depigmenting activity of raspberry ketone from *Rheum officinale*. *International Journal of Molecular Sciences* 12, 4819–4835.
Lomas, A., Leonardi-Bee, J. and Bath-Hextall, F. (2012) A systematic review of worldwide incidence of nonmelanoma skin cancer. *British Journal of Dermatology*166(5), 1069–1080.
Lyle, N. (2009) Stress modulating antioxidant effect of *Nardostachys jatamansi*. *Indian Journal of Biochemistry and Biophysics* 46, 93–98.
Maleki, M. (2017) An overview of skin lesions adapted to cutaneous leishmaniasis in Persian medicine. *Electronic Physician* 9(11), 5854–5862.

Mansouri, P., Khademi, A., Pahlevan, D., Memariani, Z., Aliasl, J. *et al.* (2016) Review of medicinal remedies on hand eczema based on Iranian traditional medicine. *Iran Journal Public Health* 45(8), 986–996.

Mardani, H., Khalili, B., Saberi, S., Lorigooini, Z., Ghaderi, M. *et al.* (2020) Effect of hydroalcoholic extracts of flower and fruit peel of *Punica granatum* on *Leishmania major* promastigotes *in vitro*. *Advanced Herbal Medicine* 6(2), 32–48.

Mohammadian Rasanan, F., Kenari, H.M., Ghassemi, M., Sabbagh, A.J., Aliasl, J. *et al.*((2022) Non-pharmacological treatments of psoriasis in Persian medicine a narrative review. *Journal of Pharmacopuncture* 25(1), 1–6.

Mojtabaee, M., Mokaberinejad, R., Adhami, S., Mansouri, P. and Rahbar, M. (2016) Nutritional advice for patients with melasma in Iranian traditional medicine. *Journal of Skin Stem Cell* 3(1), e42722.

Naseri, M., Mojab, F., Khodadoost, M., Kamalinejad, M., Davati, A.,. *et al.* (2012) The study of anti-inflammatory activity of oil-based dill (*Anethum graveolens* L.) extract used topically in formalin-induced inflammation male rat paw. *Iranian Journal of Pharmaceutical Research* 11(4), 1169–1174.

Nasir, A., Siddiqui, M.A., Ahmad, M.A. and Fatma, G. (2016) Descriptions and management of eczema (Nar-e-farsi) in modern and Unani medicine. *International Journal of Advances in Pharmacy Medicine and Bioallied Sciences* 4(1), 29–33.

Niiho, Y., Yamazaki, T., Nakajima, Y., Yamamoto, T., Ando, H. *et al.* (2005) Gastroprotective effects of bitter principles isolated from Gentian root and Swertia herb on experimentally-induced gastric lesions in rat. *Journal of Natural Medicines* 60(1), 82–88.

Niu, Y., Wang, B., Zhou, L., Ma, C., Waterhouse, G.I.N. *et al.* (2021) *Nigella sativa*: a dietary supplement as an immune-modulator on the basis of bioactive components. *Frontiers in Nutrition* 8, 722813.

NLM (2014) Skin Cancer as a Major Public Health Problem. Available at: https://www.ncbi.nlm.nih.gov/books/NBK247164/ (accessed April 2022).

Oliveira, M.L.M. de, Nunes-Pinheiro, D.C.S., Bezerra, B.M.O., Leite, L.O., Tomé, A.R. *et al.* (2013) Topical anti-inflammatory potential of pumpkin (*Cucurbita pepo* L.) seed oil on acute and chronic skin inflammation in mice. *Acta Scientiae Veterinariae* 41, 1168.

O'Sullivan, R.L., Lipper, G. and Lerner, E.A.(1988) The neuro-immuno-cutaneous-endocrine network: relationship of mind and skin. *Archives of Dermatology* 134(11), 1431–1435.

Parvizi, M.M., Fatehi, N., Jaladat, A.M., Gholampour, Z., Shahriarirad, R. *et al.* (2021) Epidemiological factors in patients with dermatologic conditions referring to clinic of traditional Persian medicine. A cross-sectional study. *International Journal of Clinical Practice* 75(11), e14788.

Patil Saraswati, B. (2008) Estrogenic activity of petroleum ether extract of seeds of *Citrus medica* on immature albino rats. *International Journal of Green Pharmacy* 2(2), 91–94.

Rafieian Kopaei, R., Khajegir, A. and Kiani, S. (2016) The association between dystemperament and prevention of diseases: a systematic review. *Journal of Clinical and Diagnostic Research* 10(9), YE01–YE06.

Roh, J.S., Han, J.Y., Kim, J.H. and Hwang, J.K. (2004) Inhibitory effects of active compounds isolated from safflower (*Carthamus tinctorius* L.) seeds for melanogenesis. *Biological and Pharmaceutical Bulletin* 27(12), 1976–1978.

Sai Ram, M., Neetu, D., Deepti, P., Vandana, M., Ilavazhagan, G. *et al.* (2003) Cytoprotective activity of amla (*Emblica officinalis*) against chromium (VI) induced oxidative injury in murine macrophages. *Phytotherapy Research* 17(4), 430–433.

Saher, S., Zulkifle, M., Ahmad, W. and Khan, T., (2021) The concept of black bile and its importance in maintenance of health. *International Journal of Health Sciences and Research* 11(4), 106–113.

Saleem, M., Alam, A. and Sultana, S. (2001a) Asafoetida inhibits early events of carcinogenesis – a chemopreventive study. *Life Science* 68(16), 1913–1921.

Saleem, M., Alam, A. and Sultana, S. (2001b) *Cheiranthus cheiri* abrogates phorbol ester induced tumor promotion response in murine skin. *Pharmaceutical Biology* 39(1), 25–34.

Saleem, M., Naseer, F., Ahmad, S., Baig, K. and Irshad, I. (2015) *In vivo* cytotoxic effects of methanol extract of *Convolvulus arvensis* on 7-12-dimethyl benz(a)antheracene (DMBA) induced skin carcinogenesis. *African Journal of Pharmacy and Pharmacology* 9(12), 397–404.

Sayyah, M., Hadidi, N. and Kamalinejad, M. (2004) Analgesic and anti-inflammatory activity of *Lactuca sativa* seed extract in rats. *Journal of Ethnopharmacology* 92(2–3), 325–329.

Schelz, Z., Csovszki, I., Bozsity, N., Hohmann, J. and Zupko, I. (2016) Antiproliferative effects of various furanoacridones isolated from *Ruta graveolens* on human breast cancer cell lines. *Anticancer Research* 36, 2751–2758.

Searle, T., Al-Niami, F. and Ali, F.R. (2021) Vitiligo: an update on systemic treatments. *Clinical and Experimental Dermatology* 46(2), 248–258.

Sepehrikia, S., Kolangi, F., Zarvandl, M. and Qaraaty, M. (2022) Types of digestion from the perspective of Persian medicine. *Journal of Current Oncology and Medical Sciences* 2(1), 144–152.

Shah, P.A., Parmar, M.Y., Thakkar, V.T. and Gandhi, T.R. (2009) Protective effect of *Hordeum vulgare* Linn. on acetaminophen-induced liver damage. *Pharmacology* 1, 336–340.

Shahouzehi, B., Sepehri G., Sadeghiyan, S. and Yaser Masoomi-Ardakani, Y. (2018) Effect of *Pistacia atlantica* resin oil on anti-oxidant, hydroxyprolin and VEGF changes in experimentally-induced skin burn in rat. *World Journal of Plastic Surgery* 7(3), 357–363.

Shakibamehr, J., Rad, J., Akram, R. and Rad, M. (2019) Effectiveness of tragacanth gel cushions in prevention of pressure ulcer in traumatic patients: a randomized controlled trial. *Journal of Caring Sciences* 8(1), 45–49.

Sharifi-Rad, J., Dey, A., Koirala, N., Shaheen, S., El Omari, N. *et al.* (2021) *Cinnamomum* species: bridging phytochemistry knowledge, pharmacological properties and toxicological safety for health benefit. *Frontiers in Pharmacology* 12, 600139.

Shin, J.Y., Che, D.N., Cho, B.O., Kang, H.J., Kim, J. *et al.* (2019) *Commiphora myrrha* inhibits itch-associated histamine and IL-31 production in stimulated mast cells. *Experimental and Therapeutic Medicine* 18(3), 1914–1920.

Shirbeigi, L., Rahbar, M.R., Mansouri, P., Salmanian, M. and Ghobadi, A. (2020) Acne in traditional Persian medicine. *Iran Journal of Public Health* 49(3), 579–587.

Singla, C., Sharma, A. and Dhiman, A. (2021) An update on phytochemistry and therapeutic properties of *Ipomoea carnea*. *Journal of Pharmacognosy and Pharmacology* 10(1), 1–6.

Soubh, A.A., Abdallah, D.M. and El-Abhar, H.S. (2015) Geraniol ameliorates TNBS-induced colitis: involvement of Wnt/β-catenin, p38MAPK, NFκB, and PPARγ signaling pathways. *Life Sciences* 136, 142–150.

Soumyanath, A., Venkatasamy, R., Joshi, M., Faas, L., Adejuyigbe, B. *et al.* (2006) UV irradiation affects melanocyte stimulatory activity and protein binding of piperine. *Photochemistry and Photobiology* 82, 1541–1548.

Steinhoff, M. (2016) Inflammation, immunology and allergy. In: Griffiths, C.E.M., Barker, J., Bleiker, T., Chalmers, R. and Creamer, D. (eds) *Rook's Textbook of Dermatology*, 9th edn (e-book). Wiley, Chichester, UK, pp. 8.1–8.2.

Taghizadieh, A., Mohammadinasab, R., Ghazi-Sha'rbaf, J., Michaleas, S.N., Vrachatis, D. *et al.* (2020) Theriac in the Persian traditional medicine. *Erciyes Medical Journal* 42(2), 235–238.

Tavakoli, A. (2021) Effect of *Plantago major* leaf gel compared to placebo on prevention of acute radiation dermatitis in breast cancer patients: a double blind, randomized clinical trial. *Iranian Registry of Clinical Trials*, registration no. IRCT20150830023823N4. Available at: https://en.irct.ir/trial/56244 (accessed 13 April 2023).

Teles, H.L. (2005) Cytotoxic lignans from the stems of *Styrax camporum. Natural Product Research* 19(4), 319–323.

Türkoğlu, M., Kılıç, S., Erkin Pekmezci, E. and Kartal, M. (2018) Evaluating antiinflammatory and antiandrogenic effects of garden cress (*Lepidium sativum* L.) in HaCaT cells. *Records of Natural Products* 12(6), 595–601.

Van Geel, N. and Speeckaert, R. (2016) Acquired pigmentary disorders. In: Griffiths, C.E.M., Barker, J., Bleiker, T., Chalmers, R. and Creamer, D. (eds) *Rook's Textbook of Dermatology*, 9th edn (e-book). Wiley, Chichester, UK, pp. 88.34–88.36.

Walzem, R.L. Dillard, C.J. and German, J.B. (2002) Whey components: millennia of evolution create functionalities for mammalian nutrition: what we know and what we may be overlooking. *Critical Reviews in Food Science and Nutrition* 42, 353–375.

Wedler, J., Weston, A., Rausenberger, J. and Butterweck, V. (2016) *In vitro* modulation of inflammatory target gene expression by a polyphenol-enriched fraction of rose oil distillation waste water. *Fitoterapia* 114, 56–62.

Werthmann, P.G., Saltzwedel, G. and Kienle, G.S. (2017) Minor regression and long-time survival (56 months) in a patient with malignant pleural mesothelioma under *Viscum album* and *Helleborus niger* extracts – a case report. *Journal of Thoracic Disease* 9(12), E1064–E1070.

Wikhe, M., Zade, V., Dabadkar, D. and Patil, U. (2013) Evaluation of the abortifacient and estrogenic activity of *Cicer arietinum* on female albino rat. *Journal of Biological Innovation* 2(3), 105–113.

Yucel, I. and Guzin, G. (2008) Topical henna for capecitabine induced hand–foot syndrome. *Investigational New Drugs* 26, 189–192.

Zakeri, S., Gorji, N., Akhtari, M. and Moeini, R. (2018) Splenectomy may have more complications than currently proven. *Medical Hypotheses* 112, 43–46.

Zargar, S. (2014) Protective effect of *Trigonella foenum-graecum* on thioacetamide induced hepatotoxicity in rats. *Saudi Journal of Biological Sciences* 21(2), 139–145.

Zou, M., Chen, Y., Sun-Waterhouse, D., Zhang, Y. and Li, F. (2018) Immunomodulatory acidic polysaccharides from *Zizyphus jujuba* cv. Huizao: insights into their chemical characteristics and modes of action. *Food Chemistry* 258, 35–42.

12 Use of Certain Herbal Drugs to Cure Various Gynecological Disorders

Seyyed Ahmad Askari[1], Shakila Hajizadeh[2], Mojgan Tansaz[3], Bahia Namavar Jahromi[4,5], and Sayyedeh Fatemeh Askari[6,7]*

[1]Traditional Medicine and Materia Medica Research Center, Shahid Beheshti University of Medical Sciences, Tehran, Iran; [2]Department of Traditional Pharmacy, School of Traditional Medicine, Shahid Beheshti University of Medical Sciences, Tehran, Iran; [3]Department of Traditional Medicine, School of Traditional Medicine, Shahid Beheshti University of Medical Science, Tehran, Iran; [4]Infertility Research Center, Shiraz University of Medical Sciences, Shiraz, Iran; [5]Department of Obstetrics & Gynecology, School of Medicine, Shiraz University of Medical Sciences, Shiraz, Iran; [6]Cardiovascular Diseases Research Center, Birjand University of Medical Sciences, Birjand, Iran; [7]Department of Pharmacognosy and Traditional Pharmacy, School of Pharmacy, Birjand University of Medical Sciences, Birjand, Iran

Abstract

In his book *Canon of Medicine*, Avicenna refers in Chapter 21 (Volume 3) to various types of gynecologic disorders such as abortion, pelvic pain, amenorrhea, cervical stenosis, uterine displacement, dysmenorrhea, false pregnancy, hysteria, libido disorders, vaginal dryness, infertility, obstetric dystocia, oligomenorrhea, polymenorrhea, postpartum hemorrhage, uterine rupture, uterine abscess, uterine dysperiment, uterine pain, uterine prolapse, and vaginitis. He also discusses methods of contraception and diagnoses of pregnancy. In addition, Avicenna lists specific vaginal prescriptions and forms of administration for each ailment, such as *forzajah* and *ḥamoul* (cotton ball), *fetilah* (wick), *shiāf* (pessary), *ābzan* and *jalous* (sitz bath), *ḥoqnah* (enema), and *bakhour* (fumigation). The ingredients of the formulations come either from animal sources (mare's milk, duck fat, chicken skin, bone marrow), plants (cinnamon, saffron, camomile, olive oil), or minerals such as sulfur, petroleum, and *koḥl* ("kohl" is a black powder, usually antimony trisulfide or lead sulfide, that was used as eye makeup, especially in Eastern countries). This chapter deals specifically with the pathology of vaginitis and its treatment in the light of Avicenna's *Canon of Medicine*.

*Email: sfaskari98@gmail.com

(eds. H. Schulz, Seyed Ahmad Emami and Farsad Nadjafi)
DOI: 10.1079/9781800621671.0012

Introduction

Vaginal complaints are one of the most common reasons forcing women to go to healthcare providers (Mills, 2017). Surveys have counted that more than 70% of females have had a vaginal problem and have used topical products to treat infections (Palmeira-de-Oliveira *et al.*, 2015). Also, vaginitis has been connected with other complications of the female reproductive organs, like increased susceptibility to and transmission of the human immunodeficiency virus (HIV) infection (Cu-Uvin *et al.*, 2002), infertility, and poor pregnancy yields (Powell and Nyirjesy, 2014). Vaginitis is described as a condition with symptoms of abnormal vaginal discharge, odor, itching, irritation, and burning (Paladine and Desai, 2018). Uncomplicated infections such as candidiasis, bacterial vaginosis (BV), or trichomoniasis are easy to diagnose and treat. However, about 8% of infected females have a more complicated condition because of failure to respond to medicines or rapid recurrence of manifestations (Mills, 2017). Different types of vaginitis are categorized based on the type of microbial agent. BV occurs when *Lactobacillus* spp., the main species in vaginal flora, are replaced by anaerobic bacteria, such as *Gardnerella*, *Mobiluncus*, *Atopobium*, *Prevotella*, *Bifidobacterium*, *Sneathia*, *Mycoplasma*, *Leptotrichia*, and some novel bacteria from the order Clostridiales referred to as BV-associated bacteria (BVAB) 1–3. Lactobacilli are supposed to supply defense against infection by preserving an acidic pH in the vagina and providing hydrogen peroxide (H_2O_2). Globally, it is estimated that 20–30% of reproductive-age women attending sexually transmitted infection clinics suffer from BV (van Schalkwyk *et al.*, 2015; Bautista *et al.*, 2016; Chen *et al.*, 2021). Multiple sexual partners, tobacco use, herpes simplex virus or HIV infection, douching, high-fat diets, age 15–44 years, alcohol intake, and stress are factors that increase the risk of BV. In contrast, hormonal contraceptive use, estrogen replacement therapy, and condom use are risk-decreasing factors (Bagnall and Rizzolo, 2017). The gold standard of BV diagnosis is the Nugent score to assess a Gram-stained vaginal discharge smear (Tam *et al.*, 1998). Another diagnostic method is Amsel, including four criteria: grayish-white vaginal discharge, pH > 4.5, clue cells on saline wet mount, and positive whiff-amine test (Amsel *et al.*, 1983). Recommended treatment for symptomatic women (even pregnant) diagnosed with BV is:

- oral metronidazole 500 mg twice daily for 7 days; or
- 0.75% w/w metronidazole gel, one applicator intravaginally every night for 5 nights; or
- 2% w/w clindamycin vaginal cream, one applicator intravaginally every night for 7 nights (Workowski *et al.*, 2015).

Trichomonas vaginalis as a causative agent of trichomoniasis infected 10–25% of pregnant women or those attending clinics (Ifeanyi *et al.*, 2018). Risk-increasing factors are a history of delinquency, approved marijuana use, multiple sexual partners, increased age, lower education, poverty, and douching (Sutton *et al.*, 2007). Symptomatic women suffer from diffuse yellow-green vaginal discharge, stinky odor, pruritus, irritation, edema, erythema, and dyspareunia. The strawberry cervix is a characteristic clinical sign detected by colposcopy and infrequently during a typical examination (Wølner-Hanssen *et al.*, 1989). Trichomoniasis is linked with ailments such as cervicitis, urethritis, pelvic inflammatory disease, ectopic pregnancy, infertility, pelvic pain, neonatal death, premature delivery, and severe disability in infants (Lossick, 1990; Ifeanyi *et al.*, 2018). The most typical test is the visualization of the motile protozoans in a saline preparation of the vaginal discharge, but the gold standard is culture (Krieger *et al.*, 1988; Hobbs and Seña, 2013). Drugs of choice for this infection are a 2-g dose of either oral tinidazole or metronidazole (Muzny and Schwebke, 2013; Workowski *et al.*, 2015). The chief complaint of BV patients is recurrence (Homayouni *et al.*, 2014). A 67% recurrence rate occurs within the first month, 76% within 6 months, and 69% during 12 months of treatment (Hay, 2009) because of antibiotic resistance or negative impacts of antibiotics on healthy vaginal microflora (Reiter and Kellogg Spadt, 2019). Therefore, attention is directed to other therapies like treating sexual partners, probiotic therapy, thermoplastic polyurethane intravaginal rings, activated charcoal, and herbal medicine (Javed *et al.*, 2019). Moreover, application of the recommended medicines is associated with adverse effects such as dryness, irritation, candidiasis, gastrointestinal disorders, nausea, and unpleasant metallic taste that leads to poor acceptance by patients (Mahboubi, 2018). Women also suffer from many other vaginal complaints like candidiasis, viral vaginitis (herpes), and chlamydia or gonorrhea. They have no definite treatments, and therefore finding new approaches and medications to treat or manage them is crucial. Traditional and complementary methods are the sources to find new treatments. With a deep historical background from the ancient era, Traditional Persian Medicine (TPM) is one of the leading traditional medical systems.

Vaginitis in View of Avicenna

Leukorrhea is excessive vaginal discharge. It is physiologic when related to different phases of the menstrual cycle. Whenever it turns into a pathologic condition, in TPM the vaginitis caused is called *sayalān-e raḥem* and was used by Avicenna and some other Persian traditional practitioners to describe this condition. Traditional medicine scholars believed that vaginitis causative factors are uterine and extrauterine. Uterine factors are related to the weakness of *qowat-e-ghaḏiah*, which feeds the uterus (based on TPM, the body performs its duties by some *qowat* or powers). This *qowat* consists of four subsets called *qowat-e-jāḏebah* (absorptive faculty), *qowat-e-māsekah* (retentive faculty), *qowat-e-hāḏemah* (digestive faculty), and *qowat-e-dāfe-ʿah* (expulsive faculty). The "absorptive faculty" absorbs the organ's proper nutrition from the blood. The "retentive faculty" maintains the absorbed food for digestion. The "digestive faculty" digests that food, and the "expulsive faculty" disposes the waste material from the organ. The debility of *qowat-e-ghaḏiah* due to the following causes increases discharges and leads to *sayalān-e raḥem*. Debility of the "digestive faculty" leads to poor digestion and increases waste. Debility of the "retentive faculty" causes the uterus to not hold in normal secretions; thus, the level of vaginal discharge increases. Debility of the "expulsive faculty" causes the accumulation of waste material in the uterus and expels it in the form of abnormal vaginal discharge (Shirooye *et al.*, 2016; Adhami *et al.*, 2017).

Extrauterine factors are related to other organs' illness that affects the healthy uterus. Production of discharge with an extrauterine origin occurs because of the body's *emtelā*. *Emtelā* means the body is full of waste materials. According to TPM, the body's health depends on the balance between the four *khelt* (humors), i.e., *dam* (sanguine), *balgham* (phlegm), *ṣafrā* (yellow bile), and *sawdā* (black bile). If any humor gets out of quantitative or qualitative balance, it becomes abnormal, and the body seeks to eliminate it. One of the disposal ways is to put waste material into the uterus and repel it through the discharge of the vagina. Therefore, if the waste material cannot pass through urine, stool, or sweat, vaginal discharge will increase and continue. Extrauterine origin of discharge can be from the four humors. If "sanguine" overcomes, thirst, blurred urine, and reddish vaginal discharge manifest; if "phlegm" overcomes, slight thirst, white urine, whitish discharge, and obesity manifest; if "yellow bile" overcomes, excessive thirst, yellow urine, and yellowish discharge manifest; and if "black bile" overcomes, dry skin, blurred face color, and dark vaginal discharge manifest (Dehdari and Hajimehdipoor, 2016; Sehar and Ansari, 2016; Meyari *et al.*, 2017).

Treatment of Vaginitis in View of Avicenna

Avicenna mentioned specific vaginal formulations and dosage forms for each condition, especially vaginitis (Askari *et al.*, 2018). The first step is purifying unbalanced humors (mentioned in the previous section) from the body by phlebotomy or purgative agents, then purifying the uterus with special purgative and drier compounds like orris and wild leek (*Allium ampeloprasum*) boiled for eliminating the pathogens. The third step is massaging the legs with emollient oils like camel grass (*Cymbopogon schoenanthus*) oil and hot herbal medicines with akarkara (*Anacyclus pyrethrum*) and black pepper (*Piper nigrum*). Then using, intravaginally, uterus tonic and astrinent agents like boiled aqueous extracts of pomegranate, akarkara, myrtle, oak gall, and *jolnār* finally to prevent a recurrence of vaginitis.

Scientific Evaluation of Medicinal Plants for Management of Vaginitis Mentioned in *Canon of Medicine*

Most of the herbs mentioned by Avicenna in the *Canon* were confirmed by *in vitro* studies as excellent antimicrobial agents, i.e., antibacterial, antifungal, and antiprotozoal activity. This section provides the latest scientific evidence on medicinal plants mentioned in the *Canon of Medicine* for vaginitis regarding their ability to ameliorate gynecologic complications, emphasizing vaginitis. The common English name, Arabic name, botanical family, scientific name, and synonym of the plants are presented in Table 12.1. Tables 12.2, 12.3, and 12.4 respectively describe some *in vitro*, *in vivo*, and clinical interventions of the individual plant species against vaginitis.

Acacia nilotica (L.) Willd. ex Delile; basionym: *Mimosa nilotica* L.

Recently, Ramalingam and Amaechi tested a mixed herbal powder extract of *Acacia arabica* (a synonym of *Acacia nilotica*) (barks) and *Terminalia chebula* (fruits) against *Streptococcus mutans, Lacticaseibacillus casei*, *Actinomyces viscosus*, and *Candida albicans* by the agar diffusion method. The mixed extract showed high antimicrobial activities (Ramalingam and Amaechi, 2020).

Artemisia absinthium L.

There are many documented data approving the antimicrobial effects of *Artemisia absinthium*. For example, davanone or silphiperfolane derivatives, components of essential oils, were found to be active against *Candida* by the paper disk diffusion method (Obistioiu *et al.*, 2014). Because of the ethanol extract's good antifungal and antibacterial activity, this valuable herb would be a candidate for gastrointestinal ailments (Ivanov *et al.*, 2021) and even vaginitis. Moreover, different nanoparticles of *A. absinthium* were tested against various microorganisms. Silver nanoparticles of *A. absinthium* showed good activity against *Pseudomonas aeruginosa*, *Escherichia coli*, *Staphylococcus aureus*, and *C. albicans* via agar well diffusion assays (Aabed and Mohammed, 2021). The copper nanoparticles of essential oil could inhibit the growth of *P. aeruginosa*, *Klebsiella pneumoniae*, *Proteus mirabilis*, *Staphylococcus epidermidis*, *Acinetobacter baumannii*, *S. aureus*, *Enterobacter aerogenes*, *C. albicans*, and *E. coli* by the disk diffusion method (Celebi *et al.*, 2021). The effect of a combination of Islamic traditional medicine drugs, including *A. absinthium* and *Quercus infectoria*, on the treatment of trichomonal vaginitis was investigated for 10 days and compared to metronidazole. Results proved the traditional combination to be as effective as metronidazole (Ahmad *et al.*, 2011).

Ceratonia siliqua L.

In an *in vivo* study, the effect of carob pods on growth performance, antioxidant activities, carcass and cecal characteristics, and equilibrium modification of the cecal microbiota population was investigated. For this purpose, 80 weaned rabbits were allocated into four dietary groups. It was found that carob pods caused an increase in daily and final body weights, glutathione peroxidase (GPx), glutathione *S*-transferase (GST), catalase (CAT), and superoxide dismutase (SOD) activities, as well as *Lactobacillus* and *Bacillus*

Table 12.1. Medicinal plants used for management of vaginitis in TPM.

Scientific name	Synonym	Common English name	Arabic name	Botanical family
Acacia nilotica (L.) Delile	*Acacia arabica* (Lam.) Willd.	Gum arabic	*Aqāqiā*	*Fabaceae*
Artemisia absinthium L.		Wormwood	*Afsantin*	*Asteraceae*
Ceratonia siliqua L.		Carob	*Kharnub*	*Fabaceae*
Cinnamomum verum J. Presl	*Cinnamomum zeylanicum* Blume	Cinnamon	*Dārṣini*	*Lauraceae*
Citrullus colocynthis (L.) Schrad.		Bitter apple	*Ḥanẓal*	*Cucurbitaceae*
Corylus avellana L.		Hazel	*Jollawaz*	*Betulaceae*
Crocus sativus L.		Saffron	*Zʿafarān*	*Iridaceae*
Iris spp.		Iris	*Irsā*	*Iridaceae*
Juglans regia L.		Walnut	*Jawz*	*Juglandaceae*
Lepidium sativum L.		Cress	*Ḥorf, habb-o-rreshād*	*Brassicaceae*
Myristica fragrans Houtt.		Nutmeg	*Jaw-e-zboā*	*Myristicaceae*
Myrtus communis L.		Myrtle	*Ās*	*Myrtaceae*
Ocimum basilicum L.		Basil	*Bādruj*	*Lamiaceae*
Olea europaea L.		Olive	*Zaitun*	*Oleaceae*
Phoenix dactylifera L.		Date palm	*Tamr*	*Arecaceae*
Pimpinella anisum L.		Anise	*Anisun*	*Apiaceae*
Pistacia lentiscus L.		Lentisk pistache, mastic	*Maṣṭakā*	*Anacardiaceae*
Portulaca oleracea L.		Purslane	*Baqlato-lḥamqā'*	*Portulacaceae*
Punica granatum L.		Pomegranate	*Jolnār*	*Lythraceae*
Quercus infectoria G. Olivier		Oak gall, oak fruit	*Afeṣ, baluṭ*	*Fagaceae*
Rubus fruticosus L.		Blackberry	*ʿOllaiq*	*Rosaceae*
Rumex spp.		Dock	*Ḥommāḍ*	*Polygonaceae*
Tamarix spp.		French tamarisk	*Ṭarfā*	*Tamaricaceae*
Thymbra capitata (L.) Cav.	*Thymus capitatus* (L.) Hoffmanns. & Link	Persian hyssop	*Ḥāshā*	*Lamiaceae*

counts, and a decrease in total cholesterol (TC), low-density lipoprotein cholesterol (LDL), high-density lipoprotein cholesterol (HDL), and triglyceride (TG) levels (Abu Hafsa *et al.*, 2017). As we know, the increment in *Lactobacillus*, normal microflora of the vagina, is a good point for treatment of vaginitis.

Cinnamomum verum J. Presl; synonym: *Cinnamomum zeylanicum* Blume

Miller *et al.* evaluated the antifungal and antibacterial activity of essential oils isolated from 59 plant species against *Lactobacillus acidophilus*, *S. aureus*, *C. albicans*, *S. mutans*, and *E. coli*. Plant samples were gathered, frozen, and steam-distilled to obtain the individual essential oils. It was identified that the essential oils from *Citrus aurantifolia* (Christm.) Swingle (*Rutaceae*), *Cinnamomum zeylanicum* (*Lauraceae*), *Origanum vulgare* L. (*Lamiaceae*), and *Lippia graveolens* Kunth (*Verbenaceae*) show significant or moderate antimicrobial activity against all abovementioned microorganisms (Miller *et al.*, 2015). The impact of cinnamon leaf and bark essential oils against *C. albicans* and *Candida auris* was investigated in another study. The antifungal activity of the essential oils against selected fungi strains (*C. albicans* ATCC 2091, *C. albicans* ATCC 10231, and *C. auris* NCPF 8971) was determined using the broth microdilution and disk diffusion

Table 12.2. Some *in vitro* studies on medicinal plants claimed to be effective in vaginitis in TPM.

Plant	Part/extract	Result	Reference
Acacia arabica	Chewing sticks/ aqueous extract	Antimicrobial effect on *S. fecalis*, by ditch plate method	Almas (2001)
Cinnamomum zeylanicum	Essential oil	Antibacterial (against Gram-positive and -negative bacterial strains), antifungal (against *Candida* strains), and antiviral (against H1N1 and HSV-1) activity, by broth microdilution method	Brochot *et al.* (2017)
	Bark oil	Antifungal activity against *Candida albicans*, by MIC test	Shahina *et al.* (2018)
Olea europaea	Leaves/methanol extract	High strength for controlling bacterial biofilms in food and related products and anti-*Candida* strains, by a broth microdilution method	Edziri *et al.* (2019)
	Leaves/acetone extract	Antimicrobial activity against oral pathogens especially *C. albicans*, by agar disk diffusion	Karygianni *et al.* (2014)
	Leaves/aqueous extract	Potential antimicrobial activity against *C. albicans* and *Escherichia coli*, by agar disk diffusion	Markin *et al.* (2003)
	Leaves/aqueous extract	Antimicrobial effect against *Bacillus cereus* and *C. albicans*, by macro broth dilution	Pereira *et al.* (2007a)
Phoenix dactylifera	Fruit/aqueous extract	Antimicrobial activity against *Staphylococcus aureus*, *E. coli*, and *C. albicans*, by zone inhibition assay	Zaheer (2018)
	Bunches/ethanol extract, butanol fraction	High activity against *C. albicans* and moderate activity against *Bacillus subtilis*, by zone inhibition assay	Fathy *et al.* (2021)
Rumex hymenosepalus	Root	Growth inhibition against *E. coli* and *C. albicans*, dose-dependently	Villalobos-Noriega *et al.* (2021)
	Root	Gram-positive bacteria, Gram-negative bacteria, and yeast (*C. albicans*) growth inhibition by synthesized silver nanoparticles, by well diffusion method	Rodríguez-León *et al.* (2018)
Rumex nervosus	Aerial parts/ aqueous extract	Antimicrobial activity against Gram-positive bacteria	Al-Asmari *et al.* (2015)
Tamarix africana	Leaves and stems/ hydroalcoholic extract	Antifungal effect against *C. albicans*, by agar disk diffusion	Chekroun-Bechlaghem *et al.* (2019)
Tamarix aphylla	Aerial parts/aqueous ethanol extract	Growth inhibition of *C. albicans*, *E. coli*, *Enterococcus faecalis*, *Klebsiella pneumoniae*, *Pseudomonas aeruginosa*, and MRSA, by agar disk diffusion	Al-Omar *et al.* (2020)

H1N1, a subtype of influenza A virus; HSV-1, herpes simplex virus serotype 1.

Table 12.3. *In vivo* studies on medicinal plants claimed to be effective in vaginitis in TPM.

Model	Species	Plant	Part	Results	Reference
Post-weaning	Piglet	*Camellia sinensis* *Punica granatum*	Leaves Fruit	↑ Average daily gain; ↑ gut health and microbial ecology; ↓ severity of an *Escherichia coli* challenge	Bontempo *et al*. (2014)
Weaned rabbits	Rabbit	*Ceratonia siliqua*	Pods	↑ Daily and final body weights; ↓ TC, LDL, HDL, and TG levels; ↑ GPx, GST, CAT, and SOD activities; ↑ *Lactobacillus* and *Bacillus* counts	Abu Hafsa *et al*. (2017)
DHEA-induced PCOS	Mice	*Cinnamomum zeylanicum*	Bark	↓ Insulin, IGF-1, and LH; improving cyclicity and ovary morphology; ↑ serum FSH level	Dou *et al*. (2018)
Doxorubicin-induced cardiotoxicity and nephrotoxicity	Rat	*Phoenix dactylifera*	Chitosan nanoparticles of extract	↓ Heart and kidney apoptosis; ↑ PD-1	Sahyon and Al-Harbi (2020)
Cyclophosphamide-induced immunosuppressed and then infected	Mice	*Punica granatum*	Peel	↓ Mortality of *Candida albicans*, moderate-activity molds	Akroum (2017)
Cyclosporine and hydrocortisone acetate-induced immunosuppressed	Rat	*Punica granatum*	Peel	↓ Growth of *C. albicans*; 100% cure in all doses after 15 days of treatment	Bassiri-Jahromi *et al*. (2018)
Citrobacter rodentium-infected	Mice	*Punica granatum*	Peel	↓ Pathogenicity of *C. rodentium* infection; ↑ abundance of proteobacteria and verrucomicrobia; ↓ actinobacteria	George *et al*. (2019)
Fat-containing diet	Broilers	*Punica granatum*	Seed	↑ Fat-containing diets digestion; ↑ crude fat digestibility and cecum *Lactobacillus* count	Rezvani *et al*. (2018)

Table 12.4. Clinical studies on medicinal plants claimed to be effective in vaginitis in TPM.

Study design	No. of patients	Disease	Plant	Part of plant	Control group	Duration of treatment	Results	Reference
Controlled, randomized, single-blind clinical trial	60	Trichomonal vaginitis	*Artemisia absinthium* *Quercus infectoria*	Aerial parts Gall	Metronidazole	10 days	As efficient as metronidazole	Ahmad *et al.* (2011)
Prospective observational pilot study	60	Vulvovaginal candidiasis	Combination of HT and other components: tea tree oil, tabebuia, *Juglans regia*, and copper	–	–	3 months	Reduction in vaginal signs (pruritus, burning, and vulvar erythema)	Zullo *et al.* (2020)
Prospective cohort study	190	Recurrent UTIs	Combination of *Olea europaea*, D-mannose, cranberry, bearberry, inulin, *Orthosiphon*, *Lactobacillus acidophilus*	Leaves	Not using supplement	10 days/month for 1 year	Fewer UTI relapses and less pain	Mainini *et al.* (2020)
Nonrandomized clinical trial	447	Infertility	*Myristica fragrans*	Mace (aril)	Routine medication for IUI	From the 7th day of menstruation until the day before performing IUI	Higher pregnancy rate but no significant difference	Kavousi *et al.* (2019)
Randomized clinical trial	120	Bacterial vaginitis	*Myrtus communis*	Leaves	Metronidazole	5 days	Efficiency better than metronidazole without any relapse	Masoudi *et al.* (2016)
Double-blind RCT	60	Candidiasis associated with denture stomatitis	*Punica granatum*	Peel	Miconazole	15 days	Similar laboratory outcome to miconazole	César de Souza Vasconcelos *et al.* (2003)

Randomized controlled, double-blind	84	Bacterial vaginitis	*Quercus infectoria*	Gall	Metronidazole	5 days	As efficient as metronidazole	Afzali *et al*. (2020)
Randomized controlled, double-blind	120	Vaginitis	*Quercus infectoria* *Myrtus communis*	Gall Leaves	Metronidazole and placebo	7 days	Improvement in clinical examination, whiff test specially for trichomoniasis	Askari *et al*. (2020)
Double-blind, randomized, placebo-controlled	92	Vulvovaginal candidiasis	*Rosa damascena* *Punica granatum* *Querqus infectoria* *Myrtus communis* *Nardostachys jatamansi*	Petals Peel Fruit Fruit Rhizome	Placebo	7 days	Reduced clinical symptoms (itching, irritation, and vaginal discharge)	Khalilzadeh *et al*. (2020)

(direct and vapor) methods. Both essential oils reduced hyphae development and caused damage to the cell membrane. According to the hemolysin production assay, cinnamon leaf and bark essential oils may inhibit hemolysin production in the studied *C. albicans* and *C. auris* strains. It was concluded that essential oils have significant antihemolytic and antifungal effects *in vitro* against *C. auris* and *C. albicans* at low doses (Tran *et al.*, 2020). Effects of essential oil and aqueous extract of cinnamon bark on biofilm formation, growth, and adherence properties of *C. albicans* were also evaluated. Cinnamon bark oil was found to be effective against preformed *C. albicans* biofilm. It has also demonstrated intense antifungal activity. Therefore, both cinnamon fractions may be promising agents for treating oral infections involving *C. albicans* due to inhibiting *C. albicans* growth, biofilm formation, and adherence (Veilleux and Grenier, 2019). Rangel *et al.* evaluated the antifungal activity of essential oil of *C. zeylanicum* leaves (prepared via the hydrodistillation method) on *Candida* spp. biofilm involved in oral infection. According to the phytochemical analysis, eugenol was found to be the major component of the *C. zeylanicum* essential oil (68.96% v/v). It was found that *C. zeylanicum* essential oil has antifungal properties through its effect on the yeast cell wall and toxicity effect on *Candida* spp. biofilms (Rangel *et al.*, 2018). The effect and mechanism of *C. zeylanicum* on the gynecological disorder polycystic ovary syndrome (PCOS) were studied by using a dehydroepiandrosterone (DHEA)-induced PCOS mouse model. It was determined that cinnamon would lead to a decrease in insulin, insulin-like growth factor 1 (IGF-1), and luteinizing hormone (LH), an increase in serum follicle-stimulating hormone (FSH) level, and improved cyclicity and ovary morphology (Dou *et al.*, 2018).

Citrullus colocynthis (L.) Schrad.; basionym: *Cucumis colocynthis* L.

Marzouk *et al.* investigated the anticandidal and antibacterial activity of aqueous and diluted acetone extracts of *Citrullus colocynthis* (from the plant's leaves, stems, roots, and three maturation stages of its seeds and fruit) against Gram-positive bacteria (*S. aureus* and *Enterococcus faecalis*), various *Candida* spp. (*Candida glabrata*, *Candida parapsilosis*, *C. albicans*, and *Candida kreusei*), and Gram-negative bacteria (*E. coli* and *P. aeruginosa*). It was identified that *C. colocynthis* has anticandidal and antibacterial activity against all strains applied in the study (Marzouk *et al.*, 2009). In another research, magnetic iron oxide nanoparticles (MNPs) were prepared with improved antibacterial activity by using hot and cold aqueous extracts of pulp and seeds of *C. colocynth*. The antimicrobial activity of these MNPs was tested against two Gram-negative (*P. aeruginosa* and *E. coli*), two Gram-positive (*S. aureus* and *Bacillus subtilis*) bacteria, and *C. albicans*. Both MNPs generated via pulp extract and cold seed extract presented comparable antimicrobial potential against the tested microorganisms. However, *C. albicans* was the microorganism that was least affected (Farouk *et al.*, 2020). Al-Ardi used *Citrullus spinosa* and *C. colocynthis* nano-extracts for *T. vaginalis* treatment *in vitro*. In that study, the plants' fruit pulps were used, and aqueous extracts with the concentrations of 100, 250, and 500 ppm were prepared. It was identified that each plant could be used as a substitute for metronidazole after completing animal and human tests (Al-Ardi, 2021).

Corylus avellana L.

Oliveira *et al.* (2008) investigated the chemical composition, antimicrobial and antioxidant activities of three hazelnut (*Corylus avellana*) cultivars. Throughout their research, these cultivars from Portugal were studied for their individual chemical composition, antioxidant capacity, and antibacterial activity. The predominant ingredient of fruits was fat, ranging from 56 to 61% w/w, with a nutritional value of roughly 650 kcal per 100 g of fruits. Oleic acid, which ranged from 80.67% v/v, was the most abundant fatty acid, followed by palmitic, linoleic, and stearic acids. To analyze samples, gas chromatography/flame ionization detector (GC/FID) was used. The antimicrobial capacity of three hazelnut cultivars was assessed against Gram-negative bacteria (*P. aeruginosa*, *E. coli*, *K. pneumoniae*), fungi (*C. albicans*, *Cryptococcus neoformans*), and Gram-positive bacteria (*Bacillus cereus*, *B. subtilis*, *S. aureus*). The antioxidant activity of aqueous hazelnut extracts was concentration-dependent, with similar results for all cultivars. Generally, all hazelnut extracts demonstrated excellent antibacterial activity against Gram-positive bacteria, indicating that these fruits showed high bioactivity (Oliveira *et al.*, 2008).

Crocus sativus L.

Carradori *et al.* investigated the anti-*Candida* effects of active agents obtained from *Crocus sativus* stigmas (called crocin-1 and safranal) and semisynthetic derivatives of safranal. These two active agents could prevent *C. albicans* germ tube production, an essential virulence component. In addition to this, more potent and promising compounds were made via chemical modification, suggesting that they could be employed as lead compounds in developing new anti-*Candida* agents (Carradori *et al.*, 2016). By using the agar well diffusion method, the antimicrobial activities of methanol and petroleum ether extracts of saffron stigmas were tested against fungi (*C. albicans*, *Aspergillus fumigatus*, and *Aspergillus niger*) and various bacterial strains (*P. aeruginosa*, *K. pneumoniae*, *S. aureus*, *E. coli*, and *Proteus vulgaris*). The results demonstrated that the methanol and petroleum ether extracts of saffron stigmas possess a high antibacterial and antifungal activity against the bacteria and fungi utilized as test organisms. Moreover, the extracts were found to have a much stronger bactericidal than fungicidal activity, according to the findings of several antimicrobial tests (Muzaffar *et al.*, 2016).

Iris species

Leaves of *Iris* × *germanica* L. were used to carry out various tests to investigate the antibacterial and antifungal activity of plant extracts obtained with five different solvents (including n-butanol, ethyl acetate, chloroform, hexane, and water). The disk diffusion assay was used to evaluate the antibacterial activity of *I.* × *germanica* against eight pathogenic bacteria (*S. aureus*, *B. subtilis*, *Bacillus atrophaeus*, *E. coli*, *P. aeruginosa*, *Erwinia carotovara*, *K. pneumoniae*, *Salmonella typhi*), and one fungal species (*C. albicans*). The research results demonstrated that the ethyl acetate and butanol fractions were more effective in controlling the growth of pathogens than the chloroform, hexane, and aqueous fractions. *C. albicans*, *S. aureus*, *E. carotovara*, *B. atrophaeus*, and *E. coli*, which were more susceptible than other microorganisms, were inhibited by all *I.* × *germanica* extracts. These findings suggest that different solvent-extracted samples from *I.* × *germanica* leaves have different antibacterial activity against various microorganisms, which might be valuable sources of antibiotics to treat fungal and bacterial infections (Uzair *et al.*, 2016). Other research investigated the anti-biofilm activity and chemical composition of 15 methanol extracts of *Iris* species against both mono- (*S. aureus* and *P. aeruginosa*) and multi-species oral biofilms (*Fusobacterium nucleatum* subsp. *nucleatum*, *Streptococcus gordonii*, *Actinomyces naeslundii*, and *Veillonella parvula*). The leaf extract of *Iris pallida* was the most efficient one in disrupting existing biofilms and inhibiting biofilm development compared to the root and rhizome extracts (Hoang *et al.*, 2020).

Juglans regia L.

The antimicrobial effects of different parts of walnut tree were investigated in several studies. In this context, comparatively high antifungal (*C. glabrata* and *C. parapsilosis*) properties of peel ethyl acetate extract against oral *Candida* strains could be observed (Noumi *et al.*, 2010). Furthermore, leaves and bark aqueous extracts inhibited the growth of Gram-positive (*S. aureus* and *S. mutans*), Gram-negative (*E. coli* and *P. aeruginosa*) bacteria species, and *C. albicans*. Another study was performed by the agar streak dilution method based on radial diffusion (Pereira *et al.*, 2007b). Ethanol extract of walnut root disrupted the cell wall and cell membrane of *Candida*, applying MIC_{90} (minimum inhibitory concentration at which ≥90% of strains within a test population are inhibited) and spot assays (Raja *et al.*, 2017). Studies performed by other research groups also presented growth inhibitory effects for bark extract, crude methanol extract, and ethyl acetate fractions against *S. aureus*, *B. subtilis*, *P. aeruginosa*, *E. coli*, and *K. pneumoniae* by disk diffusion (Bakht *et al.*, 2017). Another study approved anti-*C. albicans* and anti-herpes simplex virus (HSV) effects of walnut pellicle ethanol extract by the microdilution method (D'Angeli *et al.*, 2021). Zullo *et al.* asserted that patients affected by vulvovaginal candidiasis showed a reduction in vaginal signs such as pruritus, burning, and vulvar erythema after applying a combination of hydroxytyrosol (HT) and other components containing tea tree oil, *Tabebuia* species, *J. regia*, and copper for a duration of 3 months (Zullo *et al.*, 2020).

Lepidium sativum L.

A research study evaluated the antifungal activity of *Lepidium sativum* seeds using various hexanic and alkaloidal extracts to determine the growth of *C. albicans*. The authors noticed that there is a significant link between inhibition of crucial enzymes involved in cellular metabolism and invasion mechanisms, which then inhibits the growth of *C. albicans* (Gacemi *et al.*, 2020).

Myristica fragrans Houtt.

Setty *et al.* (2020) investigated *Myristica fragrans* (nutmeg) antimicrobial effects on primary tooth pathogens. *M. fragrans* essential oil was prepared via hydrodistillation by Clavenger's method and thin-layer chromatography (TLC), gas chromatography–mass spectrometry (GC–MS) analysis, and high-performance TLC were used to determine the individual phytoconstituents. It was deduced that *M. fragrans* active components, including myristicin, myristic acid, trimyristin, elemicin, and safrole, are effective against *E. coli*, *S. aureus*, *E. faecalis*, *S. mutans*, *C. albicans*, *L. casei*, *A. viscosus*, *Prevotella intermedia*, and *Porphyromonas gingivalis*. Thus, the essential oil of nutmeg fruits shows good antimicrobial activity (Setty *et al.*, 2020). Therefore, extracts obtained from *M. fragrans* fruits represent potent inhibitors of *S. mutans* already at a very low inhibitory concentration of 3.9 μg/ml, which is significantly lower than for other natural anticariogenic agents (Chung *et al.*, 2006). Another research detected nutmeg seed extract's antifungal activity against oral candidal infection (*C. albicans*). Compared to nystatin as a positive control, the nutmeg extract showed antifungal activity with an effective zone of inhibition ranging from 18.0 to 12.0 mm (Iyer *et al.*, 2017). It has been investigated whether using honey and 1% w/w extract of *M. fragrans* as a vaginal tablet affects the extent of success of intrauterine insemination (IUI) or not. In a non-randomized clinical trial, 447 women with infertility problems participated. They used the vaginal tablet from the seventh day of menstruation until the day before performing IUI. It was observed at study end that the use of this vaginal product for a more extended period and across several menses cycles before IUI might produce more positive results (Kavousi *et al.*, 2019).

Myrtus communis L.

Different studies show that myrtle (*Myrtus communis*) leaves have a promising pharmacologic efficacy. Aqueous and ethanol crude extracts demonstrate antimicrobial effects against Gram-positive (*S. aureus* and *E. fecalis*) and Gram-negative (*E. coli*, *P. vulgaris*, *Serratia marcescens*, *A. baumannii*, *K. pneumoniae*, *P. aeruginosa*) bacteria (Mun'im R. *et al.*, 2009). All crude extracts, including methanol, ethanol, ethyl acetate, and acetone extracts, showed good antifungal activity against *C. albicans* (Erdogan *et al.*, 2014). Many other studies confirmed antibacterial (Mansouri *et al.*, 2001; Al-Saimary *et al.*, 2002; Alem *et al.*, 2008), antifungal (Cannas *et al.*, 2014), antiviral (Jassim and Naji, 2003), and antiprotozoal (Azadbakht *et al.*, 2003; Mahdi *et al.*, 2006) effects of myrtle. In another study, myrtucommulone obtained from myrtle leaves exhibited potent anti-inflammatory effects *in vivo* when administered to mice intraperitoneally (Rossi *et al.*, 2009). Using an *in vitro* model (lipopolysaccharide-stimulated macrophages) demonstrated that myrtle essential oil has anti-inflammatory properties. Two studies about ethanol extracts showed the anti-inflammatory effect of myrtle leaves in rats (Al-Hindawi *et al.*, 1989; Sen *et al.*, 2017). Five-day use of *M. communis* and *Berberis vulgaris* has been reported to be more effective than metronidazole in patients with bacterial vaginitis without relapse (Masoudi *et al.*, 2016).

Ocimum basilicum L.

A study was conducted to investigate the effect of mastic essential oil from *Ocimum basilicum* and *Pistacia lentiscus* on trophozoites of *T. vaginalis*. At several time intervals (after 24, 48, 72, and 96 h), the authors determined the effects of different concentrations of *P. lentiscus* mastic (15, 10, and 5 mg/ml) and *O. basilicum* essential oil (30, 20, and 10 mg/ml) on trophozoite proliferation. According to the results obtained, both *O. basilicum* and *P. lentiscus* mastic inhibited the growth of *T. vaginalis* trophozoites via extensive vacuolization of the cytoplasm and causing damage to its membrane. Thus, it could be concluded that phytotherapeutic agents such as *O. basilicum* and *P. lentiscus* mastic oil may be effective in treating trichomoniasis (Ezz Eldin and Badawy, 2015). Another study investigated the antimicrobial activity of five different plant varieties against a wide range of foodborne Gram-negative bacteria, Gram-positive bacteria, molds, and yeasts. An agar well diffusion method was used to test essential oils of sweet basil (*O. basilicum*) obtained by hydrodistillation. Except for *Pseudomonas* species and *Flavimonas oryzihabitans*, all five different basil essential oils demonstrated antimicrobial activity against most of the tested organisms (Lachowicz *et al.*, 1998).

Olea europaea L.

Various researchers have examined olive leaves for their antimicrobial effects on oral pathogens like *C. albicans*, food-related bacteria, *E. coli*, and *B. cereus*, by macro and micro broth dilution and agar disk diffusion. Aqueous, methanol, and acetone extracts showed potent inhibitory effects (Markin *et al.*, 2003; Pereira *et al.*, 2007a; Karygianni *et al.*, 2014; Edziri *et al.*, 2019). A prospective cohort study investigated the prevention and treatment of menopausal cystitis by a nutraceutical product containing D-mannose, inulin, cranberry, bearberry, olive oil, *Orthosiphon* leaves, and *L. acidophilus*. In that study, 190 patients with recurrent urinary tract infection (UTI) used this nutraceutical product 10 days per month for 1 year. At study end, it was observed that users had fewer UTI relapses and less pain (Mainini *et al.*, 2020).

Phoenix dactylifera L.

Recently, Sahyon and Al-Harbi (2020) provided a novel combination of heart of *Phoenix dactylifera* (HP) extract loaded on chitosan nanoparticles and assessed its antibacterial, antioxidant, anticancer, and free radical scavenging potential (*in vitro*). The HP extract and HP extract loaded on chitosan nanoparticles (HP–ChNPs) were tested against two Gram-negative bacteria, *E. coli* and *K. pneumoniae*, and three Gram-positive bacteria, i.e., *S. aureus*, *B. subtilis*, and *B. cereus*. In addition to this, the antifungal activity of the HP extract and HP–ChNPs was tested against *C. albicans*. The agar disk diffusion method was used to evaluate the antibacterial activity of the HP–ChNPs and the HP extract (Sahyon and Al-Harbi, 2020). Soxhlet extraction (70°C, 6 h) was used to prepare a hydroalcoholic (2:8, v/v) extract of the *P. dactylifera* fruit pulp. Silver nanoparticles were synthesized using an aqueous solution (20% v/v) of the extract (AgNPs). The disk diffusion experiment revealed that the plant–nanosilver broth had effective antimicrobial action against *E. coli* ATCC 8739, *S. aureus* ATCC 6538, and *C. albicans* ATCC 10231. However, it was found that the plant extract has no antibacterial action (Shaikh *et al.*, 2018). Other researchers used a root extract of *P. dactylifera* to produce silver nanoparticles (AgNPs) to evaluate their anticancer and antimicrobial potential (against *C. albicans* and *E. coli*). According to the results obtained, it was concluded that biologically generated AgNPs have multifunctional capabilities and might be employed to treat human cancer and other infectious diseases (Oves *et al.*, 2018). The antimicrobial, anticancer, and antioxidant activities of chitosan nanoparticles loaded with an extract of *P. dactylifera* tree (HP–ChNPs) were studied in rats receiving doxorubicin-induced cardiotoxicity and nephrotoxicity. Result showed that the HP–ChNPs reduced cardiac and renal apoptosis and increased the level of programmed cell death protein-1 (PD-1) (Sahyon and Al-Harbi, 2020).

***Pimpinella anisum* L.**

Vieira *et al.* found that several essential oils isolated from *Apiaceae* species, especially anise seeds, have a potent inhibitory effect against *Candida* spp., as detected by the broth microdilution technique (Vieira *et al.*, 2019).

***Pistacia lentiscus* L.**

Milia *et al.* (2020) screened activity of *P. lentiscus* essential oil, obtained from wild plant leaves grown in North Sardinia (Italy), against *Candida* spp. and periodontal bacteria. It was found that *P. lentiscus* essential oil has a broad spectrum of activity against various *Candida* fungi species and periodontal bacteria. In this context, the authors also found a dual inhibitory effect with regard to the two enzymes cyclooxygenase (COX-1/2) and lipoxygenase (Milia *et al.*, 2020).

***Portulaca oleracea* L.**

In a study performed by Ojah *et al.* (2021), root extracts of *Portulaca oleracea* were tested for their antibacterial activity. Generally, this widespread plant species is rich in *n*-3 fatty acids, alkaloids, oxalic acid, coumarins, cardiac glycosides, flavonoids, and anthraquinone glycosides. Throughout the research, the agar well diffusion method was carried out to evaluate the antibacterial activity of different root extracts against *E. coli*, *B. subtilis*, *Enterobacter cloacae*, *C. albicans*, *K. pneumoniae*, *Micrococcus luteus*, *S. typhi*, *P. aeruginosa*, *Shigella dysenteriae*, *S. aureus*, and *Streptococcus agalactiae*, which are responsible for the spread of common illnesses in Nigeria. Except for *P. aeruginosa*, *K. pneumoniae*, and *M. luteus*, all extracts demonstrated a high MIC against the pathogens. For the first time, *P. oleracea* was found as an effective antibacterial agent, confirming the plant's ethnomedicinal usage. Another study investigated the antimicrobial activity of liquid carbon dioxide (CO_2) extracts of *P. oleracea* grown in the territory of Kazakhstan. The antimicrobial activity was assessed against *S. aureus* ATCC 6538-P, *B. subtilis* ATCC 6633, *E. coli* ATCC 8739, and *C. albicans* ATCC 10231 by the disk diffusion method and micro method of serial dilution. It was concluded that the CO_2 extracts of *P. oleracea* contain several bioactive components with a considerable antibacterial activity (Tleubayeva *et al.*, 2021). In a study aimed to investigate anti-*Candida* activities of several medicinal plants, to choose a herbal medicine for use as an effective antibacterial in humans, ethanol extracts of seven different plants such as *P. oleracea* (*baq'lah*), *Avicennia marina* (*qurm*), *Fagonia indica* (*shoka'a*), *Salvadora persica* (*souwak*), *Ziziphus spina-christi* (*sidr*), *Lawsonia inermis* (*henna*), and *Asphodelus tenuifolius* (*kufer*) were used. It was found that *L. inermis* and *P. oleracea* extracts are likely to be promising drugs for treating multidrug-resistant (MDR) bacteria and *C. albicans* in humans (Soliman *et al.*, 2017).

***Punica granatum* L.**

Both fruit and peels of *Punica granatum* (pomegranate) exhibited significant antimicrobial activity. Microparticles synthesized from ethanol extract of peel show antifungal activity against *C. albicans* by the broth microdilution reference procedure of the Clinical and Laboratory Standards Institute (CLSI) (Endo *et al.*, 2012). Anibal and co-workers confirmed that ethanol extract of whole pomegranate fruit has excellent antimicrobial activity against *C. parapsilosis*, *Candida utilis*, *Candida lusitaniae*, and *C. glabrata* through the M27-A2 broth microdilution reference procedure of CLSI (Anibal *et al.*, 2013). In another study, antifungal activity of methanol and ethanol extracts of pomegranate bark and root against *C. albicans* and *C. glabrata* could be proven via the broth dilution antifungal susceptibility test (Lavaee *et al.*, 2018). Furthermore, methanol extracts of pomegranate peel demonstrated inhibitory effect against a range of microorganisms such as *S. aureus*, *K. pneumoniae*, *B. subtilis*, *M. luteus*, *E. coli*, *C. albicans*, and *A. niger* by applying the well diffusion method of the National Committee for Clinical Laboratory Standards (NCCLS) (Qabaha, 2013). Aqueous extracts obtained from the mesocarp of pomegranate fruits could also be used as suitable agents against *S. mutans*, *Streptococcus mitis*, and *C. albicans* (Mehta *et al.*, 2014). The *in vivo* antifungal activity of acetone extracts of *P. granatum*, *Quercus suber*, and *Vicia faba* against *C. albicans* and *Trichophyton mentagrophytes* was tested in another study using immunosuppressed mice as test animals. There, a decreased mortality of *C. albicans* and moderate-activity molds was observed (Akroum, 2017). Evaluating the *in vivo* antifungal activity of the pomegranate peel extract against oral candidiasis in Wistar rats and comparing it to nystatin were aims of a study in which rats were immunosuppressed by cyclosporine and hydrocortisone acetate, got infected, and then were cured. Not only did that study present a decreasing growth of *C. albicans* but the rats were completely cured in all doses after 15 days of treatment (Bassiri-Jahromi, 2018). Accordingly, pomegranate extracts alter the mouse microbiome and dysbiosis caused by *Citrobacter rodentium* infection by decreasing the pathogenicity of *C. rodentium* infections and actinobacteria and increasing the abundance of proteobacteria and verrucomicrobia (George *et al.*, 2019). The novel plant extract product (Grazix™) can affect weaned piglets' performance and gut health when challenged with *E. coli*. This product is a standardized mixture of green tea leaves (*Camellia sinensis*) and pomegranate fruit obtained using the LiveXtract™ process. Including this product in the diet of post-weaning piglets would increase the average daily gain, gut health, and microbial ecology as well as decrease the severity of an *E. coli* challenge (Bontempo *et al.*, 2014). The effect of adding pomegranate seed extracts to fat-containing diets on nutrient digestibility, intestinal microflora, and performance of broilers has been investigated and reported to increase the digestion of fat-containing diets, crude fat digestibility, and cecum *Lactobacillus* count (Rezvani *et al.*, 2018). A similar laboratory outcome to miconazole was reported when using *P. granatum* as an antifungal agent for 15 days against candidosis associated with denture stomatitis (César De Souza Vasconcelos *et al.*, 2003). In a single-arm clinical trial, the effect of pomegranate extract on stool microbial composition was studied. Here, participants were 20 healthy volunteers, and the study duration was 4 weeks. In this context it was demonstrated that consumption of pomegranates may have health benefits due to the formation of ellagitannin metabolites, which can be attributed to changes in the microbiota (Li *et al.*, 2015).

Quercus infectoria G. Olivier

Chusri and Voravuthikunchai (2011) were able to show the antimicrobial activity of *Q. infectoria* gall against *S. aureus* and methicillin-resistant *Staphylococcus aureus* (MRSA) by applying ethanol extracts due to hypersensitivity to low and high osmotic pressure using MIC/minimum bactericidal concentration (MBC). Also, methanol and aqueous extracts of gall displayed substantial activity against *C. albicans*, *Candida krusei*, *C. glabrata*, *C. parapsilosis*, and *Candida tropicalis* using the twofold serial dilution technique (Baharuddin *et al.*, 2015). The efficacy of *Q. infectoria* and *Mimusops elengi* on gingivitis inhibition has been also reported in a study in which 20 patients with the problem of chronic gingivitis participated. The intervention group used this combination for 7 days, while the control group used Listerine mouthwash. At the end of this double-blind, randomized controlled trial (RCT), it could be shown that the herbal preparation was efficient but less so than Listerine (Choudhary *et al.*, 2015). Another RCT investigated the effect of taking myrtle and oak gall as vaginal suppositories for vaginitis. That study involved a total of 120 patients in three groups, including a placebo control group taking metronidazole and an intervention group. It lasted 7 days, and it turned out that myrtle and oak gall consumption led to improvement in clinical examination, whiff test, especially for trichomoniasis, and resulted in similar treatment efficacy to metronidazole (Askari *et al.*, 2020). In an RCT performed on bacterial vaginitis, oak gall was found to be even more efficient than metronidazole. The duration of that study was 5 days, and 84 patients took part (Afzali *et al.*, 2020). In another RCT against vulvovaginal candidiasis, 92 patients were divided into two groups (a placebo group and an intervention group) in which the patients took a vaginal tablet containing *Rosa* × *damascena*, *P. granatum*, *Q. infectoria*, *M. communis*, and *Nardostachys jatamansi* for 7 days. It was shown that the use of this vaginal tablet significantly reduced clinical symptoms such as itching, irritation, and vaginal discharge (Khalilzadeh *et al.*, 2020).

***Rubus fruticosus* L.**

Recently, synthesized silver nanoparticles of an aqueous decoction of *Rubus fruticosus* fruit showed an inhibitory effect against *E. faecalis* and *C. albicans* by the microdilution method. This activity may be related to the high ellagic acid content of *R. fruticosus* berries (Ekrikaya *et al.*, 2021).

***Rumex* species**

Different species of *the Rumex* genus possess antimicrobial activity. For example, roots and aerial parts of *Rumex crispus* (acetone and hexane extracts) showed good antibacterial activity against *Mycobacterium smegmatis*, *B. subtilis*, and *B. subtilis* var. *niger* and antifungal efficacy against *C. albicans* via the agar well diffusion method (Ulukanli *et al.*, 2005). Alcoholic extracts of *Rumex dentatus* expressed growth-inhibitory effects on *C. albicans* and *Acremonium* spp. by agar disk diffusion (Humeera *et al.*, 2013). Synthesized nanoparticles of *Rumex hymenosepalus* root played a significant role in growth inhibition against *E. coli*, *C. albicans*, Gram-positive, and Gram-negative bacteria, dose-dependently, when applying the well diffusion method (Rodríguez-León *et al.*, 2018; Villalobos-Noriega *et al.*, 2021). Furthermore, it was proven that aqueous extract of *Rumex nervosus* aerial parts was able to inhibit the growth of Gram-positive bacteria (Al-Asmari *et al.*, 2015).

***Tamarix* species**

Both *Tamarix africana* and *Tamarix aphylla* (L.) H. Karst. showed antifungal effects against *C. albicans* when hydroalcoholic extracts of the aerial parts were applied using the agar disk diffusion method. In addition, *T. aphylla* was found to have inhibitory effects against *E. coli*, *E. faecalis*, *K. pneumoniae*, *P. aeruginosa*, and MRSA (Chekroun-Bechlaghem *et al.*, 2019; Al-Omar *et al.*, 2020).

***Thymbra capitata* (L.) Cav.; basionym: *Satureja capitata* L.**

Pinna *et al.* (2019) aimed to determine the antimicrobial effects of *Citrus limon* var. *pompia* Camarda extract and *Thymbra capitata* essential oil, either incorporated in vesicular nanocarriers or as raw extracts, against *C. albicans* and *S. mutans*. Both nanovesicles and raw extracts demonstrated no activity against human gingival fibroblasts at all dosages examined (1, 10, and 100 μg/ml). The highest antimicrobial activity of *T. capitata* against *S. mutans* and *C. albicans* was demonstrated by applying the time–kill assay, disk diffusion method, MIC/MBC/minimum fungicidal concentration (MFC) assay, and transmission electron microscopy (TEM) (Pinna *et al.*, 2019). Another study screened biofilm inhibition via loading of essential oils on biocompatible poly(ε-caprolactone) nanocapsules. Essential oils distilled from *O. vulgare* and *T. capitata* (thymol and carvacrol chemotype) were encapsulated in poly(ε-caprolactone) nanocapsules, which were biocompatible. The mentioned nanosystems demonstrated antifungal, antibiofilm, and antibacterial activities against *C. albicans*, *E. coli*, and *S. aureus*. It was revealed that the polymeric nanocapsules loaded with thyme and oregano essential oils could inhibit microbial and biofilm formation, which might be an ecological alternative in developing novel antimicrobial strategies (Kapustová *et al.*, 2021). Further research evaluated the combined effects of conventional drugs with *C. verum* and *T. capitata* essential oils against elucidation of the molecular mechanism of action and biofilm formation in *C. albicans*. It was identified that *T. capitata* and *C. verum* essential oils were significantly effective against biofilm formation in *C. albicans* with 80.6 and 85.57% inhibition, respectively, at half MIC values. Moreover, essential oils reduced the synthesis of secreted aspartic proteases by more than 90% at 0.75 × MIC. Essential oils reduced the expression of adhesion and secreted aspartyl proteinase virulence factor genes when used in conjunction with amphotericin B. As a result, *C. albicans* biofilm on medical devices could be effectively treated when using *C. verum* and *T. capitata* essential oils in conjunction with amphotericin B (Essid *et al.*, 2019). The primary purpose of another study was to characterize, extract, and formulate the essential oil obtained from *T. capitata* in phospholipid vesicles: glycerosomes, liposomes, and penetration enhancer-containing vesicles (PEVs). Here, carvacrol was the major component of the steam-distilled essential oil that was found to have antibacterial activity against *L. acidophilus*, *Streptococcus sanguinis*, and *S. mutans*. Therefore, various *Thymus* essential oil formulations may be helpful for the treatment of oral cavity diseases due to the combination of antibacterial and antioxidant activities (Manconi *et al.*, 2018).

Conclusion

Vaginosis is one of the most typically encountered problems by a gynecologist. Many women frequently self-treat with over-the-counter (OTC) medicines and may present to their healthcare provider after treatment failure. Candidiasis (20 to 25% of cases), BV (40 to 50% cases), and trichomoniasis (15 to 20% cases) may occur as repetitive episodes and are therefore associated with significant treatment costs and morbidity (Paladine and Desai, 2018). Because of treatment failures and recurrent infections, there is not any completely satisfactory treatment for vaginitis in current medicine. Therefore, finding new sources of drugs is necessary to develop new medications for these conditions. In TPM, vaginitis is named *sayalān-e raḥem*, which means excessive vaginal discharge. As a prominent physician of TPM, Avicenna recommended several medicinal plants for the treatment of vaginitis (Table 12.1). Natural remedies were always a potential source of new medicines. Therefore, in this chapter, the medicinal plants recommended by Avicenna in his *Canon of Medicine* have been presented, along with current scientific evidence-based results. According to TPM the first step in the treatment of vaginosis is the elimination of the pathogen. Most of the herbs mentioned here have been confirmed in *in vitro* studies as excellent antimicrobial agents and the postulated antibacterial, antifungal, and antiprotozoal effects could be confirmed. In the second step, Avicenna emphasized the use of uterus tonics after treatment to prevent recurrence of the disease. Some of the plants studied in this context could play a very prominent role in strengthening normal vaginal flora, such as pomegranate (*Punica granatum*).

The highest level of evidence came from *in vitro* studies, but generally there are still too few animal studies or clinical trials. More research is therefore needed to better utilize the existing options of TPM in the development and administration of new pharmaceutical formulations also in the treatment of vaginosis.

References

Aabed, K. and Mohammed, A.E. (2021) Synergistic and antagonistic effects of biogenic silver nanoparticles in combination with antibiotics against some pathogenic microbes. *Frontiers in Bioengineering and Biotechnology* 9, 652362. DOI: 10.3389/fbioe.2021.652362.

Abu Hafsa, S.H., Ibrahim, S.A. and Hassan, A.A. (2017) Carob pods (*Ceratonia siliqua* L.) improve growth performance, antioxidant status and caecal characteristics in growing rabbits. *Journal of Animal Physiology and Animal Nutrition* 101(6), 1307–1315. DOI: 10.1111/jpn.12651.

Adhami, S., Tansaz, M., Malehi, A.S. and Javadnoori, M. (2017) The relationship between uterine temperament and vaginitis from Iranian traditional medicine point of view. *Indo American Journal Pharmaceutical Sciences* 4(10), 3589–3595. DOI: 10.5281/zenodo.1011051.

Afzali, E., Siahposh, A., Haghighi-Zadeh, M.H., Farajzadeh, A. and Abbaspoor, Z. (2020) The effect of Quercus (Oak Gal) vaginal cream versus metronidazole vaginal gel on bacterial vaginosis: a double-blind randomized controlled trial. *Complementary Therapies in Medicine* 52, 102497. DOI: 10.1016/j.ctim.2020.102497.

Ahmad, W. Hasan, A. Abdullah, A. and Tarannum, T. (2011) Efficacy of a combination of Unani drugs in patients of trichomonal vaginitis. *Indian Journal of Traditional Knowledge* 10(4), 727–730.

Akroum, S. (2017) Antifungal activity of acetone extracts from *Punica granatum* L., *Quercus suber* L. and *Vicia faba* L. *Journal de Mycologie Medicale* 27(1), 83–89. DOI: 10.1016/j.mycmed.2016.10.004.

Al-Ardi, M.H. (2021) Anti-parasitic activity of nano *Citrullus colocynthis* and nano *Capparis spinose* against *Trichomonas vaginalis in vitro*. *Journal of Parasitic Diseases* 45(3), 845–850. DOI: 10.1007/s12639-021-01371-4.

Al-Asmari, A.R., Siddiqui, Y.M., Athar, M.T., Al-Buraidi, A., Al-Eid, A.S. *et al.* (2015) Antimicrobial activity of aqueous and organic extracts of a Saudi medicinal plant: *Rumex nervosus*. *Journal of Pharmacy & Bioallied Sciences* 7(4), 300–303. DOI: 10.4103/0975-7406.168031

Alem, G., Mekonnen, Y., Tiruneh, M. and Mulu, A. (2008) *In vitro* antibacterial activity of crude preparation of myrtle (*Myrtus communis*) on common human pathogens. *Ethiopian Medical Journal* 46(1), 63–69.

Al-Hindawi, M.K., Al-Deen, I.H., Nabi, M.H. and Ismail, M.A. (1989) Anti-inflammatory activity of some Iraqi plants using intact rats. *Journal of Ethnopharmacology* 26(2), 163–168. DOI: 10.1016/0378-8741(89)90063-9

Almas, K. (2001) The antimicrobial effects of seven different types of Asian chewing sticks. *Odonto-Stomatologie Tropicale = Tropical Dental Journal* 24(96), 17–20.

Al-Omar, M.S., Mohammed, H.A., Mohammed, S., Abd-Elmoniem, E., Kandil, Y.I. *et al.* (2020) Anti-microbial, anti-oxidant, and α-amylase inhibitory activity of traditionally-used medicinal herbs: a comparative analysis of pharmacology, and phytoconstituents of regional halophytic plants' diaspora. *Molecules* 25(22), 5457. DOI: 10.3390/molecules25225457.

Al-Saimary, I.E., Bakr, S.S., Jaffar, T., Al-Saimary, A.E., Salim, H. *et al.* (2002) Effects of some plant extracts and antibiotics on *Pseudomonas aeruginosa* isolated from various burn cases. *Saudi Medical Journal* 23(7), 802–805.

Amsel, R., Totten, P.A., Spiegel, C.A., Chen, K.C., Eschenbach, D. *et al.* (1983) Nonspecific vaginitis. Diagnostic criteria and microbial and epidemiologic associations. *The American Journal of Medicine* 74(1), 14–22. DOI: 10.1016/0002-9343(83)91112-9.

Anibal, P.C., Peixoto, I.T., Foglio, M.A. and Höfling, J.F. (2013) Antifungal activity of the ethanolic extracts of *Punica granatum* L. and evaluation of the morphological and structural modifications of its compounds upon the cells of *Candida* spp. *Brazilian Journal of Microbiology* 44(3), 839–848. DOI: 10.1590/S1517-83822013005000060.

Askari, S.F., Jahromi, B.N., Dehghanian, A., Zarei, A., Tansaz, M. *et al.* (2020) Effect of a novel herbal vaginal suppository containing myrtle and oak gall in the treatment of vaginitis: a randomized clinical trial. *DARU Journal of Pharmaceutical Sciences* 28(2), 603–614. DOI: 10.1007/s40199-020-00365-6.

Askari, S.F., Mohagheghzadeh, A., Azadi, A., Jahromi, B.N., Tansaz, M. *et al.* (2018) A brief review on vaginal drug delivery in traditional Persian medicine. *Traditional and Integrative Medicine* 3(4), 223–229.

Azadbakht, M., Ziai, H., Abdollahi, F. and Shabankhani, B. (2003) Effect of essential oils of *Artemisia*, *Zataria* and *Myrtus* on *Trichomonas vaginalis*. *Journal of Medicinal Plants* 2(8), 35–40.

Bagnall, P. and Rizzolo, D. (2017) Bacterial vaginosis: a practical review. *Journal of the American Academy of Physician Assistants* 30(12), 15–21. DOI: 10.1097/01.JAA.0000526770.60197.fa

Baharuddin, N.S., Abdullah, H. and Abdul Wahab, W.N. (2015) Anti-*Candida* activity of *Quercus infectoria* gall extracts against *Candida* species. *Journal of Pharmacy & Bioallied Sciences* 7(1), 15–20. DOI: 10.4103/0975-7406.148742.

Bakht, J., Khan, S., Shafi, M. and Iqbal, A. (2017) Fractionation of crude extracts from controlled dried and commercially available stem bark of *Juglans regia* and their antimicrobial effects. *Pakistan Journal of Pharmaceutical Sciences* 30(5), 1581–1588.

Bassiri-Jahromi, S., Pourshafie, M.R., Mirabzade Ardakani, E., Ehsani, A.H., Doostkam, A. *et al.* (2018) *In vivo* comparative evaluation of the pomegranate (*Punica granatum*) peel extract as an alternative agent to nystatin against oral candidiasis. *Iranian Journal of Medical Sciences* 43(3), 296–304. DOI: 10.30476/IJMS.2018.40542.

Bautista, C.T., Wurapa, E., Sateren, W.B., Morris, S., Hollingsworth, B. *et al.* (2016) Bacterial vaginosis: a synthesis of the literature on etiology, prevalence, risk factors, and relationship with chlamydia and gonorrhea infections. *Military Medical Research* 3(4), 4. DOI: 10.1186/s40779-016-0074-5.

Bontempo, V., Jiang, X.R., Cheli, F., Lo Verso, L., Mantovani, G. *et al.* (2014) Administration of a novel plant extract product via drinking water to post-weaning piglets: effects on performance and gut health. *Animal* 8(5), 721–730. DOI: 10.1017/S175173111400041X.

Brochot, A., Guilbot, A., Haddioui, L. and Roques, C. (2017) Antibacterial, antifungal, and antiviral effects of three essential oil blends. *MicrobiologyOpen* 6(4), e00459. DOI: 10.1002/mbo3.459.

Cannas, S., Molicotti, P., Usai, D., Maxia, A. and Zanetti, S. (2014) Antifungal, anti-biofilm and adhesion activity of the essential oil of *Myrtus communis* L. against *Candida* species. *Natural Product Research* 28(23), 2173–2177. DOI: 10.1080/14786419.2014.925892.

Carradori, S., Chimenti, P., Fazzari, M., Granese, A. and Angiolella, L. (2016) Antimicrobial activity, synergism and inhibition of germ tube formation by *Crocus sativus*-derived compounds against *Candida* spp. *Journal of Enzyme Inhibition and Medicinal Chemistry* 31 (Suppl.2), 189–193. DOI: 10.1080/14756366.2016.1180596.

Celebi, O., Cinisli, K.T. and Celebi, D. (2021) Antimicrobial activity of the combination (Nano-Bio) of *Artemisia absinthium* with copper nanoparticles. *Materials Today: Proceedings* 45, 3809–3813. DOI: 10.1016/j.matpr.2021.01.824.

César de Souza Vasconcelos, L., Sampaio, M.C., Sampaio, F.C. and Higino, J.S. (2003) Use of *Punica granatum* as an antifungal agent against candidosis associated with denture stomatitis. *Mycoses* 46(5–6), 192–196. DOI: 10.1046/j.1439-0507.2003.00884.x.

Chekroun-Bechlaghem, N., Belyagoubi-Benhammou, N., Belyagoubi, L., Mansour, S., Djebli, N. *et al.* (2019) Antimicrobial and anti-inflammatory activities of three halophyte plants from Algeria and detection of some biomolecules by HPLC-DAD. *Natural Product Research* 35(12), 2107–2111. DOI: 10.1080/14786419.2019.1655413.

Chen, X., Lu, Y., Chen, T. and Li, R. (2021) The female vaginal microbiome in health and bacterial vaginosis. *Frontiers in Cellular and Infection Microbiology* 11, 631972. DOI: 10.3389/fcimb.2021.631972.

Choudhary, A., Smitha, C., Suresh, D. and Basu, S. (2015) Clinical evaluation of efficacy of *Quercus infectoria* and *Mimusops elengi* Linn. herbal preparation in inhibition of gingivitis. *Advances in Human Biology* 4(3), 68–76.

Chung, J.Y., Choo, J.H., Lee, M.H. and Hwang, J.K. (2006) Anticariogenic activity of macelignan isolated from *Myristica fragrans* (nutmeg) against *Streptococcus mutans*. *Phytomedicine* 13(4), 261–266. DOI: 10.1016/j.phymed.2004.04.007.

Chusri, S. and Voravuthikunchai, S.P. (2011) Damage of staphylococcal cytoplasmic membrane by *Quercus infectoria* G. Olivier and its components. *Letters in Applied Microbiology* 52(6), 565–572. DOI: 10.1111/j.1472-765X.2011.03041.x.

Cu-Uvin, S., Hyejin, K., Jamieson, D.J., Hogan, J.W., Schuman, P. *et al.* (2002) Prevalence, incidence, and persistence or recurrence of trichomoniasis among human immunodeficiency virus (HIV)-positive women and among HIV-negative women at high risk for HIV infection. *Clinical Infectious Diseases* 34(10), 1406–1411. DOI: 10.1086/340264.

D'Angeli, F., Malfa, G.A., Garozzo, A., Li Volti, G., Genovese, C. *et al.* (2021) Antimicrobial, antioxidant, and cytotoxic activities of *Juglans regia* L. pellicle extract. *Antibiotics (Basel, Switzerland)* 10(2), 10020159. DOI: 10.3390/antibiotics10020159.

Dehdari, S. and Hajimehdipoor, H. (2016) Herbal medicines for leucorrhea according to Iranian traditional medicine. *Iranian Journal of Medical Sciences* 41(3 Suppl.), S36.

Dou, L., Zheng, Y., Li, L., Gui, X., Chen, Y. *et al.* (2018) The effect of cinnamon on polycystic ovary syndrome in a mouse model. *Reproductive Biology and Endocrinology* 16(1), 99. DOI: 10.1186/s12958-018-0418-y.

Edziri, H., Jaziri, R., Chehab, H., Verschaeve, L., Flamini, G. *et al.* (2019) A comparative study on chemical composition, antibiofilm and biological activities of leaves extracts of four Tunisian olive cultivars. *Heliyon* 5(5), e01604. DOI: 10.1016/j.heliyon.2019.e01604.

Ekrikaya, S., Yilmaz, E., Celik, C., Demirbuga, S., Ildiz, N. *et al.* (2021) Investigation of ellagic acid rich-berry extracts directed silver nanoparticles synthesis and their antimicrobial properties with potential mechanisms towards *Enterococcus faecalis* and *Candida albicans*. *Journal of Biotechnology* 341, 155–162. DOI: 10.1016/j.jbiotec.2021.09.020.

Endo, E.H., Ueda-Nakamura, T., Nakamura, C.V. and Filho, B.P. (2012) Activity of spray-dried microparticles containing pomegranate peel extract against *Candida albicans*. *Molecules* 17(9), 10094–10107. DOI: 10.3390/molecules170910094.

Erdogan, E.A., Goksen, G. and Everest, A. (2014) Anti-candidal activities of some *Myrtus communis* L. extracts obtained using accelerated solvent extraction (ASE). *Journal of Applied Biology and Biotechnology* 2(5), 12–14. DOI: 10.7324/JABB.2014.2503.

Essid, R., Gharbi, D., Abid, G., Karkouch, I., Hamouda, T.B. *et al.* (2019) Combined effect of *Thymus capitatus* and *Cinnamomum verum* essential oils with conventional drugs against *Candida albicans* biofilm formation and elucidation of the molecular mechanism of action. *Industrial Crops and Products* 140, 111720. DOI: 10.1016/j.indcrop.2019.111720.

Ezz Eldin, H.M. and Badawy, A.F. (2015) *In vitro* anti-*Trichomonas vaginalis* activity of *Pistacia lentiscus* mastic and *Ocimum basilicum* essential oil. *Journal of Parasitic Diseases* 39(3), 465–473. DOI: 10.1007/s12639-013-0374-6.

Farouk, F., Abdelmageed, M., Azam Ansari, M. and Azzazy, H. (2020) Synthesis of magnetic iron oxide nanoparticles using pulp and seed aqueous extract of *Citrullus colocynth* and evaluation of their antimicrobial activity. *Biotechnology Letters* 42(2), 231–240. DOI: 10.1007/s10529-019-02762-7.

Fathy, H.M., Ibrahim, R.S., El-Hawiet, A. and Omar, A.A. (2021) Chemical constituents of date palm (*Phoenix dactylifera* L.) fruit-free bunches and their biological activities. *Pharmaceutical Chemistry Journal* 55(4), 384–391. DOI: 10.1007/s11094-021-02432-x.

Gacemi, S., Benarous, K., Imperial, S. and Yousfi, M. (2020) Lepidine B & E as new target inhibitors from *Lepidium sativum* seeds against four enzymes of the pathogen *Candida albicans*: *in vitro* and *in silico* studies. *Endocrine, Metabolic & Immune Disorders Drug Targets* 20(1), 127–138. DOI: 10.2174/1871530319666190415141520.

George, N.S., Cheung, L., Luthria, D.L., Santin, M., Dawson, H.D. *et al.* (2019) Pomegranate peel extract alters the microbiome in mice and dysbiosis caused by *Citrobacter rodentium* infection. *Food Science & Nutrition* 7(8), 2565–2576. DOI: 10.1002/fsn3.1106.

Hay, P. (2009) Recurrent bacterial vaginosis. *Current Opinion in Infectious Diseases* 22(1), 82–86. DOI: 10.1097/QCO.0b013e32832180c6.

Hoang, L., Beneš, F., Fenclová, M., Kronusová, O., Švarcová, V. *et al.* (2020) Phytochemical composition and *in vitro* biological activity of *Iris* spp. (Iridaceae): a new source of bioactive constituents for the inhibition of oral bacterial biofilms. *Antibiotics (Basel, Switzerland)* 9(7), 403. DOI: 10.3390/antibiotics9070403.

Hobbs, M.M. and Seña, A.C. (2013) Modern diagnosis of *Trichomonas vaginalis* infection. *Sexually Transmitted Infections* 89(6), 434–438. DOI: 10.1136/sextrans-2013-051057.

Homayouni, A., Bastani, P., Ziyadi, S., Mohammad-Alizadeh-Charandabi, S., Ghalibaf, M. *et al.* (2014) Effects of probiotics on the recurrence of bacterial vaginosis: a review. *Journal of Lower Genital Tract Disease* 18(1), 79–86. DOI: 10.1097/LGT.0b013e31829156ec.

Humeera, N., Kamili, A.N., Bandh, S.A., Amin, S.U., Lone, B.A. *et al.* (2013) Antimicrobial and antioxidant activities of alcoholic extracts of *Rumex dentatus* L. *Microbial Pathogenesis* 57, 17–20. DOI: 10.1016/j.micpath.2013.02.001.

Ifeanyi, O.E., Chinedum, O.K. and Chijioke, U.O. (2018) *Trichomonas vaginalis*: complications and treatment. *International Journal of Current Research in Medical Sciences* 4(5), 76–89. DOI: 10.22192/ijcrms.2018.04.05.012.

Ivanov, M., Gašić, U., Stojković, D., Kostić, M., Mišić, D. *et al.* (2021) New evidence for *Artemisia absinthium* L. application in gastrointestinal ailments: ethnopharmacology, antimicrobial capacity, cytotoxicity, and phenolic profile. *Evidence-Based Complementary and Alternative Medicine* 2021, 9961089. DOI: 10.1155/2021/9961089.

Iyer, M., Gujjari, A.K., Gowda, V. and Angadi, S. (2017) Antifungal response of oral-associated candidal reference strains (American Type Culture Collection) by supercritical fluid extract of nutmeg seeds for geriatric denture wearers: an *in vitro* screening study. *Journal of Indian Prosthodontic Society* 17(3), 267–272. DOI: 10.4103/jips.jips_10_17.

Jassim, S.A. and Naji, M.A. (2003) Novel antiviral agents: a medicinal plant perspective. *Journal of Applied Microbiology* 95(3), 412–427. DOI: 10.1046/j.1365-2672.2003.02026.x.

Javed, A., Parvaiz, F. and Manzoor, S. (2019) Bacterial vaginosis: an insight into the prevalence, alternative treatments regimen and its associated resistance patterns. *Microbial Pathogenesis* 127, 21–30. DOI: 10.1016/j.micpath.2018.11.046.

Kapustová, M., Puškárová, A., Bučková, M., Granata, G., Napoli, E. *et al.* (2021) Biofilm inhibition by biocompatible poly(ε-caprolactone) nanocapsules loaded with essential oils and their cyto/genotoxicity to human keratinocyte cell line. *International Journal of Pharmaceutics* 606, 120846. DOI: 10.1016/j.ijpharm.2021.120846.

Karygianni, L., Cecere, M., Skaltsounis, A.L., Argyropoulou, A., Hellwig, E. *et al.* (2014) High-level antimicrobial efficacy of representative Mediterranean natural plant extracts against oral microorganisms. *BioMed Research International* 2014, 839019. DOI: 10.1155/2014/839019.

Kavousi, M., Khadem Ghaebi, N., Najaf Najafi, M., Mokaberinejad, R., Feyzabadi, Z. *et al.* (2019) The effect of a natural vaginal product based on honey on the success of intrauterine insemination (IUI) in infertility treatment. *Avicenna Journal of Phytomedicine* 9(4), 310–321. DOI: 10.22038/AJP.2019.12588.

Khalilzadeh, S., Eftkhar, T., Shirbeigi, L., Tabarrai, M., Toliyat, T. *et al.* (2020) Efficacy of a vaginal tablet as a Persian medicine product on vulvovaginal candidiasis: a double-blind, randomised, placebo-controlled trial. *Pharmaceutical Biology* 58(1), 574–580. DOI: 10.1080/13880209.2020.1784236.

Krieger, J.N., Tam, M.R., Stevens, C.E., Nielsen, I.O., Hale, J. *et al.* (1988) Diagnosis of trichomoniasis. Comparison of conventional wet-mount examination with cytologic studies, cultures, and monoclonal antibody staining of direct specimens. *JAMA* 259(8), 1223–1227. DOI: 10.1001/jama.259.8.1223.

Lachowicz, K.J., Jones, G.P., Briggs, D.R., Bienvenu, F.E., Wan, J. *et al.* (1998) The synergistic preservative effects of the essential oils of sweet basil (*Ocimum basilicum* L.) against acid-tolerant food microflora. *Letters in Applied Microbiology* 26(3), 209–214. DOI: 10.1046/j.1472-765x.1998.00321.x.

Lavaee, F., Motaghi, D., Jassbi, A.R., Jafarian, H., Ghasemi, F. *et al.* (2018) Antifungal effect of the bark and root extracts of *Punica granatum* on oral *Candida* isolates. *Current Medical Mycology* 4(4), 20–24. DOI: 10.18502/cmm.4.4.382.

Li, Z., Henning, S.M., Lee, R.P., Lu, Q.Y., Summanen, P.H. *et al.* (2015) Pomegranate extract induces ellagitannin metabolite formation and changes stool microbiota in healthy volunteers. *Food & Function* 6(8), 2487–2495. DOI: 10.1039/c5fo00669d.

Lossick, J.G. (1990) Epidemiology of urogenital trichomoniasis. In: Honigberg, B.M. (ed.) *Trichomonads Parasitic in Humans*. Springer, New York, pp. 311–323. DOI: 10.1007/978-1-4612-3224-7_17.

Mahboubi, M. (2018) Systematic review: the potency of *Zataria multiflora* Boiss. in treatment of vaginal infections. *Infectio* 22(2), 76–83.

Mahdi, N.K., Gany, Z.H. and Sharief, M. (2006) Alternative drugs against *Trichomonas vaginalis*. *Eastern Mediterranean Health Journal* 12(5), 679–684.

Mainini, G., Passaro, M., Schiattarella, A., Franciscis, P., Donna, M. *et al.* (2020) Prevention and treatment of cystitis during menopause: efficacy of a nutraceutical containing D-mannose, inulin, cranberry, bearberry, *Olea europaea*, *Orthosiphon and Lactobacillus acidophilus*. *Menopause Review* 19(3), 130–134. DOI: 10.5114/pm.2020.99567.

Manconi, M., Petretto, G., D'hallewin, G., Escribano, E., Milia, E. *et al.* (2018) *Thymus* essential oil extraction, characterization and incorporation in phospholipid vesicles for the antioxidant/antibacterial treatment of oral cavity diseases. *Colloids and Surfaces. B, Biointerfaces* 171, 115–122. DOI: 10.1016/j.colsurfb.2018.07.021.

Mansouri, S., Foroumadi, A., Ghaneie, T. and Najar, A.G. (2001) Antibacterial activity of the crude extracts and fractionated constituents of *Myrtus communis*. *Pharmaceutical Biology* 39(5), 399–401. DOI: 10.1076/phbi.39.5.399.5889.

Markin, D., Duek, L. and Berdicevsky, I. (2003) *In vitro* antimicrobial activity of olive leaves. *Mycoses* 46(3-4), 132–136. DOI: 10.1046/j.1439-0507.2003.00859.x.

Marzouk, B., Marzouk, Z., Décor, R., Edziri, H., Haloui, E. *et al.* (2009) Antibacterial and anticandidal screening of Tunisian *Citrullus colocynthis* Schrad. from Medenine. *Journal of Ethnopharmacology* 125(2), 344–349. DOI: 10.1016/j.jep.2009.04.025.

Masoudi, M., Miraj, S. and Rafieian-Kopaei, M. (2016) Comparison of the effects of *Myrtus communis* L, *Berberis vulgaris* and metronidazole vaginal gel alone for the treatment of bacterial vaginosis. *Journal of Clinical and Diagnostic Research* 10(3), QC04–QC07. DOI: 10.7860/JCDR/2016/17211.7392.
Mehta, V.V., Rajesh, G., Rao, A., Shenoy, R. and Mithun Pai, B.H. (2014) Antimicrobial efficacy of *Punica granatum* mesocarp, *Nelumbo nucifera* leaf, *Psidium guajava* leaf and *Coffea canephora* extract on common oral pathogens: an *in-vitro* study. *Journal of Clinical and Diagnostic Research* 8(7), ZC65–ZC68. DOI: 10.7860/JCDR/2014/9122.4629.
Meyari, A., Tansaz, M., Yavari, M., Bahman, M., Memarzadeh Zavareh, H. *et al.* (2017) Nutritional recommendations for patients with uterine discharge in Iranian traditional medicine. *The Iranian Journal of Obstetrics, Gynecology and Infertility* 20, 68–77. DOI: 10.22038/IJOGI.2017.9836.
Milia, E., Usai, M., Szotáková, B., Elstnerová, M., Králová, V. *et al.* (2020) The pharmaceutical ability of *Pistacia lentiscus* L. leaves essential oil against periodontal bacteria and *Candida* sp. and its anti-inflammatory potential. *Antibiotics (Basel, Switzerland)* 9(6), 281. DOI: 10.3390/antibiotics9060281.
Miller, A.B., Cates, R.G., Lawrence, M., Soria, J.A., Espinoza, L.V. *et al.* (2015) The antibacterial and antifungal activity of essential oils extracted from Guatemalan medicinal plants. *Pharmaceutical Biology* 53(4), 548–554. DOI: 10.3109/13880209.2014.932391.
Mills, B.B. (2017) Vaginitis: beyond the basics. *Obstetrics and Gynecology Clinics of North America* 44(2), 159–177. DOI: 10.1016/j.ogc.2017.02.010.
Mun'im R., A., Hussin, M.S., Kadum, M.M., Kadum, Y.A. and Hamza, E.H. (2009) *In vitro* antimicrobial activities of *Myrtus communis* L. and *Cyperus rotundus* L. extracts. *Al-Mustansiriyah Journal of Science* 20(1), 1–13.
Muzaffar, S., Rather, S.A. and Khan, K.Z. (2016) *In vitro* bactericidal and fungicidal activities of various extracts of saffron (*Crocus sativus* L.) stigmas from Jammu & Kashmir, India. *Cogent Food & Agriculture* 2(1), 1158999. DOI: 10.1080/23311932.2016.1158999.
Muzny, C.A. and Schwebke, J.R. (2013) The clinical spectrum of *Trichomonas vaginalis* infection and challenges to management. *Sexually Transmitted Infections* 89(6), 423–425. DOI: 10.1136/sextrans-2012-050893.
Noumi, E., Snoussi, M., Hajlaoui, H., Valentin, E. and Bakhrouf, A. (2010) Antifungal properties of *Salvadora persica* and *Juglans regia* L. extracts against oral *Candida* strains. *European Journal of Clinical Microbiology & Infectious Diseases* 29(1), 81–88. DOI: 10.1007/s10096-009-0824-3.
Obistioiu, D., Cristina, R.T., Schmerold, I., Chizzola, R., Stolze, K. *et al.* (2014) Chemical characterization by GC–MS and *in vitro* activity against *Candida albicans* of volatile fractions prepared from *Artemisia dracunculus*, *Artemisia abrotanum*, *Artemisia absinthium* and *Artemisia vulgaris*. *Chemistry Central Journal* 8(1), 6. DOI: 10.1186/1752-153X-8-6.
Ojah, E.O., Oladele, E.O. and Chukwuemeka, P. (2021) Phytochemical and antibacterial properties of root extracts from *Portulaca oleracea* Linn. (purslane) utilized in the management of diseases in Nigeria. *Journal of Medicinal Plants for Economic Development* 5(1), a103. DOI: 10.4102/jomped.v5i1.103.
Oliveira, I., Sousa, A., Morais, J.S., Ferreira, I.C., Bento, A. *et al.* (2008) Chemical composition, and antioxidant and antimicrobial activities of three hazelnut (*Corylus avellana* L.) cultivars. *Food and Chemical Toxicology* 46(5), 1801–1807. DOI: 10.1016/j.fct.2008.01.026.
Oves, M., Aslam, M., Rauf, M.A., Qayyum, S., Qari, H.A. *et al.* (2018) Antimicrobial and anticancer activities of silver nanoparticles synthesized from the root hair extract of *Phoenix dactylifera*. *Materials Science & Engineering. C, Materials for Biological Applications* 89, 429–443. DOI: 10.1016/j.msec.2018.03.035.
Paladine, H.L. and Desai, U.A. (2018) Vaginitis: diagnosis and treatment. *American Family Physician* 97(5), 321–329.
Palmeira-de-Oliveira, R., Palmeira-de-Oliveira, A. and Martinez-de-Oliveira, J. (2015) New strategies for local treatment of vaginal infections. *Advanced Drug Delivery Reviews* 92, 105–122. DOI: 10.1016/j.addr.2015.06.008.
Pereira, A.P., Ferreira, I.C., Marcelino, F., Valentão, P., Andrade, P.B. *et al.* (2007a) Phenolic compounds and antimicrobial activity of olive (*Olea europaea* L. cv. Cobrançosa) leaves. *Molecules* 12(5), 1153–1162. DOI: 10.3390/12051153.
Pereira, J.A., Oliveira, I., Sousa, A., Valentão, P., Andrade, P.B. *et al.* (2007b) Walnut (*Juglans regia* L.) leaves: phenolic compounds, antibacterial activity and antioxidant potential of different cultivars. *Food and Chemical Toxicology* 45(11), 2287–2295. DOI: 10.1016/j.fct.2007.06.004.
Pinna, R., Filigheddu, E., Juliano, C., Palmieri, A., Manconi, M. *et al.* (2019) Antimicrobial effect of *Thymus capitatus* and *Citruslimon* var. *pompia* as raw extracts and nanovesicles. *Pharmaceutics* 11(5), 11050234. DOI: 10.3390/pharmaceutics11050234.

Powell, A.M. and Nyirjesy, P. (2014) Recurrent vulvovaginitis. *Best Practice & Research. Clinical Obstetrics & Gynaecology* 28(7), 967–976. DOI: 10.1016/j.bpobgyn.2014.07.006.

Qabaha, K.I. (2013) Antimicrobial and free radical scavenging activities of five Palestinian medicinal plants. *African Journal of Traditional, Complementary, and Alternative Medicines* 10(4), 101–108. DOI: 10.4314/ajtcam.v10i4.17.

Raja, V., Ahmad, S.I., Irshad, M., Wani, W.A., Siddiqi, W.A. *et al.* (2017) Anticandidal activity of ethanolic root extract of *Juglans regia* (L.): effect on growth, cell morphology, and key virulence factors. *Journal de Mycologie Medicale* 27(4), 476–486. DOI: 10.1016/j.mycmed.2017.07.002.

Ramalingam, K. and Amaechi, B.T. (2020) Antimicrobial effect of herbal extract of *Acacia arabica* with triphala on the biofilm forming cariogenic microorganisms. *Journal of Ayurveda and Integrative Medicine* 11(3), 322–328. DOI: 10.1016/j.jaim.2018.01.005.

Rangel, M.L., de Aquino, S.G., de Lima, J.M., Castellano, L.R. and de Castro, R.D. (2018) *In vitro* effect of *Cinnamomum zeylanicum* Blume essential oil on *Candida* spp. involved in oral infections. *Evidence-Based Complementary and Alternative Medicine 2018*, 4045013. DOI: 10.1155/2018/4045013.

Reiter, S. and Kellogg Spadt, S. (2019) Bacterial vaginosis: a primer for clinicians. *Postgraduate Medicine* 131(1), 8–18. DOI: 10.1080/00325481.2019.1546534.

Rezvani, M.R., Sayadpour, N. and Saemi, F. (2018) The effect of adding pomegranate seed extract to fat-containing diets on nutrients digestibility, intestinal microflora and growth performance of broilers. *Iranian Journal of Veterinary Research* 19(4), 310–317. DOI: 10.22099/IJVR.2018.5061.

Rodríguez-León, E., Íñiguez-Palomares, R.A., Navarro, R.E., Rodríguez-Beas, C., Larios-Rodríguez, E. *et al.* (2018) Silver nanoparticles synthesized with *Rumex hymenosepalus* extracts: effective broad-spectrum microbicidal agents and cytotoxicity study. *Artificial Cells, Nanomedicine, and Biotechnology* 46(6), 1194–1206. DOI: 10.1080/21691401.2017.1366332.

Rossi, A., Di Paola, R., Mazzon, E., Genovese, T., Caminiti, R. *et al.* (2009) Myrtucommulone from *Myrtus communis* exhibits potent anti-inflammatory effectiveness *in vivo*. *The Journal of Pharmacology and Experimental Therapeutics* 329(1), 76–86. DOI: 10.1124/jpet.108.143214.

Sahyon, H.A. and Al-Harbi, S.A. (2020) Antimicrobial, anticancer and antioxidant activities of nano-heart of *Phoenix dactylifera* tree extract loaded chitosan nanoparticles: *in vitro* and *in vivo* study. *International Journal of Biological Macromolecules* 160, 1230–1241. DOI: 10.1016/j.ijbiomac.2020.05.224.

Sehar, N. and Ansari, K.B. (2016) Concept and management of leucorrhoea in unani system of medicine. *IOSR Journal of Pharmacy* 6(6), 36–40.

Sen, A., Yuksel, M., Bulut, G., Bitis, L., Ercan, F. *et al.* (2017) Therapeutic potential of *Myrtus communis* subsp. *communis* extract against acetic acid-induced colonic inflammation in rats. *Journal of Food Biochemistry* 41(1), e12297. DOI: 10.1111/jfbc.12297.

Setty, J.V., Srinivasan, I., Sathiesh, R.T., Kale, M., Shetty, V.V. *et al.* (2020) *In vitro* evaluation of antimicrobial effect of *Myristica fragrans* on common endodontic pathogens. *Journal of the Indian Society of Pedodontics and Preventive Dentistry* 38(2), 145–151. DOI: 10.4103/JISPPD.JISPPD_214_20.

Shahina, Z., El-Ganiny, A.M., Minion, J., Whiteway, M., Sultana, T. *et al.* (2018) *Cinnamomum zeylanicum* bark essential oil induces cell wall remodelling and spindle defects in *Candida albicans*. *Fungal Biology and Biotechnology* 5, 3. DOI: 10.1186/s40694-018-0046-5.

Shaikh, A.E., Satardekar, K.V., Khan, R.R., Tarte, N.A. and Barve, S.S. (2018) Silver nanoparticles: green synthesis using *Phoenix dactylifera* fruit extract, characterization, and anti-oxidant and anti-microbial activities. *Applied Nanoscience* 8(3), 407–415. DOI: 10.1007/s13204-018-0682-3.

Shirooye, P., Yavari, M., Naiemi, S.A., Bahman, M., Meyari, A. *et al.* (2016) Pathophysiology of vaginal discharge from Iranian Traditional Medicine (ITM) point of view compared with common medicine. *The Iranian Journal of Obstetrics, Gynecology and Infertility* 19(37), 75–84.

Soliman, S., Semreen, M.H., El-Keblawy, A.A., Abdullah, A., Uppuluri, P. *et al.* (2017) Assessment of herbal drugs for promising anti-*Candida* activity. *BMC Complementary and Alternative Medicine* 17(1), 257. DOI: 10.1186/s12906-017-1760-x.

Sutton, M., Sternberg, M., Koumans, E.H., McQuillan, G., Berman, S. *et al.* (2007) The prevalence of *Trichomonas vaginalis* infection among reproductive-age women in the United States, 2001–2004. *Clinical Infectious Diseases* 45(10), 1319–1326. DOI: 10.1086/522532.

Tam, M.T., Yungbluth, M. and Myles, T. (1998) Gram stain method shows better sensitivity than clinical criteria for detection of bacterial vaginosis in surveillance of pregnant, low-income women in a clinical setting. *Infectious Diseases in Obstetrics and Gynecology* 6(5), 204–208. DOI: 10.1002/(SICI)1098-0997(1998)6:5<204::AID-IDOG3>3.0.CO;2-R.

Tleubayeva, M.I., Datkhayev, U.M., Alimzhanova, M., Ishmuratova, M.Y., Korotetskaya, N.V. *et al.* (2021) Component composition and antimicrobial activity of CO_2 extract of *Portulaca oleracea*, growing in the territory of Kazakhstan. *The Scientific World Journal* 2021, 5434525. DOI: 10.1155/2021/5434525.
Tran, H., Graham, L. and Adukwu, E.C. (2020) *In vitro* antifungal activity of *Cinnamomum zeylanicum* bark and leaf essential oils against *Candida albicans* and *Candida auris*. *Applied Microbiology and Biotechnology* 104(20), 8911–8924. DOI: 10.1007/s00253-020-10829-z.
Ulukanli, Z., Ulukanli, S., Ozbay, H., Ilcim, A. and Tuzcu, M. (2005) Antimicrobial activities of some plants from the eastern Anatolia region of Turkey. *Pharmaceutical Biology* 43(4), 334–339. DOI: 10.1080/13880200590951757.
Uzair, A., Bakht, J., Iqbal, A., Naveed, K. and Ali, N. (2016) *In vitro* antimicrobial activities of different solvent extracted samples from *Iris germinica*. *Pakistan Journal of Pharmaceutical Sciences* 29(1), 145–150.
van Schalkwyk, J., Yudin, M.H. and Infectious Disease Committee (2015) Vulvovaginitis: screening for and management of trichomoniasis, vulvovaginal candidiasis, and bacterial vaginosis. *Journal of Obstetrics and Gynaecology Canada* 37(3), 266–274. DOI: 10.1016/S1701-2163(15)30316-9.
Veilleux, M.P. and Grenier, D. (2019) Determination of the effects of cinnamon bark fractions on *Candida albicans* and oral epithelial cells. *BMC Complementary and Alternative Medicine* 19(1), 303. DOI: 10.1186/s12906-019-2730-2.
Vieira, J.N., Gonçalves, C.L., Villarreal, J., Gonçalves, V.M., Lund, R.G. *et al.* (2019) Chemical composition of essential oils from the Apiaceae family, cytotoxicity, and their antifungal activity *in vitro* against *Candida* species from oral cavity. *Brazilian Journal of Biology* 79(3), 432–437. DOI: 10.1590/1519-6984.182206.
Villalobos-Noriega, J., Rodríguez-León, E., Rodríguez-Beas, C., Larios-Rodríguez, E., Plascencia-Jatomea, M. *et al.* (2021) Au@Ag core@shell nanoparticles synthesized with *Rumex hymenosepalus* as antimicrobial agent. *Nanoscale Research Letters* 16(1), 118. DOI: 10.1186/s11671-021-03572-5.
Wølner-Hanssen, P., Krieger, J.N., Stevens, C.E., Kiviat, N.B., Koutsky, L. *et al.* (1989) Clinical manifestations of vaginal trichomoniasis. *JAMA* 261(4), 571–576. DOI: 10.1001/jama.1989.03420040109029.
Workowski, K.A., Bolan, G.A. and Centers for Disease Control and Prevention (2015) Sexually transmitted diseases treatment guidelines, 2015. *MMWR Recommendations and Reports* 64(RR-03), 1–137.
Zaheer, Z. (2018) Biogenic synthesis, optical, catalytic, and *in vitro* antimicrobial potential of Ag-nanoparticles prepared using palm date fruit extract. *Journal of Photochemistry and Photobiology. B, Biology* 178, 584–592. DOI: 10.1016/j.jphotobiol.2017.12.002.
Zullo, M.A., Schiavi, M.C., Di Pinto, A., Prata, G., Scudo, M. *et al.* (2020) Efficacy and safety of oral administration of a product based on hydroxytyrosol as preventive therapy for recurrent vulvo-vaginal candidosis: a prospective observational pilot study. *European Review for Medical and Pharmacological Sciences* 24(13), 7427–7432. DOI: 10.26355/eurrev_202007_21911.

13 Medicinal Plants for Anal Fissure in the *Canon* of Avicenna: An Evidence-Based Review

Ghazaleh Mosleh[1]*, Abdolali Mohagheghzadeh[2,3], and Meysam Zaeri[2]
[1]*Phytopharmaceutical Technology and Traditional Medicine Incubator, Shiraz University of Medical Sciences, Shiraz, Iran;* [2]*Department of Phytopharmaceuticals, Faculty of Pharmacy, Shiraz University of Medical Sciences, Shiraz, Iran;* [3]*Pharmaceutical Sciences Research Center, Shiraz University of Medical Sciences, Shiraz, Iran*

Abstract

Gastrointestinal disorders are dysfunctions affecting the digestive tract. Medicinal plants are commonly consumed as affordable and easily accessible drugs to treat various kinds of gastrointestinal ailments. Gastrointestinal disorders and their specific natural medications are described in the *Canon* of Avicenna in five distinct sections, differentiated from esophageal to anal disorders. In this regard, esophagus diseases are categorized into five general groups including dysphagia, esophageal hemorrhage, obstruction, swelling, and ulcer. Anal diseases are categorized into seven main subjects including anal fissure, fistula-like lesion, hemorrhoid, laxity, pruritus, swelling, and rectal prolapse. Today nonsurgical treatment methods of anal fissure (AF), especially herbal medicines, have been noticed. The medicinal plants recommended by Avicenna for treatment of this disease are presented in this chapter.

Introduction

Gastrointestinal disorders are dysfunctions affecting the digestive tract. Medicinal plants are commonly consumed as affordable and easily accessible drugs to treat various kinds of gastrointestinal ailments (Rokaya *et al.*, 2014). Avicenna, the reputed Persian physician (AD 980–1038), has described gastrointestinal disorders and their specific natural medications in his book called *The Canon of Medicine* in five distinct sections, differentiated from esophageal to anal disorders. In this regard, esophagus diseases are categorized into five general groups including dysphagia, esophageal hemorrhage, obstruction, swelling, and ulcer. Then Avicenna has cited stomach diseases in 18 main topics including abnormal gastric mucosa, abnormal stomach size, abdominal blood clots, decreased or lack of appetite, dystemperament, eructation, flatulence, gastric weakness, hiccups, hypersensitivity, maldigestion, nausea, pain, peptic ulcer, rash, swelling, thirst, and vomiting. Furthermore, liver diseases could be categorized into 12 major disorders according to

*Email: moslehgh@sums.ac.ir

 Medicinal Plants Used in Traditional Persian Medicine (eds. H. Schulz, Seyed Ahmad Emami and Farsad Nadjafi)
DOI: 10.1079/9781800621671.0013

Avicenna's point of view. They comprise abscess, ascites, atrophy, liver damage trauma, discharges, dystemperament, dropsy, weakness, obstruction, pain, swelling, and trauma. Moreover, yellow and black jaundices are introduced as gallbladder disorders in the *Canon*. Then, spleen diseases are categorized into five main groups including dystemperament, obstruction, pain, swelling, and spleen wind. Thereafter, intestinal diseases are explained and categorized into 11 main disorders including abnormal bowel movements, borborygmi, colic (*qoulanj*), diarrhea, dysentery (*zahir*), enteric ulcers, gripe (*maghs*), ileus, increased or decreased defecation, intestinal worms, and pain. Finally, Avicenna has explained anal diseases, which are categorized into seven main subjects including anal fissure (AF), fistula-like lesion (*nāṣoor*), hemorrhoid, laxity, pruritus, swelling, and rectal prolapse.

Present interest in nonsurgical operations and using medical herbs for management of AF (Derakhshan, 2017) inspired the authors to identify the medicinal plants recommended for AF in the *Canon* and suggest them as suitable candidates for further investigations. AF, a linear ulcer usually situated in the posterior midline anoderm, triggers defecation pain and burning sensation, as well as the appearance of small spots of blood on toilet paper (Madalinski and Kalinowski, 2009; van Meegdenburg *et al.*, 2016). It is one of the most common conditions among proctologic diseases. However, a great deal of research on the etiology of AF has yet been in vain (Lund, 2006). The accepted theory is that spasm of the internal anal sphincter followed by local trauma and pain can increase the maximum basal pressure responsible for diminished wound healing and poor blood flow at the AF site (van Meegdenburg *et al.*, 2016). In addition, repeated mechanical lesions and inflammation in the anoderm can provoke this painful disorder (van Outryve, 2006). Most AFs are acute, so they heal even spontaneously in less than 14 days (Derakhshan, 2017). Almost 235,000 new cases of acute fissures in the anus are diagnosed each year in the USA, 40% of which become chronic (Mosleh *et al.*, 2020). Several therapeutic strategies are recommended for the management of AF in current medicine; lifestyle modification, increased intake of liquids, using stool softeners, and topical medication are helpful strategies to accelerate healing of acute AF. Furthermore, internal sphincterotomy is recognized as the best therapeutic strategy for chronic fissures that fail the conservative care (Derakhshan, 2017). However, the surgical treatment may also have its own risks and complications (Mosleh *et al.*, 2020).

Anal Fissure and Its Common Treatments in the View of Avicenna

A disease entitled *sheqāq-e-maqʿad* in Traditional Persian Medicine (TPM) is denoted as AF. According to Avicenna's point of view, there are several underlying factors associated with *sheqāq-e-maqʿad* incidence. In this regard, dry and hot dystemperament of the anus followed by production of dense stools is traditionally considered as the most prevalent cause of AF. Furthermore, warm swelling in the anus, constipation, hemorrhoid, and fullness of rectal arteries are described as other probable causes of *sheqāq-e-maqʿad* (Avicenna, 1038; Derakhshan, 2017; Mosleh *et al.*, 2019). Avicenna believed that the first step in management of AF is to find and fix the main pathologic cause. Then topical medication is prescribed to stimulate the local wound healing. Generally, four types of topical drugs for *sheqāq-e-maqʿad* are cited in the *Canon*: (i) wound healing agents (*modamel*); (ii) emollients and wettings (*molayen* and *moraṭeb*); (iii) anti-inflammatory drugs (*moʿālejat-al-waram*); and (iv) anti-AF agents by attribute (*ḏo-al-khāsiah*) (Avicenna, 1038; Mosleh *et al.*, 2019).

Scientific Evaluation of the Medicinal Herbs Recommended for Anal Fissure in the *Canon*

The single medicinal plants indicated in the *Canon* for topical medication of AF are provided and the relevant scientific evidences supporting their efficacy are expressed in this section. All collected data are shown in Table 13.1.

Table 13.1. Medicinal plants mentioned in the *Canon* of Avicenna for the treatment of AF and their relevant *in vitro, in vivo*, and clinical studies.

Scientific name (family)	Common name (English/Arabic)	Part used in TPM (administration method)	*In vitro*/*in vivo*/clinical studies, extract, and effect	Reference
Aloe vera (L.) Burm.f. (*Xanthorrhoeaceae*)	Aloe/*Ṣabr*	Dried extract of the ground and squeezed leaves (topical solution of aloe dissolved in sweet wine)	*In vitro*: • Gel: ↑ macrophage cytokine production, wound healing, bFGF, TGF-β1, angiogenic effect	Zhang and Tizard (1996); Moon *et al*. (1999); Choi *et al*. (2001); Moriyama *et al*. (2016); Hormozi *et al*. (2017)
			• Different aloe extracts: ↑ wound healing, cell proliferation, EGF, blood vessels, IL-6, TGF-β1, anti-inflammatory effect; ↓TNF-α, IL-1β, IL-6, COX-2, thromboxane A2 synthase	Yagi *et al*. (2002); Fox *et al*. (2017); Li *et al*. (2017); Wahedi *et al*. (2017); Babu and Noor (2019)
			In vivo: • Gel: ↑ wound healing, wound closure, anti-inflammatory effects, epithelialization, fibroblast numbers, TGF-β gene expression, MMP-3, TIMP-2, KGF-1, VEGF, and type I collagen expressions	Jettanacheawchankit *et al*. (2009); Haritha Yadav *et al*. (2012); Desai *et al*. (2014); Tabandeh *et al*. (2014); Takzaree *et al*. (2015, 2016); Oryan *et al*. (2016); Malik Hussein (2019)
			• Different aloe extracts: ↑ wound healing and epithelialization; ↓ NO, inflammatory mediators, AST, ALT, TNF-α, IL-6, sPLA2	Sarkar *et al*. (2005); Rajput *et al*. (2009); Oryan *et al*. (2010); Rezaie *et al*. (2012); Babu and Noor (2019)
			Clinical: • Gel: ↑ wound healing; ↓ postoperative pain and pressure ulcers	Eshghi *et al*. (2010); Rahmani *et al*. (2014); Hekmatpou *et al*. (2018a)
Astragalus gummifer Labill. (*Fabaceae*)	Tragacanth gum/*Kathirā*	Gum (mucilaginous dressings made from the gum)	*In vivo*: • Mucilage/gel: ↑ wound healing	Moghbel *et al*. (2005); Fayazzadeh *et al*. (2014)
			Clinical: • Gel: ↓ pressure ulcers	Shakibamehr *et al*. (2019)
Commiphora mukul (Hook. ex Stocks) Engl. (*Burseraceae*)	Gum guggul/*Moql*	Gum resin (ointment of gum resin with kernel oil or tallow)	*In vitro* :• Gum and/or resin: ↑ HO-1, Nrf2 transcriptional activity, ARE-binding activity, Akt phosphorylation, anti-inflammatory effect, inhibiting COX-1 and COX-2	Manjula *et al*. (2006); Ahn and Youn (2008); Mencarelli *et al*. (2009; Almazari *et al*. (2012); Madasu *et al*. (2017)

Continued

Table 13.1. Continued.

Scientific name (family)	Common name (English/Arabic)	Part used in TPM (administration method)	*In vitro*/*in vivo*/clinical studies, extract, and effect	Reference
			In vivo: • Gum and/or resin: ↑ anti-inflammatory effect	Sosa *et al*. (1993); Kimura *et al*. (2001); Mencarelli *et al*. (2009)
Dracaena cinnabari Balf.f. (*Asparagaceae*)	Dragon's blood/*Damm-ol-akhwain*	Resin (compounded in other formulations)	*In vitro*: • Resin: ↓ TNF-α, IL-6, nitrite production	Gupta *et al*. (2014)
			In vivo: • Resin: ↑ anti-inflammatory effect	Alwashli *et al*. (2012); Gupta *et al*. (2014)
Erysimum × *cheiri* (L.) Crantz (*Brassicaceae*)	Common wallflower/*Kheiri*	Flower (dried powder as a dressing/TPM wallflower oil)	*In vitro*: • Different extracts of the herb: ↑ anti-inflammatory effect	Mosleh *et al*. (2021a)
			Clinical: • TPM oil in ointment: ↓ pain, itching, inflammation	Mosleh *et al*. (2020)
Lawsonia inermis L. (*Lythraceae*)	Henna/*Henā*	Leaf (topical cerate of henna leaves and TPM damask rose oil)	*In vitro*: • Leaf, stem, flower and seed: ↑ anti-inflammatory effect	Liou *et al*. (2013); Wiem *et al*. (2014); Bouhlali *et al*. (2016); Manuja *et al*. (2021)
			In vivo: • Different extracts of the herb: ↑ wound healing, epithelialization, and anti-inflammatory effects; ↓ ROS-induced damages	Nayak *et al*. (2007); Jridi *et al*. (2017); Salih *et al*. (2017); Daemi *et al*. (2019); Rekik *et al*. (2019); Ali *et al*. (2021)
			Clinical: • Henna oil: healing the grade 1 pressure ulcer • Herb/leaf mixture: ↓ decubitus ulcers and pressure ulcer	Hekmatpou *et al*. (2018b); Poursadra *et al*. (2019); Rafiei *et al*. (2019)
Olea europaea L. (*Oleaceae*)	Olive/*Zeitoon*	Oil (sitz bath of olive oil, water, and an astringent agent/enema of olive oil and verjuice)	*In vitro*: • Olive oil: ↑ TNF-α and anti-inflammatory effect; ↓ COX-2 and NO production	Miles *et al*. (2005); Rosignoli *et al*. (2013); Cardeno *et al*. (2014); Cárdeno *et al*. (2014); Takeda *et al*. (2014); Abdallah *et al*. (2018); Yonezawa *et al*. (2018); Aparicio-Soto *et al*. (2019); Elaasser *et al*. (2020); Fernández-Prior *et al*. (2021)

			Clinical: • Olive oil: ↑ wound healing; ↓ pressure ulcers and dermatitis	Amani *et al*. (2015); Nasiri *et al*. (2015); Lupiañez-Perez *et al*. (2015, 2017); Díaz-Valenzuela *et al*. (2019); Miraj *et al*. (2020)
Prunus armeniaca L. (*Rosaceae*)	Common apricot/ *Meshmesh*	Kernel oil (topical kernel oil, either singly or in combination with guggul gum resin)	*In vitro*: • Seed: ↑ anti-inflammatory effect	Chang *et al*. (2005)
			In vivo: • Seed kernel extracts: ↓ inflammatory pain, inflammation, and colitis	Hwang *et al*. (2008); Minaiyan *et al*. (2014); Karaboğa *et al*. (2018)
Quercus infectoria G. Olivier (*Fagaceae*)	Aleppo oak/*Afs*	Gall (hydroalcoholic solution of powdered nut galls suspended in a mixture of wine and water/decoction of galls in concentrated grape wine)	*In vitro*: • Gall: ↓ Set7, p65, NF-κB activity, IL-1β, IL-6, CCL-2/MCP-1, CCL-7/MCP-3; ↑ caspase-3 activity; scavenging NO, ameliorating LPS-stimulated PGE_2, PMA-stimulated superoxide production; regulating Bcl-2, Bax, p53, pAkt production	Kaur *et al*. (2004); Yoon *et al*. (2013); Chokpaisarn *et al*. (2017b)
			In vivo: • Gall/oak fruit: ↑ wound healing, anti-inflammatory effect; ↓ IL-1β, TNF-α	Jalalpure *et al*. (2002); Kaur *et al*. (2004); Choudhary (2007); Yousif *et al*. (2011); Bonab and Farahpour (2017); Chokpaisarn *et al*. (2017a,b)
			Clinical: • Gall solution: ↑ wound healing	Chokpaisarn *et al*. (2020)
Rosa × damascena Herrm. (*Rosaceae*)	Damask rose/*Ward*	Fixed oil from petals (topical TPM oil)	*In vitro*: • Polyphenol-enriched fraction obtained from rose oil distillation: ↓ IL-1β, IL-6, IL-8, RANTES, and MCP-1	Wedler *et al*. (2016)
			In vivo: • Hydroalcoholic extract of the plant: ↑ wound healing	Mansouri *et al*. (2016)
			Clinical: • TPM oil in ointment: ↓ pain, itching, inflammation	Mosleh *et al*. (2020)

Continued

Table 13.1. Continued.

Scientific name (family)	Common name (English/Arabic)	Part used in TPM (administration method)	*In vitro*/*in vivo*/clinical studies, extract, and effect	Reference
Taraxacum campylodes G.E. Haglund (*Asteraceae*)	Common dandelion/ *Khas-ol-ḥemār*	All parts (cerate of dandelion in damask rose oil)	*In vitro*: • Different extracts of the herb: ↑ anti-inflammatory effect; ↓ NO, TNF-α, NLRP3, IL-1β	Koh *et al.* (2010); Park *et al.* (2010, 2011, 2014); Zhang *et al.* (2012); Jeon *et al.* (2017); Xue *et al.* (2017); Lai *et al.* (2020); Yang *et al.* (2021)
			In vivo: • Different extracts of root/plant: ↑ anticolitis and anti-inflammatory effects; ↓ COX-2 expression, TNF-α, IL6	Jeon *et al.* (2008; Wang, *et al.* (2017); Ding and Wen (2018); Razak *et al.* (2020)
Vitex agnus-castus L. (*Lamiaceae*)	Lilac chaste/*Athlaq*	All parts, especially the seeds (powder)	*In vitro*: • Aerial part/fruit extracts: ↑ anti-inflammatory effect	Choudhary *et al.* (2009); Ahmad *et al.* (2010); Ibrahim *et al.* (2021)
			In vivo: • Seed essential oil/fruit extracts: ↑ anti-inflammatory effect	Ramezani *et al.* (2010); Aboul-Enein *et al.* (2017)

NLRP3, nod-like receptor (NLR) family pyrin domain containing 3.

Aloe vera (L.) Burm.f.; basionym: *Aloe perfoliata* var. *vera* L.

The phytochemical compounds in *Aloe vera* such as acemannan, aloesin, β-sitosterol, and aloin have *in vitro* anti-inflammatory and wound healing activities through various mechanisms of action (Zhang and Tizard, 1996; Moon *et al.*, 1999; Yagi *et al.*, 2002). Aloe gel has increasing effects on cytokine production in RAW 264.7 cells (Zhang and Tizard, 1996). Also, it has shown wound healing effects in the monolayer of human keratinocytes, primary human epidermal keratinocytes (HPEKs), and EpiDerm-FT (Choi *et al.*, 2001; Moriyama *et al.*, 2016). Aloe gel increases basic fibroblast growth factor (bFGF) and transforming growth factor-β1 (TGF-β1) in mouse embryonic fibroblasts (Hormozi *et al.*, 2017). Furthermore, it has angiogenic effects on the chorioallantoic membrane of chick embryos *in vitro* (Moon *et al.*, 1999). Besides, aloe leaf has shown wound healing effects in HaCaT keratinocyte cells (Fox *et al.*, 2017). The peptide and polypeptide fraction of the plant showed anti-inflammatory effects on RAW 264.7 cells (Babu and Noor, 2019). Aloesin of *A. vera* increases interleukin (IL)-6 and TGF-β1 while decreasing IL-1β and tumor necrosis factor-α (TNF-α) in skin cell lines, RAW 264.7, and HaCaT cells (Wahedi *et al.*, 2017). In addition, aloin increased cell proliferation, epidermal growth factor (EGF), and blood vessels in endothelial cells (Li *et al.*, 2017). Also, aloesin derivatives reduced cyclooxygenase-2 (COX-2) and thromboxane A2 synthase *in vitro* (Yagi *et al.*, 2002). On the other hand, accelerating wound healing and anti-inflammatory activities of *A. vera* are reported in various *in vivo* pharmacological studies. Aloe gel increased fibroblast numbers, TGF-β gene expression, and wound healing effects in square-shaped wounds in rats (Takzaree *et al.*, 2015, 2016). The mannose-rich polysaccharides of aloe gel increased matrix metalloproteinase-3 (MMP-3), tissue inhibitor of metalloproteinase-2 (TIMP-2), and wound closure in full-thickness skin wounds (Tabandeh *et al.*, 2014), as well as acemannan-stimulated keratinocyte growth factor-1 (KGF-1), vascular endothelial growth factor (VEGF), fibroblast proliferation, and type I collagen expressions in rats (Jettanacheawchankit *et al.*, 2009). Glucomannan induced protective effects against colitis in dextran sodium sulfate (DSS)-induced colitis murine models (Zhang *et al.*, 2019). Also, a glycoprotein fraction G1G1M1DI2 accelerated wound healing and cell proliferation in mouse models (Choi *et al.*, 2001). Furthermore, aloin has shown wound healing effects in rats (Li *et al.*, 2017), and aloesin accelerated wound closure while increasing Smad and mitogen-activated protein kinase (MAPK) signaling proteins (Wahedi *et al.*, 2017). Aloe extract accelerated epithelialization and wound contraction in the full-thickness wound in rats (Rezaie *et al.*, 2012). The leafy exudate decreased nitric oxide (NO) and inflammatory mediators in carrageenan/dextran-induced edema models in rats (Sarkar *et al.*, 2005). Peptides and polypeptide fractions of the plant decreased aspartate aminotransferase (AST), alanine aminotransferase (ALT), TNF-α, IL-6, and secretory phospholipase A2 (sPLA2) in complete Freund's adjuvant-induced paw edema models in rats (Babu and Noor, 2019). On the other hand, in a randomized, double-blind, placebo-controlled study, a cream formulation containing aloe gel decreased the anal pain and wound repair time in 49 patients via 28 days of treatment after hemorrhoidectomy (Rahmani *et al.*, 2014). Besides, aloe gel has shown preventing effects on pressure ulcers in a randomized, triple-blind study on 80 hospitalized patients after 10 days of treatment (Hekmatpou *et al.*, 2018b).

According to Avicenna's opinion, a topical solution containing aloe juice dissolved in sweet wine (the alcoholic extract) possesses anti-inflammatory and wound healing properties in the AF site (Avicenna, 1038).

Astragalus gummifer Labill.

Tragacanth mucilage presented healing properties in full-thickness wound models in albino rats and rabbits (Moghbel *et al.*, 2005; Fayazzadeh *et al.*, 2014). Furthermore, in a triple-blind randomized clinical trial on 94 patients, pad and foam formulations of tragacanth gel showed delaying effects on the onset of erythema and preventive effects on pressure ulcers in patients admitted to intensive care units (ICUs) (Shakibamehr *et al.*, 2019).

According to the *Canon*, mucilaginous dressings made from gum tragacanth, either singly or compounded with other medicaments, are recommended for the management of AF (Avicenna, 1038).

Commiphora mukul (Hook. ex Stocks) Engl.; basionym: *Balsamodendrum mukul* Hook. ex Stocks

Despite the lack of relevant clinical trials, the active constituents of *Commiphora mukul* have been studied in several experiments. Myrrhanone A of the gum resin showed anti-inflammatory effects on phorbol-12-myristate-13-acetate (PMA)/lipopolysaccharide (LPS)-activated human monocyte U937 cells (Madasu *et al.*, 2017). Guggulsterone showed anti-inflammatory activities *in vitro* via decreasing nuclear factor-κB (NF-κB) activation and COX-2 expression (Ahn and Youn, 2008). Also, it showed stimulatory effects on nuclear factor erythroid 2-related factor 2 (Nrf2) and heme oxygenase-1 (HO-1) transcriptional activity, antioxidant response element (ARE)-binding activity, and protein kinase B (Akt) phosphorylation in MCF10A epithelial cell line (Almazari *et al.*, 2012). Additionally, guggulsterone showed *in vitro* anti-inflammatory effects in $CD4^+$ cells of the intestinal lamina propria (Mencarelli *et al.*, 2009). *In vivo* studies on air-pouch granulomas models in mice indicate anti-inflammatory potential of myrrhanol A (Kimura *et al.*, 2001). Also, *E*-guggulsterone and *Z*-guggulsterone showed anti-inflammatory effects in murine models of inflammatory bowel disease and croton oil-induced ear edema models in mice (Sosa *et al.*, 1993; Mencarelli *et al.*, 2009).

In the *Canon* the ointment made by mixing *C. mukul* gum resin with tallow or kernel oil is considered to be effective for the treatment of AF (Avicenna, 1038).

Dracaena cinnabari Balf.f.

The alcoholic extract of dragon's blood resin containing 4′-hydroxy-7,8-methylenedioxyhomoisoflavan as an active constituent has shown *in vitro* pro-inflammatory activities via reducing TNF-α, IL-6, and nitrite production in macrophages (Gupta *et al.*, 2014). Also, it has shown *in vivo* anti-inflammatory effects in rat paw edema models (Alwashli *et al.*, 2012; Gupta *et al.*, 2014).

Avicenna has introduced dragon's blood resin in his *Canon* as a good astringent substance in AF formulations (Avicenna, 1038).

Erysimum × *cheiri* (L.) Crantz; basionym: *Cheiranthus* × *cheiri* L.

The hydroalcoholic extracts of wallflower flowers, aerial parts, roots, and seeds showed *in vitro* anti-inflammatory effects by inhibiting egg albumin denaturation (Mosleh *et al.*, 2021a). Noticing the several cardenolide compounds in different wallflower organs, safety evaluations and dose adjustments of variouis extracts of this herb would be necessary (Mosleh *et al.*, 2021a,b). Besides, in a randomized clinical trial on AF patients, a traditional cerate formulation containing wallflower oil exerted comparable healing effects to diltiazem 2% gel after 14 days of treatment in 64 patients (Mosleh *et al.*, 2020).

In the *Canon*, dried powder of wallflower flower is considered to be a useful dressing for AF. Furthermore, wallflower oil is reported to be a suitable topical medication for fissures induced by dry dystemperament of the anus (Avicenna, 1038).

Lawsonia inermis L.

Henna leaf extract showed anti-inflammatory activities via membrane stabilization potential and inhibition of protein denaturation *in vitro* (Bouhlali *et al.*, 2016). Also, lawsochylin A, lawsonaphthoate A, luteolin, apigenin, 4*S*-4-hydroxy-α-tetralone, and

2-butoxysuccinic acid available in leaf and stem have presented *in vitro* anti-inflammatory effects (Liou *et al.*, 2013). Different henna extracts have shown *in vivo* anti-inflammatory and wound healing properties via various mechanisms of action. The aqueous extract of henna decreased damage induced by reactive oxygen species (ROS) and ulcers in rat excision wound models (Jridi *et al.*, 2017). The oily extract presented wound healing and full reepithelialization properties in circular excision wound models in rats (Rekik *et al.*, 2019). The suspension of henna with normal saline showed anti-inflammatory and healing effects on full-thickness skin wounds in mice (Ali *et al.*, 2021). In a randomized clinical study on 108 patients, 7 days' application of topical henna oil made from leaves and stems of *L. inermis* showed wound repairing effects on grade 1 pressure ulcers (Poursadra *et al.*, 2019). Another randomized clinical trial indicated the healing effects of the herb mixture on pressure ulcers in 72 patients after 7 days of treatment (Rafiei *et al.*, 2019). In a randomized clinical study on 80 patients in critical care units, powdered henna leaves suspended in distilled water exerted preventive effects on decubitus ulcers (Hekmatpou *et al.*, 2018b).

Avicenna introduced henna in the *Canon* as an important plant for various medial indications such as inflammations, wounds, burns, bone fractures, and AF. In this regard, the cerate made from henna leaves and damask rose oil is recommended for topical use on AF (Avicenna, 1038).

***Olea europaea* L.**

Phenolic compounds including vanillic, *p*-coumaric, syringic, homovanillic and caffeic acids, as well as kaempferol, oleuropein glycoside, tyrosol, hydroxytyrosol, and an unsaponifiable fraction of the oil have shown *in vitro* anti-inflammatory effects via various mechanisms of action such as increasing TNF-α and reducing COX-2 and NO production (Miles *et al.*, 2005; Rosignoli *et al.*, 2013; Cardeno *et al.*, 2014; Takeda *et al.*, 2014; Yonezawa *et al.*, 2018). In addition, the beneficial effect of topical olive oil (fixed oil) on wound healing has been observed in several clinical studies. Topical application of olive oil for 10 days showed positive wound healing effects in a randomized, controlled clinical trial on 90 patients (Amani *et al.*, 2015). In a triple-blind, parallel, multicenter, randomized clinical trial on 831 patients, olive oil reduced pressure ulcers after 112 days of treatment (Lupiañez-Perez *et al.*, 2015, 2017). Also, according to a randomized controlled clinical trial on 173 patients, the indication of olive oil cream for 28 days decreased dermatitis successfully (Kiechl-Kohlendorfer *et al.*, 2008).

The *Canon* mentioned a mixture of olive oil and water with an astringent ingredient as a therapeutic sitz bath for AF. Moreover, the enema made from olive oil and verjuice is recommended for curing internal ulcers of the anus (Avicenna, 1038).

***Prunus armeniaca* L.**

In vitro studies of apricot kernels on mouse BV2 microglial cells have demonstrated anti-inflammatory properties (Chang *et al.*, 2005). Furthermore, apricot kernel oil (fixed oil) has shown anti-inflammatory activities in ethanol-induced gastric ulcers in rats (Karaboğa *et al.*, 2018). Kernel extract and oil showed ameliorating effects on colitis in trinitrobenzene sulfonic acid-induced colitis in rats (Minaiyan *et al.*, 2014). Besides, amygdalin isolated from apricot kernels reduced inflammatory pain in an *in vivo* study on rats via histological examination and RT-PCR (Hwang *et al.*, 2008).

According to Avicenna's viewpoint in the *Canon*, topical indication of kernel oil, either singly or in combination with *C. mukul* gum resin, has been considered useful for medical treatment of AF (Avicenna, 1038).

***Quercus infectoria* G. Olivier**

As an active ingredient in gall nuts and oak bark, gallic acid has shown potent pro-inflammatory activities *in vitro*. It increases caspase-3 activity, regulates Bcl2-associated-X protein (Bax), -cell lymphoma 2 (Bcl-2), p53 protein, and phosphorylated Akt (pAkt) production, and decreases IL-6, IL-1β, chemokine ligand (CCL)-2/monocyte chemoattractant protein (MCP)-1, and CCL-7/MCP-3 in fibroblast-like synoviocytes (Yoon *et al.*, 2013). Also, the gall presented inhibitory effects on NF-κB, Set7, and p65 activity in bone marrow-derived macrophages (Chokpaisarn *et al.*, 2017b). In addition, the alcoholic extract of gall has shown scavenging effects on NO, and ameliorating effects on LPS-stimulated prostaglandin E_2 (PGE_2) and PMA-stimulated superoxide production in a rat peritoneal macrophage model, *in vitro* (Kaur *et al.*, 2004). On the other hand, the alcoholic extract of gall presented anti-inflammatory effects on PMA-induced ear edema in mice (Kaur *et al.*, 2004). The crude aqueous extract of gall, as well as its ethanol extract, showed wound healing properties in rats (Jalalpure *et al.*, 2002; Choudhary, 2007). The aqueous extract of the internal layer of fruit containing polyphenolic compounds and tannins decreases NO and MDA, while increasing pH in rat gastric ulcer induced by stress (Jafari Barmak *et al.*, 2018). The aqueous extract of the external layer of oak fruit ameliorates the biochemical alteration; it increases glutathione (GSH) and decreases malondialdehyde (MDA) and NO in acetic acid-induced ulcerative colitis models in rats (Jafari *et al.*, 2015). Besides, in a multicenter randomized controlled trial on 51 diabetic patients with ulcers, a topical solution made from *Quercus infectoria* galls showed accelerated wound healing effects (Chokpaisarn *et al.*, 2020).

In Avicenna's *Canon*, powdered nut galls suspended in a mixture of wine and water (hydroalcoholic solution) is considered a useful topical suspension for healing of AF. Also, decoction of galls in concentrated grape wine is recommended for inflammation-induced fissures in the anus (Avicenna, 1038).

Rosa × *damascena* Herrm.

According to an *in vitro* study, a polyphenol fraction obtained from damask rose flowers was found to modify the inflammatory target gene expression via various mechanisms of action, including decreased IL-6, IL-1β, IL-8, RANTES (regulated on activation, normal T cell expressed and secreted), and MCP-1 (Wedler *et al.*, 2016). A topical lotion prepared from a hydroalcoholic extract of damask rose enhanced the healing process of full-thickness wounds in rat models (Mansouri *et al.*, 2016). A herbal cerate containing damask rose oil and wallflower oil showed potential healing effects on patients suffering from acute AF (Mosleh *et al.*, 2020). Furthermore, damask rose oil has been commercialized for the treatment of hemorrhoids in Iran (Afsari Sardari *et al.*, 2019).

In the *Canon* Avicenna has frequently prescribed damask rose oil as a suitable oily vehicle in various AF formulations. This popular traditional essential oil has significant analgesic and wound healing properties according to TPM (Avicenna, 1038; Afsari Sardari *et al.*, 2019).

Taraxacum campylodes G.E. Haglund

The leaf extract of dandelion containing luteolin and chicoric acid reduced inflammation in LPS-stimulated RAW 264.7 cells (Park *et al.*, 2010). The crude extract of dandelion has shown decreasing effects on NO and TNF-α in RAW 264.7 cells (Park *et al.*, 2014). In addition, the methanol extract of the herb with active constituents of protocatechuic acid, chlorogenic acid, caffeic acid, *p*-coumaric acid, and ferulic acid has led to anti-inflammatory effects in LPS-stimulated human umbilical vein endothelial cells (Jeon *et al.*, 2017). Also, the leaf and root containing taraxinic acid β-D-glucopyranosyl ester have induced anti-inflammatory effects in Huh7 cells (Esatbeyoglu *et al.*, 2017). On the other hand, the aqueous extract of dandelion root has anticolitis and anti-inflammatory effects in DSS-induced ulcerative colitis rats (Ding and Wen, 2018). Besides, the alcoholic extract of aerial parts, as well as taraxasterol, have shown *in vivo* anti-inflammatory effects via various mechanisms of action, including reduction of TNF-α, IL-6, and COX-2 expression

(Jeon *et al.*, 2008; Wang *et al.*, 2017; Razak *et al.*, 2020).

In the Canon, a cerate formulation containing dandelion in the vehicle of traditional damask rose oil is considered to be useful in the healing of AF (Avicenna, 1038).

Vitex agnus-castus L.

Aqueous alcoholic extracts obtained from fruits and methanolic extracts from the aerial parts of *Vitex agnus-castus* showed *in vitro* anti-inflammatory activities (Choudhary *et al.*, 2009; Ahmad *et al.*, 2010; Ibrahim *et al.*, 2021). Additionally, the essential oil of seeds and the hydroalcoholic extract of fruits have shown anti-inflammatory effects in previous *in vivo* studies (Ramezani *et al.*, 2010; Aboul-Enein *et al.*, 2017).

According to Avicenna's point of view explained in the *Canon*, all parts of *V. agnus-castus* herb, and especially the seeds, show healing effects on AF (Avicenna, 1038).

References

Abdallah, M., Marzocco, S., Adesso, S., Zarrouk, M. and Guerfel, M. (2018) Olive oil polyphenols extracts inhibit inflammatory markers in J774A.1 murine macrophages and scavenge free radicals. *Acta Histochemica* 120, 1–10.

Aboul-Enein, H.Y., Habbab, A., Sekkoum, K., Belboukhari, N. and Abdelkrim, C. (2017) Analgesic and anti-inflammatory effects of essential oils of *Vitex agnuscastus* L. from South-West of Algeria. *Current Bioactive Compounds* 13, 165–169.

Afsari Sardari, F., Mosleh, G., Azadi, A., Mohagheghzadeh, A. and Badr, P. (2019) Traditional and recent evidence on five phytopharmaceuticals from *Rosa damascena* Herrm. *Research Journal of Pharmacognosy* 6, 77–84.

Ahmad, B., Azam, S., Bashir, S., Khan, I., Adhikari, A. *et al.* (2010) Anti-inflammatory and enzyme inhibitory activities of a crude extract and a pterocarpan isolated from the aerial parts of *Vitex agnus-castus*. *Biotechnology Journal* 5, 1207–1215.

Ahn, S. and Youn, H.S. (2008) Guggulsterone suppresses the activation of NF-κB and expression of COX-2 induced by toll-like receptor 2, 3, and 4 agonists. *Food Science and Biotechnology* 17, 1294–1298.

Ali, K., Marif, H., Raoof, H., Hassan, S. and Ramzi, D. (2021) Assessment the effects of mastic gum resin, *Lawsonia inermis* and *Quercus brantii* on cutaneous wound healing in BALB/c mice. *Indian Journal of Forensic Medicine and Toxicology* 15, 2012–2023.

Almazari, I., Park, J., Park, S., Suh, J., Na, H. *et al.* (2012) Guggulsterone induces heme oxygenase-1 expression through activation of Nrf2 in human mammary epithelial cells: PTEN as a putative target. *Carcinogenesis* 33, 368–376.

Alwashli, A., Sobarry, M.A. and Alaoui, Y. (2012) Anti-inflammatory and analgesic effects of ethanol extract of *Dracaena cinnabari* balf, as endemic plant in Yemen. *International Journal of Pharma and Bio Sciences* 3, 96–100.

Amani, R., Kariman, N., Mojab, F., Majd, H. and Majidi, S. (2015) Comparison of the effects of cold compress with gel packs and topical olive oil on episiotomy wound healing. *Journal of Babol University of Medical Sciences* 17, 14–19.

Aparicio-Soto, M., Redhu, D., Sánchez-Hidalgo, M., Fernández-Bolaños, J.G., Alarcón-de-la-Lastra, C. *et al.* (2019) Olive-oil-derived polyphenols effectively attenuate inflammatory responses of human keratinocytes by interfering with the NF-κB pathway. *Molecular Nutrition and Food Research* 63, e1900019.

Avicenna, H. (1038) *The Canon of Medicine* (*Qanoon-Fil-Tib*), researched by A. Foad Basha and F. Jab-Allah. Misr University for Science and Technology, Giza, Egypt.

Babu, S. and Noor, A. (2019) *Aloe barbadensis* Miller peptide/polypeptide fraction alleviates inflammation through inhibition of proinflammatory cytokines and mediators *in vitro* and in rats with Freund's adjuvant-induced hind paw edema. *Asian Pacific Journal of Tropical Biomedicine* 9, 524–530.

Bonab, F.S. and Farahpour, M.R. (2017) Topical co-administration of *Pistacia atlantica* hull and *Quercus infectoria* gall hydroethanolic extract improves wound-healing process. *Comparative Clinical Pathology* 26, 885–892.

Bouhlali, E.T., Sellam, K., Bammou, M., Alem, C. and Filali-Zehzouti, Y. (2016) *In vitro* antioxidant and anti-inflammatory properties of selected Moroccan medicinal plants. *Journal of Applied Pharmaceutical Science* 6, 156–162.

Cardeno, A., Sanchez-Hidalgo, M., Aparicio-Soto, M. and Alarcón-de-la-Lastra, C. (2014) Unsaponifiable fraction from extra virgin olive oil inhibits the inflammatory response in LPS-activated murine macrophages. *Food Chemistry* 147, 117–123.
Cárdeno, A., Sánchez-Hidalgo, M., Aparicio-Soto, M., Sánchez-Fidalgo, S. and Alarcón-de-la-Lastra, C. (2014) Extra virgin olive oil polyphenolic extracts downregulate inflammatory responses in LPS-activated murine peritoneal macrophages suppressing NFκB and MAPK signalling pathways. *Food and Function* 5, 1270–1277.
Chang, H.K., Yang, H.Y., Lee, T.H., Shin, M.C., Lee, M.H. *et al.* (2005) Armeniacae semen extract suppresses lipopolysaccharide-induced expressions of cycloosygenase-2 and inducible nitric oxide synthase in mouse BV2 microglial cells. *Biological and Pharmaceutical Bulletin* 28, 449–454.
Choi, S.W., Son, B.W., Son, Y.S., Park, Y.I., Lee, S.K. *et al.* (2001) The wound-healing effect of a glycoprotein fraction isolated from *Aloe vera*. *British Journal of Dermatology* 145, 535–545.
Chokpaisarn, J., Chusri, S., Amnuaikit, T., Udomuksorn, W. and Voravuthikunchai, S.P. (2017a) Potential wound healing activity of *Quercus infectoria* formulation in diabetic rats. *Peer Journal* 5, e3608. DOI: 10.7717/peerj.3608.
Chokpaisarn, J., Urao, N., Voravuthikunchai, S.P. and Koh, T.J. (2017b) *Quercus infectoria* inhibits Set7/NF-κB inflammatory pathway in macrophages exposed to a diabetic environment. *Cytokine* 94, 29–36.
Chokpaisarn, J., Chusri, S. and Voravuthikunchai, S.P. (2020) Clinical randomized trial of topical *Quercus infectoria* ethanolic extract for the treatment of chronic diabetic ulcers. *Journal of Herbal Medicine* 21, 100301.
Choudhary, G.P. (2007) Evaluation of wound healing activity of galls of *Quercus infectoria*. *Asian Journal of Chemistry* 19, 3271–3273.
Choudhary, M.I., Azizuddin, Jalil, S., Nawaz, S.A., Khan, K.M., Tareen, R.B. *et al.* (2009) Antiinflammatory and lipoxygenase inhibitory compounds from *Vitex agnus-castus*. *Phytotherapy Research* 23, 1336–1339.
Daemi, A., Farahpour, M.R., Oryan, A., Karimzadeh, S. and Tajer, E. (2019) Topical administration of hydroethanolic extract of *Lawsonia inermis* (henna) accelerates excisional wound healing process by reducing tissue inflammation and amplifying glucose uptake. *Kaohsiung Journal of Medical Sciences* 35, 24–32.
Derakhshan, A.R. (2017) Natural treatments for fissure in ano used by traditional Persian scholars, Razi (Rhazes) and Ibn Sina (Avicenna). *Journal of Evidence-Based Complementary and Alternative Medicine* 22, 324–333.
Desai, S.K., Borkar, V.S., Bachchewar, R.S. and Patil, C.N. (2014) Wound healing and antimicrobial activity of *Aloe vera* gel. *International Journal of Pharmaceutical Sciences Review and Research* 27, 256–260.
Díaz-Valenzuela, A., García-Fernández, F.P., Carmona Fernández, P., Valle Cañete, M.J. and Pancorbo-Hidalgo, P.L.(2019) Effectiveness and safety of olive oil preparation for topical use in pressure ulcer prevention: multicentre, controlled, randomised, and double-blinded clinical trial. *International Wound Journal* 16, 1314–1322.
Ding, A. and Wen, X. (2018) Dandelion root extract protects NCM460 colonic cells and relieves experimental mouse colitis. *Journal of Natural Medicines* 72, 857–866.
Elaasser, M.M., Morsi, M.K., Galal, S.M., Abd El-Rahman, M.K. and Katry, M.A. (2020) Antioxidant, anti-inflammatory and cytotoxic activities of the unsaponifiable fraction of extra virgin olive oil. *Grasas Aceites* 71, e386.
Esatbeyoglu, T., Obermair, B., Dorn, T., Siems, K., Rimbach, G. *et al.* (2017) Sesquiterpene lactone composition and cellular Nrf2 induction of *Taraxacum officinale* leaves and roots and taraxinic acid β-D-glucopyranosyl ester. *Journal of Medicinal Food* 20, 71–78.
Eshghi, F., Hosseinimehr, S.J., Rahmani, N., Khademloo, M., Norozi, M.S. *et al.* (2010) Effects of *Aloe vera* cream on posthemorrhoidectomy pain and wound healing: results of a randomized, blind, placebo-control study. *Journal of Alternative and Complementary Medicine* 16, 647–650.
Fayazzadeh, E., Rahimpour, S., Ahmadi, S.M., Farzampour, S., Sotoudeh Anvari, M. *et al.* (2014) Acceleration of skin wound healing with tragacanth (*Astragalus*) preparation: an experimental pilot study in rats. *Acta Medica Iranica* 52, 3–8.
Fernández-Prior, Á., Bermúdez-Oria, A., Millán-Linares, M.D.C., Fernández-Bolaños, J., Espejo-Calvo, J.A. *et al.* (2021) Anti-inflammatory and antioxidant activity of hydroxytyrosol and 3,4-dihydroxyphenyglycol purified from table olive effluents. *Foods* 10, 227.
Fox, L.T., Mazumder, A., Dwivedi, A., Gerber, M., du Plessis, J. *et al.* (2017) *In vitro* wound healing and cytotoxic activity of the gel and whole-leaf materials from selected aloe species. *Journal of Ethnopharmacology* 200, 1–7.

Gupta, D., Verma, N., Das, H. and Gupta, R. (2014) Evaluation of anti-inflammatory activity of *Dracaena cinnabari* Balf.f. resin. *Indian Journal of Natural Products and Resources* 5, 215–222.

Hekmatpou, D., Ahmadian, F., Eghbali, M. and Farsaei, S. (2018a) Henna (*Lawsonia inermis*) as an inexpensive method to prevent decubitus ulcers in critical care units: a randomized clinical trial. *Journal of Evidence-Based Integrative Medicine* 23, 2515690X18772807.

Hekmatpou, D., Mehrabi, F., Rahzani, K. and Aminiyan, A. (2018b) The effect of *Aloe vera* gel on prevention of pressure ulcers in patients hospitalized in the orthopedic wards: a randomized triple-blind clinical trial. *BMC Complementary and Alternative Medicine* 18, 264.

Hormozi, M., Assaei, R. and Boroujeni, M.B. (2017) The effect of *Aloe vera* on the expression of wound healing factors (TGFβ1 and bFGF) in mouse embryonic fibroblast cell: *in vitro* study. *Biomedicine and Pharmacotherapy* 88, 610–616.

Hwang, H.J., Kim, P., Kim, C.J., Lee, H.J., Shim, I. *et al.* (2008) Antinociceptive effect of amygdalin isolated from *Prunus armeniaca* on formalin-induced pain in rats. *Biological and Pharmaceutical Bulletin* 31, 1559–1564.

Ibrahim, F.M., Ibrahim, A.Y., El-Newary, S.A., Hendawy, S.F. and Mahomoodally, M.F. (2021) *Vitex agnus-castus* L. (chasteberry) extracts shows *in vitro* and *in vivo* anti-inflammatory and anti-tumor propensities via reduction of cyclooxygenase-2 activity and oxidative stress complications. *South African Journal of Botany* 143, 363–373.

Jafari, K., Mirzaee, A. and Ghavamizadeh, M. (2015) Protective effect of jaft extract on acetic acid induced ulcerativecolitis in rats. *Research Journal of Pharmaceutical, Biological and Chemical Sciences* 6, 1211–1216.

Jafari Barmak, M., Dehghan Menshadi, A., Mahmoudi, R. and Bardania, H. (2018) Evaluation of the effect of the internal layer of oak fruit (jaft) extract on the prevention of gastric ulcers caused by stress in male rats. *Journal of Medicine and Life Science* 11, 225–230.

Jalalpure S.S., Patil, M.B. and Alagawadi, K.R. (2002) Wound healing activity of the galls of *Quercus infectoria* Olivier. *Journal of Natural Remedies* 2, 54–58.

Jeon, D., Kim, S.J. and Kim, H.S. (2017) Anti-inflammatory evaluation of the methanolic extract of *Taraxacum officinale* in LPS-stimulated human umbilical vein endothelial cells. *BMC Complementary and Alternative Medicine* 17, 508.

Jeon, H.J., Kang, H.J., Jung, H.J., Kang, Y.S., Lim, C.J. *et al.* (2008) Anti-inflammatory activity of *Taraxacum officinale*. *Journal of Ethnopharmacology* 115, 82–88.

Jettanacheawchankit, S., Sasithanasate, S., Sangvanich, P., Banlunara, W. and Thunyakitpisal, P. (2009) Acemannan stimulates gingival fibroblast proliferation; expressions of keratinocyte growth factor-1, vascular endothelial growth factor, and type I collagen; and wound healing. *Journal of Pharmacological Sciences* 109, 525–531.

Jridi, M., Sellimi, S., BenLassoued, K., Beltaief, S., Souissi, N. *et al.* (2017) Wound healing activity of cuttlefish gelatin gels and films enriched by henna (*Lawsonia inermis*) extract. *Colloids and Surfaces A: Physicochemical and Engineering Aspects* 512, 71–79.

Karaboğa, I., Ovalı, M.A., Yılmaz, A. and Alpaslan, M. (2018) Gastroprotective effect of apricot kernel oil in ethanol-induced gastric mucosal injury in rats. *Biotechnic and Histochemistry* 93, 601–607.

Kaur, G., Hamid, H., Ali, A., Alam, M.S. and Athar, M. (2004) Antiinflammatory evaluation of alcoholic extract of galls of *Quercus infectoria*. *Journal of Ethnopharmacology* 90, 285–292.

Kiechl-Kohlendorfer, U., Berger, C. and Inzinger, R. (2008) The effect of daily treatment with an olive oil/lanolin emollient on skin integrity in preterm infants: a randomized controlled trial. *Pediatric Dermatology* 25, 174–178.

Kimura, I., Yoshikawa, M., Kobayashi, S., Sugihara, Y., Suzuki, M. *et al.* (2001) New triterpenes, myrrhanol A and myrrhanone A, from guggul-gum resins, and their potent anti-inflammatory effect on adjuvant-induced air-pouch granuloma of mice. *Bioorganic and Medicinal Chemistry Letters* 11, 985–989.

Koh, Y.J., Cha, D.S., Ko, J.S., Park, H.J. and Choi, H.D. (2010) Anti-inflammatory effect of *Taraxacum officinale* leaves on lipopolysaccharide-induced inflammatory responses in RAW 264.7 cells. *Journal of Medicinal Food* 13, 870–878.

Lai, J., Peng, L., Chen, H., Li, R., Liu, L. *et al.* (2020) Amino acid sequence identification and anti-inflammatory potency evaluation of dandelion oligopeptides in LPS-induced RAW264.7 macrophage cells. *Journal of Molecular Structure* 1216, 128280.

Li, L.J., Gao, S.Q., Peng, L.H., Wang, X.R., Zhang, Y. *et al.* (2017) Evaluation of efficacy of aloin in treating acute trauma *in vitro* and *in vivo*. *Biomedicine and Pharmacotherapy* 88, 1211–1219.

Liou, J.R., El-Shazly, M., Du, Y.C., Tseng, C.N., Hwang, T.L. *et al.* (2013) 1,5-Diphenylpent-3-en-1-ynes and methyl naphthalene carboxylates from *Lawsonia inermis* and their anti-inflammatory activity. *Phytochemistry* 88, 67–73.

Lund, J.N. (2006) Nitric oxide deficiency in the internal anal sphincter of patients with chronic anal fissure. *International Journal of Colorectal Disease* 21, 673–675.

Lupiañez-Perez, I., Uttumchandani, S.K., Morilla-Herrera, J.C., Martin-Santos, F.J., Fernandez-Gallego, M.C. *et al.* (2015) Topical olive oil is not inferior to hyperoxygenated fatty aids to prevent pressure ulcers in high-risk immobilised patients in home care. Results of a multicentre randomised triple-blind controlled non-inferiority trial. *PLoS One* 10, e0122238.

Lupiañez-Pérez, I., Morilla-Herrera, J.C., Kaknani-Uttumchandani, S., Lupiañez-Perez, Y., Cuevas-Fernandez-Gallego, M. *et al.* (2017) A cost minimization analysis of olive oil vs. hyperoxygenated fatty acid treatment for the prevention of pressure ulcers in primary healthcare: a randomized controlled trial. *Wound Repair and Regeneration* 25, 846–851.

Madalinski, M. and Kalinowski, L. (2009) Novel options for the pharmacological treatment of chronic anal fissure – role of botulin toxin. *Current Clinal Pharmacology* 4, 47–52.

Madasu, C., Gudem S., Sistla R. and Uppuluri V.M. (2017) Synthesis and anti-inflammatory activity of some novel pyrimidine hybrids of myrrhanone A, a bicyclic triterpene of *Commiphora mukul* gum resin. *Monatshefte für Chemie* 148, 2183–2193.

Malik Hussein, S. (2019) Study of the effects of the *Aloe vera* gel on soft tissue injury regeneration: an experimental study on rabbits. *Plant Archives* 19, 490–495.

Manjula, N., Gayathri, B., Vinaykumar, K.S., Shankernarayanan, N.P., Vishwakarma, R.A. *et al.* (2006) Inhibition of MAP kinases by crude extract and pure compound isolated from *Commiphora mukul* leads to down regulation of TNF-α, IL-1β and IL-2. *International Immunopharmacology* 6, 122–132.

Mansouri, E., Hardani, A., Afzalzadeh, M.R., AmirZargar, A. and Meamar, Z. (2016) Combined effects of retinoic acid and hydro-alcoholic extract of *Rosa damascena* Mill on wound in diabetic rats. *Iranian Journal of Pharmaceutical Research* 15, 583–589.

Manuja, A., Rathore, N., Choudhary, S. and Kumar, B. (2021) Phytochemical screening, cytotoxicity and anti-inflammatory activities of the leaf extracts from *Lawsonia inermis* of Indian origin to explore its potential for medicinal uses. *Medicinal Chemistry* 17, 576–586.

Mencarelli, A., Renga, B., Palladino, G., Distrutti, E. and Fiorucci, S. (2009) The plant sterol guggulsterone attenuates inflammation and immune dysfunction in murine models of inflammatory bowel disease. *Biochemical Pharmacology* 78, 1214–1223.

Miles, E.A., Zoubouli, P. and Calder, P.C. (2005) Differential anti-inflammatory effects of phenolic compounds from extra virgin olive oil identified in human whole blood cultures. *Nutrition* 21, 389–394.

Minaiyan, M., Ghannadi, A., Asadi, M., Etemad, M. and Mahzouni, P. (2014) Anti-inflammatory effect of *Prunus armeniaca* L. (apricot) extracts ameliorates TNBS-induced ulcerative colitis in rats. *Research in Pharmaceutical Sciences* 9, 225–231.

Miraj, S., Pourafzali, S., Ahmadabadi, Z.V. and Rafiei, Z. (2020) Effect of olive oil in preventing the development of pressure ulcer grade one in intensive care unit patients. *International Journal of Preventive Medicine* 11, 23.

Moghbel, A.A.H., Hemmati, A.A., Agheli, H., Amraee, K.H. and Rashidi, I. (2005) The effect of tragacanth mucilage on the healing of full-thickness wound in rabbit. *Archives of Iranian Medicine* 8, 257–262.

Moon, E.J., Lee, Y.M., Lee, O.H., Lee, M.J., Lee, S.K. *et al.* (1999) A novel angiogenic factor derived from *Aloe vera* gel: β-sitosterol, a plant sterol. *Angiogenesis* 3, 117–123.

Moriyama, M., Moriyama, H., Uda, J., Kubo, H., Nakajima, Y. *et al.* (2016) Beneficial effects of the genus *Aloe* on wound healing, cell proliferation, and differentiation of epidermal keratinocytes. *PLoS One* 11, e0164799.

Mosleh, G., Azadi, A., Khademian, S., Heidari, R. and Mohagheghzadeh, A. (2021a) Anti-inflammatory activity and quality control of *Erysimum cheiri* (L.) Crantz. *BioMed Research International 2021*, 5526644.

Mosleh, G., Azadi, A. and Mohagheghzadeh, A. (2021b) Characterization and chromatographic fingerprint analysis of traditional wallflower oil. *Research Journal of Pharmacognosy* 8, 31–36.

Mosleh, G., Nimrouzi, M., Badr, P., Abolhasanzadeh, Z., Azadi, A. *et al.* (2019) The approach of traditional Persian medicine to treatment of anal fissure. *Traditional and Integrative Medicine* 4, 78–83.

Mosleh, G., Hosseini, S.V., Azadi, A., Bahrami, F., Ghahramani, L. *et al.* (2020) *Erysimum cheiri* and *Rosa × damascena* cerate vs. diltiazem 2% gel in the treatment of acute anal fissure: a randomized, controlled clinical trial. *European Journal of Integrative Medicine* 40, 101230.

Nasiri, M., Fayazi, S., Jahani, S., Yazdanpanah, L. and Haghighizadeh, M.H. (2015) The effect of topical olive oil on the healing of foot ulcer in patients with type 2 diabetes: a double-blind randomized clinical trial study in Iran. *Journal of Diabetes and Metabolic Disorders* 14, 38.

Nayak, B.S., Isitor, G., Davis, E.M. and Pillai, G.K. (2007) The evidence based wound healing activity of *Lawsonia inermis* Linn. *Phytotherapy Research* 21, 827–831.
Oryan, A., Naeini, A.T., Nikahval, B. and Gorjlan, E. (2010) Effect of aqueous extract of *Aloe vera* on experimental cutaneous wound healing in rat. *Veterinarski Arhiv* 80, 509–522.
Oryan, A., Mohammadalipour, A., Moshiri, A. and Tabandeh, M.R. (2016) Topical application of *Aloe vera* accelerated wound healing, modeling, and remodeling: an experimental study. *Annals of Plastic Surgery* 77, 37–46.
Park, C.M., Park, J.Y. and Song, Y.S. (2010) Luteolin and chicoric acid, two major constituents of dandelion leaf, inhibit nitric oxide and lipid peroxide formation in lipopolysaccharide-stimulated raw 264.7 cells. *Journal of Food Sciences and Nutrition* 15, 92–97.
Park, C.M., Park, J.Y., Noh, K.H., Shin, J.H. and Song, Y.S. (2011) *Taraxacum officinale* Weber extracts inhibit LPS-induced oxidative stress and nitric oxide production via the NF-κB modulation in RAW 264.7 cells. *Journal of Ethnopharmacology* 133, 834–842.
Park, C.M., Cho, C.W. and Song, Y.S. (2014) TOP 1 and 2, polysaccharides from *Taraxacum officinale*, inhibit NFκB-mediated inflammation and accelerate Nrf2-induced antioxidative potential through the modulation of PI3K–Akt signaling pathway in RAW 264.7 cells. *Food and Chemical Toxicology* 66, 56–64.
Poursadra, E., Anvari-Tafti, M., Dehghani A., Eghbali-Babadi, M. and Rafiei, Z. (2019) Comparing the effect of henna oil and olive oil on pressure ulcer grade one in intensive care units patients. *Advanced Biomedical Research* 8, 68.
Rafiei, Z., Mazaheri, M., Eghbali-Babadi, M. and Yazdannik, A. (2019) The effect of henna (*Lawsonia Inermis*) on preventing the development of pressure ulcer grade one in intensive care unit patients. *International Journal of Preventive Medicine* 10, 26.
Rahmani, N., Khademloo, M., Vosoughi, K. and Assadpour, S. (2014) Effects of *Aloe vera* cream on chronic anal fissure pain, wound healing and hemorrhaging upon defection: a prospective double blind clinical trial. *European Review for Medical and Pharmacological Sciences* 18, 1078–1084.
Rajput S.S., Soni, K.K. and Saxena, R.C. (2009) Pharmacology and phytochemistry of saponin isolated from *Aloe vera* for wound healing activity. *Asian Journal of Chemistry* 21, 1029–1032.
Ramezani, M., Amin, G. and Jalili, E. (2010) Antinociceptive and anti-inflammatory effects of hydroalcohol extract of *Vitex agnus-castus* fruit. *World Academy of Science, Engineering and Technology* 40, 624–626.
Razak, S., Afsar, T., Al-Disi, D., Almajwal, A., Arshad, M. *et al.* (2020) GCMS fingerprinting, *in vitro* pharmacological activities, and *in vivo* anti-inflammatory and hepatoprotective effect of selected edible herbs from Kashmir valley. *Journal of King Saud University –Science* 32, 2868–2879.
Rekik, D.M., Ben Khedir, S., Daoud, A., Ksouda Moalla, K., Rebai, T. *et al.* (2019) Wound healing effect of *Lawsonia inermis*. *Skin Pharmacology and Physiology* 32, 295–306.
Rezaie, A., Ebadi, A., Nazeri, M., Rezaie, S. and Elmi, F. (2012) Effect of *Aloe vera* on healing of the experimental skin wounds on rats and its comparison with zinc oxide: a geometry and histopathologic study. *Journal of Animal and Veterinary Advances* 11, 2445–2452.
Rokaya, M.B., Uprety, Y., Poudel, R.C., Timsina, B., Münzbergová, Z. *et al.* (2014) Traditional uses of medicinal plants in gastrointestinal disorders in Nepal. *Journal of Ethnopharmacology* 158 Pt A, 221–229.
Rosignoli, P., Fuccelli, R., Fabiani, R., Servili, M. and Morozzi, G.(2013) Effect of olive oil phenols on the production of inflammatory mediators in freshly isolated human monocytes. *Journal of Nutritional Biochemistry* 24, 1513–1519.
Salih, A.M., Kakamad, F.H., Salih, R.Q., Hussein, D.A., Hassan, H.A. *et al.* (2017) Effect of *Lawsonia inermis* (henna) on wound healing in Sprague–Dawley rats: a pilot study. *Wound Medicine* 18, 41–42.
Sarkar, D., Dutta, A., Das, M., Sarkar, K., Mandal, C. *et al.* (2005) Effect of *Aloe vera* on nitric oxide production by macrophages during inflammation. *Indian Journal of Pharmacology* 37, 371–375.
Shakibamehr, J., Rad, M., Akrami, R. and Rad, M. (2019) Effectiveness of tragacanth gel cushions in prevention of pressure ulcer in traumatic patients: a randomized controlled trial. *Journal of Caring Sciences* 8, 45–49.
Sosa, S., Tubaro, A., DellaLoggia, R. and Bombardelli, E. (1993) Anti-inflammatory activity of *Commiphora mukul* extracts. *Pharmacological Research* 27, 89–90.
Tabandeh, M.R., Oryan, A. and Mohammadalipour, A. (2014) Polysaccharides of *Aloe vera* induce MMP-3 and TIMP-2 gene expression during the skin wound repair of rat. *International Journal of Biological Macromolecules* 65, 424–430.

Takeda, Y., Bui, V.N., Iwasaki, K., Kobayashi, T., Ogawa, H. *et al.* (2014) Influence of olive-derived hydroxytyrosol on the toll-like receptor 4-dependent inflammatory response of mouse peritoneal macrophages. *Biochemical and Biophysical Research Communications* 446, 1225–1230.

Takzaree, N., Hassanzadeh, G., Rouini, M.R., Keshtkar, A., Manayi, A. *et al.* (2015) Histological study of wound repair with topical *Aloe vera* gel in rat. *Tehran University Medical Journal* 73, 660–667.

Takzaree, N., Hadjiakhondi, A., Hassanzadeh, G., Rouini, M.R., Manayi, A. *et al.* (2016) Transforming growth factor-β (TGF-β) activation in cutaneous wounds after topical application of *Aloe vera* gel. *Canadian Journal of Physiology and Pharmacology* 94, 1285–1290.

van Meegdenburg, M.M., Trzpis, M., Heineman, E. and Broens, P.M. (2016) Increased anal basal pressure in chronic anal fissures may be caused by overreaction of the anal-external sphincter continence reflex. *Medical Hypotheses* 94, 25–29.

Van Outryve, M. (2006) Physiopathology of the anal fissure. *Acta chirurgica Belgica* 106, 517–518.

Wahedi, H.M., Jeong, M., Chae, J.K., Do, S.G., Yoon, H. *et al.* (2017) Aloesin from *Aloe vera* accelerates skin wound healing by modulating MAPK/Rho and Smad signaling pathways *in vitro* and *in vivo*. *Phytomedicine* 28, 19–26.

Wang, Y., Li, G.H., Liu, X.Y., Xu, L., Wang, S.S. *et al.* (2017) *In vivo* anti-inflammatory effects of taraxasterol against animal models. *African Journal of Traditional, Complementary and Alternative Medicines* 14, 43–51.

Wedler, J., Weston, A., Rausenberger, J. and Butterweck, V. (2016) *In vitro* modulation of inflammatory target gene expression by a polyphenol-enriched fraction of rose oil distillation waste water. *Fitoterapia* 114, 56–62.

Wiem, A., Smail, A., Wissem, M., Faleiro, M. and Miguel, M. (2014) Antioxidant, anti-inflammatory and anti-acetylcholinesterase activities of leaf, flower and seed aqueous extracts of *Lawsonia inermis* from Tunisia. *International Journal of Pharmacy and Pharmaceutical Sciences* 6, 445–452.

Xue, Y., Zhanga, S., Dub, M. and Zhu, M.J. (2017) Dandelion extract suppresses reactive oxidative species and inflammasome in intestinal epithelial cells. *Journal of Functional Foods* 29, 10–18.

Yadav, K.C.H., Kumar, J.R., Basha, S.I., Deshmukh, G.R., Gujjula, R. *et al.* (2012) Wound healing activity of topical application of *Aloe vera* gel in experimental animal models. *International Journal of Pharma and Bio Sciences* 3, 63–72.

Yagi, A., Kabash, A., Okamura, N., Haraguchi, H., Moustafa, S.M. *et al.* (2002) Antioxidant, free radical scavenging and anti-inflammatory effects of aloesin derivatives in *Aloe vera*. *Planta Medica* 68, 957–960.

Yang, F., Ye, X.J., Chen, M.Y., Li, H.C., Wang, Y.F. *et al.* (2021) Inhibition of NLRP3 inflammasome activation and pyroptosis in macrophages by taraxasterol is associated with its regulation on mTOR signaling. *Frontiers in Immunology* 12, 632606.

Yonezawa, Y., Miyashita, T., Nejishima, H., Takeda, Y., Imai, K. *et al.* (2018) Anti-inflammatory effects of olive-derived hydroxytyrosol on lipopolysaccharide-induced inflammation in RAW264.7 cells. *European Journal of Clinical Nutrition* 80, 1801–1807.

Yoon, C.H., Chung, S.J., Lee, S.W., Park, Y.B., Lee, S.K. *et al.* (2013) Gallic acid, a natural polyphenolic acid, induces apoptosis and inhibits proinflammatory gene expressions in rheumatoid arthritis fibroblast-like synoviocytes. *Joint Bone Spine* 80, 274–279.

Yousif, M.F., Haider, M. and Sleem, A.A. (2011) Formulation and evaluation of two anti-inflammatory herbal gels. *Journal of Biologically Active Products from Nature* 1, 200–209.

Zhang, L. and Tizard, I.R. (1996) Activation of a mouse macrophage cell line by acemannan: the major carbohydrate fraction from *Aloe vera* gel. *Immunopharmacology* 35, 119–128.

Zhang, L.J., Huang, X.J., Shi, X.D., Chen, H.H., Cui, S.W. *et al.* (2019) Protective effect of three glucomannans from different plants against DSS induced colitis in female BALB/c mice. *Food and Function* 10, 1928–1939.

Zhang, X., Xiong, H. and Liu, L. (2012) Effects of taraxasterol on inflammatory responses in lipopolysaccharide-induced RAW 264.7 macrophages. *Journal of Ethnopharmacology* 141, 206–211.

14 An Evidence-Based Review of Medicinal Plants for Sexual Dysfunction Management Cited by Avicenna

Elaheh Zibaee[1*], Seyed Majid Ghazanfari[2], Lida Jarahi[3], Mina Borhani[4], and Seyed Ahmad Emami[1,4]

[1]Department of Traditional Pharmacy, School of Pharmacy, Mashhad University of Medical Sciences, Mashhad, Iran; [2]Iranian Research Center on Healthy Aging (IRCHA), Sabzevar University of Medical Sciences, Sabzevar, Iran; [3]Department of Community Medicine, Faculty of Medicine, Mashhad University of Medical Sciences, Mashhad, Iran; [4]Department of Pharmacognosy, School of Pharmacy, Mashhad University of Medical Sciences, Mashhad, Iran

Abstract

Sexual health is one of the significant aspects of health for men and women. Sexual disorders (SD) have been a common problem in all societies at all times and affect the quality of life of individuals and their spouses. Impotence or decreased sexual desire is introduced as a problem in different societies, which can be attributed to diet, mental disorders, various diseases, and medications, as well as lifestyle. The term "aphrodisiac" refers to substances, foods, medicines, and factors that increase sexual desire in both females and males. Aphrodisiacs are used to treat three types of disorder: libido (sexual desire, arousal), sexual potency (effectiveness of erection), and sexual pleasure. Traditional Persian Medicine (TPM) has paid special attention to the treatment of sexual impotence. In this chapter, aphrodisiac herbs and herbal therapeutic methods introduced by Avicenna for treatment of these disorders are shown.

Introduction

Sexual health is one of the significant aspects of health for men and women. Sexual disorders (SD) have been a common problem in all societies at all times and affect the quality of life of individuals and their spouses. Impotence or decreased sexual desire is introduced as a problem in different societies, which can be attributed to diet, mental disorders, various diseases, and medications, as well as lifestyle (Bosch *et al.*, 1991; Rosen and Ashton, 1993; Monga, 1999). The levels of sex hormones, including testosterone and progesterone, are the most important determinants of sexual desire in both sexes. SD include decreased or no erection, premature or delayed or no ejaculation, sexual dissatisfaction, and decreased secretion of sexual fluids that occur in reducing sexual desire (Bosch *et al.*, 1991). The

*Email: zibaeeelah@gmail.com

DOI: 10.1079/9781800621671.0014

term "aphrodisiac" refers to substances, foods, medicines, and factors that increase sexual desire in both females and males (Bosch *et al.*, 1991). Aphrodisiacs are used to treat three types of disorder: libido (sexual desire, arousal), sexual potency (effectiveness of erection), and sexual pleasure (Malviya *et al.*, 2011). Traditional Persian Medicine (TPM) is one of the most critical medical doctrines with an ancient history that can help us to recommend effective medicinal herbs in this context. The Persian physician Avicenna has devoted one chapter in his *Canon of Medicine* (Ibn Sinā, 2015) especially to male SD. In this chapter, Avicenna's perspective on male SD is reviewed, along with a brief discussion of pathology, holistic treatment methods, and the principles by which herbal medications work.

Sexual Disorders in View of Avicenna

SD in men are known in TPM as *ḍaʿf -e qow-e bah* (weakness of aphrodisiac) and *enzal-e zood-ras* (early ejaculation). Avicenna has classified male SD in 24 categories. One critical class includes low sexual desire (*noqṣān-o-al-bah*).

Anatomy of the genital system

The male reproductive system includes the testes, ducts, and penis:

1. *Testes*: Produce semen and are surrounded by the scrotum and many veins and arteries.
2. *Ducts*: Two ducts (vas deferens) come from the testes carrying semen to the penis.
3. *Penis*: A hollow ligament covered on both sides by muscles that make it firm. Three ducts in the penis lead to one valve in the trunk. It drains urine from the bladder and semen from the testes.

Semen, from the Traditional Persian Medicine perspective

Semen is the fourth digestion product. Therefore, intake of foods has a very influential role in semen production. The characteristics of semen-producing foods include high nutrition value and moderate warm nature. After the first digestion in the stomach, the food goes to the liver, and the second digestion takes place in the liver. The derived blood from these two digestions goes to the arteries and passes the third digestion. Then blood reaches the organs to perform the fourth digestion (tissue digestion). Thus, semen is produced in the fourth digestion of food.

Reasons for low sexual desire

Causes of low sexual desire in the view of Avicenna are:

1. Structural and dystemperament (*su-e-mizāj*) disorders in the organs that produce and transmit semen (testes, semen ducts, and penis).
2. Disorders in other essential body organs (weakness of chief organs, *ḍaʿf -e aʿḍā-e-raeisah*).
3. Semen disorders (in quantity, quality, and movement).
4. Psychological and mental disorders (*aʿrāḍ-e-nafsāni*).
5. Other causes including gas (*riḥ*) deficiency, drugs, and prolonged lack of intercourse.

Structural and dystemperament (*su-e-mizāj*) disorders in the organs that produce and transmit semen (testes, semen ducts, and penis)

Any defect in the testes, arteries, and semen ducts, including excessive laxity and the various dystemperaments in the testes and genital system, affects the production, quality, and quantity of semen.

Main dystemperament (*su-e-mizāj*) disorders and their treatment principles are as follows.

Dry dystemperament

Dry dystemperament on the genital system makes semen very thick, compact, and challenging to move.

TREATMENT. The use of various soups, milk, dairy products, fruits, and topical cold and moist oils such as lettuce and pumpkin, is helpful in these patients.

Cold dystemperament

In cold dystemperament on the genital system, patients' semen is thick and usually has difficulty coming out.

TREATMENT. These patients recover by consuming warm foods and taking moderate exercise. Foods with a warm nature such as pistachios, almonds, cinnamon, saffron, carrot seeds, nettle seeds, ginger jam, broth with warm spices, cooked peas, sparrow meat, turnips, coconut, and carrot jam, and topical warm oils such as clove and garlic are helpful for these patients. These people should avoid sour buttermilk and yogurt. Exercise is very suitable for these patients.

Warm dystemperament

In the warm dystemperament on the genital system, the patient's skin is hot, the urine color is bright yellow, and the semen color is yellow. Rapid ejaculation, enlarged testes, testicular vessels bulging, and a burning sensation among the ejaculation are other symptoms of these patients.

TREATMENT. These people tend to use cold foods and drinks. *Sekanjabin* (oxymel), lemon syrup, buttermilk, pomegranate and its paste, milk, cucumber, purslane, barberry, plum, rice, spinach, topical violet oil, and avoiding hot foods are helpful for these patients.

Wet dystemperament

In the wet dystemperament on the genital system, the semen is thin, loose, and has a large volume.

TREATMENT. Foods with a hot and dry nature, such as sparrow and pigeon meat, barbecue, partridge, cinnamon, cloves, cardamom, jujube, thyme, dill, bananas, walnuts, eggs, ginger jam, shrimp, leek, horseradish, watercress, artichoke, garlic, cumin, and topical warm oils are recommended, whereas the consumption of moisturizing foods such as beer, pumpkin, cucumber, yogurt, and buttermilk should be avoided.

Cold and dry dystemperament

According to Avicenna's *Canon*, cold and dry dystemperament on the genital system prevents the gas (*riḥ*) production, and the hot dystemperament destroys the gas (*riḥ*). Therefore, penile flaccidity can be caused by penile dystemperament.

TREATMENT. For the cold dystemperament symptoms, hot spices such as onion, leek, fenugreek, carrot seed, and hot temperament oil are recommended.

If the dry dystemperament symptoms are observed, warm and moist foods such as egg yolk, cow's milk, almond milk, milk with rice, pumpkin stew, carrot jam, and wet temperament oils such as lettuce and pumpkin oil are suitable.

Phlegmatic syndrome

Phlegmatic dystemperament on the genital system causes the weakness of the penile nerves and leads to erectile dysfunction.

TREATMENT. Hot and dry nature foods, exercise, avoiding cold and wet nature foods such as buttermilk and yogurt, and topical warm oils are helpful for these patients.

Disorders in other essential body organs (weakness of chief organs, *ḍaʿf -e aʿḍā-e-raeisah*)

Weakness in brain function (ḍaʿf -e demagh)

The brain, as the body's central organ, controls other organs. So, any disturbance in its function can affect the other organs' function. These patients' complaints are lack of pleasure during intercourse, decreased sexual desire, and penile flaccidity.

TREATMENT. General recommendations for brain health include avoiding exposure to bad smells, too many mental activities, excessive sleep and wakefulness, excessive intercourse, overeating, and consuming very sour foods. Foods that strengthen the brain include peeled almonds with candies, bird meat with chickpeas, saffron, cinnamon, and cloves, meat extract, lemongrass (*Cymbopogon jwarancusa* subsp. *olivieri* (Boiss.) Soenarko), chamomile (*Chamaemelum nobile* (L.) All.), basil (*Ocimum basilicum* L.), lavender (*Lavandula stoechas* L.), figs, apples, rose water, grapes, currants, pistachios, hazelnuts, and eggs. Treatment for all kinds of brain dystemperament disorders may be needed.

Weakness in heart function (*ḍaʿf -e qalb*)

The heart delivers the necessary blood to the organs and produces vital spirit (*ruḥ-e-ḥayawāni*). Heart weakness due to decreased spirit and gas (*riḥ*) is associated with symptoms such as intercourse reluctance, exertional dyspnea during intercourse, erectile dysfunction and ejaculation without an erection, pulse weakness, and decreased body temperature.

TREATMENT. Participating in joyous gatherings, avoiding sadness, applying good-smelling perfumes, using delicate and heart-strengthening foods such as apples, beets, and their juice, sweet pomegranates, grapes, rose water, bitter gourd, spring orange juice, saffron, bird's eggs, and barberry and plum juice (in hot temperament) are general recommendations for heart health. Treatment of heart dystemperament may be necessary.

Weakness in liver function (*ḍaʿf -e kabed*)

From the TPM perspective, the liver produces four main humors (*khelṭ*) in the body: yellow bile (*ṣafrā*), black bile (*saudā*), phlegm (*balgham*), and blood (*dam*), and produces natural spirit (*ruḥ-e-ṭabieʿi*). Therefore, liver dysfunction can lead to infertility due to the abnormal production of semen.

TREATMENT. General recommendations for liver health maintenance are avoiding overeating, excessive consumption of very cold or very hot foods, drinking icy water, water during fasting or after bathing, intercourse and exercise while sleeping, extreme thirst, eating foods that are excessively sweet, eating sticky foods and consuming dense foods, and using natural and mild sweets after meals. Foods that strengthen the liver are coriander, beer, pistachios, almonds, raisins, hazelnuts, chicory, apple and pomegranate juice, and fast-digesting meats such as poultry and mutton. In the presence of liver hot dystemperament symptoms, purslane, chicory, tamarind, lemon, plum, cucumber, pumpkin, beetroot, apple, pomegranate, and pear should be consumed. If the liver cold dystemperament symptoms are obtained, currants, cloves, cinnamon, and saffron are helpful.

Weakness in stomach function (*ḍaʿf -e maʿdah*)

The stomach is an essential organ in the body; the digestion of food begins here, and the gastric digestion product is delivered to the liver. Any food maldigestion in the stomach can cause many disorders in the liver, leading to an abnormal main humor production. Stomach weakness has symptoms such as belching, bloating, stomach heaviness after eating, and defecation disorders (constipation or diarrhea).

TREATMENT. The general principles of stomach health include avoiding extreme hunger and overeating, cramming, eating before the stomach evacuation, drinking a lot of hot or very cold water, consuming drinks with food, consuming thick and slow-digesting foods, not chewing food, and eating too fast. In order to strengthen a cold stomach, warm temperament foods such as ginger and its jam, beet jam, apple, carrot, pistachio, almond, mint, oregano, cardamom, saffron, cinnamon, cumin, and cloves are recommended.

Weakness in kidney function (*ḍaʿf -e kolyah*)

These patients' complaint is impotence and ejaculation without lust.

TREATMENT. Suitable recommendations for this group are sheep and camel milk, raisins, carrot jam, apple juice, pear, pomegranate and its paste, quail meat, and eggs.

Disorders in the middle organs between the chief organs (aʿḍā-e-raeisah) and the genital system

Included in these disorders is obstruction of the ducts between the origin and the genitals.

Disorders in particular adjacent organs

The disorders in adjacent organs include amputation of hemorrhoids and pain in the anal area due to the nerve endings in this region.

Systemic disorders

Systemic disorders are general weakness or systemic diseases of the body.

TREATMENT. In the treatment of body weakness, fortified foods such as eggs, milk, almond, pistachios, and green beans are the appropriate recommendations.

Semen disorders

Semen disorders include quantity, quality, and movement disorders.

Psychological and mental disorders (aʿrāḍ-e-nafsāni)

Conditions such as extreme fear, depression, anxiety, and sudden sadness are effective triggers in infertility.

Other causes including gas (*riḥ*) deficiency, drugs, and prolonged lack of intercourse

Gas (*riḥ*) deficiency may be due to cold or hot dystemperament in the distal region of the body. Although people believe using drugs is an effective way to cure ejaculation, narcotics can decrease the seminal fluid movement and disrupt the movement of sperm. Ejaculation is complex, and the discharge is thick. A long-time reduction in sexual activity impairs sperm production. These patients' complaints are poor erection and decreased sexual desire.

TREATMENT. If the cause of gas (*riḥ*) deficiency in the bottoms is due to extreme cold dystemperament, proper treatments are soft rubbing, grains in the food such as kidney beans and chickpeas, and onions with salt. And if the reason for the gas (*riḥ*) deficiency is hot dystemperament, use of cold temperament oils and foods such as pear and milk is helpful. Suitable recommendations for the drug-use group are giving up drugs, eating Golghand (a product obtained by mixing rose petals and sugar) with rose water and lemon balm every morning, fresh cow's milk with cantaloupe, rice with milk, carrot jam, ginger jam, egg yolk with cinnamon, birds meat, saffron, cinnamon, and chickpea.

Table 14.1 summarizes the reasons for low sexual desire according to Avicenna's *Canon of Medicine* and compares them to modern medicine.

Low Sexual Desire from the Perspective of Modern Medicine

Sexual desire (libido) is a type of motivation, need, and desire for sexual activity that is created by visual, auditory, external tactile stimuli as well as internal psychological stimuli in cortical and subcortical areas of the brain such as the limbic system and temporal lobe.

Sexual desire has biologic, psychological, and social components. Regarding the biologic aspect, sexual desire, like other instincts, is a response to a biologic need with biologic factors such as anatomy, physiology, genes, hormonal status, general health, nutrition, lifestyle, and age influencing it. The psychological aspect refers to neuropsychological issues and interpersonal relationships, whereas the social aspect refers to cultural and value issues of society and individuals (Nimbi *et al.*, 2020).

Table 14.1. Reasons for low sexual desire mentioned in Avicenna's *Canon of Medicine* compared to the results of modern medicine.

Causes of low sexual desire	Avicenna's *Canon of Medicine*		Modern medicine
	Main causes	Minor causes	
Structural and dystemperament (*su-e-mizāj*) disorders in the organs that produce and transmit semen (testes, semen ducts, and penis)	Various dystemperaments in the testes and genital system Excessive laxity	Cold, severe cold, dry, and severe cold with dryness dystemperaments	Trauma Cryptorchidism Varicocele Contact with gonadotoxins Orchitis Testicular torsion Lack of androgen Chromosomal causes Idiopathic Ejaculation dysfunction and impotence Hypospadias
Disorders in other important body organs (weakness of chief organs, *ḍaʿf -e aʿḍā-e-raeisah*)	Brain	Weakness in brain function (*ḍaʿf -e demagh*) Brain trauma	Hypothalamic disorders Pituitary disorders Thyroid disorder Brain trauma
	Heart	Weakness in heart function (*ḍaʿf -e qalb*)	Heart failure
	Liver	Weakness in liver function (*ḍaʿf -e kabed*)	Liver failure
	Stomach	Weakness in stomach function (*ḍaʿf -e maʿdah*)	Digestion disorders
	Kidney	Weakness in kidney function (*ḍaʿf -e kolyah*)	Kidney failure
	Middle organs	Obstruction of ducts between the chief organs (*aʿḍā-e-raeisah*) and the genital system	
	Special adjacent organs	Amputation of hemorrhoids Pain in the anal area	
	Systemic disorder	General weakness of the body	Systemic diseases of the body Diabetes
Semen disorders	Changes in quantity, quality, and movement	Various dystemperaments in the testes and genital system	Hormonal causes Pharmacologic causes Gonadotoxins
Psychological and mental disorders (*aʿrāḍ-e-nafsāni*)	*Aʿrāḍ -e-nafsāni*		Psychological and mental disorders
Other causes	Systemic diseases	General weakness	Chronic systemic diseases
	Gas (*riḥ*) deficiency	In the whole body Around the body	Erectile dysfunction
	Intercourse time	Long-term lack of intercourse	Time and frequency of proximity
	Drugs		Alcohol Contact with oxidative stress

Causes of low sexual desire

Biologic causes

Androgen deficiency, hypogonadism, hormonal disorders, nervous system diseases, vascular system diseases, heavy alcohol drinking, smoking, diabetes, obesity, eating disorders, sedentary lifestyle, cancer, prostate disorder, chronic illness, pelvic infections, fracture or surgery, and medications (like hormonal therapy, chemotherapy, blood pressure medications, statins and fibrates, antidepressant, antipsychotics, benzodiazepines, anticonvulsants, H2 blockers, corticosteroids) can be listed as biologic causes of low sexual desire.

Among the hormonal causes, free testosterone levels are an important factor on sexual desire. Free testosterone is crucial as it affects key receptors both centrally and peripherally to promote a normal sexual response and desire. Among other effective biologic compounds, nitric oxide as a chemotactic agent can indirectly affect sexual desire. It is a potent vasodilator throughout the body and facilitates blood flow and lubrication to the sexual organs. Neurotransmitters such as dopamine, norepinephrine, and serotonin are three important centrally acting neurotransmitters of the limbic system, associated with sexual desire.

Psychosocial causes

Psychological disorders, posttraumatic stress disorder, relationship problems, stress and exhaustion, and low self-esteem are the main psychosocial causes of low sexual desire (Jayne *et al.*, 2017).

Treatment of low sexual desire

The choice of treatment depends on the cause of the decreased libido. Sometimes a change in lifestyle, a discontinuation of a particular medication, or a change in medication (such as a reduction in dosage) can solve the problem. In general, two approaches to treatment are suggested: medication and psychotherapy. Different treatment options for biomedical approaches (e.g., hormonal replacement therapy or hormonal treatments, testosterone therapy) and psychological approaches (e.g., psychotherapy, cognitive behavioral therapy, couple psychotherapy) are available. Pro-erection medications and the use of testosterone injections or gels are the first line of treatment for male sexual desire problems. There is evidence that central dopamine neurotransmission may be important in mediating the central nervous system's regulation of sexual desire and erections. Treatments with dopamine receptor agonists (e.g., bromocriptine, pergolide) may also stimulate erections and sexual behavior (McCarthy and McDonald, 2008).

Common drugs for erectile dysfunction are known as phosphodiesterase inhibitors. The enzyme phosphodiesterase (PDE) is responsible for the degradation of cGMP and cAMP. The isoform in penile tissue is phosphodiesterase type 5 (PDE5), which exclusively degrades cGMP and reduces cGMP levels. Thus, inhibition of PDE5 increases cGMP levels, reduces intracellular calcium, relaxes smooth muscles, and dilates blood vessels.

The use of certain drugs such as bupropion and dopamine that have other main uses has also been suggested to increase libido. Bupropion is a norepinephrine and dopamine reuptake inhibitor that improves sexual desire in depressed patients as well as in patients with hypoactive sexual desire disorder (HSDD) induced by use of serotonin specific reuptake inhibitor antidepressants. Sildenafil is a PDE5 inhibitor. Inhibition of PDE5 promotes vasodilation via the sustained effects of nitric oxide in the vascular system. Furthermore, buspirone acts as an antianxiety agent that binds to both serotonin and dopamine receptors in the brain. It has been studied for its potential to affect sexual function and possible off-label use in HSDD with some promise (Engel *et al.*, 2019).

Pharmacologic Aspects of Medicinal Plants for Treating Low Sexual Desire Cited by Avicenna

This section presents scientific evidence backing up medicinal plants mentioned in

Avicenna's *Canon of Medicine* for treating low sexual desire. However, only around 30 out of more than 75 mentioned plants have been studied for this purpose. Table 14.2 presents the aphrodisiac herbs cited in the *Canon of Medicine*.

Acorus calamus (*Acoraceae*)

Ethanol extracts obtained from roots of *Acorus calamus* are enriched with α- and β-asarone, acorenone, calamenone, acoradin, and shyobunone. The extract demonstrated a PDE5-inhibiting effect in *in vitro* research studies (Ganapathy *et al.*, 2021).

Allium cepa (*Amaryllidaceae*)

It was found that extracts obtained from the bulb of *Allium cepa* raise the blood total testosterone level in rats when fed fresh onion juice (Nimrouzi *et al.*, 2020).

Apium graveolens (*Apiaceae*)

Eighty women participated in a clinical trial where they consumed 500 mg of celery (*Apium graveolens*) seeds and a placebo. The Female Sexual Function Index (FSFI) questionnaire was used to evaluate the women's sexual function. The study results showed an increase in total FSFI score and improvement in sexual desire, arousal, lubrication, and pain (Hessami *et al.*, 2021).

Alpinia officinarum (*Zingiberaceae*)

Twenty-four male rats were randomly assigned into separate groups (control, diabetic), a single dose (60 mg/kg) of streptozotocin was injected intraperitoneally to induce diabetes in the experimental group, and the diabetic rats were treated orally with a hydroalcoholic extract of *Alpinia officinarum* (200 and 500 mg/kg, respectively). After 8 weeks of treatment, blood samples, testis, and cauda epididymis were excised to evaluate specific hormonal changes, sperm parameters, and testis morphology. The studies showed ameliorated sperm damage and improved sperm morphology besides improving histological damage in the testis in diabetic rats. Testosterone levels were significantly elevated (Heidari *et al.*, 2021).

Brassica rapa (*Brassicaceae*)

In studies performed with male mice, the application of *Brassica rapa* and *Prunus amygdalus* (explained below) presented various benefits such as increases in weight of seminal vesicles, testosterone and dihydrotesterone (DHT) levels, sperm count, and sperm motility (Qureshi *et al.*, 1989).

Cicer arietinum (*Fabaceae*)

Seventy-two male albino mice were divided equally into four groups that were daily fed a diet containing 0, 20, 30, and 50% w/w of *Cicer arietinum* seeds, respectively. At day 7, 14, and 21 from the start of the experiment, the mice were anesthetized and euthanized to collect blood, testes, epididymis, and seminal vesicles. Enhancement of testicular function and elevation of serum testosterone and luteinizing hormone (LH) levels, sperm concentration, sperm motility as well as testicular levels of antioxidants including glutathione (GSH), glutathione peroxidase (GPx), and catalase (CAT) were reported (Sayed *et al.*, 2018).

Cinnamomum zeylanicum (*Lauraceae*)

Cinnamomum zeylanicum contains comparatively high amounts of cinnamaldehyde, which is mainly found in the bark of the tree. This active component enhances testosterone secretion from Leydig cells and improves sperm parameters (Alahmadi, 2020).

Crocus sativus (*Iridaceae*)

The most important valuable secondary substances occurring in the stigma of *Crocus*

Table 14.2. Plants with aphrodisiac properties cited in Avicenna's *Canon of Medicine*.

No.	Scientific name	Common name	No.	Scientific name	Common name
1	*Acorus calamus* L.	Sweet flag	36	*Moringa pterygosperma* Gaertn. Synonym: *Moringa aptera* Gaertn.	Ben oil tree
2	*Allium cepa* L.	Onion	37	*Musa* × *paradisiaca* L.	Banana
3	*Allium hirtifolium* Boiss.	Persian shallot	38	*Myristica fragrans* Houtt.	Nutmeg tree
4	*Allium porrum* L.	Leek	39	*Gundelia tournefortii* L.	Tumble thistle
5	*Alpinia officinarum* Hance	Lesser galanal	40	*Himantoglossum hircinum* (L.) Spreng.	Lizard orchid
6	*Althaea cannabina* L.	Egyptian hemp	41	*Lagoecia cuminoides* L.	Common wild cumin
7	*Androsace lactea* L.	Milkwhite rock jasmine	42	*Lepidium coronopus* (L.) Al-Shehbaz Basionym: *Cochlearia coronopus* L.	Swine's cress
8	*Apium graveolens* L.	Celery	43	*Limonium vulgare* Mill.	Sea lavender
9	*Aristolochia fontanesii* Boiss. & Reut.	Long aristolochia	44	*Linum usitatissimum* L.	Common flax
10	*Arum maculatum* L.	Snakeshead	45	*Lupinus termis* Forssk. *Lupinus albus* L.	Egyptian lupine
11	*Asarum europaeum* L.	Asarabacca	46	*Medicago sativa* L.	Alfalfa
12	*Asparagus officinalis* L.	Garden asparagus	47	*Papaver somniferum* L.	Opium poppy
13	*Astragalus tragacantha* L.	Gum tragacanth plant	48	*Pastinaca sativa* L.	Parsnip
14	*Brassica nigra* (L.) W.D.J. Koch	Black mustard	49	*Peucedanum oreoselinum* (L.) Moench. Basionym: *Athamanta oreoselinum* L.	Mountain parsley
15	*Brassica oleracea* L.	Cabbage	50	*Phaseolus vulgaris* L.	Common bean
16	*Brassica rapa* L. subsp. *rapa*	Common turnip, field mustard	51	*Phyllanthus emblica* L.	Emblic myrobalan
17	*Capparis spinosa* L.	Caper bush	52	*Pimpinella anisum* L.	Anise
18	*Centaurea behen* L.	White behen	53	*Pinus gerardiana* Wall. ex D. Don	Chilgoza pine
19	*Cheilocostus speciosus* (J. Koenig) C.D. Specht Basionym: *Banksia speciosa* J. Koenig	Crepe-ginger	54	*Piper cubeba* L.	Tailed pepper
20	*Cicer arietinum* L.	Chickpea	55	*Piper longum* L.	Long pepper
21	*Cinnamomum verum* J. Presl	Common cinnamon tree	56	*Piper nigrum* L.	Black pepper
22	*Cocos nucifera* L.	Coconut	57	*Pistacia terebinthus* L.	Turpentine tree, terebinth
23	*Colchicum autumnale* L.	Colchicum, meadow saffron, autumn crocus	58	*Pistacia vera* L.	Pistachio
24	*Corylus avellana* L.	Common hazel	59	*Prunus persica* (L.) Batsch	Peach
25	*Crocus sativus* L.	Saffron crocus	60	*Raphanus sativus* L.	Common cultivated radish

Continued

Table 14.2. Continued.

No.	Scientific name	Common name	No.	Scientific name	Common name
26	*Cucumis melo* L.	Muskmelon	61	*Roccella tinctoria* DC.	Orchil
27	*Cyperus esculentus* L.	Tiger nut	62	*Sesamum indicum* L.	Sesame
28	*Daucus carota* L.	Carrot	63	*Sisymbrium officinale* (L.) Scop.	Hedge mustard
29	*Elettaria cardamomum* (L.) Maton Basionym: *Amomum cardamomum* L.	Small cardamom	64	*Taxus baccata* L.	Common yew
30	*Eruca vesicaria* (L.) Cav. Basionym: *Brassica vesicaria* L. Synonym: *Eruca sativa* Mill.	Arugula	65	*Tribulus terrestris* L.	Caltrop
31	*Ferula assa-foetida* L.	Asafoetida	66	*Trigonella foenum-graecum* L.	Fenugreek
32	*Ferula szowitziana* DC.	Persian giant fennel	67	*Urtica pilulifera* L.	Roman nettle
33	*Fraxinus excelsior* L.	Ash	68	*Vicia faba* L.	Broad bean
34	*Glossostemon bruguieri* Desf.	Moghat	69	*Withania somnifera* (L.) Dunal	Ashwagandha
35	*Mentha* spp.	Mint	70	*Zingiber officinale* Roscoe	Ginger

sativus are crocin, picocrocin, and safranol. Studies performed with aqueous and ethanol extracts of saffron stigma in rats showed improved erectile dysfunction and sperm parameters. In a clinical trial, male patients underwent the nocturnal penile tumescence (NPT) test. Use of the International Erectile Function Index Questionnaire (IIEF-15) revealed an increase in the number and duration of erectile events from the beginning to the end of the 10-day study (Shamsa *et al*., 2009). In another research, the effect of saffron stigma, its valuable ingredients, and aqueous extracts obtained from them on sexual behavior was investigated. In that study, the aqueous extract, crocin, safranal, and sildenafil were administered intraperitoneally to male rats and their sexual behavior was investigated. It was reported that the frequency of copulation and erection increased (Hosseinzadeh *et al*., 2008).

Cyperus esculentus (*Cyperaceae*)

Cyperus esculentus tubers with the common name of "tiger nut" contain high levels of quercetin, vitamin C, vitamin E, and the mineral zinc. Two sets of sexually active male rats including intensely and moderately active were treated with raw tiger nut powder. Copulatory behavior and serum hormonal levels were measured. The powder could stimulate sexual motivation and performance in active rats, reducing mount and intromission latencies, intromission frequency, and ratio. It also increased serum testosterone levels (Allouh *et al*., 2015).

Daucus carota (*Apiaceae*)

In a randomized clinical trial conducted on women with sexual dysfunction, carrot (*Daucus carota*) improved desire, arousal, lubrication, orgasm, satisfaction, and pain (Molkara *et al*., 2018).

Eruca sativa (*Brassicaceae*)

The efficacy of ethanol extracts obtained from arugula (*Eruca sativa*) seeds in male rats with acrylamide-induced hypogonadism showed an antiapoptotic effect indicated by elevation of both B-cell lymphoma 2 (Bcl-2) and Bcl2-associated-X protein (Bax) expression and protected testis against acrylamide-induced testicular toxicity by normalizing testicular steroidogenesis, keeping Leydig cells, and improving oxidative stress status (Abd-Elsalam *et al*., 2021).

Ferula assa-foetida (*Apiaceae*)

Application of ethanol extracts from *Ferula assa-foetida* seeds in combination with root hydroethanolic extracts from the same plant caused improvement of sperm count, libido, and erectile function after 2 months of treatment (Nimrouzi *et al*., 2020).

Linum usitatissimum (*Linaceae*)

Hydroalcoholic extracts of *Linum usitatissimum* seeds were used in an experiment performed on mice. Two sets of mice containing 32 aged and 32 young ones were divided into four groups. The control group remained untreated, while the experimental groups were fed with seed extract. After 3 weeks, the brain was removed, and blood samples were collected to measure the individual sex hormone levels by ELISA. Generally, an increase in sex hormone levels was observed (Bahmanpour and Kamali, 2016).

Myristica fragrans (*Myristicaceae*)

According to studies performed with ethanol extracts from nutmeg (*Myristica fragrans*), mating efficiency and mounting behavior of mice and rats could be improved (Nimrouzi *et al*., 2020). It is known that sterols of this plant have nerve-stimulating properties (Kotta *et al*., 2013). When the general mating behavior, libido, and potency of albino rats (male and female) were evaluated and compared to sildenafil, an increase in sexual activity was noted. Thus, in addition to its use as a spice, nutmeg has been found to have aphrodisiac, nerve-stimulant, hyperlipidemic, antithrombotic, antifungal, antidysenteric, and anti-inflammatory effects, thus making a valuable contribution in the treatment of SD in men (Nimrouzi *et al.*, 2020).

Mentha species (*Lamiaceae*)

Biochemical and physical evaluation was conducted on Wistar albino rats treated with *Mentha piperita* and *Mentha spicata* tea, showing an increase in LH and follicle-stimulating hormone (FSH) levels, as well as in the diameter of the testicular tubules (Sharma and Jocob, 2001; Akdogan *et al.*, 2004).

Musa × paradisiaca (*Musaceae*)

Dessert banana (*Musa × paradisiaca*) had androgenic and anabolic effects in sexually mature male albino rats receiving root extract and was able to increase testicular/body weight ratio and decrease LH and FSH levels (Yakubu *et al.*, 2013).

Pastinaca sativa (*Apiaceae*)

According to literature, the roots, leaves, and seeds of parsnip (*Pastinaca sativa*) contain 5α-androstenone, which enhances libido and has a remarkable impact on sexual activity (Kenari *et al.*, 2021).

Papaver somniferum (*Papaveraceae*)

The findings of a clinical trial involving 100 impotent male patients showed that papaverine combined with prostaglandin E_1 (PGE_1) was more effective than PGE_1 alone and therefore can be a useful supplement in the treatment of male impotence (Drewes *et al.*, 2003).

Phyllanthus emblica (*Phyllanthaceae*)

Fruits of the amla tree (*Phyllanthus emblica*) were used as a part of a combination therapy with herbal medicines in a clinical trial. All male patients received one tablet daily of a combination of tryptophan, *Satureja montana*, *Tribulus terrestris*, and *P. emblica* extracts for 3 months, after which all patients underwent follow-up visits with the same investigations. Improvement of premature ejaculation was completely remarkable (Sansalone *et al.*, 2016).

Piper longum (*Piperaceae*)

Ethanol extracts of long pepper (*Piper longum*) fruits were administered to mice in acute (24 h) and chronic (90 days) oral toxicity studies. All external morphologic, hematologic, and spermatogenic changes, in addition to body weight and vital organ weights, were recorded. In the study, a significant increase in reproductive organ weights, sperm motility, and sperm count, and failure to illicit any spermatotoxic effect were reported (Khushbu *et al.*, 2011).

Piper nigrum (*Piperaceae*)

Ethanol and aqueous extracts of black pepper (*Piper nigrum*) offered to male mice as pellets once daily for 90 days were found to be potentially effective to improve sexual drive (Sutyarso *et al.*, 2015). In another study, black pepper fruits showed the compensation of normal estrogen level and antioxidant activity (Zodape and Gaikwad, 2020).

Pistacia terebinthus (*Anacardiaceae*)

The fruits of the so-called "turpentine tree" (*Pistacia terebinthus*) are known for their aphrodisiac effect, which has not yet been proven by objective scientific studies. However, the tree resin and the fruits of this plant are associated with an aphrodisiac effect (Bozorgi *et al.*, 2013).

Prunus amygdalus (*Rosaceae*)

Seeds of almond (*Prunus amygdalus*) were applied to male mice and effects of the seeds on the epidydimal spermatozoa, quantity and motility of sperm, and weight of testis and vesicles were investigated. An increase in weight of seminal vesicles, testosterone, DHT, sperm count, and motility were found to be the particular benefits of this plant (Qureshi *et al.*, 1989).

Sesamum indicum (*Pedaliaceae*)

Ethanol extracts of sesame seeds (*Sesamum indicum*) in male Wistar rats showed a significant improvement in sperm count, life/death ratio of sperm cells, and sperm motility when compared to a placebo group (Tahvilzadeh *et al.*, 2016).

Syzygium aromaticum (*Myrtaceae*)

Ethanol and hexane extracts of clove buds (*Syzygium aromaticum*) contain various active ingredients that increased libido and erection, mounting frequency, intromission frequency, and latency in studies performed with rats and mice. Δ^5-3β- and 17β-Hydroxysteroid dehydrogenase and testosterone levels, mating behavior, and mating performance were also increased (Nimrouzi *et al.*, 2020).

Tribulus terrestris (*Zygophyllaceae*)

Caltrop (*Tribulus terrestris*) is a famous plant species known for its aphrodisiac properties and many types of research have been conducted to investigate the individual effects of this plant. Fruits, seeds, and flowers of *T. terrestris* were used in various research studies (Ştefănescu *et al.*, 2020). Increase in testosterone level, influencing spermatogenesis and improving sexual behavior in male Swiss albino mice, reducing the level of nuclear factor-κB (NF-κB), and increasing the levels of nuclear factor erythroid 2-related factor 2 (Nrf2) and heme oxygenase-1 (HO-1) were reported in male rats by comparing treated male rats' sexual behavior and biochemical properties with a sildenafil-treated group as control (Qureshi *et al.*, 2014; Sahin *et al.*, 2016). Improvement in total FSFI score, increase in DHEA level, and decrease in serum testosterone and free testosterone levels were reported in a clinical trial on female patients of reproductive age and presenting sexual dysfunction. In rabbits, weight of the prostate, mount and intromission frequencies, and penile erection index were increased; but in chronic use, mount, intromission, ejaculatory latency, and serum levels of testosterone were decreased (Gama *et al.*, 2014).

Trigonella foenum-graecum (*Fabaceae*)

Fenugreek seeds (*Trigonella foenum-graecum*) were used in a short-term, double-blind, randomized, placebo-controlled study of 80 women aged 20 to 49 years. In this context biochemical and behavioral properties were evaluated. Generally, an increasing sexual arousal and desire in women treated with seeds of this plant was mentioned (Rao *et al.*, 2015). In another trial, 60 healthy males aged between 25 and 52 years, without erectile dysfunction, were randomly treated with an oral dose of the active treatment or placebo for 6 weeks. The primary outcome measure was the Derogatis Interview for Sexual Functioning (DISF-SR) self-administered quality of life (QOL) total score, with the specific QOL metrics being the secondary outcomes. The study outcomes were improvements in the subdomains of sexual arousal and orgasm, an increase in the physiological element of libido and QOL in self-reported satisfaction, and maintenance of normal, healthy testosterone levels (Steels *et al.*, 2011).

Urtica pilulifera (*Urticaceae*)

Different doses of *Urtica pilulifera* extracts were examined for PDE inhibitory (PDEI) activity in a clinical experiment versus control groups that received sildenafil and a placebo. The results revealed a notable dose-dependent cGMP-PDEI impact (Khanavi *et al.*, 2012).

Withania somnifera (*Solanaceae*)

Withania somnifera is another herbal plant that is known for its aphrodisiac activity. Aqueous extract of the root used in rats caused an increment in spermatogenesis and testosterone level (Bhattacharya and Muruganandam, 2003; Ambiye *et al.*, 2013). Methanol extracts of the root applied to male rats improved their libido, penile erection, and semen quality (Ilayperuma *et al.*, 2002).

Zingiber officinale (*Zingiberaceae*)

Administration of an aqueous extract of ginger (*Zingiber officinale*) to rats increased testicular weight, blood testosterone, and testicular cholesterol levels (Kamtchouing *et al.*, 2002). Furthermore, *Z. officinale* showed numerous beneficial effects and was also able to protect against the negative effects of various toxic chemicals on the sperm properties of the epididymis in male diabetic rats treated with ginger root extract. Phenols present in this plant also act as potent antioxidants (Mohammad and Hamed, 2012; Alahmadi, 2020; Nimrouzi *et al.*, 2020).

Table 14.2 shows all aphrodisiac herbs mentioned in Avicenna's *Canon* for making some recommendations for future research, given that only a few plants have been studied so far.

References

Abd-Elsalam, R.M., El Badawy, S.A., Ogaly, H.A., Ibrahim, F.M., Farag, O.M. *et al.* (2021) *Eruca sativa* seed extract modulates oxidative stress and apoptosis and up-regulates the expression of Bcl-2 and Bax genes in acrylamide-induced testicular dysfunction in rats. *Environmental Science and Pollution Research* 28(38), 53249–53266. DOI: 10.1007/s11356-021-14532-y.

Akdogan, M., Ozguner, M., Kocak, A., Oncu, M. and Cicek, E. (2004) Effects of peppermint teas on plasma testosterone, follicle-stimulating hormone, and luteinizing hormone levels and testicular tissue in rats. *Urology* 64(2), 394–398. DOI: 10.1016/j.urology.2004.03.046.

Alahmadi, B.A. (2020) Effect of herbal medicine on fertility potential in experimental animals – an update review. *Materia Socio-Medica* 32(2), 140. DOI: 10.5455/msm.2020.32.140-147.

Allouh, M.Z., Daradka, H.M. and Ghaida, J.H.A. (2015) Influence of *Cyperus esculentus* tubers (tiger nut) on male rat copulatory behavior. *BMC Complementary and Alternative Medicine* 15(1), 331. DOI: 10.1186/s12906-015-0851-9.

Ambiye, V.R., Langade, D., Dongre, S., Aptikar, P., Kulkarni, M. *et al.* (2013) Clinical evaluation of the spermatogenic activity of the root extract of ashwagandha (*Withania somnifera*) in oligospermic males: a pilot study. *Evidence-Based Complementary and Alternative Medicine* 2013, 571420. DOI: 10.1155/2013/571420.

Bahmanpour, S. and Kamali, M. (2016) The effect of flax seed (*Linum usitatissimum*) hydroalcoholic extract on brain, weight and plasma sexual hormone levels in aged and young mice. *Iranian Journal of Medical Sciences* 41(3 Suppl.), S12.

Bhattacharya, S. and Muruganandam, A. (2003) Adaptogenic activity of *Withania somnifera*: an experimental study using a rat model of chronic stress. *Pharmacology, Biochemistry, and Behavior* 75(3), 547–555. DOI: 10.1016/s0091-3057(03)00110-2.

Bosch, R.J., Benard, F., Aboseif, S.R., Stief, C.G., Lue T.F. *et al.* (1991) Penile detumescence: characterization of three phases. *The Journal of Urology* 146(3), 867–871. DOI: 10.1016/s0022-5347(17)37950-8.

Bozorgi, M., Memariani, Z., Mobli, M., Salehi Surmaghi, M. H., Shams-Ardekani M.R. *et al.* (2013) Five *Pistacia* species (*P. vera*, *P. atlantica*, *P. terebinthus*, *P. khinjuk*, and *P. lentiscus*): a review of their traditional uses, phytochemistry, and pharmacology. *The Scientific World Journal* 2013, 219815. DOI: 10.1155/2013/219815.

Drewes, S.E., George J. and Khan, F. (2003) Recent findings on natural products with erectile-dysfunction activity. *Phytochemistry* 62(7), 1019–1025. DOI: 10.1016/s0031-9422(02)00621-0.

Engel, J., Veit, M., Sinke, C., Heitland, I., Kneer, J. *et al.* (2019) Same same but different: A clinical characterization of men with hypersexual disorder in the Sex@Brain study. *Journal of Clinical Medicine* 8(2), 157. DOI: 10.3390/jcm8020157.

Gama, C.R., Lasmar, R., Gama, G.F., Abreu, C.S., Nunes, C.P. *et al.* (2014) Clinical assessment of *Tribulus terrestris* extract in the treatment of female sexual dysfunction. *Clinical Medicine Insights Women's Health* 7(7), 45–50. DOI: 10.4137/CMWH.S17853.
Ganapathy, A.A., Priya V.H. and Kumaran, A. (2021) Medicinal plants as a potential source of phosphodiesterase-5 inhibitors: a review. *Journal of Ethnopharmacology* 267, 113536. DOI: 10.1016/j.jep.2020.113536.
Heidari, H., Abdollahi, M., Khani, S., Nojavan F. and Khani, S. (2021) Effect of *Alpinia officinarum* extract on reproductive damages in streptozotocin induced diabetic male rats. *Journal of Diabetes & Metabolic Disorders* 20(1), 77–85. DOI: 10.1007/s40200-020-00711-0.
Hessami, K., Rahnavard, T., Hosseinkhani, A., Azima, S., Sayadi, M. *et al.* (2021) Treatment of women's sexual dysfunction using *Apium graveolens* L. fruit (celery seed): a double-blind, randomized, placebo-controlled clinical trial. *Journal of Ethnopharmacology* 264, 113400. DOI: 10.1016/j.jep.2020.113400.
Hosseinzadeh, H., Ziaee, T. and Sadeghi, A. (2008) The effect of saffron, *Crocus sativus* stigma, extract and its constituents, safranal and crocin on sexual behaviors in normal male rats. *Phytomedicine* 15 (6–7), 491–495. DOI: 10.1016/j.phymed.2007.09.020.
Ibn Sinā, H.A. (2015) *Al-Qanun fi'l-Tibb (The Canon of Medicine)*, Vol. 3. Egypt University of Sciences and Technology, Cairo, pp. 1049–1087 (in Arabic).
Ilayperuma, I., Ratnasooriya, W. and Weerasooriya, T. (2002) Effect of *Withania somnifera* root extract on the sexual behaviour of male rats. *Asian Journal of Andrology* 4(4), 295–298.
Jayne, C.J., Heard, M.J., Zubair, S. and Johnson, D.L. (2017) New developments in the treatment of hypoactive sexual desire disorder – a focus on flibanserin. *International Journal of Women's Health* 9, 171–178. DOI: 10.2147/IJWH.S125356.
Kamtchouing, P., Fandio, G. M., Dimo, T. and Jatsa, H.B. (2002) Evaluation of androgenic activity of *Zingiber officinale* and *Pentadiplandra brazzeana* in male rats. *Asian Journal of Andrology* 4(4), 299–302.
Kenari, H.M., Kordafshari, G., Moghimi, M., Eghbalian, F. and TaherKhani, D. (2021). Review of pharmacological properties and chemical constituents of *Pastinaca sativa*. *Journal of Pharmacopuncture* 24(1), 14–23. DOI: 10.3831/KPI.2021.24.1.14.
Khanavi, M., Azimi, H., Ghiasi, S., Hassani, S., Rahimi, R. *et al.* (2012) Specifying human platelet cAMP and cGMP phosphodiesterase inhibitory activity of the plants used in traditional Iranian medicine for the purpose of erectile dysfunction. *International Journal of Pharmacology* 8(3), 161–168. DOI: 10.3923/ijp.2012.161.168.
Khushbu, C., Roshni, S., Anar, P., Carol, M. and Mayuree, P. (2011) Phytochemical and therapeutic potential of *Piper longum* Linn. A review. *International Journal of Research in Ayurveda and Pharmacy* 2(1), 157–161.
Kotta, S., Ansari, S.H. and Ali, J. (2013) Exploring scientifically proven herbal aphrodisiacs. *Pharmacognosy Reviews* 7(13), 1–10. DOI: 10.4103/0973-7847.112832.
Malviya, N., Jain, S., Gupta, V.B. and Vyas, S. (2011) Recent studies on aphrodisiac herbs for the management of male sexual dysfunction – a review. *Acta Poloniae Pharmaceutica* 68(1), 3–8.
McCarthy, B. and McDonald, D. (2008) Assessment, treatment, and relapse prevention: male hypoactive sexual desire disorder. *Journal of Sex & Marital Therapy* 35(1), 58–67. DOI: 10.1080/00926230802525653.
Mohammad, S.M. and Hamed, H.K. (2012) Ginger (*Zingiber officinale*): a review. *Journal of Medicinal Plants Research* 6(26), 4255–4258. DOI: 10.5897/JMPR011.787.
Molkara, T., Akhlaghi, F., Ramezani, M.A., Salari, R., Vakili, V. *et al.* (2018) Effects of a food product (based on *Daucus carota*) and education based on traditional Persian medicine on female sexual dysfunction: a randomized clinical trial. *Electronic Physician* 10(4), 6577–6587. DOI: 10.19082/6577.
Monga, M. (1999) The aging penis: erectile dysfunction. *Geriatric Nephrology and Urology* 9(1), 27–37. DOI: 10.1023/a:1008340506011.
Nimbi, F.M., Tripodi, F., Rossi, R., Navarro-Cremades, F. and Simonelli, C. (2020) Male sexual desire: an overview of biological, psychological, sexual, relational, and cultural factors influencing desire. *Sexual Medicine Reviews* 8(1), 59–91. DOI: 10.1016/j.sxmr.2018.12.002.
Nimrouzi, M., Jaladat, A.-M. and Zarshenas, M.M. (2020) A panoramic view of medicinal plants traditionally applied for impotence and erectile dysfunction in Persian medicine. *Journal of Traditional and Complementary Medicine* 10(1), 7–12. DOI: 10.1016/j.jtcme.2017.08.008.
Qureshi, A., Naughton, D.P. and Petroczi, A. (2014) A systematic review on the herbal extract *Tribulus terrestris* and the roots of its putative aphrodisiac and performance enhancing effect. *Journal of Dietary Supplements* 11(1), 64–79. DOI: 10.3109/19390211.2014.887602.

Qureshi, S., Shah, A., Tariq, M. and Ageel, A. (1989) Studies on herbal aphrodisiacs used in Arab system of medicine. *The American Journal of Chinese Medicine* 17(1–2), 57–63. DOI: 10.1142/S0192415X89000103.

Rao, A., Steels, E., Beccaria, G., Inder, W.J. and Vitetta, L. (2015) Influence of a specialized *Trigonella foenum-graecum* seed extract (Libifem), on testosterone, estradiol and sexual function in healthy menstruating women, a randomised placebo controlled study. *Phytotherapy Research* 29(8), 1123–1130. DOI: 10.1002/ptr.5355.

Rosen, R.C. and Ashton, A.K. (1993) Prosexual drugs: empirical status of the "new aphrodisiacs". *Archives of Sexual Behavior* 22(6), 521–543. DOI: 10.1007/BF01543299.

Sahin, K., Orhan, C., Akdemir, F., Tuzcu, M., Gencoglu, H. *et al.* (2016) Comparative evaluation of the sexual functions and NF-κB and Nrf2 pathways of some aphrodisiac herbal extracts in male rats. *BMC Complementary and Alternative Medicine* 16(1), 318. DOI: 10.1186/s12906-016-1303-x.

Sansalone, S., Russo, G. I., Mondaini, N., Cantiello, F., Antonini, G. *et al.* (2016) A combination of tryptophan, *Satureja montana, Tribulus terrestris, Phyllanthus emblica* extracts is able to improve sexual quality of life in patient with premature ejaculation. *Archivio Italiano di Urologia e Andrologia* 88(3), 171–176. DOI: 10.4081/aiua.2016.3.171.

Sayed, A.A., Ali, A.A. and Mohamed, H.R. (2018) Fertility enhancing efficacy of *Cicer arietinu*m in male albino mice. *Cellular and Molecular Biology* 64(4), 29–38.

Shamsa, A., Hosseinzadeh, H., Molaei, M., Shakeri, M.T. and Rajabi, O. (2009) Evaluation of *Crocus sativus* L. (saffron) on male erectile dysfunction: a pilot study. *Phytomedicine* 16(8), 690–693. DOI: 10.1016/j.phymed.2009.03.008.

Sharma, N. and Jocob, D. (2001) Antifertility investigation and toxicological screening of the petroleum ether extract of the leaves of *Mentha arvensi*s L. in male albino mice. *Journal of Ethnopharmacology* 75(1), 5–12. DOI: 10.1016/s0378-8741(00)00362-7.

Steels, E., Rao, A. and Vitetta, L. (2011) Physiological aspects of male libido enhanced by standardized *Trigonella foenum-graecum* extract and mineral formulation. *Phytotherapy Research* 25(9), 1294–1300. DOI: 10.1002/ptr.3360.

Ştefănescu, R., Tero-Vescan, A., Negroiu, A., Aurică, E. and Vari, C.-E. (2020) A comprehensive review of the phytochemical, pharmacological, and toxicological properties of *Tribulus terrestris* L. *Biomolecules* 10(5), 752. DOI: 10.3390/biom10050752.

Sutyarso, Kanedi, S.M. and Rosa, E. (2015) Effects of black pepper (*Piper nigrum* Linn.) extract on sexual drive in male mice. *Research Journal of Medicinal Plants* 9(1), 42–47. DOI: 10.3923/rjmp.2015.42.47.

Tahvilzadeh, M., Hajimahmoodi, M., Toliyat, T., Karimi, M. and Rahimi, R. (2016) An evidence-based approach to medicinal plants for the treatment of sperm abnormalities in traditional Persian medicine. *Andrologia* 48(8), 860–879. DOI: 10.1111/and.12676.

Yakubu, M.T., Oyeyipo, T.O., Quadri, A.L. and Akanji, M.A. (2013) Effects of aqueous extract of *Musa paradisiaca* root on testicular function parameters of male rats. *Journal of Basic and Clinical Physiology and Pharmacology* 24(2), 151–157. DOI: 10.1515/jbcpp-2012-0059.

Zodape, G. and Gaikwad, V. (2020) Effect of *Piper nigrum* (Linn.) on infertility induced by ethionamide and para amino salicylic acid in female Sprague–Dawley rats. *Biomedical and Pharmacology Journal* 13(2), 1029–1036. DOI: 10.13005/bpj/1972.

15 The Current Situation of Medicinal and Aromatic Plants' Industry, Research, Education, and Standardization in Iran. How Can Traditional Persian Medicine Products be Successfully Adapted to the Needs of the European Market?

Farsad Nadjafi[1,2]* and Hartwig Schulz[3]

[1]*Shahid Beheshti University, Medicinal Plants and Drugs Research Institute Tehran, Iran;* [2]*MateriaMed GmbH, Vienna, Austria;* [3]*Consulting and Project Management for Medicinal and Aromatic Plants, Stahnsdorf, Germany*

Abstract

The use of medicinal plants and natural products in the treatment of human and animal diseases has a long history in Iran. This may be due to the vast geographical area of this country and its climatic diversity, rich plant diversity, cultural diversity, and also its history as one of the first cradles of human civilization. The existence of the Silk Road and the extensive commercial connections of this country with other civilizations have played a role in the richness of this knowledge and its development. As in other parts of the world, this rich knowledge was neglected in Iran for many years after modernization, and the use of medicinal plants was limited to traditional herbal healers in rural areas. In recent years, Iran has followed the global trend and traditional and alternative medicine, cultivation of herbs, related industries as well as research and education in this field have been greatly developed. In this chapter we show briefly the situation of medicinal and aromatic plants in Iran from different aspects as well as the potentials and also the requirements to introduce this rich knowledge to international markets like Europe.

Current Situation of Medicinal and Aromatic Plants in Iran

Cultivation and wild collection

Iran, with an area of 1,648,195 million km^2, is the 18th largest country in the world and situated in three areas of Asia (West, Central, and South) in the Middle East. It has about 33% of its area under cultivation, 14 million km^2 of meadows, 60 million km^2 of steppes, and 16 million km^2 of deserts. Because of the particular climatic importance (with 11 climate zones out of 13 climate

*Email: md@materiamed.at

 Medicinal Plants Used in Traditional Persian Medicine
(eds. H. Schulz, Seyed Ahmad Emami and Farsad Nadjafi)
DOI: 10.1079/9781800621671.0015

zones worldwide), Iran is a rich source of medicinal plants, some of which have been used in traditional medicine for centuries. The flora of Iran is very rich with almost 8000 species and, together with Turkey, these countries have the highest plant diversity and percentage of endemic plants (more than 2000 species) in the Middle East and West Asia (Ghahreman and Attar, 1999). It is estimated that nearly 13% of Iran's plant flora (850–1000 species) are used as herbs in the traditional and ethno-medicine of the country (Bagheri and Reghan, 1994). Cultivation of herbs has a long history in Iran and the cultivation of some of them, like saffron, goes back to 3000 years ago (Koocheki, 2013). Cultivation of herbs in Iran is mostly done by local smallholders. According to the statistics of the Iranian Ministry of Agriculture, the cultivation area of medicinal plants in 2015 was about 171,463 ha (Salehi Sardoei, 2022). Various herbs and spices are cultivated in different parts of the country such as coriander (*Coriandrum sativum*), tarragon (*Artemisia dracunculus*), fennel (*Foeniculum vulgare*), dill (*Anethum graveolens*), cumin (*Cuminum cyminum*), black cumin (*Nigella sativa*), damask rose (*Rosa* × *damascena*), saffron (*Crocus sativus*), blue mallow (*Malva sylvestris*), savory (*Satureja hortensis*), anise (*Pimpinella anisum*), spearmint (*Mentha spicata*), and henna (*Lawsonia inermis*). The farmers used their own selected propagation materials and the vast geographical distribution and different climatic conditions resulted in a rich genetic diversity of land races with different growth and quality characteristics. The cultivation and postharvest methods in these agricultural systems are based on traditional knowledge and are very labor-intensive (Fig. 15.1). Women have an important role in the collection and processing of herbs. The products are mostly sold in local markets but rarely used by the pharmaceutical and food industries, as they often do not meet the required standards. The reason is that the farmers use traditional processing methods like natural drying that may cause hygiene problems.

In recent years, the development of medicinal plant industries and export markets has significantly increased the demand for standardized herbal raw materials. These increased demands have led to the industrial cultivation of some aromatic and medicinal plant species such as cumin, peppermint, lemon balm (*Melissa officinalis*), coriander, damask rose, fennel, saffron, and tarragon, as well as introduced some new herbs such as *Stevia rebaudiana* and *Aloe vera* into the country's agricultural systems (Fig. 15.2). As acreage expanded, so did the need for mechanization and improved varieties, especially in terms of yield, resistances against drought, pests, and diseases, salt tolerance as well as quality.

Among the medicinal and spice plants, saffron has the largest area with almost 100,000 ha (Salehi Sardoei, 2022). Although the cultivation and harvesting of saffron require a lot of labor, the farmers are very interested in growing this plant species. The reason is its low water requirement and high economic income to be obtained from this crop. In 2014, 135 t of saffron were exported from Iran and it is estimated that the production volume in 2022 will reach more than 300 t (Salehi Sardoei, 2022). Damask rose is another plant that is extensively cultivated in Iran. Together with Turkey and Bulgaria, Iran is the largest producer of damask rose in the world. This species had a cultivation area of 18,000 ha in 2016. More than 85% of the cultivation area is located in Fars, Kerman, Isfahan, and East Azerbaijan provinces (Salehi Sardoei, 2022). The livelihood of many farmers in these regions is completely related to the cultivation and processing of this plant. Despite the large number of processing companies in the country, industrial production is mainly focused on rose water, which is a cheap product, while valuable products like essential rose oil, rose concrete, and rose absolute are produced only in very low amounts. This is mainly due to traditional concepts of the industry and low investment in new distillation and extraction technologies in recent years. Cumin is another spice plant cultivated mostly in dryland agricultural systems of the country. The area of this plant under cultivation and production in Iran has significantly increased in recent years. In 2020, the cultivation

Fig. 15.1. Harvesting of blue mallow flowers by local farmers in Iran. (By Farsad Nadjafi.)

Fig. 15.2. A mechanized peppermint farm in Iran. (By Farsad Nadjafi.)

area was nearly 18,000 ha with more than 12,000 ha cultivated in Razavi Khorasan Province alone (Sefidkon, 2021). Coriander is also produced in Iran mostly for seed production and the spice industry; the cultivation area was nearly 7000 ha in 2021 and located mainly in Hamadan Province. Ninety-five percent of the production is exported to countries such as Pakistan, Germany, and the United Arab Emirates (Salehi Sardoei, 2022).

In Iran only 54 species are cultivated, while most of the medicinal plants (>95%) used in the country are collected from the wild (Mashayekhan *et al.*, 2016). It is estimated that 40% of rural families in Iran live in poverty and rely on medicinal plant collection for their livelihood. Medicinal plants can provide a significant source of income for rural people in Iran especially through the sale of wild collected material, which contributes 10% to the total income of poorer households in the northeastern part of Iran (Mashayekhan *et al.*, 2016). Some wild collected medicinal plants of the country that are well known in the international market are tragacanth (*Astragalus gummifer*), galbanum gum (*Ferula gummosa*), asafoetida (*Ferula assa-foetida*), and licorice (*Glycyrrhiza glabra*). Due to increased harvesting and unsustainable harvesting methods, many medicinal plant species are currently threatened in the country. The plants are mostly exported and introduced in the markets as raw and unprocessed materials. There exist some licorice extraction companies with a production capacity of 18,350 t in 2021, of which 95% is exported (Salehi Sardoei, 2022).

Research and education

With the development of the medicinal plant industry, the need for scientific information and expert personnel in this field had increased. Research and education related to medicinal plants and natural products in Iran are managed by the following three ministries: the Ministry of Science, Research and Technology; the Ministry of Health; and the Ministry of Agriculture. The Ministry of Health is responsible for education at graduate level and research related to traditional medicine and pharmacognosy. The Ministry of Science, Research and Technology is responsible for establishing scientific and technological bases in the field of medicinal and aromatic plants in areas such as agriculture, phytochemistry, biology and process engineering, etc. and for training experts in these fields. This Ministry has also created a National Network for Research and Technology of Medicinal Plants of Iran to improve the collaboration between laboratories, promote cooperation between industry and research institutes, and organize scientific and technical conferences and workshops related to medicinal plants (https://en.mpnet.ir/, accessed 19 April 2023).

Figure 15.3 shows an overview of the situation of education and research in the field of medicinal plants in 2014 and 2015. The number of scientific papers, the number of companies in science and technology parks, the number of journals that published articles on medicinal plants, and the research in universities increased from 2014 to 2015. Since standards in the field of medicinal plants were largely lacking in Iran, the responsible organizations developed appropriate standards in cooperation with universities, research institutes, and industrial centers.

One of these standards, which was of particular importance to the country, is the national standard GAP (Good Agricultural Practices of Medicinal and Aromatic Plants). In addition, the number of companies associated to universities has been increased, reference laboratories have been installed, and more entrepreneurial offerings for medicinal plants have been developed (Fig. 15.4).

The revival of Traditional Iranian/Persian Medicine (TIM/TPM) and its use in the country's healthcare system was followed in recent years in medical universities. Figure 15.5 shows the situation of TIM/TPM and pharmacognosy in the universities during 2014 and 2015. Development of traditional medicine was not only pursued in universities, but also specialized clinics were established to serve the public. The number of these clinics that serve TIM/TPM was seven in the year 2015 (unpublished data of the National Council for Science and Technology Development of Medicinal Plants and Traditional Medicine).

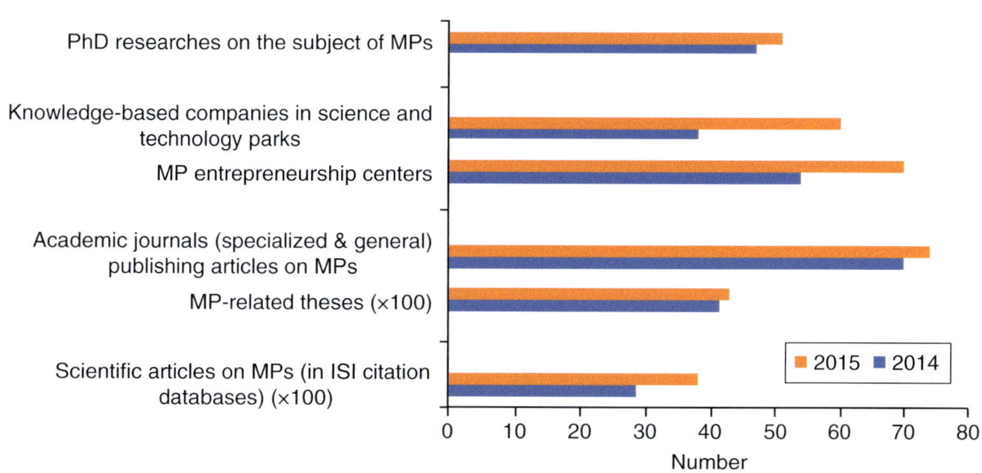

Fig. 15.3. Education and research situation in the field of medicinal plants (MPs) in Iran in 2014 and 2015. (Unpublished data of the National Council for Science and Technology Development of Medicinal Plants and Traditional Medicine.)

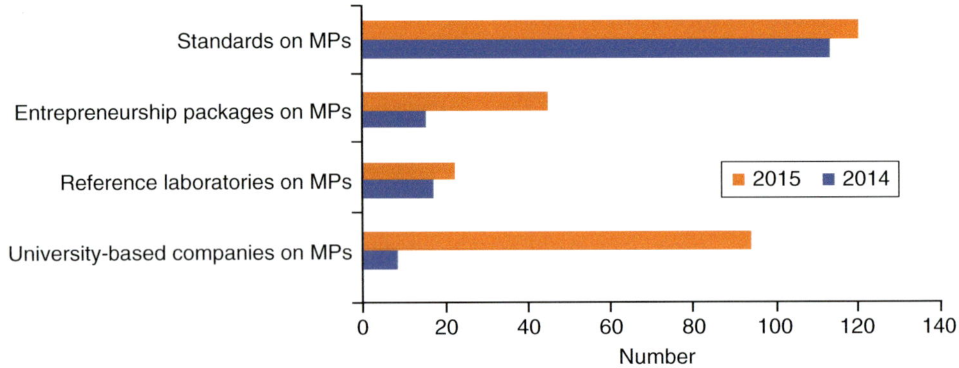

Fig. 15.4. Innovation and technology of medicinal plants (MPs) in Iran during 2014 and 2015. (Unpublished data of the National Council for Science and Technology Development of Medicinal Plants and Traditional Medicine.)

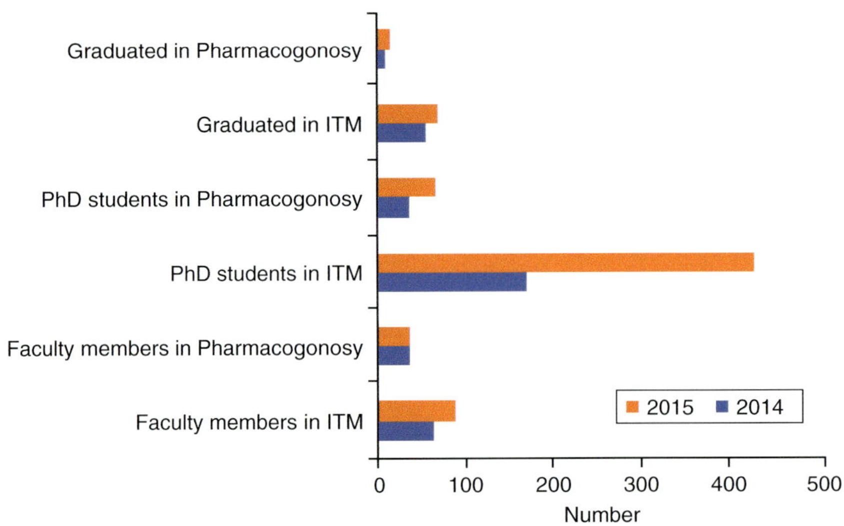

Fig. 15.5. The situation of Traditional Iranian Medicine (TIM) and pharmacogonosy fields in the medical universities in Iran during 2014 and 2015. (Unpublished data of the National Council for Science and Technology Development of Medicinal Plants and Traditional Medicine.)

Industry

The trend toward herbal medicine, medicinal plants as well as the use of traditional medicine medical services has led to the continous growth of herbal products industries in the country. In 2015, already 350 medicinal plant companies existed in Iran and from them, 183 companies produced herbal waters, plant extracts, and essential oils. It should be said that most of them are selling herbal waters due to the high national consumption and good markets in Arab countries (unpublished data of the National Council for Science and Technology Development of Medicinal Plants and Traditional Medicine). Figure 15.6 shows the number of herbal products and licenses issued to manufacture herbal medicines and other plant-based products during 2014 and 2015. To support the medicinal plant industries, the government placed 56 herbal drugs under insurance coverage.

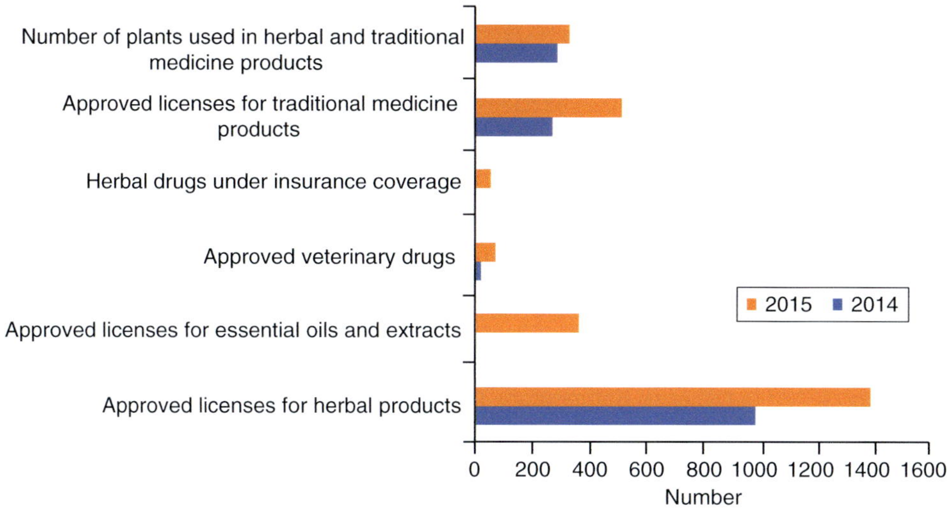

Fig. 15.6. The situation of herbal products in Iran during 2014 and 2015. (Unpublished data of the National Council for Science and Technology Development of Medicinal Plants and Traditional Medicine.)

National policy-making processes aiming to improve the market for medicinal plants

Due to the large number of government organizations involved in the issue of medicinal plants, in order to improve coordination among these organizations and ministries and to define national strategies, the National Council for Science and Technology Development of Medicinal Plants and Traditional Medicine was established at the Vice Presidency of Science and Technology in Iran. This Council has prepared the national document on medicinal plants and traditional medicine, which defines the country's strategies in this area. All ministries and governmental organizations are obligated to perform their duties within the framework of these strategies.

Adaptation of Traditional Persian Medicine Products to European Market Needs

Using natural products as medicines must have been an enormous challenge for early humans and it is very likely that they often ingested poisonous plants in search of food, resulting in vomiting, diarrhea, coma, or other toxic reactions. In this way, however, early humans were able to develop a huge knowledge about edible materials and natural medicines. Subsequently, humans invented fire, learned how to make alcohol, developed religions, achieved technological breakthroughs, and learned how to develop new medicines. All the different forms of Traditional Medicine such as Traditional Chinese Medicine (TCM), Ayurveda, Kampo, Traditional Korean Medicine (TKM), Traditional Persian Medicine (TPM), and Unani use natural products and have been practiced around the world for hundreds or even thousands of years, and they have developed into properly regulated systems of medicine. In their various manifestations, they have certain shortcomings, but they are still a valuable part of human knowledge (Fabricant and Farnsworth, 2001; Alves and Rosa, 2007).

Persia was a center of academic knowledge in ancient times and became the first world empire in history, stretching at those times from Bulgaria to Pakistan and from Egypt to the Caucasus. The experience and knowledge of Greece, Egypt, Babylon, India, and also China could thus enrich TPM and develop it over the last 4000 years.

During the Middle Ages and the following centuries in Europe, mainly monks, experienced women, and later pharmacists practiced the art of healing by means of naturally obtained extracts and preparations. This Traditional European Medicine (TEM) was enriched from the 15th century onward by the discovery of America and India with new plants and healing methods. Today, multicomponent mixtures and their natural substances are still in the focus of modern pharmaceutical research. In this context, scientists specialized in various disciplines are working closely together to find new applications for plant constituents in complementary medicine or to rediscover traditional knowledge with the techniques and methods available today on the basis of current studies (Bunse *et al.*, 2022).

Thus, in the Western world, phytotherapy became the culmination point for the emergence of today's classical medicine. Today, nearly a quarter of all modern medicines are derived from natural products, which thus represent a resource for primary care, but also have become very important for innovation and discovery of new active substances. As can be demonstrated by the discovery of artemisinin to combat malaria, the extensive knowledge of natural products is very promising for the future, and it is therefore expected that modern analytical techniques will be a very important tool to identify other pharmaceutically relevant active compounds in plants endemic to Iran (Wu, 2009).

In order to reorganize and support the worldwide application of Traditional Medicine systems, the World Health Organization (WHO) has published various policy papers aiming to strengthen quality assurance, safety, proper use, and efficacy, in particular through the regulation of traditional medical products. In this context, the importance of training, skill development, services, and reliable therapies is also clearly emphasized (WHO, 2015). More than 100 million Europeans are currently users of TCM, with one-fifth using TCM regularly and preferring healthcare including TCM. There are many more TCM users in Africa, Asia, Australia, and North America.

In modern times, it is often forgotten that TEM and TPM have a common heritage based on the four elements doctrine, the so-called "humoral medicine". Various medicinal plants known in Europe, such as sage (*Salvia officinalis*), fennel (*Foeniculum vulgare*), and lovage (*Levisticum officinale*), had already been mentioned by Avicenna and are later found in various herbal books on herbal remedies of monastic medicine. In Europe, Persian physicians were long regarded as role models, and many elements of TPM were incorporated into European healing practice in the 13th century. Avicenna in particular is considered the "father of modern medicine" because he was the first who introduced clinical trials to experimentally test medicines for their individual effectiveness of natural healing substances. Important approaches to health prevention can also be taken from various Islamic writings such as the *Taqwim al-Sihha* (*Tacuinum sanitatis*) written by Ibn Butlan, in which individual food materials were assigned to the four elements (warm, cold, moist, dry) and in this way the different health effects of a food were presented for the first time.

In recent years, medical systems such as TCM and Ayurveda are attracting more and more attention in the Western world, where the demand for inner balance and a holistic view of body and mind has increased significantly. In this context, numerous vitality and wellness organizations have emerged that have developed individual concepts from these two traditional systems and adapted them to the Western way of life. In addition to the respective herbal therapies, one of the main goals is a change of the nutritional pattern based on Avicenna's approach that a balanced diet using valuable food components has a significant impact on human health. Today, it is widely recognized among healthcare providers that access to a healthy diet, optimally adapted to our living conditions, significantly strengthens our immune system and reduces the risk of diabetes and cardiovascular disease, including inflammation, elevated cholesterol, high blood pressure, and poor sleep. The main emphasis of this "food as medicine" concept is put on the consumption of a variety of herbs and spices, legumes, nuts and seeds, fruits and vegetables, while the concept of conventional medicine is based primarily on technological medical advances to treat diseases with various synthetic drugs. However, the concept of "food as medicine" does not always work equally well for all people, which is why the individual case must be carefully examined. In many cases, however, it is advisable to carry out the change in diet with the help of dietary supplements and in conjunction with appropriate conventional medical treatment.

The marketing of phytopharmaceuticals based on endemic species is difficult because these plant species must first be approved by the European authorities, and this process is usually time-consuming and costly. To enable inclusion in the European Pharmacopoeia (Ph.Eur.), it may therefore be advantageous to first cultivate the relevant species in Europe in order to increase awareness in individual European countries. Corresponding studies were conducted by Bomme *et al.* (2006) on the introduction of traditional Chinese medicinal plants to the European market, for example to compare possible quality differences of plants from experimental cultivation in Germany and corresponding imported plants from China. Detailed botanical species identification using DNA sequence and fingerprint analyses was performed in the cultivation of 16 selected Chinese medicinal plant species. In addition, the harvesting technique and various active ingredients were determined. Through this controlled and documented cultivation, the relevant data for inclusion in the Ph.Eur. with regard to drug safety and general quality requirements could be collected (Bomme, 2001).

Before the introduction of Western medicine in the 16th century, TCM was the predominant form of medical care in China (Dong, 2013). Like TPM, TCM is based on thousands of years of medical practice and experience, from which its efficacy and effectiveness are derived. It is generally noteworthy that today traditional medical systems and modern medicine are increasingly converging, and it has become possible to determine the pharmacology as well as the mechanisms of action of numerous endemic medicinal plants in TPM and TCM with the help of modern technology. With the understanding of life sciences, it was thus possible to make the traditional medical systems more understandable for modern medicine (Tu *et al.*, 2002; Zhang and Li, 2011; Zhang *et al.*, 2012; Dong, 2013; Haidan *et al.*, 2016).

Quality assessment

Reliable quality control of medicinal plants, which have been collected and used worldwide for the pharmaceutical industry, has always been one of the main concerns of the WHO. Therefore, internationally accepted guidelines for assessing the quality of the individual plant raw materials were established that can be used in each country to set up individual quality specifications of the medicinal drugs. These guidelines were developed to promote the proper use and development of traditional medicine and, in this context, to ensure an objective description of the quality of individual pharmaceutical products. According to the WHO (2000), the following specific objectives represent the most important guidelines:

- harmonization of recognized and most important concepts applied in Traditional Medicine;
- elaboration of key issues in the development of methods for research and evaluation in Traditional Medicine;
- improving quality and importance of research in the field of Traditional Medicine; and
- provision of appropriate evaluation methods to facilitate regulatory development and registration in Traditional Medicine.

In order to ensure high quality, safety, and efficacy of medicinal plants, most of which come from wild collections, it is first important to reliably identify the respective plant species. In this context, particular attention should be paid to the current botanical names (genus and species), but also synonyms and vernacular names, of the plant parts used for the individual preparations, as well as detailed instructions for agricultural production and collection conditions according to the good agricultural practice of the respective country. For quality control, different sources can be used, such as the individual WHO quality control methods (WHO, 1998b) for medicinal drugs and the WHO monographs of selected medicinal plants (WHO, 1999).

When researching and evaluating herbal medicinal products for which there is no long history of use or that have not been researched at all, the "WHO research guidelines for the evaluation of safety and efficacy" should be regarded in detail. In the case of well-documented traditional uses of herbal medicinal products, the procedure can be simplified and it is mostly sufficient to follow a reduced number of procedures for conducting research and evaluating safety and efficacy according to the relevant WHO guidelines (WHO, 1993, 1998a).

Whereas in the past the term "quality" was described primarily as product quality (efficacy, composition, taste, shelf life, etc.), for some years we must also ensure that suitability value, ecological quality, social and sociocultural value, and nutritional convenience quality are guaranteed.

Furthermore, particularly in the food sector including food supplements, there has been an enormous increase in labels with regard to aspects of fair trade, regionality, organic cultivation, and sustainability.

The individual quality of medicinal and aromatic plants offered on the market, as well as extracts and distillates isolated from them, is usually specified by different parameters, using different physical, chemical, microbial, and sensory test methods (described by the International Organization for Standardization (ISO) and various national/international pharmacopoeia). For the respective products (e.g., teas, drugs, spices, essential oils, and solvent extracts) usually also company-internal specifications exist. In addition to the classical sum parameters such as density, refractive index, optical rotation as well as various titrimetrically determined key parameters, nowadays mainly chromatographic methods such as gas chromatography (GC) and high-performance liquid chromatography (HPLC) are used for qualitative and quantitative analyses. In addition, vibrational spectroscopic methods are increasingly applied especially for rapid and mostly nondestructive characterization of plants and the extracts obtained from them. In most cases, it is possible to determine the botanical origin of the plant material (species, chemotype) under investigation and to detect reliably possible adulterations and impurities.

Contaminants

The European Union has created a very complex set of regulations for medicinal products, including in particular various regulations, guidance documents of the European Medicines Agency (EMA), monographs, pharmacopoeias, and other specific regulations of the national health authorities. All these legal provisions also apply to herbal medicinal products, which must therefore meet the same quality, safety, and efficacy requirements as all other medicinal products (Steinhoff, 2019).

However, there are some specific things to consider due to the natural origin of herbal raw materials that mainly relate to their production and quality control. In addition, since medicinal plants can be exposed to various environmental influences, special legal requirements for their purity must be observed. The Ph.Eur. therefore mentions tests for pesticides, heavy metals, and microbial impurities, as well as mycotoxins, in the general monographs for herbal drugs and herbal drug extracts under defined conditions. In addition to these general monographs, individual monographs for a variety of herbal drugs and for some extracts may also contain individual rules for the absence or limitation of certain contaminants. The individual methods for determining these contaminants are also described in more detail in specific chapters of the Ph.Eur. The analytical methods used in quality control of herbal plants and the related products such as essential oils must generally be validated according to individual criteria that are also described in the Ph.Eur. in detail.

Currently, the Ph.Eur. contains a list of about 70 plant protection products (PPPs) that are considered relevant in phytopharmaceutical raw materials and an individual limit value is provided for each of these substances. Generally, with consistent application of the *Guideline on Good Agricultural and Collection Practice (GACP) for Starting Materials of Herbal Origin* (EMA, 2006), the responsible health authorities may permit waiving the cost-intensive pesticide analyses.

The list of PPPs is continuously updated depending on the individual approval status and adapted to the respective risk assessments made by the European Food Safety Authority (EFSA). Detailed information on the substances and their permitted maximum residue levels in the respective products can be found on the website of the European Commission (https://food.ec.europa.eu/system/files/2021-12/pesticides_ppp_app-proc_guide_phys-chem-ana_chem-subst.pdf, accessed 27 April 2023).

Use of plant protection products

In case of suspicion of other PPPs not listed, the Ph.Eur. refers to the EU Regulation No. 396/2005 on pesticide residues in food (European Parliament and Council of the European Union, 2005). Provided that the individual batches as well as the standards of GACP (EMA, 2006) can be fully documented, the responsible health authorities will waive the usally mandatory testing of PPPs.

The respective analytical methods applied to determine PPPs must also be documented, according to the documents published by the European Commission. In practice, several methods may be used, but they must be derived from the official "S19 method of the German Research Society (DFG)". According to guidance of the EMA's Committee for Herbal Medicinal Products (HMPC) for preparation of the application dossier, it is sufficient to cite these official methods validated on different herbal matrices (EMA, 2016a).

Heavy metals

The general monograph entitled *Herbal Drugs* of the Ph.Eur. (Council of Europe, 2016) establishes the individual limits for cadmium (1.0 ppm), lead (5.0 ppm), and mercury (0.1 ppm). These limits are mandatory "unless otherwise specified in an individual monograph or unless otherwise justified and approved." Because certain herbal drugs may have naturally occurring higher levels of certain heavy metals, there are several exemptions in the respective individual monographs with higher limits for cadmium (e.g., tormentil and willow bark) and for lead (Iceland moss and stinging nettle root). The Ph.Eur. also states that "herbal drugs used for the preparation of extracts may, in justified cases, exceed the limits for heavy metals specified in the monograph provided that the resulting extract meets the requirements for heavy metals" (Council of Europe, 2016).

In addition, the pharmacopoeia presents modern techniques for the determination of heavy metals in herbal drugs and herbal drug preparations, such as atomic absorption spectrometry (AAS), inductively coupled plasma atomic emission spectrometry (ICP-AES), or inductively coupled plasma mass spectrometry (ICP-MS).

Polycyclic aromatic hydrocarbons

Medicinal plants may be contaminated with polyaromatic hydrocarbons (PAHs) taken up from contaminated soil or air. In addition, PAHs can also enter herbal drugs during drying or roasting processes by direct application of open fire. For the food sector, Commission Regulation No. 1881/200619 (Commission of the European Communities, 2006) sets the following maximum levels for PAHs in food supplements, dried herbs, and dried spices (except cardamom and smoked capsicum): benzo(a)pyrene, 10.0 μg/kg; sum of the four PAHs (benzo(a)pyrene, benz(a)anthracene, benzo(b)fluoranthene, chrysene), 50.0 μg/kg. The currently available information on possible sources of PAH occurrence in herbal preparations, such as smoking, roasting, drying, or environmental effects, has been presented in a reflection paper published by the EMA's HMPC (EMA, 2016b). This paper describes not only the effects of PAHs on human health, but it also evaluates potential sources of PAH occurrence in herbal preparations. Although the potential contribution of herbal medicinal products to overall PAH exposure is considered to be very low, possible sources of contamination should be identified and their individual impact on product quality should be carefully investigated.

Official methods for the determination of PAHs in herbal drugs or herbal preparations do not yet exist. The reflection paper published by the HMPC provides an overview of various methods useful for the extraction of PAHs from different plant matrices. Furthermore, various recommendations for the separation of individual PAHs by GC and HPLC and the identification of individual substances applying mass spectrometry, ultraviolet absorption, and fluorescence are described in detail (EMA, 2016b).

Pyrrolizidine alkaloids

Since the occurrence of pyrrolizidine alkaloids (PAs) in various herbal and medicinal teas has become known over the years (BfR, 2013), a possible contamination of medicinal plants with various PA-containing weeds, such as different *Senecio* species, has been discussed. In order to counteract possible PA contamination, a great deal of research has been conducted focused on this topic and as a result, a "Code of Practice" has been developed in collaboration with herb growers (Dittrich *et al.*, 2016). This document provides a framework for the implementation of individual measures in pharmaceutical companies along the entire process chain from cultivation to control and release of the finished product. Since complete prevention of PA contamination is not possible with the current state of the art, the main focus has been on preventive measures.

Although contamination of medicinal plants with PAs is still a challenge for growers and manufacturers today, a remarkable decrease in PA contamination in herbal drugs and extracts used for the production of herbal teas and phytomedicinal products has been observed in recent years. However, it has also become clear that the problem of PA contamination will not be completely solved in the near future and it can therefore be assumed that the limit value for PAs of 1.0 μg/day for herbal medicinal products will continue to be maintained.

Protection of genetic resources/benefit sharing

Growers and collectors of medicinal plants/herbal substances must ensure that they avoid damage to existing wildlife habitats. CITES (Convention on International Trade in Endangered species of Wild Fauna and Flora) must be adhered to. Furthermore, batches of collected medicinal plant material should be unambiguously and unmistakeably traceable to their individual sources. Therefore, appropriate labeling and batch assignment should be carried out as early as possible. In this context, it should also be possible to clearly distinguish cultivated medicinal plants from wild-collected medicinal plants.

Seeds used for the cultivation of medicinal plants should be clearly identified and traceable in terms of genus, species, variety, or chemotype (the same applies to vegetatively propagated medicinal plants). For the production of medicinal plants intended for organic production, not only must an appropriate certificate be available, but also the starting material should be free from pests and diseases to ensure healthy plant growth. If possible, species that are naturally resistant or tolerant to various plant diseases should preferably be used.

Generally, the use of pesticides should be kept to a minimum also in the conventional cultivation of medicinal plants. The application of PPPs should be carried out only by qualified personnel with approved equipment. In the case of harvested plants, the applicable regulations on maximum residue limits provided in the Ph.Eur., various European directives, the Codex Alimentarius, etc. must be observed.

The Nagoya Protocol on Access to Genetic Resources and the Fair and Equitable Sharing of Benefits Arising from their

Utilization, signed by 50 countries in 2014, aims to:

> create conditions to encourage and support research that contributes to the conservation and sustainable use of biodiversity, particularly in developing countries, including through facilitated measures for access for non-commercial research purposes, taking into account the need to consider changing the intent for such research.

The main aim is to effectively reduce biopiracy, which is often still practiced by developing countries today, and to ensure a binding participation under international law in the products manufactured with the plant raw materials (e.g., phytopharmaceuticals) and the profits generated thereby.

Evidence-based practices in Traditional Iranian/Persian Medicine

Evidence-based medicine (EBM) has become increasingly important during the last 20 years. Generally, this discipline integrates clinical experience and patient values with the best available research information. The main objective is to further improve the use of high-quality clinical research in clinical decision-making.

In this context, there have also been numerous efforts to confirm valid scientific evidence for the individual health efficacies described in TIM/TPM. For example, a study performed by Arabshahi *et al.* (2015) described the concept of knowledge-based practice in TIM/TPM from the perspective of experts. That qualitative study was conducted in the form of face-to-face in-depth interviews to obtain the opinions of TIM/TPM experts about the concept of evidence-based practice. The study not only addressed the limitations of evidence-based practice in Iranian traditional medicine, but also presented the differences between conducting research in modern and Iranian traditional medicine. However, in general, the authors conclude that evidence-based practice and the use of commonly accepted references are essential in TIM/TPM to objectify its efficacy and effectiveness.

Markets for herbal teas, medicinal plants, and phytopharmacy

The Islamic Republic of Iran is the second largest economy in the Middle East and North Africa after Saudi Arabia, with an estimated gross domestic product (GDP) of US$447.7 billion in 2017 and a population of approximately 80.6 million (The World Bank, 2019). In the years after the Islamic Revolution, Iranian pharmaceutical policy was developed based on the mandatory generic drug policy (Cheraghali, 2017).

References

Alves, R.R. and Rosa, I.M. (2007) Biodiversity, traditional medicine and public health: where do they meet? *Journal of Ethnobiology and Ethnomedicine* 3, 14. DOI:10.1186/1746-4269-3-14.

Arabshahi, S.K.S., Kenari, H.M., Kordafshari, G., Ardakani, M.R.S. and Shoaleh Bigdeli, S. (2015) Criteria for evidence-based practice in iranian traditional medicine. *Acta Medica Iranica* 53(7), 419–424.

Bagheri, M. and Reghan, M.S. (1994) Study the status and use of medicinal plants in Iran and in the world. *Forest and Rangelands Journal* 33, 15–19 (in Farsi).

BfR (2013) Pyrrolizidine alkaloids in herb teas and teas. Opinion no. 018/2013 of the German Federal Institute for Risk Assessment (BfR) of 05 July 2013. Available at: https://www.bfr.bund.de/cm/349/pyrrolizidine-alkaloids-in-herbal-teas-and-teas.pdf (accessed 27 April 2023).

Bomme, U. (2001) Anbauversuche mit chinesischen Arzneipflanzen. *Deutsche Apotheker Zeitung* 37, 67.

Bomme, U., Friedl, F. and Hummelsberger, J. (2006) Anbau Chinesischer Arzneipflanzen in Bayern – ein Pilotprojekt. *Deutsche Zeitschrift für Akupunktur* 49, 44–45.

Bunse, M., Daniels, R., Gründemann, C., Heilmann, J., Kammerer, D.R. *et al.* (2022) Essential oils as multicomponent mixtures and their potential for human health and well-being. *Frontiers in Pharmacology* 13, 956541. DOI: 10.3389/fphar.2022.956541.

Cheraghali, A.M. (2017) Trends in Iran pharmaceutical market. *Iranian Journal of Pharmaceutical Research* 1, 1–7.

Commission of the European Communities (2006) Commission Regulation (EC) No 1881/2006 of 19 December 2006 setting maximum levels for certain contaminants in foodstuffs. *Official Journal of the European Union* 49, 20.12.2006, L 364/5–L364/14.

Council of Europe (2016) *Herbal Drugs*. General monograph no. 1433, Ph.Eur. 9th edn. Council of Europe, Strasbourg, France.

Dittrich, H., Hösel, K., Sievers, H., Klier, B., Waimer, F. *et al.* (2016) Code of practice zur Vermeidung und Verringerung von Kontaminationen pflanzlicher Arzneimittel mit Pyrrolizidinalkaloiden. *Pharmaceutical Industry* 78, 836–845.

Dong, J.C. (2013) The relationship between traditional chinese medicine and modern medicine. *Evidence-Based Complementary Alternative Medicine* 2013, 153148. DOI: 10.1155/2013/153148.

EMA (2006) *Guideline on Good Agricultural and Collection Practice (GACP) for Starting Materials of Herbal Origin.* EMEA/HMPC/246816/2005 of 20 February 2006. European Medicines Agency, Committee on Herbal Medicinal Products, London.

EMA (2016a) *Guideline on the use of the CTD format in the preparation of a registration application for traditional herbal medicinal products: Final – revision 2*. EMA/HMPC/71049/2007 Rev. 2 of 5 April 2016. European Medicines Agency, Committee on Herbal Medicinal Products, London.

EMA (2016b) *Reflection paper on Polycyclic Aromatic Hydrocarbons (PAH) in herbal medicinal products/ traditional herbal medicinal products*. EMA/HMPC/300551/2015 of 31 May 2016. European Medicines Agency, Committee on Herbal Medicinal Products, London

European Parliament and Council of the European Union (2005) Regulation (EC) 396/2005 of the European Parliament and of the Council of 23 February 2005 on maximum residue levels of pesticides in or on food and feed of plant and animal origin and amending Council Directive 91/414/EEC. *Official Journal of the European Union* 48, 16.03.2005, L 70/1–L 70/16.

Fabricant, D.S. and Farnsworth, N.R. (2001) The value of plants used in traditional medicine for drug discovery. *Environmental Health Perspectectives* 109, 69–75.

Ghahreman, A. and Attar, F. (1999) *Biodiversity of Plant Species in Iran*, Vol. 1. Tehran University Publication, Tehran.

Haidan, Y., Qianqian, M., Li, Y. and Guangchun, P. (2016) The traditional medicine and modern medicine from natural products. *Molecules* 21(5), 559. DOI:10.3390/molecules21050559.

Koocheki, A. (2013) Research on production of Saffron in Iran: past trend and future prospects. *Saffron Agronomy & Technology* 1(1), 3–21.

Mashayekhan, A., Pourmajidian, M.R., Jalilvand, H., Gholami, M.R. and Teimouri, M.S. (2016) Economic importance and GIS mapping of medicinal plants in Iran. *Journal of Applied Sciences and Environmental Management* 20(3) 646–650.

Salehi Sardoei, S.A. (2022) Iranian medicinal plants: from economically to ethnomedicine studies. *International Journal of Advanced Biological and Biomedical Research* 10(2), 98–116. DOI: 10.22034/ ijabbr.2022.540138.1369.

Sefidkon, F. (2021) National approach to make a transformation in the economic value of medicinal plants. *Iranian Nature Journal* 6(1), 135–143.

Steinhoff, B. (2019) Challenges in the quality of herbal medicinal products with a specific focus on contaminants. *Phytochemical Analysis* 32(2), 117–123. DOI: 10.1002/pca.2879.

The World Bank (2019) Islamic Republic of Iran; An Overview. Available at: https://www.worldbank.org/en/ country/iran/overview (accessed 20 January 2023).

Tu, P.F., Guo, H.Z. and Guo, D.A. (2002) Researches on active constituents of natural and traditional medicine and development of new drugs. *Journal of Peking University. Health Sciences* 34, 513–518.

WHO (1993) *Research Guidelines for Evaluating the Safety and Efficacy of Herbal Medicines*. World Health Organization Regional Office for the Western Pacific, Manila.

WHO (1998a) *Regulatory Situation of Herbal Medicines: A Worldwide Review*. WHO/TRM/98.1. World Health Organization, Geneva, Switzerland.

WHO (1998b) *Quality Control Methods for Medicinal Plant Materials*. World Health Organization, Geneva, Switzerland.

WHO (1999) *WHO Monographs on Selected Medicinal Plants*, Vol. I. World Health Organization, Geneva, Switzerland.

WHO (2000) *General Guidelines for Methodologies on Research and Evaluation of Traditional Medicine.* WHO/EDM/TRM/2000.1. World Health Organization, Geneva, Switzerland.

WHO (2015) *Global Health History Seminar on Traditional Medicine and Ayurveda*. World Health Organization, Geneva, Switzerland.

Wu, Y.L. (2009) Artemisinin – the revelation of the history and reality. *Chemical Progress in Chemistry* 21, 2365–2371.

Zhang, L.H. and Li, J. (2011) Current situation and developing trends of modernization of traditional Chinese medicine. *Journal of Zhejiang University (Medical Science)* 40, 349–353.

Zhang, L., Yan, J., Liu, X., Ye, Z., Yang, X. *et al.* (2012) Pharmacovigilance practice and risk control of Traditional Chinese Medicine drugs in China: current status and future perspective. *Journal of Ethnopharmacology* 140, 519–525.

Index